Biotechnology in Agriculture and Food Processing

Opportunities and Challenges

Biotechnology in Agriculture and Food Processing

Opportunities and Challenges

Edited by
Parmjit S. Panesar
Satwinder S. Marwaha

CRC Press is an imprint of the
Taylor & Francis Group, an informa business

CRC Press
Taylor & Francis Group
6000 Broken Sound Parkway NW, Suite 300
Boca Raton, FL 33487-2742

First issued in paperback 2017

Version Date: 20130521

ISBN 13: 978-1-138-07326-5 (pbk)
ISBN 13: 978-1-4398-8836-0 (hbk)

Library of Congress Cataloging-in-Publication Data

Panesar, P. S. (Parmjit Singh)
Biotechnology in agriculture and food processing : opportunities and challenges / authors, P.S. Panesar, S.S. Marwaha.
pages cm
"A CRC title."
Summary: "An instructive and comprehensive overview of the use of biotechnology in various agriculture and food applications, this book discusses the current status of applications for improving quality and productivity of agriculture and food products. It discusses plant tissue culture techniques, genetic engineering in food ingredient production, bio-fertilizers, bio-pesticides, enzymes, genetically modified foods etc. The book is divided into four parts: fundamental concepts, biotechnology in agriculture, biotechnology in food processing and biotechnological management of crop residues and waste/by products of agro industries"-- Provided by publisher.
Includes bibliographical references and index.
ISBN 978-1-4398-8836-0 (hardback)
1. Food--Biotechnology. 2. Crops--Genetic engineering. 3. Agricultural processing. 4. Biotechnology--Industrial applications. I. Marwaha, S. S. II. Title.

TP248.65.F66P36 2013
664--dc23 2013020077

Visit the Taylor & Francis Web site at
http://www.taylorandfrancis.com

and the CRC Press Web site at
http://www.crcpress.com

Contents

Preface....vii
Acknowledgments....ix
Editors....xi
Contributors....xiii

Part I Fundamental Concepts

1. **Biotechnology and Its Role in Agriculture and Food Processing**....3
   Parmjit S. Panesar and Satwinder S. Marwaha

2. **Genomics in Agriculture and Food Processing**....45
   Peter C. McKeown, Channa Keshavaiah, Antoine Fort, Reetu Tuteja, Manash Chatterjee, Rajeev K. Varshney, and Charles Spillane

Part II Biotechnology in Agriculture

3. **Plant Cell and Tissue Culture Techniques in Crop Improvement**....73
   Dinesh K. Srivastava, Geetika Gambhir, and Poornima Sharma

4. **Genetic Transformation and Crop Improvement**....133
   Kailash C. Bansal and Dipnarayan Saha

5. **Production of Biofertilizers**....169
   Dinesh Goyal and Santosh K. Goyal

6. **Production of Biopesticides**....191
   Surinder Kaur, Gurpreet S. Dhillon, Satinder K. Brar, and Ramesh Chand

Part III Biotechnology in Food Processing

7. **Production of Fermented Foods**....219
   Sanjeev K. Soni, Raman Soni, and Chetna Janveja

8. **Functional Foods**....279
   Kulwant S. Sandhu and Sarabjeet K. Sra

9. **Enzymes in Food Processing** ... 329
Ivana G. Sandri, Luciani T. Piemolini-Barreto, and Roselei C. Fontana

10. **Production of Polysaccharides** ... 355
Gordon A. Morris and Stephen E. Harding

11. **Production of Sweeteners** ... 387
Pedro Fernandes and Joaquim M.S. Cabral

12. **Production of Biocolors** ... 417
Nuthathai Sutthiwong, Yanis Caro, Philippe Laurent, Mireille Fouillaud, Alain Valla, and Laurent Dufossé

13. **Production of Bioflavors** ... 447
Jyothi Ramamohan, Harish K. Chopra, and Shubhneet Kaur

14. **Genetically Modified Foods** ... 479
Konstantin Skryabin and Victor Tutelyan

Part IV Biotechnological Management of Crop Residues and By-Products of Agro-Industries

15. **Production of Mushrooms** ... 509
Pardeep K. Khanna and Shivani Sharma

16. **Value Addition of Agro-Industrial Wastes and Residues** ... 557
Parmjit S. Panesar, Satwinder S. Marwaha, and Reeba Panesar

Index ... 607

Preface

Biotechnology has a long history of use in agriculture and food processing. It is a key technology and has played a significant role in these areas. It represents a combination of both traditional and latest techniques based on molecular biology and recombinant DNA technology. It has not only been helpful in boosting the yield of agricultural crops but has also led to the generation of novel varieties, which can play a significant role in overcoming the problem of hunger and malnutrition worldwide. Biotechnological techniques in food processing offer the prospect of improving existing processes and developing novel products. The applications of biotechnology range from the production of fermented foods to the development of biosensors and food safety kits, which can help ensure the quality and safety of food products.

Agriculture and food processing are integrated disciplines, with the latter having its roots in the former. The role of biotechnology in these areas reflects the true picture of its range of applications from "farm to fork." Keeping this in view, the compilation of this volume is expected to provide comprehensive information to the readers on the subject of biotechnology. The objective of this volume is to provide a perspective on the important biotechnological issues that have direct relevance to agriculture and food processing. Even with restricted perspective, the potential range of the topics is enormous, and it is not possible to review the whole field comprehensively in a single volume. Hence, a careful selection of chapters has been made to provide readers with both an in-depth study and a broad perception.

This volume is divided into four parts. Part I explains the fundamental concepts of the role of biotechnology and genomics in agriculture and food processing. Part II focuses on specific applications of biotechnology in agriculture and includes chapters on plant cell and tissue culture techniques, and genetic transformation in crop improvement, besides production of biofertilizers and biopesticides. Part III comprises of chapters that deal with different aspects of biotechnology in food processing such as production of fermented foods, functional foods, enzymes in food processing, production of polysaccharides, production of sweeteners, biocolors and bioflavors, and genetically modified foods. Part IV focuses on chapters on the management of crop residues and by-products of agro-industries, comprising mushroom production and value addition to agro-industrial wastes and residues.

We hope that this volume will contribute to understanding the significance of biotechnology applications in agriculture and food processing. We also hope that it will help in recognizing the role of microorganisms, enzymes, and biotechnological techniques in both disciplines, which can give further understanding of mechanisms contributing to the improvement of the overall quality of agriculture and food products.

In summary, this volume, *Biotechnology in Agriculture and Food Processing: Opportunities and Challenges,* is an instructive and comprehensive overview of current knowledge about the interventions of biotechnology in agriculture and food applications. In each chapter, a brief introduction, fundamentals, and applications of the topic are provided, together with future prospects and recent references. Efforts have been made to present each chapter in a well-documented and illustrated manner with suitable tables and figures. To make each chapter complete in itself, some information may be similar to that in other chapters, but, considering the subject matter as a whole, should not be considered as repetitive.

In conclusion, this volume is an indispensable treatise that offers a holistic view of all the aspects related to the subject and will be a useful source not only to academicians but also to people working in industry. It is also hoped that both students and professionals will equally benefit from this volume.

Parmjit S. Panesar
Satwinder S. Marwaha

Acknowledgments

We are indeed grateful to all the contributors for sharing their expertise and for their cooperation and promptness in revising the manuscript. We acknowledge the various sources of illustrations used in this volume, which have been presented in the form of figures and tables and also in the form of references given at the end of every chapter. We also appreciate the support rendered by the research scholars in compiling this volume.

We are especially indebted to Dr. Gagandeep Singh, commissioning editor, and Marsha Pronin, project coordinator, CRC Press, Taylor & Francis Group, for their support in bringing out this volume. We also extend our thanks to the production team of Taylor & Francis Group for their efforts in the production of this volume.

Last but not least, we acknowledge the support of our family members in the preparation of this volume.

Editors

Parmjit S. Panesar is a doctorate in biotechnology from Punjabi University, Patiala, India, with more than 15 years experience in the area of food biotechnology and enzyme/fermentation technology. He now works as professor, Biotechnology Research Laboratory, Department of Food Engineering and Technology, Sant Longowal Institute of Engineering and Technology (deemed to be university; established by the Government of India), Longowal, Punjab, India. In 2005, he was awarded BOYSCAST (Better Opportunities for Young Scientists in Chosen Areas of Science and Technology) fellowship by the Department of Science and Technology (DST), Ministry of Science and Technology, Government of India, to carry out advance research in industrial biotechnology at Chembiotech Laboratories, University of Birmingham Research Park, Birmingham, United Kingdom. His postdoctoral research focused on the development of immobilized cell technology for the production of lactic acid from whey. Besides, he has also been involved in U.S.- and Australian-sponsored projects on the scale-up of a commercial polysaccharide derivative in biomedical applications and critical evaluation of a commercial test kit for the detection of trace levels of explosives. In 1999, Dr. Panesar was awarded the Young Scientist Fellowship by the Punjab State Council for Science and Technology, Punjab. He has undergone advance training from different leading institutes/labs, including International Centre for Genetic Engineering and Biotechnology, Trieste, Italy; Institute of Microbial Technology, Chandigarh, India; Central Food Technological Research Institute, Mysore, Karnataka, India; and National Dairy Research Institute, Karnal, India.

Dr. Panesar has published more than 80 international/national scientific papers, 50 book reviews in peer-reviewed journals, and 20 chapters and has authored/edited 4 books. He is a member of the editorial advisory boards of national/international journals and of various national and international professional bodies and committees, including scientific panel on "genetically modified organisms and foods" constituted by the Food Safety and Standards Authority of India. He is also a member of curriculum development committee for designing syllabi for the food technology and biotechnology courses of various institutes/universities of India. He has successfully handled four major research projects sponsored by different funding agencies, such as Council of Scientific and Industrial Research, Ministry of Human Resources Development, New Delhi. He is a life member of the International Forum on Industrial Bioprocesses, Biotechnology Research Society of India, Association of Microbiologists of India, Association of Food Scientists, Indian Science Congress Association, and Technologists and Indian Society for Technical Education. He is also on the reviewer panel of many national/international journals related to food and biotechnology. His research interests include microbial fermentation, application of immobilized cells/enzymes in different bioprocesses, and synthesis of prebiotics.

Satwinder S. Marwaha received his doctorate in microbiology from Punjab Agricultural University, Ludhiana, India, in 1981 and is now the chief executive officer of Punjab Biotechnology Incubator (PBTI), an important component of cluster of agri-food biotechnology being developed in the Knowledge City, SAS Nagar, Mohali, India. He was director (biotechnology) of the Punjab State Council for Science and Technology from 1997 to 2006 and professor and chairman, Department of Biotechnology, Punjabi University, Patiala, India, from 1994 to 1997. He now works as visiting professor and adjunct professor at the leading central universities of India.

Dr. Marwaha has completed many research projects for the scale-up of laboratory-developed processes to technologies for transfer to field and industry. He has been a member of academic councils/board of studies and committees of state- and national-level institutions/organizations/corporate as well as a member of the board of directors of a number of national biotech/life sciences industries. He also serves as a member of the State Expert Appraisal Committee (SEAC) constituted by the Ministry of Environment and Forest, Government of India, Scientific Panel on GM Crops and Foods of Food Safety and Standards Authority of India (FSSAI), and member of the Board of Directors of Export Inspection Council (EIC) of India.

In recognition of his professional contribution, he was conferred with the State Award by the Government of Punjab (India) in 2003 for promoting science and technology. Professor Marwaha was also a commonwealth fellow in 1982–1983 at the University of Birmingham, United Kingdom, and pursued his research activities in the area of fermentation technology. He has published more than 120 research papers and review articles in international/national journals and has contributed more than 30 chapters in books published by national and international publishing houses. He has also published eight books in the areas of agriculture, including agro-processing biotechnology, food biotechnology, and environmental biotechnology.

Contributors

Kailash C. Bansal
National Bureau of Plant Genetic Resources
Indian Council of Agricultural Research
New Delhi, India

Satinder K. Brar
Centre for Water, Earth and Environment (ETE)
National Institute of Scientific Research (INRS)
University of Quebec
Quebec, Canada

Joaquim M.S. Cabral
Department of Bioengineering
Higher Technical Institute
Technical University of Lisbon
and
Centre for Biological and Chemical Engineering
Higher Technical Institute
Institute for Biotechnology and Bioengineering
Lisboa, Portugal

Yanis Caro
Laboratory of Chemistry, Natural Products and Food Science
Food Engineering Department
University of Réunion Island
La Réunion, France

Ramesh Chand
Department of Mycology & Plant Pathology
Institute of Agricultural Sciences
Banaras Hindu University
Varanasi, India

Manash Chatterjee
BenchBio Private Limited
Jai Research Foundation
Valsad, India

Harish K. Chopra
Department of Chemistry
Sant Longowal Institute of Engineering and Technology
Longowal, India

Gurpreet S. Dhillon
Centre for Water, Earth and Environment (ETE)
National Institute of Scientific Research (INRS)
University of Quebec
Quebec, Canada

Laurent Dufossé
Laboratory of Chemistry, Natural Products and Food Science
Food Engineering Department
University of Réunion Island
La Réunion, France

Pedro Fernandes
Faculty of Engineering
Lusophone University of Humanities and Technologies
and
Department of Bioengineering
Higher Technical Institute
Technical University of Lisbon
and
Centre for Biological and Chemical Engineering
Higher Technical Institute
Institute for Biotechnology and Bioengineering
Lisboa, Portugal

Roselei C. Fontana
Institute of Biotechnology
University of Caxias do Sul
Caxias do Sul, Brazil

Antoine Fort
Plant & AgriBiosciences Research Centre
Botany and Plant Science
School of Natural Sciences
National University of Ireland
Galway, Ireland

Mireille Fouillaud
Laboratory of Chemistry, Natural Products and Food Science
Food Engineering Department
University of Réunion Island
La Réunion, France

Geetika Gambhir
Department of Biotechnology
Dr. Y.S. Parmar University of Horticulture & Forestry
Solan, India

Dinesh Goyal
Department of Biotechnology & Environmental Sciences
Thapar University
Patiala, India

Santosh K. Goyal
Centre for Conservation and Utilization of Blue Green Algae
Indian Agricultural Research Institute
New Delhi, India

Stephen E. Harding
National Centre for Macromolecular Hydrodynamics
School of Biosciences
University of Nottingham
Nottingham, United Kingdom

Chetna Janveja
Department of Microbiology
Punjab University
Chandigarh, India

Shubhneet Kaur
Department of Food Engineering and Technology
Sant Longowal Institute of Engineering and Technology
Longowal, India

Surinder Kaur
Department of Mycology & Plant Pathology
Institute of Agricultural Sciences
Banaras Hindu University
Varanasi, India

Channa Keshavaiah
Faculty of Science
University of South Bohemia
Branisovska
Ceske Budejovice, Czech Republic

Pardeep K. Khanna
Department of Microbiology
Punjab Agricultural University
Ludhiana, India

Philippe Laurent
Laboratory of Chemistry, Natural Products and Food Science
Food Engineering Department
University of Réunion Island
La Réunion, France

Satwinder S. Marwaha
Punjab Biotechnology Incubator
Mohali, India

Peter C. McKeown
Plant & AgriBiosciences Research Centre
Botany and Plant Science
School of Natural Sciences
National University of Ireland
Galway, Ireland

Gordon A. Morris
Department of Chemical & Biological Sciences
School of Applied Sciences
University of Huddersfield
Huddersfield, United Kingdom

Parmjit S. Panesar
Department of Food Engineering and Technology
Sant Longowal Institute of Engineering and Technology
Longowal, India

Reeba Panesar
Department of Food Engineering and Technology
Sant Longowal Institute of Engineering and Technology
Longowal, India

Luciani T. Piemolini-Barreto
Center for Science and Technology
Department of Food Engineering
University of Caxias do Sul
Caxias do Sul, Brazil

Jyothi Ramamohan
JRM Natural Products Private Limited
Chennai, India

Dipnarayan Saha
National Bureau of Plant Genetic Resources
Indian Council of Agricultural Research
New Delhi, India

Kulwant S. Sandhu
Department of Food Science and Technology
Punjab Agricultural University
Ludhiana, India

Ivana G. Sandri
Center for Science and Technology
Department of Food Engineering
University of Caxias do Sul
Caxias do Sul, Brazil

Poornima Sharma
Department of Biotechnology
Dr. Y.S. Parmar University of Horticulture and Forestry
Solan, India

Shivani Sharma
Department of Microbiology
Punjab Agricultural University
Ludhiana, India

Konstantin Skryabin
Centre 'Bioengineering'
Russian Academy of Sciences
Moscow, Russia

Raman Soni
Department of Biotechnology
D.A.V. College
Chandigarh, India

Sanjeev K. Soni
Department of Microbiology
Punjab University
Chandigarh, India

Charles Spillane
Plant & AgriBiosciences Research Centre
Botany and Plant Science
School of Natural Sciences
National University of Ireland
Galway, Ireland

Sarabjeet K. Sra
Punjab Horticultural Postharvest Technology Centre
Punjab Agricultural University
Ludhiana, India

Dinesh K. Srivastava
Department of Biotechnology
Dr. Y.S. Parmar University of Horticulture and Forestry
Solan, India

Nuthathai Sutthiwong
Department of Agricultural Technology
Thailand Institute of Scientific and Technological Research
Pathum Thani, Thailand

Reetu Tuteja
Plant & AgriBiosciences Research Centre
Botany and Plant Science
School of Natural Sciences
National University of Ireland
Galway, Ireland

Victor Tutelyan
Institute of Nutrition
Russian Academy of Medical Sciences
Moscow, Russia

Alain Valla
National Center for Scientific Research
Chemistry and Biology of Natural Products
Quimper, France

Rajeev K. Varshney
Centre of Excellence in Genomics
International Crops Research Institute for the Semi-Arid Tropics
Patancheru, India

Part I

Fundamental Concepts

1

Biotechnology and Its Role in Agriculture and Food Processing

Parmjit S. Panesar and Satwinder S. Marwaha

CONTENTS

1.1 Concept of Biotechnology 4
1.2 Historical Developments 5
1.3 Interdisciplinary Nature of Biotechnology 6
1.4 Biotechnology in Agriculture 7
  1.4.1 Biotechnological Techniques in Agriculture 8
    1.4.1.1 Plant Cell and Tissue Culture 8
    1.4.1.2 Genetically Modified Crops 9
    1.4.1.3 Biocontrol Agents 13
1.5 Biotechnology in Food Processing 15
  1.5.1 Milk Processing 16
  1.5.2 Cereal Processing 18
  1.5.3 Fruit and Vegetable Processing 19
  1.5.4 Meat and Fish Processing 20
  1.5.5 Production of Single Cell Protein 21
  1.5.6 Production of Food Additives 22
    1.5.6.1 Production of Biopigments 22
    1.5.6.2 Production of Bioflavors 25
    1.5.6.3 Production of Organic Acids 25
    1.5.6.4 Production of Gums 27
    1.5.6.5 Production of Amino Acids 28
    1.5.6.6 Production of Vitamins 28
    1.5.6.7 Production of Bacteriocins 29
    1.5.6.8 Production of Biosurfactants 30
    1.5.6.9 Production of Oils and Fats 30
  1.5.7 Production of Alcoholic Beverages 31
  1.5.8 Food Industry Waste Management 32
1.6 Impact of Genomics on Agriculture and Food Processing 32
1.7 Summary and Future Prospects 34
References 35

1.1 Concept of Biotechnology

Man has exploited traditional techniques of biotechnology for thousands of years in various activities and its origin can be traced back to prehistoric times. However, the discovery of genetic engineering is primarily responsible for the current "biotechnology boom" and the much publicity of biotechnology. It is likely to be the key technology of twenty-first century and is expected to play an important role in societal transformations through its wide range of applications. Biotechnology includes a wide range of diverse technologies and in recent years tremendous advances are seen in agriculture and food biotechnology, including transgenic crops and genetically modified (GM) organisms/foods. Biotechnological techniques can not only improve the existing processes and products, but these can also result in the development of totally new products with desired characteristics. The applications of biotechnology transcend all the aspects related to agriculture and food processing beginning right from their production to consumption or in other words "farm to fork."

Biotechnology as the word indicates is the interaction between science of biology and technology. The word "bio" in biotechnology means life and refers to microbes and other living cells including animal and plant cells. The word "technology" involves growth of living cells under defined conditions to achieve products/processes at optimum efficiency (Arora et al., 2000). True to its name, it deals with the exploitation of biological agents or their components for generating useful products. As evident from the wide range of applications of biotechnology and divergent range of techniques involved, a precise definition of the subject is relatively difficult.

Biotechnology may be defined as "the use of living organisms in systems or processes for the manufacture of useful products; it may involve algae, bacteria, fungi, yeast, cells of higher plants and animals, or subsystems of any of these or isolated components from living matter" (Gibbs and Greenhalgh, 1983). Biotechnology comprises the "controlled and deliberate application of simple biological agents—living or dead, cells or cell components—in technically useful operations, either of productive manufacture or as service operations" (Bulock, 1987). U.S. National Science Foundation has defined biotechnology as the "controlled use of biological agents such as microorganisms or cellular components for beneficial use" (Singh 2010). European Federation of Biotechnology has defined biotechnology as "the integrated use of biochemistry, microbiology, and engineering sciences in order to achieve technological applications of the capabilities of microorganisms, cultured tissue cells, and parts thereof" (Singh 2010). A simpler approach defines biotechnology as the use of biological organisms or processes in any technological application (Riley and Hoffman, 1999).

Biotechnological tools can help in the sustainable development of agriculture as well as food processing, improving the production, availability, and nutritional value of foods (Pérez-Magariño and González-Sanjosé, 2003). It not only deals with traditional tools and techniques but also considers the new molecular biology and recombinant DNA techniques for the improvement of agricultural practices and food processing. This introductory chapter provides a complete overview of historical developments and the range of established biotechnological activities in agriculture and food processing.

1.2 Historical Developments

As stated earlier, origin of biotechnology discipline can be traced back to prehistoric times, though the term biotechnology seems to be of recent origin. The traditional or old biotechnology refers to the conventional techniques used for the production of wine, vinegar, curd, etc., using microorganisms in early 5000 BC. But the microorganisms were first observed under microscope during the seventeenth century. The eighteenth and nineteenth centuries proved to be the milestones in the development of biotechnology. During this period, G.H. Mendel gave the law of genetics, and later on pasteurization technique and microbial fermentation processes were established by Louis Pasteur. The production of alcoholic beverages constitutes a very important segment of biotechnology, which has seen significant growth from the time of Pasteur. In 1917, Karl Ereky, a Hungarian engineer coined the term biotechnology. In late nineteenth century, A. Jost gave the term genetic engineering, which led to the development of GM crops/organisms. The discoveries related to the production of monoclonal antibodies, cloning of plants/animals, and development of gene transfer methods have also been accomplished during this period. These developments resulted in the development of new biotechnology. Thus, the historical developments in biotechnology discipline can be summarized in four phases: (1) production of foods and beverages by natural fermentation process, (2) production of biomass, solvents (ethanol, acetic acid, butanol, etc.), and waste treatment processes under non-sterile conditions, (3) use of pure single culture under controlled septic condition, and finally (4) the development of applied genetics and recombinant DNA technology (Smith, 2009). The comprehensive information on the milestones in the historical developments of biotechnology has been encapsulated in Figure 1.1.

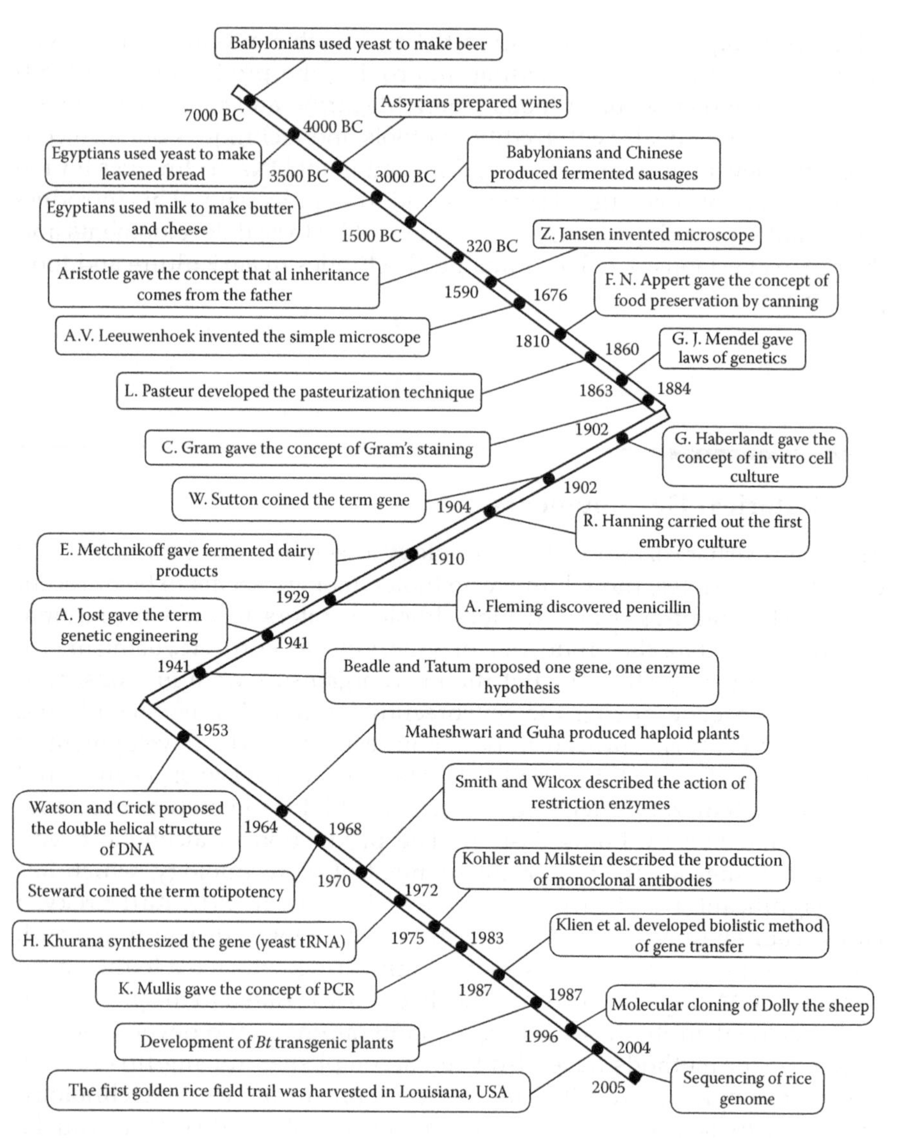

FIGURE 1.1
Historical developments of biotechnology.

1.3 Interdisciplinary Nature of Biotechnology

Biotechnology is truly multidisciplinary in nature and it encompasses several disciplines of basic sciences and engineering. The science discipline from which biotechnology draws heavily are microbiology, chemistry, biochemistry, genetics, molecular biology, immunology, cell and tissue culture,

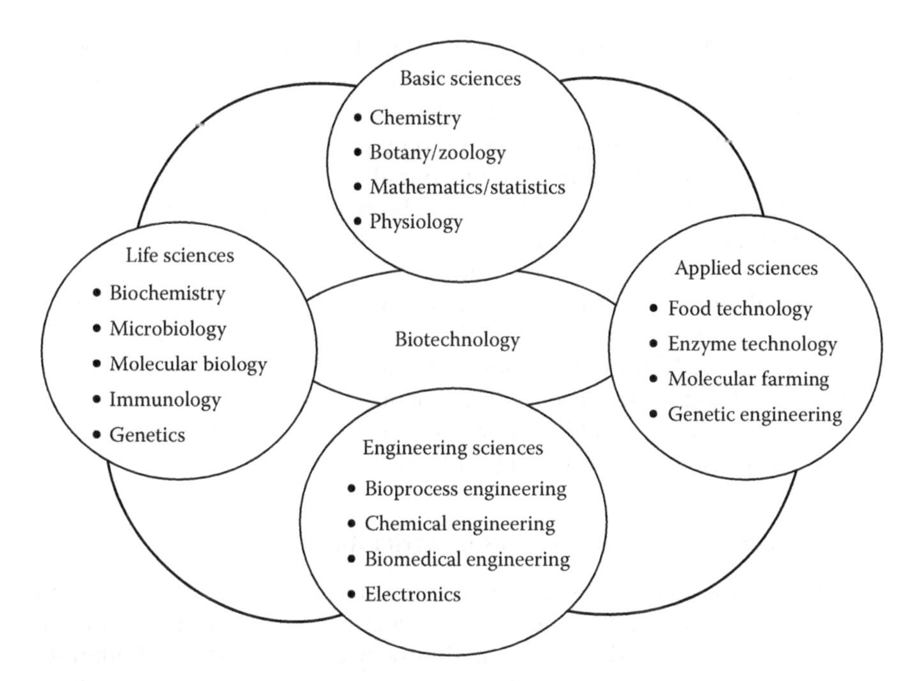

FIGURE 1.2
Interdisciplinary nature of biotechnology.

and physiology. On the engineering side, it leans heavily on process, chemical, and biochemical engineering regarding large-scale cultivation of microorganisms and cells and their downstream processing. Biotechnology has diverse range of applications in agricultural, pharmaceutical, food, medical, and environmental fields (Figure 1.2).

1.4 Biotechnology in Agriculture

Since the beginning of agriculture, farmers and researchers have been trying to alter the genetic makeup of crops through conventional breeding programs for the enhancement of crop yield. But, with the rapid rise in population and limited sources of natural resources, the ever-increasing need of crop-derived products and yield can be better helped out by employing modern biotechnological techniques instead of traditional breeding methods (Ozor and Igbokwe, 2007). With the development and understanding of new biotechnological techniques, plant breeders and researchers have now relied more on genetic modifications and recombinant techniques for upgrading the production and trait improvement in different crops. Biotechnology has opened many new doors for researchers to understand the genetic makeup of different crops by which they can add, delete, modify, and silence a single or group of genes to obtain the desirable results. It allows the researchers to isolate a single/multiple gene from plant having the desired trait and then transfer from one plant to another (Wieczorek, 2003). This process has not only eliminated the repeated trials with 1000 genes of unknown functions but also helped to cope up with the technical hurdles faced during traditional breeding programs (Raney, 2004). Thus, biotechnology can play an effective role in designing of crops that can perform better in the changing environment.

Biotechnology has also enabled the use of crops as bioreactors for the production of biological products in the form of vaccines, biocolorants, antibodies, drugs, etc. Tissue culture has been employed for the mass production of endangered and useful crops in terms of quality and quantity. Further, the remarkable development of bioinformatics and genomics has strengthened the pursuits of biotechnology in agriculture.

1.4.1 Biotechnological Techniques in Agriculture

Biotechnological techniques, including conventional plant breeding, plant tissue culture, and modern genetic engineering, play important roles in improving the agro-products both qualitatively (nutritionally) and quantitatively (product yield and productivity). Micropropagation, embryo culture, somatic embryogenesis, and somaclonal variations can be achieved through tissue culture technique. Genetic engineering is mainly employed when a new trait is to be introduced, which is not present in the germplasm of the crop or it is very difficult to improve the existing trait by conventional breeding programs. Although there are many complex and diverse techniques used for genetic modification of plants, this process mainly involves four steps: (1) isolation of desired gene, (2) cloning of the gene in suitable vector, (3) transfer of gene in suitable host, and (4) selection of recombinant. The gene can be transferred into the cells by different methods, which involve *Agrobacterium*-mediated transformation, particle gun method, electroporation, or by protoplast fusion method. Transformed cells are then cultured under controlled conditions, which are further induced to develop into small plants containing the inserted gene (Suslow et al., 2002; Singh, 2010). To determine the integrity and authenticity of the inserted gene into the host plants, different biotechnological methods like polymerase chain reaction, probe hybridization, microarrays, and DNA fingerprinting have been developed.

The following sections provide a comprehensive view of important biotechnological tools/techniques applied in agriculture.

1.4.1.1 Plant Cell and Tissue Culture

Tissue culture, which involves the generation of useful products from plant cells, tissues, and organs, has played a significant role in crop improvement since the 1940s. The explants are continuously maintained under in vitro conditions or passed through different phases for the regeneration of the complete plant. Thus, plant tissue culture forms an integral part of any biotechnology activity. It involves the achievement of rapid clonal multiplication (micropropagation), virus elimination, cell culturing as well as in vitro pollination, somaclonal variations, gene transfer, and development of molecular marker–assisted breeding (Suslow et al., 2002).

Plant tissue culture has played a considerable role in the propagation of plants that are difficult to propagate through vegetative means or in which there is a need to propagate a plant tissue having the desired trait (Thiart, 2003; Singh and Shetty, 2011). Many ornamental crops, vegetable and fruit crops, as well as food crops have been propagated in vitro by micropropagation techniques (Sangwan et al., 1997).

In embryo culture, the immature embryos are cultured under controlled conditions to produce hybrid plants for the genetic improvement of crops and also to rescue them from wide crosses. Similarly, protoplast fusion is another alternative method for developing hybrids with new genetic variations in laboratories under controlled conditions. In protoplast fusion, the protoplasts from two genetically different plants are allowed to fuse with each other that results in the integration of DNA of both the plants, and the fused cell is then induced to develop into a complete plant having the traits derived from parents (Suslow et al., 2002).

Artificial seeds are produced by encapsulating a plant part in a matrix, which can be used to grow into a whole plant. Plant tissue may generally consist of somatic embryos that have been grown aseptically in tissue culture to develop them into the complete individual plant (Redenbaugh et al., 1988; Paunescu, 2009). Artificial seeds have proved to be promising in the propagation of transgenic plants, seedless plants, and plant lines having problems in seed propagation. The cloning nature of artificial seeds can not only reduce the time period of selection procedure but also bring forward the potentials of biotechnology in agriculture expansion in a cost-effective manner (Saiprasad, 2001).

1.4.1.2 Genetically Modified Crops

In the last few decades, modern biotechnology has been increasingly applied to various crops to increase the yield and nutritional content. Genetic modification generally involves the manipulation or designing of existing and new crops to achieve the desired traits. Thus, the most widespread application of biotechnology in agriculture by far is GM crops, having the potential to impart desirable characteristics in crops (Table 1.1). GM crops involve the genetic modification of the nuclear DNA of the plant having the desired trait through the permanent integration of recombinant DNA sequences (Morrin, 2008; Cowan, 2011). The introduction of first GM crop for human consumption began in 1994, when the federal Food and Drug Administration (FDA) approved the first whole GM food for commercialization, i.e., Flavr Savr

TABLE 1.1
List of GM Crops along with Their Modified Genes and Benefits

Crop	Gene Introduced	Benefits	References
Brinjal	A gene (*Cry 1Ac*) is inserted from the soil bacterium *Bacillus thuringiensis*	Plant becomes resistant against lepidopteran insects	Kumar et al. (2011)
Canola	Herbicide-tolerant genes have been introduced	Resistant to glyphosate, glufosinate ammonium, and bromoxynil herbicides	Demeke et al. (2002)
Corn	*Bt* crystal protein gene transferred from *B. thuringiensis*	Resistant to glyphosate or glufosinate herbicides and insect resistance	Moellenbeck et al. (2001)
Cotton	*Bt* crystal protein gene transferred from soil bacterium	Protects the plants from bollworm	Morin (2008)
Hawaiian papaya	Papaya ring spot virus resistance genes have been introduced	Variety is resistant to the papaya ring spot virus	Gonsalves (2004)
Potatoes	1. Antisense copy of the granule-bound starch synthase (*GBSS*) gene is introduced	High amylopectin ratio	Hofvander et al. (2004)
	2. *B1* gene isolated from the skin of tree frogs, *Phyllomedusa bicolor,* has been inserted	*B1* inhibited the growth of fungi and bacteria that cause blackleg in potato plants	Osusky et al. (2005)
Rice	1. Three new genes introduced in golden rice—two from daffodils and one from a bacterium	Vitamin A deficiency can be treated	Bagwan et al. (2010)
	2. Anthranilate synthase alpha subunit 1 (*ASA1*) gene from garlic plants has been introduced into indica rice	*ASA1* gene from garlic plants makes it susceptible to sap sucking pests	Yarasi et al. (2008)
	3. Cholera toxin B subunit (*CTB*) gene of *Vibrio cholerae* was modified to make it suitable form for rice seeds	Rice-based CTB vaccine remained stable at room temperature and maintains immunogenicity	Nochi et al. (2007)

TABLE 1.1 (continued)

List of GM Crops along with Their Modified Genes and Benefits

Crop	Gene Introduced	Benefits	References
Soybean	1. Enolpyruvylshikimate phosphate synthase (*EPSPS*) gene from a soil bacterium is introduced	Herbicide resistance	Tengel et al. (2001)
	2. Diacylglycerol acyltransferase 2A from the soil fungus *Umbelopsis ramanniana* is introduced	Transgenic soybeans have much higher content of stored lipids	Lardizabal et al. (2008)
Sugar beet	Herbicide-resistant genes are introduced	Resistance to glyphosate and glufosinate herbicides	Whitman (2000)
Sugarcane	Pesticide resistance gene is introduced	Resistance to pesticides, high-sucrose cane	Bagwan et al. (2010)

tomato, which was soon followed by the introduction of GM crops of cotton, maize, soybean, sugar beet, canola, papaya, banana, and golden rice (Ruse and Castle, 2002).

1.4.1.2.1 Benefits of GM Crops

GM crops usually contain a gene encoding a desired trait like herbicide tolerance, insect tolerance, endurance to biotic/abiotic stress, etc. (Figure 1.3). Genetically engineered crops also facilitate agriculturalists to meet the problems of malnutrition (Table 1.1) by improving the quality and quantity of various crops (golden rice, GM soybean, GM canola, GM maize, etc.) by keeping in pace with declining sources of water and land (Kishore and Shewmaker, 1999; Sharma et al., 2002).

Farmers generally employ large amount of pesticides and herbicides annually to protect their crops from harmful pests and weeds. Due to increase in knowledge and awareness, consumers are not willing to eat food that has been treated with chemical pesticides because of potential health hazards and environmental concerns (Bagwan, 2010). The concomitant progress in molecular genetics made it possible to incorporate resistance genes from unrelated organisms into susceptible crops to make them herbicide and pesticide resistant.

Many different types of insects not only damage the different types of crops but also reduce the yield and quality of crop-derived products. Several species of bacteria produce proteins, which can kill the insect larvae that ingest these bacteria with their food. All of the commercially available insect-tolerant plants contain a gene from the soil bacterium *Bacillus thuringiensis* (*Bt*). *Bt* toxins or "cry" genes are highly effective for controlling many pest

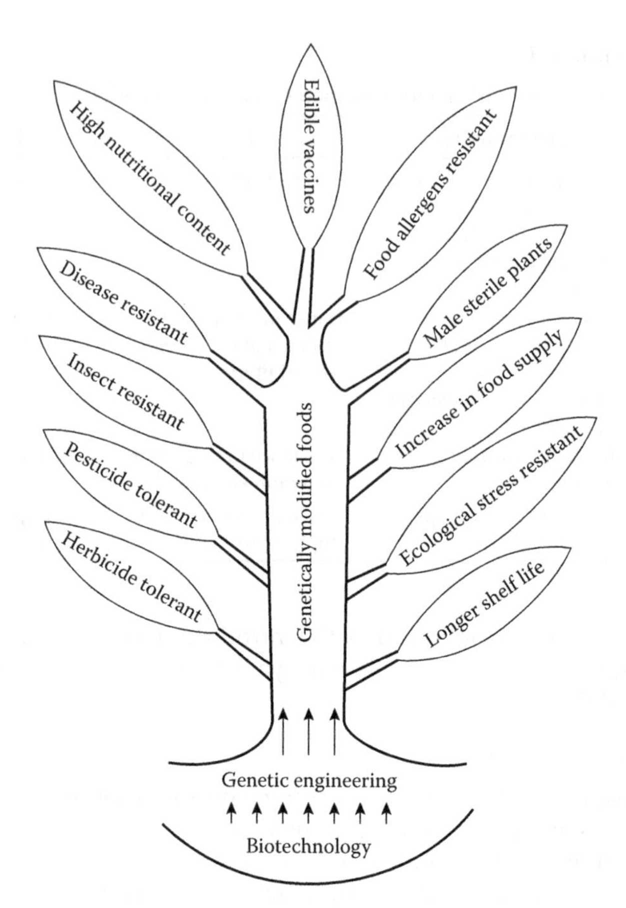

FIGURE 1.3
Potential benefits of genetically modified crops.

organisms, like beetles and moth larva (Table 1.1), but are not toxic to mammals and most other nontarget organisms. Genetic engineering has also led to the development of male sterile plants, in which the "barnase" gene isolated from *Bacillus amyloliquefaciens* has been transformed in crop species like tobacco and rapeseed oil (Mariani et al., 1990; Daniell, 2002).

1.4.1.2.2 Crops as Edible Vaccines

A new approach for delivering vaccine antigen is the use of inexpensive, oral vaccines derived from transgenic food crops, i.e., edible vaccines. These vaccines offer exciting possibilities in reducing the trouble caused by diseases like hepatitis and diarrhea by using transgenic fruits and vegetables, particularly in the developing world where storing and administering vaccines are major problems (Chaitanya and Kumar, 2006). Edible vaccines are mainly developed by using the genetic engineering technique in which the selected genes are introduced into the desired plants and the transgenic plant is further induced to express the recombinant protein. This protein triggers the immune response to produce antigen-specific antibodies and functions as an oral vaccine to treat the specific disease, when the recombinant plant is taken by an organism as food. Some of the potential benefits of plant-derived vaccines include their heat stability, easy administration, easy storage and transportation, and generation of systemic/mucosal immunity, and more importantly they are free of pathogens, contaminants, and toxins (Mishra et al., 2008).

A rice-based mucosal vaccine (needle-free vaccination) has been developed in which the cholera toxin B (*CTB*) gene of *Vibrio cholerae* was modified to make it suitable for rice grains, which is easy to store and administer (Nochi et al., 2007). Alzheimer's disease (a neurodegenerative disease) can also be treated by reducing the accumulation of β-amyloid with the help of transgenic tomato (Youm et al., 2008).

1.4.1.2.3 Safety Concerns of GM Foods

The level of risk to accept a food by community is greatly influenced by past experience and knowledge about particular food. The safety assessment of a GM food has been achieved by the comparison between GM food and its conventional counterpart. This is generally referred to as the concept of substantial equivalence (Kuiper et al., 2001; FAO/WHO, 2003). The basic idea is that novel GM organism (GMO)-derived food products should be at least as safe as the traditional products. Today, there is an increase in the number of new and modified varieties of food and food additives that do not have any history, so these foods need to be carefully assessed before they reach the market for commercialization and ensure that these are safe for human consumption. Many factors are taken into account during the safety assessment of GM foods, such as identity, source, composition, transformation process, and protein expression products of the novel DNA and their effects in the future. Some of the problems related to safety concerns of GM foods involve unintended harm to other organisms and economical and religious concerns (Abbas et al., 2010).

1.4.1.3 Biocontrol Agents

The consumer awareness about the harmful health effects of chemical fertilizers, pesticides, insecticides, herbicides, and environmental concerns is pushing scientists and researchers to develop and use biocontrol agents derived from different microbial sources for the control of pests, pathogens, or weeds in agriculture system. Many biocontrol agents produced by companies through the fermentation process are now available in market worldwide and providing a choice to the farmers to use an alternative to chemical agents currently being used (Ahmed et al., 2002) to control insect pests, weeds, and various harmful diseases. Biocontrol agents protect the environment and the crops from the damage being done by the continuous and intensive use of chemical insecticides/pesticides as well as check the increase in the number of pesticide/insecticide/herbicide-resistant insects. Antibiosis, competition, suppression, parasitism, induced resistance, predation, and hypovirulence are some of the common mechanisms for the biological control of various plant diseases (Hoggag and Mohamed, 2007; Khokhar et al., 2012).

1.4.1.3.1 Biopesticides

Different types of microbial sources (viruses, bacteria, fungi) are used for the control of various insects, but only few of them are commercially applicable. *B. thuringiensis*, a spore-forming bacterium, is widely used for the control of certain pests in crops as it produces crystal proteins (δ-endotoxins) that are processed into toxins by the proteases present in an insect's midgut, which ultimately results in its death. Several different types of crystal proteins are produced by different bacterial strains having specific insect targets (Rossas Garcia, 2009). The hazelnut leaf holer, a serious pest of hazelnut and oat trees, has been controlled with four different *Bacillus* spp. (Demir et al., 2002). Baculoviruses are regarded as safe and selective bioinsecticides and are being used for the insect control on a commercial scale (Trang and Chaudhari, 2002). The nucleopolyhedrosis virus (NPV) and granulosis virus (GV) have been commercially used for the control of *Helicoverpa armigera* and *Spodoptera litura*, respectively (Singh, 2010). Similarly, entomophagous fungi (*Metarhizium anisopliae* and *Beauveria bassiana*) are used to control brown plant hoppers, rice bugs, and lepidopterans at a global scale (Chi et al., 2005), whereas *Hirsutella thompsonii* and *Verticillium lecanii* are used to control citrus mites and aphids, respectively (Singh, 2010).

1.4.1.3.2 Bioherbicides

The presence of undesirable weeds along with cultivated crops causes significant loss annually not only in terms of quantity but also from the ecological perspective. Bioherbicides are phytopathogenic or microbial toxins used to control harmful weeds biologically. Among different microbial sources, fungal pathogens are found to be attractive as biocontrol agents for weed control in various crops. A fungus, *Myrothecium verrucaria*, has been reported to be used as bioherbicide to control sicklepod and various other herbaceous weeds (Walker and Tilley, 1997; Hoagland et al., 2007). A wood rotting basidiomycetes, *Chondrostereum purpureum*, is popular in the Netherlands for the control of *Prunus* and *Populus* spp. (De Jong, 2000). *M. verrucaria* in combination with glyphosate has been observed to be effective in the control of kudzu (*Pueraria lobata*), redvine (*Brunnichia ovata*), and trumpet creeper (*Campsis radicans*) weeds (Boyette et al., 2006).

1.4.1.3.3 Disease Control Agents

Crop diseases need to be controlled by biocontrol agents to maintain the yield and quality of crops. Among bacteria, the strains of *Pantoea*, *Bacillus*, and *Pseudomonas* are found to be effective biocontrol agents. But fungal-based biocontrol agents have gained more importance as compared to bacterial sources, because of their wide spectrum use in disease control and retaining the crop yield. *Trichoderma* spp. is an efficient biocontrol agent that is commercially produced to control various soilborne fungal pathogens (Pandya and Saraf, 2010).

The occurrence of small wounds or cuts in fruits during the postharvest processing of crops can result in great economic losses due to the attack of harmful pathogens on them. Lactic acid bacteria (LAB) isolated from fresh fruits and vegetables were found to be effective as a biocontrol agent against harmful bacteria and fungi that are mainly responsible for the spoilage of various fruits and vegetables (Trias et al., 2008).

1.4.1.3.4 Biofertilizers

The acquisition of essential nutrients from soil by crops is one of the major constraints in agriculture, particularly in areas prone to various biotic and abiotic stresses (Maiti, 2010). Biofertilizers are the best alternative to tackle this problem as they constitute the live formulations of agriculturally beneficial microorganisms, which can not only mobilize the availability of nutrients to seeds, roots, and soil by their biological activity but also help to improve the fertility of soil (Pandya and Saraf, 2010). Many species of *Rhizobium* are capable of forming root nodules in legumes. *Cyanobacteria, Azolla,* and *Azotobacter* play an important role in agriculture by biological nitrogen fixation. Similarly, *Bacillus* and *Mycorrhizal* spp. help in increasing the availability of phosphorous (Singh, 2010). Some *Mycorrhizal* spp. also result in the improvement of growth, nodulation, and nutrient uptake by different crops as compared to the control, whereas *Aspergillus niger* and *Aspergillus tubingensis* were found to be effective in solubilizing rock phosphates and improving the growth of maize plants in rock phosphate–rich soils (Pandya and Saraf, 2010).

1.5 Biotechnology in Food Processing

The food-processing industry is the oldest and largest industry using biotechnological processes and techniques. The old biotechnological processes used in the food-processing sector are based on fermentation such as bread production, brewing of beer, production of wine, yogurt, cheese, etc. There have been continuous and consistent efforts in exploiting microorganisms for the production of various value-added products such as food, feed, enzymes, amino acids, vitamins, single cell proteins (SCP), polysaccharides, organic acids, beverages, etc. (Figure 1.4). Biotechnology in the food-processing sector has not only provided quality foods that are tasty, nutritious, convenient, and safe but is also focusing further on improving processes to provide even

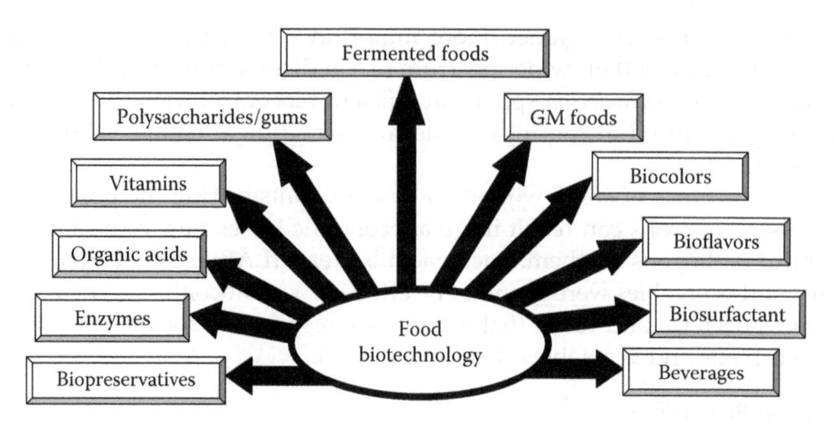

FIGURE 1.4
Role of biotechnology in food processing.

more nutritious, palatable, and stable food (John Innes Centre, 1998). Further development of the food-processing sector based on biotechnology depends upon the refinement of existing processes, such as fermentation, applications of immobilized biocatalyst technology, production of additives, and processing aids, as well as the development of other novel processes and techniques for use in food processing (Knorr and Sinske, 1985).

The significant role of biotechnology in different sectors of food processing is discussed in the following sections.

1.5.1 Milk Processing

The application of biotechnology in making dairy products can be traced back thousands of years. The role of fermented milk in human nutrition is well documented and the virtues of these products were known to man even during the ancient days of civilization. Fermented milk products have long been an important component of nutritional diet. The consumer's interest in fermented milk products is gaining momentum due to the development of new food-processing techniques, changing social attitudes, and scientific evidence of health benefits of certain fermented product ingredients (Gandhi, 2000). In recent years, the demand of fermented milk products has considerably increased because of their distinct benefits such as increased digestibility, nutritive value, reduced lactose content, increased calcium and iron absorption, protection against gastrointestinal infection, immune system stimulation, and lowering of serum cholesterol (Grunewald, 1992; Arora et al., 2000). Several types of fermented milk products are popular all over the world (Table 1.2). The wide degree of variety seen in fermented dairy products comes from differences in the type and condition of the milk used, bacterial strains employed, and the methods of fermentation, processing, storage, and ripening (Angold et al., 1989). Historically, fermentation process

TABLE 1.2

Some Important Fermented Milk Products

S. No.	Name	Type of Milk	Microorganisms Involved
1	Curd	Buffalo's or cow's milk	*Lactococcus lactis* subsp. *lactis* *Lactobacillus delbrueckii* subsp. *bulgaricus* *Lactobacillus plantarum* *Streptococcus lactis* *Streptococcus thermophilus* *Streptococcus cremoris*
2	Yogurt	Cow's milk	*Lactobacillus acidophilus* *Streptococcus thermophilus* *Lactobacillus bulgaricus*
3	Cultured butter milk	Buffalo's or cow's milk	*Streptococcus lactis* subsp. *diacetylactis* *Streptococcus cremoris*
4	Lassi	Buffalo's or cow's milk	*Lactobacillus bulgaricus*
5	Acidophilus milk	Cow's milk	*Lactobacillus acidophilus*
6	Bulgarian butter milk	Cow's milk	*Lactobacillus delbrueckii* subsp. *bulgaricus*
7	Shrikhand	Buffalo's or cow's milk	*Streptococcus thermophilus* *Lactobacillus bulgaricus*
8	Kumiss	Mare's, camel's, or ass's milk	*Lactobacillus acidophilus* *Lactobacillus bulgaricus* *Saccharomyces* spp. *Micrococci*
9	Kefir	Sheep's, cow's, goat's, or mixed milk	*Streptococcus lactis* *Streptococcus cremoris* *Saccharomyces* spp. *Micrococci*
10	Leben	Goat's or sheep's milk	*Streptococcus lactis* *Streptococcus thermophilus* *Lactobacillus bulgaricus* Lactose fermenting yeast
11	Cheese	Cow's, Buffalo's, or goat's milk, sheep milk	*Lactococcus lactis* subsp. *lactis, Lactococcus lactis* subsp. *cremoris, Lactococcus lactis* subsp. *diacetylactis, Streptococcus thermophilus, Lactobacillus delbueckii* subsp. *bulgaricus* *Propionibacterium shermanii, Penicillium roqueforti*, etc.

involved unpredictable and slow souring of milk caused by the organisms inherently present in the milk. However, modern techniques involve the use of specific lactic microorganisms to carry out specific fermentation under controlled conditions to produce products of superior physical, chemical, nutritional, and sensory qualities (Gandhi, 2000).

The increasing awareness of consumers toward healthy and therapeutic products has resulted in the incorporation of probiotic cultures in different milk products, since these bacterial cultures have been claimed to exert beneficial effect on the human body. Therefore, bacterial cultures can play a dual role in transforming milk into a diverse range of fermented dairy products (yogurt, cheese, kefir, etc.) and contributing to the important role of colonizing bacteria. Dairy products represent a major portion of the total European functional foods market and are at the forefront of probiotic developments. The probiotic bacteria used in commercial products are mainly members of the genera *Lactobacillus* and *Bifidobacterium* (Reuter, 1997; Heller, 2001; Dardir, 2012). In the dairy markets, yogurt with its existing health image is well positioned to capitalize on the growth of healthy foods (Stanton et al., 2001). Many cultured dairy foods such as bioghurt, yakult, actimel, activia, danito, chamyto, etc. are already marketed as therapeutic and dietetic products (Siro et al., 2008; Granato et al., 2010).

New biotechnological techniques (genetic techniques) have also been used to manipulate and improve the properties of bacteria that are important to the dairy industry (Coffey et al., 1994). Both classical and molecular approaches have been employed to improve strains involved in yogurt and cheese production for increasing the efficiency of substrate conversion, regulating the production of flavor-enhancing metabolites, and developing starter cultures resistant to bacteriophage and bacteriocin attack.

1.5.2 Cereal Processing

Cereal grains constitute one of the most important sources of dietary proteins, carbohydrates, vitamins, minerals, and fiber for people all over the world. However, the nutritional quality of cereals and the sensorial properties of their products are sometimes inferior or poor in comparison to dairy products, due to lower protein content, deficiency of certain essential amino acids (lysine), presence of antinutrients, low starch availability, etc. (Chavan et al., 1989).

Fermentation is the most simple and economical way of improving the nutritional value, sensory properties, and functional qualities of cereals (Blandino et al., 2003). It leads to general improvement in the shelf life, texture, taste, and aroma of the final product. Several volatile compounds are formed during cereal fermentations, which contribute to a complex blend of flavors in the final products (Chavan et al., 1989). Bread making, the most important ancient craft, is one of the oldest applications of biotechnology in cereal processing. From the biotechnological context, besides fermentation, the naturally occurring enzymes or enzymes added in the flour play a very important role in the process (Angold et al., 1989). Enzymes may be added in the dough in the form of constituents of other natural materials such as malt or soybean flour or as special enzyme preparations. Carbohydrates are the important functional enzymes in baking, α-amylase and β-amylase being the major catalysts in the conversion of starch into fermentable sugars. Besides these, other groups of enzymes such as proteases, lipases, lipoxidases, and glucose oxidases can also influence the baking process.

Biotechnology has significantly contributed to baking in the selection and improvement of yeast strains besides the production of efficient enzymes. It can improve the baking process with improvements in cereal grains and starter culture through recombinant DNA technology, use of enzymes as processing aids, and application of advanced batch and continuous fermentation technologies (Linko et al., 1997).

Traditional fermented cereal food products prepared from most common types of cereals such as rice, ragi, maize, wheat, corn, or sorghum are well known in many parts of the world. The popular examples of these food products include idli, dosa, dhokla, soy sauce, kishk, tarhana, ogi, kenkey, pozol, ijera, kisra, boza, etc. (Blandino et al., 2003; Yegin and Fernández-Lahore, 2012). The microbiology of many of these products is quite complex and in most of these products, the microflora responsible for fermentation is indigenous and includes strains of LAB, yeast, and fungi. The common fermenting bacteria are species of *Leuconostoc, Lactobacillus, Streptococcus, Pediococcus, Micrococcus, Bacillus,* and *Flavobacterium.* The fungal genera *Aspergillus, Paecilomyces, Cladosporium, Fusarium, Penicillium,* and *Trichothecium* are most frequently found in certain products. The common fermenting yeasts such as *Saccharomyces, Trichosporon, Candida,* and *Kluyveromyces* are associated with traditional fermentations of cereals (Steinkraus, 1998; Soni and Arora, 2000).

With the newer developments in functional foods, some new cereal-based fermented foods such as yosa that are considered as probiotic products have been developed (Wood, 1997). Yosa is a new snack food made from cooked oat bran pudding and fermented with LAB and *Bifidobacteria.* After fermentation, it is flavored with sucrose or fructose and fruit jam (Salminen and Von Wright, 1998). It is mainly consumed in Finland and other Scandinavian countries. Other traditional cereal-based fermented foods have been modified to aid the control of some diseases.

Besides production of fermented cereal products, biotechnology has played significant role in the starch hydrolysis industry through application of enzymes such as α-amylase, β-amylase, glucoamylase, and hemicellulases (Arora et al., 2000). A number of enzymes are presently being employed in the food industry to hydrolyze starch slurries to produce glucose, maltose syrups, and maltodextrins. The production of high fructose corn syrup (HFCS) is one of the most important commercial applications of enzymes.

1.5.3 Fruit and Vegetable Processing

Fruits and vegetables are highly perishable commodities and need to be used within short duration of time period after harvesting, thus needing proper postharvest care. Therefore, there is need to avoid the wastage of this highly nutritive and perishable commodity especially during the glut season using suitable processing technologies. Fruits, which are naturally rich in juices and sugars, are a suitable medium for growth of yeasts as well as alcoholic fermentation (Montet et al., 1999). Vegetables have a low sugar content and neutral pH and provide a natural medium for growth of microflora. Lactic acid is empirically found to stabilize such flora and has been used for this purpose since ancient times. Therefore, the preservation of fruits and vegetables through fermentation involves positive action of microflora, and thus biotechnological processes are potentially of interest for the further developments and improvements with new techniques (Angold et al., 1989). The commonly used fermented vegetable products include sauerkraut, olives, and pickles.

Generally, the sequence of prevalence of microflora during fermentation of vegetables includes initial dominance of salt-resistant LAB, heterofermentative *Leuconostoc* spp., and *Lactobacillus brevis* followed by homofermentative *Lactobacillus plantarum* and *Pediococcus* spp., which result in acid production and lowering of pH, thereby favoring the growth of yeasts. Fermentation by the aforementioned microflora results in the conversion of sugars present in vegetables to lactic acid, acetic acid, carbon dioxide, and several other metabolites, which inhibit the growth of undesirable microorganisms, thereby ensuring a long shelf life of the fermented products (Soni and Arora, 2000).

Microorganisms during fermentation are helpful in removal or detoxification of naturally occurring toxins and anti-nutritional compounds present in many fruits and vegetables (Battcock and Azam-Ali, 1998). Fermentation can also enhance the nutritional value of a food product through increased vitamin levels in the final product, besides flavor and appearance of the final product. Microorganisms contain certain enzymes, such as cellulases, which are incapable of being synthesized by humans. Microbial cellulases hydrolyze cellulose into sugars which are then readily digestible by humans. Similarly, pectinases soften the texture of foods and liberate sugars for digestion. Thus, fermentation food products have improved digestibility as compared to unfermented foods (Parades-Lopez, 1992; Kovac and Raspor, 1997).

Biotechnology has also greatly benefited fruit juice industry through the application of enzymes. The enzymes like pectinases, cellulases, hemicellulases, amylases, and proteinases have contributed to the improvement of long-established processes, giving increased yields and widening the range of the available products (Angold et al., 1989). Another major application of enzymes is removal of bitterness from citrus fruits through use of naringinase preparations, which is being tried world over (Arora et al., 2000). The hydrolysis of limonin (causing delayed bitterness) using limonin dehydrogenase is also being tried. Furthermore, the latest biotechnological techniques, such as recombinant DNA technology, have allowed the selection and production of enzymes with improved properties, high concentration, and specificity.

1.5.4 Meat and Fish Processing

Biotechnology has played a significant role in meat and fish processing and preservation through fermentation and enzymes. Meat and fish being highly perishable, their preservation through fermentation has been practiced since ancient times. Meat fermentation results in unique and distinctive meat properties such as flavor and palatability, color, microbiological safety, tenderness, and desirable attributes (Ockerman and Basu, 2007). Fermentation in case of meat mainly involves the action of bacteria, which results in lowering of pH, thereby suppressing pathogenic as well as spoilage microflora (Soni and Arora, 2000). Among fermented meat products, fermented sausages are the most popular and important products. The technology of fermentation in meat products has generally remained restricted to either semidry or dry sausages. Some popular examples include genoa, salami, cervelat, thuringer, pepperoni, frischwurst, soudjouk, etc. (Schut, 1978; Wani and Sharma, 1999).

The microflora of fresh refrigerated meat primarily includes psychotropic pseudomonads and small numbers of LAB. The primary genera of bacteria that have effectively been used as meat starter cultures are *Lactobacillus, Pediococcus,* and *Micrococcus* (Bacus, 1986). The primary criteria for the selection of these cultures are their salt and nitrite tolerance, flavor production, acidification rate, and absence of toxicity. Among the non-LAB, *Staphylococcus carnosus* is the most important component of starter preparations (Hammes and Knauf, 1994). *Staphylococcus xylosus* has also been successfully employed in the fermentation of sausages (Stahnke, 1995). Yeasts encourage color development and improve aroma of dry sausages. Yeast growth is sometimes regarded as desirable and surface inoculation of dry sausages with starter cultures consisting of mainly *Debaryomyces* yeast is also carried out (Bacus, 1986; Lücke, 2000). *Debaryomyces hansenii* is the most common species of yeast in the fermented meat products. *Candida formata* as a starter preparation grows mainly on the surface and on the outer part of the sausages (Hammes and Knauf, 1994). Molds also contribute to the characteristic aroma and surface appearance of mold ripened dry type sausages. Generally, *Penicillium chrysogenum,* a nontoxic strain, is applied as mold starter cultures (Wani and Sharma, 1999)

Besides production of fermented meat sausages, biotechnology is playing an important role in meat tenderization through application of enzymes especially proteases. Papain, bromelain, and ficin are being used on a commercial scale for meat tenderization (Lawrie, 1998; Grzonka et al., 2007). Furthermore, proteases are also being employed in bone cleaning and flavor formation in meat industry (Lantto et al., 2009). Lipase can be used for flavor development in sausage production.

1.5.5 Production of Single Cell Protein

Microorganisms have been used by people world over for centuries, but primarily in the preparation of fermented food and beverages. However, the direct use of microorganisms as protein sources was certainly an interesting perception that led to the development of SCP. SCP is the dried cells of microorganisms such as algae, actinomycetes, bacteria, yeasts, molds, and higher fungi grown on a large scale for use as protein sources in human foods or animal feeds (Litchfield, 1983). These are excellent source of proteins that are comparable to other sources of proteins. The algae of genus *Spirulina* is the most popular example of SCP. Extensive research has been conducted into fermentation science and technology for SCP production (Solomons and Litchfield, 1983). Many raw materials have been tested as carbon and energy sources for SCP production. A number of microorganisms (bacteria, yeast, and molds) have been used for the production of SCPs though each has its own advantages and disadvantages.

Microorganisms are capable of using a wide range of substrates both from renewable and nonrenewable sources. To improve the economics of the process, the alternative sources like cellulosic materials, food-processing waste, or other low-cost materials are being encouraged for SCP production. The SCP production processes utilizing different raw materials have been carried out on a commercial scale using various organisms. The substrates used and producer organisms include sulfite waste liquor (*Paecilomyces variotii, Candida utilis*), molasses (*Saccharomyces cerevisiae*), and cheese whey (*Kluyveromyces fragilis*), while the Symba process developed in Sweden utilizes starchy wastes, for example, wastewater from potato-processing plant, by combining two yeasts, *Endomycopsis fibuligera* and *C. utilis* (Skogman, 1976). High capital and operating costs and the need for extensive nutritional and toxicological assessments have limited the development and commercialization of new processes for SCP production (Litchfield, 1983). The concentration of nucleic acid in SCP is a major concern for its intake as dietary food. Some of these problems may be solved through possible pathways employing genetic engineering as well as using suitable nucleic acid reduction processes (Goldberg, 1988; Martinez et al., 1990; Abou-Zeid et al., 1995).

1.5.6 Production of Food Additives

Due to consumer awareness about health and environment, the use of chemical additives in the food and pharmaceutical industries is the cause of concern. The use of biological sources for the extraction and production of food additives has shown encouraging response to meet the consumer expectations. The development of biotechnological processes is an initiative in the direction of production of functional food additives having better nutritional and shelf life as compared to synthetic ones. The major food additives that can be produced through biotechnological routes (Table 1.3) are discussed in brief later and in detail in Chapters 8, 10, 12, and 13.

1.5.6.1 Production of Biopigments

The use of food colorants (pigments) in the food products is a significant and decisive factor for both food manufacturers and consumers in determining the acceptability of such food products in the market (Spears, 1988). A variety of plants, animals, and microorganisms have been reported as the

TABLE 1.3

Food Additives and Their Microbial Sources

Food Additive	Microorganism(s) Involved
Biopigments	
Yellow, orange, red	*Monascus pilosus*, *M. purpureus*, *M. ruber*
Astaxanthin	*Xanthophyllomyces dendrorhous*
Arpink red	*Penicillium oxalicum*
β-Carotene	*Blakeslea trispora*, *Mucor circinelloides*, *Phycomyces blakesleeanus*, *Eremothecium ashbyii*
Monascorubramine	*Monascus purpious*
Lycopene	*Flavobacterium sporotrichioides*, *Bradyrhizobium* spp.
Zeaxanthin	*Flavobacterium* spp.
Isorenieratene	*Brevibacterium aurartiacum*
Bioflavors	
Monosodium glutamate	*Micrococcus glutamicus*
Glutamic acid	*Corynebacterium glutamicum*
Terpenes	*Ceratocystis* spp., *Ascoidea hylecoeti*, *Cronartium fusiforme*, *Phellinus* spp., *Pleurotus euosmus*
Esters	*Saccharomyces* spp., *Hansenula* spp., *Candida utilis*
Lactones	*Sporobolomyces odorus*, *Trichoderma viride*
Pyrazines	*Bacillus subtilis*, *Corynebacterium glutamicum*
Vanillin	*Aspergillus niger*
Organic acids	
Lactic acid	*Lactobacillus* spp.
Acetic acid, gluconic acid	*Acetobacter* spp.
Citric acid	*Aspergillus* spp.
Benzoic acid	*Streptomyces maritimus*
Propionic acid	*Propionibacterium freudenreichii*
Gums	
Xanthan	*Xanthomonas campestris*
Pullulan	*Aureobasidium pullulans*
Dextran	*Leuconostoc mesenteroides*, *L. dextranicum*
	Pseudomonas spp., *Azotobacter* spp.
Alginates	*Sphingomonas paucimobilis*, *Pseudomonas elodea*
Gellan gum	*Sphingomonas paucimobilis*, *Sphingomonas elodea*
Amino acids	
Glutamic acid	*Corynebacterium glutamicum*, *Arthrobacter*
Lysine	*Brevibacterium lactofermentum*
Threonine	*Escherichia coli*
Methionine	*Corynebacterium glutamicum*
Aspartic acid	*Corynebacterium* spp.
Phenylalanine	*Corynebacterium glutamicum*
Tryptophan	Recombinant *E. coli* and *Corynebacterium glutamicum*

(continued)

TABLE 1.3 (continued)

Food Additives and Their Microbial Sources

Food Additive	Microorganism(s) Involved
Biopreservatives	
Nisin	*Lactococcus lactis*
Lacticin 3147	*Lactococcus lactis*
Pediocin	*Pediococcus acidilactici*
Biosurfactants	
Rhamnolipids	*Pseudomonas* spp.
Sophonolipids	*Torulopsis* spp.
Trehaloses	*Mycobacterium* spp., *Rhodococcus* spp.
Phospholipids	*Arthobacter* spp., *Corynebacterium* spp.
Fatty acids	*Candida* spp., *Pseudomonas* spp., *Micrococcus* spp., *Candida* spp., *Penicillium* spp., *Aspergillus* spp.
Lipopeptides	*Bacillus* spp., *Streptomyces* spp., *Corynebacterium* spp., *Mycobacterium* spp.
Vitamins	
Riboflavin	*Bacillus subtilis, Eremothecium ashbyii, Ashbya gossypii*
Cobalamin	*Propionibacterium shermanii, P. freudenreichii*
Ascorbic acid	*Gluconobacter oxydans, Corynebacterium* spp.
Carotenoids	*Blakeslea trispora, Dunaliella* spp.
L-Carnitine	*Agrobacterium* spp., *Sporobolomyces salmonicolor*
Vitamin K_2	*Flavobacterium* spp.
Dihomo-γ-linolenic acid	*Mortierella alpine*
Arachidonic acid	*Mortierella alpine*
Eicosapentaenoic acid	*Mortierella alpine*

natural source of biocolorants. Natural colorants have been derived from plants such as pepper, red beet, grapes, and saffron (FDA/IFIC, 1993; Bridle and Timberlake, 1997). Nowadays, different microorganisms are being used for the production of food grade pigments such as red color from *Monascus* spp., astaxanthin from *Xanthophyllomyces dendrorhous*, Arpink red color from *Penicillium oxalicum*, riboflavin from *Ashbya gossypii*, β-carotene from *Blakeslea trispora*, and carotenoids from *Rhodotorula glutinis* and *Phaffia rhodozyma* (Dufosse, 2006; Frengova and Beshkova, 2009). Other explored fungal sources for the production of β-carotene are *Mucor circinelloides* and *Phycomyces blakesleeanus*. Riboflavin (vitamin B_2) is a yellow food colorant and can also be produced through fermentation using fungi *Eremothecium ashbyii* and *A. gossypii* (Jacobson and Wasileski, 1994; Stahmann et al., 2000). Besides the aforementioned microorganisms, other microorganisms such as *Serratia*, *Streptomyces*, and *Sporidiobolus* can also be employed to produce biopigments in good amounts (Kim et al., 1997; Valduga et al., 2009). Interested readers may get detailed information from review articles published by Dufosse (2006) and Chattopadhyay et al. (2008) on the production of biopigments in Chapter 12.

1.5.6.2 Production of Bioflavors

Flavors and fragrances find a wide range of applications in food, feed, cosmetic, chemical, and pharmaceutical industries. Different biotechnological routes such as microorganisms, enzymatic, and tissue culture techniques have been employed for the production of flavors (Janssens et al., 1992; Vandamme and Soetaert, 2002; Dubal et al., 2008). Microbial fermentation imparts particular flavor to many of the fermented foods due to the action of microbial enzymes on the food constituents. Fermented dairy products such as curd, yogurt, and cheese are common examples. This property of microbial fermentation has been further exploited for the production of flavor compounds (Janssens et al., 1992). The best known microbial product is monosodium glutamate (MSG) produced from *Micrococcus glutamicus* (Angold et al., 1989).

Flavoring compounds such as esters, diacetyl, pyrazines, lactones, terpenes, etc. have been produced through microbial technology (Singhal and Kulkarni, 1999). In addition, some 5′ nucleotides are produced by the action of *Penicillium citrinum* on nucleic acid. The large-scale propagation and fermentation of glutamic acid production by *Corynebacterium glutamicum* have been well studied. The microorganisms also offer an opportunity to investigate the possibility of coproducing glutamate and tetramethylpyrazine in a fermentation process (Demain et al., 1967). Yeasts also have the potential to produce de novo flavors. *Sporobolomyces odorus* (now *Sporidiobolus salmonicolor*) is the typical example, which produces a high variety of lactones such as γ-decalactone or δ-lactones (Tahara et al., 1972). *Kluyveromyces lactis* and *Williopsis saturnus* also have the potential to synthesize large amounts of terpenes and fruity ester flavors, respectively. *K. lactis* produces de novo fruity, floral flavor terpenes such as citronellol, linalool, and geranio (Vandamme and Soetaert, 2002; Vandamme, 2003). Solid-state fermentation (SSF) and biotechnological tools like immobilization have also been applied in production of flavoring compounds and similar biotransformations.

1.5.6.3 Production of Organic Acids

Organic acids constitute important biochemicals, which are widely used in the food, pharmaceutical, and chemical industries. Several microorganisms possess the ability to convert carbohydrates to organic acids with high yields. For example, *Lactobacillus* spp. produce lactic acid, *Acetobacter* spp. produce acetic acid and gluconic acid, and *Aspergillus* spp. produce citric acid. Microbial fermentation has the advantage that, by choosing a strain of LAB producing only one of the isomers, an optically pure product can be obtained, whereas synthetic production always results in a racemic mixture of lactic acid (Litchfield, 1996).

The choice of an organism primarily depends on the nature of carbohydrate to be used for fermentation. *Lactobacillus delbrueckii* subsp. *delbrueckii* is capable of fermenting sucrose, whereas *L. delbrueckii* subsp. *bulgaricus* is able to use lactose. *Lactobacillus helveticus* has the ability to utilize both lactose (Gasser, 1970) and galactose. *Lactobacillus amylophilus* is able to ferment starch (Nakamura and Crowell, 1979) and *Lactococcus* (*Lc.*) *lactis* can ferment glucose, sucrose, and galactose (Narayanan et al., 2004).

Citric acid is produced either in an anhydrous form or as the monohydrate at a commercial level. Generally, citric acid production is carried out using submerged fermentation of starch or sucrose-based media by filamentous fungus *A. niger* (Vandenberghe et al., 2000). SSF has been an alternative method for citric acid production using agro-industrial residues (Prado, 2002). Among different microorganisms, *A. niger* is the most popular microbial source for citric acid production (Grewal and Kalra, 1996). Although *A. niger* is the traditional producer of citric acid, other microorganisms have also been used for citric acid production.

During the past decades, the use of yeasts as citric acid producers has attracted the attention of researchers (Arzumanov et al., 2000; Roukas, 2006). Yeasts have some advantages as compared to *A. niger* strains such as the fermentation time is short (half the time of *A. niger* fermentation) and thus productivity is better. Moreover, yeast strains are insensitive to variations in molasses and can be used for developing a continuous process (Kapoor et al., 1983; Milsom, 1987). A variety of yeasts have been tested for the citric acid production (Kapoor et al., 1983). Among all these, strains of *Candida* are widely used for the production of citric acid. Some other yeast strains belonging to genera *Yarrowia, Hansenula, Pichia,* and *Torula* have also been tested for citric acid production.

Many fermentative bacteria are known for the production of acetic acid; however, acetic acid bacteria are the popular bacteria for its commercial production. Examples of commercially used strains are *Acetobacter aceti, Acetobacter pasteurianus, Acetobacter peroxidans, Gluconobacter oxydans,* etc. (Ghose and Bhadra, 1985). Gluconic acid is a noncorrosive, nonvolatile, mild organic acid that finds applications in the food, feed, beverage, textile, pharmaceutical, and construction industries (Singh and Kumar, 2007). Among various microbial fermentation processes, the method utilizing the fungus *A. niger* is one of the most widely used. The carbohydrate source for gluconate production is either glucose crystals or dextrose syrup (Blom et al., 1952; Hatcher, 1972). Besides fungal species, bacterial sources such as *Pseudomonas, Acetobacter, Gluconobacter,* etc. have also been explored for gluconic acid production (Ramachandran et al., 2006; Singh and Kumar, 2007).

Propionic acid is a naturally occurring carboxylic acid and is used in the production of artificial fruit flavors, pharmaceuticals, cellulose acetate propionate, and preservatives for food and animal feed. Propionic acid bacteria since long have been used in the dairy industry. These bacteria play important roles in the development of the characteristic flavor in Swiss type cheeses. The microorganisms used for the production of propionic acid are *Propionibacterium* spp., *Clostridium propionicum,* and *Megasphaera elsdenii* (Sethi and Maini, 1999). Different substrates such as glucose, maltose, sucrose, and whey lactose have been employed for propionic acid production using *Propionibacteria* spp. (Colomban et al., 1993; Quesada-Chanto et al., 1994).

1.5.6.4 Production of Gums

Microbial exopolysaccharides (gums) have attracted worldwide attention due to their novel and unique physical properties. These have multifarious applications in industries, especially in food and pharmaceutical industries as emulsifiers, stabilizers, binders, gelling agents, lubricants, film formers, and thickening and suspending agents (Sutherland, 1998; Marwaha and Arora, 1999). Among the biopolymers that have been the subject of extensive studies are xanthan, pullulan, alginates, and gellan.

Xanthan is one microbial polysaccharide that is being produced commercially on a very large scale. It is an extracellular heteropolysaccharide that is produced by the *Xanthomonas campestris* commonly from glucose or sucrose through fermentation (Marwaha and Arora, 1999; Garcia-Ochoa et al., 2000). Besides macronutrients such as carbon and nitrogen, *X. campestris* needs several nutrients, including micronutrients (e.g., potassium, iron, and calcium salts), to produce xanthan gum. It has also been observed that nitrogen, phosphorous, and magnesium influence the growth, whereas nitrogen, phosphorous, and sulfur influence the production of xanthan by *X. campestris* (Garcia-Ochoa et al., 1992).

Pullulan is one of the few neutral, water-soluble microbial polysaccharides that can be produced in large quantities through fermentation (Pollock et al., 1992). It has a wide range of commercial and industrial applications in many areas such as food, healthcare, pharmaceuticals, and even lithography (Singh et al., 2008). This polymer is produced by strains of *Aureobasidium pullulans* from different substrates (Marwaha and Arora, 1999). The quantity of polymer produced by fungi has been affected by both type and concentration of nitrogen sources of the medium (Auer and Seviour, 1990).

Dextran is a D-glucose polymer, which is commercially produced using *Leuconostoc mesenteroides* and *Leuconostoc dextranicum* with sucrose as substrate (Marwaha and Arora, 1999). This polymer is synthesized extracellularly and the enzyme involved in its synthesis is extracellular dextran sucrase. Microbial alginates have physical properties quite similar to those of algal alginates and can be used for the same applications as algal alginates. The microbial alginates possess *O*-acetyl groups attached to D-mannuronic acid, which is absent in algal alginates and is produced by *Pseudomonas* and *Azotobacter* species (Pena et al., 2006). The process for the production of heteropolysaccharides consisting of D-mannuronic acid and L-guluronic acid from *A. vinelandii* has been commercialized (Marwaha and Arora, 1999).

Gellan gum is also gaining much importance in food, pharmaceutical, and chemical industries due to its novel properties. It is a comparatively new gum, which can be produced by fermentation using *Sphingomonas paucimobilis* and *Pseudomonas elodea* (Giavasis et al., 2000; Bajaj et al., 2007). Different strains of *S. paucimobilis* have been employed for gellan gum production (Kang et al., 1982; Bajaj et al., 2006). The media used for production of gellan gum contains carbon source, nitrogen source, and inorganic salts. The exact quantity of carbon utilization depends in part upon the other ingredients of the medium (Kang et al., 1982). Carbohydrates such as glucose, fructose, maltose, sucrose, and mannitol can be used either alone or in combination as carbon source.

1.5.6.5 Production of Amino Acids

The interest in microbial production of amino acids has increased considerably because of their increased use in food, feed, flavorings, seasonings, cosmetics, and medicines and as intermediates in chemical industry. Microbial and enzymatic biosyntheses have advantages over chemical synthesis as the former yield biologically active isomers, while chemical synthesis produces a racemic product having equal amounts of D- and L-isomers, which are difficult to separate (Marwaha and Arora, 1999). The main examples of amino acids produced through fermentation are glutamic acid, produced by *C. glutamicum* and *Arthrobacter* spp., and lysine from *Brevibacterium lactofermentum* and *C. glutamicum*. The presence of biotin in the medium is the significant factor in glutamic acid fermentation, which is an essential growth factor for glutamic acid bacteria. Lysine is produced from aspartate, which also serves as a starting material for the manufacture of homoserine, methonine, threonine, and L-isoleucine. Cane molasses is commonly used as a carbon source in the production of lysine, though other carbohydrate materials can also be used (Hermann, 2003). Different inorganic pure ammonia and ammonium salts like ammonium sulfate and complex organic components like peptone hydrolysates or corn steep liquor can be used as nitrogen sources. Besides these, inorganic nutrients and essential elements are also required for growth of microorganisms for the production of lysine (Coello et al., 2000). The history of strain improvement for lysine production demonstrates the development of genetic techniques from simple selection, through progressive mutagenesis, to protoplast fusion and recombinant DNA technology (Angold et al., 1989; Ikeda, 2003).

Aspartic acid is also gaining importance as it is the key constituent in the production of aspartame (sweetener). The production of L-aspartic acid is being carried out by the enzymatic (using aspartase) addition of ammonia to fumaric acid (Yokote et al., 1978; Marwaha and Arora, 1999).

1.5.6.6 Production of Vitamins

Vitamins are essential dietary nutrients for human beings and their deficiency can lead to certain diseases. Keeping in view their importance, different biotechnological processes (i.e., fermentation and microbial/enzymatic transformation) for their production have been developed. Biotechnological processes have the advantage of being fully biological, single-step methods as compared to chemical multistep processes. Different vitamins have been produced through biotechnological means. The most important vitamins that are produced commercially through fermentation processes are riboflavin and vitamin B_{12}.

The organisms used for riboflavin production are bacterium *Bacillus subtilis*, yeast *Candida famata*, and the fungi *Eremothecium ashbyii* and *A. gossypii* (Crueger and Crueger, 1989; Lim et al., 2001). *B. subtilis* and *C. famata* utilize glucose, but the fungus *A. gossypii* requires plant oil as its sole carbon source. The supplementation of guanosine triphosphate, which is a process-limiting precursor, is required for the improved riboflavin production.

Vitamin B_{12} (cobalamin) is currently being synthesized through microbial bioprocesses (Rehm and Reed, 1986). Today, cobalamin is exclusively produced by fermentation processes, using selected microorganisms. Although a number of microorganisms from different genera are known for vitamin B_{12} synthesis, mainly the two genera, *Propionibacterium* (*Propionibacterium shermanii* and *Propionibacterium freudenreichii*) and *Pseudomonas* (*Pseudomonas denitrificans*), are used for industrial production of vitamin B_{12} because of their rapid growth and high productivity of selected mutants (Marwaha and Sethi, 1984; Marwaha and Arora, 1999).

L-Ascorbic acid is another important vitamin that is also synthesized through biotechnological routes. The preferred microorganisms for production of ascorbic acid are bacteria and fungi. *G. oxydans* and *Corynebacterium* spp. are organisms used for the production of vitamin C on a commercial scale (Boudrant, 1990). Carotenoids have also been produced through microbial fermentation, and among the different microorganisms, high yields have been obtained with the fungus *B. trispora* and the alga *Dunaliella* (Ninet and Renaut, 1979).

1.5.6.7 Production of Bacteriocins

The bacteriocins are ribosomally synthesized antimicrobial proteinaceous compounds. These are produced by almost all microorganisms and mostly act against closely related species (Klaenhammer, 1993). Although chemical preservatives are being employed in the preservation of food products, increasing awareness among consumers about the potential health risks associated with these substances has stimulated researchers to explore the possibility of using bacteriocins as biopreservatives in food-processing industry.

Bacteriocins have been produced from different microbial sources. The major research interest is shown globally toward the production of bacteriocins from LAB (Stiles, 1996), as they are predominant microflora of many foods and are safe to consume. Many representatives of this group have been consumed for thousands of years and have never reported to pose any health risk to humans and are designated as generally recognized as safe (GRAS) organisms. The popular bacteriocin-producing LAB belong to genera of *Lactococcus* (nisin), *Pediococcus* (pediocin), *Lactobacillus* (lactocin), *Enterococcus* (enterocin), etc.

Bacteriocins of gram-positive bacteria are abundant and more diverse as that of gram-negative bacteria (Tagg et al., 1976; Jack et al., 1995) and therefore bacteriocins produced by the former group of bacteria have been largely studied and also biochemically and genetically characterized (Navaratna et al., 1998). Most of the known ribosomally synthesized antimicrobial peptides produced by bacteria have been identified and those produced by LAB, in particular the lantibiotics, are the most extensively studied (Papagianni and Anastasiadou, 2009). Nisin, the lantibiotic produced by *Lc. lactis* strains, is undoubtedly the most well-known and studied bacteriocin. Nisin is a classical example of a bacteriocin that is currently approved for food use in the United States and has been successfully used for several decades as a food preservative in more than 50 countries (Chen and Hoover, 2003). The other most promising bacteriocins are the pediocins, which are produced by *Pediococcus* spp. Besides this, many recombinant bacteriocins have also been produced and purified from gram-negative bacteria (Riley, 1993; Braun et al., 1994).

1.5.6.8 Production of Biosurfactants

Biosurfactants are natural surface-active compounds, which have a wide range of applications in the petroleum, pharmaceuticals, biomedical, and food-processing industries (Desai and Banat, 1997; Banat et al., 2000). These are amphiphilic molecules containing hydrophobic and hydrophilic moieties, produced by a wide variety of microorganisms, which either adhere to cell surface or are excreted extracellularly in the growth medium (Muthusamy et al., 2008). The widely used microorganisms for biosurfactant production are *Pseudomonas aeruginosa, B. subtilis,* and *Candida bombicola* (Desai and Banat, 1997; Banat et al., 2000). The type and proportion of the biosurfactants produced depends on the microbial strain, the carbon source used, and the culture conditions. Some examples of biosurfactants are rhamnolipids (*P. aeruginosa*), surfactin (*B. subtilis*), emulsan (*Acinetobacter calcoaceticus*), and sophorolipids (*C. bombicola*). Other cultural conditions such as C:N ratio, nitrogen source, and presence of other nutrients influence the production of biosurfactants. Besides synthetic substrates, agro-industrial by-products have also been used for the production of biosurfactants (Makkar and Cameotra, 2002).

1.5.6.9 Production of Oils and Fats

The oils and fats are practically all derived from animal or vegetable sources. The production of fat from microbial sources to meet the demand is certainly a novel idea. A number of oleaginous microorganisms have been isolated and characterized. Microorganisms have the potential to produce lipids, i.e., single cell oil, which are highly specific (Ratledge, 1991). Microorganisms that accumulate more than 20% of their biomass are known as the oleaginous species. Some of the examples include *R. glutinis, Lipomyces starkeyi, Candida* spp., *Mortierella isabellina, Rhizopus arrhizus*, etc. (Wynn and Ratledge, 2005). Much of the potential is being seen in unusual polysaturated fatty acids due to medical benefits (Angold et al., 1989).

1.5.7 Production of Alcoholic Beverages

The history of alcoholic beverages is as old as the history of man and is indeed the oldest application of biotechnology. In economic terms, wine was the first alcoholic beverage to be made. Examples of non-distilled alcoholic beverages are wine and beer, and distilled beverages include whisky, brandy, gin, rum, etc. Currently, yeast (*S. cerevisiae*) is the major ethanol-producing microorganism used all over the world (Saigal, 1993). In the preparation of alcoholic beverages, different raw materials and methods are used and the efficiency of ethanol production is mainly dependent on the nature of yeast strain. The essential step in all the fermentation processes is the conversion of glucose into alcohol by yeast. The enzymes present in yeast catalyze the breakdown of glucose (Esser and Karsch, 1984; Panesar et al., 2000).

Wine is the oldest of the alcoholic beverages made from complete or partial fermentation of grape or any other fruit juice by yeast (*S. cerevisiae* var. *ellipsoideus*) and a subsequent aging process (Amerine et al., 1980). Wine fermentation may be "natural" or "artificial" according to whether the natural yeast flora or an artificially grown yeast culture is used.

Beer basically refers to a fermented beverage, which involves the extraction of malted barley alone or with other cereals or carbohydrate-rich materials, boiling the extracted material with hops, cooling the extract, fermenting it using yeast, and the final product is then clarified and filled in an effervescent condition (Briggs et al., 1981). The hops (*Humulus lupulus*) are added to the mash to provide pleasant bitterness and flavor and to increase the shelf life of beer (Reed, 1987). There are two major types of beers mainly distinguished on the basis of fermentation used for their production. In bottom fermented beers, as the fermentation subsides, bottom yeast (*Saccharomyces uvarum*) tends to flocculate and settle. In top fermented beers, top yeast (*S. cerevisiae*) rises to surface during fermentation where it is recovered by skimming.

Whisky may be defined as the potable spirit obtained by distillation of aqueous infusion of malted barley and other cereals that have been fermented with yeast (*S. cerevisiae*) and normally matured in oak barrels (Russell and Stewart, 1999). This alcoholic beverage has evolved to a refined flavorful beverage primarily due to advancements in technology (Andrews, 1987). The various types of whiskies differ primarily in the nature and proportion of raw materials used in addition to malted barley.

The existing strains available with the alcohol industries have several limitations such as low temperature and alcohol tolerance and limited substrate range (Rogers et al., 1984). Therefore, efforts are going on to overcome these hindrances and to find suitable alternate organisms with better efficiencies. Attempts have been made to improve the existing technology and raw materials and also to have better strains for higher ethanol production through genetic manipulations. Immobilization cell technology has also been applied in the production of alcoholic beverages. Among various ethanologens and other microbes, *Zymomonas*, a gram-negative bacterium, has emerged as a potential organism for ethanol production (Panesar et al., 2000, 2006). Although this microbe has a number of advantages over yeast, it has not been used for ethanol production on a commercial scale due to some technical hindrances.

1.5.8 Food Industry Waste Management

The significant quantity of wastes generated by food-processing industry is rich in biodegradable materials, which makes them suitable substrates for the production of value-added products using biotechnological innovations. Biotechnological techniques have played a significant role in the production of different value-added products using agro-industrial wastes as raw materials. The bioutilization of waste materials through microbial fermentation will not only help in the reduction of environmental pollution but also result in the preparation of bioproducts. Significant reductions of biochemical oxygen demand (BOD) with concomitant production of useful bioproducts such as lactic acid, citric acid, ethanol, enzymes, SCP, etc. have been achieved. Recently, novel products such as exopolysaccharides, biofuels, glycerol, poly-3-hydroxybutyric acid, biosurfactants, and bacteriocins have been generated using food industry wastes through biotechnological means (Kosseva, 2011; Panesar and Kennedy, 2012). See Chapter 16 for detailed information regarding the aforementioned subject.

1.6 Impact of Genomics on Agriculture and Food Processing

Genomics is mainly divided into three broad categories, i.e., structural, comparative, and functional genomics. Structural genomics deals with the three-dimensional structures of proteins on the basis of their amino-acid sequence, whereas comparative genomics is mainly concerned with the analysis and comparison of genomes from different species in order to understand the evolution of species and to further determine the functions of genes and intergenic regions. Functional genomics, the most widely used genomics, deals with the development and application of experimental approaches to assess gene function by global transcript profiling, reverse genetics, and map-based cloning (Gutterson and Zhang, 2004). The main motive of genomics is to find out the genes responsible for the Mendelian and complex traits of any species so that they can be further used in fundamental and applied sciences for the development of products with desired traits (Machuka, 2004).

The remarkable development in genomic strategies allowed the researchers to sequence the complete genome sequence of many species that are further deposited in data banks, many of which are freely accessible on the Internet. These databases facilitate the identification and selection of genes and proteins involved with different traits, development of quantitative trait loci (QTLs) and molecular markers for selection of various traits in different crop species, and also development of transgenic crops (Gutterson and Zhang, 2004). The databases involved with the genome sequence of food-related microorganisms and plants have opened new doors for their functional genomic approaches including transcriptome, proteome, and metabelome analysis along with their structural analysis (Saito and Matsuda, 2010). Many advanced food and pharmaceutical industries have relied upon these databases to identify novel proteins and drug targets. Genomics in food biotechnology has also helped in identifying the resistance mechanism of pathogens and in the identification of probiotic strains (Kuipers, 1991).

One of the best biotechnological tool that allows the researchers for further improvement in the quality and quantity of crop yield is molecular marker–assisted selection (MAS), which involves the selection of plants carrying genomic regions that are involved in the expression of desired trait (Choudhary et al., 2008). These markers are generally near the sequence of desired gene. Thus, genetic linkage between genes and molecular markers will help the researchers in prediction, selection, and transfer of desired gene from one crop to another or between the same crop species and also important in mapping of these markers on different chromosomes, which will further help in identifying the location and distance of important genes from markers, respectively (Hoisington, 2001). Molecular markers have played important role to deal with the underlying concepts of quantitative traits which are the key limitations in classical genetic programs. MAS has become possible for the selection of traits governed by major genes as well as QTLs. Molecular markers that are widely employed to study the genetic pattern in different crops are amplified fragment length polymorphism (AFLP), random amplified polymorphic DNA (RAPD), simple sequence repeats (SSR), restriction fragment length polymorphism (RFLP), and single nucleotide polymorphism (SNP). These molecular markers have their own advantages and disadvantages, thus need careful evaluation before being effectively used for genetic analysis (Saini et al., 2004). Similarly, another approach known as advanced backcross QTL (AB-QTL) has been used in agriculture system for simultaneous detection and transfer of useful QTLs from the wild crop species to the relative popular cultivars for the improvement of trait (Tanksley and Nelson, 1996). This technique has been successfully utilized in many crops like rice, wheat, barley, etc. Further, marker-assisted recurrent selection (MARS) and genome-wide selection (GWS) techniques have been proposed to overcome the problems associated with MAS and AB-QTL. The former approaches are very helpful to get desirable alleles from many QTLs for highly polygenic traits (Semagn et al., 2010).

New genomic approaches that rely upon gene knockout, gene silencing, and gene overexpression technologies are being used to stop the functioning of those genes that are not desirable or to overexpress the desirable gene, respectively, in many crop species (Angaji et al., 2010). These approaches can be helpful to cope up with the economic constraints of crop species that is not possible by traditional breeding programs.

1.7 Summary and Future Prospects

Recent biotechnological techniques have a distinct impact on agriculture systems and food processing and have opened up newer possibilities for rapid improvement in the quantity and quality of foods. These interventions can have a significant impact on the quality, nutritional value, and safety of the plant and animal products that are the basis of the food-processing industry. Almost every ingredient used in food production has living organisms in its source, which could be plant, animal, or microorganism. It is established fact that both plant and animal food sources available to early humans have evolved through natural selection. Nowadays, biotechnology is playing a significant role in improvement of traits of not only microorganisms but also plants and animals. Exciting opportunities in unique ingredients, new product development, cost reductions, and novel processing methods will occur by application of new biotechnological strategies. Therefore, it is believed that biotechnology in future may become a major force for human existence.

Besides the important contributions of biotechnological techniques in agriculture and food processing, it is also important to consider any potential human health or environmental risks. Despite the widespread beneficial applications of genetic engineering in agriculture and food industry, the technology is surrounded by controversy and there is a critical need to examine the safety of the technology in view of any potential human health or environmental risks. Moreover, improvements are needed not only in the characterization of food materials, and processing methods, but also in the quality/safety of processed food products and utilization of agro-industrial waste/by-products.

References

Abbas, K.A., Lasekan, O., and Khalil, S.K. 2010. Recent advances, advantages and limitations of genetically modified foods: A review. *Journal of Food, Agriculture and Environment* 8: 232–236.

Abou-Zeid, A.Z.A., Khan, J.A., and Abulanaja, K.O. 1995. On methods for reduction of nucleic acid contents in a single cell protein from gas oils. *Bioresource Technology* 52: 21–24.

Ahmed, S., Saleem, A.M., and Rauf, I. 2002. Field efficacy of some bioinsecticides against maize and jowar stem borer, *Chilo partellus* (Pyralidae: Lepidoptera). *International Journal of Agriculture and Biology* 4: 332–334.

Amerine, M.A., Kunkee, K.E., Ough, C.S., Singleton, V.L., and Webb, A.D. 1980. *The Technology of Wine Making*, 4th edition. AVI Publishing Co. Inc., Westport, CT.

Andrews, S. 1994. *Food and Beverage Service—Training Manual*. Tata McGraw-Hill Publishing Co. Ltd., New Delhi, India, pp. 87–121.

Angaji, S.A., Hedayati, S.S., Hosein Poor, R., Samad Poor, S., Shiravi, S., and Madani, S. 2010. Applications of RNA interference in plants. *Plant Omics Journal* 3: 77–84.

Angold, R., Beech, G., and Taggart, J. 1989. *Food Biotechnology*. Cambridge University Press, New York.

Arora, J.K., Marwaha, S.S., and Bakshi, A. 2000. Biotechnological advancement in food processing. In: *Food Processing: Biotechnological Applications*, eds. S.S. Marwaha and J.K. Arora, pp. 1–24. Asiatech Publishers Inc., New Delhi, India.

Arzumanov, T.E., Sidorov, I.A., Shishkanova, N.V., and Finogenova, T.V. 2000. Mathematical modeling of citric acid production by repeated batch culture. *Enzyme and Microbial Technology* 26: 826–833.

Auer, D.P.F. and Seviour, R.J. 1990. Influence of varying nitrogen sources on polysaccharide production by *Aureobasidium pullulans* in batch culture. *Applied Microbiology and Biotechnology* 32: 637–644.

Bacus, J.N. 1986. Fermented meat and poultry products. *Advances in Meat Research* 2: 123–164.

Bagwan, D.J., Patil, J.S., Mane, S.A., Kadam, V.V., and Vichare, S. 2010. Genetically modified crops: Food of the future (review). *International Journal of Advanced Biotechnology and Research* 1: 21–30.

Bajaj, I.B., Saudagar, P.S., Singhal, R.S., and Pandey, A. 2006. Statistical approach to optimization of fermentative production of gellan gum from *Sphingomonas paucimobilis* ATCC 31461. *Journal of Biosciences and Bioengineering* 102: 150–156.

Bajaj, I.B., Survase, S.A., Saudagar, P.S., and Singhal, R.S. 2007. Gellan gum: Fermentative production, downstream processing and applications. *Food Technology and Biotechnology* 45: 341–354.

Banat, I.M., Makkar, R.S., and Cameotra, S.S. 2000. Potential commercial applications of microbial surfactants. *Applied Microbiology and Biotechnology* 53: 495–508.

Battcock, M. and Azam-Ali, S. 1998. Fermented fruits and vegetables: A global perspective. FAO Agricultural Services Bulletin No. 134, Food and Agriculture Organization of the United Nations, Rome, Italy.

Blandino, A., Al-Aseeri, M.E., Pandiella, S.S., Cantero, D., and Webb, C. 2003. Cereal-based fermented foods and beverages. *Food Research International* 36: 527–543.

Blom, R.H., Pfeifer, V.F., Moyer, A.J., Traufler, D.H., and Conway, H.F. 1952. Sodium gluconate production–fermentation with *Aspergillus niger*. *Industrial and Engineering Chemistry* 44: 435–440.

Boudrant, J. 1990. Microbial processes for ascorbic acid biosynthesis: A review. *Enzyme and Microbial Technology* 12: 322–329.

Boyette, C.D., Reddy, N.K., and Hoagland, R.E. 2006. Glyphosate and bioherbicide interaction for controlling kudzu (*Pueraria lobata*), redvine (*Brunnichia ovata*), and trumpetcreeper (*Campsis radicans*). *Biocontrol Science and Technology* 16: 1067–1077.

Braun, V., Pilsl, H., and Gross, P. 1994. Colicins: Structures, modes of actions, transfer through membranes and evolution. *Archives of Microbiology* 161: 199–206.

Bridle, P. and Timberlake, C.F. 1997. Anthocyanins as natural food colour-selected aspects. *Food Chemistry* 58: 103–109.

Briggs, D.E., Wadison, A., Statham, R., and Taylor, J.F. 1981. The use of extruded barley, wheat and maize as adjunct in mashing. *Journal of Institute of Brewing* 92: 468–474.

Bulock, J.D. 1987. Introduction to basic biotechnology. In: *Basic Biotechnology*, eds. J.D. Bulock and B. Kristiansen, pp. 3–10. Academic Press, London.

Chaitanya, K.V. and Kumar, U.J. 2006. Edible vaccines. *Sri Ramachandra Journal of Medicine* 1: 33–36.

Chattopadhyay, P., Chatterjee, S., and Sukanta, S.K. 2008. Biotechnological potential of natural food grade biocolourants. *African Journal of Biotechnology* 7: 2972–2985.

Chavan, J.K., Kadam, S.S., and Beuchat, L.R. 1989. Nutritional improvement of cereals by fermentation. *Critical Reviews in Food Science and Nutrition* 28: 349–400.

Chen, H. and Hoover, D.G. 2003. Bacteriocins and their food application. *Comprehensive Reviews in Food Science and Food Safety* 2: 82–100.

Chi, B.T.V., Hung, Q.P., Nhan, T.N., Thanh, D.N., Hong, B.T.T., and Loc, T.N. 2005. Economic performance by using bioinsecticides and chemical insecticides to control rice insect pests. *Omonrice* 13: 63–68.

Choudhary, K., Choudhary, O.P., and Shekhawat, N.S. 2008. Marker assisted selection: A novel approach for crop improvement. *American–Eurasian Journal of Agronomy* 1: 26–30.

Coello, N., Brito, L., and Nonus, M. 2000. Biosynthesis of L-lysine by *Corynebacterium glutamicum* grown on fish silage. *Bioresource Technology* 73: 221–225.

Coffey, A.G., Daly, C., and Fitzgerald, G. 1994. The impact of biotechnology on the dairy industry. *Biotechnology Advances* 12: 625–633.

Colomban, A., Roger, L., and Boyaval, P. 1993. Production of propionic acid from whey permeate by sequential fermentation, ultrafiltration, and cell recycling. *Biotechnology and Bioengineering* 42: 1091–1098.

Cowan, T. 2011. *Agricultural Biotechnology: Background and Recent Issues*. Congressional Research Service, Washington, DC.

Crueger, W. and Crueger, A. 1989. *Biotechnology*, 2nd edition. Sinauer Associates, Inc., Sunderland, MA, pp. 219–228.

Daniell, H. 2002. Molecular strategies for gene containment in transgenic crops. *Nature Biotechnology* 20: 581–586.

Dardir, H.A. 2012. In vitro evaluation of probiotic activities of lactic acid bacteria strains isolated from novel probiotic dairy products. *Global Veterinaria* 8: 190–196.

De Jong, M.D. 2000. The BioChon story: Deployment of *Chondrostereum purpureum* to suppress stump sprouting in hard woods. *Mycologist* 14: 58–62.

Demain, A.L. 2000. Small bugs, big business: The economic power of the microbe. *Biotechnology Advances* 18: 499–514.

Demain, A.L., Jackson, M., and Trenner, N.R. 1967. Thiamine-dependent accumulation of tetramethylpyrazine accompanying a mutation in the isoleucine–valine pathway. *Journal of Bacteriology* 94: 323–326.

Demeke, T., Giroux, R.W., Reitmeier, S., and Simon, S.L. 2002. Development of a polymerase chain reaction assay for detection of three canola transgenes. *Journal of American Oil Chemists Society* 79: 1015–1019.

Demir, I., Sezen, K., and Demibrag, Z. 2002. The first study on bacterial flora and biological control agent of *Anoplus roboris* (Sufr. Coleoptera). *The Journal of Microbiology* 40: 104–108.

Desai, J.D. and Banat, I.M. 1997. Microbial production of surfactants and their commercial potential. *Microbiological Molecular Reviews* 61: 47–64.

Dubal, S.A., Tilkari, Y., Momin, S.A., and Borkar, I.V. 2008. Biotechnological routes in flavour industries. *Advances in Biotechnology* 3: 20–31.

Dufosse, L. 2006. Microbial production of food grade pigments. *Food Technology and Biotechnology* 44: 313–321.

Esser, K. and Karsch, T. 1984. Bacterial ethanol production: Advantage and disadvantages. *Process Biochemistry* 17: 116–121.

FAO/WHO. 2003. Report on safety assessment of foods derived from genetically modified animals, including fish. Rome, Italy, November 17–21.

FDA/IFIC. 1993. Regulation of color additives. Food and Drug Administration. FDA/IFIC Brochure.

Frengova, G.I. and Beshkova, D.M. 2009. Carotenoids from *Rhodotorula* and *Phaffia*: Yeasts of biotechnological importance. *Journal of Industrial Microbiology and Biotechnology* 36: 163–180.

Gandhi, D.N. 2000. Fermented dairy products and their role in controlling food borne diseases. In *Food Processing: Biotechnological Applications*, eds. S.S. Marwaha and J.K. Arora, pp. 209–220. Asiatech Publishers Inc., New Delhi, India.

Garcia-Ochoa, F., Santos, V.E., Fritsch, A.P. 1992. Nutritional study of *Xanthomonas campestris* in xanthan gum production by factorial design of experiments. *Enzyme and Microbial Technology* 14: 991–996.

Gasser, F. 1970. Electrophoretic characterization of lactate dehydrogenases in the genus *Lactobacillus*. *Journal of General Microbiology* 14: 233–252.

Ghose, T.K. and Bhadra, A. 1985. Acetic acid. In: *Comprehensive Biotechnology*, ed. M. Moo-Young, p. 724. Pergamon, New York.

Giavasis, I., Harvey, L.M., and McNeil, B. 2000. Gellan gum. *Critical Reviews in Biotechnology* 20: 177–211.

Gibbs, D.F. and Greenhalgh, M.E. 1983. *Biotechnology, Chemical Feedstocks and Energy Utilization*. Francis Pinter (Publ.), London.

Goldberg, I. 1988. Future prospects of genetically engineered single cell protein. *Trends in Biotechnology* 6: 32–38.

Gonsalves, D. 2004. Transgenic papaya in Hawaii and beyond. *AgBioForum* 7: 36–40.

Granato, D., Branco, G.F., Cruz, A.G., Faria, J.D.A.F., and Shah, N.P. 2010. Probiotic dairy products as functional foods. *Comprehensive Reviews in Food Science and Food Safety* 9: 455–470.

Grewal, H.S. and Kalra, K.L. 1996. Fungal production of citric acid. *Biotechnology Advances* 13: 209–234.

Grunewald, K.K. 1992. Serum cholesterol levels in rats fed skim milk fermented by *Lactobacillus acidophilus*. *Journal of Food Science* 47: 2078–2079.

Grzonka, Z., Kasprzykowski, F., and Wiczk, W. 2007. Cysteine proteases. In: *Industrial Enzymes: Structure Function and Applications*, eds. J. Ploaina and A.P. MacCabe, pp. 181–195. Springer, Dordrecht, the Netherlands.

Gutterson, N. and Zhang, J.Z. 2004. Genomic applications to biotech traits: A revolution in progress? *Current Opinion in Plant Biology* 7: 226–230.

Hammes, W.P. and Knauf, H.I. 1994. Starters in the processing of meat products. *Meat Science* 36: 155.

Hatcher, J.H. 1972. Gluconic acid production. U.S. Patent 3,669,840.

Heller, K.J. 2001. Probiotic bacteria in fermented foods: Product characteristics and starter organisms. *American Journal of Clinical Nutrition* 73: 374S–379S.

Hermann, T. 2003. Industrial production of amino acids by *Coryneform* bacteria. *Journal of Biotechnology* 104: 155–172.

Hoagland, R.E., Weaver, M.A., and Boyette, C.D. 2007. *Myrothecium verrucaria* fungus. A bioherbicide and strategy to reduce its non target risk. *Allelopathy Journal* 19: 179–192.

Hofvander, P., Persson, T.P., Tallberg, A., and Wikstrom, O. 2004. Genetically engineered modification of potato to form amylopectin-type starch. U.S. Patent 6,784,338, B1.

Hoggag, M.W. and Mohamed, A.H. 2007. Biotechnological aspects of microorganisms used in plant biological control. *World Journal of Agriculture Sciences* 3: 771–776.

Hoisington, D. 2001. Application of biotechnology to maize improvement. Past, present and future prospects. *Seventh Eastern and Southern Africa Regional Maize Conference*, February 11–15, pp. 7–11, Nairobi, Kenya: CIMMYT (International Maize and Wheat Improvement Centre) and KARI (Kenya Agricultural Research Institute).

Ikeda, M. 2003. Amino acid production processes. *Advances in Biochemical Engineering and Biotechnology* 79: 1–35.

Jack R.W., Bierbaum, G., Hiedrich, C., and Sahl, H.G. 1995. The genetics of lantibiotic biosynthesis. *Bioessays* 17: 793–802.

Jacobson, G. and Wasileski, J. 1994. Production of food colorants by fermentation. In: *Bioprocess Production of Flavor, Fragrance, and Color Ingredients*, ed. A. Gabelman, pp. 205–237. John Wiley & Sons, New York.

Janssens, L., DePooter, H.L., Schamp, N.M., and Vandamme, E.J. 1992. Production of flavours by microorganisms. *Process Biochemistry* 27: 195.

John Innes Centre. 1998. *Biotech Bytes: Food Biotechnology*. http://www.jic.bbsrc.ac.uk/exhibitions/bio-future/

Kang, K.S., Colegrove, G.T., and Veeder, G.T. 1982. Deacetylated polysaccharide S-60. U.S. Patent 4,385,123.

Kapoor, K.K., Chaudhary, K., and Tauro, P. 1983. Citric acid. In: *Prescott and Dunn's Industrial Microbiology*, ed. G. Reed, pp. 709–747. MacMillan Publishers Ltd., U.K.

Khokhar, M.K., Gupta, R., and Sharma, R. 2012. Biological control of plant pathogens using biotechnological aspects: A review. *Open Access Scientific Reports* 1: 277. doi:10.4172/scientificreports.277.

Kim, S. W, Seo, W.T., and Park, Y.H. 1997. Enhanced production of β-carotene from *Blakeslea trispora* with span 20. *Biotechnology Letters* 19: 561–562.

Kishore, G.M. and Shewmaker, C. 1999. Biotechnology: Enhancing human nutrition in developing and developed worlds. *Proceedings of National Academy of Sciences U S A*, 96: 5968–5972.

Klaenhammer, T.R. 1993. Genetics of bacteriocins produced by lactic acid bacteria. *FEMS Microbiology Reviews* 12: 39–85.
Knorr, D. and Sinskey, A. 1985. Biotechnology in food production and processing. *Science* 229: 1224–1229.
Kosseva, M.R. 2011. Management and processing of food wastes. In: *Comprehensive Biotechnology*, 2nd edition, Vol. 6, ed. M. Moo-Young, pp. 557–593. Elsevier, New York.
Kovac, B. and Raspor, P. 1997. The use of the mould *Rhizopus oligosporus* in food production. *Food Technology and Biotechnology* 35: 69–73.
Kuiper, H.A., Kleter, G.A., Noteborn, H.P.J.M., and Kok, E.J. 2001. Assessment of the food safety issues related to genetically modified foods. *The Plant Journal* 27: 503–528.
Kuipers, P.O. 1999. Genomics for food biotechnology: Prospects of the use of high-throughput technologies for the improvement of food microorganisms. *Current Opinion in Biotechnology* 10: 511–516.
Kumar, S., Prasanna, L.P.A., and Wankhade, S. 2011. Potential benefits of *Bt* brinjal in India—An economic assessment. *Agricultural Economics Research Review* 24: 83–90.
Lantto, R., Kruus, K., Puolanne, E., Honkapää, K., Roininen, K., and Buchert, J. 2009. Enzymes in meat processing. In: *Enzymes in Food Technology*, eds. R.J. Whitehurst and M. van Oort, pp. 264–289. Wiley-Blackwell, Oxford, U.K..
Lardizabal, K., Effertz, R., Levering, C. et al. 2008. Expression of *Umbelopsis ramanniana* DGAT2A in seed increases oil in soybean. *Plant Physiology* 148: 89–96.
Lawrie, R.A. 1998. *Lawrie's Meat Science*, 6th edition. Woodhead Publishing Ltd., Cambridge, U.K.
Lim, S.H., Choi, J.S., and Park, E.Y. 2001. Microbial production of riboflavin using riboflavin overproducers, *Ashbya gossypii*, *Bacillus subtilis*, and *Candida famate*: An overview. *Biotechnology and Bioprocess Engineering* 6: 75–88.
Linko, Y.Y., Javanainen, P., and Linko, S. 1997. Biotechnology of bread baking. *Trends in Food Science and Technology* 8: 339–344.
Litchfield, J.H. 1983. Single cell proteins. *Science* 219: 740–746.
Litchfield, J.H. 1996. Microbial production of lactic acid. *Advances in Applied Microbiology* 42: 45–95.
Lücke, F.K. 2000. Utilization of microbes to process and preserve meat. *Meat Science* 56: 105–115.
Machuka, J. 2004. Agriculture genomics and sustainable development: Perspectives and prospects for Africa. *African Journal of Biotechnology* 3: 127–135.
Maiti, D. 2010. Improving activity of native arbuscular mycorrhizal fungi (AMF) for mycorrhizal benefits in agriculture: Status and prospect. *Journal of Biofertilizers and Biopesticides* 2: 113.
Makkar, R.S. and Cameotra, S.S. 2002. An update on the use of unconventional substrates for biosurfactant production and their new applications. *Applied Microbiology and Biotechnology* 58: 428–434.
Mariani, C., De Beuckeleer, M., Truettner, J., Leemans, J., and Goldberg, R.B. 1990. Induction of male sterility in plants by a chimaeric ribonuclease gene. *Nature* 347: 737–741.
Martinez, M.C., Sanchez-Montero, J.M., Sinisterra, J.V., and Ballesteros, A. 1990. New insolubilized derivatives of ribonuclease and endonuclease for elimination of nucleic acids in single cell protein concentrate. *Biotechnology and Applied Biochemistry* 12: 643–652.

Marwaha, S.S. and Arora, J.K. 1999. Production of gums, amino acids and vitamins. In: *Biotechnology: Food Fermentation*, eds. V.K. Joshi and A. Pandey, pp. 1231–1258. Educational Publishers and Distributors, New Delhi, India.

Marwaha, S.S. and Sethi, R.P. 1984. Utilization of dairy waste for vitamin B_{12} fermentation. *Agriculture Wastes* 9: 111–130.

Milsom, P.E. 1987. Organic acids by fermentation, especially citric acid. In: *Food Biotechnology*, eds. R.D. King and P.S.J. Cheetham, pp. 273–308. Elsevier Applied Science, London, U.K.

Mishra, N., Gupta, P.N., Khatri, K., Goyal, K.A., and Vyas, P.S. 2008. Edible vaccines: A new approach to oral immunization. *Indian Journal of Biotechnology* 7: 283–294.

Moellenbeck, D.J., Peters, M.L., Bing, J.W. et al. 2001. Insecticidal proteins from *Bacillus thuringiensis* protect corn from corn rootworms. *Nature Biotechnology* 19: 668–672.

Montet, D., Loiseau, G., Zakhia, N., and Mouquet, C. 1999. Fermented fruits and vegetables. In: *Biotechnology: Food Fermentation*, Vol. 2, eds. V.K. Joshi and A. Pandey, pp. 951–969. Educational Publishers & Distributors, New Delhi, India.

Morin, K.X. 2008. Genetically modified food from crops: Progress, pawns, and possibilities. *Analytical and Bioanalytical Chemistry* 392: 333–340.

Muthusamy, K., Gopalkrishnan, S., Ravi, T.K., and Sivachidambaram, P. 2008. Biosurfactants: Properties, commercial production and application. *Current Science* 94: 736–747.

Nakamura, L.K. and Crowell, C.D. 1979. *Lactobacillus amylophilus*, a new starch hydrolyzing species from spices waste-corn fermentation. *Developments in Industrial Microbiology* 20: 531–540.

Narayanan, N., Roychoudhury, P.K., and Srivastava, P. 2004. L(+) lactic acid fermentation and its product polymerization. *Electronic Journal of Biotechnology* 7: 167–179.

Navaratna, M.A., Sahl, H.G., and Tagg, J.R. 1998. Two components anti *Staphylococcus aureus* lantibiotic activity produced by *Staphylococcus aureus* C55. *Applied and Environmental Microbiology* 64: 4803–4808.

Ninet, L. and Renaut, J. 1979 Carotenoids. In: *Microbial Technology*, eds. H.J. Peppler, and D. Perlman, pp. 529–544. Academic Press, New York.

Nochi, T., Takagi, H., Yuki, Y. et al. 2007. Rice-based mucosal vaccine as a global strategy for cold-chain- and needle-free vaccination. *Proceedings of National Academy of Sciences USA* 104: 10986–10991.

Ockerman, H.W. and Basu, L. 2007. Production and consumption of fermented meat products. In: *Handbook of Fermented Meat and Poultry*, ed. F. Toldrá, pp. 9–16. Blackwell Publishing, Oxford, U.K.

Osusky, M., Osuska, L., Kay, W., and Misra, S. 2005. Genetic modification of potato against microbial diseases: In vitro and in planta activity of a dermaseptin B1 derivative, MsrA2. *Theoretical and Applied Genetics* 111: 711–722.

Ozor, N. and Igbokwe, E.M. 2007. Roles of agricultural biotechnology in ensuring adequate food security in developing societies. *African Journal of Biotechnology* 6: 1597–1602.

Pandya, U. and Saraf, M. 2010. Application of fungi as biocontrol agent and their biofertilizer potential in agriculture. *Journal of Advance in Developmental Research* 1: 90–99.

Panesar, P.S., Chopra, H., Marwaha, S.S., and Joshi, V.K. 2000. Technologies for the production of alcoholic beverages. In: *Food Processing: Biotechnological Applications*, eds. S.S. Marwaha and J.K. Arora, pp. 191–208. Asiatech Publishers, New Delhi, India.

Panesar, P.S. and Kennedy, J.F. 2012. Biotechnological approaches for the value addition of whey. *Critical Reviews in Biotechnology* 32: 327–348.

Panesar, P.S., Marwaha, S.S., and Kennedy, J.F. 2006. *Zymomonas mobilis*—An alternative ethanol producer. *Journal of Chemical Technology and Biotechnology* 81: 623–635.

Papagianni, M. and Anastasiadou, S. 2009. Pediocins: The bacteriocins of Pediococci. Sources, production, properties and applications. *Microbial Cell Factories* 8: 3.

Parades-Lopez, O. 1992. Nutrition and safety considerations. In: *Applications of Biotechnology to Traditional Fermented Foods*. Report of an Ad Hoc Panel of the Board on Science and Technology for International Development, pp. 153–158. National Academy Press, Washington, DC.

Paunescu, A. 2009. Biotechnology for endangered plant conservation: A critical overview. *Romanian Biotechnological Letters* 14: 4095–4103.

Pena, C., Hernandez, L., and Galindo, E. 2006. Manipulation of the acetylation degree of *Azotobacter vinelandii* alginate by supplementing the culture medium with 3-(*N*-morpholino)-propane-sulfonic acid. *Letters in Applied Microbiology* 43: 200–204.

Pérez-Magariño, S. and González-Sanjosé, M.L. 2003. Biotechnology in food production. In: *Encyclopedia of Food Sciences and Nutrition*, 2nd edn., eds. B. Caballero, P. Finglas, and L. Trugo, pp. 500–506. Academic Press, London, U.K.

Pollock, T.J., Thorne, I., and Armentrout, R.W. 1992. Isolation of new *Aureobasidium* strains that produce high-molecular-weight pullulan with reduced pigmentation. *Applied Environment and Microbiology* 58: 877–883.

Prado, F.C. 2002. Desenvolvimento de Bioprocesso em escala semipiloto para produção de ácido cítrico por fermentação no estado sólido a partir do bagaço de mandioca, MSc Thesis. Universidade Federal do Paraná, Brazil.

Quesada-Chanto, A., Afschar, A.S., and Wagner, F. 1994. Microbial production of propionic acid and vitamin B12 using molasses or sugar. *Applied Microbiology and Biotechnology* 41: 378–383.

Ramachandran, S., Fontanille, P., Pandey, A., and Larroche, C. 2006. Gluconic acid: Properties, applications and microbial production: A review. *Food Technology and Biotechnology* 44: 185–195.

Raney, T. 2004. Agricultural biotechnology for developing countries. *Health Policy and Development* 2: 122–130.

Ratledge, C. 1991. Microorganisms for lipids. *Acta Biotechnologica* 11: 429–438.

Redenbaugh, K., Slade, D., Viss, P., and Kossler, M. 1988. Artificial seeds: Encapsulation of somatic embryos. In: *Forest and Crop Biotechnology, Progress and Prospects*, ed. F.A. Valentine, pp. 400–419, Springer Verlag, New York.

Reed, G. 1987. *Prescott and Dunn's Industrial Microbiology*, 4th edn. CBS Publishers, New Delhi, India.

Rehm, H.J. and Reed, G. 1988. *Biotechnology: A Comprehensive Treatise*, Vol. 4, Verlage Chemiew, Weinheim, Germany.

Reuter, G. 1997. Present and future of probiotics in Germany and in Central Europe. *Bioscience Microflora* 16: 43–51.

Riley, M.A. 1993. Molecular mechanisms of colicin evolution. *Molecular Biology Evolution* 10: 1380–1395.

Riley, P.A. and Hoffman, L. 1999. Value-enhanced crops: Biotechnology's next stage. *Agricultural Outlook*, pp. 18–23. USDA/Economic Research Service, Washington, DC.

Rogers, P.L., Lee, K.J., Scotnicki, M.L., and Lee, J.H. 1984. Recent developments in the *Zymomonas* process for ethanol production. *Critical Reviews in Biotechnology* 1: 272–288.

Rossas-Garcia, M.N. 2009. Biopesticide production from *Bacillus thuringiensis*: An environment friendly alternative. *Recent Patents on Biotechnology* 3: 28–36.

Roukas, T. 2006. Biotechnology of citric acid production. In: *Food Biotechnology*, eds. K. Shetty, G. Paliyath, A. Pometto, and R.E. Levin, pp. 351–405. Taylor & Francis Group, Boca Raton, FL.

Ruse, M. and Castle, D. 2002. *Genetically Modified Foods: Debating Biotechnology*. Prometheus Books, Amherst, New York.

Russell, I. and Stewart, R. 1999. Cereal based alcoholic beverages. In: *Biotechnology: Food Fermentation*, Vol. 2, eds. V.K. Joshi and A. Pandey, pp. 745–780. Educational Publishers and Distributors, New Delhi, India.

Saigal, D. 1993. Yeast strain development for ethanol production. *Indian Journal of Microbiology* 33:159–168.

Saini, N., Jain, N., Jain, S., and Jain, K.R. 2004. Assessment of genetic diversity within and among basmati and non-basmati rice varieties using AFLP, ISSR and SSR markers. *Euphytica* 140: 133–146.

Saiprasad, G.V.S. 2001. Artificial seeds and their applications. *Resonance* 6: 39–47.

Saito, K. and Matsuda, F. 2010. Metabolomics for functional genomics, systems biology, and biotechnology. *Annual Review of Plant Biology* 61:463–489.

Salminen, S. and Von Wright, A. 1998. Safety of probiotic bacteria: Current perspectives. In: *Functional food research in Europe*, eds. T. Mattila-Sandholm and T. Kauppila, pp. 105–106. Julkaisija-Utgivare, Finland.

Sangwan, S.R., Sangwan Norreel, S.B., and Harada, H. 1997. In vitro techniques and plant morphogenesis: Fundamental aspects and practical applications. *Plant Biotechnology* 14: 93–100.

Schut, J. 1978. The European sausage industry. In: *Proceedings of the 31st Annual Reciprocal Meat Conference*, Vol. 31, 5, June 18–22. American Meat Sciences Association, Chicago, IL.

Semagn, K., Bjernstad, A., and Xu, Y. 2010. The genetic dissection of quantitative traits in crops. *Electronic Journal of Biotechnology* 13 : 1–45.

Sethi, V. and Maini, S.B. 1999. Production of organic acids. In: *Biotechnology: Food Fermentation*, Vol. 2, eds. V.K. Joshi and A. Pandey, pp. 1259–1290. Educational Publishers & Distributors, New Delhi, India.

Sharma, H.C., Crouch, J.H., Sharma, K.K., Seetharama, N., and Hash, C.T. 2002. Applications of biotechnology for crop improvement: Prospects and constraints. *Plant Sciences* 163: 381–395.

Singh, B.D. 2010. General and industrial microbiology. In: *Biotechnology: Expanding Horizons*, 3rd edn. Kalyani Publishers, New Delhi, India.

Singh, B.D. 2010. Introduction to biotechnology. In: *Biotechnology: Expanding Horizons*, 3rd edn., pp. 1–12. Kalyani Publishers, New Delhi, India.

Singh, O.V. and Kumar, R. 2007. Biotechnological production of gluconic acid: Future implications. *Applied Microbiology and Biotechnology* 75: 713–722.

Singh, R.S., Saini, G.K., and Kennedy, J.F. 2008. Pullulan: Microbial sources, production and applications. *Carbohydrate Polymers* 73: 515–531.

Singh, G. and Shetty, S. 2011. Impact of tissue culture on agriculture in India. *Biotechnology, Bioinformatics and Bioengineering* 1: 279–288.

Singhal, R.S. and Kulkarni, P.R. 1999. Production of food additives by fermentation. In: *Biotechnology: Food Fermentation*, eds. V.K. Joshi and A. Pandey, pp. 1145–1200. Educational Publishers and Distributors, New Delhi, India.

Siro, I., Kapolna, E., Kapolna, B., and Lugasi, A. 2008. Functional food: Product development, marketing and consumer acceptance—A review. *Appetite* 51: 456–467.

Skogman, H. 1976. Production of Symba yeast from potato waste. In: *Food from Waste*, eds. G.C. Birch, K.J. Parker, and J.T. Worgan, pp. 167–179. Applied Science Publication, London, U.K.

Smith, J.E. 2009. *Biotechnology*, 5th edn. Cambridge University Press, Cambridge, U.K.

Solomons, G.L. and Litchfield, J.H. 1983. Single cell protein. *Critical Reviews in Biotechnology* 1: 21–58.

Soni, S.K. and Arora, J.K. 2000. Indian fermented foods: Biotechnological approaches. In: *Food Processing: Biotechnological Applications*, eds. S.S. Marwaha and J.K. Arora, pp. 143–190. Asiatech Publishers Inc., New Delhi, India.

Spears, K. 1988. Developments in food colourings: The natural alternatives. *Trends in Biotechnology* 6: 283–288.

Stahmann, K.P., Revuelta, J.L., and Seulberger, H. 2000. Three biotechnical processes using *Ashbya gossypii*, *Candida famata*, or *Bacillus subtilis* compete with chemical riboflavin production. *Applied Microbiology and Biotechnology* 53: 509–516.

Stahnke, L.H. 1995. Dried sausage fermented with *Staphylococcus xylosus* at different temperature and with different ingredient levels. Part I—Chemical and bacteriological data. *Meat Science* 41: 179.

Stanton, C., Gardiner, G., Meehan, H., Collins, K., Fitzgerald, G., Brendan Lynch, P., and Paul Ross, R. 2001. Market potential for probiotics. *American Journal of Clinical Nutrition* 73: 476S–483S.

Steinkraus, K.H. 1998. Bio-enrichment: Production of vitamins in fermented foods. In: *Microbiology of Fermented Foods*, ed. J.B. Wood, pp. 603–619. Blackie Academic and Professional, London, U.K.

Stiles, M.E. 1996. Biopreservation by lactic acid bacteria. *Antonie Van Leeuwenhoek* 70: 331–345.

Suslow, V.T., Thomas, R.B., and Bradford, J.K. 2002. *Biotechnology Provides New Tools for Plant Breeding*. Agricultural Biotechnology in California Series, Seed Biotechnology Centre, UC Davis.

Sutherland, I.W. 1998. Novel and established applications of microbial polysaccharides. *Trends in Biotechnology* 16: 41–46.

Tagg, J.R., Dajani, A.S., and Wannamaker, L.W. 1976. Bacteriocins of gram-positive bacteria. *Bacteriology Reviews* 40: 722–756.

Tahara, S., Fujiwara, K., Ishizaka, H., Mizutani, J., and Obata, Y. 1972. Gamma decalactone: One of constituents of volatiles in cultured broth of *Sporobolomyces odorus*. *Agriculture and Biological Chemistry* 36: 2585–2587.

Tanksley, S.D. and Nelson, J.C. 1996. Advanced backcross QTL analysis: A method for the simultaneous discovery and transfer of valuable QTLs from unadapted germplasm into elite breeding lines. *Theoretical and Applied Genetics* 92: 191–203.

Tengel, C., Schubler, P., Setzke, E., Balles, J., and Haubels, S.M. 2001. PCR based detection of genetically modified soybean and maize in raw and highly processed foodstuffs. *Biotechniques* 31: 426–429.

Thiart, S. 2003. Manipulation of growth by using tissue culture techniques. *Combined Proceedings International Plant Propagator's Society* 53: 61–67.

Trang, K.T.T. and Chaudhari, S. 2002. Bioassay of nuclear polyhedrosis virus (npv) and in combination with insecticide on *Spodoptera litura* (Fab). *Omonrice* 10: 45–53.

Trias, R., Baneras, L., Montesinos, E., and Badosa, E. 2008. Lactic acid bacteria from fresh fruit and vegetables as biocontrol agents of phytopathogenic bacteria and fungi. *International Microbiology* 11: 231–236.

Valduga, E., Valério, A., Tatsch, P.O., Treichel, H., Furigo, J.A., and Luccio, M.D. 2009. Optimization of the production of total carotenoids by *Sporidiobolus salmonicolor* (CBS 2636) using response surface technique. *Food Bioprocess Technology* 2: 415–421.

Vandamme, E.J. 2003. Bioflavours and fragrances via fungi and their enzymes. *Fungal Diversity* 13: 153–166.

Vandamme, E.J. and Soetaert, W. 2002. Bioflavors and fragrances via fermentation and biocatalysis. *Journal of Chemical Technology and Biotechnology* 77: 1323–1332.

Vandenberghe, L.P.S., Soccol, C.R., Pandey, A., and Lebeault, J.M. 2000. On-line monitoring of citric acid production using respirometry in solid state fermentation with cassava bagasse. In: *Proceedings of International Symposium on the Bioconversion of Renewable Raw Materials*, Hannover, Germany.

Walker, H.L. and Tilley, A.M. 1997. Evaluation of an isolate of *Myrothecium verrucaria* from sicklepod (*Senna obtusifolia*) as a potential mycoherbicide agent. *Biological Control* 10: 104–112.

Wani, S.A. and Sharma, S.D. 1999. Fermented meat products. In: *Biotechnology: Food Fermentation*, eds. V.K. Joshi and A. Pandey, pp. 971–1002. Educational Publishers and Distributors, New Delhi, India.

Whitman, B.D. 2000. Genetically modified foods: Harmful or helpful? *CSA Discovery Guides*. http://www.csa.com/discoveryguides/gmfood/overview.php

Wieczorek, A. 2003. Use of biotechnology in agriculture—Benefits and risks. University of Hawaii, Honolulu, HI. 6 pp. (Biotechnology; BIO-3).

Wood, P.J. 1997. Functional foods for health: Opportunities for novel cereal processes and products. In: *Cereals: Novel Uses and Processes*, eds. G.M. Campbell, S.L. McKee, and C. Webb, pp. 233–239. Plenum Press, New York.

Wynn, J.P. and Ratledge, C. 2005. Oils from microorganisms. In: *Bailey's Industrial Oil and Fat Products*, 6th edn., ed. F. Shahidi, pp. 121–153. John Wiley & Sons, New York.

Yarasi, B., Sadumpati, V., Immanni, P.C., Vudem, R.D., and Khareedu, R.V. 2008. Transgenic rice expressing *Allium sativum* leaf agglutinin (ASAL) exhibit high level resistance against major sap sucking pests. *BMC Plant Biology* 8: 102–106.

Yegin, S. and Fernández-Lahore, M. 2012. Boza: A traditional cereal-based, fermented Turkish beverage. In: *Handbook of Plant-Based Fermented Food and Beverage Technology*, 2nd edn., eds. Y.H. Hui and E. Özgül Evranuz, pp. 533–542. CRC Press, Boca Raton, FL.

Yokote, Y., Maeda, S., Yabushita, H., Noguchi, S., Kimura, K., and Samejima, H. 1978. Production of L-aspartic acid by *E. coli* aspartase immobilized on phenol-formaldehyde resin. *Journal of Solid-Phase Biochemistry* 3: 247–261.

Youm, W.J., Jeon, H.J., Kim, H. et al. 2008. Transgenic tomatoes expressing human beta-amyloid for use as a vaccine against Alzheimer's disease. *Biotechnology Letters* 30: 1839–1845.

2

Genomics in Agriculture and Food Processing

Peter C. McKeown, Channa Keshavaiah, Antoine Fort, Reetu Tuteja, Manash Chatterjee, Rajeev K. Varshney, and Charles Spillane

CONTENTS

2.1 Introduction....45
2.2 Harnessing Genomics in Crop Improvement Programs....48
2.2.1 Genomics and Marker-Assisted Selection in Plant Breeding....48
2.2.2 Next Generation Sequencing-Facilitated Crop Improvement....48
2.2.3 Next Generation Sequencing Techniques for Identifying Genomic Markers....51
2.2.4 Genotyping by Sequencing....52
2.2.5 Mapping by Genome-Wide Association Studies....53
2.2.6 Genomic Selection....54
2.3 Improvement of Nutritional Traits in Staple Food Crops through Gene Modification....55
2.3.1 Humanitarian Golden Rice Initiative....55
2.3.2 GMOs within the Biological Species Concept....56
2.3.3 Improving Nutritional Qualities of Crop Varieties by Mutagenesis....57
2.3.4 Targeting Induced Local Lesions in Genomes....58
2.3.5 Genome Editing....59
2.3.6 Genomics and the Food-Processing Industry....59
2.3.7 Genomic Analysis of Nutrient Availability during Food Processing and Storage....59
2.3.8 Microbial Genomics for Efficient Food Processing....60
2.3.9 Microbial Genomics in Food Safety....61
2.4 Summary and Future Prospects....62
References....63

2.1 Introduction

Micronutrient malnutrition (MNM), which has been described as "hidden hunger," is caused by insufficient dietary nutrients and is endemic in many parts of the developing world. Key contributing factors to hidden hunger malnutrition include insufficient nutrient levels within staple crops and loss of nutrients during storage and processing of plant-derived foods. Therefore, although a key aim of modern crop biotechnology is to meet rising calorific requirements (Tester and Langridge, 2010), it will also be essential that increased food production is nutritionally optimized, especially in developing countries (Ribaut et al., 2010). Much of the world's population suffers from malnutrition caused by lack of protein, vitamins, and minerals. Recent advances in genomics have already facilitated progress toward improved food, feed, fiber, fuel, and biomass production (Mittler and Blumwald, 2010). This chapter is focused on the technologies that have allowed these advances and on their future development, as summarized in Figure 2.1.

Tackling malnutrition (undernutrition and overnutrition) in both developing and developed countries will clearly require major multidisciplinary efforts and include both improvements to plant-based food production and maximizing the bioavailability of nutrients in existing foods (Beddington, 2010; Hirschi, 2009). Micronutrient enrichment could also provide tools for tackling this, an approach that has been termed "biomedical agriculture," which will not be further elaborated in this chapter. Readers who are interested to know more on this subject can refer to the previous review published on this subject (Thompson and Thompson, 2009). Targets for improving seed quality have also recently been

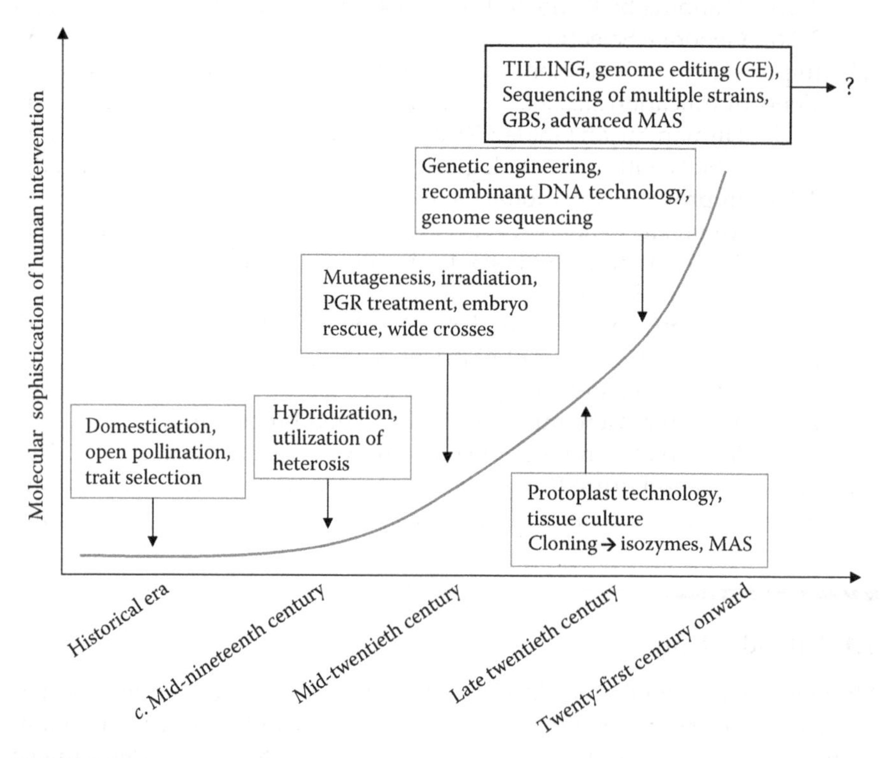

FIGURE 2.1
Development of crop biotechnology.

reviewed (Ligterink et al., 2012), suggesting the need for a "generalized genetical genomics approach" based on the analysis of recombinant inbred lines (RILs) by phenotypic- and omics-based approaches. Further challenges include improvement of postharvest processing to ensure that food retains its nutritional properties in a form accessible for human consumption. This also implies that the crop produce will be protected from spoilage, decay, or other loss of beneficial properties. Genomic resources can therefore provide assistance both by identifying plant-breeding targets for improved postharvest characteristics and through increased understanding of the pathogenic microbes that may spoil/contaminate foods or lead to reductions in its nutrient bioavailability. Some of the points at which biotechnology adds value to the crop and food-processing supply chain are summarized in Figure 2.2. There is also a pressing need to reduce the inputs required to produce staple crops (Edwards and Batley, 2010) while simultaneously improving their nutritional potential (White and Brown, 2010) and storage capabilities (Dahmani-Mardas et al., 2010).

The development of genomic resources already allows the limited use of marker-assisted selection (MAS) to be used in crop-breeding programs (Varshney et al., 2006). MAS will become more widely applicable due to ongoing sequencing of plant genomes and the application of next-generation sequencing (NGS) approaches to crop improvement. The technological approaches underlying recent developments are discussed in the following

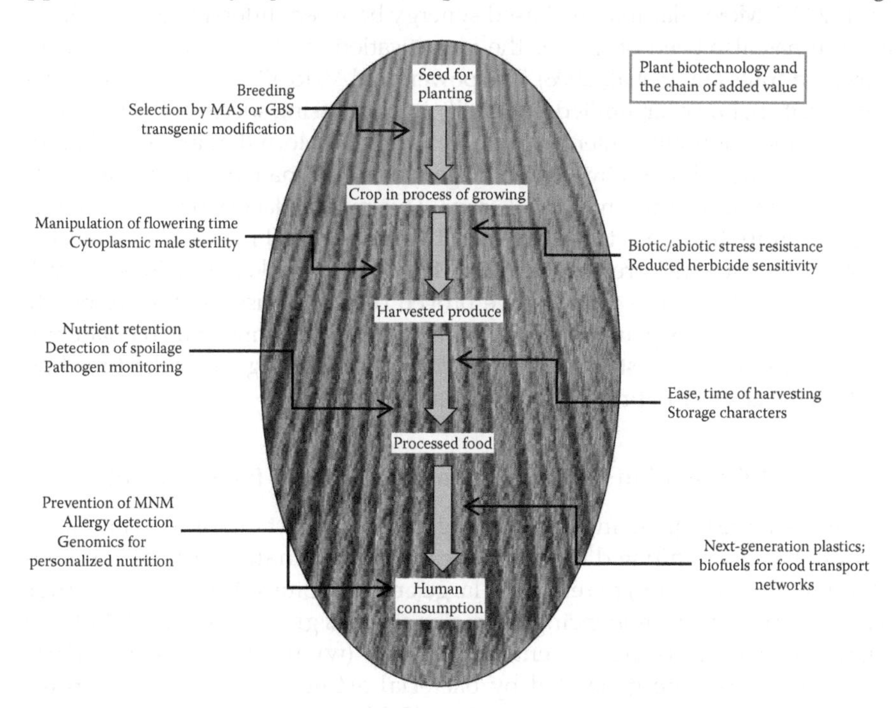

FIGURE 2.2
Biotechnology and added value in agriculture and food processing.

section, together with how these have been used in the identification of markers for use in breeding programs. Emerging techniques to harness natural variation by transgenic and genome-editing technology are also elaborated.

2.2 Harnessing Genomics in Crop Improvement Programs

2.2.1 Genomics and Marker-Assisted Selection in Plant Breeding

Genomic tools have the potential to revolutionize plant-breeding programs on a scale similar or greater than the development of hybrid maize or the Green Revolution (Moose and Mumm, 2008). In the context of more nutritious crops, much attention has been focused on the production of biofortified plant-derived foods (Mayer et al., 2008). Using molecular markers, genetic maps are developed based on the mapping population segregating for the target trait in a given crop species. Analysis of genetic mapping and phenotyping data for the target trait segregating within the mapping population can provide molecular markers associated with the trait. Such molecular markers can subsequently be used to select superior lines for the target traits from the segregating populations, a process termed marker-assisted selection or MAS (Ganal et al., 2009; Varshney et al., 2006). Molecular marker–based synergy between different species within a taxon has also been of use for the identification of markers associated with agri-food traits (Tang et al., 2008). The power of MAS to allow selection for polygenic traits is, however, limited, especially in outcrossing species, which usually include micronutrient content (Jannink et al., 2010; Morrell et al., 2012). Earlier approaches were limited by the lack of genomic information that was available especially for minor and noncereal crops. This made it laborious and challenging to identify the genes that were causative for a trait of interest or to develop markers for these (Morrell et al., 2012). Many of the markers previously developed in species such as tomato have proved of limited use as they are poorly characterized across genotypes or are insufficiently polymorphic (Foolad and Panthee, 2012). This illustrates the limitations of developing molecular-breeding tools in the absence of proper genome characterization.

2.2.2 Next Generation Sequencing-Facilitated Crop Improvement

Whole genome sequencing of many organisms important for food and agriculture has been initiated in recent decades, from small microbial genomes (Markowitz et al., 2010) to relatively large crop genomes (Feuillet et al., 2011). The current state of sequencing for major crops is given in Table 2.1, but still many other projects are currently underway (www.phytozome.net). Early plant genomes were generated by bacterial artificial chromosomes (BAC) clone-by-clone sequencing approaches (Table 2.1), but continuing reductions in cost/unit of sequence data are expected to further accelerate the

TABLE 2.1

List of Selected Plants with Available Completed and Partial Genome Sequences

Class	Species	Remarks	References
Model dicots	*Arabidopsis lyrata*	207 Mb determined by 8.3× dideoxy sequence coverage; 32,670 protein-coding loci	Hu et al. (2011)
	Arabidopsis thaliana (accession Col-0)	157 Mb as 5 chromosomes; BAC-by-BAC sequencing, additional 454 pyrosequencing; 27,416 protein-coding loci	(*Arabidopsis*, 2000); latest release TAIR10 www.arabidopsis.org/doc/news/breaking_news/140
	A. thaliana (1001 accessions)	Whole genome shotgun (WGS) sequencing performed for 80 accessions; genetic maps generated across 1001	See www.1001genomes.org Weigel and Mott (2009)
	Brassica (crops; used as *A. thaliana* relatives)	283 Mb arranged as 40,367 scaffolds 26,374 protein-coding loci	v1.2 of *Brassica rapa* available at phytozome; see brassicadb.org/brad (Cheng et al., 2011)
Dicot crops	*Cajanus cajan*	606 Mb (sequenced by Illumina), arranged as scaffolds; 48,680 predicted protein-coding loci	Not yet available at phytozome (Varshney et al., 2012)
	Carica papaya	4114 contigs covering 135 Mb 27,332 protein-coding loci	asgpb.mhpcc.hawaii.edu/papaya. The ASGPB Hawaii Papaya Genome Project
	Citrus clementina	1128 scaffolds covering 296 Mb	International Citrus Genome Consortium
		25,385 protein-coding loci	v0.9 available at JGI; unpublished data
	Citrus sinensis	12,574 scaffolds covering 319 Mb, from WGS of 2 Gb sequence; 25,376 protein-coding loci	International Citrus Genome Consortium v1.0 JGI; unpublished data
	Glycine max	Chromosome-level assembly, 975 Mb as 20 chromosomes; performed by WGS; 54,175 protein-coding loci	v1.1, "Glyma1," currently available Schmutz et al. (2010)
	Gossypium raimondii	764 Mb, combination of Sanger sequencing, 454 sequencing, and Illumina sequencing	Unpublished data, v1.0 of the D genome Available at JGI
	Linum usitatissimum	Seven libraries covering 350 Mb, WGS sequencing (Illumina); 44,453 protein-coding loci	Unpublished data, Beijing Genome Institute Deyholos group, University of Alberta, Canada

(*continued*)

TABLE 2.1 (continued)

List of Selected Plants with Available Completed and Partial Genome Sequences

Class	Species	Remarks	References
	Manihot esculentum	454 WGS; 12,977 scaffolds covering 533 Mb (of 760 Mb); 30,666 protein-coding loci	Cassava4 assembly, as reported in Prochnik et al. (2012)
	Populus trichocarpa (timber crop)	403 Mb arranged as 19 pseudochromosomes 40,668 protein-coding loci	v2 currently available (Tuskan et al., 2006)
	Prunus persica	8 scaffolds (as pseudochromosomes), size unclear ~35,000 protein-coding loci	Unpublished data from the International Peach Genome Initiative; v1.0 at JGI
	Ricinus communis	25,878 scaffolds covering 400 Mb 31,221 protein-coding loci	v1.0 made available at JCVI castorbean.jcvi.org/index.php
	Solanum lycopersicum	WGS: 760 Mb arranged as 91 scaffolds assigned across 12 chromosomes; ~34,727 protein-coding loci	Not yet available at phytozome; see Tomato Sequencing Consortium (Consortium, 2012)
	Solanum tuberosum	WGS: 727 Mb mostly arranged as super-scaffolds; ~39,031 protein-coding loci	Potato Sequencing Consortium (Consortium, 2011)
Cereals	*Brachypodium distachyon* (2× relative of wheat)	272 Mb arranged 5 chromosomes + 78 unmapped scaffolds 26,522 protein-coding loci	v1.0 available at JGI; www.brachypodium.org (Vogel et al., 2010)
	Oryza sativa subsp. *indica*	372 Mb arranged as 12 chromosomes 55,986 protein-coding loci	Yu et al. (2002); rice.plantbiology.msu.edu
	O. sativa subsp. *japonica*	372 Mb arranged as 12 chromosomes 55,986 protein-coding loci	MSU release 7.0 available (Ouyang et al., 2007)
	Setaria italica	WGS sequencing, 406 Mb (out of 515) arranged as 9 pseudochromosomes; 35,471 protein-coding loci	v2.1 available from JGI www.phytozome.net/foxtailmillet.php
	Sorghum bicolor	Large-scale shotgun sequencing, 10 chromosomes, 698 Mb 34,496 protein-coding loci	v1.0 available from JGI (Paterson et al., 2009)
	Zea mays	Genome of accession B73, 454 sequencing of Mo17.2500 Mb; 65,291 cDNAs	AGPv2 available; www.maizesequence.org/
Other	*Vitis vinifera*	487 Mb arranged as 19 chromosomes, WGS sequencing 26,346 protein-coding loci	French–Italian Public Consortium for Grapevine Genome Characterization (Jaillon et al., 2007)

use of genomics in plant breeding (Thudi et al., 2012). Recent advances in sequencing technologies arising from the advent of NGS techniques have transformed biology by making it possible to sequence millions of base pairs in one run (Egan et al., 2012; Varshney et al., 2009). Several NGS technologies, such as Illumina systems (www.illumina.com), 454/Roche systems (www.454.com), and ABI Solid systems (www.appliedbiosystems.com), are currently referred to as second-generation sequencing technologies (SGS) and have accelerated DNA sequencing by reducing both time and cost (Delseny et al., 2010). This has made possible the resequencing of multiple individuals of the same species including *Arabidopsis*, rice, sorghum, and maize (Table 2.1). Current versions of SGS suffer from certain technical disadvantages such as short sequence reads, expense of reagents, read biases, and, in some instances, high error rates. Such constraints are being mitigated by the so-called third-generation sequencing (TGS) technologies, which generate longer sequence reads in a shorter time at even lower cost per instrument run. These include the Single-Molecule Real-Time (SMRT™) Sequencer (www.pacificbiosciences.com), Heliscope™ Single Molecule Sequencer (www.helicosbio.com), and the Ion Personal Genome Sequencer (Thudi et al., 2012). Apart from the benefit of longer read lengths (expected to be >1 kb), the majority of TGS technologies do not require cloning and amplification, thereby saving costs compared to earlier sequencing technologies.

Multiplexed second- and third-generation sequencing has the potential to massively accelerate crop-breeding programs for improving traits related to food processing and nutrient availability in two ways. In the first instance, MAS and quantitative trait locus (QTL) analysis may be accelerated by the identification of far larger numbers of genetic markers than were previously available (Xu et al., 2012). For example, it may become possible to apply QTL analysis to interbred populations such as the multi-parent advanced generation intercross (MAGIC), nested association (NAM), and recombinant inbred advanced intercross lines (RIAIL), which have been claimed to allow up to single-nucleotide resolution (Cavanagh et al., 2008; Morrell et al., 2012; Varshney and Dubey, 2009). Massively increased identification of single-nucleotide polymorphisms (SNPs) at high density allows the use of entirely new breeding methods that were not feasible previously, such as genomic selection (GS). The sequencing of many individuals or subpopulations can identify causal sources of variation for traits of interest, which can be accessed directly (see Section 2.3). Recent progress in each of these areas and their prospects for improving nutrient bioavailability content of crops is discussed in the following section.

2.2.3 Next Generation Sequencing Techniques for Identifying Genomic Markers

One major potential of NGS technologies is the generation of markers for widely studied crops that can be used across many cultivars such as in rice (Yamamoto et al., 2010). The identification of markers for use in nonmodel species has also attracted considerable attention (Ganal et al., 2009). Recent developments such as reduced-representation sequencing (Davey et al., 2011; Garvin et al., 2010) have been developed to allow scientists to sequence only the genetically informative regions of a genome, which contain many unique sequences and are gene-rich, as opposed to less informative regions that contain many paralogous sequences and are gene-poor. Although the procedures involved in these techniques vary, all typically involve digestion of DNA with restriction enzymes to reduce the complexity of the mixture and enrich the genomically informative unique sequences. Reduced-representation sequencing was pioneered by the use of the "complexity reduction of polymorphic sequences" technique (van Orsouw et al., 2007), which identified large numbers of SNPs in maize by applying NGS to AFLP-digested genomic fragments.

A further advance has been the use of reduced-representation libraries (Van Tassell et al., 2008), which allowed deep NGS for SNP recovery in soybean (Hyten et al., 2010a). Similar techniques have been adapted for use in beans (Hyten et al., 2010b) and chickpea (Gujaria et al., 2011) for which little genomic data were available and also potato (Hamilton et al., 2011) in which few molecular markers were available despite the recent sequencing of its genome (Consortium, 2011). Similar alternative approaches include restriction site–associated DNA sequencing (RAD-seq), which sequences only DNA surrounding restriction sites and has proven of particular use in barley by using existing expressed sequence tag (EST) databases as a reference (Chutimanitsakun et al., 2011). The recently developed restriction enzyme sequence comparative analysis (RESCAN) has been used to detect natural and induced SNPs in *Arabidopsis* and rice (Monson-Miller et al., 2012). An additional approach that was used in *Aegilops tauschii*, a wild progenitor of wheat, was to use Roche 454 sequencing to produce a low genome coverage dataset and use this as a reference for subsequent deeper coverage sequencing (You et al., 2011). The optimal techniques for applying NGS for the purpose of SNP identification in any given genome currently remain a matter for empirical determination. Nonetheless, it is clear that the techniques established in the last few years now allow molecular markers to be identified at far higher density than before, even in nonmodel crop species where widespread polyploidy and high transposable element content can be a challenge.

2.2.4 Genotyping by Sequencing

A further technique using the power of NGS technologies for marker identification is the development of genotyping by sequencing (GBS). GBS has the advantages of being high throughput, high resolution, and inexpensive and can be used for genotyping populations even of species with large, unsequenced genomes (Hamilton and Robin Buell, 2012). The use of GBS has been demonstrated for recombinant inbred populations in barley, wheat, and maize (Elshire et al., 2011), using methylation-sensitive restriction enzymes to reduce complexity while avoiding repetitive regions. Detailed guides have been developed for adapting GBS for use in other crops. In species of barley, which lacks a complete genome sequence, the advantage is that a reference map need only be developed around the restriction sites, effectively allowing the consensus of the clusters to be used in place of a reference. This was put into practice in the analysis of biparental barley and wheat by use of a dual-enzyme restriction approach (Poland et al., 2012). Published GBS data has so far only been derived from a few other species such as eucalyptus (Grattapaglia et al., 2011), but its use is currently being widely explored in maize, grapevine, switchgrass (*Panicum virgatum*), citrus fruits, and others (www.intlpag.org).

2.2.5 Mapping by Genome-Wide Association Studies

Genome-wide association studies (GWAS), also termed linkage disequilibrium association mapping (LDAM), is another technique that allows the identification of loci linked with favorable traits. Such traits could include improved nutrient content or ease of processing, and GWAS can be extended for the study of polygenic traits with the correct experimental design (Hall et al., 2010). GWAS differs from GS in that the SNPs used as markers are derived from an analysis of individuals spanning the diversity of the species rather than within subpopulations used for breeding purposes. GWAS facilitates mining for novel alleles for nutritional improvement that can be targeted through crop improvement or by gene transfer, assuming that causative variation can be distinguished from correlated features. GWAS holds particular appeal in plants as patterns of linkage disequilibrium (LD) are more favorable in domesticated crops than they typically are in animals (Hamblin et al., 2011). Research in the model plant *Arabidopsis thaliana* (*Arabidopsis*, thale cress) has had particular success in identifying loci associated with traits of environmental and agronomic interest (Atwell et al., 2010), and the power of GWAS may be increased still further by use of haplotypes containing multiple SNPs (Hamblin and Jannink, 2011). The most comprehensive GWAS-based program to date is that of Huang et al. (2010) who are currently sequencing 517 rice landraces with the aim of investigating 14 economically important traits. GWAS may also be possible with fewer markers than are required for GS, provided that LD is sufficiently large: As collections of large numbers of markers are becoming more available, this is not such a great advantage as in the past. GWAS also shares an important drawback with traditional QTL studies as it does not retain the use of all markers as predictors. This makes GWAS less amenable to the detection of small effect loci that are likely to occur in outcrossing species (Morrell et al., 2012) or in species derived from wild progenitors with large effective population sizes such as maize and barley (Hamblin et al., 2011).

Finally, it should be noted that NGS techniques are as applicable to ribonucleic acid (RNA) transcriptome sequencing as they are to DNA sequencing including SNP discovery via transcript sequencing. For example, this has recently been done in maize (Cronn et al., 2012; Hansey et al., 2012) where it has been concluded that polymerase chain reaction (PCR)-based strategies are only reasonable for identifying small genomic regions (≤50 kb). However, larger targets require the use of transcriptomic approaches. Such approaches have been widely applied in well-studied species but are also applicable to nonmodels, for example, fruits (Ward et al., 2012). These may be useful for crop improvement geared toward nutrient enrichment as well as increasing yield. In conclusion, rapid advances in sequencing technology promise massive advances in conventional plant breeding for increased bioavailable nutritional content in plants.

2.2.6 Genomic Selection

In many studies, complex traits such as yield seem to be controlled by many QTLs of small effect. MAS in such cases is quite challenging due to requirement of unmanageable population size to pyramid a large number of QTLs (Heffner et al., 2009). As a result, MAS is considered a good approach for improving one or a few traits. Availability of NGS technology has facilitated use of a new breeding methodology referred to as genomic selection (GS) or genome-wide selection (GWS). Instead of identifying markers associated with a trait and then using them for improving the trait, GS identifies breeding lines with higher performance based on parameters called "genomic estimated breeding values" (GEBVs) for using such lines in breeding crosses (Hamblin et al., 2011; Meuwissen et al., 2001). For measuring GEBVs, a population comprised of breeding lines referred to as a "training population" is genotyped with genome-wide markers using GBS approaches or large-scale SNP markers. In the majority of the cases, breeders already have extensive phenotyping data available on the training population. The genotyping data and phenotyping data are used to develop models for calculating GEBVs. Subsequently, in any population (referred to as the test population), segregating lines or breeding lines can be used for genotyping with the same set of markers used for developing the models. Applying these models to the genotyping data of the test population provides GEBVs for these lines. The superior lines based on higher GEBVs are then used for further crosses. The same procedure is performed in the next generations. After a few generations, lines showing indication of improved traits in terms of GEBVs are taken to field trials and the best performing lines for the target complex traits can be advanced further for varietal release.

The application of GS to agriculture was pioneered in animal breeding (Goddard and Hayes, 2007) and is currently being optimized for plants (Heffner et al., 2009). Issues remain with regards to choice of models for assessing genotype × environment (G × E) interactions and variation not captured by markers (Crossa et al., 2010). Simple ridge regression models have proved robust in maize (Piepho, 2009) and even produce good results when applied to populations of sibling plants (Jannink et al., 2010). GS holds promise for crop improvement as the GEBVs that it predicts can correlate with observed phenotypes with a value of 0.84, which may be high enough to allow selection based on markers alone (Heffner et al., 2009). Jannink et al. (2010) have, however, cautioned that some researchers may be attempting to apply GS before the theoretical understanding needed to ensure robust experimental designs has been developed. Nevertheless, trial programs targeting grain quality in wheat, a crop with a challenging genome, found GS to be significantly more accurate than MAS (Heffner et al., 2011), providing the first confirmation of its utility for plant improvement. GS has also been proposed as a key emerging tool for forestry (Grattapaglia and Resende, 2011).

2.3 Improvement of Nutritional Traits in Staple Food Crops through Gene Modification

Earlier sections discussed the advances in sequencing technology and their use for identifying genetic markers for use in plant breeding. However, such approaches can also be used to identify loci (genes) causal for traits of interest that can be harnessed directly via genetic modification (Moose and Mumm, 2008). Many transgenic approaches have been employed for improving nutritional composition of model plants and food plants (Davies, 2007; Hirschi, 2009). However, the heavy regulatory burdens that regulators have placed on genetically modified (GM) crops have generated major cost barriers to entry and limited commercialization/deployment of GM crops for a small number of highly profitable traits (Peng, 2011). Although many regulatory systems attempt to distinguish between "GM" and "non-GM" crops, over time ongoing technological advances will make this distinction increasingly difficult for regulators to sustain in a scientifically valid manner (Herman et al., 2009; Morris and Spillane, 2008).

2.3.1 Humanitarian Golden Rice Initiative

Arguably, the most famous example of nutritional enrichment in a crop has come from the ongoing development of golden rice, the grains of which have massively increased levels of β-carotene. The first strain of golden rice was unveiled over 10 years ago (Ye et al., 2000) as an explicit means of addressing provitamin A deficiency, although its classification as a heavily regulated GM crop has greatly delayed its humanitarian deployment to the benefit of the poor (Potrykus, 2010). The first golden rice was transformed with β-carotene pathway genes from *Narcissus* and a bacterial phytoene desaturase under the control of an endosperm-specific promoter (Beyer et al., 2002). Consumption of golden rice can increase retinol (vitamin A) levels (Tang et al., 2009). Varieties of golden rice that were developed later accumulate levels of provitamin A >20-fold higher than in the original variety and could deliver up to 50% of a child's daily vitamin A requirements due to the replacement of the *Narcissus* gene promoter with a more efficient one derived from maize (Paine et al., 2005). Golden rice could not have been achieved via conventional breeding, even aided by extensive marker populations, as β-carotene content in rice endosperm is both negligible and invariant (Potrykus, 2010). However, once the transgene cassette for elevated β-carotene is present within rice, introgression into non-GM varieties can be used to produce new lines with elevated β-carotene content (Datta et al., 2007).

A range of other crops including maize and potatoes have been targeted for β-carotene enrichment using manipulation of similar pathways (Aluru et al., 2008; Giuliano et al., 2008). This highlights that such metabolic engineering is applicable to many species, not just seed crops. Other staple crops from the developing world such as banana/plantain and cassava are currently being developed with elevated micronutrient compositions. There has also been progress in generating crops enriched in mineral ions (White and Broadley, 2009) and for enriching multiple limiting nutrients within a single food source, as demonstrated in maize hyperaccumulating folate, ascorbate, and β-carotene (Naqvi et al., 2009). The Consultative Group on International Agricultural Research (CGIAR)'s Harvest Plus program is one of the major international humanitarian initiatives, which is developing micronutrient dense staple crops through both conventional and transgenic approaches (depending on which is the most efficient route for the crop and micronutrient in question). Harvest Plus is currently focused on developing high iron (beans and pearl millet), high vitamin A (maize and sweet potato), and high zinc (rice and wheat) varieties of the staple crops in developing countries (www.harvestplus.org).

2.3.2 GMOs within the Biological Species Concept

Crops that are classed as GM are considered as Genetically Modified Organisms (GMOs) and face significant regulatory hurdles and public perception issues, particularly in relation to regulatory and risk-assessment systems within European member states. Techniques that are not classed as GM are therefore attractive targets for biofortification. Two related classes of crop improvement represent "gray areas" between GM and non-GM plants. The terms cisgenesis and intragenesis refer to a special case of transgenic technology involving transformation of an organism with genetic material from another member of its own gene pool, that is to say, its own species where interbreeding between individuals of the species is possible. Cisgenic and intragenic transformation approaches involve the transfer of a gene from one species to another with which it is sexually compatible, that is, within its own species according to the biological species concept (Rommens, 2004; Rommens et al., 2007; Schouten and Jacobsen, 2008; Schouten et al., 2006). In these cases, the transfer of DNA by GM technology only accelerates what would have been possible in a less precise way via traditional breeding programs and introgression. At the current time, crops produced in this way are still classed as GM according to process-based regulatory systems such as those used by the European Union (Davies, 2007; Rommens et al., 2007).

Genetic modification is a particular important tool for traits regulated by loci that have previously been subject to selective sweeps, as such loci are likely to have been so heavily selected that very little natural variation exists between domestic varieties. In this instance, variation cannot be selected for and needs to be actively reintroduced (Morrell et al., 2012). Nutritional and processing characters desirable to humans may be highly disadvantageous in the wild, such as softer seed coats (Hamblin et al., 2011). Such reintroduction of genetic variation at key pathway genes can potentially be achieved by use of traditional crosses to landraces and/or crop wild relatives, but in many cases such introgression is highly laborious and is linked to undesirable characters; thus, intragenic/cisgenic type approaches hold clear promise in this regard. Rapid analysis of crop near-relatives, for example, wild species within *Vitis* genus, has been demonstrated via NGS techniques (Myles et al., 2010) and can identify sources of suitable material for such cisgenic/intragenic approaches.

2.3.3 Improving Nutritional Qualities of Crop Varieties by Mutagenesis

In contrast to the recent development of GM technology, chemical- and radiation-based mutagenesis approaches for crop improvement have a long history of improving nutritional qualities of everyday food crops (Takeda and Matsuoka, 2008). However, despite major genetic and genomic changes, such mutants are not classified as genetically modified and hence have a less burdensome regulatory route than is currently the case for transgenic crops and foods especially within the European Union member states (Gómez-Galera et al., 2012). Such mutants are typically selected from forward genetic screens for desirable phenotypes and, if the mutation displays simple Mendelian inheritance, introgressed by backcrossing into elite varieties. The backcrossing process removes unwanted secondary mutations, which may have occurred at other regions of the genome. In some genetic backgrounds, the introgressed mutant allele may have superior or inferior qualities due to gene × genotype interactions and an empirical approach to determine this has still to be employed. Depending on the locus, a small genetic change generating a mutant allele can lead to major changes to phenotype, especially if the gene mutated is rate-determining for a biochemical pathway. Mutagenesis has therefore proved valuable in introducing variation for nutritional and anti-nutritional properties. One example is the generation of a variety of pea with reduced phytate, an indigestible phosphorous source that chelates zinc and iron ions and prevents them being efficiently absorbed from food (Warkentin et al., 2012). Modification of seed storage proteins to minimize their concentration of anti-nutritional compounds has also been achieved in soybean (Lee et al., 2011).

2.3.4 Targeting Induced Local Lesions in Genomes

Transgenic approaches represent a powerful means to improve crops through introduction of novel genes or changing the expression patterns and levels of existing genes. However, transgenic approaches may not be available for all crops or varieties. An alternative technique, termed targeting induced local lesions in genomes (TILLING), has been developed as an application of reverse genetics to facilitate mutagenesis (McCallum et al., 2000a) and has been used in many crops (Table 2.2). TILLING combines traditional chemical

TABLE 2.2

Applications of TILLING, Eco-TILLING, and GE in Different Crop Species

Technique	Species Applied	References	Remarks
TILLING	Species with sequenced genomes:		
	Soybean	Stephenson et al. (2010)	
	Brassica (various)	Weil (2009)	
	Cereals (various)	Parry et al. (2009)	
	Tomato	Minoia et al. (2010)	Included analysis of storage traits
	Cucumber	Minoia et al. (2010)	
	Nonsequenced genomes:		
	Peanut, sunflower, lettuce, melons	Minoia et al. (2010)	
	Wheat	Uauy et al. (2009)	275 verified alleles developed
Eco-TILLING	Bananas	Till et al. (2010)	80 clonal cultivars, 800 markers developed
	Brassica	Wang et al. (2010)	Altered erucic acid content
	Teff, *Eragrostis tef*	Esfeld et al. (2009)	Harvesting traits were targeted
Genome editing	Zn finger nucleases (ZFNs):		
	Model organisms and culture cells	Urnov et al. (2010)	
	Maize	Weinthal et al. (2010)	
	Soybean	Curtin et al. (2011)	Demonstration that paralogs could be distinguished
	TALE nucleases (TALENS):	Mahfouz et al. (2011)	
	Arabidopsis thaliana	Cermak et al. (2011)	
	Culture cells	Mussolino et al. (2011)	Reduced risk of cytotoxicity claimed in this report

mutagenesis with molecular screening approaches to discover induced point mutations in genes that are known to be controlling important traits in crops whose genome sequence is known. TILLING can add new alleles to existing genomes and faces minimal regulatory hurdles in most countries except Canada as TILLING is typically classified as non-GM. TILLING has therefore been used in the improvement of domesticated crops based on prior knowledge of genes of interest for trait improvements (Kurowska et al., 2011; McCallum et al., 2000b; Slade and Knauf, 2005).

2.3.5 Genome Editing

Genome editing (GE) is a loosely defined term coined to refer to a suite of emerging techniques that permit mutagenesis directly at a locus of interest through the use of sequence-specific DNA-modifying enzymes (Marton et al., 2010). Unlike TILLING, the mutagenesis process used in GE is selected from specific loci that have been identified as *a priori* candidates, whereas TILLING involves generating mutations across the genome followed by detection of mutations in the gene/locus of interest (Table 2.2). As with TILLING, GE also poses major challenges for simplistic regulatory frameworks that attempt to effect process-driven distinctions between "GM" and "non-GM" (Morris and Spillane, 2008). Despite the "proofs of principle" shown for applications of GE in different species (Table 2.2), GE remains another technique that is essentially exploratory, albeit of great promise for food crop improvement in the future.

2.3.6 Genomics and the Food-Processing Industry

The ongoing genomics revolution is now beginning to impact heavily on crop improvement strategies, with many new agri-food products under development. Postharvest processing of crops is important for ensuring that plant nutrients are retained for human consumption. In particular, genomics provides immense amounts of data on the plants, microorganisms, and biochemical/enzymatic pathways required to produce novel and improved foods by fermentation and other biochemical conversion processes. Future food security relies not only upon the development of crops with metabolite portfolios optimal for human health but also on their effective storage; for example, ~26% of all consumer food was lost due to postharvest perishing in the United States in 2008 (Hodges et al., 2011).

2.3.7 Genomic Analysis of Nutrient Availability during Food Processing and Storage

Functional genomics of cooked rice has been assessed with respect to bioactive content (Heuberger et al., 2010), and SNPs associated with altered production of free phenolics and vitamin E have been identified in both coding regions and untranslated regions (UTRs) of genes involved in their biosynthesis. Analysis of the effects of processing have also been essential for demonstrating that consumption of golden rice leads to the expected increases in retinol (Pillay et al., 2011). Similar approaches have assessed the effects of cooking on provitamin A–enriched maize (Li et al., 2007) and iron-enriched soybean (Hoppler et al., 2008). Breeding programs for nutrient bioavailability have had some success in increasing the bioavailability of essential amino acids in maize, as a development of basic research on transgenic mechanisms for altering amino-acid content (Frizzi et al., 2008). Such approaches were instrumental in the development of quality protein maize (Nuss and Tanumihardjo, 2011). Maize is relatively rich in niacin, but it must be processed correctly to make it available in the diet, which is essential for preventing pellagra (Nuss and Tanumihardjo, 2010). The effects of typical cooking techniques on protein availability from enriched sorghum have also been assessed in a similar manner (da Silva et al., 2011; Taylor and Taylor, 2011). Development of suitable makers for bioavailability therefore remains a priority. While better postharvest storage and handling techniques drastically increase the shelf life of food, plant genomics can also be harnessed to improve postharvest qualities (Han and Korban, 2011). This has been applied to vegetables (Silva Dias and Ortiz, 2011) and especially cash fruits through TILLING (Dahmani-Mardas et al., 2010) and transgenic approaches (Meli et al., 2010). Other examples include the application of genomics for improved understanding of the mechanisms of postharvest losses by disease or premature ripening. These have led to new preservation techniques such as treatment with salicylic acid (Asghari and Aghdam, 2010).

2.3.8 Microbial Genomics for Efficient Food Processing

A major challenge for the food industry is to produce safe foods with the desired functionality by minimal processing technologies, many of which rely upon microbe-mediated and enzyme-mediated processing techniques. The emergence of genomics has opened up new avenues for the systematic analysis of microbial metabolism and microbial responses to environment (Canchaya et al., 2006; Sieuwerts et al., 2008, 2010). Rational development of novel food-processing techniques will therefore be aided by advances in microbial genomics. Many such genomics-assisted approaches have been taken to assist major food-processing industries such as cheese (Irlinger and Mounier, 2009), beer (Smid and Hugenholtz, 2010), and vinegar (Raspor and Goranovič, 2008). As little prior genomic knowledge is needed to identify different bacterial strains, it is possible to apply these techniques to specialized environments involved in traditional forms of food processing worldwide. The roles of *Lactobacillus plantarum*, *Lactobacillus fermentum*, *Leuconostoc pseudomesenteroides*, and *Enterococcus casseliflavus* in cocoa fermentation have been identified by genetic approaches (Camu et al., 2007), as have the roles of fungi associated with the fermentation of Puer tea (Abe et al., 2008) and acetobacteria used in artisanal balsamic vinegar manufacture (Gullo et al., 2009). However, the importance of microbial fermentation for maximizing nutrient bioavailability has not been thoroughly researched. Studies that have examined this area include reports on Acetobacteraceae diversity, which may assist their use in the production of precursors to various nutrients, including vitamin C (Raspor and Goranovič, 2008).

2.3.9 Microbial Genomics in Food Safety

Microbial genomics is increasingly being used in toxicological assessments to ensure food safety (Brown and van der Ouderaa, 2007). Genome sequences are now available for many microbes that cause food-borne diseases, for example, *Escherichia coli, Listeria monocytogenes, Yersinia enterocolitica, Clostridium botulinum, Campylobacter jejuni,* and various *Salmonella* species. Analysis of methods for preventing food-borne diseases has suggested that early-stage identification and typing of pathogenic and spoilage microorganisms are crucial (Quested et al., 2010). Well-established pre-genomics era methods for such molecular typing include plasmid typing, pulsed-field gel electrophoresis, ribotyping, and RAPD analysis (Kathariou, 2002). The ability of genomic information to provide rapidly screened strain-specific genetic signatures for detecting microbial contamination is a key advancement (Fang et al., 2010). For example, comparison of the genome sequence data of 97 strains of *C. jejuni* was needed to identify rare examples of strain-specific genomic regions (Taboada et al., 2004). The complete genome sequences of different strains of *L. monocytogenes* and the closely related nonpathogenic strain *L. innocua* have been determined (Gilmour et al., 2010; Glaser et al., 2001), and as gene order is largely conserved, regions that have undergone insertions or deletions between their chromosomes are the best possible regions for use as pathogenicity markers (Nelson et al., 2004).

In some instances, marker identification has already led to industrial applications (Rasooly and Herold, 2008). For example, DNA chip–based technologies have made use of the identification of markers for *Bacillus subtilis* sporulation to allow spoilage risk to be determined (Brul et al., 2006). A PCR-based screen has been developed to allow identification of bacterial genomic regions associated with pathogenicity in situ and its efficacy demonstrated against *Salmonella* (Lermo et al., 2007). Furthermore, an array-based method has been reported that is able to distinguish between 11 agents of major food-borne diseases through the use of 70 unique nucleotide sequences identified from their genomes (Kim et al., 2008). In most instances, these technological advances are relatively simple and easy to use within the food-processing industry, suggesting that the limiting scientific factor in the application of microbial genomics to food processing remains the identification of diagnostic sequences. In addition to benefiting from the advances in NGS technology described in previous sections, the study of microbes has also made major progress through the field of metagenomics. Metagenomics describes the sampling of genome sequences of a community of organisms inhabiting a common environment and can be broadly defined as any type of analysis of DNA obtained directly from the environment (Hugenholtz and Tyson, 2008; Streit and Schmitz, 2004) and may improve understanding of how complex microbial communities function and their members interact within their niches (Valenzuela et al., 2006). For microbes that are difficult to culture or detect, the advent of metagenomics promises dramatic advances (Gilbert and Dupont, 2011), although the diversity of substrates in which microbes occur in crop and food processing may prove an obstacle to immediate application of metagenomics to food-borne microbes (Justé et al., 2008).

2.4 Summary and Future Prospects

A major challenge for global development is to ensure that the world's burgeoning population has sufficient access to essential dietary nutrients in a form accessible to human digestion. This target still appears distant because many staple plant-based foodstuffs are lacking in essential nutrients, or contain them in a form that is lost or inaccessible when consumed. The genome sequences for many cereals and other staple crops are now available (Table 2.1), and so are the genome sequences of many of the main microbes associated with food processing and loss. Some of the opportunities that this wealth of genomic data offers for improving the nutrient content of plant-derived food have been described in this chapter. Genomics has already revolutionized breeding methodology through mutagenesis and MAS and by identifying functions of genes. A wide range of prototype varieties with improved nutritional traits have been developed, but few have reached the markets (particularly transgenic varieties), meaning that their integration into sustainable agricultural systems remains a challenge for the future (Figure 2.2). In contrast to the situation in crops, many genetics and genomics-based applications have been developed for understanding the role of microbial pathogens in fermentation and food processing and in combating food-borne pathogens. On the other hand, genomics has not been majorly applied to the improvement of nutrient bioavailability during food processing. To conclude, advances in genomics in food and agriculture will be essential for helping to prevent malnutrition and allowing the development of diets personalized for health via nutrigenomics targeted at individuals or populations.

References

Abe, M., Takaoka, N., Idemoto, Y., Takagi, C., Imai, T., and Nakasaki, K. 2008. Characteristic fungi observed in the fermentation process for Puer tea. *International Journal of Food Microbiology* 124: 199–203.

Aluru, M., Xu, Y., Guo, R. et al. 2008. Generation of transgenic maize with enhanced provitamin A content. *Journal of Experimental Botany* 59: 3551–3562.

Asghari, M. and Aghdam, M.S. 2010. Impact of salicylic acid on post-harvest physiology of horticultural crops. *Trends in Food Science & Technology* 21: 502–509.

Atwell, S., Huang, Y.S., Vilhjalmsson, B.J. et al. 2010. Genome-wide association study of 107 phenotypes in *Arabidopsis thaliana* inbred lines. *Nature* 465: 627–631.

Beddington, J. 2010. Food security: Contributions from science to a new and greener revolution. *Philosophical Transactions of the Royal Society B: Biological Sciences* 365: 61–71.

Beyer, P., Al-Babili, S., Ye, X. et al. 2002. Golden rice: Introducing the β-carotene biosynthesis pathway into rice endosperm by genetic engineering to defeat vitamin A deficiency. *The Journal of Nutrition* 132: 506S–510S.

Brown, L. and van der Ouderaa, F. 2007. Nutritional genomics: Food industry applications from farm to fork. *British Journal of Nutrition* 97: 1027–1035.

Brul, S., Schuren, F., Montijn, R., Keijser, B.J.F., van der Spek, H., and Oomes, S.J.C.M. 2006. The impact of functional genomics on microbiological food quality and safety. *International Journal of Food Microbiology* 112: 195–199.

Camu, N., De Winter, T., Verbrugghe, K. et al. 2007. Dynamics and biodiversity of populations of lactic acid bacteria and acetic acid bacteria involved in spontaneous heap fermentation of cocoa beans in Ghana. *Applied and Environmental Microbiology* 73: 1809–1824.

Canchaya, C., Claesson, M.J., Fitzgerald, G.F., van Sinderen, D., and O'Toole, P.W. 2006. Diversity of the genus *Lactobacillus* revealed by comparative genomics of five species. *Microbiology* 152: 3185–3196.

Cavanagh, C., Morell, M., Mackay, I., and Powell, W. 2008. From mutations to MAGIC: Resources for gene discovery, validation and delivery in crop plants. *Current Opinion in Plant Biology* 11: 215–221.

Cermak, T., Doyle, E.L., Christian, M. et al. 2011. Efficient design and assembly of custom TALEN and other TAL effector-based constructs for DNA targeting. *Nucleic Acids Research* 39: e82.

Cheng, F., Liu, S., Wu, J. et al. 2011. BRAD, the genetics and genomics database for *Brassica* plants. *BMC Plant Biology* 11: 136.

Chutimanitsakun, Y., Nipper, R., Cuesta-Marcos, A. et al. 2011. Construction and application for QTL analysis of a Restriction Site Associated DNA (RAD) linkage map in barley. *BMC Genomics* 12: 4.

Consortium. 2011. Genome sequence and analysis of the tuber crop potato. *Nature* 475: 189–195.

Consortium. 2012. The tomato genome sequence provides insights into fleshy fruit evolution. *Nature* 485: 635–641.

Cronn, R., Knaus, B.J., Liston, A. et al. 2012. Targeted enrichment strategies for next-generation plant biology. *American Journal of Botany* 99: 291–311.

Crossa, J., Campos, G.d.l., Pérez, P. et al. 2010. Prediction of genetic values of quantitative traits in plant breeding using pedigree and molecular markers. *Genetics* 186: 713–724.

Curtin, S.J., Zhang, F., Sander, J.D. et al. 2011. Targeted mutagenesis of duplicated genes in soybean with zinc-finger nucleases. *Plant Physiology* 156: 466–473.

Dahmani-Mardas, F., Troadec, C., Boualem, A. et al. 2010. Engineering melon plants with improved fruit shelf life using the TILLING approach. *PLoS ONE* 5: e15776.

Datta, S., Datta, K., Parkhi, V. et al. 2007. Golden rice: Introgression, breeding, and field evaluation. *Euphytica* 154: 271–278.

Davey, J.W., Hohenlohe, P.A., Etter, P.D., Boone, J.Q., Catchen, J.M., and Blaxter, M.L. 2011. Genome-wide genetic marker discovery and genotyping using next-generation sequencing. *Nature Reviews Genetics* 12: 499–510.

Davies, K.M. 2007. Genetic modification of plant metabolism for human health benefits. *Mutation Research/Fundamental and Molecular Mechanisms of Mutagenesis* 622: 122–137.

Delseny, M., Han, B., and Hsing, Y.I. 2010. High throughput DNA sequencing: The new sequencing revolution. *Plant Science* 179: 407–422.

Edwards, D. and Batley, J. 2010. Plant genome sequencing: Applications for crop improvement. *Plant Biotechnology Journal* 8: 2–9.

Egan, A.N., Schlueter, J., and Spooner, D.M. 2012. Applications of next-generation sequencing in plant biology. *American Journal of Botany* 99: 175–185.

Elshire, R.J., Glaubitz, J.C., Sun, Q. et al. 2011. A robust, simple genotyping-by-sequencing (GBS) approach for high diversity species. *PLoS ONE* 6: e19379.

Esfeld, K., Plaza, S., and Tadele, Z. 2009. Bringing high-throughput techniques to orphan crop of Africa: Highlights from the tef TILLING project. *Gene Conserve* 33: 804–807.

Fang, H., Xu, J., Ding, D. et al. 2010. An FDA bioinformatics tool for microbial genomics research on molecular characterization of bacterial foodborne pathogens using microarrays. *BMC Bioinformatics* 11(Suppl 6): S4.

Feuillet, C., Leach, J.E., Rogers, J., Schnable, P.S., and Eversole, K. 2011. Crop genome sequencing: Lessons and rationales. *Trends in Plant Science* 16: 77–88.

Foolad, M.R. and Panthee, D.R. 2012. Marker-assisted selection in tomato breeding. *Critical Reviews in Plant Sciences* 31: 93–123.

Frizzi, A., Huang, S., Gilbertson, L.A., Armstrong, T.A., Luethy, M.H., and Malvar, T.M. 2008. Modifying lysine biosynthesis and catabolism in corn with a single bifunctional expression/silencing transgene cassette. *Plant Biotechnology Journal* 6: 13–21.

Ganal, M.W., Altmann, T., and Röder, M.S. 2009. SNP identification in crop plants. *Current Opinion in Plant Biology* 12: 211–217.

Garvin, M.R., Saitoh, K., and Gharrett, A.J. 2010. Application of single nucleotide polymorphisms to non-model species: A technical review. *Molecular Ecology Resources* 10: 915–934.

Gilbert, J.A. and Dupont, C.L. 2011. Microbial metagenomics: Beyond the genome. *Annual Review of Marine Science* 3: 347–371.

Gilmour, M., Graham, M., Van Domselaar, G. et al. 2010. High-throughput genome sequencing of two *Listeria monocytogenes* clinical isolates during a large foodborne outbreak. *BMC Genomics* 11: e120.

Giuliano, G., Tavazza, R., Diretto, G., Beyer, P., and Taylor, M.A. 2008. Metabolic engineering of carotenoid biosynthesis in plants. *Trends in Biotechnology* 26: 139–145.

Glaser, P., Frangeul, L., Buchrieser, C. et al. 2001. Comparative genomics of *Listeria* species. *Science* 294: 849–852.

Goddard, M.E. and Hayes, B.J. 2007. Genomic selection. *Journal of Animal Breeding and Genetics* 124: 323–330.

Gómez-Galera, S., Twyman, R.M., Sparrow, P.A.C. et al. 2012. Field trials and tribulations—Making sense of the regulations for experimental field trials of transgenic crops in Europe. *Plant Biotechnology Journal* 10: 511–523.

Grattapaglia, D., de Alencar, S., and Pappas, G. 2011. Genome-wide genotyping and SNP discovery by ultra-deep Restriction-Associated DNA (RAD) tag sequencing of pooled samples of *E. grandis* and *E. globulus*. *BMC Proceedings* 5: P45.

Grattapaglia, D. and Resende, M. 2011. Genomic selection in forest tree breeding. *Tree Genetics & Genomes* 7: 241–255.

Gujaria, N., Kumar, A., Dauthal, P. et al. 2011. Development and use of genic molecular markers (GMMs) for construction of a transcript map of chickpea (*Cicer arietinum* L.). *Theoretical and Applied Genetics* 122: 1577–1589.

Gullo, M., De Vero, L., and Giudici, P. 2009. Succession of selected strains of *Acetobacter pasteurianus* and other acetic acid bacteria in traditional balsamic vinegar. *Applied and Environmental Microbiology* 75: 2585–2589.

Hall, D., Tegström, C., and Ingvarsson, P.K. 2010. Using association mapping to dissect the genetic basis of complex traits in plants. *Briefings in Functional Genomics* 9: 157–165.

Hamblin, M.T., Buckler, E.S., and Jannink, J.L. 2011. Population genetics of genomics-based crop improvement methods. *Trends in Genetics* 27: 98–106.

Hamblin, M.T. and Jannink, J.L. 2011. Factors affecting the power of haplotype markers in association studies. *Plant Genetics* 4: 145–153.

Hamilton, J., Hansey, C., Whitty, B. et al. 2011. Single nucleotide polymorphism discovery in Elite North American potato germplasm. *BMC Genomics* 12: 302.

Hamilton, J.P. and Robin Buell, C. 2012. Advances in plant genome sequencing. *The Plant Journal* 70: 177–190.

Han, Y. and Korban, S.S. 2011 Transgenic approaches to improve fruit quality. In: *Breeding for Fruit Quality*, pp. 151–171. John Wiley & Sons, Inc., New York.

Hansey, C.N., Vaillancourt, B., Sekhon, R.S., de Leon, N., Kaeppler, S.M., and Buell, C.R. 2012. Maize (*Zea mays* L.) genome diversity as revealed by RNA-sequencing. *PLoS ONE* 7: e33071.

Heffner, E.L., Jannink, J.-L., Iwata, H., Souza, E., and Sorrells, M.E. 2011. Genomic selection accuracy for grain quality traits in biparental wheat populations. *Crop Science* 51: 2597–2606.

Heffner, E.L., Sorrells, M.E., and Jannink, J.L. 2009. Genomic selection for crop improvement. *Crop Science* 49: 1–12.

Herman, R.A., Chassy, B.M., and Parrott, W. 2009. Compositional assessment of transgenic crops: An idea whose time has passed. *Trends in Biotechnology* 27: 555–557.

Heuberger, A.L., Lewis, M.R., Chen, M.H., Brick, M.A., Leach, J.E., and Ryan, E.P. 2010. Metabolomic and functional genomic analyses reveal varietal differences in bioactive compounds of cooked rice. *PLoS ONE* 5: e12915.

Hirschi, K.D. 2009. Nutrient biofortification of food crops. *Annual Review of Nutrition* 29: 401–421.

Hodges, R.J., Buzby, J., and Bennett, B. 2011. Postharvest losses and waste in developed and less developed countries: Opportunities to improve resource use. *The Journal of Agricultural Science* 149: 37–45.

Hoppler, M., Schönbächler, A., Meile, L., Hurrell, R.F., and Walczyk, T. 2008. Ferritin-iron is released during boiling and in vitro gastric digestion. *The Journal of Nutrition* 138: 878–884.

Hu, T.T., Pattyn, P., Bakker, E.G. et al. 2011. The *Arabidopsis lyrata* genome sequence and the basis of rapid genome size change. *Nature Genetics* 43: 476–481.

Huang, X., Wei, X., Sang, T. et al. 2010. Genome-wide association studies of 14 agronomic traits in rice landraces. *Nature Genetics* 42: 961–967.

Hugenholtz, P. and Tyson, G.W. 2008. Microbiology: Metagenomics. *Nature* 455: 481–483.

Hyten, D., Cannon, S., Song, Q. et al. 2010a. High-throughput SNP discovery through deep resequencing of a reduced representation library to anchor and orient scaffolds in the soybean whole genome sequence. *BMC Genomics* 11: e38.

Hyten, D., Song, Q., Fickus, E. et al. 2010b. High-throughput SNP discovery and assay development in common bean. *BMC Genomics* 11: 475.

Irlinger, F. and Mounier, J. 2009. Microbial interactions in cheese: Implications for cheese quality and safety. *Current Opinion in Biotechnology* 20: 142–148.

Jaillon, O., Aury, J.M., Noel, B. et al. 2007. The grapevine genome sequence suggests ancestral hexaploidization in major angiosperm phyla. *Nature* 449: 463–467.

Jannink, J.-L., Lorenz, A.J., and Iwata, H. 2010. Genomic selection in plant breeding: From theory to practice. *Briefings in Functional Genomics* 9: 166–177.

Justé, A., Thomma, B.P.H.J., and Lievens, B. 2008. Recent advances in molecular techniques to study microbial communities in food-associated matrices and processes. *Food Microbiology* 25: 745–761.

Kathariou, S. 2002. *Listeria monocytogenes* virulence and pathogenicity, a food safety perspective. *Journal of Food Protection* 65: 1811–1829.

Kim, H.-J., Park, S.-H., Lee, T.-H., Nahm, B.-H., Kim, Y.-R., and Kim, H.-Y. 2008. Microarray detection of food-borne pathogens using specific probes prepared by comparative genomics. *Biosensors and Bioelectronics* 24: 238–246.

Kurowska, M., Daszkowska-Golec, A., Gruszka, D. et al. 2011. TILLING—A shortcut in functional genomics. *Journal of Applied Genetics* 52: 371–390.

Lee, K.J., Kim, J.-B., Kim, S.H. et al. 2011. Alteration of seed storage protein composition in soybean [*Glycine max* (L.) Merrill] mutant lines induced by γ-irradiation mutagenesis. *Journal of Agricultural and Food Chemistry* 59: 12405–12410.

Lermo, A., Campoy, S., Barbé, J., Hernández, S., Alegret, S., and Pividori, M.I. 2007. In situ DNA amplification with magnetic primers for the electrochemical detection of food pathogens. *Biosensors and Bioelectronics* 22: 2010–2017.

Li, S., Tayie, F.A.K., Young, M.F., Rocheford, T., and White, W.S. 2007. Retention of provitamin A carotenoids in high β-carotene maize (*Zea mays*) during traditional African household processing. *Journal of Agricultural and Food Chemistry* 55: 10744–10750.

Ligterink, W., Joosen, R.V.L., and Hilhorst, H.W.M. 2012. Unravelling the complex trait of seed quality: Using natural variation through a combination of physiology, genetics and -omics technologies. *Seed Science Research* 22: S45–S52.

Mahfouz, M.M., Li, L., Shamimuzzaman, M., Wibowo, A., Fang, X., and Zhu, J.-K. 2011. De novo-engineered transcription activator-like effector (TALE) hybrid nuclease with novel DNA binding specificity creates double-strand breaks. *Proceedings of the National Academy of Sciences U S A* 108: 2623–2628.

Markowitz, V.M., Chen, I.-M.A., Palaniappan, K. et al. 2010. The integrated microbial genomes system: An expanding comparative analysis resource. *Nucleic Acids Research* 38: D382–D390.

Marton, I., Zuker, A., Shklarman, E. et al. 2010. Nontransgenic genome modification in plant cells. *Plant Physiology* 154: 1079–1087.

Mayer, J.E., Pfeiffer, W.H., and Beyer, P. 2008. Biofortified crops to alleviate micronutrient malnutrition. *Current Opinion in Plant Biology* 11: 166–170.

McCallum, C.M., Comai, L., Greene, E.A., and Henikoff, S. 2000a. Targeted screening for induced mutations. *Nature Biotechnology* 18: 455–457.

McCallum, C.M., Comai, L., Greene, E.A., and Henikoff, S. 2000b. Targeting Induced Local Lesions IN Genomes (TILLING) for plant functional genomics. *Plant Physiology* 123: 439–442.

Meli, V.S., Ghosh, S., Prabha, T.N., Chakraborty, N., Chakraborty, S., and Datta, A. 2010. Enhancement of fruit shelf life by suppressing *N*-glycan processing enzymes. *Proceedings of the National Academy of Sciences U S A* 107: 2413–2418.

Meuwissen, T.H.E., Hayes, B.J., and Goddard, M.E. 2001. Prediction of total genetic value using genome-wide dense marker maps. *Genetics* 157: 1819–1829.

Minoia, S., Petrozza, A., D'Onofrio, O. et al. 2010. A new mutant genetic resource for tomato crop improvement by TILLING technology. *BMC Research Notes* 3: e69.

Mittler, R. and Blumwald, E. 2010. Genetic engineering for modern agriculture: Challenges and perspectives. *Annual Review of Plant Biology* 61: 443–462.

Monson-Miller, J., Sanchez-Mendez, D., Fass, J., Henry, I., Tai, T., and Comai, L. 2012. Reference genome-independent assessment of mutation density using restriction enzyme-phased sequencing. *BMC Genomics* 13: e72.

Moose, S.P. and Mumm, R.H. 2008. Molecular plant breeding as the foundation for 21st century crop improvement. *Plant Physiology* 147: 969–977.

Morrell, P.L., Buckler, E.S., and Ross-Ibarra, J. 2012. Crop genomics: Advances and applications. *Nature Reviews Genetics* 13: 85–96.

Morris, S.H. and Spillane, C. 2008. GM directive deficiencies in the European Union. *EMBO Reports* 9: 500–504.

Mussolino, C., Morbitzer, R., Lütge, F., Dannemann, N., Lahaye, T., and Cathomen, T. 2011. A novel TALE nuclease scaffold enables high genome editing activity in combination with low toxicity. *Nucleic Acids Research* 39: 9283–9293.

Myles, S., Chia, J.-M., Hurwitz, B. et al. 2010. Rapid genomic characterization of the genus *Vitis*. *PLoS ONE* 5: e8219.

Naqvi, S., Zhu, C., Farre, G. et al. 2009. Transgenic multivitamin corn through biofortification of endosperm with three vitamins representing three distinct metabolic pathways. *Proceedings of the National Academy of Sciences USA* 106: 7762–7767.

Nelson, K.E., Fouts, D.E., Mongodin, E.F. et al. 2004. Whole genome comparisons of serotype 4b and 1/2a strains of the food-borne pathogen *Listeria monocytogenes* reveal new insights into the core genome components of this species. *Nucleic Acids Research* 32: 2386–2395.

Nuss, E.T. and Tanumihardjo, S.A. 2010. Maize: A paramount staple crop in the context of global nutrition. *Comprehensive Reviews in Food Science and Food Safety* 9: 417–436.

Nuss, E.T. and Tanumihardjo, S.A. 2011. Quality protein maize for Africa: Closing the protein inadequacy gap in vulnerable populations. *Advances in Nutrition: An International Review Journal* 2: 217–224.

van Orsouw, N.J., Hogers, R.C.J., Janssen, A. et al. 2007. Complexity Reduction of Polymorphic Sequences (CRoPS™): A novel approach for large-scale polymorphism discovery in complex genomes. *PLoS ONE* 2: e1172.

Ouyang, S., Zhu, W., Hamilton, J. et al. 2007. The TIGR Rice Genome Annotation Resource: Improvements and new features. *Nucleic Acids Research* 35: D883–D887.

Paine, J.A., Shipton, C.A., Chaggar, S. et al. 2005. Improving the nutritional value of golden rice through increased pro-vitamin A content. *Nature Biotechnology* 23: 482–487.

Parry, M.A.J., Madgwick, P.J., Bayon, C. et al. 2009. Mutation discovery for crop improvement. *Journal of Experimental Botany* 60: 2817–2825.

Paterson, A.H., Bowers, J.E., Bruggmann, R. et al. 2009. The *Sorghum bicolor* genome and the diversification of grasses. *Nature* 457: 551–556.

Peng, W. 2011. GM crop cultivation surges, but novel traits languish. *Nature Biotechnology* 29: 302–302.

Piepho, H.P. 2009. Ridge regression and extensions for genomewide selection in maize. *Crop Science* 49: 1165–1176.

Pillay, K., Siwela, M., Derera, J., and Veldman, F. 2011. Provitamin A carotenoids in biofortified maize and their retention during processing and preparation of South African maize foods. *Journal of Food Science and Technology* 10: 1007.

Poland, J.A., Brown, P.J., Sorrells, M.E., and Jannink, J.-L. 2012. Development of high-density genetic maps for barley and wheat using a novel two-enzyme genotyping-by-sequencing approach. *PLoS ONE* 7: e32253.

Potrykus, I. 2010. Regulation must be revolutionized. *Nature* 466: 561–561.

Prochnik, S., Marri, P.R., Desany, B., Rabinowicz, P.D., Kodira, C., Mohiuddin, M., Rodriguez, F., Fauquet, C., Tohme, J., and Harkins, T. 2012. The cassava genome: Current progress, future directions. *Tropical Plant Biology* 5: 88–94.

Quested, T.E., Cook, P.E., Gorris, L.G.M., and Cole, M.B. 2010. Trends in technology, trade and consumption likely to impact on microbial food safety. *International Journal of Food Microbiology* 139(Suppl): S29–S42.

Rasooly, A. and Herold, K.E. 2008. Food microbial pathogen detection and analysis using DNA microarray technologies. *Foodborne Pathogens and Disease* 5: 531–550.

Raspor, P. and Goranovič, D. 2008. Biotechnological applications of acetic acid bacteria. *Critical Reviews in Biotechnology* 28: 101–124.

Ribaut, J.M., de Vicente, M.C., and Delannay, X. 2010. Molecular breeding in developing countries: Challenges and perspectives. *Current Opinion in Plant Biology* 13: 213–218.

Rommens, C.M. 2004. All-native DNA transformation: A new approach to plant genetic engineering. *Trends in Plant Science* 9: 457–464.

Rommens, C.M., Haring, M.A., Swords, K., Davies, H.V., and Belknap, W.R. 2007. The intragenic approach as a new extension to traditional plant breeding. *Trends in Plant Science* 12: 397–403.

Schmutz, J., Cannon, S.B., Schlueter, J. et al. 2010. Genome sequence of the palaeopolyploid soybean. *Nature* 463: 178–183.

Schouten, H.J. and Jacobsen, E. 2008. Cisgenesis and intragenesis, sisters in innovative plant breeding. *Trends in Plant Science* 13: 260–261.

Schouten, H.J., Krens, F.A., and Jacobsen, E. 2006. Do cisgenic plants warrant less stringent oversight? *Nature Biotechnology* 24: 753–753.

Sieuwerts, S., de Bok, F.A.M., Hugenholtz, J., and Vlieg, J.E.T.V. 2008. Unraveling microbial interactions in food fermentations: From classical to genomics approaches. *Applied and Environmental Microbiology* 74: 4997–5007.

Sieuwerts, S., Molenaar, D., van Hijum, S.A.F.T. et al. 2010. Mixed-culture transcriptome analysis reveals the molecular basis of mixed-culture growth in *Streptococcus thermophilus* and *Lactobacillus bulgaricus*. *Applied and Environmental Microbiology* 76: 7775–7784.

da Silva, L.S., Taylor, J., and Taylor, J.R.N. 2011. Transgenic sorghum with altered kafirin synthesis: Kafirin solubility, polymerization, and protein digestion. *Journal of Agricultural and Food Chemistry* 59: 9265–9270.

Silva Dias, J. and Ortiz, R. 2011 Transgenic vegetable crops: Progress, potentials, and prospects. In: *Plant Breeding Reviews*, ed. J. Janick, pp. 151–246. John Wiley & Sons, Inc., Hoboken, NY.

Slade, A.J. and Knauf, V.C. 2005. TILLING moves beyond functional genomics into crop improvement. *Transgenic Research* 14: 109–115.

Smid, E.J. and Hugenholtz, J. 2010. Functional genomics for food fermentation processes. *Annual Review of Food Science and Technology* 1: 497–519.

Stephenson, P., Baker, D., Girin, T. et al. 2010. A rich TILLING resource for studying gene function in *Brassica rapa*. *BMC Plant Biology* 10: e62.

Streit, W.R. and Schmitz, R.A. 2004. Metagenomics—The key to the uncultured microbes. *Current Opinion in Microbiology* 7: 492–498.

Taboada, E.N., Acedillo, R.R., Carrillo, C.D. et al. 2004. Large-scale comparative genomics meta-analysis of *Campylobacter jejuni* isolates reveals low level of genome plasticity. *Journal of Clinical Microbiology* 42: 4566–4576.

Takeda, S. and Matsuoka, M. 2008. Genetic approaches to crop improvement: Responding to environmental and population changes. *Nature Reviews Genetics* 9: 444–457.

Tang, H., Bowers, J.E., Wang, X., Ming, R., Alam, M., and Paterson, A.H. 2008. Synteny and collinearity in plant genomes. *Science* 320: 486–488.

Tang, G., Qin, J., Dolnikowski, G.G., Russell, R.M., and Grusak, M.A. 2009. Golden rice is an effective source of vitamin A. *The American Journal of Clinical Nutrition* 89: 1776–1783.

Taylor, J. and Taylor, J.R.N. 2011. Protein biofortified sorghum: Effect of processing into traditional African foods on their protein quality. *Journal of Agricultural and Food Chemistry* 59: 2386–2392.

Tester, M. and Langridge, P. 2010. Breeding technologies to increase crop production in a changing world. *Science* 327: 818–822.

Thompson, M.D. and Thompson, H.J. 2009. Biomedical agriculture: A systematic approach to food crop improvement for chronic disease prevention. In: *Advances in Agronomy*, ed. S. Donald, pp. 1–54. Academic Press, New York.

Thudi, M., Li, Y., Jackson, S.A., May, G.D., and Varshney, R.K. 2012. Current state-of-art of sequencing technologies for plant genomics research. *Briefings in Functional Genomics* 11: 3–11.

Till, B., Jankowicz-Cieslak, J., Sági, L. et al. 2010. Discovery of nucleotide polymorphisms in the *Musa* gene pool by Ecotilling. *Theoretical and Applied Genetics* 121: 1381–1389.

Tuskan, G.A., Difazio, S., Jansson, S. et al. 2006. The genome of black cottonwood, *Populus trichocarpa* (Torr. & Gray). *Science* 313: 1596–1604.

Uauy, C., Paraiso, F., Colasuonno, P. et al. 2009. A modified TILLING approach to detect induced mutations in tetraploid and hexaploid wheat. *BMC Plant Biology* 9: 1–14.

Urnov, F.D., Rebar, E.J., Holmes, M.C., Zhang, H.S., and Gregory, P.D. 2010. Genome editing with engineered zinc finger nucleases. *Nature Reviews Genetics* 11: 636–646.

Valenzuela, L., Chi, A., Beard, S. et al. 2006. Genomics, metagenomics and proteomics in biomining microorganisms. *Biotechnological Advances* 24: 197–211.

Van Tassell, C.P., Smith, T.P.L., Matukumalli, L.K. et al. 2008. SNP discovery and allele frequency estimation by deep sequencing of reduced representation libraries. *Nature Methods* 5: 247–252.

Varshney, R.K., Chen, W., Li, Y. et al. 2012. Draft genome sequence of pigeon pea (*Cajanus cajan*), an orphan legume crop of resource-poor farmers. *Nature Biotechnology* 30: 83–89.

Varshney, R.K. and Dubey, A. 2009. Novel genomic tools and modern genetic and breeding approaches for crop improvement. *Journal of Plant Biochemistry and Biotechnology* 18: 127–138.

Varshney, R.K., Hoisington, D.A., and Tyagi, A.K. 2006. Advances in cereal genomics and applications in crop breeding. *Trends in Biotechnology* 24: 490–499.

Varshney, R.K., Nayak, S.N., May, G.D., and Jackson, S.A. 2009. Next-generation sequencing technologies and their implications for crop genetics and breeding. *Trends in Biotechnology* 27: 522–530.

Vogel, J.P., Garvin, D.F., Mockler, T.C., Schmutz, J., Rokhsar, J., and Bevan, M.W. 2010. Genome sequencing and analysis of the model grass *Brachypodium distachyon. Nature* 463: 763–768.

Wang, N., Shi, L., Tian, F. et al. 2010. Assessment of *FAE1* polymorphisms in three *Brassica* species using EcoTILLING and their association with differences in seed erucic acid contents. *BMC Plant Biology* 10: e137.

Ward, J.A., Ponnala, L., and Weber, C.A. 2012. Strategies for transcriptome analysis in nonmodel plants. *American Journal of Botany* 99: 267–276.

Warkentin, T.D., Delgerjav, O., Arganosa, G. et al. 2012. Development and characterization of low-phytate pea. *Crop Science* 52: 74–78.

Weigel, D. and Mott, R. 2009. The 1001 Genomes Project for *Arabidopsis thaliana. Genome Biology* 10: 1–5.

Weil, C.F. 2009. TILLING in grass species. *Plant Physiology* 149: 158–164.

Weinthal, D., Tovkach, A., Zeevi, V., and Tzfira, T. 2010. Genome editing in plant cells by zinc finger nucleases. *Trends in Plant Science* 15: 308–321.

White, P.J. and Broadley, M.R. 2009. Biofortification of crops with seven mineral elements often lacking in human diets—Iron, zinc, copper, calcium, magnesium, selenium and iodine. *New Phytologist* 182: 49–84.

White, P.J. and Brown, P.H. 2010. Plant nutrition for sustainable development and global health. *Annals of Botany* 105: 1073–1080.

Xu, Y., Lu, Y., Xie, C., Gao, S., Wan, J., and Prasanna, B. 2012. Whole-genome strategies for marker-assisted plant breeding. *Molecular Breeding* 29: 833–854.

Yamamoto, T., Nagasaki, H., Yonemaru, J. et al. 2010. Fine definition of the pedigree haplotypes of closely related rice cultivars by means of genome-wide discovery of single-nucleotide polymorphisms. *BMC Genomics* 11: 267.

Ye, X., Al-Babili, S., Klöti, A. et al. 2000. Engineering the provitamin A (β-carotene) biosynthetic pathway into (carotenoid-free) rice endosperm. *Science* 287: 303–305.

You, F., Huo, N., Deal, K. et al. 2011. Annotation-based genome-wide SNP discovery in the large and complex *Aegilops tauschii* genome using next-generation sequencing without a reference genome sequence. *BMC Genomics* 12: e59.

Yu, J., Hu, S., Wang, J. et al. 2002. A draft sequence of the rice genome (*Oryza sativa* L. ssp. *indica*). *Science* 296: 79–92.

Part II

Biotechnology in Agriculture

3

Plant Cell and Tissue Culture Techniques in Crop Improvement

Dinesh K. Srivastava, Geetika Gambhir, and Poornima Sharma

CONTENTS

3.1 Introduction....75
3.2 Plant Cell and Tissue Culture....76
3.2.1 Plant Cell and Tissue Culture Nutrient Medium....76
3.2.2 Techniques of Plant Cell Culture....77
3.2.3 Applications of Cell Culture....78
3.2.3.1 Mutant Selection....78
3.2.3.2 Production of Secondary Metabolites....78
3.2.3.3 Biotransformation....79
3.2.4 Successful Case Studies....79
3.3 Meristem Culture....82
3.3.1 Methods for Virus Elimination....83
3.3.1.1 Thermotherapy/Heat Treatment....83
3.3.1.2 Chemotherapy....84
3.3.1.3 Other In Vitro Methods....84
3.3.2 Virus Indexing....85
3.4 Micropropagation....85
3.4.1 Micropropagation Systems....86
3.4.1.1 Single-Node Culture....86
3.4.1.2 Axillary Branching....86
3.4.1.3 Regeneration of Adventitious Buds/Shoots....87
3.4.2 Stages of Micropropagation....87
3.4.2.1 Preparation and Pretreatment of the Explants....87
3.4.2.2 Establishment of Aseptic Culture....87
3.4.2.3 Multiplication of Shoots....88
3.4.2.4 Root Regeneration in In Vitro Developed Shoots....90
3.4.2.5 Hardening for Subsequent Field Planting....92
3.4.2.6 Molecular Analysis of Genetic Stability in Micropropagated Plants....92
3.4.3 Progress in Micropropagation of Horticultural and Forest Tree Species....93
3.4.4 Successful Case Study....93

3.5 Anther Culture and Haploid Production 95
3.5.1 Techniques of Anther Culture and Isolated Pollen Culture 96
3.5.2 Culture Media and Nutritional Requirements 96
3.5.2.1 Direct Androgenesis 98
3.5.2.2 Indirect Androgenesis 98
3.5.3 Factors Affecting Androgenesis 99
3.5.4 Applications of Haploid Plants 99
3.5.4.1 Development of Pure Homozygous Lines 99
3.5.4.2 Hybrid Development 99
3.5.4.3 Induction of Mutations 99
3.5.4.4 Significance of Early Release of Varieties 100
3.5.5 Successful Case Studies 100
3.6 Embryo Culture 101
3.6.1 Types of Embryo Culture 102
3.6.1.1 Mature Embryo Culture 102
3.6.1.2 Immature Embryo Culture 102
3.6.2 Factors Affecting Embryo Culture 102
3.6.3 Applications of Embryo Culture 102
3.6.3.1 Prevention of Embryo Abortion 103
3.6.3.2 Precocious Germination 103
3.6.3.3 Seed Dormancy 103
3.6.3.4 Nutritional Requirements 103
3.6.4 Successful Case Studies 104
3.7 Protoplast Culture and Somatic Cell Hybridization 104
3.7.1 Isolation, Culture, and Regeneration of Protoplasts 105
3.7.1.1 Sequential Enzymatic Treatment (Two Steps) 105
3.7.1.2 Mixed Enzymatic Treatment (Simultaneous) 106
3.7.2 Protoplast Culture and Regeneration 106
3.7.3 Protoplast Fusion 106
3.7.4 Cybridization 108
3.7.5 Identification and Selection of Heterokaryons or Hybrid Cells 109
3.7.5.1 Auxotrophic Mutants 109
3.7.5.2 Drug Sensitivity 109
3.7.5.3 Density Gradient Centrifugation 109
3.7.5.4 Flow Cytometry 110
3.7.6 Characterization of Somatic Hybrid Plants 110
3.7.7 Application of Somatic Hybrids 110
3.7.7.1 Somatic Hybrids for Gene Transfer 110
3.7.7.2 Successful Transfer of Disease Resistance Genes by Protoplast Fusion 111
3.7.7.3 Transfer of Male Sterility 111
3.7.8 Successful Case Studies 112

3.8 Somaclonal Variations.......... 112
3.8.1 Factors Affecting Somaclonal Variation.......... 114
3.8.1.1 Physiological Causes.......... 114
3.8.1.2 Genetic Causes.......... 114
3.8.1.3 Biochemical Causes.......... 115
3.8.2 Detection and Isolation of Somaclonal Variations.......... 115
3.8.2.1 Screening.......... 115
3.8.2.2 In Vitro Cell Selection.......... 115
3.8.3 Applications of Somaclonal Variations.......... 116
3.8.4 Successful Case Studies.......... 117
3.9 Summary and Future Prospects.......... 117
References.......... 119

3.1 Introduction

Plant cell and tissue culture offers a number of nonconventional approaches that may supplement the conventional method of crop improvement. The improvement of agricultural crops has until recently been largely confined to conventional breeding approaches. Such programs rely on interspecific sexual hybridization of plants that have desirable heritable characteristics and on naturally or artificially induced random mutation. Plant tissue culture is also a way to study the mechanism by which cells differentiate, thereby providing an experimental approach to link genotype with phenotype. Organ redifferentiation can be manipulated in dedifferentiated tissue (callus) by subjecting it to the interacting influence of a whole host of plant growth regulators and nutrient constituents. Plant tissue culture permeates plant biotechnology and cements together its various aspects; to a large extent, the tissue culture revolution has occurred because of the needs of this new plant biotechnology. The past few years have witnessed an ever-increasing interest in plant cell and tissue culture and its application of in vitro techniques to plant breeding, clonal propagation, and disease eradication.

The in vitro plant cell culture techniques were developed initially to demonstrate the totipotency of plant cells predicted by German botanist Gottlieb Haberlandt (1902). Totipotency is the ability of a plant cell to divide and differentiate into a complete plantlet. G. Haberlandt was the first to culture isolated, fully differentiated cells on a nutrient medium containing glucose, peptone, and Knop's salt solutions. The cells remained alive but failed to divide. They could be induced to divide in culture only after several decades. Efforts to demonstrate totipotency led to the development of techniques for culturing of plant cells under defined conditions. This was made possible by brilliant contributions of Gauthret (1934) and White (1939). Most of the modern tissue culture media have been derived from the research work of Skoog and coworkers during the 1950s and 1960s. During the second half of the last century, the progress in this area has been so spectacular that cultured plant cells are being used in all areas of biology for a variety of purposes.

3.2 Plant Cell and Tissue Culture

Plant cell culture is a technique to isolate plant cells (from various sources), culture, and regenerate whole plantlet from it on an appropriate nutrient medium, or to grow the cells in large volume (suspension cultures) for the production of native plant constituents, i.e., secondary metabolites. The progress in this field has been so spectacular that it is possible to isolate secondary metabolites from a number of plant species under in vitro conditions such as *Aloe saponaria, Catharanthus roseus, Coffea arabica, Digitalis purpurea, Gentiana macrophylla,* and *Podophyllum hexandrum* (Karuppusamy, 2009). Plant physiologist and plant biochemists have recognized the merits of single cell system over intact organs and whole plants for studying cell metabolism and the effect of various substances on cellular responses. The cloning of single cells permits crop improvement through the extension of the techniques of microbial genetics to higher plants.

3.2.1 Plant Cell and Tissue Culture Nutrient Medium

All the plant cell and tissue culture media are synthetic and are chemically defined. A variety of synthetic media have been developed such as MS medium (Murashige and Skoog, 1962), White's medium (White, 1963), B5 medium (Gamborg et al., 1968), and Woody plants medium (Lloyd and McCown, 1980), which are widely used all over the world in research and commercial plant cell and tissue culture laboratories.

Plant tissue culture medium consists of inorganic nutrients (macronutrients [N, P, K, Ca, S, Mg] and micronutrients [Fe, Zn, Mn, Cu, B, Mo]), vitamins, carbon source, and plant growth regulators. Agar agar is used as the gelling agent for most of the tissue culture experiments except for the cell suspension or some specific experiments. The pH of the medium is usually adjusted between 5.6 and 5.8.

A suitable plant cell culture medium can be devised for a new plant system in several ways. It would be better to start with a well-known basal medium such as MS medium or B5 medium. The most variable factors in plant tissue media are plant growth regulators. A simple approach is to first test several concentrations of auxins and cytokinins in combination or to identify a suitable combination of the two that gives best callus formation/shoot regeneration.

The best selected concentration and combination of auxins and cytokinins are then tried with different auxins and cytokinins at that concentration. Using these concentrations of plant growth regulators, different standard media compositions may be evaluated. Further, it would be worthwhile to check MS salt concentration with the best combinations of growth regulators as well as different sucrose concentrations (2%–6%) to decide its optimal level.

3.2.2 Techniques of Plant Cell Culture

A number of techniques are being used for the isolation and culture of plant cells. Single cells can be isolated from plant organs, particularly leaf, either by mechanical or enzymatic means. During mechanical isolation, leaves are cut into small pieces, ground in a suitable medium, the resulting homogenate is filtered through muslin cloth or sieves of different pore sizes, and finally cultured on the appropriate nutrient medium (Gnanam and Kulandaivelu, 1969; Srivastava et al., 1990). Mechanical isolation of cells has at least two distinct advantages over the enzymatic method: (i) it eliminates the exposure of cells to the harmful effect(s) of enzymes and (ii) the cells need not be plasmolyzed, which is often desirable in physiological and biochemical studies. In the enzymatic method, the lower epidermis of leaves is peeled off and the leaves are cut into moderate pieces, which are incubated in a macrozyme or pectinase solution. Enzymatic method for the isolation of single cells has been found convenient as it is possible to obtain high yields from preparations of spongy parenchyma with minimum damage or injury to the cells. This can be accomplished by providing osmotic protection to the cells while the enzyme macerozyme degrades the middle lamella and the cells are separated.

The most widely applied approach is to obtain a single cell system from cultured tissues. The explants are cultured on the nutrient medium with suitable proportions of auxins and cytokinins. Explants on such a medium exhibit callusing at the cut ends. The callus may be separated from the explants and transferred to the fresh nutrient medium for further proliferation. Pieces of undifferentiated calli are transferred to liquid medium, which is continuously agitated to obtain a suspension culture (Srivastava et al., 1987). Agitation allows to break down the callus into smaller clumps/single cells and also helps to maintain uniform distribution of cells/cell clumps in the medium. The viability of the cells can be determined by Evan's blue and fluorescein diacetate (FDA) (Bhojwani and Razdhan, 2004). Plant cells can be cultured using a number of techniques such as (i) filter paper raft-nurse tissue technique, (ii) micro-chamber technique, (iii) microdrop method, (iv) thin layer liquid method, and (v) Bergmann's cell plating technique. Bergmann's cell plating technique is widely used for cell culturing in which the cells remain embedded in the soft agar medium and are observable under the microscope. The macroscopic colonies after they develop are isolated and cultured separately.

3.2.3 Applications of Cell Culture

The cell culture technique has a number of applications such as mutant selection, production of secondary metabolites, and biotransformation.

3.2.3.1 Mutant Selection

Since selection of mutation will be exercised at cellular level, no chimeras will be obtained, which is sometimes a drawback of mutation breeding methods, where mutants are selected at the level of whole plants. Using this strategy, cell lines resistant to amino acid analogs, antibiotics, herbicides, and fungal toxin have actually been isolated.

3.2.3.2 Production of Secondary Metabolites

Many higher plants are major sources of a large variety of biochemicals, which are metabolites of both primary and secondary metabolism including various natural products having antimicrobial, insecticidal, molluscicidal, and hormonal properties and valuable pharmaceutical and pharmacological activities, in addition to their use as agrochemicals, food additives, flavor and fragrance ingredients (Balandrin and Klocke, 1988). In recent years, plant cell suspension cultures and immobilized cells are being utilized for the production of the chemicals on a commercial scale (Figure 3.1) due to the advantages over extraction from whole plant such as cell culture gives better yield and quality of the product because it is not influenced by the environment and

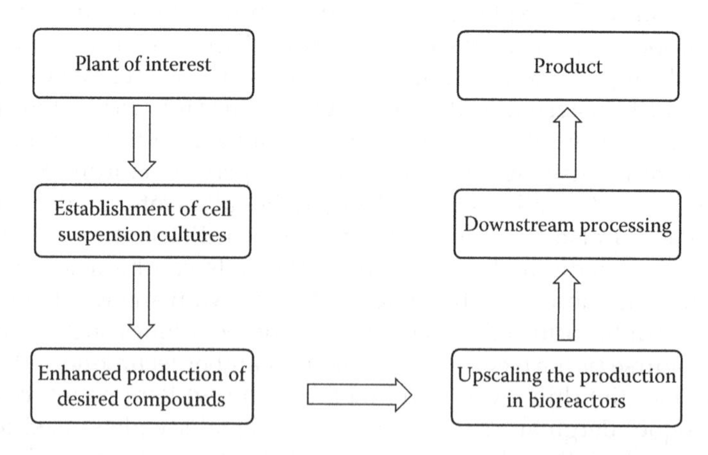

FIGURE 3.1
Flowchart showing the production of secondary metabolites.

the production schedule can be predicted and controlled in the laboratory. In many cases, the chemical synthesis of metabolites is not possible or economically feasible. The desirable medicinal compounds from plants can be produced at a commercial level under in vitro conditions using plant cell and tissue culture techniques (Ramachandra Rao and Ravishankar, 2002).

The three commercially viable products of secondary metabolism are ginseng saponins, shikonin, and berberine, and each product has diversified uses, including medicinal applications. Ginseng is produced in large-scale root cultures, whereas the other two products are produced in cell cultures. Intensive activity has centered on production of natural drugs or chemoprotective compounds from plant cell culture. Some of the most prominent pharmaceutical products in the latter category include ajmalicine (a drug for circulatory problems) from *C. roseus* and taxol (a phytochemical effective in treatment of ovarian cancer) from *Taxus* species. In vitro production of secondary metabolites from higher plants is surveyed and summarized in Table 3.1.

Transgenic hairy root cultures have revolutionized the role of plant tissue culture in secondary metabolite production. They are unique in their genetic and biosynthetic stability, faster in growth, and more easily maintained. Using this methodology, a wide range of chemical compounds has been synthesized (Giri and Narasu, 2000).

3.2.3.3 Biotransformation

Plant cell cultures and suspension cultures are also being utilized, for producing valuable products including secondary metabolites through biotransformation. In this technique, low-cost precursors are used as a substrate and transformed into value-added high-cost products. Suspension cultures of *Digitalis lantana* can convert digitoxin or methyl digitoxin into medically important metabolites, which are used for treatment of heart diseases. *Datura* cell cultures possess the ability to convert hydroquinone into arbutin (used as diuretic and urinary antiseptic) through glycosylation.

3.2.4 Successful Case Studies

Taxol (plaxitaxol), a complex diterpene alkaloid found in the bark of the *Taxus* tree, is one of the most promising anticancer agents known due to its unique mode of action on the microtubular cell system. The first study on the production of taxol (paclitaxel) by *Taxus* cell cultures has been reported by Christen et al. (1989). Paclitaxel was found to accumulate at high yields (1.5 mg/L) exclusively in the second phase of growth. A similar level of paclitaxel from *Taxus brevifolia* cell suspension cultures following 10 days in culture with optimized medium containing 6% fructose has been established (Kim et al., 1995).

TABLE 3.1

In Vitro Production of Secondary Metabolites from Plant Cell, Tissue, and Organ Cultures

Plant Name	Active Ingredient	Culture Medium and Plant Growth Regulator(s)	Culture Type	References
Aconitum heterophyllum	Aconites	MS + 2,4-D + Kinetin	Hairy root	Giri et al. (1997)
Anchusa officinalis	Rosamarinic acid	B5 + 2,4-D	Suspension	De-Eknamkul and Ellis (1985)
Agave amaniensis	Saponins	MS + Kinetin	Callus	Andrijany et al. (1999)
Artemisia annua	Artimisinin	MS + NAA + Kinetin	Callus	Baldi and Dixit (2008)
Allium sativum	Allin	MS + IAA + Kinetin	Callus	Malpathak and David (1986)
Aloe saponaria	Glucosides	MS + 2,4-D + Kinetin	Suspension	Yagi et al. (1983)
Azadirachta indica	Azadirachtin	MS + 2,4-D	Suspension	Sujanya et al. (2008)
Arachis hypogaea	Resveratrol	G5 + 2,4-D + Kinetin	Hairy root	Kim et al. (2008)
Artemisia absinthum	Essential oil	MS + NAA + BAP	Hairy root	Nin et al. (1997)
Angelica gigas	Deoursin	Liquid MS + 2,4-D + GA_3	Hairy root	Xu et al. (2008)
Beta vulgaris	Betalain pigments	MS + IAA	Hairy root	Taya et al. (1992)
Brucea javanica	Cathin	MS + IAA + GA_3	Suspension	Wagiah et al. (2008)
Brugmansia candida	Tropane	MS + 2,4-D + IAA	Hairy root	Marconi et al. (2008)
Bupleurum falcatum	Saikosaponin	LS+ 2,4-D	Callus	Wang and Huang (1982)
Camellia chinensis	Flavones	MS + 2,4-D + NAA	Callus	Nikolaeva et al. (2009)
Capsicum annuum	Capsiacin	MS + 2,4-D + Kinetin	Callus	Umamaheswari and Lalitha (2007)
Cassia obtusifolia	Anthraquinone	MS + TDZ + IAA	Hairy root	Ko et al. (1995)
Catharanthus roseus	Indole alkaloid	MS + 2,4-D + GA_3 + Vanadium	Suspension	Tallevi and Dicosmo (1988)
Cayratia trifoliata	Stilbenes	MS + IAA + GA_3	Suspension	Roat and Ramawat (2009)
Centella asiatica	Asiaticoside	MS + 2,4-D	Hairy root	Kim et al. (2007)

TABLE 3.1 (continued)

In Vitro Production of Secondary Metabolites from Plant Cell, Tissue, and Organ Cultures

Plant Name	Active Ingredient	Culture Medium and Plant Growth Regulator(s)	Culture Type	References
Chrysanthemum cinerariaefolium	Pyrithrins	MS + 2.4-D + Kinetin	Callus	Rajasekaran et al. (1991)
Coffea arabica	Caffeine	MS + 2,4-D + Kinetin	Callus	Waller et al. (1983)
Corydylis terminalis	Corydalin	MS + 2,4-D + BAP	Callus	Taha et al. (2008)
Coscinium fenustratum	Berberin	MS + 2,4-D + BAP	Callus	Khan et al. (2008)
Crataegus sinaica	Flavonoid	MS + 2,4-D + NAA + BAP	Callus	Maharik et al. (2009)
Cymbopogon citrates	Essential oil	MS + IAA + GA_3	Shoot	Quiala et al. (2006)
Digitalis purpurea	Cardioactive glycosides	MS + 2,4-D + BA	Hairy root	Saito et al. (1990)
Gentiana macrophylla	Glucoside	MS + IAA + Kinetin	Hairy root	Tiwari et al. (2007)
Geranium thunbergii	Tannin	MS + 2,4-D + BAP	Hairy root	Ishimaru and Shimomura (1991)
Gynostemma pentaphyllum	Saponin	MS + 2,4-D + BAP	Hairy root	Fei et al. (1993)
Hypericum perforatum	Hypericin	Liquid MS + NAA + GA_3	Suspension	Hohtola et al. (2005)
Hypericum perforatum	Hyperforin	MS + 2,4-D + Leucine	Multiple shoot	Karppinen et al. (2007)
Ipomoea cairica	Lignan	MS + IAA + Kinetin	Callus	Paska et al. (1999)
Lithospermum erythrorhizon	Shikonin	MS + 2,4-D + Kinetin	Hairy root	Fukui et al. (1998)
Papaver somniferum	Codeine	LS + BA + NAA	Hairy root	Williams and Ellis (1992)
Piper solmsianum	Piperine	MS + 2,4-D + BA	Suspension	Balbuena et al. (2009)
Plumbago rosea	Plumbagin	MS + $CaCl_2$	Callus	Komaraiah et al. (2003)
Podophyllum hexandrum	Podophyllotaxin	MS + BAP + GA_3	Shoot	Li et al. (2009)
Psoralea cordifolia	Isoflavones	MS + TDZ + BAP	Multiple shoot	Shinde et al. (2009)

(continued)

TABLE 3.1 (continued)

In Vitro Production of Secondary Metabolites from Plant Cell, Tissue, and Organ Cultures

Plant Name	Active Ingredient	Culture Medium and Plant Growth Regulator(s)	Culture Type	References
Rauvolfia serpentine	Serpentine	MS + BAP + IAA	Callus	Salma et al. (2008)
Rubia akane	Anthraquinone	B5 + NAA + Kinetin	Hairy root	Park and Lee (2009)
Salvia officinalis	Terpenoids	MS + 2,4-D + BA	Callus	Santos-Gome et al. (2002)
Silybium marianum	Silymarin	MS + IAA + GA_3	Hairy root	Rahnama et al. (2008)
Solanum aculeatissi	Steroidal saponin	MS + 2,4-D	Hairy root	Ikenaga et al. (1995)
Stevia rebaudiana	Stevioside	MS + BA + NAA	Callus	Dheeranapattana et al. (2008)
Vaccinium myrtillus	Flavonoids	MS + BAP + NAA	Callus culture	Hohtola et al. (2005)
Vitis vinifera	Resveratrol	MS + IAA + GA_3 + UV	Callus	Kin and Kunter (2009)
Withania somnifera	Withanoloid A	MS + IAA + Kinetin	Hairy root	Murthy et al. (2008)
Withania somnifera	Steroidal lactone	MS + 2,4-D + BA	Callus	Mirjalili et al. (2009)
Zataria multiflora	Rosmarininc acid	MS + IAA + Kinetin	Callus	Francoise et al. (2007)

3.3 Meristem Culture

Many crops are infected by various diseases caused by fungi, bacteria, viruses, mycoplasma, and nematodes. Pathogen attack does not always lead to the death of plants, but often the infection by them considerably reduces the yield and quality of crops. Control of plant bacterial and fungal diseases may be obtained through the application of chemicals. However, it has not been possible to control viral diseases through the use of chemicals. Chemicals that can affect virus multiplication usually exert high phytotoxicity on the host plant. In addition, they are usually very expensive and once the treatment stops, the viruses may rapidly build up again. Meristem-tip culture was first carried out by Morel and Martin (1952) to get virus-free plants of dahlias. Later on, this technique has been applied widely to many

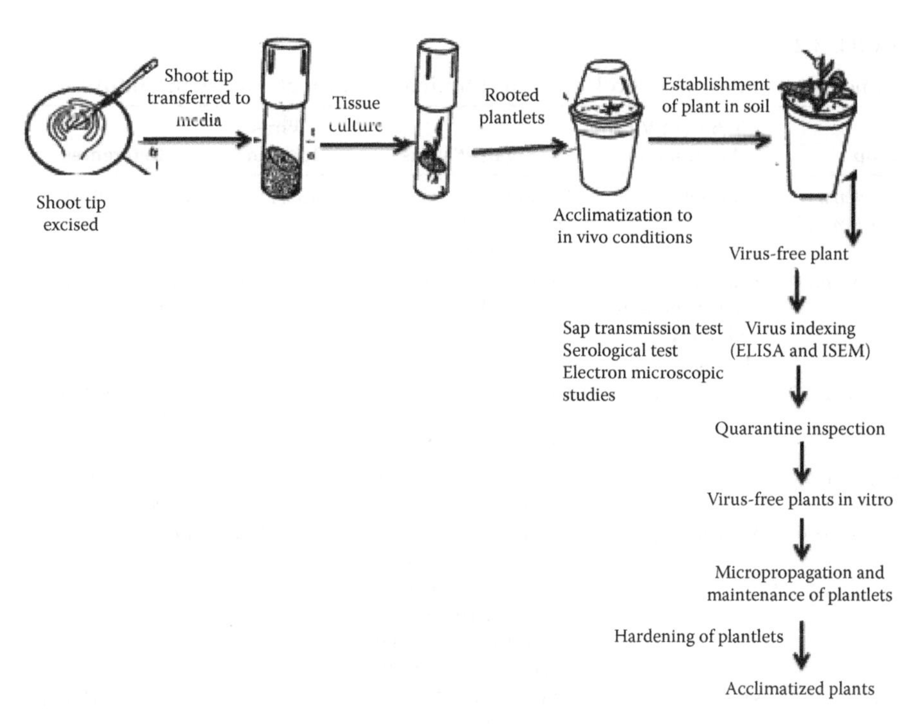

FIGURE 3.2
Technique of meristem tip culture.

vegetatively propagated crops to obtain virus-free plants from originally diseased material. The successful rapid multiplication of orchids by shoot meristem culture, demonstrated by Morel (1965), increased the interest in the application of tissue culture techniques as an alternative means of asexual propagation. Meristem culture involves the development of an already existing shoot meristem and subsequently the regeneration of adventitious roots from shoots. Shoot apical meristem lies in the "shoot tip" beyond the youngest leaf or the first leaf primordium. The meristem tip must be small enough to eradicate viruses, yet large enough to develop into a shoot. Although roots may form on the shoot directly in the same medium, often the shoot has to be transferred to another medium in order for roots to develop. Meristem-tip culture is used successfully to remove viruses from plants (Figure 3.2 and Table 3.2).

3.3.1 Methods for Virus Elimination

3.3.1.1 Thermotherapy/Heat Treatment

During the period when meristem-tip culture was not discovered, the in vivo eradication of viruses from plants was achieved by heat treatment (50°C–52°C) of whole plants (Hollings, 1965). In contrast to thermotherapy alone, meristem-tip

TABLE 3.2
Some Selected Research Work on Virus Elimination through Meristem Culture

Crop	Method of Virus Elimination	Against Virus	Virus Indexing	References
Nicotiana rustica	Thermotherapy	Cucumber mosaic virus (CMV) and alfalfa mosaic virus	ELISA	Walkey (1976)
Grapevine	Somatic embryogenesis with heat therapy	Grapevine fanleaf virus (GFLV)	ISEM and ELISA	Goussard and Wiid (1992)
Grapevine	Thermotherapy	Grapevine fanleaf virus (GFLV) and Arabis mosaic virus (AMV)	ELISA	Leonhardt et al. (1998)
Sugarcane	Meristem culture	Sugarcane yellow leaf virus (SCYLV)	RT-PCR	Chatenet et al. (2001)
Strawberry	Meristem culture	Strawberry mild yellow edge virus (SMYEV), strawberry crinkle virus (SCV), and strawberry vein banding virus (SVBV)	DAS-ELISA	Biswas et al. (2007)
Lily	Shoot meristem-tip culture and in vitro thermotherapy	Lily symptomless virus (LSV)	ELISA	Nesi et al. (2009)
Potato	Meristem-tip culture	Potato virus Y (PVY)	ELISA	Tale et al. (2011)

culture is widely applicable. Heat treatment in combination with meristem culture has been reported to increase the number of meristem tips that may be regenerated to plantlets and the percentage of virus-free plants obtained from such meristem tips (Stac-Smith and Mellor, 1968; Ng and Hahn, 1985).

3.3.1.2 Chemotherapy

Chemicals reported to inactivate or inhibit plant viruses have been reviewed (Long and Cassells, 1986). They include purine and pyrimidine analogs, amino acids, plant growth regulators, antibiotics, and other substances. By culturing meristem tips or other explants in the presence of a suitable chemical, such as ribavirin, it may be possible to maintain inhibitory conditions long enough to eliminate viruses successfully (Wambugu et al., 1985).

3.3.1.3 Other In Vitro Methods

Virus elimination through callus cultures has also been reported. Virus-free plants were obtained from the callus culture of plants infected with tobacco mosaic virus (TMV) (Murakishi and Carlson, 1976). Virus-free plants were also obtained from mesophyll protoplast culture of tobacco leaves infected with potato virus X (PVX) (Shepard, 1975).

In a number of plant species where meristem-tip culture continues to be difficult, shoot-tip grafting has been studied and found to be effective in virus elimination (Navarro et al., 1975; Huang and Millikan, 1980). Shoot tips of 0.14–0.18 mm in length isolated aseptically from a diseased plant were grafted on to young etiolated root stock seedlings grown in vitro. Shoot-tip grafting had been used to produce virus-free plants in other crops including peach and apple.

3.3.2 Virus Indexing

Sensitive and reliable methods of virus detection are crucial in obtaining disease-free germplasm. The regenerated plants are transplanted from tubes into sterile soil in pots and are kept in an isolation room for further growth and monitoring for disease expression. Virus indexing methods vary depending on the type of virus(es) that is involved and on the availability of facilities. Ideally, regenerated plants should be tested for viruses by a variety of methods and these should be repeated several times. Virus indexing may take up to 1 year.

Some methods used for the indexing of plant materials may include monitoring of plants for possible occurrence of symptoms over a long period of time, sap inoculation of test plants, grafting on suitable indicator plants, and inspections of materials under the electron microscope or immunosorbent electron microscopy (ISEM) and by immunoassays like enzyme-linked immunosorbent assay (ELISA) or dot-blot assay. Recent developments in biotechnology permit detection of viruses by means of monoclonal antibodies (Thomas et al., 1986) as well as by nucleic acid hybridization technique (Old and Primrose, 1994).

3.4 Micropropagation

A variety of plant species can be conventionally propagated through the technique of cell tissue and organ culture, which is described as clonal propagation or micropropagation. The chief objective of clonal propagation is to produce progeny plants, which are identical in genotype with their parent plant. This is achieved by the following processes that are employed for micropropagation of different plant species: (i) proliferation of axillary bud and (ii) induction of adventitious buds, bulb, and protocorms. The most significant advantage offered micropropagation over the conventional methods is that in a relatively short time and space, a large number of plants can be produced starting from a single individual. In nature, clonal propagation occurs by apomixis (seed development without meiosis and fertilization) and/or vegetative reproduction (regeneration of new plants from vegetative parts). Apomixis is restricted to only a few species; therefore, foresters have adopted the methods of vegetative reproduction for clonally multiplying selected forest plant species and cultivars. Plant propagation by this method is termed micropropagation because miniature shoots or plantlets are initially derived.

3.4.1 Micropropagation Systems

3.4.1.1 Single-Node Culture

This is the simplest, most natural and safe method (with respect to variation) used for micropropagation of plants that form a stem with leaves and bud in their axils, but it is difficult with rosette plants. The rate of propagation is strongly dependent on the number of nodes formed within a particular time interval.

3.4.1.2 Axillary Branching

Axillary buds have their dormancy broken by breaking apical dominance with cytokinin (Figure 3.3). This method has become the most important propagation method as it is being simple and safe (with respect to variation).

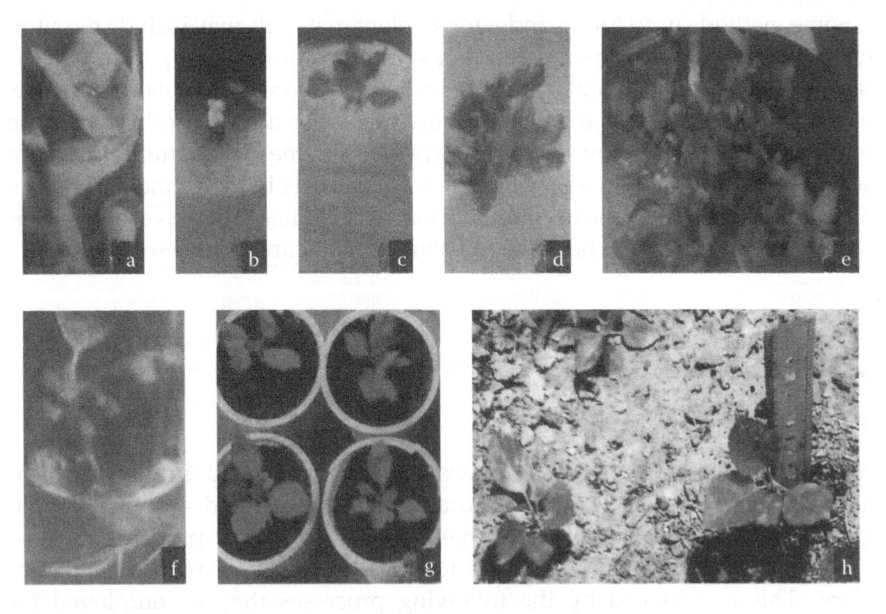

FIGURE 3.3
In vitro clonal multiplication of apple rootstock (MM111). (a) Axillary bud cultured on filter-paper bridge on MS establishment medium, (b) sprouting of axillary bud, (c) shoot regeneration, (d) shoot proliferation, (e) shoot multiplication, (f) root regeneration in in vitro developed shoots, (g) hardening of in vitro regenerated plantlets in plastic pots, and (h) in vitro regenerated plantlets in field. (Reproduced from Kaushal, N. et al., *Indian J. Exp. Biol.*, 43, 561, 2005. With permission.)

Another advantage is that the propagation rate is relatively fast and the genetic stability is usually preserved (Modgil et al., 2005).

3.4.1.3 Regeneration of Adventitious Buds/Shoots

This method includes the formation of adventitious buds/shoots on explants from leaves, petioles, and stem. However, the percentage of plant species that can regenerate adventitious buds is relatively small and is often restricted to herbaceous plants. The chances of obtaining mutations is much higher with the earlier method mentioned here, particularly with so-called chimeric plants.

3.4.2 Stages of Micropropagation

Micropropagation can be divided into six stages, which are discussed in the following sections.

3.4.2.1 Preparation and Pretreatment of the Explants

The nature of the explants to be used for in vitro propagation is to a certain extent governed by the method of shoot multiplication to be adopted. For enhanced axillary branching, only such explants are suitable that carry preformed vegetative buds. When the objective is to produce virus-free plant from an infected individual, it becomes obligatory to start with submillimeter shoot tips. However, if the stock plant is virus tested, the most suitable explant is nodal cuttings. The physiological state of the parent plant at the time of explant excision has a definite influence on the response of the buds. Explants from actively growing shoots at the beginning of the growing season generally give best results (Anderson, 1980). The seasonal fluctuations in the response of the shoot buds may be minimized by maintaining the parent plants under light and temperature conditions required for continual vegetative growth in glasshouse or growth cabinets. Standard methods for surface sterilization of plant tissue and organs should generally be adequate to achieve asepsis in cultures. Special precautions need to be taken when explants are derived from field-grown materials, which are often necessary in cleaning an elite plant. Washing of plant material in running tap water for 20–30 min prior to surface sterilization reduces the microflora population substantially (De Fossard, 1976).

3.4.2.2 Establishment of Aseptic Culture

The establishment of the explant is universally achieved on MS medium with a number of variations and concentrations of plant growth regulators. The choice of the explants is often the shoot meristem and axillary buds. The predominant reason for this is the far higher degree of stability in culture and greater plasticity. The establishment of explants in tissue culture is also affected by factors such as the season when the cuttings are taken. Spring is often found to be the best time to take shoots as they possess considerable vigor and less infection. In some species, the production of phenolic compounds by the explants has deleterious effects, resulting in browning of the medium and is toxic to the tissue. This problem can be solved by a combination of factors such as treatment with ascorbate, soaking in water for 24 h prior to culture, and frequent subculturing of the explants. Charcoal and polyvinyl pyrrolidone additives are also beneficial.

3.4.2.3 Multiplication of Shoots

This is the most crucial stage during micropropagation. Broadly, three approaches have been followed to achieve in vitro shoot multiplication, i.e., through callusing, adventitious bud formation, and enhanced axillary branching.

The potentiality of plant cells to multiply indefinitely in cultures and their totipotent nature permit a very rapid multiplication of several plant types. Differentiation of plants from cultured cells callus may occur via shoot root formation or somatic embryogenesis where applicable. A somatic embryo is an embryo derived from a somatic cell, other than zygote, usually on culture in vitro and the process is known as somatic embryogenesis. In contrast, embryos developing from zygotes are called zygotic embryos or often simply embryos, while those derived from pollen are known as pollen embryo or androgenetic embryos. Somatic embryos generally originate from single cells that divide to form a group of meristematic cells. Usually, this multicellular group becomes isolated by breaking cytoplasmic connections with the other cells around it and subsequently by cutinization of the outer walls of this differentiating cell mass. The cells of meristematic mass continue to divide to give rise to globular, heart-shaped, torpedo, and cotyledonary stages and finally develop into a plantlet. In general, the essential features of somatic embryo development especially after the globular stage are comparable to those of zygotic embryos. This is often the fastest method of shoot multiplication and has been suggested as a potential method for cloning plant species (Murashige, 1978). However, there are several drawbacks in this method, and as far as possible, it should be avoided in clonal propagation of a cultivar. The most serious objection against the use of callus culture or somatic embryogenesis for shoot multiplication is the genetic instability of their cells. Plants propagated by cell and callus cultures showed genetic variations (somaclonal variation), whereas those from bud cultures were uniformly genetically stable (Shenoy and Vasil, 1992). Buds arising from any place other than leaf axil or the shoot apex are termed adventitious buds. A number of plant species produce adventitious buds in vivo from different organs, i.e., roots, bulbs, and leaves. Another merit of in vitro technique that helps in promoting the overall rate of plant multiplication is that explants as small as 20–50 mg in weight, which fail to survive in nature, are able to produce adventitious buds in cultures. Under the influence of the appropriate combination of growth regulators, in cultures, adventitious buds can be induced on the leaf and stem segments of even those species that are normally not propagated vegetatively. Adventitious bud formation involves the risk of splitting the chimeras leading to pure type plants (Skirvin, 1978). However, there are few recent reports on the regeneration of plants through direct and indirect organogenesis (using leaf and petiole explants) and the study of their genetic stability using molecular techniques (Thakur and Srivastava, 2006; Thakur et al., 2008, 2012; Gaur, 2011; Husaini and Srivastava, 2011). The development of reliable regeneration systems from mature tissues (explants) is a prerequisite for the application of gene transfer technique to the genetic improvement of crops (Figures 3.4 through 3.6).

Axillary buds are usually present in the axil of each leaf and every bud has the potential to develop into a shoot. In nature, these buds remain dormant for various periods depending on the growth pattern of the plant. In culture, the rate of shoot multiplication by enhanced axillary branching can be substantially enhanced by growing shoots in a medium containing a suitable cytokinin at an appropriate concentration with or without auxin. Due to the continuous availability of cytokinin, the shoot formed by the

FIGURE 3.4
Plant regeneration in Himalyan poplar (*Populus ciliata* Wall; male plant) from leaf explants. (a) Leaf explants cultured on shoot regeneration medium, (b) leaf explants showing callus formation and shoot initiation, (c) shoot proliferation, (d) shoot multiplication, (e) root initiation from regenerated shoot, (f) hardened plant. (Reproduced from Thakur, A.K. and Srivastava, D.K., *In Vitro Cell. Dev. Biol. Plant*, 42, 144, 2006. With permission.)

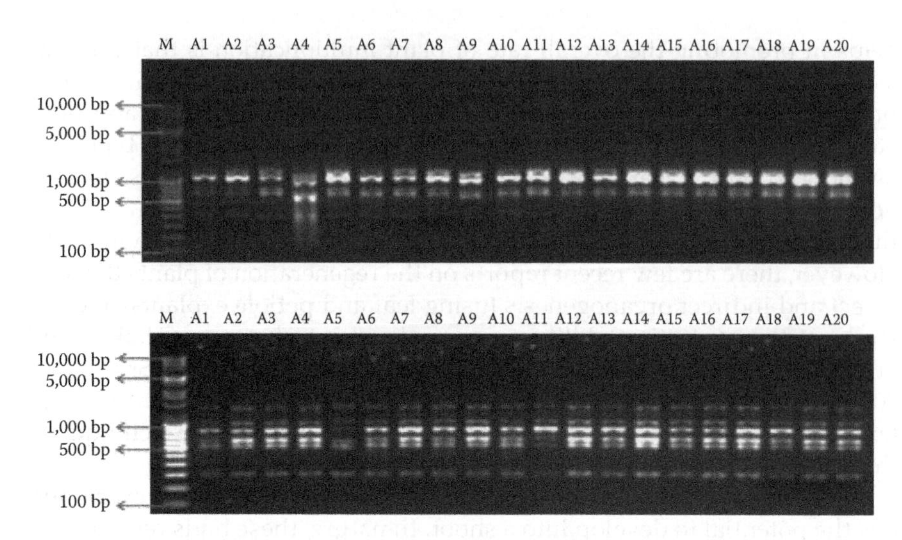

FIGURE 3.5
Genetic fidelity studies in in vitro raised plantlets of Himalayan poplar (*Populus ciliata* Wall.) by using RAPD. (Reproduced from Gaur, A., Studies on genetic fidelity of tissue culture raised plants of Himalayan poplar (*Populus ciliate* Wall.), M.Sc. Thesis, Dr. Y.S. Parmar University of Horticulture and Forestry, Nauni, Solan (H.P.), India, p. 103, 2011. With permission.)

bud already present on the explants (nodal segment or shoot-tip cutting) develops axillary bud that may grow directly into shoots. This process may be repeated several times and the initial explants are transformed into a mass of branches. There is a limit to which shoot multiplication can be achieved in a single passage after which further axillary branching stops. At this stage, however, if miniature shoots are excised and planted on a fresh medium of the same composition, the shoot multiplication cycle can be repeated. This process can go indefinitely and can be maintained throughout the year.

3.4.2.4 Root Regeneration in In Vitro Developed Shoots

Adventitious and axillary shoots developed in cultures in the presence of a cytokinin generally lack roots. To obtain full plants, the shoots must be transferred to a root regeneration medium that is different from the shoot multiplication medium as it contains low concentrations of auxins (IAA, IBA, NAA, or 2,4-D). The number of shoot multiplication cycles after which root regeneration exercise is to be started is governed by the number of plants to be produced through micropropagation and the available nursery facilities. For root regeneration treatment, individual shoots measuring about 1 cm in length are excised and transferred to the root regeneration medium. Individual rooted plants could be easily transferred to the pots

FIGURE 3.6
Plant regeneration studies in strawberry (*Fragaria annanasa* Dutch cv. chandler). (a) shoot regeneration from leaf derived callus, (b) direct shoot regeneration from petiole explants, (c) direct shoot regeneration from leaf explants, (d) shoot multiplication, (e) root regeneration in in vitro developed shoots, (f) in vitro regenerated plantlet, (g) hardening of in vitro regenerated plantlets in pots, (h) in vitro regenerated plantlets after 5 weeks of acclimatization. (Reproduced from Husaini, A.M. and Srivastava, D.K., *Phytomorphology*, 61(1–2), 55, 2011. With permission.)

for hardening. Somatic embryo carries a preformed radicle and develops directly into the plantlets. Rarely, however, the embryos enter dormancy after maturity and require special treatment for germination. In some of the plants, it has been possible to treat the regenerated shoots as minicuttings and root them out of culture. The basal cut end of the shoot is treated with standard rooting powder or different concentration of various auxins separately and planted in a potting mix. In some of the plant species, root regeneration from shoots formed in cultures is far better in vivo than in vitro. Where possible, root regeneration under nonsterile conditions should reduce the cost of plant production by cutting down a step in aseptic manipulation.

3.4.2.5 Hardening for Subsequent Field Planting

Micropropagation on a large scale can be successful only when plants after transfer from the cultures to the soil show high survival rates and the cost involved in the process is low. The transfer of plantlets from the culture vessels to the soil requires a careful stepwise procedure. The roots of the regenerated plantlets are gently washed to remove the agar medium sticking to them. The most essential requirement for successful transplantation is to maintain the plants under a very high humidity (90%–100%) for the first 10–15 days by keeping them under mist or covering them with clear plastic. Some small holes may be poked in the plastic for air circulation. After spending a few days under high humidity, the plants are moved to the greenhouse bench but are continued to be shaded for few more days. After this additional hardening, which may last for 4–6 weeks after transplantation, the plants are now ready to grow under normal field conditions.

3.4.2.6 Molecular Analysis of Genetic Stability in Micropropagated Plants

The occurrence of somaclonal variation is a potential drawback when the propagation of an elite tree is extended where clonal fidelity is required to maintain the advantages of desired elite genotypes. Micropropagated plants from the cultures of preformed structures such as shoot tips, axillary buds and from the tissues of hard wood shoot cuttings have been reported to maintain clonal fidelity (Modgil et al., 2005), but there is still a possibility that in vitro culture plantlets exhibit somaclonal variations (Rani et al., 1995). This variation is often heritable and therefore unwanted in clonal propagation (Brieman et al., 1987). Thus, screening of micropropagated plants at an early stage is essential to reduce the chances for inclusion of variable genotype.

The detection of off-types among micropropagated plants, especially of forest trees, by morphological observations and karyotype analysis of metaphase chromosome has several limitations such as extensive evaluation time needed for assessment. Isozymes provide limited information and are affected by environmental variations. On the other hand, DNA markers are an attractive means for detecting somaclonal variations, since they are more informative and are not developmentally affected. Restriction fragment length polymorphism (RFLP), though has been used for screening of tissue culture–derived plants (Vallés et al., 1993), is laborious and usually involves radioactivity and is not suited for routine application of tissue culture systems. Random amplified polymorphic DNA (RAPD) requires only small amount of starting DNA and does not require prior DNA sequence information nor involves radioactivity (Welsh and McClelland, 1990; Williams et al., 1990), and data can be generated faster with less labor than other methods like RFLP and microsatellites.

3.4.3 Progress in Micropropagation of Horticultural and Forest Tree Species

During the past many years, considerable success has been achieved with respect to micropropagation of horticultural, ornamental, and medicinal plant and forest tree species (Table 3.3). Explants that have shown positive response in culture for regeneration, however, are largely restricted to juvenile material. Clonal multiplication of woody perennials in vitro on a commercial scale requires enormous efforts directed toward establishing cultures from adult explants. Further effective treatments to induce efficient rooting in in vitro multiplied shoots and quality improvement of somatic embryos to achieve high frequency conversion to plantlets must be found. Successful micropropagation have been achieved in horticultural species (apple, strawberry, cherry, walnut, kiwi, banana, mango, peach, plum, pomegranate, guava, and date palm), forestry species (*Alnus glutinosa, Alnus nepalensis, Betulla palatphylla, Dendrocalamus hamiltonii* [bamboo], *Eucalyptus tereticornis, Pinus radiata, Pinus taeda, Populus ciliata, Populus deltoids, Populus euranericana, Populus tremula, Punica granatum, Robinia pseudoacacia, Salix* spp., and *Tectona grandis*), and ornamental species (rose, gerbera, carnation, and *Lilium*).

3.4.4 Successful Case Study

Sun Agrigenetics Plant Biotech, a Vadodra-based company (personal communication), is providing micropropagated plants of banana, sugarcane, potato, lemon, fig, cucumber, sweet potato, rose, watermelon, *Anthurium, Coccinia indica,* and *Trichosanthes dioica* of different varieties to the farmers of Gujarat and other states of India.

TABLE 3.3

Some Selected Research Work on Micropropagation Studies in Various Crop Species

Taxa	Type of Culture	Source of Explant	Result	References
Hevea brasiliensis Müll. Arg.	Callus	Callus	Embryogenesis	Ferriere et al. (1992)
Alnus nepalensis	Axillary bud	Axillary bud	Multiple shoots and complete plantlets	Kaur et al. (1992)
Colt-Cherry rootstock	Axillary bud	Axillary bud	Multiple shoots and complete plantlets	Sharma et al. (1992)
Manihot esculenta Crantz	Axillary bud culture	Node, axillary buds, meristems	Formation of shoots and multiple shoots	Konan et al. (1997)
Actinidia deliciosa Kiwi	Axillary bud	Axillary bud	Multiple shoots	Kumar et al. (1998)
Manihot esculenta Crantz	Somatic embryo	Somatic embryo	Shoot organogenesis, somatic embryogenesis	Ma (1998)
Vitis vinifera L. Grape	Axillary bud	Axillary bud	Multiple shoots and complete plantlet	Mhatre et al. (2000)
Phyllanthus coroliniensis Walter	Axillary shoot	Nodal segments	Multiple shoots, rooting initiated	Catapan et al. (2000)
Apple rootstock MM106, MM111	Axillary bud	Axillary bud	Multiple shoots and complete plantlets	Sharma et al. (2000), Kaushal et al. (2005)
Rosa hybrida cv. Baronesse	Apical bud	Apical bud	Shoot multiplication	Carelli and Echeverrigaray (2002)
Phyllanthus urinaria L.	Node and callus	Node	Callus induction, multiple shoots, rooted in vitro	Catapan et al. (2002)
Manihot esculenta Crantz	Somatic embryo culture	Somatic embryo	Somatic embryo culture Differentiated, developed, and germinated	Groll et al. (2002)
Mentha arvensis Mint	Axillary bud	Axillary bud	Shoot induction and multiplication	Dhawan et al. (2003)
Manihot esculenta Crantz	Axillary bud	Mononodal microcutting	Plants were recovered	Rommanee et al. (2003)

TABLE 3.3 (continued)

Some Selected Research Work on Micropropagation Studies in Various Crop Species

Taxa	Type of Culture	Source of Explant	Result	References
Azadirachta indica Neem	Nodal segment	Nodes and internodes	Plantlet formation	Chaturvedi et al. (2004)
Euphorbia tirucalli L.	Axillary bud	Internode	Adventitious bud proliferation, shoot regeneration	Uchida et al. (2004)
Jatropha curcas L.	Axillary bud	Leaf segments, axillary buds, nodes	Adventitious shoots, multiple shoots	Sujatha and Sailaja (2005)
Mallotus repandus (Willd.)	Node, internode	Node, internode	Shoot induction, shoot elongation, organogenesis	Prathanturarug et al. (2007)
Ginger	Axillary bud	Axillary bud	Shoot organogenesis	Jagadev et al. (2008)
Talinum portulacifolium L.	Axillary bud	Axillary bud	Well-rooted and partially acclimatized plantlets	Thangavel et al. (2008)
Blackberry	Axillary bud	Axillary bud	Shoot induction	Abadi and Hamidoghli (2009)
Apple rootstocks	Apical bud and nodal segment	Apical bud	Shoot elongation and multiplication	Dobránszki and Da Silva (2010)
Etlingera elatior (Torch ginger)	Axillary bud	Axillary bud	Shoot induction and elongation	Abdelmageed et al. (2011)

3.5 Anther Culture and Haploid Production

Haploid plants are obtained from pollen grains by culturing anthers or isolated pollen grains on a suitable nutrient medium, and this constitutes anther or pollen culture. The anthers may be taken from plants grown in the field or in pots, but ideally these plants should be grown under controlled temperature, light, and humidity. The optimum condition may differ often from species to species and the capacity for haploid production declines with age of the donor plants. Haploid plants are of special interest to the geneticist and plant breeder, because it helps in the induction of maximum genetic variability of germplasm sources to secure a wider scope for selection and introduction of better trait qualities in existing crop species. In addition, doubling the chromosome number of a haploid to produce doubled haploid results in a completely homozygous plant. Owing to the great theoretical and applied value of haploids, many methods have been tried to induce their occurrence, but they have extremely low frequency with which they occur in nature. Ever since A.D. Bergner discovered haploid plants in *Datura stramonium* in 1921, plant breeders have worked intensively to obtain haploids either in vivo or in vitro. Various techniques for in vivo haploid production, such as treatment with irradiated pollen, delayed pollination, utilization of alien cytoplasm or pollinator, distant hybridization, and polyembryony, have been employed, which result in their production in a small number. Anther culture was first demonstrated by two Indian scientists Guha and Maheshwari (1964) to develop haploid plants in *Datura inoxia* Mill. Subsequently, Bourgin and Nitsch (1967) obtained first haploid plants from isolated anthers of *Nicotiana*. To date, androgenic haploids have been reported in more than 170 species and hybrids distributed within 25 families. Haploid production is of great use and its advance has attracted the attention of geneticists world over. The techniques of anther culture, culture medium, mode of androgenesis, and factor affecting androgenesis are discussed later.

3.5.1 Techniques of Anther Culture and Isolated Pollen Culture

The technique for the excision and culture of anther is relatively simple and efficient. Closed flower buds, which have anther containing uninucleate microspores, are most suitable for the induction of androgenesis. The surface-sterilized unopened buds are used for isolation of microspores. The microspores are extracted from the anthers at the end of the inductive period when microspores with two identical nuclei can be seen. In case of large pollen, the pollen grains are then squeezed out of the anther by pressing them against the side of the beaker with the piston of syringe. Anther tissue debris is removed by filtering the suspension through nylon sieve. The pollen suspension is centrifuged at low speed. The supernatant containing fine debris is discarded and the pellet of the pollen resuspended in fresh medium and washed twice. Then the pollens are mixed with appropriate culture medium for regeneration.

3.5.2 Culture Media and Nutritional Requirements

Basal media of White (1934), Murashige and Skoog (1962), and Nitsch and Nitsch (1969), with slight modifications and addition of growth regulators, have been used for the anther culture. The normal level of sucrose is 2%–4% but anthers have been observed to grow better on media with 6%–12% sucrose, it may be due to the osmotic effect rather than a need for a higher carbohydrate level. The nutrient requirement of the excised anthers is much simpler than those of isolated microspores. In the isolated microspores, it is obvious that certain factors responsible for the induction of androgenesis, which might have been provided by the anther, are missing and these have to be provided through medium. The requirement for auxins and cytokinins depends on their endogenous level in the anther. Media rich in growth regulators encourage the proliferation of tissues other than microspores and should be avoided because in such cases a mixed calli with cells of different ploidy levels are obtained (Nitsch and Nitsch, 1969). In culture, microspores undergo various modes of androgenesis (Figure 3.7), which leads to

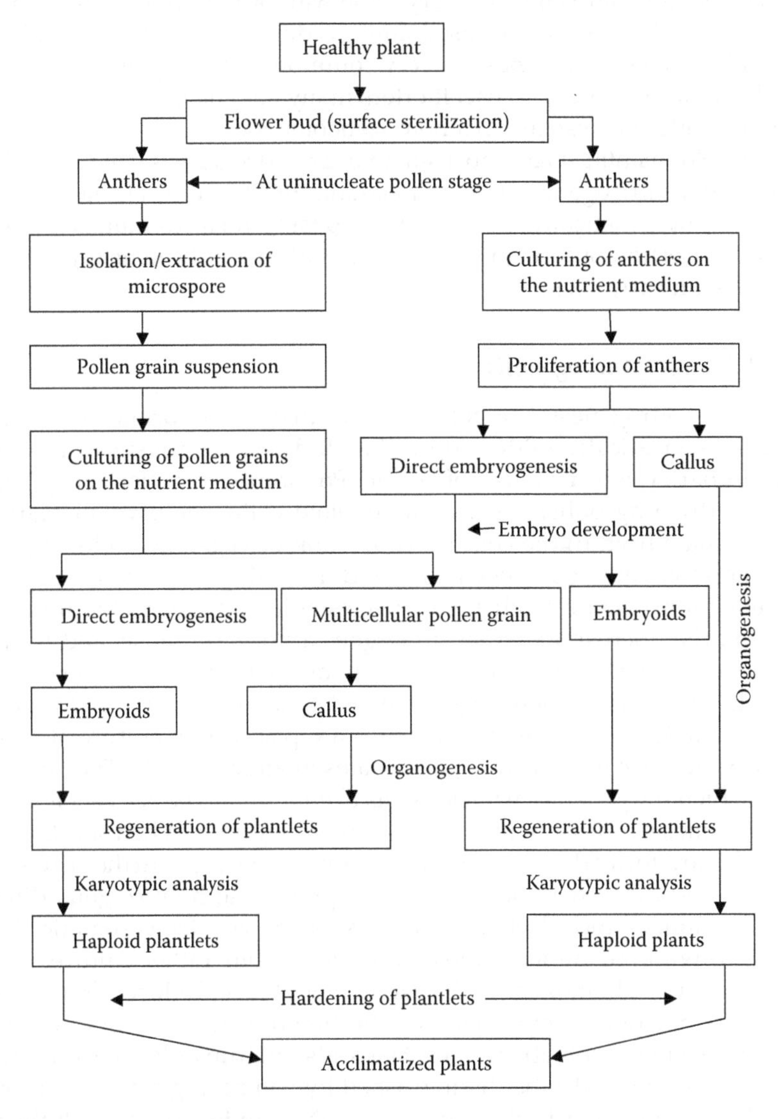

FIGURE 3.7
Various modes of androgenesis.

the formation of haploids either by direct androgenesis (embryogenesis) or indirect androgenesis (via callus formation).

3.5.2.1 Direct Androgenesis

Direct androgenesis is also called pollen-derived embryogenesis. In this case, pollen directly acts as a zygote and therefore passes through various embryogenic stages similar to zygotic embryogenesis. When pollen grains reached the globular stage of embryo, the wall of the pollen is broken and the embryo released. The released embryo develops cotyledons and then the plants. Direct androgenesis is very common in many plants of the family Solanaceae and Brassicaceae. It offers many advantages such as (i) it is a haploid, single cell system; (ii) a homogeneous population of pollen grains at the developmental stages most suitable for androgenesis can be obtained by gradient centrifugation (Kyo and Harada, 1990); and (iii) isolated microspores can be genetically modified by exposing them to mutagenic treatments or insertion of foreign genes before culture and the new genotypes selected at an early stage.

3.5.2.2 Indirect Androgenesis

In indirect androgenesis the pollen grains, instead of normal embryogenesis, divide erratically to develop callus. Indirect androgenesis has been found in barley, wheat, *Vitis*, coffee, etc. Possibility of pollen morphology from the dividing pollen varies. The haploid callus, embryo, or plantlets may originate from (i) the continued division of vegetative cells of pollen (the generative cells soon degenerate) and (ii) multiple divisions of generative cells (nonvegetative cells). For species cultured during the uninucleate stage, the microspore either undergoes a normal mitosis and forms a vegetative and a generative nucleus or divides to form two similar-looking nuclei. In those cases where vegetative and generative nuclei are formed in culture or where binucleate microspores are placed into culture, it is usually the vegetative nucleus that participates in androgenesis. The only species in which the generative nucleus has been found to be actively involved in androgenesis is black henbane (*Hyoscyamus niger* L.). When similar looking nuclei are formed, one or both nuclei may undergo further divisions. In some cases, the two nuclei will fuse, producing homozygous diploid plants or callus. Since diploid callus may also arise from somatic tissue associated with the anther, diploids produced from anther culture cannot be assumed to be homozygous. To verify that plants produced from anther culture are haploid, chromosome counts should be made from root tips or other meristematic somatic tissues. Haploids and diploids recovered from anther culture may also be distinguished by comparing size of cells, particularly stomatal guard cells, karyotype, DNA content or through the use of flow cytometry.

3.5.3 Factors Affecting Androgenesis

Reports of haploid production through anther culture have been steadily increasing. This has been possible by carefully monitoring a number of factors that influence androgenesis in vitro. The various factors that effect androgenesis are (i) genotype of donor plants, (ii) age and physiological condition of donor plants, (iii) stage of microspore or pollen development, (iv) culture medium, and (v) temperature and light.

3.5.4 Applications of Haploid Plants

The conventional method to produce homozygous plants is lengthy and laborious, requiring 6–8 recurrent cycles of inbreeding. Moreover, this approach is impractical for self-incompatible and male sterile plants and tree species. Through rapid achievement of homozygous traits in double haploids, pollen-derived haploid plants have been used in breeding and improvement of crop species. Some of the applications of this technique are discussed in the following sections.

3.5.4.1 Development of Pure Homozygous Lines

Homozygosity is achieved in the quickest possible way making genetic and breeding research much easier through anther/microspore culture methods. Homozygosity is still more important for those plants that have a very long juvenile phase such as fruit trees, bulbous plants, and forestry trees. Even if repeated self-pollination is possible, achievement of homozygosity in this group of plants is an extremely long process.

3.5.4.2 Hybrid Development

As a result of complete homozygosity obtained from diploidization of haploids, one can rapidly fix traits in the homozygous conditions. Pure homozygous lines can be used for the production of pure F1 hybrids.

3.5.4.3 Induction of Mutations

Haploid cell cultures are useful material for studying somatic cell genetics, especially for mutation and cell modification. The majority of mutations induced are recessive and therefore are not expressed in the diploid cells because of the presence of dominant allele. Single cells and isolated pollens have the advantage over the entire plant in that they can be plated and screened in large numbers. Mutants that are resistant to antibiotics, herbicides, and toxins have been isolated in a number of plants species. By subjecting haploid *Nicotiana tabacum* cells to methionine sulfoximine, Carlson (1973) regenerated mutant plant that showed a considerably lower level of infection by *Pseudomonas tabaci*.

TABLE 3.4

Some Selected Research Work on Anther Culture in Various Crop Species

Plant	Type of Haploid Production	References
Cyclamen persicum	Androgenesis (direct pollen embryogenesis)	Ishizaka and Uematsu (1993)
Brassica vegetables	Androgenesis (anther culture)	Cao et al. (1996)
Capsicum annuum	Androgenesis (pollen culture)	Supena et al. (2006)
Rice	Androgenesis (pollen culture)	Raina and Irfan (1998)
Triticum turgidum	Anther and microspore culture	Touraev et al. (2009)
Secale cereal	Anther culture	Basu et al. (2010)
Wheat	Anther culture	Basu et al. (2011)

3.5.4.4 Significance of Early Release of Varieties

Based on anther culture, many varieties have been released. In Japan, a tobacco variety F211 resistant to bacterial wilt has been obtained through anther culture (Matsuda and Ohashi, 1973). In sugarcane, selection among anther culture–derived haploids led to the development of superior lines with tall stem and higher sugar content. In bell peppers, dihaploid lines exhibited all shades of color ranging from dark green to light green. These reports have encouraged many plant breeders to incorporate anther culture in breeding methods (Table 3.4).

3.5.5 Successful Case Studies

A successful protocol for double haploid production in Indonesian hot pepper (*Capsicum annuum* L.) has been reported (Supena et al., 2006). The critical factors of the protocol are selection of flower buds with more than 50% late unicellular microspores, pretreatment of the buds at 4°C for 1 day, followed by culture of the anthers in double-layer medium system for 1 week at 9°C and thereafter at 28°C in continuous darkness. The medium contained Nitsch components and 2% maltose, with 1% activated charcoal in the solid under layer and 2.5 μm zeatin and 5 μm indole-3-acetic acid in the liquid upper layer. All the 10 genotypes of hot pepper tested, responded to this protocol. This protocol can be used as a potential tool for producing double haploid plants for hot pepper breeding.

The great achievements being made in the field of haploid production technology in some selected crop species and its implication to modern agriculture and in crop development programs have been established (Basu et al., 2011). This technique also has useful applications in genetic transformation for generating polyploidy wheat. Wheat cultivars developed from double haploids from both anther culture and maize induction systems have been released for cultivation in all the major continents. Several techniques have been adapted for the production of haploid plants such as anther culture and isolated microspore culture. With the ability to increase the yield of haploids in bread wheat and durum wheat, the haploidy technique may play an ever-increasing role in basic cytogenetic, genetic, and genomic research as well as in applied plant breeding.

3.6 Embryo Culture

Embryo culture, also known as embryo rescue, is one of the earliest and successful forms of in vitro culture techniques that is used to assist in the development of plant embryos and is defined as a sterile isolation and growth of immature embryo in vitro, with the goal of obtaining a viable plant (Sage et al., 2010). Plant embryos are multicellular structures that have the potential to develop into a new plant. The most widely used embryo rescue procedure involves excising plant embryos and placing them onto culture media (Miyajuma, 2006). Embryo rescue was first documented in the eighteenth century when Charles Bonnet excised *Phaseolus* and *Fagopyrum* embryos and planted them in soil and the cross resulted in dwarf plants (Sharma et al., 1996). Soon after this, scientists started placing the embryos in various nutrient media. During the period of 1890–1904, systems for embryo rescue became systematic by applying nutrient solutions that contained salts and sugars and applying aseptic technique (Amanate-Bordeos et al., 1992). The first successful in vitro embryo culture was performed by Hanning in 1904; he, however, described problems with precocious embryos that resulted in small, weak, and often inviable embryo (Mehetre and Aher, 2004).

Embryo culture is also used in propagation of orchids, as orchids are difficult to propagate since their seeds lack any stored food and the embryo is virtually naked. In many orchids, embryo development is incomplete at the time their seed mature. Young or mature orchid embryos are removed from seeds and placed on suitable nutrient media. The embryos develop into seedlings either directly or through protocorm callus formation. Embryo rescue is most often used to create interspecific and intergeneric crosses that would normally produce seeds that are aborted. Interspecific incompatibility in plants can occur for many reasons, but most often embryo abortion occurs (Sharma et al., 1996). In plant breeding, wide hybridization crosses can result in small shrunken seeds that indicate that fertilization has occurred, however, the seed fails to develop. Many times, remote hybridizations will fail to undergo normal sexual reproduction, thus embryo rescue can assist in circumventing this problem (Bridgen, 1994). Liabach (1929) demonstrated the practical application of this technique by isolating and growing the embryos of interspecific cross *Linum perenne* and *Linum austriacum* that aborted in vivo.

3.6.1 Types of Embryo Culture

Depending on the organ cultured, it may be referred to as embryo, ovule, or ovary culture. Ovule culture or in ovolo embryo culture is a modified technique of embryo rescue whereby embryos are cultured while still inside their ovules to prevent damage during the excision process (Cisneros and Zur, 2010). Ovary or pod culture employs the use of an entire ovary into culture. It becomes necessary to excise the entire small ovary and ovule to prevent early embryo abortion. However, it is technically difficult to isolate the tiny intact embryos, so often ovaries with young embryos or entire fertilized ovules are used (Ikeda et al., 2003).

3.6.1.1 Mature Embryo Culture

This type of culture involves the mature embryos derived from ripe seeds and is carried out when embryos do not survive in vivo or become dormant for long periods to eliminate the inhibition of germination. Excision of embryos from testa bypasses the seed dormancy caused by chemical inhibitors or mechanical resistance present in these structures covering the embryo. Embryos excised from the developing seed stages are capable of growing on a simple inorganic media supplemented with certain growth supplements.

3.6.1.2 Immature Embryo Culture

This is done to avoid the embryo abortion and production of viable plant. When the individuals from two different species of same genus or different genera are crossed, this often lead to failure due to various pre- and postfertilization barriers that prevent the successful gene transfer from wild to cultivated species. The basic principle of embryo rescue technique is the aseptic isolation of embryo and its transfer to a suitable medium under optimum culture conditions.

3.6.2 Factors Affecting Embryo Culture

The main factors that influence success of embryo culture include the time of culture, the composition of the medium, and temperature and light. These various factors should be carefully monitored while carrying out embryo culture.

3.6.3 Applications of Embryo Culture

Embryo culture is one of the earliest and well-established practical forms of tissue culture technique with proven value to the plant breeders (Dunwell, 1986). Its major application in plant breeding has been for interspecific hybridization, haploid production, precocious germination studies, shortening of breeding cycles, and overcoming seed dormancy. However, many unsuccessful crosses occur due to embryo abortion.

3.6.3.1 Prevention of Embryo Abortion

Early embryo abortion occurs primarily because the endosperm fails to develop properly (Hu and Wang, 1986). With interspecific crosses, intergeneric crosses, and crosses between diploids and tetraploids, the endosperm often develops poorly or not at all. Several successful cases have been documented with embryos arising from interspecific hybrids and intergeneric hybrids by aseptically culturing the embryo in a nutrient medium, which then possesses the potential for initiating development by avoiding postzygotic barriers within the mother plant (Williams and De Lautour, 1980; Williams et al., 1982; Ramming, 1990; Koba et al., 1991; Morgan and Thomas, 1991).

3.6.3.2 Precocious Germination

Precocious germination is defined as the germination of embryos before the completion of normal embryo development. Usually, it results in formation of weak seedlings. Embryos can be cultured under various conditions to determine what simulates embryological development and to understand the factors that regulate the orderly development of embryos in nature. Precocious germination occurs because inhibitors are lost when the testa is removed or because the negative osmotic potential is a higher value in vivo. Precocious germination has been prevented in *Prunus* through ovule culture, where the integument acts as a natural inhibitor (Ramming, 1985).

3.6.3.3 Seed Dormancy

By excising embryos, germination occurred without delay at all stages of seed maturity. Small or young embryos that abort at early stages of development are often difficult to isolate. In such situations, it may be possible to rescue embryos by ovary or ovule culture methods (Rangan, 1984). Ovaries excised (calyx, corolla, and stamens removed) is surface-sterilized and cultured with the cut end of the pedicel inserted into the nutrient medium. The ovary then develops into a fruit with fully developed seeds if it all goes well. For ovule culture, the sterilized ovary is opened and the fertilized ovules are scooped out and transferred to the surface of the culture medium. The reasons for the successful recovery of hybrids from ovary or ovule culture rather than through embryo culture are probably related to nutritional and physical factors and protection of the embryo by the maternal or sporophytic tissues.

3.6.3.4 Nutritional Requirements

Embryo culture is also useful in basic studies. Growing embryos outside the ovule (ex ovulo) is an excellent way to study the nutrition and metabolism of the embryos at various stages of development. The technique can also be used to examine the effects of plant growth regulators and environmental conditions

TABLE 3.5

Embryo Culture of Some Interspecific Crosses for Embryo Rescue and Resistance Development

Crossing Species	Importance of Embryo Culture	References
Actinidia deliciosa × *Actinidia eriantha*	Overcome nonviability	Mu et al. (1990)
Actinidia deliciosa × *Actinidia arguta*	Overcome nonviability	Mu et al. (1991)
Brassica napus × *Brassica oleracea*	Resistance to triazine	Ayotte et al. (1989)
Linum perenne × *Linum austriacum*	Seed germination	Abraham and Ramachandran (1960)
Hordeum vulgare × *Hordeum bulbosum*	Resistant to powdery mildew	Snape et al. (1989)
Triticum aestivum × *Thinopyrum scirpeum*	Salt tolerance	Farooq et al. (1993)
Vitis vinifera × *Vitis rotundifolia*	Yield increase and seedless progeny	Goldy et al. (1988)

on zygotic embryogenesis, and the regeneration potentials of whole embryos and their segments (Yeung et al., 1981). Embryo culture can be used to localize sites of germination promoters and inhibitors, for studies of embryogenesis, and for cryopreservation (Grout, 1986). Some of the selected research work on embryo culture of few interspecific crosses have been cited in Table 3.5.

3.6.4 Successful Case Studies

The immature embryos of apple rootstocks *Malus prunifolia* (Marubakaido) and *Malus pumila* (M9) have been rescued after 40–60 days of pollination (Dantas et al., 2006). The culture was put into MS culture media supplemented with agar (6 g/L) and casein hydrolysate (500 mg/L). Embryos originated from interspecific crosses and open pollination showed differences in the in vitro responses, depending on the female parent, the developmental stage of the embryo, and the culture medium composition. Embryos of the *M. pumila* rootstock, rescued within 40 days after pollination and put in culture medium supplemented with indoleacetic acid (IAA), gibberellic acid (GA_3), kinetin, and maltose, resulted in a normal development of plantlets. The crossing of responsive species and the use of the technique of embryo culture provided a rapid and uniform germination and, consequently, the development of fully normal seedlings.

3.7 Protoplast Culture and Somatic Cell Hybridization

Protoplast fusion or somatic hybridization is one of the most important uses of protoplast culture. This is particularly significant for hybridization between species or genera that cannot be made to cross by conventional method of sexual hybridization. A nonconventional genetic procedure involving fusion between isolated somatic protoplasts of the two different species (which are sexually incompatible) under in vitro condition and subsequently development of their fused product (heterokaryon) to a hybrid plant is known as somatic hybridization. The fusion of protoplasts has become an important tool in plant somatic cell genetics and crop improvement, especially for the production of interspecific hybrids. Of all the possible starting points for plant genetic manipulation, only protoplasts offer the opportunity to take advantage of all the technologies now available. This technology offers great promise for achieving wide crosses between species with a hope to develop new varieties. While offering great promise, this technique has to date been successfully applied to a number of plant species. Since the first successful isolation of protoplasts by Cocking (1960), substantial progress has been made toward improving the technology. Attempts have also been made to isolate protoplasts from several crop species and protoplast-based plant regeneration systems are made available for a great number of species (Maheshwari et al., 1986). The improvements that have occurred include modification of protoplast isolation procedures, media composition (Kao and Michayluk, 1975), preconditioning of protoplast donor tissues (Shahin, 1984), utilization of conditioned media or feeder cells (Bellincampi and Morpugo, 1987; Kyozuka et al., 1987; Lee et al., 1990), and manipulation of culture environment. Somatic cell fusion, thus, offers new ground to achieve novel genetic changes in plants.

3.7.1 Isolation, Culture, and Regeneration of Protoplasts

Protoplasts (cell without cell wall) are the biologically active and most significant materials of cells. When the cell wall is mechanically or enzymatically removed, the isolated protoplast is known as "naked plant cell" on which most of recent researches are based. The plant cell wall acts as the physical barrier and protects cytoplasm from microbial invasion and environmental stress. It consists of a complex mixture of cellulose, hemicellulose, pectin, lignin, lipids, protein, etc. For dissolution of different components of the cell wall, it is essential to have the respective enzymes. Protoplasts can be isolated from a range of plant tissues: leaves, stems, roots, anthers, and even pollens. The method for protoplast isolation can be classified into two main groups.

3.7.1.1 Sequential Enzymatic Treatment (Two Steps)

This approach involves initial incubation of macerated plant tissues with pectinase, which, in turn, are then converted into protoplasts by a cellulase treatment. A sequential or two-step method for isolating mesophyll protoplasts using commercial preparations of enzymes has been employed (Takebe et al., 1968).

3.7.1.2 Mixed Enzymatic Treatment (Simultaneous)

In this approach, plant tissues are plasmolyzed in the presence of a mixture of pectinase and cellulase, thus inducing concomitant separation of cells and degradation of their walls to release the protoplasts directly. By using two enzymes together, protoplast has been isolated in one step. The protoplast can also be isolated from cell suspension using this approach (Srivastava et al., 1987). The isolated protoplasts should be healthy and viable in order to undergo sustained division and regeneration. The viability of the protoplasts can be checked by Evan's blue and FDA methods.

3.7.2 Protoplast Culture and Regeneration

Isolated protoplasts are cultured by feeder layer technique, microdrop culture, or Bergmann's cell plating technique. During isolation and culture, the protoplasts require osmotic protection until they regenerate cell wall. Inclusion of an osmoticum in both isolation and culture media prevents rupture of protoplasts. The first step in the protoplast culture is the development of cell wall around the membrane of isolated protoplasts. This is followed by induction of divisions in the protoplast-derived new cell giving rise to a small cell colony. By manipulating the nutritional and physiological conditions in the nutrient media, cell colonies are induced to grow a callus and regenerate into complete plantlets (Figure 3.8).

3.7.3 Protoplast Fusion

Protoplast fusion involves mixing of protoplasts of two different genomes. It can be achieved by either spontaneous or induced fusion methods. The isolated protoplasts are devoid of cell wall, which makes them easy tool for undergoing in vitro fusion. An important aspect has been that incompatibility barriers do not exist during the cell fusion process at interspecific or intergeneric levels. Thus, plant protoplast represents the finest single cell system that could offer exciting possibilities in the field of somatic cell genetics and crop improvement. To achieve induced fusion, a suitable agent is generally necessary. These agents are known as fusogens. A variety of fusogens (sodium nitrate, calcium ions at high pH, polyethylene glycol [PEG] treatment, and electrofusion) have successfully been used to fuse protoplasts and produce somatic hybrids (Figure 3.9).

The protoplast fusion results into two types of somatic hybrids, i.e., symmetric hybrids and asymmetric hybrids. Symmetric hybrids consist of complete set of chromosomes from both the parents. Earlier efforts were made to obtain somatic hybrids among closely related and cross-compatible species resulting in somatic hybrids that resembled the sexual hybrids. For instance, the somatic hybrids among *Brassica campestris* and *Brassica oleracea* resembled the *B. napus* with the refinements in the technique of protoplast isolation, fusion, and culture.

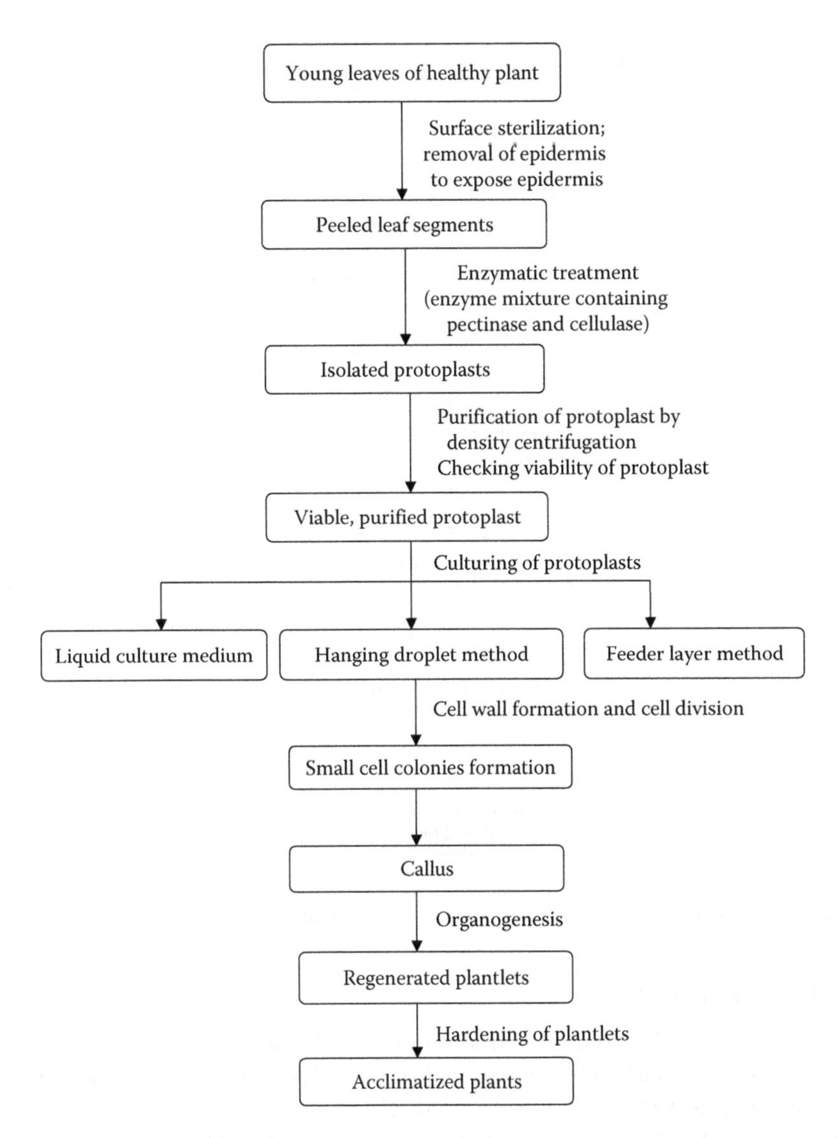

FIGURE 3.8
Culture of isolated protoplasts and regeneration of plants.

Asymmetric hybrids result from preferential or unidirectional loss of chromosomes of one fusion partner. Asymmetrization occurs spontaneously or it can be induced artificially. The final product of protoplast fusion among phylogenetically remote species is usually asymmetric combination of two genomes with parts of one or both genomes being lost during the in vitro passage. The extent direction of asymmetrization is largely random and hence unpredictable. Now growing interest lies in the artificial production of asymmetric cytoplasmic hybrids for single-step transfer of useful cytoplasmic traits.

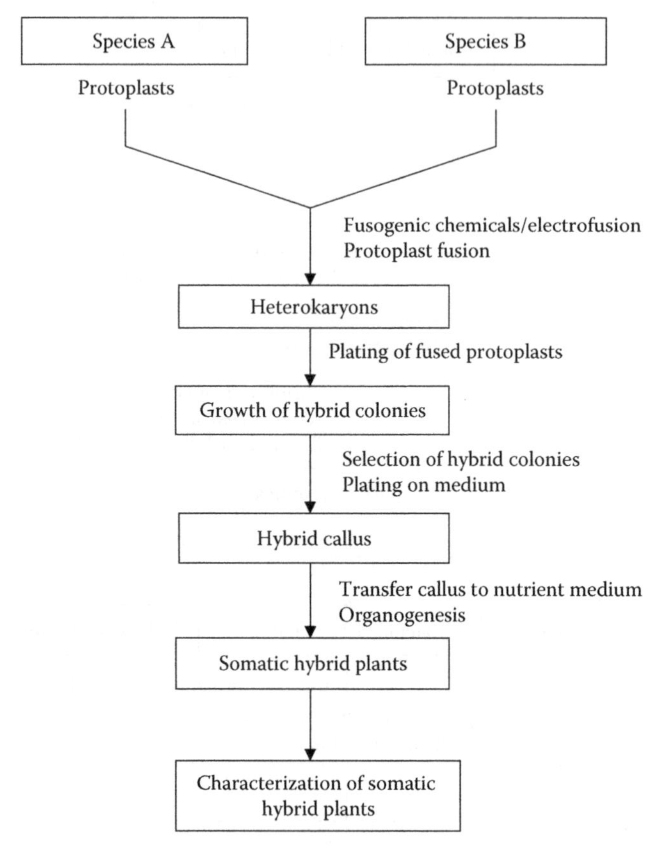

FIGURE 3.9
Fusion of protoplasts and production of somatic hybrid plants.

3.7.4 Cybridization

In somatic hybridization, combination of cytoplasmic organelles from both the parent takes place and recombination of mitochondrial and chloroplast genome occurs frequently. Fusion products with nucleus of one parent and extranuclear cytoplasmic genomes of the other parent are referred to as cybrid and the process to obtain cells or plants with such genetic combination/s is called cybridization. Cybrids may be produced in relatively high frequency by (i) irradiating (with x-rays or gamma rays or iodoacetamide treatment) the protoplasts of one species prior to fusion in order to inactivate their nucleus or (ii) preparing enucleate protoplasts (cytoplasts) of one species and fusing them with normal protoplasts of other species.

The most significant aspects of the use of protoplast fusion are that it can produce nuclear cytoplasmic combinations that are difficult or impossible to obtain by conventional plant breeding techniques. Because angiosperms usually have cytoplasmic genomes that are maternally inherited, new cytoplasmic nuclear combinations can be produced sexually only by back crossing, which is both time consuming and unidirectional. Although organelle DNA represents only 10^{-3} to 10^{-5} of the total DNA of higher plants, it carries important genes that control chlorophyll content, herbicide resistance, and cytoplasmic male sterility. The establishment of new nuclear cytoplasmic genome combination is important both for fundamental and molecular studies of plant cells. Another way of transferring cytoplasmic genes is the introduction of chloroplasts, mitochondria, or their respective DNAs into normal protoplasts followed by culture and plant regeneration.

3.7.5 Identification and Selection of Heterokaryons or Hybrid Cells

Following fusion treatment, the fused protoplasts in culture medium regenerate cell walls and undergo mitosis resulting in a mixed population of parental cells and heterokaryotic fusion products or hybrids. Heterokaryons or hybrid cells must be distinguished from the other cells. Development of methods to select heterokaryons from nonfused or homoplasmically fused protoplast is essential, which can be achieved through the following procedures.

3.7.5.1 Auxotrophic Mutants

The selection of heterokaryons (somatic hybrid cells) as a result of complementation by auxotrophic mutants may be useful as only the hybrid cells are expected to survive in the minimal medium. The selection of somatic hybrids by utilizing protoplasts of nitrate reductase–deficient and chlorate-resistant mutant lines of tobacco isolated by Muller and Grafe has been done (Glimelius et al., 1978; Wallin et al., 1979).

3.7.5.2 Drug Sensitivity

Drug sensitivity is also used for the selection of heterokaryons. The differential sensitivity of protoplasts isolated from *Petunia parodii* and *Petunia hybrida* to the antibiotic actinomycin D has been utilized and it has been observed that the addition of actinomycin D to the culture medium apparently has little effect on the regeneration potential of *P. parodii* protoplasts, but those of *P. hybrida* fail to divide (Power et al., 1976). Heterokaryons, however, are able to grow despite the presence of drug and ultimately differentiate into somatic hybrid plants. A similar procedure was adopted in the selection of somatic hybrids between *Nicotiana sylvestris* and *Nicotiana knightiana* (Maliga et al., 1977).

3.7.5.3 Density Gradient Centrifugation

Differences in density for vacuolated, evacuolated, and homoplasmically fused protoplasts have been used in the separation of heterokaryons by density gradient centrifugation (Hampp and Steingraber, 1985).

3.7.5.4 Flow Cytometry

The technique is very useful for sorting or concentrating the heterokaryons of normal cells that lack genetic markers. In the technique, the protoplast has been labeled with fluorescein isothiocyanate (FITC) or rhodamine isothiocyanate (RITC). Heterokaryons were sorted by means of the fluorescence of these dyes. Heterokaryon has been selected manually or by flow cytometry from fusion products between *B. oleracea* and *B. campestris* (Sunderberg et al., 1987).

3.7.6 Characterization of Somatic Hybrid Plants

A number of techniques are used to characterize the somatic hybrid plants such as isozyme patterns, ribulose-1,5-bisphosphate carboxylase oxygenase (RuBisCO) subunit polypeptides, restriction pattern of nuclear DNA coding for ribosomal RNA (rDNA), and restriction pattern of mitochondrial DNA and chloroplast DNA. Electrophoretic analysis of the isozyme pattern is a rapid method used to detect hybrid-like callus colonies or plants. This method is also used to differentiate cybrids from nuclear hybrids because most enzymes are encoded on the nuclear genome. A close correlation between chloroplast hybridity determined by ribulose-1,5-bisphosphate carboxylase large subunit analysis and the restriction pattern of cpDNA has been reported (Asahi et al., 1988). In several plant species combinations, the restriction enzyme patterns of rDNA differ and their nuclear DNA can be distinguished, after digestion with restriction enzyme and subsequent hybridization with a labeled heterologous rDNA or rRNA fragment as a probe. Due to restriction site polymorphism in mitochondrial DNA and chloroplast DNA, the electrophoretic patterns of the fragment DNA after digestion with appropriate restriction enzymes are characteristics of the cytoplasmic type. It has proved useful for confirming the cytoplasmic hybridity of the hybrids or cybrids.

3.7.7 Application of Somatic Hybrids

Sexual hybridization between closely related species has been used for years to improve cultivated crops. Unfortunately, sexual hybridization is limited in most cases particularly to cultivars within a species or at the best to a few wild species closely to the cultivated crops. Cell engineering of higher plants can be affected by fusion of somatic cells to enable partial or complete transfer of organelles between different genotypes. Therefore, the gene flow across sexual barriers can be achieved and unique nuclear cytoplasmic combinations can be selected and properly evaluated in breeding programs.

3.7.7.1 Somatic Hybrids for Gene Transfer

Protoplast has already been used to produce new species combinations that overcome barriers of conventional breeding. Wide interspecies and intergeneric hybrids have been produced in several laboratories and new gene combinations particularly for mixture of nuclear and cytoplasmic gene have been recovered in somatic hybrids. Between 1981 and 1984, emphasis shifted to the transfer of agriculturally useful traits resulting in a list of hybrid plants expressing useful traits. Agriculturally useful traits present in divergent wild species should be identified and procedures outlined to permit transfer by protoplast fusion into a cultivated species. Other traits that have been suggested as prospects for transfer by protoplast fusion include the disease resistance in *Arachis*, alkaloid content in *Solanum*, drought tolerance in *Lilium*, and altered linolenic acid content in *Brassica*.

3.7.7.2 Successful Transfer of Disease Resistance Genes by Protoplast Fusion

Transfer of disease resistance by protoplast fusion between sexually incompatible plant species has been carried out due to the development of this technology. *Solanum brevidens*, a non-tuber-bearing wild species, has resistance to potato leaf roll virus (PLRV) and to frost. But it is difficult to sexually cross this species with the cultivated potato, i.e., *Solanum tuberosum*. The protoplast from diploid *S. brevidens* and diploid *S. tuberosum* has been fused to obtain somatic hybrids from which plants were regenerated (Austin et al., 1985). Hybridity was verified by the plant morphology and by cytological observation. Interestingly, somatic hybrid plants show resistance to PLRV. A hexaploid somatic hybrid by protoplast fusion of a diploid *S. brevidens* carrying resistance to PLRV and tetraploid *S. tuberosum* carrying late blight resistance has been developed (Healgeson et al., 1986). All the hybrid plants showed late blight resistance and most of the hybrid plants were resistance to PLRV. These tetraploid and hexaploid hybrids of potato have been shown to have reasonable levels of female fertility. The tubers of hexaploid hybrid potatoes did not decay in storage and found to be resistant to the soft rot (Austin et al., 1988). Several progenies of the hybrids are resistant to the soft rot indicating that this trait is genetically stable.

3.7.7.3 Transfer of Male Sterility

Cytoplasmic male sterility, an important agricultural characteristic, is defined as the inhibition of male gametophyte development specified by cytoplasmically inherited factors. The finding of alloplasmic cytoplasmic male sterility, which was produced by interspecific crosses, has led to the supposition that cytoplasmic male sterility is the result of the incompatibility between the nuclear genomes of one species and cytoplasmic mitochondrial genome of the other. Cytoplasmic male sterility traits can be successfully transferred by protoplast fusion. Somatic fusion in which one partner is cytoplasmic male sterile and other fertile does not however necessarily yield only male sterile hybrids. Both male sterile and male fertile plants have been reported for *Nicotiana* (Galun et al., 1982), *Petunia* (Clark et al., 1986), and *Brassica* (Chetrit et al., 1985).

3.7.8 Successful Case Studies

The feasibility of introgression of genes by means of somatic hybridization has been evaluated to exploit new sources of disease resistance in rose breeding (Schum et al., 2002). They had established protocol for (i) regeneration of protoplasts in a variety of rose genotypes, (ii) PEG-mediated fusion, and (iii) preferential regeneration of heterologous fusion products. Protoplasts isolated from nonembryogenic source material gave rise to callus in *Rosa canina, Rosa caudata, Rosa corymbifera* "Laxa," *Rosa multiflora* (two accessions), *Rosa roxburghii, Rosa spinosissima, Rosa wichuraiana* (two accessions), as well as in *Rosa hybrida* "Elina" and "Pariser Charme." Protoplasts isolated from embryogenic cell suspensions of "Heckenzauber" and "Pariser Charme," as well as from nonembryogenic suspensions of the hybrid *Rosa persica* × *Rosa xanthina* were regenerated into plants. In order to suppress sustained cell divisions of nonfused protoplasts or of homologous fusion products, protoplasts were pretreated with either rhodamine 6G (0.1 m mol) or iodoacetate (0.5–1 mmol) for 15 min as well as with x-rays (300 Gy) at a dose rate of 3 Gy/min. Induced defects were mostly complementary, thus indicating that corresponding pretreatment of protoplasts prior to fusion allows preferential regeneration of the heterologous fusion products aimed at. Specific genotypes of different wild species within the genus *Rosa* were identified to carry resistance genes against *Diplocarpon rosae*, the causal agent of black spot (von Malek-Podjaski, 1999). Putative somatic hybrid callus lines were obtained from "Heckenzauber" + *R. wichuraiana* or *R. multiflora* as well as from "Pariser Charme" + *R. wichuraiana, R. multiflora*, or *R. roxburghii*, respectively. Shoots were regenerated from the combination of "Pariser Charme" and *R. wichuraiana*. The hybrid character of some selected regenerates was exemplarily confirmed by flow cytometry and AFLP analysis.

3.8 Somaclonal Variations

The regeneration of plants from callus, leaf explants, or plant protoplasts in culture is frequently a source of variability and such genetic variations found in the in vitro cultured cells are collectively referred to as somaclonal variation and the plants derived from such cells are called as "somaclones." It has been observed that the long-term callus and cell suspension culture and plants regenerated from such cultures are often associated with chromosomal variations. Variants selected in tissue cultures have been referred to as calliclones from callus cultures (Skirvin, 1978) or protoclones from protoplast cultures. Larkin and Scowcroft (1981), working at the Division of Plant Industry, C.S.I.R.O., Australia, gave the term "somaclones" for plant variants obtained from cultures of somatic tissues, although Evans et al. (1984) preferred the term gametoclonal variations for variant clones specifically raised from gametic and gametophytic cells. The term "somaclonal variation" refers to the variation that has been observed in plants regenerated from cell and tissue cultures, but in general not from axillary bud or shoot-tip cultures. Somaclonal variation is disadvantageous for the production of a clone by micropropagation, but it can be a source of valuable traits and several new varieties have been obtained by this way in some ornamental and crop species. The mutations can be detected after plant regeneration, on the field or in laboratory, but the application of selection pressures at the cell level increases the proportion of modified cells and plants. Such a selection is applicable for tolerance to physical or chemical environmental factors such as cold, drought, salt, or toxic ions. Modified plants have been regenerated in several plant species after in vitro culture and selection; for some clones or seed progenies, the physiological and genetical bases of the stress tolerance have been elucidated. As pointed out by Karp (1989), somaclonal variation is a phenomenon of broad taxonomic occurrence, reported for species of different ploidy levels and for outcrossing and inbreeding, vegetatively and seed propagated, and cultivated and noncultivated plants. During the last 10 years, somaclonal variation has been used in the laboratory of cytogenetics, in order to produce mutated lines in several crops. Some of these lines were then analyzed by genetical, physiological, and biochemical methods to elucidate the nature of the mutations where the genetic variations with desired or improved characters are introduced into the plants and new varieties are created that can exhibit disease resistance, improved quality, and yield in plants like cereals, legumes, oil seeds, tuber crops, etc.

Chromosomal rearrangements are an important source of this variation. The variations can be genotypic or phenotypic, which in the latter case can be either genetic or epigenetic in origin. Typical genetic alterations are changes in the number of chromosome (polyploidy and aneuploidy), change in the structure of chromosome (translocations, deletions, insertions, and duplications), and change in the DNA sequence (base mutations). Typical epigenetic alteration (nonheritable) related events are gene amplification and gene methylation.

The cumulative evidence from large number of scientific publications shows that somaclonal variation in cultured cells is more the rule than the exception. The usefulness of this variability in crop improvement program was first demonstrated through the recovery of disease-resistant plants in potato (resistance against late blight and early blight) and sugarcane (resistance against eyespot, Fiji disease, and downy mildew). Regenerated plants with altered chromosomal changes often show changes in leaf shape and color, growth rate and habit, and sexual fertility. Such changes are sometimes seen in regenerated plants with apparently normal chromosome constitution, implying that somaclonal variation may extend to the level of individual genes.

3.8.1 Factors Affecting Somaclonal Variation

The development of new crop varieties through somaclonal variation is largely influenced by physiological, genetic, and biochemical causes, which are discussed in the following sections.

3.8.1.1 Physiological Causes

These include genotype, explant source, duration of culture, and plant growth hormone effects. The genotype influences both the frequency of regeneration and somaclonal variation. The source of explant is also a critical variable for somaclonal variations due to different selective pressures exerted against different explants sources in cultures, resulting in a spectrum of somaclonal variation among regenerated plants. Chromosomal variations have been recorded due to several factors including nuclear fragmentation associated with first cell division of callus initiation and abnormalities of the mitotic process leading to aneuploidy and endoreduplication, occurring during cell culture initiation. It has been stated that variation increases with increasing duration of culture. The composition of culture medium also influences the frequency of karyotypic alterations in culture cells. The growth regulators, auxins and cytokinins, have been considered most frequently responsible for chromosomal variability.

3.8.1.2 Genetic Causes

At the genetic level, evidence has been presented for the involvement of many phenomena, including gross alterations in chromosome number and structure, point mutations, mitotic recombination, and the amplification, deamplification, deletion, transposition, or methylation of DNA sequences in nuclear, mitochondrial, or chloroplast genomes (Larkin, 1987; Evans, 1989; Karp, 1989; Brown, 1991; van den Bulk, 1991). Repetitive sequences may have a general role in somaclonal variation. Anomalies that seem peculiar to somaclonal populations are the occasional occurrence of homozygous variants (perhaps due to mitotic crossing over), and instances of directed changes resulting in population shifts (Larkin, 1987; Karp, 1989). In general though, the frequency of variants displaying a desired trait is very low. The chromosomal or molecular changes may result in stable alterations that are transmitted to sexual progeny, but variants frequently are not stable, particularly through meiosis. Some of the unstable changes may be due to the activation of transposable elements, while some may have a basis in DNA methylation (Karp, 1989). Many instances of selection of variant cell lines concern epigenetic variation—adaptations to selection pressure that are reversed when the pressure is removed.

3.8.1.3 Biochemical Causes

The biochemical causes involved in somaclonal variations include lack of photosynthetic ability due to alteration in carbon metabolism, biosynthesis of starch via carotenoid pathway, nitrogen metabolism, and antibiotic resistance.

3.8.2 Detection and Isolation of Somaclonal Variations

Isolation of somaclonal variants can be grouped into two broad categories, which are discussed in the following sections.

3.8.2.1 Screening

This method involves the observation of large number of cells/plants from tissue culture and detection of variants. In general, R1 progeny (progeny of regenerated R0 plants) are used for the identification of variant plants. R2 progeny (progeny of R1 plants) are used for confirmation. This screening procedure has been employed for a number of plants. Computer-based automated cell-sorting devices have been used to screen as many as 1000–2000 cells/s from which variant cells can be automatically separated. These variant cells are further regenerated to produce complete plantlets. This approach is widely used for the isolation of variants that produce high yield and desirable traits and is also used to obtain cell clones that produce higher quantities of certain biochemical agents.

3.8.2.2 In Vitro Cell Selection

In this technique, an appropriate selection pressure is applied under in vitro conditions, which permits the survival/growth of variant cells only during culture. When the selection pressure allows only the variant (mutagenic) cells to survive, it is called positive selection. In negative selection, the selection pressure allows only the wild-type cells to survive. These wild-type cells are later killed by counter-selection pressure. The variant cells are rescued by the removal of counter-selection agents. Positive selection approach may be further subdivided into four categories:

1. *Direct selection*: In this method, the selection agent kills the wild-type cells. The mutant (variant) cells remain unaffected. The mutant cells continue growing and dividing in the culture medium. This is the most common method employed to obtain variants that are resistant to toxins, herbicides, antibiotics, high salt concentrations, etc.
2. *Rescue selection*: In this method, the selection agent kills the wild-type cells. The mutant (variant) cells remain alive but do not divide due to unfavorable environment created by the selection agent. The selection agent is then removed to recover the variants. This method is used to obtain variants that are resistant to aluminum, cold temperature, etc.

3. *Stepwise selection*: In this method, the concentration of selection agent is increased in a stepwise manner, e.g., to obtain variants that are resistant to high salt concentration. First, low concentration of salt/toxin is added to the medium. Those cells that survive are then subjected to higher salt/toxin concentration and those cells that survive at this concentration are further subjected to higher salt/toxin concentration and so on.
4. *Double selection*: In this method, variants with two traits are selected simultaneously with the same selection agent, e.g., selection of a variant, which shows antibiotic resistance (streptomycin resistance) and development of chlorophyll. Here, streptomycin resistance is the first trait and development of chlorophyll in the cells is the second trait.

3.8.3 Applications of Somaclonal Variations

Somaclonal variations represent useful sources of introducing genetic variations that could be of value to plant breeders. Somaclonal variation methodologies are simpler and easier as compared to recombinant DNA technology. It helps in the development and production of plants tolerant to biotic (disease resistance) and abiotic stress. Salt tolerance in tobacco, maize and aluminum tolerance in carrot, and herbicide resistance in tobacco (resistant to sulfonylurea) have been developed. *Lathyrus sativa* (*Lathyrus* Bio L 212) having seeds with low content of neurotoxin has been developed. A somaclonal variant of *Citronella java* (with 37% more oil and 39% more citronellon), a medicinal plant, has been released as Bio-13 for commercial cultivation by Central Institute for Medicinal and Aromatic Plants (CIMAP), Lucknow, India. Supertomatoes-Heinz Co. and DNA Plant Technology Laboratories (United States) developed supertomatoes with high solid component by screening somaclones that helped in reducing the shipping and processing costs.

Several cell lines resistant to herbicides have been isolated and among the important achievements are tobacco, soybean, wheat, and maize resistant to various herbicides such as glyphosate, sulfonyl urea, and imidazolinone. Efforts have been made to follow in vitro approaches to obtain plants expressing tolerance to herbicides. Exposure of cell and protoplast cultures to various herbicides induces mutations that may yield tolerant cell lines, which ultimately regenerate into plants. The greatest contribution of variation, induced with respect to resistance to disease, was well-reviewed by Van den Bulk (1991). Using the in vitro cell selection technique, cell lines resistant to various pathogens (biotic stresses) have been developed in alfalfa (Acrioni et al., 1987), potato (Behnke, 1979), barley (Chawla and Wenzel, 1987), rice (Vidhyasekaran et al., 1990), and pea (Kumar et al., 1996). One of the major constraints in the development of agriculture is salt stress (abiotic stress). Since the first report on in vitro regeneration of sodium chloride–tolerant tobacco plants (Nabors et al., 1980), scientists have developed a number of cell lines and plants in several plant species, namely, alfalfa, rice, maize, sorghum, *Solanum nigrum, Brassica juncea, Brunus, Citrus,* and tomato that can withstand the rigors of salinity. In some cases, somaclones displaying tolerance stable through subsequent sexual generations have been identified (Spiegel-Roy and Ben-Hayyim, 1985; Jain et al., 1991; Winicov, 1991; Srivastava et al., 1995, 1996). Mandal et al. (1999) developed a salt-tolerant somaclone BTS24 from indigenous rice cultivar Pokkali. This somaclone yielded 36.6 q/ha under normal soil. Some of the somaclones developed for disease and herbicide resistance are cited in Table 3.6.

3.8.4 Successful Case Studies

Two strategies have been studied for the production of herbicide (imazapyr)-tolerant sugarcane, namely, (i) screening populations from breeding crosses for naturally occurring tolerant genotypes and (ii) producing tolerant genotypes through in vitro cell mutagenesis (Koch et al., 2009). In the first, over 11,000 seedlings were sprayed with 0.1–1.5 L/ha Arsenal (250 g/L active ingredient—imazapyr), after which 1.25 L/ha Arsenal was selected to test 12,000 seedlings. The second approach exploited the regeneration of herbicide-tolerant plants through induced somaclonal variation. The regenerated plants were acclimatized in the greenhouse and sprayed with Arsenal to confirm tolerance.

3.9 Summary and Future Prospects

Plant cell and tissue culture techniques have vast potential in crop improvement. The technique has now reached a level of sophistication at which it is possible to get most plant species into culture. The degree of understanding of cell biology of cultured cells, although low in comparison to some other system, is also steadily increasing. Plant cell and tissue culture hold great promise for controlled production of useful secondary metabolites on demand. Discoveries of cell culture capable of producing specific medicinal compounds at rate similar or superior to that of intact plants have accelerated in last few years. Plant cell suspension cultures are also being used for producing valuable products including secondary metabolites through biotransformation. In this technique, low-cost precursors are used as a substrate and transformed into value-added high-cost products. Plant tissue culture has also opened a new vista in the field of plant genetic engineering, as the development of reliable plant regeneration protocol is a prerequisite for the application of gene transfer technique to the genetic improvement of crops. Micropropagation techniques are becoming increasingly popular as an

TABLE 3.6

Somaclonal Variants for Disease Resistance and Herbicide Resistance

S. No.	Crop	Pathogen or Disease or Herbicide	References/Sources
1.	Rice	*Helminthosporium oryzae, Xanthomonas oryzae*	
2.	Wheat	*Helminthosporium sativum, Pseudomonas syringae*	
3.	Barley	*Rhynchosporium secalis, Helminthosporium sativum*	
4.	Oats	*Helminthosporium victoriae*	
5.	Maize	*Helminthosporium maydis*	
6.	Sugarcane	*Helminthosporium sacchari, Sclerospora sacchari, Ustilago scitaminea, Puccinia melanocephala,* Fiji disease	
7.	Tobacco	*Phytophthora parasitica, Pseudomonas syringae, P. solanacearum,* tobacco mosaic virus	
8.	Eggplant	*Verticillium dahlie,* little leaf disease	
9.	Potato	*Alternaria solani, Phytophthora infestans, Streptomyces scabies,* potato virus X, potato virus Y, potato leaf roll virus	
10.	Tomato	*Fusarium oxysporum, Pseudomonas solanacearum*	
11.	Hop	Tomato mosaic virus	
12.	Alfalfa	*Verticillium albo-atrum, Fusarium solani, F. oxysporum*	
13.	Celery	*Fusarium oxysporum, Septoria apii, Cercospora apii*	
14.	Lettuce	*Pseudomonas cichorii*	
15.	Rape	*Bremia lactucae,* lettuce mosaic virus	
16.	Banana	*Phoma lingam, Alternaria brassicicola*	
17.	Peach	*Fusarium oxysporum, Xanthomonas campestris* pv. *pruni*	
18.	Poplar	*Xanthomonas campestris, Pseudomonas syringae* *Septoria musiva, Melampsora medusa*	
19.	*Zea mays*	Glyphosate	Forlani et al. (1992)
20.	*Datura innoxia*	Chlorosulfuron	Rathinasabapathi and King (1992)
21.	*Glycine max*	Imazethapyr	Tareghyan et al. (1995)
22.	*Nicotiana tobacum*	Glyphosate	Stefanov et al. (1996)
23.	*Gossypium hirsutum*	Sulfonyl urea	Rajasekaran et al. (1996)
24.	*Zea mays*	Cycloxydim	Landes et al. (1996)
25.	*Triticum aestivum*	Difenzoquat	Bozorgipour and Snape (1997)
26.	*Beta vulgaris*	Imidazolinone	Wright et al. (1998)
27.	Sugarcane	Imazapyr	Koch et al. (2009)

alternative means of plant vegetative propagation and a considerable success has been achieved in case of micropropagation of horticultural, ornamental, and medicinal plant and forest tree species. Virus elimination through meristem culture and callus culture is an important aspect of plant cell and tissue culture and has also been reported in a number of plant species. The developments in the field of biotechnology permit detection of viruses for the testing of these plants using molecular techniques. The ultimate benefit resulting from the use of such techniques is that these are more rapid, efficient, and safe for use in the transfer of vegetative material.

Anther culture techniques hold great promise for the production of haploid plants and homozygous lines that have been used in plant breeding and improvement of crop species. Embryo culture or embryo rescue is one of the earliest and successful forms of in vitro culture technique for developing rare hybrids, where embryo aborts at an early stage due to certain incompatible barriers. Somatic cell hybridization and cybridization are few of the most important significances of protoplast culture and protoplast fusion for hybridization between species that are sexually incompatible. Using this technique, the cell engineering of higher plants can be affected by fusion of somatic cells to enable partial or complete transfer of organelles between different genotypes. Thus, gene flow across sexual barriers can be achieved and unique nuclear cytoplasmic combinations can be selected and properly evaluated in breeding programs. Somaclonal variation and gametoclonal variation represent useful sources of introducing genetic variation that could be of value to plant breeders. In vitro cell selection for abiotic and biotic stresses has also been effective in the development of certain agronomically useful mutations.

References

Abadi, A.J.N. and Hamidoghli, Y. 2009. Micropropagation of thornless trailing blackberry (*Rubus* sp.) by axillary bud explants. *Australian Journal of Crop Science* 3(4): 191–194.

Abdelmageed, A.H.A., Faridah, Q.Z., Norhana, F.M.A., Julia, A.A., and Kadir, M.A. 2011. Micropropagation of *Etlingera elatior* (Zingiberaceae) by using axillary bud explants. *Journal of Medicinal Plants Research* 5(18): 4465–4469.

Abraham, A. and Ramachandran, K. 1960. Growing *Colocasia* embryos in culture. *Current Science* 29: 342–343.

Acrioni, S., Pezzolli, M., and Damiani, F. 1987. In vitro selection of alfalfa plants resistant to *Fusarium oxysporum* f. sp. *medicagenesis. Theoretical and Applied Genetics* 74: 700–705.

Amanate-Bordeos, A.D., Nelson, R.J., Oliva, N.P., Dalmacio, R.D., Leung, H., and Sitch, L.A. 1992. Transfer of blast and bacterial blight resistance from the tetraploid wild rice *Oryza minuta* to the cultivated rice, *O. sativa. Theoretical and Applied Genetics* 84: 345–354.

Anderson, W.C. 1980. Tissue culture propagation of red raspberries. In: *Proceedings of Conference on Nursery Production of Fruit Plants through Tissue Culture—Applications and Feasibility*. Agric. Res. Sci. Educ. Admin., U.S.D.A., Beltsville, MD.

Andrijany, V.S., Indrayanto, G., and Soehono, L.D. 1999. Simultaneous effect of calcium, magnesium, copper and cobalt on sapogenin steroids content in callus cultures of *Agave amaniensis*. *Plant Cell Tissue and Organ Culture* 55: 103–108.

Asahi, T., Kumashiro, T., and Kubo, T. 1988. Constitutional of mitochondrial and chloroplast genome in male sterile tobacco obtained by protoplast fusion of *Nicotiana tabcum* and *N. debneyi*. *Plant Cell Physiology* 29: 43–49.

Austin, S., Bear, M.A., and Helegson, J.P. 1985. Transfer of resistance to potato leaf roll virus from *Solanum brevidense* into *Solanum tuberosum*. *Plant Science* 39: 75–82.

Austin, S., Lojkowska, E., Ehlenfeldt, M.K., Kelman, A., and Helgeson, J.P. 1988. Fertile interspecific somatic hybrids of *Solanum*: A novel source of resistant to *Erwinia* soft rot. *Phytopathology* 24: 51–53.

Ayotte, R., Harney, P.M., and Machado, V.S. 1989. The transfer of triazine resistance from *Brassica napus* L. to *Brassica oleraceae* L. IV second and third backcross to *Brassica oleraceae* and recovery of an 18-chromosome, triazine resistant BC3. *Euphytica* 40: 15–19.

Balandrin, M.J. and J.A. Klocke. 1988. Medicinal, aromatic and industrial materials from plants. In: *Biotechnology in Agriculture and Forestry. Medicinal and Aromatic Plant*, ed. Y.P.S. Bajaj, 1–36. Springer-Verlag, Berlin, Germany.

Balbuena, T.S., Santa-Catarina, C., Silvera, V., Kato, M.J., and Floh, E.I.S. 2009. In vitro morphogenesis and cell suspension culture establishment in *Piper solmsianum* DC. (Piperaceae). *Acta Botanica Brasilica* 23: 229–236.

Baldi, A. and Dixit, V.K. 2008. Enhanced artemisinin production by cell cultures of *Artemisia annua*. *Current Trends in Biotechnology and Pharmacology* 2: 341–348.

Basu, S.K., Datta, M., Sharma, M., and Kumar, A. 2011. Haploid production technology in wheat and some selected higher plants. *Australian Journal of Crop Science* 5(9): 1087–1093.

Basu, S.K., Eudes, F., and Kovalchuk, I. 2010. Role of *recA/RAD51* gene family in homologous recombination repair and genetic engineering of transgenic plants. In: *Applications of Plant Biotechnology: In vitro Propagation, Plant Transformation and Secondary Metabolite Production*, eds. A. Kumar and S. Sopory, 231–255. I.K. International Publishing House Pvt. Ltd., New Delhi, India.

Behnke, M. 1979. Selection of potato callus for resistance to culture filtrate of *Phytophthora infestans* and regeneration of resistant plants. *Theoretical and Applied Genetics* 55: 69–71.

Bellincampi, D. and Morpugo, G. 1987. Conditioning factor affecting growth in plant cells in culture. *Plant Science* 51: 83–91.

Bhojwani, S.S. and Razdan, M.K. 2004. *Plant Tissue Culture: Theory and Practice*. An Imprint of Elsevier, North Holland, the Netherlands.

Biswas, M.K., Hossain, M., and Islam, R. 2007. Virus free plantlets of strawberry through meristem culture. *World Journal of Agricultural Sciences* 3(6): 757–763.

Bourgin, J.P. and Nitsch, J.P. 1967. Obtaining *de Nicotiana* haploides a partir d'étamines cultiivées. *In Vitro Annals of Physiology and Vegetables* 9: 377–382.

Bozorgipour, R. and Snape, J.W. 1997. An assessment of somaclonal variation as a breeding tool for generating tolerant genotypes in wheat (*Triticum aestivum* L.). *Euphytica* 94: 335–340.

Bridgen, M.P. 1994. Review of plant embryo culture. *Horticultural Science* 29: 1243–1246.

Brieman, A., Ritem, A.D., Barp, A., and Shaskin, H. 1987. Heritable somaclonal variation in wild barley (*Hordeum spontaneum*). *Theoretical and Applied Genetics* 74: 104–112.

Brown, P.T.H. 1991. The spectrum of molecular changes associated with somaclonal variation. *IAPTC Letter* 66: 14–15.

Cao, M.Q., Li, Y., Liu, F., Jiang, T., and Liu, G.S. 1996. Application of anther culture and isolated microspore culture to vegetable crop improvement. *Acta Horticulturae* 392: 29–32.

Carelli, B.P. and Echeverrigaray, S. 2002. An improved system for the in vitro propagation of rose cultivars. *Scientia Horticulturae* 92:69–74.

Carlson, P.S. 1973. Methionine sulfoximine-resistant mutants of tobacco. *Science* 180: 1366–1368.

Catapan, E., Fleith, O.M., and Maria, V.A. 2000. In vitro culture of *Phyllanthus caroliniensis* (Euphorbiaceae). *Plant Cell Tissue and Organ Culture* 62: 195–202.

Catapan, E., Luis, M., da Silva, B., Moreno, F.N., and Viana, A.M. 2002. Micropropagation, callus and root culture of *Phyllanthus urinaria* (Euphorbiaceae). *Plant Cell Tissue and Organ Culture* 70: 301–309.

Chatenet, M., Delage, C., Ripolles, M., Irey, M., Lockhart, B.E.L., and Rott, P. 2001. Detection of sugarcane yellow leaf virus in quarantine and production of virus-free sugarcane by apical meristem culture. *Plant Disease* 85: 1177–1180.

Chaturvedi, R., Razadan, M.K., and Bhojwani, S.S. 2004. In vitro clonal propagation of a tree of neem (*Azadirachta indica* A. Juss.) by forced axillary branching. *Plant Science* 166: 501.

Chawla, H.S. and Wenzel, G. 1987. In vitro selection of barley and wheat for resistance against *Helminthosporium sativum*. *Theoretical and Applied Genetics* 74: 841–845.

Chetrit, P., Mathieu, C., Vedel, F., Pelletier, G., and Primard, C. 1985. Mitochondrial DNA polymorphism induced by protoplast fusion in *Crucifera*. *Theoretical and Applied Genetics* 69: 361–363.

Christen, A.A., Bland, J., and Gibson, D.M. 1989. Cell cultures as a means to produce Taxol. *Proceedings of the American Association for Cancer Research* 30: 566.

Cisneros, A. and Zur, T. 2010. Embryo rescue and plant regeneration following inter-specific crosses in the genus *Hylocereus* (Cactaceae). *Euphytica* 174: 73–82.

Clark, E.M., Schnabelrauch, L., Hanson, M.R., and Sink, K.C. 1986. Differential fate of plastid and mitochondrial genomes in *Petunia* somatic hybrids. *Theoretical and Applied Genetics* 72: 748–755.

Cocking, E.C. 1960. A method for the isolation of plant protoplasts and vacuoles. *Nature* 187: 927–929.

Dantas, A.C.M., Bonetti, J.I., Nodari, R.O., and Guerra, M.P. 2006. Embryo rescue from interspecific crosses in apple rootstocks. *Pesquisa Agropecuaria Brasileira* 41(6): 122–127.

De Fossard, R.A. 1976. *Tissue Culture for Plant Propagators*. University of New England Press, Armidale, Australia.

De-Eknamkul, D. and Ellis, B.E. 1985. Effects of macronutrients of growth and rosmarine acid formation in cell suspension cultures of *Anchusa officinalis*. *Plant Cell Reports* 4: 46–49.

Dhawan, S., Shasany, A.K., Naqvi, A.A., Kumar, S., and Khanuja, S.P.S. 2003. Menthol tolerant clones of *Mentha arvensis*: Approach for in vitro selection of menthol rich genotypes. *Plant Cell Tissue and Organ Cultures* 75: 87–94.

Dheeranapattana, S., Wangprapa, M., and Jatisatienr, A. 2008. Effect of sodium acetate on stevioside production of *Stevia rebaudiana*. *Acta Horticulturae (ISHS)* 786: 269–272.

Dobránszki, J. and Da Silva, J.A. 2010. Micropropagation of apple—A review. *Biotechnology Advances* 28(4): 462–488.

Dunwell, J.M. 1986. Pollen, ovule and embryo culture as tools in plant breeding. In: *Plant Tissue Culture and its Agricultural Applications,* eds. L.A. Withers and P.G. Alderson, 375–404. Butterworths, London.

Evans, D.A. 1989. Somaclonal variation: Genetic basis and breeding applications. *Trends in Genetics* 5(2): 46–50.

Evans, D.A., Sharp, W.R., and Medina-Filho, H.P. 1984. Somaclonal and gametoclonal variation. *American Journal of Botany* 71: 759–774.

Farooq, S., Iqbal, N., Shah, T.M., and Asghar, M. 1993. Intergeneric hybridization for wheat improvement. VIII. Transfer of salt tolerance from *Tinopyrum scirpeum* into wheat. *Journal of Genetics and Breeding* 47: 191–198.

Fei, H.M., Mei, K.F., Shen, X., Ye, Y.M., Lin, Z.P., and Peng, L.H. 1993. Transformation of *Gynostemma pentaphyllum* by *Agrobacterium rhizogenes* saponin production in hairy root cultures. *Acta Botanica Sinica* 35: 626–631.

Ferriere, N.M., Grout, H., and Carron, M.P. 1992. Origin and ontogenesis of somatic embryos in *Hevea brasiliensis* (Euphorbiaceae). *American Journal of Botany* 79: 174–180.

Forlani, G., Nielsen, E., and Racchi, M.L. 1992. A glyphosate resistant 5-enol pyruvyl shikimate 3 phosphate synthase confers tolerance to a maize cell line. *Plant Science Limeric* 85: 9–15.

Francoise, B., Hossein, S., Halimeh, H., and Zahra, N.F. 2007. Growth optimization of *Zataria multiflora* Boiss. tissue cultures and rosmarinic acid production improvement. *Pakistan Journal of Biological Sciences* 10: 3395–3399.

Fukui, H., Feroj, H.A.F.M., Ueoka, T., and Kyo, M. 1998. Formation and secretion of a new benzoquinone by hairy root cultures of *Lithospermum erythrorhizon. Phytochemistry* 47: 1037–1039.

Galun, E., Arzee-Gonen, P., Fluhr, R., Edelman, M., and Aviv, D. 1982. Cytoplasmic hybridization in *Nicotiana*: Mitochondrial DNA analysis in progenies resulting from fusion between protoplast having different organelle constitutions. *Molecular and General Genetics* 186: 50–56.

Gamborg, O.L., Miller, R.A., and Ojima, K. 1968. Nutrient requirements of suspension cultures of soybean root cells. *Experimental Cell Research* 50: 151–158.

Gaur, A. 2011. Studies on genetic fidelity of tissue culture raised plants of Himalayan poplar (*Populus ciliate* Wall.), M.Sc. Thesis, p. 103. Dr. Y.S. Parmar University of Horticulture and Forestry, Nauni, Solan (H.P.), India.

Gauthret, R.J. 1934. Culture du tissue cambial. *Comptes Rendus de I' Academie des Sciences* 198: 2195–2196.

Giri, A., Banerjee, S., Ahuja, P.S., and Giri, C.C. 1997. Production of hairy roots in *Aconitum heterophyllum* Wall. using *Agrobacterium rhizogenes. In Vitro Cellular and Developmental Biology-Plant* 33: 280–284.

Giri, A. and Narasu, M.L. 2000. Transgenic hairy roots: Recent trends and applications. *Biotechnology Advances* 18: 1–22.

Glimelius, K., Erikssan, T., Grafe, R., and Muller, A.J. 1978. Somatic hybridization of nitrate-reductase deficient mutants of *Nicotiana tabacum* by protoplast fusion. *Plant Physiology* 44: 273–277.

Gnanam, A. and Kulandaivelu, G. 1969. Photosynthetic studies with leaf cells suspensions from higher plants. *Plant Physiology* 44: 1451–1456.

Goldy, R.G., Emershad, R.L., Ramming, D.W., and Chaparro, J.X. 1988. Embryo culture as a means of introgressing seedlessness from *Vitis vinifera* to *Vitis rotundifolia. Horticultural Science* 23: 886–889.

Goussard, P.G. and Wiid, J. 1992. The elimination of fanleaf virus from grapevines using in vitro somatic embryogenesis combined with heat therapy. *South African Journal of Enology and Viticultures* 13(2): 81–83.

Groll, J., Mycock, D.J., and Gray, V.M. 2002. Effect of medium salt concentration on differentiation and maturation of somatic embryos of cassava (*Manihot esculenta* Crantz). *Annals Botany* 89: 645–648.

Grout, B.W.W. 1986. Embryo culture and cryopreservation for the conservation of genetic resources of species with recalcitrant seed. In: *Plant Tissue Culture and its Agricultural Applications*, eds. L.A. Withers and P.G. Alderson, 303–309. Butterworths, London.

Guha, S. and Maheshwari, S.C. 1964. In vitro production of embryos from anthers of *Datura. Nature* 204: 497–498.

Haberlandt, G. 1902. Kulturversuche mit isollierten pflanzenzellen. Sitzungsber. Akad. Wiss. Wien, Math-Naturwiss. KI, Abt. 1. 111: 69–92.

Hampp, R. and Steingraber, M. 1985. Electric field induced fusion of evacuolated mesophyll protoplasts of oat. *Naturwissenschaften* 72: 91–92.

Helgeson, J.P., Hunt, G.J., Haberlach, G.T., and Austin, S. 1986. Somatic hybrid between *Solanum brevidens* and *Solanum tuberosum*: Expression of a late blight resistance gene and potato leaf roll resistance. *Plant Cell Reports* 3: 212–214.

Hohtola, A., Jalonen, J., Tolnen, A. et al. 2005. Natural product formation by plants, enhancement, analysis, processing and testing. In: *Sustainable Use Renewable Natural Resources—From Principles to Practices*, eds. A. Jalkanen and P. Nygren, pp. 34–69. University of Helsinki Publication, Finland.

Hollings, M. 1965. Disease control through virus stock. *Annual Review of Phytopathology* 3: 367–396.

Hu, C. and Wang, P. 1986. Embryo culture: Technique and application. In: *Handbook of Plant Cell Culture*, ed. D.A. Evans, W.R. Sharp, and P.V. Ammirato, pp. 43–96. Macmillan, New York.

Huang, S.C. and Millikan, D.F. 1980. In vitro micrografting of apple shoot tips. *Horticulture Science* 15: 741–742.

Husaini, A.M. and Srivastava, D.K. 2011. Efficient plant regeneration from leaf and petiole explants of strawberry (*Fragaria* × *ananassa* Duch.). *Phytomorphology* 61(1–2): 55–62.

Ikeda, N., Niimi, Y., and Han, D. 2003. Production of seedling from ovules excised at the zygote stage in *Lilium* spp. *Plant Cell Tissue and Organ Cultures* 73(2): 159–166.

Ikenaga, T., Oyama, T., and Muranaka, T. 1995. Growth and steroidal saponin production in hairy root cultures of *Solanum aculeatissi. Plant Cell Reports* 14: 413–417.

Ishimaru, K. and Shimomura, K. 1991. Tannin production in hairy root cultures of *Geranium thunbergii. Phytochemistry* 30: 825–828.

Ishizaka, H. and Uematsu, J. 1993. Production of plants from pollen in *Cyclamen persicum* Mill. through anther culture. *Japanese Journal of Breeding* 43: 207–218.

Jagadev, P.N., Panda, K.N., and Beura, S. 2008. A fast protocol for in vitro propagation of ginger (*Zingiber officinale* Rosc.) of a tribial district of India. *Acta Horticulturae* 765: 101–108.

Jain, S., Nainawatee, H.S., Jain, R.K., and Chowdhury, J.B. 1991. Proline status of genetically stable salt-tolerant *Brassica juncea* L. somaclones and their parent cv. Prakash. *Plant Cell Reports* 9: 684–687.

Kao, K.N. and Michayluk, M.R. 1975. Nutritional requirements for growth of *Vicia hajastana* cells and protoplasts at a very low population density in liquid media. *Planta* 126: 105–110.

Karp, A. 1989. Can genetic instability be controlled in plant tissue cultures? *IAPTC Newsletter* 58: 2–11.

Karppinen, K., Hokkanen, J., Tolonen, A., Maltila, S., and Hohtola, A. 2007. Biosynthesis of hyperforin and adhyperforin from amino acid precursors in shoot cultures of *Hypericum perforatum*. *Phytochemistry* 68: 1038–1045.

Karuppusamy, S. 2009. A review on trends in production of secondary metabolites from higher plants by in vitro tissue, organ and cell cultures. *Journal of Medicinal Plants Research* 3: 1222–1239.

Kaur, R., Srivastava, D.K., and Sharma, D.R. 1992. Micropropagation of *Alnus nepalensis* Don.—A multipurpose forest tree. In: *Proceedings of "World Congress on Cell and Tissue Culture,"* Washington, DC.

Kaushal, N., Modgil, M., Thakur, M., and Sharma, D.R. 2005. In vitro clonal multiplication of an apple rootstock by culture of shoot apices and axillary buds. *Indian Journal of Experimental Biology* 43: 561–565.

Khan, T., Krupadanam, D., and Anwar, Y. 2008. The role of phytohormone on the production of berberine in the calli culture of an endangered medicinal plant, turmeric (*Coscinium fenustratum* L.). *African Journal of Biotechnology* 7: 3244–3246.

Kim, O.T., Bang, K.H., Shin, Y.S. et al. 2007. Enhanced production of asiaticoside from hairy root cultures of *Centella asiatica* (L.) Urban elicited by methyl jasmonate. *Plant Cell Reports* 26: 1914–1949.

Kim, J.S., Lee, S.Y., and Park, S.U. 2008. Resveratrol production in hairy root culture of peanut, *Arachis hypogaea* L. transformed with different *Agrobacterium rhizogenes* strains. *African Journal of Biotechnology* 7: 3788–3790.

Kim, J.H., Yun, J.H., Hwang, Y.S., Byun, S.Y., and Kim, D.I. 1995. Production of taxol and related taxanes in *Taxus brevifolia* cell cultures: Effect of sugar. *Biotechnology Letters* 17(1): 101–106.

Kin, N. and Kunter, B. 2009. The effect of callus age, VU radiation and incubation time on *trans*-resveratrol production in grapevine callus culture. *Tarim Bilimleri Dergisi* 15: 9–13.

Ko, K.S., Ebizuka, Y., Noguchi, H., and Sankawa, U. 1995. Production of polypeptide pigments in hairy root cultures of *Cassia* plants. *Chemical & Pharmaceutical Bulletin* 43: 274–278.

Koba, T., Handa, M., and Shimada, T. 1991. Efficient production of wheat–barley hybrids and preferential elimination of barley chromosomes. *Theoretical and Applied Genetics* 81: 285–292.

Koch, A.C., Ramgareeb, S., Snyman, S.J., Watt, M.P., and Rutherford, R.S. 2009. Pursuing herbicide tolerance in sugarcane: Screening germplasm and induction through mutagenesis. *Proceedings of South African Sugar Technologists Association* 82: 629–632.

Komaraiah, P., Ramakrishna, S.V., Reddanna, P., and Kavikishore, P.B. 2003. Enhanced production of plumbagin in immobilized cells of *Plumbago rosea* by elicitation and in situ adsorption. *Journal of Biotechnology* 10: 181–187.

Konan, N.K., Schöpke, C., Carcamo, R., Beachy, R.N., and Fauquet, C. 1997. An efficient mass propagation system for cassava (*Manihot esculenta* Crantz) based on nodal explants and axillary bud derived meristems. *Plant Cell Reports* 16: 444–449.

Kumar, S., Chander, S., Gupta, H., and Sharma, D.R. 1998. Micropropagation of *Actinidia deliciosa* from axillary buds. *Phytomorphology* 48(3): 303–307.

Kumar, A., Srivastava, D.K., Gupta, V.K., and Kohli, U.K. 1996. Cellular selection and characterization of pea lines resistant to culture filtrate of *Fusarium oxysporum*. In: *Trends in Plant Tissue Culture and Biotechnology*, eds. L.K. Pareek and P.L. Swarankar, pp. 162–168. Agro Botanical Publisher, India.

Kyo, M. and Harada, H. 1990. Control of the developmental pathway of tobacco pollen in vitro. *Planta* 168: 427–432.

Kyozuka, J., Hayashi, Y., and Shinomoto, K. 1987. High frequency plant regeneration from rice protoplasts by novel nurse culture methods. *Molecular & General Genetics* 206: 408–413.

Landes, M., Walther, H., Gerber, M., Auxier, B., and Brown, H. 1996. New possibilities for post-emergence grass weed control with cycloxydim and sethoxydim in herbicide tolerant corn hybrids. In: *Proceedings of the Second International Weed Control Congress*, 869–874, Denmark.

Larkin, P.J. 1987. Somaclonal variation, history, method and meaning. *Iowa State Journal of Research* 61: 393–434.

Larkin, P.J. and Scowcroft, W.R. 1981. Somaclonal variation: Novel source of variability from cell cultures for plant improvement. *Theoretical and Applied Genetics* 60: 197–214.

Lee, L., Schroll, R.E., Grimes, H.D., and Hodges, T.K. 1990. Plant regeneration from indica rice (*Oryza sativa* L.) protoplasts. *Planta* 178: 325–333.

Leonhardt, W., Wawrosch, Ch., Auer, A., and Kopp, B. 1998. Monitoring of virus diseases in Austrian grapevine varieties and virus elimination using in vitro thermotherapy. *Plant Cell, Tissue and Organ Culture* 52: 71–74.

Li, W., Li, M., Yang, D.L., Xu, R., and Zhang, Y. 2009. Production of podophyllotaxin by root culture of *Podophyllum hexandrum* Royle. *Electronic Journal of Biology* 5: 34–39.

Liabach, F. 1929. Ectogenesis in plants: Methods and genetic possibilities of propagating embryos otherwise dying in the seed. *Journal of Heredity* 20: 201–208.

Llyod, G.B. and Mc Cown, B.H. 1980. Commerically feasible micropropagation on mountain laurel, *Kalmia latifolia*, by use of shoot tip culture. *Combined Proceedings of International Plant Propagation Society* 30: 421–437.

Long, R.D. and Cassells, A.C. 1986. Elimination of viruses from tissue cultures in the presence of antivirus chemicals. In: *Plant Tissue Culture and its Agricultural Applications*, eds. L.A. Withers and P.G. Alderson, pp. 239–248. Butterworths, London, UK.

Ma, G., 1998. Effects of cytokinins and auxins on cassava shoot organogenesis and somatic embryogenesis from somatic embryo explants. *Plant Cell Tissue and Organ Culture* 54: 1–7.

Maharik, N., Elgengaihi, S., and Taha, H. 2009. Anthocyanin production in callus cultures of *Crataegus sinaica* Bioss. *International Journal of Academic Research* 1: 30–34.

Maheshwari, S.C., Gill, R., Maheshwari, N., and Gharyal, P.K. 1986. The isolation and culture of protoplasts. In: *Differentiation of Protoplasts and Transformed Plant Cells*, J. Reinert and H. Binding, eds., 20–48. Springer-Verlag, New York.

Maliga, P., Lazar, G., Joo, F., Nagy, A.H., and Menczel, L. 1977. Restoration of morphogenetic potential in *Nicotiana* by somatic hybridization. *Molecular and General Genetics* 157: 291–296.

Malpathak, N.P. and David, S.B. 1986. Flavor formation in tissue cultures of garlic (*Allium sativum* L.). *Plant Cell Reports* 5: 446–447.

Mandal, A.B., Chowdhury, B., and Sheeja, T.E. 1999. Development and characterization of salt tolerant somaclones in rice cultivar Pokkali. *Indian Journal of Experimental Biology* 38: 74–79.

Marconi, P.L., Selten, L.M., Cslcena, E.N., Alvarez, M.A., and Pitta-Alvarez, S.I. 2008. Changes in growth and tropane alkaloid production in long term culture of hairy roots of *Brugmansia candida*. *Electronic Journal of Integrative Biosciences* 3: 38–44.

Matsuda, T. and Ohashi, Y. 1973. Inheritance to resistance to bacterial wilt disease in tobacco. *Japanese Journal of Breeding* 23: 175–180.

Mehetre, S.S. and Aher, A.R. 2004. Embryo rescue: Tool to overcome incompatible interspecific hybridization in *Gossypium* Linn. A review. *Indian Journal of Biotechnology* 3: 29–36.

Mhatre, M., Salunkhe, C.K., and Rao, R.S. 2000. Micropropagation of *Vitis vinifera* L.: Towards an improved protocol. *Scientia Horticulturae* 84:357–363.

Mirjalili, M.H., Moyano, E., Bonfill, M., Cusido, R.M., and Palazon, J. 2009. Steroidal lactones from *Withania somnifera*, an antioxidant plant for novel medicine. *Molecules* 14: 2373–2393.

Miyajuma, D. 2006. Ovules that failed to form seeds in zinnia (*Zinnia violacec* Cav). *Scientia Horticulturae* 107(2): 176–182.

Modgil, M., Mahajan, K., Chakrabarti, S.K., Sharma, D.R., and Sobti, R.C. 2005. Molecular analysis of genetic stability in micropropagated apple rootstock MM106. *Scientia Horticulturae* 104: 151–160.

Morel, G.M. 1965. Clonal propagation of orchids by meristem culture. *Cymbidium Society News* 20: 3–11.

Morel, G.M. and Martin, Q.C. 1952. Guersion de dahlia attients d'une maladie a virus. *Comptes Rendues Academie de Sciences (Paris)* 235: 1324–1325.

Morgan, W.G. and Thomas, H. 1991. A study of chromosome association and chiasma formation in amphidiploids between *Lolium multiflorum* and *Festuca drymeja*. *Heredity* 67: 241–245.

Mu, X.J., Tsai, D.R., An, H.X., and Wang, W.L. 1991. Embryology and embryo rescue of interspecific hybrids in *Actinidia*. *Acta Horticulturae* 297: 93–97.

Mu, X.J., Wang, W.L., Cai, D.R., and An, H.X. 1990. Embryology and embryo rescue of interspecific cross between *Actinidia deliciosa* cv. Hayward and *A. eriantha*. *Acta Botanica Sinica* 2: 425–431.

Murakishi, H.H. and Carlson, P.S. 1976. Regeneration of virus free plant from dark-green islands of tobacco mosaic virus-infected tobacco leaves. *Phytopathology* 66: 931–932.

Murashige, T. 1978. Principles of rapid propagation. In: *Propagation of Higher Plants through Tissue Culture*, eds. K.W. Hughes et al., pp. 14–24. U.S. Department of Energy, Washington, DC.

Murashige, T. and Skoog, F. 1962. A revised medium for rapid growth and bioassays with tobacco tissue cultures. *Plant Physiology* 15(3): 473–497.

Murthy, H.N., Dijkstra, C., Anthony, P. et al. 2008. Establishment of *Withania somnifera* hairy root cultures for the production of withanolide. *Journal of Integrative Plant Biology* 50: 915–981.

Nabors, M.W., Gibbs, S.E., Bernstein, C.S., and Meis, M.E. 1980. NaCl-tolerant tobacco plants from cultured cells. *Z Pflanzenphysiol.* 97: 13–17.

Navarro, L., Roistacher, C.N., and Murashige, T. 1975. Improvement of shoot-tip grafting in vitro for virus free citrus. *Journal of the American Society for Horticultural Sciences* 100: 471–479.

Nesi, B., Trinchello, D., Lazzereschi, S., and Grassootti, A. 2009. Production of lily symptomless virus-free plants by shoot meristem tip culture and in vitro thermotherapy. *Horticultural Science* 44(1): 217–219.

Ng, S.Y.C. and Hahn, S.K. 1985. Application of tissue culture to tuber crops at IITA. In: *Proceedings, Inter-Center Seminar on Biotechnology in International Agricultural Research*, 24–40. IRRI, Los Banos, Philippines.

Nikolaeva, T.N., Zagoskina, N.V., and Zaprometov, M.N. 2009. Production of phenolic compounds in callus cultures of tea plant under the effect of 2,4-D and NAA. *Russian Journal of Plant Physiology* 56: 45–49.

Nin, S., Bennici, A., Roselli, G., Mariotti, D., Schiff, S., and Magherini, R. 1997. *Agrobacterium* mediated transformation of *Artemisia absinthum* L. (worm wood) and production of secondary metabolites. *Plant Cell Reports* 16: 725–730.

Nitsch, J.P. and Nitsch, C. 1969. Haploid plants from pollen grains. *Science* 163: 85–87.

Old, R.W. and Primrose, S.B. 1994. *Principles of Gene Manipulation: An Introduction to Genetic Engineering*. Blackwell Scientific Publications, London, UK.

Park, S.U. and Lee, S.Y. 2009. Anthraquinone production by hairy root culture of *Rubia akane* Nakai: Influence of media and auxin treatment. *Scientific Research and Essays* 4: 690–693.

Paska, C., Innocent, G., Kunvari, M., Laszlo, M., and Szilagyi, L. 1999. Lignan production by *Ipomea cairica* callus culture. *Phytochemistry* 52: 879–883.

Power, J.B., Frearson, E.M., Hayward, C. et al. 1976. Somatic hybridization of *Petunia hybrida* and *Petunia parodii*. *Nature* 263: 500–502.

Prathanturarug, S., Soonthornchareonnon, N., Chuakul, W., Phaidee, Y., and Saralamp, P. 2007. An improved protocol for micropropagation of *Mallotus repandus* (Willd.) Mull.-Arg. *In vitro Cellular Developmental Biology-Plant* 43: 275–279.

Quiala, E., Barbon, R., Jimenez, E., Feria, M.D., Chavez, M., Capote, A., and Perez, N. 2006. Biomass production of *Cymbopogon citratus* (DC.) Stapf. A medicinal plant in temporary immersion systems. *In Vitro Cellular Developmental Biology-Plant* 42: 298–300.

Rahnama, H., Hasanloo, T., Shams, M.R., and Sepehrifar, R. 2008. Silymarin production by hairy root culture of *Silybium marianum* (L.) Gaertn. *Iranian Journal of Biotechnology* 6: 113–118.

Raina, S.K. and Irfan, S.T. 1998. High-frequency embryogenesis and plantlet regeneration from isolated microspores of indica rice. *Plant Cell Reports* 17: 957–962.

Rajasekaran, K., Grula, J.W., and Anderson, D.M. 1996. Selection and characterisation of mutant cotton cell lines resistant to sulphonylurea and imidazolinone herbicides. *Plant Science* 119: 115–124.

Rajasekaran, T., Rajendran, L., Ravishankar, G.A., and Venkataraman, L.V. 1991. Influence of nutrient stress on pyrethrin production by cultured cells of pyrethrum (*Chrysanthemum cinerariaefolium*). *Current Science* 60: 705–707.

Ramming, D.W. 1985. In ovulo embryo culture of early-maturing *Prunus*. *Horticultural Science* 20: 419–420.

Ramming, D.W. 1990. The use of embryo culture in fruit breeding. *Horticultural Science* 25: 393–398.

Rangan, T.S. 1984. Culture of ovules. In: *Cell Culture and Somatic Cell Genetics of Plants*, ed. I.K. Vasil, pp. 227–231. Academic Press, New York.

Rani, V., Parida, A., and Raina, S.N. 1995. Random amplified polymorphic DNA (RAPD) markers for genetic analysis in micropropagated plants of *Populus deltoids* Marsh. *Plant Cell Reports* 14: 459–462.

Rao, R.S. and Ravishankar, G.A. 2002. Plant cell cultures: Chemical factories of secondary metabolites. *Biotechnology Advances* 20: 101–153.

Rathinasabapathi, B. and King, J. 1992. Physiological response of herbicide resistant cell variants of *Datura innoxia* to branched chain amino acids. *Plant Science Limerick* 81: 191–198.

Roat, C. and Ramawat, K.G. 2009. Elicitor induced accumulation of stilbenes in cell suspension cultures of *Cayratia trifoliata* (L.) Domin. *Plant Biotechnology Reports* 3: 135–138.

Rommanee, C., Salak, P., Wichien, Y., and Akira, S. 2003. Routine cryopreservation of in vitro grown axillary apices of cassava (*Manihot esculenta* Crantz) by vitrification: Importance of simple mononodal culture. *Scientia Horticulturae* 98: 485–492.

Sage, T.L., Strumas, F., Cole, W.W., and Barret, S. 2010. Embryo rescue and plan regeneration following interspecific crosses in the genus *Hylocereus* (Cactaceae). *Euphytica* 174: 73–82.

Saito, K., Yoshimatsu, K., and Murakoshi, T. 1990. Genetic transformation of foxglove (*Digitalis purpurea*) by chimeric foreign genes and production of cardioactive glycosides. *Plant Cell Reports* 9: 121–124.

Salma, U., Rahman, M.S.M., Islam, S. et al. 2008. The influence of different hormone concentration and combination on callus induction and regeneration of *Rauvolfia serpentina* (L.) Benth. *Pakistan Journal of Biological Sciences* 11: 1638–1641.

Santos-Gomes, P.C., Seabra, R.M., Andrade, P.B., and Fernandes-Ferreira, M.M. 2002. Phenolic antioxidant compounds produced by in vitro shoots of sage (*Salvia officinalis* L.). *Plant Science* 162: 981–987.

Schum, A., Hofmann, K., and Felten, R. 2002. Fundamentals for integration of somatic hybridization in rose breeding. *Acta Horticulturae* 572: 29–36.

Shahin, E.A. 1984. Isolation and culture of protoplasts: Tomato. In: *Cell Culture and Somatic Cell Genetics of Plants: Laboratory Procedure and their Application*, Vol. 1, ed. I.K. Vasil, 381–390. Academic Press, Orlando, FL.

Sharma, D.R., Chauhan, P.S., Kaur, R., and Srivastava, D.K. 1992. Micropropagation of Colt—A semidwarf rootstock of cherry. *Indian Journal of Horticulture* 49: 209–212.

Sharma, D.R., Kaur, R., and Kumar, K. 1996. Embryo rescue in plants: Review. *Euphytica* 89: 325–337.

Sharma, M., Modgil, M., and Sharma, D.R., 2000. Successful propagation in vitro of apple rootstock MM106 and influence of phloroglucinol. *Indian Journal of Experimental Biology* 38: 1236–1240.

Shenoy, V.B. and Vasil, I.K. 1992. Biochemical and molecular analysis of plants derived from embryogenic tissue cultures of napiergrass (*Pannisetum purpureum* K. Schum). *Theoretical and Applied Genetics*. 83: 947–955.

Shepard, J.F. 1975. Regeneration of plants from protoplasts of potato virus X-infected tobacco leaves. *Virology* 66: 492–501.

Shinde, A.N., Malpathak, N., and Fulzele, D.P. 2009. Induced high frequency shoots regeneration and enhanced isoflavones production in *Psoralea corylifolia*. *Records of Natural Products* 3: 38–45.

Skirvin, R.M. 1978. Natural and induced variation in tissue culture. *Euphytica* 27: 241–266.

Snape, J.W., Jie, X., and Parker, B.B. 1989. Gene transfer from *Hordeum bulbosum* into cultivated barley. In: *Annual Report, AFRC, Institute of Plant Science Research and John Innes Institute*, for 1988, 2–3. Norwich, UK.

Spiegel-Roy, P. and Ben-Hayyim, G. 1985. Selection and breeding for salinity tolerance in vitro. *Plant and Soil* 89: 243–252.

Srivastava, D.K., Bhatt, P.N., and Mehta, A.R. 1990. Cell division of mechanically isolated mesophyll cells of Rose. *Science, Technology & Medicine* 37–38: 1–6.

Srivastava, D.K., Desai, H.V., Bhatt, P.N., and Mehta, A.R. 1987. Cell suspension and protoplast culture in *Solanum melongena* L. In: *Plant Cell and Tissue Culture of Economically Important Plants*, ed. G.M. Reddy, 37–41.Vecon Press, India.

Srivastava, D.K., Gupta, V.K., and Sharma, D.R. 1995. Isolation and characterization of water stress tolerant callus cultures of tomato (*Lycopersicon esculentum* L. cv. Kt-I). *Indian Journal of Plant Physiology* 38: 99–104.

Srivastava, D.K., Gupta, V.K., and Sharma, D.R. 1996. Regeneration in water stress tolerant callus cultures of tomato (*Lycopersicon esculentum* L.). In: *Trends in Plant Tissue Culture and Biotechnology*, eds. L.K. Pareek and P.L. Swarankar, 178–182. Agro Botanical Publisher, India.

Stac-Smith, R. and Mellor, F.C. 1968. Eradication of potato virus X and S by thermotherapy and axillar bud culture. *Phytopathology* 53: 199–203.

Stefanov, K.L., Djilianov, D.L., Vassileva, Z.Y., Batchvarova, R.B., Atanassov, A.I., Kuleva, L.V., and Popov, S.S. 1996. Changes in lipid composition after in vitro selection for glyphosate tolerance in tobacco. *Pesticide Science* 46: 369–374.

Sujanya, S., Poornasri, D.B., and Sai, I. 2008. In vitro production of azadirachtin from cell suspension cultures of *Azadirachta indica. Journal of Biosciences* 33: 113–120.

Sujatha, M. and Sailaja, M. 2005. Stable genetic transformation of castor (*Ricinus communis* L.). via *Agrobacterium tumefaciens* mediated gene transfer using embryo axes from mature seeds. *Plant Cell Reports* 23: 803–810.

Sunderberg, E., Landgren, M., and Glimelius, K. 1987. Fertility and chromosome stability in *Brassica napus* resynthesized by protoplast fusion. *Theoretical and Applied Genetics* 75: 96–104.

Supena, E.D.J., Suharsono, S., Jacobsen, E., and Custers, J.B.M. 2006. Successful development of a shed-microspore culture protocol for doubled haploid production in Indonesian hot pepper (*Capsicum annuum* L.). *Plant Cell Reports* 25: 1–10.

Taha, H.S., El-Rahman, A., Fathalla, M., Kareem, A.E., and Aly, N.E. 2008. Successful application for enhancement and production of anthocyanin pigment from calli cultures of some ornamental plants. *Australian Journal of Basic and Applied Sciences* 2: 1148–1156.

Takebe, I., Otsuki, Y., and Aoki, S. 1968. Isolation of tobacco mesophyll cells in intact and active state. *Plant Cell Physiology* 9: 115–124.

Tale Miassar, M.Al., Hassawi, D.S., and Abu-Romman, S.M. 2011. Production of virus free potato plants using meristem culture from cultivars grown under Jordanian environment. *American–Eurasian Journal of Agriculture & Environmental Sciences* 11(4): 467–472.

Tallevi, S.G. and Dicosmo, F. 1988. Stimulation of indole alkaloid content in vanadium treated *Catharanthus roseus* suspension cultures. *Planta Medica* 54: 149–152.

Tareghyan, M.R., Collin, H.A., Putwain, P.D., and Mortimer, A.M. 1995. Characterization of somaclones of soybean resistant to imazethapyr. In: *Brighton Crop Protection Conference: Weeds. Proceedings of the International Conference*, 375–380. Brighton, UK.

Taya, M., Mine, K., Kinoka, M., Tone, S., and Ichi, T. 1992. Production and release of pigments by cultures of transformed hairy roots of red beet. *Journal of Fermentation and Bioengineering* 73: 31–36.

Thakur, A.K., Aggarwal, G., and Srivastava, D.K. 2012. Genetic modification of lignin biosynthetic pathway in *Populus ciliate* Wall. via *Agrobacterium*-mediated antisense *CAD* gene transfer for quality paper production. *National Academy Science Letters* 35(2): 79–84.

Thakur, A.K., Sharma, S., and Srivastava, D.K. 2008. Direct organ genesis and plant regeneration from petiole explant of male Himalayan poplar (*Populus ciliata* Wall.) *Phytomorphology* 58(1–2): 49–55.

Thakur, A.K. and Srivastava, D.K. 2006. High efficiency plant regeneration from leaf explants of male Himalayan poplar (*Populus ciliata* Wall.). *In Vitro Cellular and Developmental Biology-Plant* 42: 144–147.

Thangavel, K., Maridass, M., Sasikala, M., and Ganesan, V. 2008. In vitro micropropagation of *Talinum portulacifolium* L. through axillary bud culture. *Ethnobotanical Leaflets* 12: 413–418.

Thomas, J.E., Massalski, P.R., and Harrison, B.D. 1986. Production of monoclonal antibodies to African cassava mosaic virus and differences in their reactivities with other whitefly-transmitted germiniviruses. *Journal of General Virology* 67: 2739–2748.

Tiwari, K.K., Trivedi, M., Guang, Z.C., Guo, G.Q., and Zheng, G.C. 2007. Genetic transformation of *Gentiana macrophylla* with *Agrobacterium rhizogenes*: Growth and production of secoiridoid glucoside gentiopicroside in transformed hairy root cultures. *Plant Cell Reports* 26: 199–210.

Touraev, A., Foster, B.P., and Jain, S.M. 2009. *Advances in Haploid Production in Higher Plants*. Springer Science + Business Media B.V., The Netherlands.

Uchida, H., Nakayachi, O., Otani, M. et al. 2004. Plant regeneration from inter node explants of *Euphorbia tirucalli*. *Plant Biotechnology* 21: 397–399.

Umamaheswai, A. and Lalitha, V. 2007. In vitro effect of various growth hormones in *Capsicum annuum* L. on the callus induction and production of capsiacin. *Journal of Plant Science* 2: 545–551.

Vallés, M.P., Wang, Z.Y., Montavon, P., Potrykus, I., and Spangenberg, G. 1993. Analysis of genetic stability of plants regenerated from suspension cultures and protoplasts of meadow fescue (*Festuca pratensis* Huds.). *Plant Cell Reports* 12: 101–106.

Van den Bulk, R.W. 1991. Application of cell and tissue culture and in vitro selection for disease resistance breeding. *Euphytica* 56: 269–285.

Vidhyasekaran, P., Long, D.H., Borromeo, E.S., Dapata, F.J., and Mew, T.W. 1990. Selection of brown spots resistant rice plants from *Helminthosporium oryzae* toxins resistant callus. *Annals of Applied Biology* 117: 515–523.

von Malek-Podjaski, B. 1999. Erschließung von Resistenzquellen gegenüber dem Erreger des Sternrußtaus (*Diplocarpon rosae* Wolf) a Rosen unter Verwendung molekularer Marker. Ph.D. Thesis, University of Hannover.

Wagiah, M.E., Alam, G., Wiryowidagdo, S., and Attia, K. 2008. Improved production of the indole alkaloid cathin-6-one from cell suspension cultures of *Brucea javanica* (L.) Merr. *Indian Journal of Science and Technology* 1: 1–6.

Walkey, D.G.A. 1976. High temperature inactivation of cucumber and alfalfa mosaic viruses in *Nicotiana rustica* cultures. *Annals of Applied Biology* 84: 183–192.

Waller, G.R., Mac Vean, C.D., and Suzuki, T. 1983. High production of caffeine and related enzyme activities in callus cultures of *Coffea arabica* L. *Plant Cell Reports* 2: 109–112.

Wallin, A., Glimelius, K., and Eriksson, T. 1979. Formation of hybrid cells by transfer of nucleus via fusion of miniprotoplasts from cell lines of nitrate reductase deficient tobacco. *Zeitschrift fur Pflanzen-Physiologie* 91: 89–94.

Wambugu, F.M., Secor, G.A., and Gudmestad, N.C. 1985. Eradication of potato virus Y and S from potato by chemotherapy and cultured axillary bud tips. *American Potato Journal* 62: 667–672.

Wang, P.J. and Huang, C.I. 1982. Production of saikosaponins by callus and redifferentiated organs of *Bupleurum falcatum* L. In *Plant Tissue Culture*, ed. Fujiwara, pp. 71–72. Maruzen, Tokyo, Japan.

Welsh, J. and McClelland, M. 1990. Fingerprinting genomes using PCR with arbitrary primers. *Nucleic Acids Research* 12: 104–106.

White, P.R. 1939. Potentially unlimited growth of excised plant callus in artificial nutrient. *American Journal of Botany* 26: 59–64.

White, P.R. 1934. Potentially unlimited growth of excised tomato tips in a liquid medium. *Plant Physiology* 9: 585–600.

White, P.R. 1963. *The Cultivation of Animal and Plant Cells*, 2nd edition. Ronald Press, New York.

Williams, E. and Lautour, G.D. 1980. The use of embryo culture with transplanted nurse endosperm for the production of interspecific hybrids in pasture legumes. *Botanical Gazette* 141: 252–257.

Williams, E.G., Verry, I.M., and Williams, W.M. 1982. Use of embryo culture in interspecific hybridization. In: *Plant Improvement and Somatic Cell Genetics*, eds. I.K. Vasil, W.R. Scowcroft, and K.J. Frey, pp. 119–128. Academic Press, New York.

Williams, J.G.K., Kubelik, A.R., Livak, K.J., Rafalski, J.A., and Tingey, S.V. 1990. DNA polymorphisms amplified by arbitrary primers are useful as genetic markers. *Nucleic Acids Research* 18: 6531–6535.

Williams, R.D. and Ellis, B.E. 1992. Alkaloids from *Agrobacterium rhizogenes* transformed *Papaver somniferum* cultures. *Phytochemistry* 32: 719–723.

Winicov, I. 1991. Characterization of salt tolerant alfalfa (*Medicago sativa* L.) plants regenerated from salt tolerant cell lines. *Plant Cell Reports* 10: 561–564.

Wright, T.R., Bascomb, N.F., Sturner, S.F., and Penner, D. 1998. Biochemical mechanism and molecular basis of ALS-inhibiting herbicide resistance in sugarbeet somatic cell selections. *Weed Science* 46: 13–23.

Xu, H., Kim, Y.K., Suh, S.Y., Udin, M.R., Lee, S.Y., and Park, S.U. 2008. Deoursin production from hairy root culture of *Angelica gigas*. *Journal of the Korean Society for Applied Biological Chemistry* 51: 349–351.

Yagi, A., Shoyama, Y., and Nishioka, I. 1983. Formation of tetrahydroanthracene glucosides by callus tissue of *Aloe saponaria*. *Phytochemistry* 22: 1483–1484.

Yeung, E.C., Thorpe, T.A., and Jensen, C.J. 1981. In vitro fertilization and embryo culture. In: *Plant Tissue Culture: Methods and Applications in Agriculture*, ed. T.A. Thorpe, pp. 253–271. Academic Press, New York.

Zhang, X.G. and Liu, P.Y. 1998. Protoplast fusion between *Cucumis sativus*, *Cucurbitamoschata* and *Cucurbita ficifolia* (in Chinese). *Journal of Southwest Agricultural University* 20: 293–297.

4

Genetic Transformation and Crop Improvement

Kailash C. Bansal and Dipnarayan Saha

CONTENTS

4.1 Introduction 133
4.2 History of Plant Genetic Transformation 136
4.3 Importance of Plant Genetic Transformation 138
4.3.1 Understanding Basic Plant Biology 138
4.3.2 Applied Plant Biotechnology 140
4.4 Methods for Developing Transgenic Plants 141
4.4.1 Agrobacterium Transformation 142
4.4.1.1 Tissue Culture–Dependent Plant Transformation 142
4.4.1.2 Tissue Culture–Independent Plant Transformation 143
4.4.2 Biolistic Transformation and Chloroplast Genetic Engineering 144
4.4.3 Other Means of Plant Transformation 145
4.5 Post-Transformation Steps in Developing Transgenic Plants 146
4.5.1 Selection of the Modified Plant Cells and Regeneration 147
4.5.2 Evaluation and Characterization of Transgene Prior to Commercialization 148
4.5.3 Transgene Regulation for Commercialization 148
4.6 Plant Trait Modification through Genetic Transformation 149
4.6.1 Herbicide Tolerance 157
4.6.2 Insect and Disease Resistance 158
4.6.3 Quality Improvement 159
4.6.4 Abiotic Stress Tolerance 160
4.7 Conclusions and Future Prospects 161
References 162

4.1 Introduction

The continuous and steep increase in world population poses a serious challenge for agriculture to meet the demand of feeding this burgeoning population with limited available resources. Intensified agriculture and conventional breeding approaches in the twenty-first century have witnessed increase in food grain production to a great extent (Godfray et al., 2010), but this increase may not be sufficient to keep with the pace of rising world population. Additionally, the diminishing agricultural resources (land, water) and the calamity of climatic change have been considered to negatively impact the agricultural production (Swaminathan and Kesavan, 2012). Thus, the threat alarming in current scenario is the global food and nutritional insecurity, especially in the developing countries. The adoption of efficient agricultural technologies is considered one of the major approaches to meet the global food and nutritional security. Although criticized and debated widely, genetic engineering is serving as one of the important tools for crop improvement with respect to its rapidity, precision, and environment-friendly nature. These beneficial features of plant genetic engineering have helped this technique to emerge as one of the alternative tools over the traditional breeding method for genetic improvement of crops in terms of improving nutritional quality and productivity. Genetic transformation has thus acquired a status of modern-day tool for molecular breeding of crops.

The recent advancements in plant molecular biology, genomics, and plant genetic engineering have paved an efficient way to introduce genes into crop plants from distantly related plants and also from unrelated non-plant sources. This impact is quite visible in many crop species that are being feasible nowadays for genetic modifications for superior agronomic traits, such as abiotic stress tolerance, herbicide tolerance, disease and insect pest resistance, improved nutritional qualities, and several other desirable trait modifications (Vain, 2007). Thus, desirable engineering and expression of foreign genes in suitable cassettes have gained a tremendous momentum in developing myriad of transgenic plants leading to foresee a new regime of "green revolution" in terms of total food production and nutritional security. Besides being promising in crop improvement, genetic engineering also significantly contributes in understanding fundamental biological science and has commercial applications in medicines, such as production of recombinant proteins and vaccines (Fischer et al., 2004).

There are several approaches to bring about genetic improvements in field crops. All these approaches have their own advantages and limitations (Table 4.1). Genetic transformation is the key tool of genetic engineering for basic and applied plant transgenic studies, which have been the focal area of research for plant biologists in the past two decades or more. The development of transformation technologies gained momentum during the 1980s with a series of research on *Agrobacterium*-based vectors for plant transformation (Bevan, 1984; Gelvin, 2003). During the period, development in transformation technology flourished with a variety of techniques to introduce genes. Apart from *Agrobacterium*-based T-DNA, direct DNA transformation methods through microprojectile bombardment (Klien et al., 1987), electroporation, silicon carbide (Frame et al., 1994), microinjection (Neuhaus and Spangenberg, 1990), and polyethylene glycol (PEG)-mediated protoplast transformation (Hayashimoto et al., 1990)

TABLE 4.1

Comparison of Different Means of Crop Improvement

Features	**Conventional Breeding**	**Mutation Breeding**	**Molecular Breeding (Marker Assisted)**	**Genetic Engineering**
Time required	Requires very long time to develop variety	Less time for mutation but requires significant time for screening of desirable trait	Requires less time to develop variety	Rapid and precise genetic improvement possible
Approach	Genotype dependent and crossing involves random mixing of genes	Random followed by screening for desirable trait	Targeted, genotype dependent	Targeted, genotype independent
Skills required	Less	Less	Moderate	High
Technical difficulties	Minimum	Moderate	Moderate	High
Inputs	Less inputs, extensive labor required	Moderate inputs, extensive labor required	High inputs, moderate labor required	High inputs, less labor required
Species barrier	Species barrier is the limitation; applicable to only closely related species	Species barrier not involved	Moderate species barrier	No species barrier; can be employed across species and beyond gene pool
Undesirable factor	Linkage drag, i.e., higher chance of transfer of undesirable genes along with desirable gene(s)	Type and site of mutations are often not known	Influence of genetic background and environment is often very profound	No undesirable "linkage drag" but transgene silencing is often a concern
Specific requirement	Extensive phenotyping and sexual crossing is required	Needs exposure to chemical or physical mutagenic agent	Genome sequence information is often required	Requires knowledge of gene function and "breeding" with cultivars
Prediction of results before hand	Possible	Not possible	Possible	Possible to some extent

were also developed. Each of these methods has its own contributions and limitations in the successful production of transgenic plants. However, the trend in research on the development of transformation technology has slightly slowed down worldwide since the mid-1990s due to various reasons, including the lack of major improvement in the existing transformation technologies (Vain, 2007). The developments in genetic transformation in the past have been an encouraging endeavor although the major challenges that lie ahead are to workout efficient transformation strategies in all major crops, including cereals, pulses, and vegetables, which remain recalcitrant till date.

4.2 History of Plant Genetic Transformation

Plant genetic transformation has a rich history of more than 100 years (Figure 4.1), which started with the discovery of *Agrobacterium tumefaciens*, a gram-negative soil phytopathogenic bacterium that causes tumor-like crown gall disease in dicotyledonous plants (Smith and Townsend, 1907). The basis of this disease was later identified as the transfer of a specialized segment of tumor-inducing (Ti) plasmid DNA or transferred DNA (T-DNA) into the host plant genome, thereby genetically hitchhiking the host genome (Chilton et al., 1977). Thereafter, the molecular basis of genetic modification by the stable integration of T-DNA in the host plant genome has been the focal area of research for many biologists leading to all the achievements made till date. It was unfurled that the T-DNA region of the Ti plasmid consists of oncogenes, which help the *Agrobacterium* to force plant cells to produce opines in excessive quantity (Gelvin, 2003). The next steps that included efforts of the biologists to eliminate those oncogenes and replace them with any DNA without any effect on the transfer of T-DNA to plants have inspired the intriguing fact that this strategy can be used to develop a vector system for delivering desirable genes into the plant genome (Bevan, 1984). This dream became a reality and a new era of plant genetic engineering started in the 1980s when Zambryski et al. (1983) first successfully demonstrated the use of Ti plasmid as a vector system to produce transgenic plant tissues of potato, carrot, petunia, and tobacco. This was further aided by the development of vector systems (Hoekema et al., 1983) and introduction of non-plant origin antibiotic resistance gene into the T-DNA vector system for positive selection of transformed plant cells (Bevan, 1984; Herrera-Estrella et al., 1983). Stepping on the success stone of these investigations, *Agrobacterium* T-DNA-mediated genetic transformation has been worked out in several dicotyledonous plants as they were found to be a "natural host." Outside "normal host range," the first report of successful T-DNA transformation in monocot was first successfully demonstrated in maize (Chilton, 1993). The following years witnessed the development of

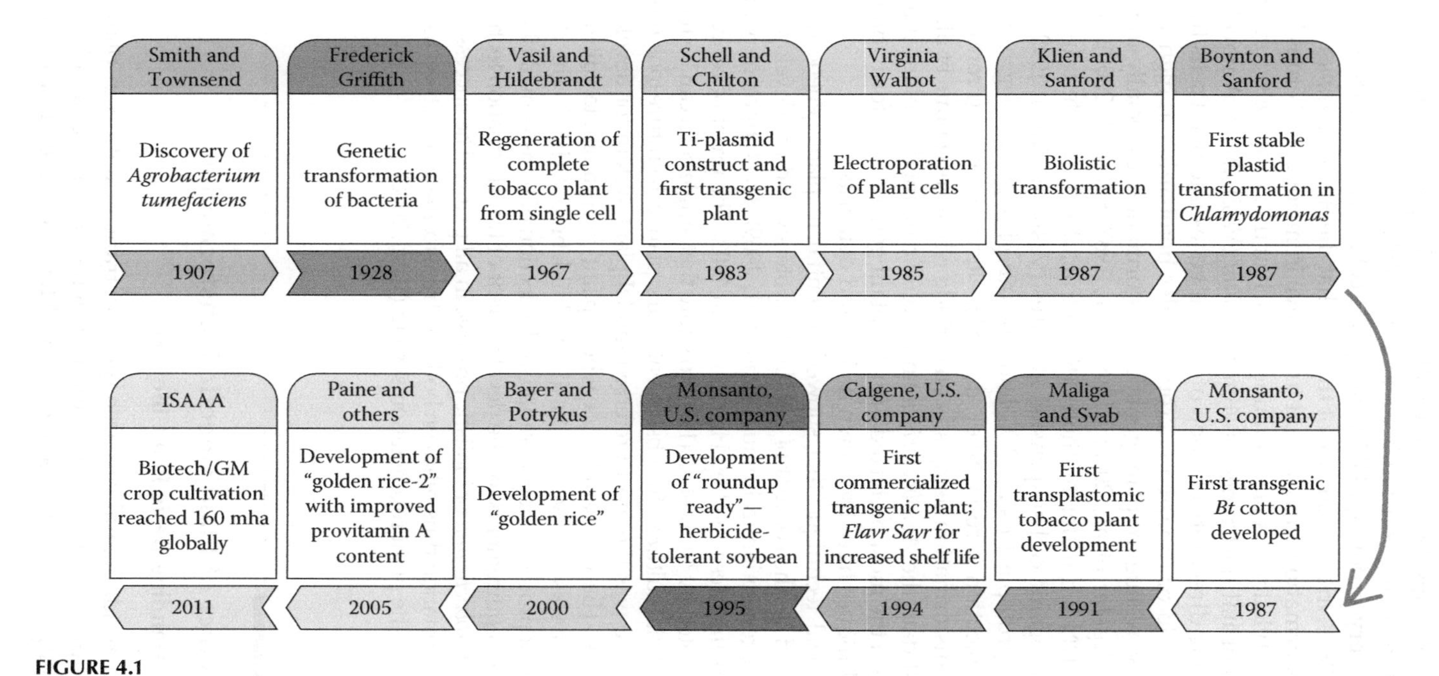

FIGURE 4.1
Major milestones in plant transformation and genetic engineering.

genetic transformation through *Agrobacterium* T-DNA in several economically important crops, like soybean, cotton, *Brassica*, potato, tomato, rice, etc., and the list is growing constantly (Valentine, 2003).

Direct delivery of foreign DNA into plant cells was also conceptualized during the 1980s, in consequence of the fact that the *Agrobacterium*-mediated genetic transformation was difficult to achieve in major cereal crops belonging to monocotyledonous plant families, which at that time were considered outside the normal host range of *Agrobacterium* (Vasil, 2005). A novel technique involving direct introduction of foreign genes into plant cells through DNA-coated gold or tungsten microprojectiles bombarded at high velocity opened a new avenue for plant genetic transformation, especially in monocot plants (Altpeter et al., 2005; Klien et al., 1987; Sanford, 2000). Several other direct DNA transfer methods (Rao et al., 2009) include transformation of embryogenic cell suspension cultures (Vasil and Vasil, 1992), transformation of protoplast using PEG and calcium phosphate (Negrutiu et al., 1987), microinjection (Crossway et al., 1986), electroporation (Fromm et al., 1986), and transformation using silicon carbide whiskers (Frame et al., 1994). However, till date the *Agrobacterium* and biolistic method of plant transformation remain the most preferred ones for plant genetic transformation, and both the techniques have their own merits and demerits.

The additional technique that showed potential in plant genetic transformation is based on the concept of cellular totipotency and the development of whole plant regeneration techniques (Rao et al., 2009). The totipotent nature of plant cell was first described by Haberlandt in 1902. The proof of concept was first demonstrated by Muir et al. (1954) to show that single cells can be regenerated from artificial cultures. In subsequent years, plant cell culture took momentum and tobacco was the most preferred model plant system for development of such studies. The regeneration of whole plant from a single cultured cell of tobacco plant was achieved by Vasil and Hildebrandt (1965). The development of efficient process of regeneration and selection of transformed plant cells became prime requirements for the development of transformation techniques, for example, *Agrobacterium* or biolistic process, in all plants. The success of plant genetic transformation thus relies on both the transformation method as well as the efficiency of the plant regeneration system (Vasil, 2008).

4.3 Importance of Plant Genetic Transformation

4.3.1 Understanding Basic Plant Biology

In basic and fundamental research, genetic transformation in plants has profound impact on the identification of genes and understanding their functional importance in plant biology. Most of these transformations involve manipulation of nuclear genome. The major approaches being the forward and reverse genetic screens in combination with functional genomics led to identification of numerous genes and regulatory elements to understand their involvement in complex biological process. Throughout the development of *Agrobacterium* as a plant genetic transformation tool, biologists were involved in understanding the molecular biology of *Agrobacterium*–host interaction process (Gelvin, 2003). Thus, the initial phase of *Agrobacterium* research was mostly to understand the *Agrobacterium* species and host range, the molecular basis of transfer of T-DNA into plant cell and its integration into plant genome, the response of plant cell to *Agrobacterium* infection, and the manipulation of the T-DNA region to utilize it for plant genetic engineering. The host range of *Agrobacterium* was found to be very wide from plant to yeast, fungi, and even HeLa cells, and sea urchin embryos. However, the susceptibility index of *Agrobacterium* transformation varies tremendously from species to species or even between cultivars (Gelvin, 2009). *Agrobacterium* transformation through forward genetic screens produced a large number of T-DNA mutagenized lines of model plant *Arabidopsis thaliana* for identification of genes involved in *Agrobacterium*-mediated transformation in plants. Further studies led to the identification of *Arabidopsis rat* and *hat* mutants that were resistant to *Arabidopsis* transformations and hypersusceptible to *Agrobacterium* transformations, respectively, and their corresponding genes (Gelvin, 2009). These forward and reverse genetic screens along with the yeast two hybrid protein–protein interactions and transcriptional genomic approaches proved to play important roles in understanding the basic knowledge of *Agrobacterium*–host interaction and the mechanism of genetic transformations (Gelvin, 2009). T-DNA-based genetic transformation in plants is increasingly being used in trapping of genes and promoter elements to understand their role in plant biology as it is considered the most reliable tool to assess gene function through gain or loss of function (Springer, 2000; Srinivasan and Saha, 2010). Functional characterization of genes, and regulatory elements through overexpression or downregulation (silencing) and promoter deletion studies, respectively, requires transformation tool to be invariably employed in model or native plant system.

Similar to nuclear transformation, the recent development in the transformation techniques enabled us to manipulate organellar genomes like chloroplast and mitochondria. The powerful technique of biolistic transformation has enabled to study the chloroplast genome–encoded information and led to a deeper understanding of the plastid gene expression (Bock, 2001; Maliga, 2004). For instance, plastid transformation in tobacco helped in understanding the basis of cytidine (C) to uridine (U) RNA editing and the evolutionary processes involved with such organelle gene expression (Bock, 2000). Plastid transformation facilitated overexpression of genes leading to the identification of the plastid-encoded *accD* gene function in the production of acetyl-CoA carboxylase to increase leaf longevity and seed yield in tobacco (Madoka et al., 2002). Overexpression of nuclear-encoded α-subunit of anthranilate synthase gene in plastids resulted into increased tryptophan production (Zhang et al., 2001a). Similarly, the plastid transformation also aided in detailed understanding of regulations in plastid mRNA transcription process (Bock, 2001; Maliga, 2004). Both plastid-encoded RNA polymerase (PEP) and the nuclear-encoded phage-type RNA polymerase (NEP) were identified to regulate the transcription process of different groups of plastid genes in knockout lines of tobacco (Maliga, 2004). Like nuclear transformation, direct introduction of foreign genes into plastids has been increasingly used in functional genomics studies, such as gene knockouts and targeted mutagenesis of plastid genes. These studies led to the significant understanding of the molecular basis of bioenergetic process within the plastid compartment (Bock, 2001). The efficiency of the homologous recombination system in plastids actually facilitated the site-directed mutagenesis for reverse genetics–based functional genomics in plastids of model plant system tobacco. The gene knockout strategies through deletion and insertional mutagenesis in *ycf* gene led to the understanding of several gene functions related to photosynthetic pigments in the tobacco plastid genome (Bock, 2001; Ruf et al., 2000). Since most of the photosynthetic genes are located on the plastid genome, plastid transformation and manipulating *rbcL* gene by replacement or relocating strategies played an important role in probing into the functions of genes involved in the formation of complex ribulose-1,5-bisphosphate carboxylase oxygenase (Rubisco) enzyme in higher plants (Whitney and Andrews, 2001). The plastid transformation system also helped to a great extent in understanding the original theory of gene transfer to chloroplasts due to endosymbiosis process and the mechanistic details of plastid genome evolution in *Chlamydomonas* and higher plants (Maliga, 2004).

4.3.2 Applied Plant Biotechnology

Genetic transformation in plants, besides exploited for understanding basic plant biology or functions of genes and regulatory elements, is also most importantly used as an efficient tool for bringing genetic improvements in crops through transgenic technology. Genetic transformation provides a powerful and rapid tool to genetically modify the plant genome through introduction of foreign genes from various origins. It results in the production of gene-specific products in transgenic plant lines that have revolutionized the modern molecular agriculture. Till date, several agricultural and horticultural crops are being genetically modified for a variety of traits, such as increase in productivity and quality, resistance to pests, diseases, and herbicides, tolerance to various abiotic stresses, and many other agronomic and commercial traits.

The research on plant biotechnology usually revolves around increasing crop production for food sufficiency through development of high yielding or better performing varieties against difficult production challenges.

Over the years, plant biotechnology exploited genetic transformation through *Agrobacterium* or other direct methods to revolutionize the agricultural production. The *Agrobacterium* mode of transformation is the most popular method for development of transgenic plants except for cereals where biolistic transformation proved more efficient in the process. However, with the recent refinement of technologies, cereals have also become amenable to genetic transformations through *Agrobacterium* (Hiei and Komari, 2008; Shrawat and Good, 2011). Chloroplast transformation through biolistic method also has a great impact on engineering agronomic traits in plants owing to its potential to produce enhanced expression of recombinant proteins in comparison to nuclear transformation (Bock, 2001; Maliga, 2004). The targeted traits for crop improvement through genetic transformations include three major areas. The first most important area is related to plant characters and metabolism under which the major emphasis is on (i) increase in the leaf photosynthesis rate to maximize carbon assimilation and crop yield (Ruan et al., 2012); (ii) increase in nitrogen use efficiency through increased nitrogen uptake and accumulation (Shrawat et al., 2008); (iii) increase in seed growth and tillering/branching pattern; (iv) plant cell structure alteration (e.g., increase in lignin biosynthesis) for enhanced biomass production (Hisano et al., 2009); (v) increase in seed oil production (Tan et al., 2011); and (vi) nutritional quality improvement (McGloughlin, 2008). The next most important target for crop improvement through genetic transformation includes genetic modification of crop plants against variety of abiotic stresses (Bhatnagar-Mathur et al., 2008; Varshney et al., 2011), such as (i) increased water use efficiency against drought stress, which is one of the priority traits; (ii) pathway modification for increased performance against salt stress; (iii) engineering transcription factor genes for temperature stress tolerance; (iv) increased submergence or flooding stress (hypoxia) tolerance; and (v) micronutrient deficiency stress tolerance, etc. Lastly, the important target areas of plant transformation include genetic improvement of crop plants for superior performance against herbicides (Mulwa and Mwanza, 2006) and insects/pests (Sharma, 2008) and resistance against nematodes and various disease-causing pathogens (Collinge et al., 2008). Apart from crop improvement, genetic transformation in plants is also targeted to enhance the production of recombinant proteins for biopharmaceutical applications, e.g., plant-derived vaccines (Boothe et al., 2010).

4.4 Methods for Developing Transgenic Plants

Genetic transformation is the principal method through which transgenic plants are developed. The success lies in the efficient introduction of exogenous DNA material through different means and its successful integration

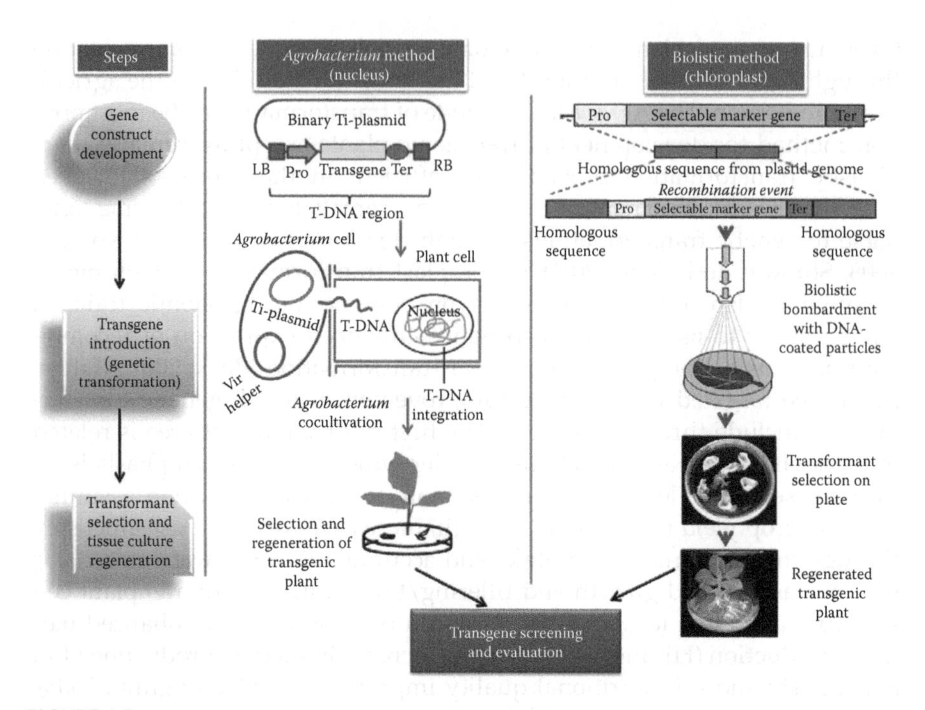

FIGURE 4.2
Different steps of genetic transformation required for developing transgenic plants. The most commonly used *Agrobacterium* (useful for nuclear transformation) and biolistic method (useful for both nuclear and chloroplast transformation) are only depicted.

with the native DNA material. Thus, the transformation events basically involve three steps: development of gene constructs, transgene introduction, followed by transformant selection and regeneration of transgenic plants (Figure 4.2). The different methods of transfer of exogenous genes are being discussed in the following sections.

4.4.1 *Agrobacterium* Transformation

4.4.1.1 Tissue Culture–Dependent Plant Transformation

A series of landmark discoveries in *Agrobacterium*-derived gene transfer mechanism in plants and advancements in designing of DNA delivery vehicle (Ti plasmid) and selectable marker gene has greatly facilitated exploitation of *Agrobacterium*-mediated genetic transformation process in a variety of crop plants (Gelvin, 2003). The *Agrobacterium* genetic transformation process primarily involves designing of appropriate vector construct for cloning of the gene of interest (GOI) and selectable marker system. Several disarmed T-DNA binary vectors were developed for this purpose (Hellens et al., 2000; Lee and Gelvin, 2008). Once the *Agrobacterium* cell consisting of the vector

T-DNA construct is used to coinfect plant tissue, the GOI is delivered into the plant cell through the natural mechanism of T-DNA transfer and integration (Tzfira and Citovsky, 2006; Zupan et al., 2000). However, the success of genetic transformation in most of the plants is dependent on the efficiency of the plant cells to respond to tissue culture for regeneration (totipotency) of transgenic plant (Rao et al., 2009; Vasil, 2008). This form of *Agrobacterium*-mediated genetic transformation in plants thus follows the tissue culture path. The success of transformation using *Agrobacterium* requires the identification of model tissue culture system with high offspring regeneration capacity. Several factors associated with plant tissue culture also affect the genetic transformation system (Alimohammadi and Bagherieh-Najjar, 2009). The genotype of explants, monocot versus dicot, and tissue types, such as leaf, stem, hypocotyls, stem, etc., determine the efficiency of plant genetic transformation using *Agrobacterium*. The other important factors for development of transgenic plants are the optimal protocol for inoculation and subsequent selection and regeneration of transformed cells. The response of different *Agrobacterium* strains to different culture conditions are also major factors influencing the genetic transformation through tissue culture. In conclusion, factors like type and age of tissue, the size of explants, the duration of cocultivation, and the growth of cells prior to inoculation are very critical points to be assessed for the successful plant genetic transformations in several dicot plants. Using *Agrobacterium*-based transformation in conjunction with organogenesis via tissue culture, to date, most dicotyledonous plants of major commercial importance, as well as many plants of academic interest, have been genetically transformed. It has now become possible to transform monocotyledonous crops such as rice, maize, and barley, even though *Agrobacterium* is not a usual pathogen to infect cereals (Shrawat and Good, 2011; Shrawat and Lorz, 2006).

4.4.1.2 Tissue Culture–Independent Plant Transformation

Besides tissue culture–dependent techniques, a unique but simple transformation by *Agrobacterium* has been developed predominantly in model plant *A. thaliana* (Bechtold and Pelletier, 1998). The technique is commonly referred to as in planta transformation that avoids the cumbersome tissue culture regeneration process (Clough and Bent, 1998). Initially, researchers attempted to use apical meristem to transform and allow regeneration into plants directly (Birch, 1997). The major advantage of the in planta transformation is to eliminate the chances of undesirable somatic mutations or epigenetic changes (DNA methylation) that take place due to culturing of cells on artificial media (Phillips et al., 1994).

The in planta transformation procedure involves preparing of *Agrobacterium* solution with engineered T-DNA, growing of young plant with undeveloped siliques, and dipping of inflorescences of the intact plants in the bacterial suspension. The plants are then allowed to imbibe

Agrobacterium for 24 h under high humidity and dark conditions before putting them under normal growth conditions. The plants continue to grow and set seed and a proportion of the next generation of seedlings are found to contain the inserted DNA in several thousands (Azpiroz-Leehan and Feldmann, 1997). The efficiency of *Agrobacterium* in planta transformation can be increased by addition of surfactant Silwet L-77 and acetosyringone in the *Agrobacterium* solution prior to "floral dip" and forcing *Agrobacterium* into plant tissues using brief vacuum infiltration (Bent, 2000). The mechanism of *Agrobacterium* in planta transformation in *Arabidopsis* has been described by Bent (2000). Through different evidences, it has been postulated that the transformation usually occurs in developing flowers after the individual gametophyte cells have formed. It has been observed that the ovule is the primary target for *Agrobacterium* transformation in developing flowers through this method (Bent, 2000). Besides *Arabidopsis,* in planta *Agrobacterium* transformation through infiltration has been reported to be promising in other plant species as well including the model plant *Medicago truncatula* (Trieu et al., 2000). Few members of *Brassica* family (*Brassica campestris* L. ssp. *chinensis, Brassica napus,* and *Brassica carinata*) have been reported to respond to the in planta transformation (Bent, 2000; Liu et al., 1998; Shanker et al., 2008). In a recent study, in planta transformation has been demonstrated in a non-Brassicaceae plant, cotton, through pistil drip inoculation of *Agrobacterium* solution. The process resulted into stable transgenic cotton plants during one growing season with a transformation efficiency of 0.46%–0.96% (TianZi et al., 2010). The advantages of this simplified *Agrobacterium* transformation method should inspire adoption of this technique to other plant species as well in the coming years.

4.4.2 Biolistic Transformation and Chloroplast Genetic Engineering

The method of direct delivery of DNA started with the development of transformation of protoplasts of plant cells (Shillito, 1999). It was developed as a result of the quest for searching of alternative transformation techniques for cereal crops, which were at that time considered not amenable to *Agrobacterium* transformation (Rao et al., 2009; Vasil, 2005). At the early phase of development of this technique, the protoplasts suspended in DNA solutions were provided with electric pulse for the transformation process followed by tissue culture–mediated selection and regeneration of transformants. Slowly, the technique of protoplast transformation in cereals was superseded by the development of *Agrobacterium*-mediated transformation of embryogenic tissues or cells using super-virulent strains of *Agrobacterium* and addition of acetosyringone (Komari and Kubo, 1999; Rao et al., 2009). However, the research on direct delivery of DNA into plant cells has led to a pioneering discovery of transforming plants by high-velocity shooting of DNA-coated gold or tungsten particles into intact plant tissues (Klien et al., 1987; Sanford, 2000). The procedure is known as microprojectile bombardment or biolistic transformation. The technique has been improved over the years to develop a biolistic particle delivery system. The basic process involves coating of DNA onto micron-sized gold or tungsten particles through precipitation with calcium chloride and spermidine. The millions of DNA-coated gold or tungsten microparticles are loaded onto a macrocarrier disk. A burst of high-pressure helium gas after rupturing a plastic disk is used to propel them to target cells, while the macrocarrier is retained by a stopping screen. The process requires evacuating the chamber of the biolistic device to subatmospheric pressures. By doing so, the frictional drag of the microparticles is reduced as they travel toward the target cells. The major advantages of biolistic transformation is that it works independent of the plant genotype, the delivered DNA can be manipulated to influence the quality and structure of the resultant transgene loci, little chances of transgene instability and silencing, and high level of transgene expression.

To date, particle bombardment is the most effective technique to achieve plastid transformation in plants and perhaps is the only method so far used to achieve mitochondrial transformation. The plastid transformation in higher plants has been established in the recent past to engineer several agronomic traits including herbicide resistance, insect and pathogen resistance, abiotic stress tolerance, increased photosynthesis, and also production of edible crops engineered to produce "biopharmaceuticals" (Daniell et al., 2002; Koop et al., 2007; Maliga and Bock, 2011). With the burst of information on chloroplast genomics, development of suitable plastid transformation vectors has become easy. This if coupled with efficient plant regeneration techniques can take plastid transformation beyond the model plant systems to major crops for fulfilling the promise of transplastomic crops in increasing agricultural production (Bansal and Saha, 2012).

4.4.3 Other Means of Plant Transformation

The development of alternative methods is always a perpetual quest to find efficient, simple, and economical means of plant transformation. Therefore, besides *Agrobacterium* and biolistic methods, several other methods have been reported to facilitate transfer of foreign genes into plant cells. The electroporation-mediated transformation requires application of strong electrical pulse field for in vitro introduction of DNA into cells and protoplasts. This technique was originally developed to transform protoplast but later found its applicability in intact plant cell transformation. Using the electroporation technique with embryogenic protoplast or cells, fertile transgenic plants were developed in rice (Shimamoto et al., 1989), barley (Salmenkallio-Marttila et al., 1995), and sugarcane (Arencibia et al., 1992). PEG is also used for genetic transformation of plant protoplasts. The technique is similar to electroporation method, where PEG is used instead of the electric pulse. This method has been demonstrated in maize and barley (Daveya et al., 2005). However, PEG-mediated genetic transformation was not found efficient because of low transformation frequency and not many species could be regenerated through protoplast culture. Similar to PEG-induced protoplast transformation, liposome-mediated transformation involves transfer of DNA through endocytosis of liposomes, which are actually circular lipid molecules with aqueous center capable of carrying DNA molecule (Rao et al., 2009). Liposome-mediated transformation involves three steps: (i) attachment of liposome onto cell surface, (ii) fusion at the site of attachment, and (iii) delivery of DNA inside the cell and nucleus. The event of integration of engulfed DNA into the host DNA is random. Except in tobacco (Dekeyse et al., 1990) and wheat (Zhu et al., 1993), despite the simplicity of the technique, liposome-mediated transformation found very little success probably due to its low efficiency of generating transformants.

Microinjection is one of the other methods used for plant genetic transformation, originally adopted from large animal cells. It involves introduction of DNA into nucleus or cytoplasm through a glass micro-capillary pipette and a manipulator. The technique involves immobilization of cells using a suction pipette followed by injection of DNA into the immobilized cell. In case of plant protoplasts instead of sucking pipette poly-L-lysine or agarose is used to attach cells to the glass surface. Microinjection transformation has been demonstrated in tobacco and *Vicia faba* (Knoblauch et al., 1999). Because of technical difficulties, microinjection has not developed as a routine transformation technique in plants. The silicon carbide fiber–mediated transformation is one of the least complicated methods. In this process, the small needlelike silicon carbide whiskers are mixed with plant cells and GOI followed by vortexing. It pierces the cells allowing the entry of the DNA into cells. This technique has often being used in a variety of plant species, such as maize, rice, wheat, and tobacco.

The other direct methods of genetic transformation in plants include DNA transfer through pollen tube pathway. In this process, the styles are cut shortly after pollination and foreign DNA is applied on the cut styles. The DNA enters through the cut end and flows down the pollen tube to reach the ovule. This technique of plant genetic transformation has been demonstrated in a number of plant species including rice, wheat, and soybean (Rao et al., 2009). Several other techniques like shoot apex method, sonication-assisted *Agrobacterium*-mediated (SAAT) transformation, electrophoresis, etc. have been attempted for the development of transgenics, but none of them could be used as efficiently as the *Agrobacterium* and biolistic technique for plant transformation.

4.5 Post-Transformation Steps in Developing Transgenic Plants

After genetic transformation event, several steps are involved in the development of transgenic plant and its commercialization (Figure 4.3). Although genetic transformation is the crucial event on which the success of

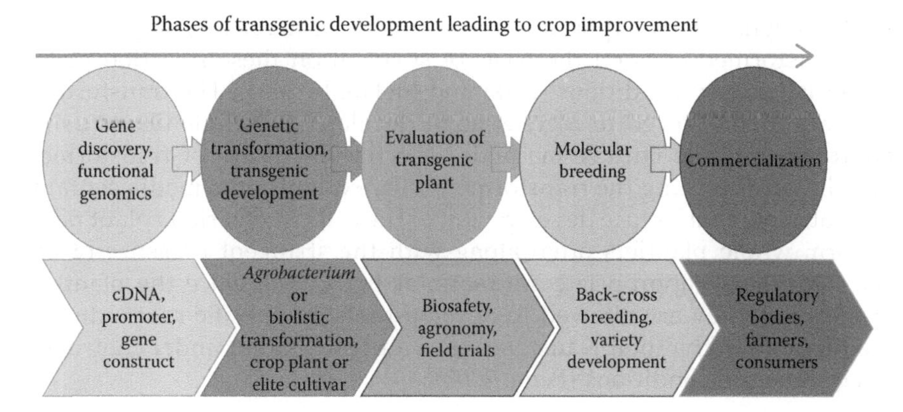

FIGURE 4.3
Different steps/phases of transgenic plant development leading to crop improvement and commercialization.

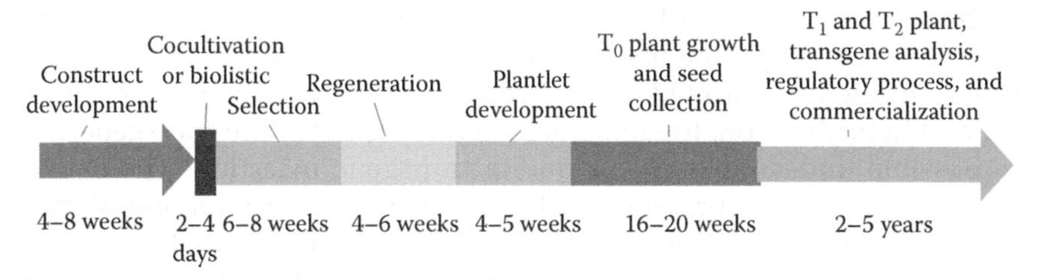

FIGURE 4.4
Average time taken for each step of plant transformation leading to crop improvement through transgenic approach.

developing transgenic plant depends, the post-transformation steps involve rigorous and time-consuming processes to commercialize it. Starting from gene construct development and transformation, it takes approximately 2–5 years or even more for developing a transgenic plant for commercialization (Figure 4.4). The different important steps of post-transformation events are described later.

4.5.1 Selection of the Modified Plant Cells and Regeneration

After transformation, the major challenge is to differentiate between the modified and unmodified plant cells, which have not incorporated the desired genes. For this purpose, most often, several selectable marker genes conferring resistance to antibiotic or herbicide are engineered along with the transgene or co-introduced along with the GOI to favor growth of only transformed cells relative to the untransformed ones. The cells harboring these transgene along with the selectable marker gene are then deliberately exposed to the antibiotic or herbicide, only to select the transformed cells on the basis of survival on these selection plates under laboratory conditions (Miki and McHugh, 2004). The transformed cells are then allowed to regenerate to develop into whole plants using appropriate tissue culture methods. The tissue culture of transformed cells involves placing the transformed cells/explants onto media containing nutrients that induce development of the cells into various plant parts to form whole plantlets often along with the stringent presence of the selection agent (appropriate antibiotic or herbicide). Once the plantlets are formed, they are allowed to develop roots on specific root-inducing media followed by their transfer to pots and incubation under controlled environmental conditions (Pena, 2005).

4.5.2 Evaluation and Characterization of Transgene Prior to Commercialization

Evaluation of plant transformation demonstrates that the introduced GOI has been integrated into the host genome and has normal inheritance pattern. Several experiments are employed to check the transgene copy number in the transformed plant through techniques like Southern blotting and hybridization, the intactness and integrity of the transgene, and the possible undesirable effects due to transgene insertion. The experiments on the gene expression of the transgene (i.e., production of messenger RNA and/or protein) are carried out to make sure that the gene is functional. The performance of the transgene in the host plant is tested by growing them in a confined greenhouse or screenhouse to evaluate the engineered trait in the modified plant without any undesirable characteristics. After that, the transgenic plants are evaluated in fields through confined trials in the open environment. As a next step, the transgenic plants are evaluated in multilocation field trials to establish the performance of the transgenic plant under multi-environmental conditions. The food and environmental safety assessment of these transgenic plants are a must. It is to be carried out in addition to the evaluation of the transgenic plant performance. Once the transgenic plant passes through all these steps, it is then considered for clearance from the regulatory body for commercial production. Often, the transgenic plants are bred with the locally adopted elite varieties through backcrossing for targeted introgression of traits and evaluated for their performances for commercial release (Visarda et al., 2009).

4.5.3 Transgene Regulation for Commercialization

Although the transgenic technology promises of a booming agricultural production and maximum economic benefits, it must have to pass through the "regulatory" system prior to entry into the public domain for commercial production and public consumption. Based on the needs and public perceptions and adaption to the technology, different countries have different regulatory frameworks for experimental and commercial purposes. For example in the United States, the three federal agencies the United States Department of Agriculture (USDA), the Environmental Protection Agency (EPA), and the Food and Drug Administration (FDA) share the responsibility of regulating biotechnology products including the transgenic plants (Chamberlin, 2010). The USDA consists of the Animal and Plant Health Inspection Service (APHIS) under which the office of Biotechnology Regulatory Service (BRS) is largely responsible for regulating GM crops and products at all levels including research and development, import, interstate movement, field trials, and commercial release and cultivation. The EPA regulates biotechnology-derived plants, microbial pesticides, or new chemical substances, whereas the FDA regulates the safety of the biotechnology-derived foods, drugs for human or animal use, biologics for human use, and medical devices.

In India, the premier organization to regulate the transgenic crops and animals is the Department of Biotechnology, Government of India. The transgenic events and approvals are governed by the Indian Environment Protection Act (EPA) 1986. Under the regulatory system, different committees or institutions are constituted, which are responsible for issuing regulatory certificates at different stages. The Institutional Biosafety Committee (IBSC) has the responsibility to monitor all the biotechnology-related experiments and their safety issues. Hierarchically, the Review Committee on Genetic Manipulation (RCGM) is responsible for chalking out the guidelines for regulatory process on genetically modified organisms (GMO), reviews and categorizes risks of all genetic engineering projects to permit experiments with high-risk categories under containments, and authorizes import of GMOs. The apex committee, Genetic Engineering Approval Committee (GEAC), is responsible for providing approval for activities involving large-scale use of biotechnology-derived products for research and industrial productions after considering safety issues from environmental and public health angles. Often, the release of transgenic products requires approval from the Ministry of Agriculture apart from being approved by the GEAC (Damodaran, 2005).

4.6 Plant Trait Modification through Genetic Transformation

Genetic transformations in plants have greatly advanced the production of transgenic plants and genetic improvement in a variety of agronomic traits (Table 4.2). The major agronomic traits targeted through genetic transformation in plants are discussed later in brief.

TABLE 4.2

Major Genetic Transformation Applications to Improve Yield and Quality of Major Crop Plants

Crop	Type of Trait Improvement	Specific Characters	Gene Employed	Genetic Transformation Methods
Rice	Herbicide tolerance	Tolerance to phosphinothricin (Basta)	*bar*	Biolistic, PEG mediated
		Tolerance to glyphosate	G6 EPSPS-encoding gene	*Agrobacterium*
	Biotic stress resistance	Tolerance to rice tungro and stripe virus	Coat protein gene	Biolistic, electroporation
		Rice yellow mottle virus resistance	*orf2*	Biolistic
		Sheath blight resistance	Chitinase-encoding gene	PEG mediated
		Fungal resistance and sheath blight resistance	*chi11*, *RC7*, and *NPR1*	*Agrobacterium*, biolistic, and PEG
		Resistance to bacterial blight	*Xa21*	Biolistic
		Tolerance to stem borer, leaf folder, and brown plant hopper (BPH)	Several *Cry* (*Bt*) genes	Biolistic, electroporation, *Agrobacterium*
	Tolerance to abiotic stresses	Enhanced tolerance to salinity	*codA*	Electroporation
		Drought and salinity tolerance	*DREB*, *TPSP*	*Agrobacterium*
	Quality improvement	Enhanced seed iron content	*ferritin* gene	*Agrobacterium*
		Higher grain filling and seed weight	*Glgc*	*Agrobacterium*
		Increased C4 photosynthesis	*PEPC*	*Agrobacterium*
		Increased provitamin A biosynthesis	*psy*, *crt*1, *icy* gene	Biolistic and *Agrobacterium*
Wheat	Herbicide tolerance	Tolerance to glyphosate (Roundup)	EPSPS-encoding gene	Biolistic
		Tolerance to phosphinothricin (Basta)	*bar*	Biolistic
	Biotic stress resistance	Increased protection against powdery mildew; protection against *Fusarium* head blight	Antifungal protein AG-AFP, chitinase and ribosome inactivating protein (RIP)	Biolistic
		Protection against wheat streak mosaic virus	Coat protein gene	Biolistic

TABLE 4.2 (continued)

Major Genetic Transformation Applications to Improve Yield and Quality of Major Crop Plants

Crop	Type of Trait Improvement	Specific Characters	Gene Employed	Genetic Transformation Methods
	Tolerance to abiotic stresses	Enhanced response to water stress	*HVA1*	Biolistic
		Enhanced drought tolerance	*DREB1* gene	Biolistic
	Quality improvement	Improved baking quality, pasting, and dough property for improved bread making quality	High-molecular-weight glutenin (*HMW-GS*) genes	Biolistic
		Increased phytate content for use of wheat as animal feed	*phyA* gene	Biolistic
		Increased photosynthesis	*PEPC* gene	Biolistic
Maize	Herbicide tolerance	Tolerance to glyphosate (Roundup)	EPSPS-encoding gene	Biolistic
		Tolerance to glufosinate	*Pat*	Biolistic
		Tolerance to phosphinothricin (Basta)	*bar*	Electroporation
		Tolerance to 2,4-D	AAD1-encoding gene	Biolistic
	Biotic stress resistance	Increased resistance to insects; especially corn borers	*Bt* (*Cry*) gene	Biolistic
		Increased resistance against maize streak virus	Mutated Rep protein gene	Biolistic
	Abiotic stress tolerance	Increased salt tolerance	*NHX1* gene	*Agrobacterium*
		Increased drought stress tolerance	*betA* gene	Biolistic
		Improved drought tolerance	*ZmPLC1* gene	*Agrobacterium*
		Drought tolerance	*TsCBF1* gene	*Agrobacterium*
	Quality improvement	Increase in free lysine content in grains	Dihydropicolinate synthase (*DHPS*) gene	Biolistic
		Increased phytic acid	*phyA2* gene	Biolistic
		Increased level of vitamins (β-carotene, ascorbic acid, and folate)	*psy1*, *crt1*, and *dhar*	Biolistic

(*continued*)

TABLE 4.2 (continued)

Major Genetic Transformation Applications to Improve Yield and Quality of Major Crop Plants

Crop	Type of Trait Improvement	Specific Characters	Gene Employed	Genetic Transformation Methods
Mustard/ canola	Herbicide tolerance	Resistance to glyphosate herbicide "Roundup" in canola	*EPSPS* gene	Biolistic
		Resistance to herbicide sulfonylurea	*crsI-I*	*Agrobacterium*
		Resistance to bromoxynil	*Bxn*	*Agrobacterium*
		Resistance to herbicide phosphinothricin (PPT) in Indian mustard	*bar* gene	*Agrobacterium*
	Biotic stress resistance	Resistance against aphids in Indian mustard	Wheat germ agglutinin (*WGA*) gene	*Agrobacterium*
		Increased resistance against insects, like diamondback moth and cabbage looper	*Bt* (*Cry*) genes	*Agrobacterium*
	Abiotic stress tolerance	Increased tolerance to salt and cold stress in Indian mustard	Choline oxidase (*codA*) gene	*Agrobacterium*
		Increased salinity stress tolerance	*AtNHX1* gene	*Agrobacterium*
	Quality improvement	Male sterility and restorer of fertility system for hybrid breeding	*barnase* and *barstar* genes	*Agrobacterium*
		Increased production of high γ-linolenic acid	δ-12 *desaturase* gene	*Agrobacterium*
		Increased production of stearate in seed oil	*Garm FatA1* gene	*Agrobacterium*
		Increased carotenoid and oleic acid level in canola seeds	Phytoene synthase (*crtB*) gene	*Agrobacterium*
		Increased polyhydroxy butyrate (PHB) synthesis for biopolymer production	*phb*A, *phb*B, and *phb*C genes	*Agrobacterium*
		Blood anticoagulant "hirudin" production	OBHIRT (oleosin–hirudin) fusion protein	*Agrobacterium*

TABLE 4.2 (continued)

Major Genetic Transformation Applications to Improve Yield and Quality of Major Crop Plants

Crop	Type of Trait Improvement	Specific Characters	Gene Employed	Genetic Transformation Methods
Cotton	Herbicide tolerance	Tolerance to 2,4-D	*tfdA* gene	*Agrobacterium*
		Glyphosate (Roundup)	*EPSPS* gene	*Agrobacterium*
		Bromoxynil (BXN)	*bxn*—nitrilase-encoding gene	*Agrobacterium*
		Sulfonylurea (SU)	Acetolactate synthase (*als*) gene	*Agrobacterium*
		Tolerance to glufosinate	*bar* and *pat* gene	*Agrobacterium*
	Biotic stress resistance	Insect resistance against cotton boll worm, cotton boll weevil, etc.	*Bt* (*Cry*) genes	*Agrobacterium*
		Resistance against Lepidopteran pest	Cowpea protease inhibitor (*CpTi*) gene	*Agrobacterium*
		Insect resistance	Choline oxidase (*Cho*) gene	*Agrobacterium*
		Enhanced resistance to root knot nematode	AtNPR1	*Agrobacterium*
		Enhanced resistance against fungal pathogens	*D4E1* gene	*Agrobacterium*
		Enhanced defense response against different plant pathogens	Harpin-encoding hpa_{xoo} gene	*Agrobacterium*
		Resistance to *Verticillium* wilt disease	Glucose oxidase gene	*Agrobacterium*
	Abiotic stress tolerance	Increased drought tolerance	*AtLOS5* gene	*Agrobacterium*
	Quality improvement	Increased number of lint fibers	*iaaM* gene	*Agrobacterium*
		Fiber quality improvement by elongation and thickening of secondary wall properties	Silencing of *GhADF1* gene	*Agrobacterium*
		Improved fiber quality by increasing thickness of cellulose secondary wall	Spinach sucrose phosphate synthase (*sps*) gene	*Agrobacterium*
		Increase in fiber strength	Keratin gene from rabbit hair	*Agrobacterium*
		Increase in fiber strength	Spider silk protein gene *ADF3*	Biolistic

(continued)

TABLE 4.2 (continued)

Major Genetic Transformation Applications to Improve Yield and Quality of Major Crop Plants

Crop	Type of Trait Improvement	Specific Characters	Gene Employed	Genetic Transformation Methods
Soybean	Herbicide tolerance	Tolerance to glyphosate (Roundup)	*EPSPS* gene	*Agrobacterium*
	Biotic stress resistance	Enhanced resistance to soybean dwarf virus	Soybean dwarf virus (*SbDV*) coat protein gene	Biolistic
		Resistant to Lepidopteran caterpillars	*Cry* protein gene	Biolistic
		Insect resistant	*pin* II gene	Biolistic
	Abiotic stress tolerance	Increased tolerance to water stress	*NTR1* gene	*Agrobacterium*
		Greater tolerance to drought stress	Δ1-Pyrroline-5-carboxylate synthase (*P5CR*) gene	*Agrobacterium*
		Increased tolerance to salinity stress	*TaNHX2* gene	*Agrobacterium rhizogenes*
	Quality improvement	Increased levels of glycinin, consisting of sulfur-containing amino acid	Glycinin-encoding gene *Gy1*	Biolistic
		Lowered linolenic acid content	Silencing of *FAD3* gene	*Agrobacterium*
		Increased oleic acid in soybean seeds	Silencing of *FAD2-1A* and *1B* gene	*Agrobacterium*
		Enhanced production of tocopherols	Homogentisate geranylgeranyl transferase (*HGGT*) gene, *VTE3* and *VTE4*	*Agrobacterium*
		Increased tocopherol content	*γ-TMT* gene	*Agrobacterium*
		Increased seed oil content	Gene encoding diacylglycerol acyltransferase (*DGAT2*)	*Agrobacterium*
Tomato	Herbicide tolerance	Resistant to sulfonylurea herbicides	Acetolactate synthase (*ALS*)-encoding gene from tobacco	*Agrobacterium*

TABLE 4.2 (continued)

Major Genetic Transformation Applications to Improve Yield and Quality of Major Crop Plants

Crop	Type of Trait Improvement	Specific Characters	Gene Employed	Genetic Transformation Methods
	Biotic stress resistance	Increased resistance to tomato fruit borer, tobacco hornworm, tomato fruit worm, and tomato pinworm	*Cry* protein genes	*Agrobacterium*
		Increased resistance to nematodes	Cysteine protease inhibitor gene	*Agrobacterium*
		Increased resistance to bacterial wilt and spot	*ceropin B* gene	*Agrobacterium*
		Resistance to tobacco mosaic virus	*tobacco N* gene	*Agrobacterium*
		Resistance to tomato spotted wilt virus	Nucleocapsid protein (*N*) gene of TSWV	*Agrobacterium*
	Abiotic stress tolerance	Increased drought tolerance	*Osmyb4* gene	*Agrobacterium*
		Increased frost tolerance	Antifreeze gene (*afa3*)	*Agrobacterium*
		Enhanced salinity stress tolerant	*AtNHX1* gene	*Agrobacterium*
		Increased tolerance to cold	Osmotin-encoding gene	*Agrobacterium*
	Quality improvement	Delayed fruit ripening due to degradation of cell wall (pectin) degrading enzyme polygalacturonase	Antisense polygalacturonase gene	*Agrobacterium*
		Delayed ripening by reduction in ethylene production	ACC synthase, ACC deaminase, and *S*-adenosylmethionine hydrolase (SAMase)	*Agrobacterium*
		Increased provitamin A	Phytoene desaturase gene	*Agrobacterium*
		Enhanced isoflavone production	Isoflavone synthase–encoding gene	*Agrobacterium*

(*continued*)

TABLE 4.2 (continued)

Major Genetic Transformation Applications to Improve Yield and Quality of Major Crop Plants

Crop	Type of Trait Improvement	Specific Characters	Gene Employed	Genetic Transformation Methods
		Reduced lycopene content for improving taste	Geraniol synthase gene	*Agrobacterium*
		Increased production of polyamines thus increasing long lasting on the vine, increased juice, and enhanced nutrition quality	*S*-Adenosylmethionine decarboxylase (*SAMdc*) gene	*Agrobacterium*
Potato	Herbicide tolerance	Herbicide bromoxynil resistance	*bxn* gene	*Agrobacterium*
		Resistance to herbicide phosphinothricin	*bar* gene	*Agrobacterium*
		Resistance to herbicides interfering with photosynthesis, such as atrazine (AT), chlortoluron (CT), and methabenzthiazuron (MT)	*CYP1A1*, *CYP2B6*, and *CYP2C19* genes	*Agrobacterium*
		Enhanced resistance to herbicide, drought, and salt tolerance	*AtDHAR1* gene	*Agrobacterium*
	Biotic stress resistance	Increased resistance to potato virus X and potato virus Y	*PVX* and *PVY* coat protein gene	*Agrobacterium*
		Potato tuber moth and other Lepidopteran insect resistance	*Bt Cry* protein genes	*Agrobacterium*
		Resistance against bacterial and fungal pathogens	*AP24*, dermaseptin, and lysozyme-encoding gene	*Agrobacterium*
	Abiotic stress tolerance	Drought and salinity tolerance	*BADH* gene	*Agrobacterium*
		Improved tolerance to multiple environmental stress	Nucleoside diphosphate kinase 2 (*AtNDPK2*) gene	*Agrobacterium*

TABLE 4.2 (continued)

Major Genetic Transformation Applications to Improve Yield and Quality of Major Crop Plants

Crop	Type of Trait Improvement	Specific Characters	Gene Employed	Genetic Transformation Methods
		Enhanced accumulation of ascorbate for improvedtolerance against various abiotic stresses	D-Galacturonic acid reductase (*GalUR*) gene	*Agrobacterium*
		Salt stress tolerance	Mannitol-1-phosphate dehydrogenase (*mtlD*) gene	*Agrobacterium*
	Quality improvement	Reduced tuber browning	Polyphenol oxidase–encoding genes	*Agrobacterium*
		Increased protein and essential amino acid content in tubers	*AmA1* gene	*Agrobacterium*
		Increased level of inulin content in tubers	*1-SST* (sucrose: sucrose 1-fructosyltransferase) and *1-FFT* (fructan: fructan 1-fructosyltransferase) genes	*Agrobacterium*

4.6.1 Herbicide Tolerance

Genetic transformation in plants is being aimed globally to bring about genetic improvement in certain traits that facilitate production of crops of economic and agricultural importance. The method of genetic transformation has emerged as a supplemental tool to increase the efficiency of crop production by way of developing transgenic plants with improved traits such as disease resistance, insect/pest resistance, herbicide tolerance, abiotic stress tolerance, nutritional quality improvement, delayed fruit ripening, and improvement in shelf life. Among these various traits, the most desirable agronomic trait targeted through genetic transformation is the herbicide tolerance. The herbicide-tolerant transgenic plants are developed with an idea to simplify weed management and reduce the associated costs of crop production (Gianessi, 2005). Commercially, the most successful herbicide-tolerant trait that has been introduced into transgenic plants through genetic transformation technology was to allow them to grow against the broad spectrum foliar herbicide glyphosate and glufosinate. These herbicides have compounds similar to amino acids and functionally block the molecular targets in the amino-acid biosynthetic pathways in plants (Slater et al., 2003). The transgenic plants with glyphosate tolerance actually contain genes coding for resistant forms of certain enzymes like 5-enolpyruvylshikimate-3-phosphate synthase (EPSPS) or glyphosate oxidoreductase enzyme that degrades the herbicide (Duke, 2005). Plants developed for resistance to glufosinate contain a *bar* gene derived from bacteria *Streptomyces hygroscopicus*, which encodes an enzyme phosphinothricin acetyl transferase (PAT) that actually detoxifies the phosphinothricin, the active component of the herbicide glufosinate (Thompson et al., 1987). Vast amount of canola today in Canada and the United States has been successfully commercialized with these transgenic traits. One of the potential benefits of herbicide-resistant plants is that they are environment and health friendly, because of the potential to minimize application of more toxic chemical herbicides and polluting the water, food, soil, and air. The other indirect benefit from the herbicide-resistant plants is the reduced or minimized tillage agriculture that prevents loss of top soil and irreversible environment damage.

4.6.2 Insect and Disease Resistance

The next most important trait modification through genetic transformation in crop plants is insect resistance usually through introduction of *Bacillus thuringiensis* (*Bt*) crystal (*Cry*) protein gene. The *Bt Cry* proteins are ingested by insects and subsequently dissolved in the alkaline midgut environment to produce truncated toxic component of the *Cry* protein. This active toxic component binds to specific receptors of the cell membrane of epithelial cells in the midgut and results into perforation of the cell membrane. The perforation of cell membrane results into osmotic lysis and imbalance of solutes and toxic substances, thus killing the insect (de Maagd et al., 2003). *Bt Cry* proteins are grouped into different subfamilies according to their selectivity against specific insect types, such as Lepidoptera, Diptera, Coleoptera, etc. (MacIntosh et al., 1990). Due to high specificity of the *Bt* toxins, the transgenic *Bt* crops possess no significant health hazard to humans as compared to the chemical pesticides (Mendelsohn et al., 2003). The commercialized cultivation of *Bt* crops has provided evidence of reduced insecticide application to the tune of 70%–80%, especially in the *Bt* cotton fields of India and China. The environmental and economic gains of pesticide control are the major benefits of *Bt* transgenics. However, the development of resistance against the *Bt* toxins is the major concern related to the use of *Bt* transgenic crops. To counter this problem, several strategies have been adopted to maintain durable resistance to insects in the field. This includes a high level of transgene expression, cultivation of different plant lines with different insect resistance genes, pyramiding of transgenes, refugia strategy by cultivating susceptible plants along with the resistant ones, and tissue-specific expression of transgenes (Roush, 1998). Among them, the most successful method adopted under the field conditions is the combination of high dose of insect-resistant protein and refuge strategy to prevent development of insect resistance.

Among the various transgenic traits to target disease resistance in crop plants, virus resistance is the most predominant trait. The concept of pathogen-triggered resistance is now proven and established in imparting resistance to many viruses in different crop species (Beachy, 1997). Coat protein (CP)-mediated resistance from mild strains of viruses has been exploited through plant genetic transformations in papaya transgenic plants in Hawaii islands against the papaya ring spot virus (PRSV) (Gonsalves et al., 2006). Another approach used for developing virus-resistant crops uses the antisense or co-suppression techniques. Technically, this involves insertion of a complementary version of the target gene in an antisense orientation to reduce the amount of viral protein when the virus infects a plant. It has been demonstrated in the development of transgenic plants showing resistance against potato leaf roll virus (PRLV). This technology has been adopted for imparting resistance against other viruses as well. Currently, transgene-mediated RNA silencing and generation of small interfering RNAs appear to be primary mechanisms that confer resistance to plant viruses (Prins et al., 2008).

4.6.3 Quality Improvement

Numerous studies have demonstrated the power of plant genetic engineering to produce transgenic plants that have enhanced nutrition and a better shelf life. The most notable example in this category is the development of "golden rice" enriched with provitamin A in which three different genes, namely, phytoene synthase from daffodil, phytoene desaturase from a bacteria, and lycopene cyclase from daffodil, were mobilized into rice (Ye et al., 2000). Vitamin A deficiency, which also interferes with the bioavailability of iron, affects 413 million children worldwide. Ferritin rice rich in iron is another example developed by introducing the ferritin gene of *Phaseolus* into rice (Goto et al., 1999; Vasconcelos et al., 2003). Iron fortification in transgenic rice can provide 30%–50% of the daily adult iron requirement. A transgenic tomato rich in lycopene, a pigment that imparts red color to tomato, has been developed by Indian scientists (Mehta et al., 2002). A gene encoding a protein with a balanced composition of all eight essential amino acids, *Ama*1 from *Amaranthus*, has been overexpressed into potato to increase their nutritive value (Chakraborty et al., 2000). Research results have shown that vitamin C and vitamin E contents can also be elevated significantly in transgenic plants.

Transgenic fruit and vegetable crops have been developed with delayed ripening to save postharvest losses that occur primarily due to overripening (Bansal et al., 2000). As a matter of fact, transgenic tomato with delayed ripening was the first product to be commercialized in the world in 1994. Since then, many public and private institutions have produced transgenic tomatoes fit for processing and with improved postharvest characteristics. Genes and promoters have been cloned to regulate the expression of ripening-related genes for improving texture and delayed ripening in tomato (Anjanasree and Bansal, 2003). Production of antigens in transgenic plants has also become a reality. These transgenic plants promise to be used as edible vaccines that have gone through clinical trials with encouraging results.

4.6.4 Abiotic Stress Tolerance

Research over the past two decades has provided a better understanding of the molecular biology of stress responses in plants. This has led to the identification of several genes and gene products that get induced upon exposure of plants to various abiotic stresses, namely, drought, salinity, and low and high temperatures (Bartels and Sunkar, 2005; Grover et al., 2003). Consequently, biotechnological tools have been applied to transfer candidate genes from diverse sources to susceptible crop plants for developing transgenics resistant to abiotic stresses (Tayal et al., 2004). Few of the notable transgenics of this category are described later. One of the reports is in a model crop tobacco by Indian scientists, who expressed two genes together in a pathway, i.e., *gly*I and *gly*II (Singla-Pareek et al., 2003). Transgenic rice with increased tolerance to drought and salinity is another such example (Garg et al., 2002; Jhang et al., 2003). These transgenic rice plants overexpressing two genes (*ots*A and *ots*B) from *Escherichia coli* produced higher amounts of trehalose, a non-reducing disaccharide form of glucose. Trehalose is known to play an important role in imparting stress tolerance to a large variety of organisms ranging from bacteria and fungi to invertebrate animals. Another notable example is the biosynthesis of glycine betaine through expression of a bacterial *cod*A gene encoding choline oxidase in transgenic rice (Mohanty et al., 2002; Sakamoto et al., 1998) and transgenic tomato (Goel et al., 2010b).

Transgenic wheat transformed with *mtl*D gene of *E. coli* for the biosynthesis of mannitol showed improved growth performance under drought and salinity stresses (Abebe et al., 2003). It is also possible to introduce abiotic stress tolerance in transgenic plants by efficient scavenging of reactive oxygen species. This was demonstrated by expressing wheat catalase gene in transgenic rice (Matsumura et al., 2002). Overexpression of AtNHX1, a vacuolar Na^+/H^+ antiport from *A. thaliana* in *B. napus*, led to improved growth of the transgenic plants under salinity stress (Zhang et al., 2001b). Similarly, transgenic tomato overexpressing the *AtNHX1* gene accumulated high sodium in leaves, but not in the fruits, and exhibited increased tolerance to salt stress (Zhang and Blumwald, 2001). Transgenic tomato ectopically expressing the *Arabidopsis CBF1* gene showed enhanced resistance to drought, low temperature, and oxidative stresses (Hsieh et al., 2002a, 2002b). Similarly, transgenic tomato and mustard plants with osmotin gene have been developed for elevated tolerance to abiotic stresses (Goel et al., 2010). Although tolerance to abiotic stresses is considered a multigenic trait, the aforementioned examples prove that single gene introduction can confer tolerance to a range of abiotic stresses. Nevertheless, efforts are needed to pyramid more than one gene to develop transgenic plants tolerant to multiple stresses, which normally occur under natural field conditions and are on the rise with the changing climate (Varshney et al., 2011).

4.7 Conclusions and Future Prospects

To match the food production trends with that of population growth, it is imperative that modern tools of biotechnology should be adopted. Intensive efforts are in progress to develop high yielding transgenic crop varieties that not only are endowed with resistance to diseases and insects/pests and are tolerant to an array of abiotic stresses such as drought, salinity, high temperature, and other environmental stresses but also are highly nutritive. Development of such varieties has now become possible (and routine in some cases) using the modern methods of plant genetic transformation. Consequently, the area under transgenic crops is increasing since the first transgenic crop was grown in field in 1996; from less than 2 mha in 1996 to about 160 mha in 2011–2012 in 29 countries. The approach of developing transgenic crops, based on rDNA technology and different methods of gene transfer to plants, is more precise and accurate than the conventional approach of developing new food crop varieties. Both *Agrobacterium*-mediated and microprojectile bombardment-based transformations are now routinely used for development of transgenic plants. Through this approach, scientists are developing foods that are wholesome, nutritious, safe, and tasty. Both developed and developing countries are making use of this opportunity to boost their food production and to speed up their economic progress. Chloroplast transformation is also fast emerging as a novel approach for developing transgenics and to prevent the transgenes flow to weedy or wild relatives (Bansal and Sharma, 2003).

Several novel techniques are also under development to develop genetically transformed plants with desired characteristics. In the present-day context, it has been considered and proven that the marker-free transgenic plants will be the demand of the future. As a requirement, more research emphasis is needed on the genomics, cloning technology, and vector design, so as to eliminate the need for residual bacterial selectable marker genes in the future. The genetically transformed plants also play an important role in understanding gene function and metabolic pathways. Numerous plant genes associated with agronomically important traits have now been identified for genetic improvement of crops against a variety of production challenges and thus usher in the revolution of food and feed production.

References

Abebe, T., Guenzi, A.C., Martin, B., and Cushman, J.C. 2003. Tolerance of mannitol-accumulating transgenic wheat to water stress and salinity. *Plant Physiology* 131: 1748–1755.

Alimohammadi, M. and Bagherieh-Najjar, M.B. 2009. *Agrobacterium*-mediated transformation of plants: Basic principles and influencing factors. *African Journal of Biotechnology* 8: 5142–5148.

Altpeter, F., Baisakh, N., Beachy, R. et al. 2005. Particle bombardment and the genetic enhancement of crops: Myths and realities. *Molecular Breeding* 15: 305–327.

Anjanasree, K.N. and Bansal, K.C. 2003. Isolation and characterization of ripening-related expansin cDNA from tomato. *Journal of Plant Biochemistry and Biotechnology* 12: 31–35.

Arencibia, A., Molina, P., Gutierrez, C. et al. 1992. Regeneration of transgenic sugarcane (*Saccharum officinarum* L.) plants from intact meristematic tissue transformed by electroporation. *Biotecnología Aplicada* 9: 156–165.

Azpiroz-Leehan, R. and Feldmann, K.A. 1997. T-DNA insertion mutagenesis in *Arabidopsis*: Going back and forth. *Trends in Genetics* 13: 152–156.

Bansal, K.C., Barthakur, S., and Roy, S.K. 2000. Post-harvest biotechnology in fruits and vegetables. In: *Biotechnology and its Application in Horticulture*, ed. S.P. Ghosh, 165–982. Narosa Publishers, New Delhi, India.

Bansal, K.C. and Saha, D. 2012. Chloroplast genomics and genetic engineering for crop improvement. *Agricultural Research* 1: 53–66.

Bansal, K.C. and Sharma, R.K. 2003. Chloroplast transformation as a tool for prevention of gene flow from GM crops to weedy or wild relatives. *Current Science* 84: 1286–1287.

Bartels, D. and Sunkar, R. 2005. Drought and salt tolerance in plants. *Critical Reviews in Plant Sciences* 24: 23–58.

Beachy, R.N. 1997. Mechanisms and applications of pathogen-derived resistance in transgenic plants. *Current Opinion in Biotechnology* 8: 215–220.

Bechtold, N. and Pelletier, G. 1998. In planta *Agrobacterium* mediated transformation of adult *Arabidopsis thaliana* plants by vacuum infiltration. *Methods in Molecular Biology* 82: 259–266.

Bent, A.F. 2000. *Arabidopsis* in planta transformation. Uses, mechanisms, and prospects for transformation of other species. *Plant Physiology* 124: 1540–1547.

Bevan, M. 1984. Binary *Agrobacterium* vectors for plant transformation. *Nucleic Acids Research* 12: 8711–8721.

Bhatnagar-Mathur, P., Vadez, V., and Sharma, K.K. 2008. Transgenic approaches for abiotic stress tolerance in plants: Retrospect and prospects. *Plant Cell Reports* 27: 411–424.

Birch, R.G. 1997. Plant transformation: Problems and strategies for practical application. *Annual Review of Plant Physiology and Plant Molecular Biology* 48: 297–326.

Bock, R. 2000. Sense from nonsense: How the genetic information of chloroplasts is altered by RNA editing. *Biochimie* 82: 549–557.

Bock, R. 2001. Transgenic plastids in basic research and plant biotechnology. *Journal of Molecular Biology* 312: 425–438.

Boothe, J., Nykiforuk, C., Shen, Y. et al. 2010. Seed-based expression systems for plant molecular farming. *Plant Biotechnology Journal* 8: 588–606.

Chakraborty, S., Chakraborty, N., and Datta, A. 2000. Increased nutritive value of transgenic potato by expressing a non allergenic seed albumin gene from *Amaranthus hypochondriacus*. *Proceedings of the National Academy of Sciences U S A* 97: 3724–3729.

Chamberlin, K.D.C. 2010. Deployment: Regulations and steps for commercialization. In: *Transgenic Crop Plants, Volume 2, Utilization and Biosafety*, eds. C. Kole, C. Michler, A.G. Abbott, and T.C. Hall, 391–410. Springer-Verlag, Heidelberg, Germany.

Chilton, M.D. 1993. *Agrobacterium* gene transfer: Progress on a "poor man's vector" for maize. *Proceedings of the National Academy of Sciences U S A* 90: 3119–3120.

Chilton, M.D., Drummond, M.H., Merlo, D.J. et al. 1977. Stable incorporation of plasmid DNA into higher plant cells: The molecular basis of crown gall tumorigenesis. *The Cell* 11: 263–271.

Clough, S.J. and Bent, A.F. 1998. Floral dip: A simplified method for *Agrobacterium*-mediated transformation of *Arabidopsis thaliana*. *The Plant Journal* 16: 735–743.

Collinge, D.B., Lund, O.S., and Christensen, H.T. 2008. What are the prospects for genetically engineered, disease resistant plants? *European Journal of Plant Pathology* 121: 217–231.

Crossway, A., Oakes, J.V., Irvinem J.M., Ward, B., Knauf, V.C., and Shewmaker, C.K. 1986. Integration of foreign DNA following microinjection of tobacco mesophyll protoplasts. *Molecular and General Genetics* 202: 179–185.

Damodaran, A. 2005. Re-engineering biosafety regulations in India: Towards a critique of policy, law and prescriptions. *Law, Environment and Development Journal* 1(1): 1–20. http://www.lead-journal.org/content/05001.pdf

Daniell, H., Khan, M.S., and Allison, L.A. 2002. Milestones in chloroplast genetic engineering: An environmentally friendly era in biotechnology. *Trends in Plant Sciences* 7: 84–91.

Daveya, M.R., Anthonya, P., Powera, J.B., and Loweb, K.C. 2005. Plant protoplasts: Status and biotechnological perspectives. *Biotechnology Advances* 23: 131–171.

de Maagd, R.A., Bravo, A., Berry, C., Crickmore, N., and Schnepf, H.E. 2003. Structure, diversity, and evolution of protein toxins from spore forming entomopathogenic bacteria. *Annual Review of Genetics* 37: 409–433.

Dekeyse, R.A., Claes, B., De Rycke, R.M.U., Habets, M.E., Van Montagu, M.C., and Caplan, A.B. 1990. Transient gene expression in intact and organized rice tissues. *The Plant Cell* 2: 591–601.

Duke, S.O. 2005. Taking stock of herbicide-resistant crops ten years after introduction. *Pest Management Science* 61: 211–218.

Fischer, R., Stoger, E., Schillberg, S., Christou, P., and Twyman, R.M. 2004. Plant based production of biopharmaceuticals. *Current Opinion in Plant Biology* 7: 152–158.

Frame, B.R., Drayton, P.R., Bagnall, S.V. et al. 1994. Production of fertile transgenic maize plants by silicon carbide whisker-mediated transformation. *The Plant Journal* 6: 941–948.

Fromm, M.E., Taylor, L.P., and Walbot, V. 1986. Stable transformation of maize after gene-transfer by electroporation. *Nature* 319: 791–793.

Garg, A.K., Kim, J.K., Owens, T.G. et al. 2002. Trehalose accumulation in rice plants confers high tolerance levels to different abiotic stresses. *Proceedings of the National Academy of Sciences U S A* 99: 15898–15903.

Gelvin, S.B. 2003. *Agrobacterium* mediated plant transformation: The biology behind the gene-jockeying tool. *Microbiology and Molecular Biology Reviews* 67: 16–37.

Gelvin, S.B. 2009. *Agrobacterium* in the genomics age. *Plant Physiology* 150: 1665–1676.

Gianessi, L.P. 2005. Economic and herbicide use impacts of glyphosate-resistant crops. *Pest Management Science* 61: 241–245.

Godfray, H.C.J., Beddington J.R., Crute, I.R. et al. 2010. Food security: The challenge of feeding 9 billion people. *Science* 327: 812–818.

Goel, D., Singh, A., Yadav, V., Babbar, S., and Bansal, K.C. 2010. Over expression of osmotin gene confers tolerance to salt and drought stress in transgenic tomato (*Solanum lycopersicum* L.). *Protoplasma* 245: 133–141.

Gonsalves, D., Vegas, A., Prasartsee, V., Drew, R., Suzuki, J.Y., and Tripathi, S. 2006. Developing papaya to control papaya ringspot virus by transgenic resistance, intergeneric hybridization, and tolerance breeding. *Plant Breeding Reviews* 26: 35–78.

Goto, F., Yoshihara, T., Shigemoto, N., Toki, S., and Takaiwa, F. 1999. Iron fortification of rice seed by the soybean ferritin gene. *Nature Biotechnology* 17: 282–286.

Grover, A., Aggarwal, P.K., Kapoor, A., Katiyar-Agarwal, S., Agarwal, M., and Chandramouli, A. 2003. Addressing abiotic stresses in agriculture through transgenic technology. *Current Science* 84: 355–376.

Hayashimoto, A., Li, Z., and Murai, N. 1990. A polyethylene glycol-mediated protoplast transformation system for production of fertile transgenic rice plants. *Plant Physiology* 93: 857–863.

Hellens, R., Mullineaux, P., and Klee, H. 2000. A guide to *Agrobacterium* binary Ti vectors. *Trends in Plant Sciences* 5: 446–451.

Herrera-Estrella, L., de Block, M., Messens, E., Hernalsteens, J.P., van Montagu, M., and Schell, J. 1983. Chimeric genes as dominant selectable markers in plant cells. *EMBO Journal* 2: 987–995.

Hiei, Y. and Komari, T. 2008. *Agrobacterium*-mediated transformation of rice using immature embryos or calli induced from mature seed. *Nature Protocols* 3: 824–834.

Hisano, H., Nandakumar, R., and Wang, Z.Y. 2009. Genetic modification of lignin biosynthesis for improved biofuel production. *In Vitro Cellular and Developmental Biology-Plant* 45: 306–313.

Hoekema, A., Hirsch, P.R., Hooykaas, P.J.J., and Schilperoort, R.A. 1983. A binary plant vector strategy based on separation of vir- and T-region of the *Agrobacterium tumefaciens* Ti plasmid. *Nature* 303: 179–180.

Hsieh, T.H., Lee, J.T., Charng, Y.Y., and Chan, M.T. 2002a. Tomato plants ectopically expressing Arabidopsis CBF1 show enhanced resistance to water deficit stress. *Plant Physiology* 130: 618–626.

Hsieh, T.H., Lee, J.T., Yang, P.T., et al. 2002b. Heterology expression of the Arabidopsis C-repeat/dehydration response element binding factor 1 gene confers elevated tolerance to chilling and oxidative stresses in transgenic tomato. *Plant Physiology* 129: 1086–1094.

Jhang, I.C., Oh, S.J., Seo, J.S. et al. 2003. Expression of a bifunctional fusion of the *Escherichia coli* genes for trehalose-6-phosphate phosphatase in transgenic rice plants increases trehalose accumulation and abiotic stress tolerance without stunting growth. *Plant Physiology* 131: 516–524.

Klein, T.M., Wolf, E.D., Wu, R., and Sanford, J.C. 1987. High velocity microprojectiles for delivering nucleic acids into living cells. *Nature* 327: 70–73.

Knoblauch, M., Hibberd, J.M., Gray, J.C., and van Bel, A.J.E. 1999. A galinstan expansion femtosyringe allows microinjection of eukaryotic organelles and prokaryotes. *Nature Biotechnology* 17: 906–909.

Komari, T. and Kubo, T. 1999. Methods of genetic transformation: *Agrobacterium tumefaciens*. In: *Advances in Cellular and Molecular Biology of Plants. Molecular Improvement of Cereal Crops*, ed. I.K. Vasil, 43–82. Kluwer, Dordrecht, the Netherlands.

Koop, H.U., Herz, S., Golds, T.J. and Nickelsen, J. 2007. The genetic transformation of plastids. *Topics in Current Genetics* 19: 457–510.

Lee, L.Y. and Gelvin, S.B. 2008. T-DNA binary vectors and systems. *Plant Physiology* 146: 325–332.

Liu, F., Cao, M.Q., Yao, L., Li, Y., Robaglia, C., and Tourneur, C. 1998. In planta transformation of pakchoi (*Brassica campestris* L. ssp. *chinensis*) by infiltration of adult plants with *Agrobacterium*. *Acta Horticulturae* 467: 187–192

MacIntosh, S.C., Stone, T.B., Sims, S.R. et al. 1990. Specificity and efficacy of purified *Bacillus thuringiensis* proteins against agronomically important insects. *Journal of Invertebrate Pathology* 56: 258–266.

Madoka, Y., Tomizawa, K.I., Mizoi, J., Nishida, I., Nagano, Y., and Sasaki, Y. 2002. Chloroplast transformation with modified *accD* operon increases acetyl-Co-A carboxylase and causes extension of leaf longevity and increase in seed yield in tobacco. *Plant and Cell Physiology* 43: 1518–1525.

Maliga, P. 2004. Plastid transformation in higher plants. *Annual Reviews in Plant Biology* 55: 289–313.

Maliga, P. and Bock, R. 2011. Plastid biotechnology: Food, fuel, and medicine for the 21st century. *Plant Physiology* 155: 1501–1510.

Matsumura, T., Tabayashi, N., Kamagat, Y., Souma, C., and Saruyam, H. 2002. Wheat catalase expressed in transgenic rice can improve tolerance against low temperature stress. *Physiologia Plantarum* 116: 317–327.

McGloughlin, M.N. 2008. Nutritionally improved agricultural crops. *Plant Physiology* 147: 939–953.

Mehta, R.A., Cassol, T., Li, N., Ali, N., Handa, A.K., and Mattoo, A.K. 2002. Engineered polyamine accumulation in tomato enhances phytonutrient content, juice quality, and vine life. *Nature Biotechnology* 20: 613–618.

Mendelsohn, M., Kough, J., Vaituzis, Z., and Matthews, K. 2003. Are *Bt* crops safe? *Nature Biotechnology* 21: 1003–1009.

Miki, B. and McHugh, S. 2004. Selectable marker genes in transgenic plants: Applications, alternatives and biosafety. *Journal of Biotechnology* 107: 193–232.

Mohanty, A., Kathuria, H., Ferjani A. et al. 2002. Transgenics of an elite indica rice variety *Pusa basmati* 1 harbouring the *codA* gene are highly tolerant to salt stress. *Theoretical and Applied Genetics* 106: 51–57.

Muir, W.H., Hildebrandt, A.C., and Riker, A.J. 1954. Plant tissue cultures produced from isolated single cells. *Science* 119: 877–878.

Mulwa, R.M.S. and Mwanza, L.M. 2006. Biotechnology approaches to developing herbicide tolerance/selectivity in crops. *African Journal of Biotechnology* 5: 396–404.

Negrutiu, I., Shillito, R.D., Potrykus, I., Biasini, G., and Sala, F. 1987. Hybrid genes in the analysis of transformation conditions. I. Setting up a simple method for direct gene transfer in plant protoplasts. *Plant Molecular Biology* 8: 363–373.

Neuhaus, G. and Spangenberg, G. 1990. Plant transformation by microinjection techniques. *Physiologia Plantarum* 79: 213–217.

Pena, L. (2005). *Transgenic Plants: Methods and Protocols. Methods in Molecular Biology*, vol. 286. Humana Press, Totowa, NJ.

Phillips, R.L., Kaeppler, S.M., and Olhoft, P. 1994. Genetic instability of plant tissue cultures: Breakdown of normal controls. *Proceedings of the National Academy of the Sciences U S A* 91: 5222–5226.

Prins, M., Laimer, M., Noris, E., Schubert, J., Wassenegger, M., and Tepfer, M. 2008. Strategies for antiviral resistance in transgenic plants. *Molecular Plant Pathology* 9: 73–83.

Rao, A.Q., Bakhsh, A., Kiani, S. et al. 2009. The myth of plant transformation. *Biotechnology Advances* 27: 753–763.

Roush, R.T. 1998. Two-toxin strategies for management of insecticidal transgenic crops: Can pyramiding succeed where pesticide mixtures have not? *Philosophical Transaction of the Royal Society B* 353: 1777–1786.

Ruan, C.J., Shao, H.B., and da Silva, J.A.T. 2012. A critical review on improvement of carbon fixation in plants using genetic engineering. *Critical Reviews in Biotechnology* 32: 1–21.

Ruf, S., Biehler, K., and Bock, R. 2000. A small chloroplast-encoded protein as a novel architectural component of the light-harvesting antenna. *Journal of Cellular Biology* 149: 369–377.

Sakamoto, A., Alia, and Murata, N. 1998. Metabolic engineering of rice lead in to biosynthesis of glycine betaine and tolerance to salt and cold. *Plant Molecular Biology* 38: 1011–1019.

Salmenkallio-Marttila, M., Aspegren, K., Kerman, S. et al. 1995. Transgenic barley (*Hordeum vulgare* L.) by electroporation of protoplasts. *Plant Cell Reports* 15: 301–304.

Sanford, J.C. 2000. The development of the biolistic process. *In Vitro Cellular and Development Biology-Plant* 36: 303–308.

Shanker, S., Chinnusamy, V., and Bansal, K.C. 2008. A simplified floral dip method for transformation of *Brassica napus* and *B. carinata*. *Journal of Plant Biochemistry and Biotechnology* 17: 197–200.

Sharma, H.C. 2008. Genetic transformation of crops for resistance to insect Pests. In: *Biotechnological Approaches for Pest Management and Ecological Sustainability*, 208–254. Taylor & Francis, Boca Raton, FL.

Shillito, R. 1999. Methods of genetic transformation: Electroporation and polyethylene glycol treatment. In: *Advances in Cellular and Molecular Biology of Plants. Molecular Improvement of Cereal Crops*, ed. I.K. Vasil, 9–20. Kluwer, Dordrecht, Netherlands.

Shimamoto, K., Terada, R., Izawa, T., and Fujimoto, H. 1989. Fertile transgenic rice plants regenerated from transformed protoplasts. *Nature* 338: 274–276.

Shrawat, A.K., Carroll, R.T., DePauw, M., Taylor, G.J., and Good, A.G. 2008. Genetic engineering of improved nitrogen use efficiency in rice by the tissue-specific expression of alanine aminotransferase. *Plant Biotechnology Journal* 6: 722–732.

Shrawat, A.K. and Good, A.G. 2011. *Agrobacterium tumefaciens*-mediated genetic transformation of cereals using immature embryos. *Methods in Molecular Biology* 710: 355–372.

Shrawat, A.K. and Lorz, H. 2006. *Agrobacterium*-mediated transformation of cereals: A promising approach crossing barriers. *Plant Biotechnology Journal* 4: 575–603.

Singla-Pareek, S.L., Reddy, M.K., and Sopory, S.K. 2003. Genetic engineering of the glyoxalase pathway in tobacco leads to enhanced salinity tolerance. *Proceedings of the National Academy of Sciences U S A* 100: 14672–14677.

Slater, A., Scott, N.W., and Fowler, M.R. 2003. *Plant Biotechnology: The Genetic Manipulation of Plants*. Oxford University Press, New York.

Smith, E.F. and Townsend, C.O. 1907. A plant-tumor of bacterial origin. *Science* 24: 671–673.

Springer, P.S. 2000. Gene traps. *The Plant Cell* 12: 1007–1020.

Srinivasan, R. and Saha, D. 2010. Promoter trapping in plants using T-DNA mutagenesis. In: *Molecular Techniques in Crop Improvement*, 2nd edition, eds. S.M. Jain and D.S. Brar, pp. 545–577. Springer, Heidelberg, Germany.

Swaminathan, M.S. and Kesavan, P.C. 2012. Agricultural research in an era of climate change. *Agricultural Research* 1: 3–11.

Tan, H., Yang, X., Zhang, F. et al. 2011. Enhanced seed oil production in canola by conditional expression of *Brassica napus* LEAFY COTYLEDON1 and LEC1-LIKE in developing seeds. *Plant Physiology* 156: 1577–1588.

Tayal, D., Srivastava, P.S., and Bansal, K.C. 2004. Transgenic crops for abiotic stress tolerance. In *Plant Biotechnology and Molecular Markers*, eds. P.S. Srivastava, A. Narula, and S. Srivastava, pp. 346–365. Kluwer Academic Publishers, the Netherlands.

Thompson, C.J., Movva, N.R., Tizard, R. et al. 1987. Characterization of the herbicide resistance gene *bar* from *Streptomyces hygroscopicus*. *EMBO Journal* 6: 2519–2523.

Tianzi, C., Shenjie, W., Jun, Z., Wangzhen, G., and Tianzhen, Z. 2010. Pistil drip following pollination: A simple in planta *Agrobacterium*-mediated transformation in cotton. *Biotechnology Letters* 32: 547–555.

Trieu, A.T., Burleigh, S.H., Kardailskym I.V. et al. 2000. Transformation of *Medicago truncatula* via infiltration of seedlings or flowering plants with *Agrobacterium*. *The Plant Journal* 22: 531–541.

Tzfira, T. and Citovsky, V. 2006. *Agrobacterium*-mediated genetic transformation of plants: Biology and biotechnology. *Current Opinion in Biotechnology* 17: 147–154.

Vain, P. 2007. Thirty years of plant transformation technology development. *Plant Biotechnology Journal* 5: 221–229.

Valentine, L. 2003. *Agrobacterium tumefaciens* and the plant: The David and Goliath of modern genetics. *Plant Physiology* 133: 948–955.

Varshney, R.K., Bansal, K.C., Aggarwal, P.K., Datta, S.K., and Craufurd, P.Q. 2011. Agricultural biotechnology for crop improvement in a variable climate: Hope or hype? *Trends in Plant Science* 16: 363–371.

Vasconcelos, M.V., Datta, K., Oliva, N. et al. 2003. Enhanced iron and zinc accumulation in transgenic rice with the ferritin gene. *Plant Science* 164: 371–378.

Vasil, I.K. 2005. The story of transgenic cereals: The challenge, the debate and the solution—A historical perspective. *In Vitro Cellular and Development Biology-Plant* 41: 577–583.

Vasil, I.K. 2008. A history of plant biotechnology: From the cell theory of Schleiden and Schwann to biotech crops. *Plant Cell Reports* 27: 1423–1440.

Vasil, V. and Hildebrandt, A.C. 1965. Differentiation of tobacco plants from single, isolated cells in microcultures. *Science* 150: 889–892.

Vasil, I.K. and Vasil, V. 1992. Advances in cereal protoplast research. *Physiologia Plantarum* 85: 279–283.

Visarada, K.B.R.S., Meena, K., Aruna, C., Srujana, S., Saikishore, N., and Seetharama, N. 2009. Transgenic breeding: Perspectives and prospects. *Crop Science* 49: 1555–1563.

Whitney, S.M. and Andrews, T.J. 2001. Plastome-encoded bacterial ribulose-1,5-bisphosphate carboxylase/oxygenase (RubisCO) supports photosynthesis and growth of tobacco. *Proceedings of the National Academy of Sciences U S A* 98: 14738–14743.

Ye, X., Al-Babili, S., Kloti, A. et al. 2000. Engineering the provitamin A (beta-carotene) biosynthetic pathway into (carotenoid-free) rice endosperm. *Science* 287: 303–305.

Zambryski, P., Joos, P.H., Genetello, C., Leemans, J., Van Montagu, M., and Schell, J. 1983. Ti plasmid vector for the introduction of DNA into plant cells without alteration of their normal regeneration capacity. *EMBO Journal* 2: 2143–2150.

Zhang, H.X. and Blumwald, E. 2001. Transgenic salt-tolerant tomato plants accumulate salt in foliage but not in fruit. *Nature Biotechnology* 19: 765–768.

Zhang, H.X., Brotherton, J.E., Widholm, J.M., and Portis, A.R. 2001a. Targeting a nuclear anthranilate synthase alpha-subunit gene to the tobacco plastid genome results in enhanced tryptophan biosynthesis. Return of a gene to its pre-endosymbiotic origin. *Plant Physiology* 127: 131–141.

Zhang, H.X., Hodson, J.N., Williams, J.P., and Blumwald, E. 2001b. Engineering salt tolerant *Brassica* plants: Characterization of yield and seeds oil quality in transgenic plants with increased vacuolar sodium accumulation. *Proceedings of the National Academic of the Sciences U S A* 98: 12832–12836.

Zhu, Z., Sun, B., Liu, C., Xiao, G., and Li, X. 1993. Transformation of wheat protoplasts mediated by cationic liposome and regeneration of transgenic plantlets. *Chinese Journal of Biotechnology* 9: 257–261.

Zupan, J., Muth, T.R., Draper, O., and Zambryski, P.C. 2000. The transfer of DNA from *Agrobacterium tumefaciens* into plants: A feast of fundamental insights. *The Plant Journal* 23: 11–28.

5

Production of Biofertilizers

Dinesh Goyal and Santosh K. Goyal

CONTENTS

5.1 Introduction 169
5.2 Microbial Cultures as Biofertilizers 170
5.2.1 Nitrogen-Fixing Microorganisms 170
5.2.1.1 Rhizobium 172
5.2.1.2 Azotobacter 173
5.2.1.3 Azospirillum 173
5.2.1.4 Blue-Green Algae 173
5.2.2 Phosphate-Solubilizing Bacteria 174
5.2.3 Vesicular Arbuscular Mycorrhiza 175
5.2.4 Plant Growth–Promoting Rhizobacteria 176
5.2.5 Microbial Consortia 176
5.3 Benefits of Biofertilizers 177
5.4 Production Technology 178
5.4.1 Raw Material and Equipment 179
5.4.2 Selection of Microbial Strain 179
5.4.3 Preparation of Mother Culture 179
5.4.4 Mass Multiplication 181
5.4.5 Mixing of Culture with Carrier Material and Packing 181
5.4.6 Quality Control 182
5.5 Global Status 183
5.6 Constraints 185
5.7 Summary and Future Prospects 186
References 187

5.1 Introduction

The initiative taken in the late 1960s and early 1970s to make the country self-reliant to meet the food requirement resulted in a shift from traditional agriculture to intensive farming involving stepped up use of chemical fertilizers and pesticides (Goyal, 2001; Tandon, 1992). The steps that brought the green revolution in the country in the long run have turned the soils barren. The runoff from fields carrying leftover fertilizers and pesticides contaminated the ground water, which gets reflected in our daily diet. These detrimental effects of current agricultural practices can be appreciably reduced by using biofertilizers in combination with organic manure and chemical fertilizers and induce long-term sustainability in agricultural production (Tandon, 1992).

Microorganisms that augment, conserve, and mobilize plant nutrients in soil are termed as biofertilizers. Biofertilizers are formulations containing beneficial and stress-compatible autochthonous microorganisms in living form, which easily colonize the soil ecosystem. Sometimes these microbial cultures are more appropriately referred to as "microbial inoculants" or "bioinoculants" (Goyal, 1982; Roy et al., 2006). Efforts by scientists during the last 50 years have culminated into the creation of a *cafeteria* of biofertilizers, which can be exploited for supplementation of organic matter, providing biologically fixed nitrogen, mobilization of insoluble phosphates, secretion of growth factors, improving the physical and chemical properties of the soil, better utilization of chemical fertilizers, bioconcentration of nutrients in the rhizosphere, and acceleration of the process of composting (Motsara et al., 1995; Roy et al., 2006; SubbaRao, 1993).

This chapter presents comprehensive information on the various attributes of biofertilizers, their production, and application for deriving agronomic advantage.

5.2 Microbial Cultures as Biofertilizers

The "cafeteria" of biofertilizers, developed (Table 5.1) during the last few decades, is crop-specific. It can be used for a variety of crops under diverse agronomic and ecological conditions and has been recognized as one of the sources of plant nutrients (Roy et al., 2006). Microbial cultures can significantly benefit crop plants and can be developed into inoculants on an industrial scale. Biofertilizer technology can be adequately exploited to improve soil fertility and production economics (Goyal, 2001; Motsara et al., 1995; SubbaRao, 1993; Tandon, 1992).

5.2.1 Nitrogen-Fixing Microorganisms

Nitrogen, in the atmosphere, is relatively inert and nonreactive, and to be of use to plants, it must be converted (fixed) into inorganic form so that it can be taken up by the plants. In industrial process, commonly called as Haber–Bosch process, nitrogen is converted into ammonia and then to fertilizers like urea and ammonium sulfate at the expense of a large amount of energy in the presence of a catalyst (Wolfe, 2001). Biological nitrogen fixation is a

TABLE 5.1

Microorganisms Being Used as Biofertilizers for Various Crops

Function	Organisms	Crops
Nitrogen fixation	*Rhizobium meliloti, R. leguminosarum R. trifolii, R. lupini, R. japonicum*	Arhar, pea, lentil, gram, green gram, rajmah, cowpea, berseem, lucerne
	Azotobacter chroococcum, A. vinelandii	Groundnut, soybean
	A. beijerinckii	Wheat, barley
	Acetobacter diazotrophicus	Maize, cotton, paddy
	Azospirillum lipoferum, A. amazonense	Sugarcane
	A. seropedica, A. americanum	Maize, sugarcane
	Cyanobacteria (blue-green algae)	Jowar, paddy, pearl millet
	Free living: *Anabaena, Nostoc, Aulosira Calothrix, Tolypothrix*	Paddy
	Symbiotic: *Azolla–Anabaena azollae*	
Phosphate solubilization	*B. polymyxa, B. megaterium P. striata, Aspergillus awamori Penicillium digitatum*	All crops
Nutrient translocation	Endomycorrhizae: *Glomus, Gigaspora Acaulospora*	Cereals, forest tree species
	Ectomycorrhizae: *Laccaria, Pisolithus*	
Organic matter decomposition	Cellulolytic:	Forest tree species
	Fungi: *Trichoderma, Chaetomium Aspergillus*	
	Bacteria: *Bacillus, Cellulomonas Cytophaga*	
	Lignolytic:	
	Fungi: *Clavaria, Cephalosporium*	
	Bacteria: *Pseudomonas, Flavobacterium*	
Growth accelerators	PGPR *Pseudomonas, Xanthomonas*	All crops

mimicry of this process accomplished inside the cells of some microorganisms known as diazotrophs in the presence of iron–molybdenum enzyme nitrogenase that grows in the absence of chemical nitrogen and contributes more than twice the amount of fixed nitrogen annually as compared to industrial and other means (Postgate, 1998). *Klebsiella pneumoniae* and *Azotobacter vinelandii* have been studied most because of their genetic traceability and fast growth. Other nitrogen-fixing bacteria are *Azotobacter, Beijerinckia* and some species of *Klebsiella, Clostridium, Desulfovibrio,* purple sulfur bacteria, purple non-sulfur bacteria, and green sulfur bacteria (Postgate, 1998). Symbiotic nitrogen-fixing bacteria are *Rhizobium* and *Frankia,* whereas *Azospirillum* and *Acetobacter* are associative symbionts. Nitrogen-fixing blue-green algae like *Nostoc, Anabaena,* and *Tolypothrix* are predominantly free-living nitrogen fixers, whereas some like *Anabaena azollae* and *Anabaena cycadeae* are symbiotic.

Surprisingly, all nitrogen-fixing microorganisms are prokaryotic and live freely either in soil or in symbiosis with plants. They convert molecular nitrogen from the air to ammonia and pass it on to the crop plants (Postgate, 1998). Presently in the United States, United Kingdom, Russia, India, and other countries, *Rhizobium* inoculants for leguminous crops have been fairly well adapted by the farmers.

5.2.1.1 Rhizobium

Bacteria belonging to this genus form symbiotic association with roots of leguminous plants like soybean, chickpea, pea, broad beans, lentil, green gram, pigeon pea, and ground nut and fix atmospheric nitrogen in root nodules. Fastidious rhizobia like *Rhizobium leguminosarum* are crop-specific and promiscuous. Forms like cowpea rhizobia are nonspecific symbionts. Based on this property of preference for forming root nodules on different legumes, rhizobia are classified into seven inoculation groups. However, based on GC content, genome size, and sequence analysis of 16S–23S intergenic spacer (ITS) region, root-nodulating rhizobia have been classified into genera including *Rhizobium, Bradyrhizobium, Mesorhizobium, Azorhizobium,* and *Sinorhizobium* (Ensifer) (Ramirez-Bahena et al., 2008). These bacteria enter the root hair, form inoculation thread, and reach the cortex. In the cortex, they form polymorphic bacteroids that are surrounded by leghemoglobin. They provide nitrogen to the plant that in turn supplies energy and hydrogen donors for the activity of nitrogenase enzyme converting nitrogen to ammonia. The plant also provides organic acids for binding ammonia into amino acids.

Studies suggest that *Rhizobium*–legume symbiosis co-opted a signaling pathway, including receptor from the more ancient mycorrhizal symbiosis to form a symbiotic interface and the two highly exocytotic vesicle-associated membrane proteins (VAMPs) required for the formation of symbiotic membrane interface in both interactions (Ivanov et al., 2012). Legumes form either indeterminate or determinate types of nodules, with these groups differing widely in nodule morphology and often in the developmental program by which rhizobia form nitrogen-fixing bacteroids (Terpolilli et al., 2012).

Biotic and abiotic factors affect nodulation efficiency, which is also governed by the presence of native rhizobial population, cultivar, and edaphic conditions. *Rhizobium* inoculation in leguminous crops leads to 15%–30% increase in crop yield. Seed inoculation (pelleting) with *Rhizobium* along with the application of recommended dose of fertilizer (20:40:40 N:P:K kg/ha) resulted in maximum grain and straw yield, protein content, and nutrient uptake by cowpea *Vigna unguiculata* L. (Dekhane et al., 2012). With combined and individual inoculation of *Rhizobium* and phosphate-solubilizing bacteria (PSB) in fenugreek (*Trigonella foenum-graecum* L.), it was found that *Rhizobium* and PSB alone gave high yield as compared to control, whereas *Rhizobium* and PSB in combination gave significantly higher yield (Mehta et al., 2011). *R. leguminosarum* bv. *trifolii* was found to enhance rice production (Yanni and Dazzo, 2010).

5.2.1.2 Azotobacter

This genus comprises free-living, nitrogen-fixing bacteria that have been shown to produce several plant growth–promoting (PGP) substances also. In addition, these bacteria induce disease resistance and flowering synchrony in plants. Pelleting of seeds with *Azotobacter* is recommended for cereal crops like paddy and wheat, vegetable crops such as potato and tomato, seed crops like mustard and rapeseed, and fiber crops like jute and cotton. The enhanced growth of bamboo (*Bambusa bamboo*) and maize (*Zea mays*) due to combined inoculation with *Azotobacter chroococcum* and PGP rhizobacteria (PGPR) was attributed to indole acetic acid (IAA) production and phosphorus solubilization. Yield of rice (Yanni and El-Fattah, 1999), cotton (Anjum et al., 2007), and wheat (Hegazi et al., 1998) increased with the application of *Azotobacter.*

5.2.1.3 Azospirillum

Azospirilli are free-living nitrogen-fixing bacteria, which form a temporary symbiotic association and potential inoculants for agriculture (Okon, 1985). They stimulate plant growth through nitrogen fixation and production of growth substances like auxins, gibberellins, and cytokinins. It is estimated that almost 10%–15% of the required nitrogen can be met by *Azospirillum* amendment. *Azospirillum*-based biofertilizers are applied in jowar, bajra, ragi, barley, oats, and forage crops. *Azospirillum*-inoculated lettuce seeds showed higher seed germination under saline conditions (Fasciglione et al., 2011). The variant strains of *Azospirillum brasilense* SMp 30 and SMΔi3-6 were more efficient surface colonizers of *Sorghum* roots as compared to wild-type strain in increasing productivity (Kochar and Shrivastava, 2012). Foliar application of micronutrients (Mn^+ Fe^+ Zn^+) together with *Azospirillum, Azotobacter,* or *Bacillus* sp. can lead to increase in yield of wheat (Eleiwa et al., 2012). PGP ability of *Azo. brasilense* has been elucidated in agriculture on commercial scale (Fibach-Paldi et al., 2012). A simple method for detecting and identifying *Azospirillum* from rhizosphere soils has been developed to ascertain colonization (Shime-Hattori et al., 2011).

5.2.1.4 Blue-Green Algae

Blue-green algae, popularly known as cyanobacteria, constitute most important input in rice cultivation (Goyal, 1982). Their profuse growth in rice fields plays a critical role in the sustenance of natural fertility of this ecosystem (Rogar and Watanabe, 1986; Stewart, 1975). Their ecological significance in the rhizosphere and endophytic colonization has also been documented (Prasanna and Nayak, 2007). Cyanobacteria are best developed as supplements to chemical fertilizers, since they are capable of fixing nitrogen (Venkataraman, 1981). Cyanobacteria form a major source of biologically fixed nitrogen in tropical rice fields. They also synthesize and liberate amino acids, vitamins, and auxins, play a vital role in reducing oxidizable matter content of the soil by providing oxygen in the submerged rhizosphere, ameliorate salinity and buffer the pH, solubilize phosphates, and increase the fertilizer use efficiency of crop plants. The waterlogged conditions, high humidity, temperature, and shade provided by the paddy crop canopy, afford optimal conditions for rapid multiplication of cyanobacteria. Algalization improves water-holding capacity of soil, increases soil aggregation, and enhances the microbial activity (Goyal, 1982; Venkataraman, 1981). In addition, cyanobacteria improve soil health, benefit plant growth, increase crop yield, add organic matter, and modify soil texture.

Commercially, algal biofertilizers can be produced using selected, region-specific, autochthonous nitrogen-fixing cyanobacteria for the production of soil-based inoculum. For the commercial production, it is necessary to search for most suitable, easily available cheap carrier material that ensures high titer value and longer shelf life. The former method is an easy way to produce algal inoculum on small scale. The algae are grown in soil-lined open-air pits, with the dried algal flakes collected and applied in the field. The classical method of producing clay-based algal biofertilizer on commercial scale has been described, in which commonly available montmorillonite clay, locally known as *multani mitti*, has been used as carrier material (Goyal et al., 1997). Extensive field trials coordinated by the Indian Agricultural Research Institute (IARI), New Delhi, India, revealed that the algalization provides 25–30 kg N/ha/season and resulted in 30% increase in yield of paddy crop (Goyal et al., 1997; Venkataraman, 1981). The water fern (*Azolla*) that harbors the nitrogen-fixing blue-green algae *An. azollae* in its leaf cavities multiplies very fast in symbiotic association and provides biologically fixed nitrogen and organic matter. *Azolla* is also suitable for human consumption, as feed supplement for a variety of animals like fish, ducks, cattle, and poultry; as adjutant in biogas production; and for controlling weeds and mosquitoes under waterlogged conditions (Raja et al., 2012). In rice cultivation, it is used both as a green manure crop and a dual crop. *Azolla* is represented by seven species, namely, *Azolla caroliniana*, *Azolla filiculoides*, *Azolla mexicana*, *Azolla microphylla*, *Azolla nilotica*, *Azolla pinnata*, and *Azolla rubra*. Among them, *A. pinnata*, endemic to India, is highly susceptible to strong light and high temperature, not a fast grower and good nitrogen fixer. Attempts are now being made to introduce more efficient strains of *A. microphylla* and *A. mexicana*.

5.2.2 Phosphate-Solubilizing Bacteria

Phosphorus (P) is one of the most essential macronutrients needed for plant growth and development (Ehrlich, 1990). This nutrient tends to become unavailable to the crop plants as it readily gets immobilized in the soil as calcium phosphate. Soil harbors several microorganisms that can mobilize this insoluble phosphate. Bacteria like *Pseudomonas* sp. and *Bacillus megaterium* var. *phosphaticum* and fungi like *Aspergillus* and *Penicillium* are important phosphate solubilizers.

These microbes can be made to provide almost 20%–25% of phosphorus requirement of plants through artificial inoculation. Bacterial strains belonging to genera *Pseudomonas, Bacillus,* and *Rhizobium* also have the ability to solubilize insoluble inorganic phosphates such as tricalcium phosphate, dicalcium phosphate, hydroxylapatite, and rock phosphate. Tricalcium phosphate and hydroxylapatite seem to be more readily degradable than rock phosphate (Banerjee et al., 2006). Production of organic acids like gluconic acid appears to be the common pathway of phosphate solubilization.

5.2.3 Vesicular Arbuscular Mycorrhiza

Mycorrhiza is a mutualistic association between roots of higher plants and some specialized soil fungi. Their mycelia ramify through the soil and bioconcentrate plant nutrients in the rhizosphere. They are not involved in nitrogen fixation. They form a mantle around the roots and arbuscules inside the cell called endomycorrhiza or vesicular arbuscular mycorrhiza (VAM) that are grouped under order Glomales (Sathiyadash et al., 2010). The diagnostic feature of arbuscular mycorrhizae (AM) is the development of a highly branched arbuscule within root cortical cells. The fungus initially grows between cortical cells, soon penetrates the host cell wall, and grows within the cell. In this association, neither the fungal cell wall nor the host cell membrane is breached. As the fungus grows, the host cell membrane invaginates and envelops the fungus, creating a new compartment where material of high molecular complexity is deposited. This apoplastic space prevents direct contact between the plant and fungus cytoplasm and allows for efficient transfer of nutrients between the symbionts. Other structures produced by some AM fungi include vesicles, auxiliary cells, and asexual spores. Vesicles are thin-walled lipid-filled structures that are usually formed in intercellular spaces. The AM type of symbiosis is very common as the fungi involved can colonize a vast range of both herbaceous and woody plants, indicating a general lack of host specificity (Ricardo et al., 2011; Sathiyadash et al., 2010).

Most plants, especially the forest trees, depend on mycorrhizal association for adequate supply and uptake of nutrients like phosphorus, zinc, and other micronutrients. Other benefits include tolerance to drought, high soil temperature, soil toxins, extreme pH levels, as well as protection against root pathogens. Forest tree species are common hosts to AM fungi, but some of them also host dark septate endophytic fungi (Ricardo et al., 2011). In agriculture, they are best used for cereal crops; however, success of symbiosis is regulated by the soil conditions in addition to the type of cultivar and associating fungus (Uma et al., 2012). VAM fungi showed maximum association with roots of Indian rice cultivars (Bhattacharjee and Sharma, 2011). Synergistic effect of co-inoculation of VAM and PSB on maize was reported (Suri et al., 2011). Combined inoculation with VAM, *Azotobacter,* and *Pseudomonas* also benefitted biomass yield and nutrient content in *Stevia rebaudiana* Bert (Das et al., 2007).

5.2.4 Plant Growth–Promoting Rhizobacteria

Beneficial soil microbes present in the rhizosphere that enhance plant growth are called PGPR. Biofortification of wheat leads to the development of micronutrient-dense staple crops through inoculation of PGPR and cyanobacteria (Rana et al., 2012). Combined inoculation of earthworms and PGPR synergistically increased the concentration of N, P, and K in the soils (Wu et al., 2012). The effect of PGPR, *Burkholderia gladii* BA-7, *Bacillus subtilis* OSU-142, *B. megaterium* M-3, and *Azospirillum brasilense* sp. 245 on vegetative development and mineral uptake by grapevine (1103 P and 41 B) root stocks revealed that A. *brasilense* sp. 245 and *B. subtilis* OSU-142 were superior than other strains (Sabir et al., 2011). PGPR in the presence of compost performed better than all other treatments and favorably influenced the microbial community in the rhizosphere of forage corn (Piromyou et al., 2011). Co-inoculation of PGPR, *B. subtilis* SU47, and *Arthrobacter* sp. SU 18 could alleviate the adverse effect of soil salinity on wheat growth (Upadhyay et al., 2011). Rhizobial isolates from root nodules of fenugreek (*Trigonella foenumgraecum*) exhibited plant growth–promotory traits with properties of biocontrol agents against *Fusarium oxysporum* (Kumar et al., 2011). Several strains of nitrogen-fixing bacilli, mainly *Bacillus* and *Paenibacillus*, were isolated from rice fields in South Brazil, which displayed important PGP characteristics that could significantly increase the crop yield (Beneduzi et al., 2008).

5.2.5 Microbial Consortia

Early work on free-living bacteria in soil indicated that certain strains, when applied to seeds or roots, may benefit crops by stimulating plant growth or by reducing the damage from soilborne plant pathogens. Free-living bacteria may also influence the symbiosis between microorganisms and plants and thereby stimulate plant growth indirectly. PGPR may increase plant growth through nitrogen fixing, increasing the availability of nutrients in the rhizosphere, positively influencing root morphology and development, and promoting other beneficial plant–microbe interactions (Vessey, 2003). A range of diazotrophic PGPR participate in interaction with C3 and C4 crop plants like rice, wheat, maize, sugarcane, and cotton and significantly increase their growth and grain yield (Kennedy et al., 2004). PGPRs increase plant growth indirectly by preventing the deleterious effects of phytopathogens and directly by providing plants with assimilable compounds and facilitating the uptake of nutrients (Glick, 1995).

IAA production by the indigenous isolates of *Azotobacter* and fluorescent *Pseudomonas* in the presence and absence of tryptophan was studied by Khan et al. (2004). They screened the strains for their intrinsic ability to produce IAA in the presence of varying amounts of L-tryptophan in terms of root elongation of germinating seeds. They also reported that inoculation of wheat seedlings with *A. brasilense* increased the number and length of lateral roots. Inoculation of canola seeds with *Pseudomonas putida* GR12-2 resulted in two- or threefold increase in the length of seedling roots.

Co-inoculation with PGPR and an AM fungus had synergistic effect on various growth parameters of wheat (Kennedy et al., 2004; Khan and Zaidi, 2007). Numerous studies have shown a substantial increase in plant growth and seed yield following inoculation with PGPR strains including PSMs and nitrogen fixers on legumes (Perveen et al., 2002). Under phosphorus deficiency, plants inoculated with AM fungi either alone or in combination with PSB increased phosphorus uptake in wheat (Raja et al., 2002) and maize (Evans and Miller, 1990). Increase in dry matter accumulation, grain yield, and phosphorus uptake by wheat plants was observed due to synergism between PSB and free-living nitrogen-fixing bacteria. Similarly, a significant increase in the dry matter yield of wheat plants with co-inoculation of rock phosphate–solubilizing fungi *Aspergillus niger, Penicillium citrinum,* and *Glomus constrictum* was reported (Omar, 1998). During interaction, the PSB increase the availability of phosphorus that in turn promotes nitrogen fixation and release of PGP substances by *Azotobacter* (Kucey et al., 1989).

The efficiency of microbial inoculants is a direct manifestation of their establishment in the niche they are placed. To monitor their colonization efficiency, the most common method employed is tracking of the phenotype of the marker gene. Antibiotic-resistant genes, such as the *nptII* gene encoding resistance to kanamycin, were the first to be used as markers. Genes encoding metabolic enzymes have also been used as nonselective markers. These include *lacZY* (encoding β-galactosidase and lactose permease) and *gusA* encoding β-glucuronidase (GUS). The enzymes encoded by *lacZ* and *gusA* cleave the uncolored substrates X-gal or X-glc, respectively, to produce blue-colored products. The difference in this method is that it does not rely on incorporation of an inhibitory compound to the agar medium. The *lacZ* has been used to study nodule infection by *Rhizobium* and root colonization by *Azospirillum* (Katipitiya et al., 1995). Genetic engineering was employed to improve the nitrogen biofertilizer potential of *Anabaena* sp. strain PCC7120, which was found to be stable, eco-friendly, and useful for environmental application as nitrogen biofertilizer in paddy fields (Chaurasia and Apte, 2011). The rapid and simple polymerase chain reaction (PCR) method was developed as a useful tool for detecting and identifying a variety of indigenous *Azospirillum* within populations of rhizosphere bacteria from agricultural samples (Kochar and Shrivastava, 2012).

5.3 Benefits of Biofertilizers

With ever-increasing population, it is necessary to increase crop productivity per unit area without causing further deterioration of soil health and maintaining crop–nutrient balance. Biofertilizers have an edge over the chemical fertilizers in providing a linear supply of nutrient as compared to the cyclic availability from the latter. They are widely accepted as cost-effective supplements and substitutes to chemical fertilizers that do not adversely impact the environment. Since the biofertilizers are healthy populations of living microbes, they proliferate in the soil and lead to "population buildup" ensuring sustained additive effect of superimposed inoculation (Roy et al., 2006; Tandon, 1992). The symbiotic association of *Rhizobium* and legume plants has been shown to have residual benefit on succeeding crop. The benefits conferred by the biofertilizers (Motsara et al., 1995; SubbaRao, 1993) are (i) reduced use of energy-intensive chemical fertilizers without affecting the overall crop production; (ii) improvement in physical and chemical properties of soil through addition of organic matter; (iii) better utilization of the applied chemical nutrients; (iv) prevention of loss of nutrients through percolation, runoff, and immobilization; (v) bringing about linear natural soil fertilization; (vi) maintenance of soil health; (vii) population buildup ensuring sustained effect of superimposed inoculation; (viii) mobilization of crop nutrients; (ix) secretion of growth regulators that enhances nutrient-utilizing efficiency of the crop plants; and (x) residual effect on the succeeding crop. Some of the biofertilizers such as *Azolla* can be produced at the farmer's field itself and serve as green manure (Singh 1989).

Several formulations of microbial inoculants using different carrier materials like agar, granular soil, porous gypsum, peat granules, kaolin and montmorillonite clay, farmyard manure granules and broth, frozen concentrates, and freeze-dried inoculants are now commercially available (Motsara et al., 1995). The choice and demand of the microbial inoculants depend upon the crop and cropping pattern of the area. Since a variety of benefits accrue from biofertilizers, the concept of using a consortia of microbial inoculants comprising compatible microbes is a more practical approach.

5.4 Production Technology

The production technology comprises product development and process optimization. While process optimization revolves around maximizing production and cutting input costs without compromising quality, product development assumes vital importance as it provides promising and domesticated microbial strains and recipe of growth medium and conditions supporting optimum production. Biofertilizer production needs to be supported by a very strong research and development unit that continues to look for super strains, acts as a depository of microbes, and provides pure and actively growing scaled-up inoculum for large-scale production (Goyal, 2001).

5.4.1 Raw Material and Equipment

Manufacturing involves selection of suitable strain, mass multiplication, mixing of the culture with carrier material, and packing in polythene/HDPE bags and storage in the dark at a low temperature. Mother cultures (*Rhizobium, Azotobacter, Pseudomonas,* etc.), carrier material (lignite, peat, or charcoal in powder form, etc.) that supports the growth of microorganisms, suitable growth media (Tables 5.2 and 5.3), and all equipments are required for any basic microbiological work including fermenter for industrial-scale production.

5.4.2 Selection of Microbial Strain

Pure cultures of various microbes employed as biofertilizers are maintained at many agricultural institutes/universities such as IARI, New Delhi; Regional Biofertilizer Development and Production Centres, Ghaziabad; and Microbial Type Culture Collection (MTCC), Chandigarh. There are international culture collection centers like American Type Culture Collection (ATCC), Texas; Nitrogen Fixation by Tropical Agricultural Legumes (NifTAL), University of Hawaii; and International Rice Research Institute (IRRI), Manila, Philippines, which also maintain the culture to be used as biofertilizers. Authentic culture of desired strain can be purchased from these identified sources. The cultures are maintained in axenic state and periodically subcultured adopting standard techniques. It is always desirable for a biofertilizer production unit to develop its own region- and crop-specific strains to ensure long-term reliability and availability of efficient cultures.

5.4.3 Preparation of Mother Culture

An actively growing pure population of any organism on agar slant or in broth is called stock culture. The mother culture is prepared from the stock culture in a conical flask of 500 or 1000 mL capacity. The conical flasks filled to half of their capacity with the appropriate medium (Jensen's broth for *Azotobacter* or Pikovskaya's broth for *Pseudomonas*) are plugged with nonabsorbent cotton and sterilized in an autoclave for 15–20 min at 15 lbs pressure at 121°C temperature. The flasks are cooled to room temperature and are aseptically inoculated with the mother culture. The flasks are incubated on a shaker for 72–90 h so as to get the optimum growth of the microbial culture. This mother culture is used for inoculating fermenter or flasks of larger capacity for mass multiplication.

TABLE 5.2

Media Composition (g/L) for Production of Microbial Inoculants

Azotobacter		*Pseudomonas*		*Rhizobium*		BGA		*Azospirillum*	
Jensen's Medium (Jensen, 1954)		**Pikovskaya Medium (Pikovskaya, 1948)**		**CRYEMA Medium (Fred et al., 1932)**		**BG-11 (+N/–N) Medium (Stanier et al., 1971)**		**Malate Medium (Okon et al., 1977)**	
Sucrose	20.0	Glucose	10.0	Mannitol	10.0	K_2HPO_4	0.04	Malic acid	5.0
K_2HPO_4	1.0	Tricalcium phosphate	5.0	K_2HPO_4	0.4	$CaCl_2 \cdot H_2O$	0.036	K_2HPO_4	0.5
$MgSO_4 \cdot 7H_2O$	0.5	$(NH_4)_2SO_4$	0.5	$MgSO_4 \cdot 7H_2O$	0.2	Citric acid	0.006	$MgSO_4 \cdot 7H_2O$	0.2
NaCl	0.2	NaCl	0.2	NaCl	0.1	Ferric ammonium citrate	0.006	NaCl	0.1
$CaCO_3$	2.0	$MgSO_4 \cdot 7H_2O$	0.1	$CaCO_3$	3.0	EDTA	0.001	$CaCl_2$	0.02
$FeSO_4 \cdot 7H_2O$	0.1	KCl	0.2	Yeast extract	0.5	Sodium carbonate	0.02	$Na_2MoO_4 \cdot 2H_2O$	0.002
		Yeast extract	0.5	Agar	15.0	Sodium nitrate	1.5	$MnSO_4 \cdot H_2O$	0.01
						Trace metal mix	1 mL	KOH	4.5
								Biotin	0.1 mg
								Fe-EDTA (1.64%)	4 mL
								Bromothymol blue	3 mL (0.5% alcoholic solution)
pH	7.5	7–7.2		7.0		7–7.2		6.8	

TABLE 5.3

Composition of the Trace Metal Mix

Constituents	Quantity (g/L)
Boric acid	2.86
Manganese chloride	1.81
Zinc sulfate	0.222
Sodium molybdate	0.039
Copper sulfate	0.079
Cobalt nitrate	0.0492

5.4.4 Mass Multiplication

Azotobacter and *Pseudomonas* can be multiplied on large scale either in a fermenter or in conical flasks. The fermenter is a vessel of varying capacity where growth conditions are controlled automatically. It is used for rapid and more efficient production of bacterial cultures on large scale under completely controlled conditions. In this method, the medium is taken in a fermenter and sterilized. The pH of the medium is adjusted to the desired level and inoculated using 1% liquid mother culture. Optimum growth is ensured by regulating temperature, nutrient concentration, pH, and oxygen supply. Alternatively, shake culture method can be used for small-scale production. The shake cultures are raised in a way similar to the mother cultures except the volume of the culture (Motsara et al., 1995).

5.4.5 Mixing of Culture with Carrier Material and Packing

The carrier materials that are used for bacterial cultures are easily available and less bulky, have high organic matter content and water-holding capacity, and support the growth of the organism. Peat is reported to be the best carrier material for the bacterial culture. In India, despite being costly and not being easily available, lignite is extensively used to prepare bacterial inoculants, as it is cheaper, has larger particle size, enables higher titer value, and confers longer shelf life. The carrier materials are autoclaved to sterilize prior to mixing with culture. The well-grown broth culture is added to powdered lignite or charcoal in 1:3 ratio and mixed thoroughly. The lignite loaded with the microbial culture is packed in polyethylene bags and sealed. The culture packets are perforated with sterilized needle for aeration to maintain them in live form as well as for the long shelf life of the inoculants. These packets are stored in a cool and dry place away from agrochemicals. If stored at 15°C–20°C, they retain the desired titer value for 6 months, and at 0°C–4°C, their shelf life can be extended up to 2 years (Motsara et al., 1995). The flowchart for the production of biofertilizer is summarized in Figure 5.1.

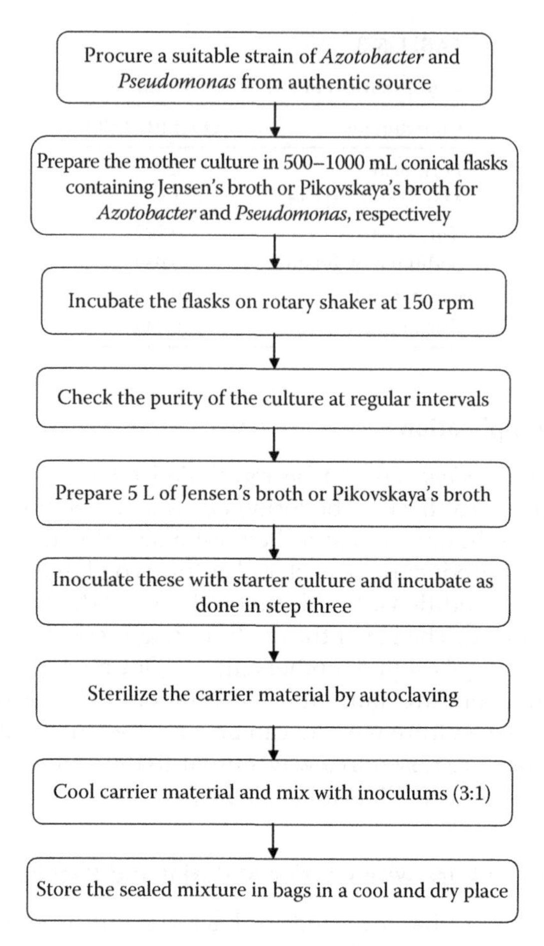

FIGURE 5.1
Flowchart of production of bacterial biofertilizer.

5.4.6 Quality Control

Quality of the microbial inoculants has to be controlled at all stages of production starting from culture or broth culture, production and mixing of broth with carrier, carrier selection, and processing, packaging, sealing, storage, distribution, and marketing to ensure their quality (Table 5.4). As per the Fertilizer Control Order 1985 No. 11-3/83-STU of the Government of India, tolerance limit for bacterial biofertilizers is 1×10^7 CFU/g of carrier material in the form of powder or granules or 5×10^7 CFU/mL in the case of liquid inoculums (Motsara et al., 1995). All over the world, product quality has been identified as the key issue and has set parameters and guidelines for their manufacturing, testing, and quality assurance including bioefficacy before supply to farmers for field application.

TABLE 5.4

Quality Specifications for Major Biofertilizers as per Fertilizer Control Order 1985 No. 11-3/83-STU of the Government of India

	Nitrogen-Fixing Bacteria			Phosphate Solubilizers
Parameters	***Azotobacter***	***Rhizobium***	***Azospirillum***	**PSB**
Base	Carrier-based in the form of moist/dry powder or granules, or liquid-based			
Particle size in case of carrier[a]-based material	All materials shall pass through 0.15–0.212 mm IS sieve.			
Viable count colony-forming units (CFU)	CFU minimum 5×10^7 cell/g of powder/granule or carrier material or 1×10^8 cells/mL of liquid			
pH	6.5–7.5 for moist/dry powder-granulated carrier-based and 5.0–7.5 for liquid-based, especially for PSB			
Contamination level	No contamination at 10^5 dilution			
Moisture % by weight (maximum in case of carrier based)	30–40			
Efficiency character	The strain should be capable of fixing at least 10 mg of nitrogen per gram of sucrose consumed	Should show effective nodulation on all the species listed on the packet	Formation of white pellicle in semisolid nitrogen-free bromothymol blue media	The strain should have phosphate-solubilizing capacity in the range of minimum 30%, when tested spectrophotometrically in terms of zone formation, minimum 5 mm solubilization zone in prescribed media having at least 3 mm thickness

[a] Type of carrier: The carrier material such as peat, lignite, peat soil, humus, wood charcoal, or similar material favoring growth of the organism.

5.5 Global Status

Biofertilizers hold promise as they ensure consistent supply of crop nutrients, nurture the soil, and are cost-effective and eco-friendly (Roy et al., 2006). Marginal saving in input cost and appreciable increase in production are a boon for the farmers. Sustained food security can be ensured only if the farmers get the cheap and easily manageable farm inputs. Biofertilizers have been shown to provide a proxy of 15%–25% nitrogenous fertilizers. On a conservative estimate, every 10% saving through the use of biofertilizers is expected to result in an annual saving of 1.0 MT of nitrogenous fertilizers (Goyal, 1982). Most of the production of biofertilizers is being done in public sector by research institutions, universities, and few state and cooperative fertilizer units. During the last 50 years, remarkable growth in the application of biofertilizers has taken place all over the world. More than 25 countries including the United States, United Kingdom, Canada, Russia, Japan, China, Brazil, Thailand, Uganda, Zimbabwe, and India have realized their potential for sustainable agriculture, food security, and environmental benefits. As a result, there are considerable investments in research, field trials, production, and applications of biofertilizers on a large scale (IPNI, 2011). Surprisingly, the involvement of private sector is extremely limited in spite of it being a low-investment and high-benefit technology. The status of usage, knowledge base, and outlook for biofertilizers in various regions of world has, however, been quite different. The market has grown steadily throughout the world in the past couple of years, with rise in the number of biofertilizer producers. Consistent decline in the quality of agricultural produce, increase in the level of contamination of ground water, and the deterioration of soil health has also lead to increase in demand for bioinoculants, and there is a growing tendency to shift from chemical-based agriculture to organic agriculture. Biofertilizer firms all over the world perceive a huge demand in the near future in the wake of massive organic farming initiative (IPNI, 2011). The utilization of fertilizers is also not uniform and seems to be directly proportional to the size of the land holdings. In India, about 68 million farmers owning 65% of the total arable land use only 30% of the total fertilizers consumed in the country. In India, there are nearly 60 production units of biofertilizers with present-day annual production estimated to be slightly more than 25,000 tons, which has grown four times in the past decade and is far below the actual demand. Consumption of biofertilizers in North Eastern states in India has risen by 50% during the last 15 years. In Brazil, around 65,000 tons of biofertilizers are utilized annually for crops such as beans, maize, rice, sugarcane, soybean, forage crops, and vegetables. In North America, Argentina, Paraguay, Bolivia, and Uruguay, most of the soybean crops are inoculated with *Bradyrhizobium* inoculants in nearly 70% of the area. PGPR such as *Azospirillum* and *Pseudomonas* have also been introduced as bioinoculants (IPNI, 2011). China has over 300 biofertilizer-producing enterprises with an annual production of half a million tons and the application area touching to 167 million hectares. However, most satisfactory usage has been of *Rhizobium* inoculants. PSB have gained more importance in phosphorus nutrition in Russia, wherein microbial consortia provide both plant protection and phosphorus nutrition to the crops (Zhigletsova et al., 2010). Some of the industries involved in production of biofertilizers in India and other countries are listed in Table 5.5.

TABLE 5.5

Commercial Scale Producers of Biofertilizers in India and Other Countries

Company Name	Product Description
Agrilife, Hyderabad, India	Develop and produce microbial fertilizer like Agri VAM® and Agri Life NITROFIX™
Bodisen Biotech Inc., Shaanxi, China	Develop and produce biofertilizers
Crop Services International, Michigan	Develop and produce microbial fertilizer like Myco Seed Treat
Cropmaster, Lake Panasoffkee	Develop and produce microbial fertilizer like NITRO-FIX™
Earthcrew Inc., Oregon	Develop and produce microbial fertilizer products like Soluble MAXX (mycorrhizal-based)
Geo Care Biotech Ltd., Chaojhou Township, Taiwan	Develop and produce microbial fertilizer like organic fermented liquid fertilizer (SuperFeed-A)
Green Planet Bio Products, Punjab, India	Develop and produce microbial fertilizer like Powerplant Grow, Powerplant Boom
Greenview Biotech, Taipei Hsien, Taiwan	Develop and produce microbial fertilizer and enzymes like Sell Enzeme
Gujarat Life Sciences Pvt. Ltd., India	Develop and produce fertilizers like Wonderlife and Superlife
Gujarat State Fertilizers & Chemicals (GSFC) Ltd., India	Develop and produce biofertilizer using *Azotobacter*, *Azospirillum* culture, etc.
Taiyen Biotech Co. Ltd., Taiwan	Develop and produce microbial fertilizer
Yan Ten Biotech Corp., Chiayi, Taiwan	Develop and produce microbial fertilizer like biological organic fertilizer and biological organic liquid fertilizer
Yuen-Foong Yu Bio-Tech Co. Ltd., Taiwan	R&D microorganism application and biofertilizer

5.6 Constraints

There exists a large consumer base, which is already aware of the biofertilizers and their advantages over the chemical fertilizers. However, the quantum of their application worldwide is negligible. An aggressive popularization campaign is required to accelerate their market penetration and preference for usage by the farmers over their chemical counterparts. This will be further facilitated if various government agencies, industrial units, and public and private research and development institutions join hands to popularize economic and ecological benefits accruing from the use of biofertilizers (IPNI, 2011). A focused approach toward users' awareness, quality control mechanisms, technical and fiscal support to small-scale businesses along with the necessary research and development support is required to promote development in biofertilizer production technology (Goyal, 2001).

Successful implementation and acceptance of biofertilizers at large scale will depend on timely availability of high-quality products with sufficient shelf life, application at proper and precise time, and use of specified equipments, supported by appropriate education and orientation of trainers, distributors, and users (IPNI, 2011). Production facilities and related technology and selection of cost-effective carrier material and additives need to be improved. More field trials are required for validation and to build the confidence of end users. The production and applications of biofertilizers can be further improved if the user's requirements are properly addressed. Since varied response of biofertilizers is observed in different regions owing to different geo-climatic conditions, there is a need to develop specific products suitable for different crops and soil types. Worldwide governmental funding agencies are continuing to provide large grants for further research and development of biofertilizer technology due to all associated constraints (IPNI, 2011).

5.7 Summary and Future Prospects

The use of microbial inoculants in agriculture has the potential of providing benefits in terms of environmental protection, conservation of nonrenewable resources, and improved food quality. Gradual shift from chemical agriculture by striking a balance between organic and inorganic inputs can restore nutrient balance in the soil, protect soil fertility, prevent soil deterioration, and ensure crop productivity and sustainable production. Since a variety of benefits accrue from biofertilizers, the concept of using consortia of microbial inoculants comprising compatible microbes will be a more practical approach.

Increasingly, it is becoming apparent that the best solution for tracking a microbe is to use more than one marker simultaneously. Molecular tagging of inoculant strains such as *Rhizobium*, *Bradyrhizobium*, *Azotobacter*, *Azospirillum*, *Bacillus polymyxa*, and *Pseudomonas striata* is essential for ecological monitoring and to assess their performance in the field.

Growing micronutrient deficiencies in soil and dwindling mineral resources have necessitated the need to look for alternative cheap source of similar carrier material that supports the growth of microorganism and simultaneously provides certain micronutrients. Fly ash, a resultant of combustion of coal at high temperature, has been regarded as a resource material for soil amendment (Jala and Goyal, 2006) and as a carrier for different types of microbial inoculants (Gaind and Gaur, 2003). In conjunction with organic manure and microbial inoculants, fly ash can enhance plant biomass production. Inherent nutrient status of soils is low and the finite nutrient resources of these soils have been exhausted at a faster rate since the ushering of the green revolution. It is a challenging task for converting intensive farming by sustainable agriculture. Concerted efforts are required to generate public awareness, strict quality control, technical and fiscal support to small-scale manufacturers, and R&D support to commercialize the concept with a focus on building farmers' confidence.

References

Anjum, M.A., Sajjad, M.R., Akhtar, N. et al. 2007. Response of cotton to plant growth promoting rhizobacteria (PGPR) inoculation under different levels of nitrogen. *Journal of Agricultural Research* 45: 135–143.

Banerjee, M.R., Yesmin, L., and Vessey, J.K. 2006. Plant growth promoting rhizobacteria as biofertilizers and biopesticides. In: *Handbook of Microbial Biofertilizers*, ed. M.K. Rai. Haworth Press, New York.

Beneduzi, A., Peres, D., Vargas, L.K., Bodanese-Zanettini, M.H., and Passaglia, L.M.P. 2008. Evaluation of genetic diversity and plant growth promoting activities of nitrogen-fixing bacilli isolated from rice fields in South Brazil. *Applied Soil Biology* 39: 311–320.

Bhattacharjee, S. and Sharma, G.D. 2011. The vesicular arbuscular mycorrhiza associated with three cultivars of rice (*Oryza sativa* L.). *Indian Journal of Microbiology* 51: 377–383.

Chaurasia, A.K. and Apte, S.K. 2011. Improved eco-friendly recombinant *Anabaena* sp. strain PCC7120 with enhanced nitrogen biofertilizer potential. *Applied Environmental Microbiology* 77: 395–399.

Das, K., Dang, R., Shivananda, T.N., and Sekeroglu, N. 2007. Influence of biofertilizers on the biomass yield and the nutrient content in *Stevia rebaundiana* Bert. grown in Indian subtropics. *Journal of Medicinal Plant Research* 1: 5–8.

Dekhane, S.S., Khafi, H.R., Raj, A.D., and Parmar, R.M. 2012. Effect of biofertilizer and fertility levels on yield, protein content and nutrient uptake of cow pea (*Vigna unguiculata* L.). *Legume Research* 34: 51–54.

Ehrlich, H.L. 1990. *Geomicrobiology*, 2nd edition. Dekker, New York.

Eleiwa, M.E., Hamed, E.R., and Shehata, H.S. 2012. Biofertilizers and/or some micronutrients role on wheat plants grown on newly reclaimed soil. *Journal of Medicinal Plants Research* 6: 3359–3369.

Evans, A.C. and Miller, M.H. 1990. The role of external mycelial network in the effect of soil disturbance upon vesicular arbuscular mycorrhizal colonization in maize. *New Phytologist* 114: 65–71.

Fasciglione, G., Casnovas, E.M., Yommi, A., Sueldo, R.J., and Barassi, C.A. 2011. *Azospirillum* improves lettuce growth and transplant under saline conditions. *Journal of the Science of Food and Agriculture* 92: 2518–2523.

Fibach-Paldi, S., Burdman, S., and Okon, Y. 2012. Key physiological properties contributing to rhizosphere adaptation and plant growth promotion abilities of *Azospirillum brasilense*. *FEMS Microbiology Letters* 326: 99–108.

Fred, E.B., Baldwin, I.L., and Mc Coy, F. 1932. Root Nodule Bacteria and Leguminous Plants with Supplement. University of Wisconsin Studies in Science, vol. 5, Madison, WI, 343 pp.

Gaind, S. and Gaur, A.C. 2003. Evaluation of fly ash as a carrier for diazotrophs and phosphobacteria. *Bioresource Technology* 95: 187–190.
Glick, B.R. 1995. The enhancement of plant growth by free-living bacteria. *Canadian Journal of Microbiology* 41: 109–117.
Goyal, S.K. 1982. Algal biofertilizer for vital soil and free nitrogen. *Proceedings of Indian National Science Academy* 59: 295–302.
Goyal, D. 2001. Biofertilizers: Exploitation potential and commercialization prospects. In: *Business Opportunities in Biotechnology*, ed. S. Tiwari, 79–87. Confederation of Indian Industry (CII), New Delhi, India.
Goyal, S.K., Singh, B.V., Nagpal, V., and Marwaha, T.S., 1997. An improved method for production of algal biofertilizer. *Indian Journal of Agricultural Science* 67: 314.
Hegazi, NA., Faye, M., Amin, G. et al. 1998. Diazotrophs associated with non-legumes grown in sandy soil. In: *Nitrogen Fixation with Non-Legumes*, eds. K.A. Malik, M.S. Mirza, and L.K. Ladha, 209–222. Kluwer, Dordrecht, the Netherlands.
IPNI Biofertilizers Report. 2011. A summary of the status of biofertilizers. www.ipni.net/article/IPNI-3215 (accessed September 12, 2012).
Ivanov, S., Fedorova, E.E., Limpens, E. et al. 2012. Rhizobium–legume symbiosis shares an exocytotic pathway required for arbuscular formation. *Proceedings of the National Academy of Sciences U S A* 109: 21.
Jala, S. and Goyal, D. 2006. Fly ash as a soil ameliorant for improving crop production—A review. *Bioresource Technology* 97: 1136–1147.
Jensen, H.L. 1954. The Azotobacteriaceae. *Bacteriological Review* 18: 195–214.
Katipitiya, S., New, P.B., Elmerich, C., and Kennedy, I.R. 1995. Improved N_2 fixation in 2,4 D treated wheat roots associated with *A. lipoferum*: Studies of colonization using reporter genes. *Soil Biology Biochemistry* 27: 447–452.
Kennedy, I.R., Choudhury, A.I.M.A., and Kecske, M.L. 2004. Non-symbiotic bacterial diazotrophs in crop-farming systems: Can their potential for plant growth promotion be better exploited? *Soil Biology Biochemistry* 36: 1229–1244.
Khan, M.S., Ahmad, F., and Ahmad, I. 2004. Indole acetic acid production by the indigenous isolates of *Azotobacter* and fluorescent *Pseudomonas* in the presence and absence of tryptophan. *Turkish Journal of Biology* 29: 29–34.
Khan, M.S. and Zaidi, A. 2007. Synergistic effect of inoculum with plant growth-promoting rhizobacteria and an arbuscular mycorrhizal fungus on the performance of wheat. *Turkish Journal of Agriculture* 31: 355–362.
Kochar, M. and Shrivastava, S. 2012. Surface colonization by *Azospirillum brasilense* SM in the indole-3-acetic acid dependent growth improvement of *Sorghum*. *Journal of Basic Microbiology* 52: 123–131.
Kucey, R.M.N., Janzen, H.H., and Leggett, M.E. 1989. Microbiology mediated increases in plant available phosphorous. *Advanced Agronomy* 42: 199–228.
Kumar, H., Dubey, R.C., and Maheshwari, D.K. 2011. Effect of plant growth promoting rhizobial on seed germination, growth promotion and suppression of *Fusarium* wilt of fenugreek (*Trigonella foenum-graecum* L.). *Crop Protection* 30: 1396–1403.
Mehta, R.S., Godara, A.S., and Meena B.S. 2011. Effect of nitrogen, phosphorus and biofertilizer levels on yield attributes, yield and economics of fenugreek (*Trigonella foenum-graecum* L.). *Progressive Horticulture* 43: 271–275.
Motsara, M.R., Bhattacharya, P., and Srivastava, B. 1995. *Biofertilizer Technology: Marketing and Usage—A Source Book-Cum-Glossary*. FDCO, New Delhi, India.
Okon, Y. 1985. *Azospirillum* as a potential inoculant for agriculture. *Trends in Biotechnology* 3: 223–228.

Okon, Y., Albrecht, S.L., and Burris, R.H. 1977. Methods for growing *Spirillum lipoferum* and for counting it in pure culture and in association with plants. *Applied Environmental Microbiology* 33: 85–87.

Omar, S.A. 1998. The role of rock-phosphate-solubilizing fungi and vascular arbuscular mycorrhiza (VAM) in growth of wheat plants fertilized with rock phosphate. *World Journal Microbiology and Biotechnology* 14: 211–218.

Perveen, S., Khan., M.S., and Zaidi, A. 2002. Effect of rhizospheric microorganisms on growth and yield of greengram (*Phaseolus radiatus*). *Indian Journal of Agricultural Science* 72: 421–423.

Pikovskaya, R.I. 1948. Mobilization of phosphorous in soil in connection with the vital activity of some microbial species. *Mikrobiologiya* 17: 362–370.

Piromyou, P., Buranabanyat, B., Tantasawat, P., Tittabutr, P., Boonkerd, N., and Teaumroong, N. 2011. Effect of plant growth promoting rhizobacteria (PGPR) inoculation on microbial community structure in rhizosphere of forage corn cultivated in Thailand. *European journal of Soil Biology* 47: 44–54.

Postgate, J. 1998. *Nitrogen Fixation*, 3rd edition. Cambridge University Press, Cambridge, UK.

Prasanna, R. and Nayak, S. 2007. Influence of diverse rice soil ecologies on cyanobacterial diversity and abundance. *Wetlands Ecology Management* 15: 127–134.

Raja, W., Rathaur, P., John, S.A., and Ramteke, P.W. 2012. *Azolla*: An aquatic pteridophyte with great potential. *International Journal of Research in Biological Sciences* 2: 68–72.

Raja, A.R., Shah, K.H., Aslam, M., and Memon, M.Y. 2002. Response of phosphobacterial and mycorrhizal inoculation in wheat. *Asian Journal of Plant Science* 1: 322–323.

Ramirez-Bahena, M.H., Garcia-Fraile, P., Peix, A. et al. 2008. Revision of the taxonomic status of the species *Rhizobium leguminosarum* (Frank 1879) Frank 1889^{AL}, *R. phaseoli* Dangeard 1926^{AL} and *R. trifolii* Dangeard 1926^{AL}. *R. trifolii* is a later synonym of *R. leguminosarum*. Reclassification of the strain *R. leguminosarum* DSM 30132(NCIMB 11478) as *R. pisi* sp. nov. *International Journal of Systematic and Evolutionary Microbiology* 58: 2484–2490.

Rana, A., Joshi, M., Prasanna, R., Shivay, Y.S., and Nain, L. 2012. Biofortification of wheat through inoculation of plant growth promoting rhizobacteria and cyanobacteria. *European Journal of Soil Biology* 50: 118–126.

Ricardo, A., Peraza, H., and Hamel, C. 2011. Soil–strain compatibility: The key to effective use of arbuscular mycorrhizal inoculants? *Mycorrhiza* 21: 183–193.

Rogar, P.A. and Watanabe, I. 1986. Technologies for utilizing biological NF in wetland rice: Potentialities, current usage and limiting factors. *Fertilizer Research* 9: 39–77.

Roy, R.N., Finck A., Blair G.J., and Tandon, H.L.S. 2006. *Plant Nutrition for Food Security: A Guide for Integrated Nutrient Management*. FAO, Rome, Italy.

Sabir, A., Yazici., Kara, Z., and Sahin, F. 2011. Growth and mineral acquisition response of grapevine rootstocks (*Vitis* spp.) to inoculation with different strains of plant growth-promoting rhizobacteria (PGPR). *Journal of the Science of Food and Agriculture* 92: 2148–2153.

Sathiyadash, K., Muthukumar, T., and Uma, E. 2010. Arbuscular mycorrhizal and dark septate fungal associations in South Indian grasses. *Symbiosis* 52: 21–32.

Shime-Hattori, S., Kobayashi, S., Ikeda, S., Asano, R., Shime, H., and Shinano T. 2011. A rapid and simple PCR method for identifying isolates of the genus *Azospirillum* within the populations of rhizosphere bacteria. *Journal of Applied Microbiology* 111: 915–924.

Singh, P.K. 1989. Use of *Azolla* in Asian agriculture. *Applied Agricultural Research* 4: 149–161.

Stanier, R.Y., Kunisawa, R., Mandal, M., and Cohen-Bazire, G. 1971. Purification and properties of unicellular BGA (order Croococcales). *Bacteroid Review* 35: 171–205.

Stewart, W.D.P. 1975. *Nitrogen Fixation by Free-Living Microorganisms*. Cambridge University Press, Cambridge.

SubbaRao, N.S. 1993. *Biofertilizers in Agriculture and Forestry*. Oxford and IBH, New Delhi, India.

Suri, V.K., Choudhary, A.K., Chander, G., Verma, T.S., Gupta, M.K., and Dutt, N. 2011. Improving phosphorus use through co-inoculation of vesicular arbuscular mycorrhizal fungi and phosphate-solubilizing bacteria in maize in an acidic alfisol. *Communication in Soil Science and Plant Analysis* 42: 2265–2273.

Tandon, H.L.S. 1992. *Fertilizers, Organic Manures, Recyclable Waste and Biofertilizers*. FDCO, New Delhi, India.

Terpolilli, J.J., Hood, G.A., and Poole, P.S. 2012. What determines the efficiency of N_2-fixing *Rhizobium*–legume symbioses? In: *Advances in Microbial Physiology*, ed. R.K. Poole, Vol. 60, 325–389. Academic Press, London.

Uma, E., Sathiyadash, K., Loganathan, J. and Muthukumar, T. 2012. Tree species as hosts for arbuscular mycorrhizal and dark septate endophyte fungi. *Journal of Forestry Research* 23: 641–649.

Upadhyay, S.K., Singh, J.S., Saxena, A.K., and Singh, D.P. 2011. Impact of PGPR inoculation on growth and antioxidant status of wheat under saline conditions. *Plant Biology* 14: 605–611.

Venkataraman, G.S. 1981. *Blue-green Algae for Rice Production—A Manual for Its Promotion*. FAO Soils Bulletin No. 46. FAO, Rome, Italy.

Vessey, J.K. 2003. Plant growth promoting rhizobacteria as biofertilizers. *Plant and Soil* 255: 571–586.

Wolfe, D.W. 2001. *Tales from the Underground: A Natural History of Subterranean Life*. Perseus Publisher, Cambridge, MA.

Wu, F., Wan, J.H.C., Wu, S., and Wong, M. 2012. Effects of earthworms and plant growth promoting rhizobacteria (PGPR) on availability of nitrogen, phosphorus, and potassium in soil. *Journal of Plant Nutrition and Soil Science* 175: 423–433.

Yanni, Y.G. and Dazzo, F.B. 2010. Enhancement of rice production using endophytic strains of *Rhizobium leguminosarum* bv. *trifolii* in extensive field inoculation trials within Egypt Nile delta. *Plant and Soil* 336: 129–142.

Yanni, Y.G. and El-Fattah, F.K.A. 1999. Towards integrated biofertilization management with free living and associative dinitrogen fixers for enhancing rice performance in the Nile delta. *Symbiosis* 27: 319–331.

Zhigletsova, S.K., Dunajtsev, I.A., and Besaeva, S.G. 2010. Possibility of application of microorganisms for solving problems of ecological and food safety. *Agrochemistry* N6: 83–96.

6

Production of Biopesticides

Surinder Kaur, Gurpreet S. Dhillon,
Satinder K. Brar, and Ramesh Chand

CONTENTS

6.1 Introduction 191
6.2 Technology of Biopesticide Production 195
6.2.1 Selection of Strain 195
6.2.2 Production Technology 199
6.2.3 Downstream Processing 200
6.3 Mode of Action of Crystal Proteins 205
6.4 Marketing Constraints/Challenges in Biopesticide Development 206
6.5 Advantages of Biopesticides 207
6.5.1 Resistance Management 208
6.5.2 Restricted-Entry Intervals 208
6.5.3 Residual Effect 208
6.6 Biopesticides and Integrated Pest Management: A Combined Approach 209
6.7 Summary and Future Prospects 209
References 210

6.1 Introduction

The advancement in science and technology especially in the field of agriculture, plant pathology, genetics, plant breeding, microbiology, and biotechnology has led to the agricultural intensification so as to feed the ever-growing population. The need to save the crops from the attack of insects/pests to ensure the food security for the growing population worldwide has resulted in the extensive use of chemical pesticides in agriculture. Despite that, the pest-induced losses are on the rise. The realization of the negative effects of these chemicals on nature and natural resources has forced the scientific community and technologist to search for alternatives of chemicals to control the pests/insects, which can damage the crops. In the current scenario, the focus is more on reliable, sustainable, and environment-friendly agents for pest control, i.e., the biopesticides.

A pesticide of biological origin, i.e., viruses, bacteria, pheromones, plant, animal compounds, and certain minerals, is termed as biopesticide. Biopesticides are gaining wide interest due to the adverse impacts of chemical pesticides. They also exist in our kitchen, for instance, canola oil and baking soda have pesticidal applications and are considered as biopesticides (www.epa.gov). They are target-specific and do not harm human beings or beneficial organisms as compared to chemical pesticides that are generally broad spectrum and pose nontarget effects on natural predators and parasites. Biopesticides are generally grouped into three major categories (www.epa.gov/oecaagct/tbio.html): (i) pesticides of microbial origin (e.g., bacterial, fungal, viral, or protozoan) that contains microorganisms as the active ingredient (a.i) (such pesticides are relatively specific for its target pest[s]), (ii) microorganism-derived molecules that act as the active ingredient against pests (the most widely used biopesticides in this category are obtained from subspecies and strains of *Bacillus thuringiensis* (popularly called *Bt.*). Different strains of *Bt.* produce a different mix of proteins that can bind to a larval gut receptor and specifically kill one or a few related species of insect larvae. Some species of *Bt.* control moth larvae found on plants, while others are specific for larvae of flies and mosquitoes), and (iii) plant-originated pesticidal substances termed as plant-incorporated protectants (PIPs). PIPs are produced by plants through genetic manipulations. For example, *Bt.* gene has been introduced in the plant's own genetic material by genetic engineering. Genetically modified plants are thereby capable of manufacturing toxic substances that are detrimental to the pests. However, such modifications are still under the heat of debate and sociocultural obligations in many countries.

The key feature of biopesticides is their eco-friendly nature and biodegradability, thereby reducing their residual effect and environmental pollution. The formulation of new legislation and the awareness regarding the side effects of chemical pesticides has resulted in their declined use, albeit at a slow pace. Therefore, alternative pest management methods are highly sought. The incorporation of biopesticides in agricultural systems is posed by many challenges, such as understanding of a biopesticide–pest interaction in the ecosystem, particularly by the end users.

According to a new technical market Business Communications Company, Inc., research report on biopesticides, the global market for biopesticides was estimated at $1.6 billion in 2009, which is expected to increase to $3.3 billion in 2014 with a 5-year compound annual growth rate (CAGR) of 15.6% (www.bccresearch.com). Similarly, owing to the intensification of agriculture, the much larger segment of the global pesticides market represented by synthetic pesticides is expected to reach nearly $48 billion in 2014, after increasing at a CAGR of 3% from the estimated value of $41.2 billion in 2009 (www.bccresearch.com). An overview of the general concepts of biopesticides, particularly *Bt.*-based biopesticides, is given in this chapter. In the view of sustainable and eco-friendly agriculture, the benefits of using biopesticides have been discussed followed by the importance of selecting an efficient and potent strain for the production of effective and reliable biopesticides.

Biocontrol measures are generally divided into three major groups: (i) importation; (ii) conservation/augmentation of natural enemies, such as predators and parasitoids; and (iii) microbial pesticides. Importation is the introduction of a foreign agent for biological regulation of populations. Conservation and augmentation allows the native agent to grow as homeostatic factor against a pest. Among the measures discussed earlier, the importation of biocontrol agents poses the greatest risks to the ecological equilibrium (Myers et al., 2000). Microbial pesticides are perhaps more widely used and cheaper than the other biocontrol methods of pest management. Several microbes used as biopesticides are listed in Table 6.1. Fungi secrete several toxic compounds, such as beauverin, lilacin, and cell wall-degrading

TABLE 6.1

Group of Microorganisms with Biocontrol Potential

Group	Microorganisms	Modes of Action	References
Fungi	*B. bassiana, M. anisopliae, N. rileyi, Aspergillus niger, Verticillium lecanii, Paecilomyces* sp., *Isaria purchasi, Fusarium* sp., *Cordyceps* sp., *Entomophthora* sp., *Trichoderma* sp., *Candida oleophila* (yeast)	Secretion of toxic compounds or enzymes	Szewczyk et al. (2006)
Nematode	*Steinernema abbasi, S. carpocapsae, S. feltiae, S. glaseri, S. neocurtillae, S. scapterisci, S. thermophilum, S. cubanum, S. scarabaei, Heterorhabditis bacteriophora, H. baujardi, H. downesi, H. zealandica, H. megidis, H. poinari*	Penetrate the insect hosts through their natural openings and then defecate symbiotic bacteria (*Xenorhabdus*/ *Photorhabdus*). The bacteria quickly kill the hosts with toxins and the nematode then reproduces in the decaying host tissues.	Brown et al. (2004), Divya and Sankar (2009)
Virus	Family: Baculoviridae; nucleopolyhedrovirus (NPVs) and granulovirus (GVs)	Widely used as expression vectors. Viruses produce a large number of occlusion bodies in infected cells (polyhedra and granules, typical polyhedra) that allow them to survive in the environment and transmit the disease from one insect to another.	Van Regenmortel et al. (2000)

enzymes, such as proteases and chitinases (Uribe and Khachatourians, 2004). Some fung acts specifically, while some have a wide spectrum of action, such as *Beauveria bassiana*, *Nomuraea rileyi*, and *Metarhizium anisopliae*, which can infect about 100 species of insects from different orders (Bidochka et al., 2001; Fang et al., 2005). New-generation fungal pesticides are engineered to enhance their biocontrol potential (Lager et al., 1996).

Among all the biopesticides discussed, *Bt.*-based biopesticides for insect pest management are most widely used (Crickmore, 2006). *Bt.* was first isolated by Ishiwata in 1903 from diseased silkworms, *Bombyx mori* (Lepidoptera: Bombycidae), and later by Berliner in 1915 from Mediterranean flour moth, in Thuringe (Germany), hence the name *thuringiensis*. *Bt.* is a gram-positive bacteria mostly isolated from the natural environment, such as soils, sericulture, granaries, and phylloplanes, and from dead insects (Smith and Couche, 1991; Kuo and Chak, 1996). To date, more than 69 serotypes have been isolated, out of which three serotypes are commercialized as microbial insecticides: (i) *Bt.* var. *kurstaki* has been used as a control agent against numerous Lepidopterans (gypsy moth, spruce budworm, vegetable pests such as diamondback moth and cabbage looper, butterflies, and moths) in agriculture and forestry; (ii) *Bt.* var. *israelensis* is used mostly against Diptera (mosquito and black fly species); and (iii) *Bt.* var. *tenebrionis* is used as a potent larvicide for some Coleoptera families, particularly for elm leaf beetle and Colorado potato beetle (Wright and Ramos, 2005).

Thousands of *Bt.* strains have been used for the production of microbial insecticides. Some of the most important strains of *Bt.* being exploited for different insects belong to the orders Lepidoptera (butterflies and moths), Diptera (mosquitoes and biting flies), and Coleoptera (beetles). Commercially available and registered *Bt.* products by the Environmental Protection Agency (EPA) include *Bt. aizawai* (Lepidoptera, used for wax moth larvae in honeycombs and for diamondback moth caterpillars), *Bt.* var. *israelensis* (Diptera, frequently used for mosquitoes and black flies), *Bt.* var. *kurstaki* (Lepidoptera, frequently used for gypsy moth, spruce budworm, and many vegetable pests, such as diamondback moth and cabbage looper, and also used for fruit pests, such as leaf rollers on apples and peach twig borer), and *Bt.* var. *tenebrionis* (Coleoptera, used for elm leaf beetle, Colorado potato beetle). *Bt.* var. *kurstaki* is the most commonly used *Bt.* formulation against many leaf-feeding larvae on vegetables, shrubs, fruit trees, and conifers. While using *Bt.*-based biopesticides, certain points are to be considered:

1. Initial *Bt.* application must be performed immediately before or just after egg hatch while the larvae are still small. Plants should be monitored periodically for worm eggs and pheromone traps should be used to determine adult moths.
2. Good spray that ensures maximum coverage of plant surface area is essential to ensure that the maximum larvae feed the *Bt.*-treated plants for satisfactory control. Spray volume, pressure, droplet size, and dilution of active ingredient (protein crystals and spores) must be optimized.

3. Target insect species should be properly identified and *Bt.* strain be selected accordingly.
4. Formulations should be prepared according to the climatic conditions of the area. Environmental factors, such as temperature, rainfall, pH, and sunlight, are some of the most important factors that require attention.

Currently, different *Bt.*-based biopesticides commercially available in the market for controlling a wide variety of agricultural and forestry pests, including disease vectors, are listed in Table 6.2.

6.2 Technology of Biopesticide Production

6.2.1 Selection of Strain

Bt.-based biopesticides have a wide scope and potential for pest management; however, it is still undergoing many advancements in formulation development, which in turn require efficient and potent strains. The development of an efficient biopesticide begins with a carefully crafted microbial screening procedure deployed by mass production protocols for the optimization of product quantity and quality. It leads to the formation of a product formulation that preserves shelf life, aids product delivery, and enhances bioactivity. Designing successful formulations requires critical considerations, such as preserving biomass viability during stabilization, drying, and rehydration. These parameters also aid in biomass/formulation delivery and target coverage. The traditional screening procedures generally screen out only those potential microbial agents that are both efficient and amenable as well as generate enhanced end product in liquid culture and increase the likelihood of selecting agents with enhanced commercial development potential (Moazami, 2007). General flow sheet for biopesticide production is described in brief in Figure 6.1.

During the scale-up of biomass production, the efficacy and the amenability of the final product and quantity must be maintained without compromising the product efficacy or formulation stability. Therefore, identifying promising and potential bacterial strains is very important in terms of industrial and agricultural applications. In other words, selection of a potent microbial strain that economically produces stable propagules and provides sustainable pest management under field conditions is an important goal of the selection process (Wraight et al., 2001).

The major objective of research in biopesticide industry is aimed at developing a formulation to obtain higher volumetric productivity (g/L/h) and production of recombinant proteins using high-cell-density cultivation (Brar et al., 2005). This implicates the need of an efficient and rapidly growing strain that contains a stable plasmid and overexpresses an easily extractable

TABLE 6.2
List of Commercially Available *Bt.* Formulations

Strains	Company	Trade Name	Active Ingredient	Target Pest
Natural strains	Valent Biosciences Corp.	Biobit	*B. thuringiensis* var. *kurstaki*	Lepidopterans
	Valent Biosciences Corp.	DiPel	*B. thuringiensis* var. *kurstaki*	Lepidopterans
	Certis	CoStar	*B. thuringiensis* var. *kurstaki*	Lepidopterans
	Bayer CropScience	Florbac	*B. thuringiensis* var. *aizawai*	Lepidopterans
	Certis	Javelin/ Delfin	*B. thuringiensis* var. *kurstaki*	Lepidopterans
	Valent Biosciences Corp.	Thuricide	*B. thuringiensis* var. *kurstaki*	Lepidopterans and certain leaf-eating worms
	Valent Biosciences Corp.	Teknar	*B. thuringiensis* var. *israelensis*	Mosquito and black fly larvae
	Valent Biosciences Corp.	Bactimos	*B. thuringiensis* var. *israelensis*	Dipterans
	Valent Biosciences Corp.	Vectolex GC	*B. sphaericus* serotype H5a5b, strain 2362	Dipterans
	TermiumPlus®	Bactospeine	*B. thuringiensis* var. *kurstaki* HD-1	Lepidopterans
	Extonetpip	Acrobe	*B. thuringiensis* var. *israelensis*	Dipterans
	Valent Biosciences Corp.	Novodor	*B. thuringiensis* var. *tenebrionis*	Coleopterans
	Trident®	Trident	*B. thuringiensis* var. *tenebrionis*	Coleopterans
	Certis	Deliver	*B. thuringiensis* var. *kurstaki*	Lepidopterans
	Certis	Jackpot WP	*B. thuringiensis* var. *kurstaki*	Lepidopterans
	Certis	Turix WP/ Agree WP	*B. thuringiensis* var. *kurstaki*	Lepidopterans
	AFA Environment Inc.	Agribac	*B. thuringiensis* var. *kurstaki*	More than 30 insect species
	Valent Biosciences Corp.	XenTari	*B. thuringiensis* var. *kurstaki*	Particularly effective against *Spodoptera* ssp. and *Plutella xilostella*

TABLE 6.2 (continued)

List of Commercially Available *Bt.* Formulations

Strains	Company	Trade Name	Active Ingredient	Target Pest
	Valent Biosciences Corp.	Vectobac	*B. thuringiensis* var. *israelensis*	Mosquito and fly larvae
	Valent Biosciences Corp.	GnatrolDG	*B. thuringiensis* var. *israelensis*	Larval stage of sciarid mushroom flies
	Valent Biosciences Corp.	Foray	*B. thuringiensis* var. *kurstaki*	Lepidopterans
Transconjugant strains	Certis	Agree WG	*B. thuringiensis* var. *aizawai*	Lepidopterans
	Certis	Condor	*B. thuringiensis* var. *kurstaki*	Lepidopterans
	Cutlass®	Cutlass	*B. thuringiensis* var. *kurstaki*	Lepidopterans
	Thermo Trilogy	Design	*B. thuringiensis* var. *aizawai*	Lepidopterans
Recombinant strains	Raven®	Raven	Cry1Ac (2×), Cry3A Cry3Bb (imported)	Lepidopterans
	Certis	CRYMAX	Cry1Ac (3×), Cry2A Cry1C (imported)	Lepidopterans
	Certis	Lepinox WDG	*B. thuringiensis* var. *kurstaki*	Lepidopterans

Source: Sanahuja, G. et al., *Plant Biotechnol. J.*, 9, 283, 2011.

protein or active agents, survival ability, and sustained efficacy after delivery to the target. The characteristics, composition, and entomotoxicity of biopesticidal formulations vary not only with the type of habitat (foliage, soil, water, and warehouse size), rheology of technical material (viscosity, particle size, and density), insect species (feeding habits, feeding niche, and life cycle), mode of action (oral/contact), host–pathogen–environment interactions (behavioral changes, resistance, and stability), mode of application (aerial and land), application rate (L/ha and kg/ha), and pathogen (type, characteristics, and regeneration mechanism) but also most importantly with the type of strain being used for biopesticide production.

The search for efficient *Bt.* biopesticides involves the isolation of novel strains from soil and larvae cadavers exhibiting different entomotoxic activities. This offers a wide range of biopesticides that could be used under different environmental conditions for pest management programs. Earlier *Bt.* strains were isolated from natural environments, such as from soils, litters, and dead insect

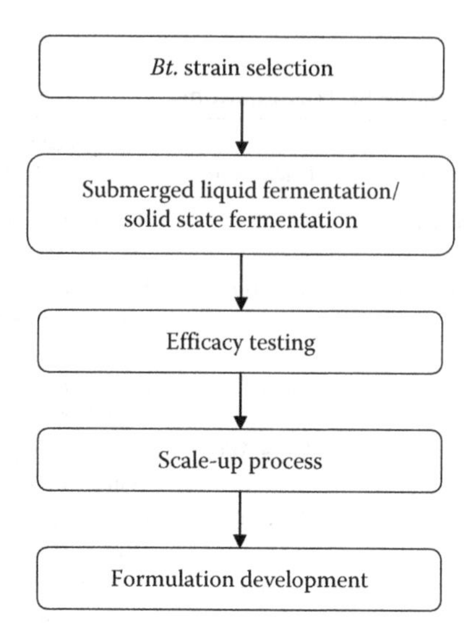

FIGURE 6.1
General flow sheet for bioinsecticide production.

larvae (Carozzi et al., 1991; Kaelin et al., 1994; Burges and Hurst, 1997). Isolation of potent *Bt.* strains from the crude gruel and fish meal (Zouari et al., 1998) and municipal and industrial wastewater sludge has also been explored (Vidyarthi et al., 2002; Yezza et al., 2004). Due to the high concentration of decomposed organic contents and other inorganic nutrients, wastewater sludge provides a natural source for microbial growth (Jain and Tyagi, 1992; Shooner et al., 1996), and hence, it has been proposed as a potential raw material for *Bt.* production (Tirado-Montiel et al., 2001). Therefore, *Bt.* isolated from wastewater sludge followed by cultivation in the same medium has higher entomotoxicity, better growth characteristics, and high sporulation due to its pre-acclimatization in sludge (Mohammedi et al., 2006).

The identification of the appropriate *Bt.* strain for development as a bioinsecticide could be a complex procedure. The selection process must evaluate the potential of *Bt.* isolate to form spores and a stable plasmid that can be economically mass-produced and acquiescent of available application technologies. It should possess enough entomotoxicity to consistently infect the target pests under their natural environment and ecological conditions (Wraight et al., 2001; Jackson and Schisler, 2002; Jaronski, 2007). Critical environmental factors, such as temperature, UV radiation, rainfall, dew, pH, and microclimate, can have a profound influence on the growth and entomotoxicity of a *Bt.* strain against the target pest (Faria and Wraight, 2001; Yeo et al., 2003). Strains collected from different environments differ in virulence, desiccation tolerance, thermal tolerance, speed of germination and infection, environmental stability, reproduction, and UV tolerance (Jackson, 1997; Vega et al., 1999). The environment in which the biopesticide is to be applied necessitates the need of selection of an efficient strain.

Different strains of *Bt.* differ in their nutritional requirements, for example, organic and inorganic nitrogen sources to amino acids especially in carbon catabolite repression, which regulate toxin (δ-endotoxin) synthesis. Therefore, in order to increase the δ-endotoxin production, it is important to overcome the repressive regulation. *Bt.* cultivation using complex substrates needs preliminary endogenous hydrolysis before the assimilation of proteins and starch-based materials (Reid et al., 1986). This is achieved by using potent hydrolytic enzymes (proteases and amylases) producing *Bt.* strain (Priest, 1977). Further, the advancement in the production technology of biopesticides along with the need of producing efficient formulations of biopesticides has been detailed.

6.2.2 Production Technology

Among several biopesticides available in the market, *Bt.* var. *kurstaki* is one of the widely used biopesticides (Prabhakar and Bishop, 2009). *Bt.* formulation contains a mixture of spores and insecticidal crystal proteins (ICPs) obtained after growth and sporulation in a bioreactor. *Bt.* biopesticides are usually marketed as solid and liquid formulations. Nutritional and environmental conditions during bacterial growth using solid-substrate (SSF) and liquid-state submerged fermentation (SmF) influence the form and efficacy of the final product (Magan, 2001; Ying and Feng, 2006). The type of formulation and the adjuvants used during the application also influences the efficacy of the biopesticide (Friesen et al., 2006; Costa et al., 2008).

There are several limiting factors for the successful development of *Bt.* biopesticides such as raw material being used. The raw materials used for the *Bt.*-based biopesticide production are a substantial part of production cost, which suggests the need for cheaper raw materials for formulation development. There is an urgent need to find high-yielding, cost-effective, easily available raw materials for the production of efficient *Bt.* formulations. Several reports have been published on cost reduction of *Bt.* production process through substitution of high-cost medium ingredients of soy flour and fish meal with complex agro-industrial wastes, such as cassava starch, maize starch, rice straw, wheat bran, corn steep liquor, sugarcane molasses, cheese whey, and coconut waste (Zouari et al., 1998; Vega, 1999; Vora and Sethna, 1999; Abdel-Hameed, 2001; Khuzhamshukurov et al., 2001); starch industry wastewater; and municipal and industrial wastewater sludge (Sachdeva et al., 2000; Tirado-Montiel et al., 2001). These low-cost agro-industrial wastes contain necessary nutritional elements for sustained growth, high sporulation, and crystal formation by *Bt.* The bioconversion of residues to value-added products preserves the sustainable agriculture and reduces the production cost of *Bt.*, which would be both socially and environmentally useful.

Wastewater sludge is considered a global problem and its sustainable management via value addition is an excellent substitute to other disposal options, such as land filling and incineration. Wastewater sludge has emerged as an economically feasible, inexpensive raw material option for the sustainable production of biopesticides (Tirado-Montiel et al., 2003; Yezza et al., 2004, 2005; Barnabe et al., 2005). Utilizing wastewater or wastewater sludge is more economical and efficient as compared to semisynthetic media currently used in commercial production. Wastewater sludge is a negative-cost raw material (Sachdeva et al., 2000; Vidyarthi et al., 2002) and possesses higher entomotoxicity (higher toxicity of crystal protein) even at low cell/spore concentration compared to soya meal medium (Yezza et al., 2004, 2005; Barnabe et al., 2005). In addition, reuse of wastewater or wastewater sludge is economical and eco-friendly approach. Wastewater and wastewater sludge have been well-researched as an alternative to costly synthetic medium for fermentation, optimization, nutrient assimilation, bench-scale and pilot plant studies, and formulations (Lachhab et al., 2001; Tirado-Montiel et al., 2001; Vidyarthi et al., 2002; Brar et al., 2004, 2005; Barnabe et al., 2005; Yezza et al., 2005). However, fermentation using wastewater or wastewater sludge poses specific problems due to the complex nature of the substrates. Preliminary studies regarding the development of liquid formulations from fermented wastewater or wastewater sludge have yielded potential for efficient formulations. Thus, *Bt.*-based biopesticide production from wastewater or wastewater sludge would be a resourceful cleanup option as well as an augmenting path for commercialization of economical biopesticides. Besides this, wastewater and wastewater sludge from starch have also been extensively researched from a process point of view comprising pretreatment (improvement of substrate complexity), medium amelioration agents (enhancing nutrient assimilation), optimizing process parameters, and scale-up and formulation development studies (Vidyarthi et al., 2002; Brar et al., 2004; Yezza et al., 2004; Barnabe et al., 2005). Recently, statistical software has been utilized for parameter optimization, such as response surface methodology (RSM) that covers large interval of the parameters to determine the optimal values of these parameters. Parameter optimization, such as moisture content, viable spores, entomotoxicity, and interactions between the operational parameters, can be done precisely by using RSM (Adjalle et al., 2011).

6.2.3 Downstream Processing

Downstream processing of biopesticides is one of the most important steps in the development of an efficient formulation. The downstream process varies significantly with the microbe used for mass production, the type of substrate used, and the required entomotoxicity of the final product. It also depends on the process throughput, the physical characteristics, and the impurities present in the substrate as well as the available infrastructure (Keller et al., 2001). In the case of *Bt.*, the final fermented broth contains spores, cell debris, inclusion bodies, enzymes, and other residual solids. These substances need to be recovered for the subsequent development of the formulation (Bernhard and Utz, 1993; Rowe and Margaritis, 2004). Commercial *Bt.* formulations contain viable spores, ICPs, an array of enzyme systems, such as proteases, chitinases, phospholipases, and many unknown virulent factors along with various inerts/adjuvants. The penultimate step in the production of biopesticides is the recovery of broth components (crystal protein, spores, vegetative insecticidal proteins, and enzyme systems, such as proteases, chitinases, and phospholipases) in an incisively active manner to develop formulations henceforth (Bernhard and Utz, 1993; Rowe and Margaritis, 2004). Literature on harvesting methods is very scarce as most of the *Bt.* commercial production technologies carried out by industries is proprietary and secure.

Earlier studies on downstream processing of biopesticides were based on lactose–acetone technique for spore and crystal toxin recovery; however, this method suffered due to high labor cost and low yields (Dulmage, 1970). In order to overcome such technicalities, centrifugation techniques with an extra step of spray drying were developed to obtain dry powder for solid formulations (Rojas et al., 1996; Teera-Arunsiri et al., 2003). Gradual advances are occurring in the field of downstream processing techniques, such as ultrafiltration (UF), microfiltration, and vacuum filtration, which are being used these days for efficient cell recovery for biopesticide development. However, the use of these techniques to separate insoluble solids' active ingredient from the soluble liquid (inert) fraction of the harvest liquor suffers because of high cost (Gulati et al., 2000; Christy and Vermant, 2002; Darnon et al., 2003), and hence, centrifugation could be used as a simple, cost-effective, and versatile technique.

There are certain issues in downstream processing, particularly from SmF owing to the thermolabile nature of the microorganisms, low concentration of the final products formed, and poor stability. In such cases, stabilizing adjuvants play a significant role in preventing spore desiccation. The liquid medium (broth or centrifuge slurry) is often contaminated by bacteria or certain fungi and therefore requires the supplementation of specific biocidal chemicals or rapid drying of the broth or centrifuge slurry (Soper and Ward, 1981). However, these methods are applicable only for small volumes of the liquid. Apart from this method, the foam flotation process has also been applied in order to recover crystal-enriched suspensions of *Bt.* In this method, the spores are selectively entrained in the foam and get separated from suspensions (Sharpe et al., 1981). This method also suffers some constraints, such as processing is laborious and requires the use of various chemicals.

Spray dying method is used to remove water from the broth. In this process, the broth slurry is passed through the heated inlet at 150°C–200°C. Subsequently, fermented liquid is thickened by centrifugation and filtration using filter aids, such as ascelite, Superfloc, among others, to reduce handling volume, and the resulting powder is collected in the spray dryer

(Taborsky, 1992). The advantage of using spray drying method is that large volumes of the medium can be handled (Zhou et al., 2004). However, the loss of bioactivity has been reported due to the continuous exposure of the bioactive components to the high temperatures (Tamez-Guerra et al., 1996). Therefore, following aspects of *Bt.* formulation (dry powder) needs to be addressed while using spray drying method:

1. Determination of optimal values of drying parameters such as inlet temperature, spraying pressure, hot airflow, and feed rate of fermented broths.
2. Evaluation of the impact of these viable parameters on spores and entomotoxicity of *Bt.* Preheated air is blown through the equipment to control flow. The hot air comes in contact with the droplets of spray-dried sample–air mixture in the "drying chamber" and gets dried instantaneously. If the contact surface of the droplets and the hot air is large, more than 90% of the moisture is evaporated in the "drying chamber." Finally, the sample that is deprived of the maximum amount of moisture is dried and sent into the "cyclone" where the vapor of moisture separates from the powder particles and is collected in a "product vessel."

The efficacy and duration of contact between the hot air and the surface of the sample droplets are one of the important parameters that need to be optimized as it determines good drying and safeguards the quality of the powder produced. The efficacy and duration of the contact between hot air and the droplets of sample are measured in terms of moisture, outlet temperature, and concentration of the viable spores of *Bt.* as well as entomotoxicity of the powder. The important parameters for the development of dry powder formulations are as follows: (i) inlet temperature that measures the degree of hot air, (ii) the rate of feed that defines the concentration of the sample and determines the water content in the air–sample mixture, (iii) the pressure of atomization of air defining the size and surface of the droplets, and (iv) the flow rate of aspiration of hot air that determines the duration of the contact between the hot air and droplets of pulverized sample–air mixture (Adjalle et al., 2011).

Disk–stack centrifuge or a rotary vacuum filter is one of the efficient methods for spore recovery. Disk–stack centrifuges are used in separation processes, where the physical characteristic of the sedimentation depends on the density gradient. Sedimentation is the physical process in which the suspensions and emulsions or a combination of both is exposed to gravitational force, either the earth's gravity or by using a centrifuge. By applying g force, the droplets and/or particles acquire settling velocity (V_g) and are sent to the bottom/wall of the container/centrifuge bowl. Settling velocity increases with the increase in particle diameter, density difference, and lower dynamic viscosity of the fluid. Settling velocity can be determined by using Stokes' law:

$$V_g = \frac{d^2(\rho_s - \rho_i) \times g}{18\mu}$$

where
 g is the acceleration of gravity (m/s^2)
 d is the particle diameter (m)
 ρ_s is the particle density (kg/m^3)
 ρ_i is the medium density (kg/m^3)
 μ is the viscosity of medium (kg/ms)

An efficient recovery of active spore–crystal complex of *Bt.* (>99%) has been reported by Rojas et al. (1996). However, the concentration of dry solids produced by filtration (31.5%) was higher to centrifugation (7.5%). Continuous centrifugation has also been used to recover *Bt.* var. *berliner* with recovery rate of 85%–90%. In this process, decrease in separation efficiency was observed with the increase in flow rates (Zamola et al., 1981). Lactose (5%) is used to supplement the final fermented broth. Lactose functions as cryoprotectant in order to prevent clumping during storage. Subsequently, lactose–acetone coprecipitation and centrifugation are done in order to achieve higher δ-endotoxin recovery efficiency (Dulmage and Rhodes, 1971). The final products (powder/suspension) are suitably formulated as aqueous (flowable) or oil concentrate, spraying powder, or granulate. The advancement in the downstream processing techniques and biotechnological advances, such as gene expression or transformation, can lead to the development of an efficient biopesticides.

Micellar-enhanced ultrafiltration (MEUF) has been used to separate dissolved organic compounds, such as thuringiensin from aqueous streams (Tzeng et al., 1999). In this process, the separation depends on the UF membrane, type, and concentration of surfactant and conditions, such as pH, ionic strength, temperature, and transmembrane pressure (Markels et al., 1995). Downstream processing involves the production of micelles by adding a surfactant to an aqueous stream containing organic solute to separate target compound for subsequent purification. It is a pressure-driven membrane-based separation process that makes use of the micellar properties of a surfactant (Fillipi et al., 1999). Tzeng et al. (1999) used a cationic cetylpyridinium chloride surfactant-based MEUF process to recover thuringiensin from the supernatant of *Bt.* fermentation broth. Two major factors responsible for the increased thuringiensin recovery (up to 94.6%) were the CPC concentration (4% w/v) and membrane pore, whereas ionic strength, pH adjustments, and micelle formation temperature were not necessarily within the temperature range studied. Finally, they developed a novel two-step MEUF processing scheme that includes microfiltration and two-step UF processes. Spray-dried thuringiensin with CPC was more effective in reducing the population of fly larvae than without CPC.

Biocatalytic conversion processes used for formulation development often face certain limitation in productivity and yield due to product inhibition or degradation (Lye and Woodley, 1999). Thus, efforts are needed to minimize such inhibitions and degradation by optimizing physiological and technological parameters. *In situ* product removal (ISPR) technique that has been used to remove proteins in various commercial processes has been successfully employed in biopesticide production (Agrawal and Burns, 1996). The retrieval of a biochemical product is done from the vicinity of a cell during active fermentation. ISPR has been developed to overcome such limitations by keeping the dissolved product concentration low in the reactor (Schugerl, 2000; Alba-Perez, 2001; Stark and von Stockar, 2003; Schugerl and Hubbuch, 2005). In this technique, the product is removed from the catalyst as soon as the product is formed in the reactor thereby reducing the number of downstream processing steps. ISPR is part of the general concept of process integration or intensification, which represents the improved coordination of upstream, reaction, and downstream technologies (von Stockar and van der Wielen, 2003; Schugerl and Hubbuch, 2005; Buque-Taboada et al., 2006). This technique depends on the properties of the target product, the expected benefits, and the biocatalyst involved. Whole bacterial cells can be removed from the reactor by ISPR techniques such as extraction, adsorption, and evaporation (Schugerl, 2000; Alba-Perez, 2001; Stark and von Stockar, 2003; Schugerl and Hubbuch, 2005). In some cases, electrodialysis, precipitation, complexation, and membrane-assisted separation techniques have also been employed (Schugerl, 2000; Alba-Perez, 2001; Fernandes et al., 2003; Stark and von Stockar, 2003; Schugerl and Hubbuch, 2005). In ISPR, the separation of the cells and product is generally achieved by employing (i) an internal configuration, where the fermentation liquid is in contact with a product-removing phase that is present in the fermenter, or (ii) an external configuration, where the fermenter liquid is via a loop in contact with a product-removing phase in an external unit (Alba-Perez, 2001; Stark and von Stockar, 2003). The choice of configuration depends on the properties of the target product and biocatalyst, the number of the relevant phases involved, and the type of recovery process employed (Buque-Taboada et al., 2006).

Microfiltration membranes function as a physical sieve and are applicable to separate larger-size particles, such as suspended solids, particulates, and microorganisms. The membranes are highly porous and have discernible pores. The membranes are made up of ceramics, Teflon, polypropylene, or other plastics (Ulbricht, 2006). The main objective of an efficient downstream processing is to reduce the number of operating units involved in the process thereby reducing overall process and validation costs while maintaining the viability of the cells. At the same time, it should be simple to handle and economical while imposing fewer burdens on the environment also. Among all the previously discussed methods, centrifugation appears to be a better method for downstream processing of biopesticides (Brar et al., 2006). It has the potential to enhance the recovery rates with further advancement in design and speed.

6.3 Mode of Action of Crystal Proteins

Bt. belongs to the *Bacillus cereus* sensu lato group. Bacteria in this group synthesize δ-endotoxins, a protein crystal having insecticidal activity. Approximately 25% of the dry weight of the bacterium is comprised of δ-endotoxins. This crystal protein specifically acts on the intestinal epithelium of susceptible insects.

The δ-endotoxins in commercial formulations function effectively after ingestion by insects. After ingestion, the crystals are dissolved in the intestinal tract, facilitated with a specific pH of the gut (usually alkaline) and the specific gut membrane structures required to bind the toxin (Figure 6.2). The high pH and reducing conditions in the guts of the susceptible insects are necessary in order to destabilize the ionic bonding and the disruption of intermolecular disulfide bridges. δ-Endotoxin binds to receptors on the surface of intestinal epithelial cells in susceptible insects after solubilization and activation (Van Rie et al., 1990). The structure of an activated δ-endotoxin, the Cry3Aa, enables it to perforate the epithelial membranes. According to Knowles (1994), the toxin acts by osmocolloidal cytolysis, which results in the formation of pores in intestinal cells. The formation of pores in the intestinal cells causes cell lysis due to the inhibition of ion exchange. Another mechanism has been proposed by Zhang et al. (2006), according to which intracellular signaling binds the toxin to a cadherin-like receptor that leads to apoptosis of the cells. Immediately after ingestion, the protein toxin damages the gut lining, leading to gut paralysis followed by destruction of the intestinal epithelium. The hemolymph and the intestinal cavity come in contact resulting in decreased intestinal pH. The bacterial spores are further allowed to germinate and multiply in the insect cadaver.

Generally, the *Bt.* spores do not spread to other insects or cause disease outbreaks on their own. The knowledge of the insect physiology and the susceptible stage of development should be followed closely for the efficient functioning of the biopesticide containing δ-endotoxins. However, it must be ensured that the bacterium is ingested by the insect in sufficient amount.

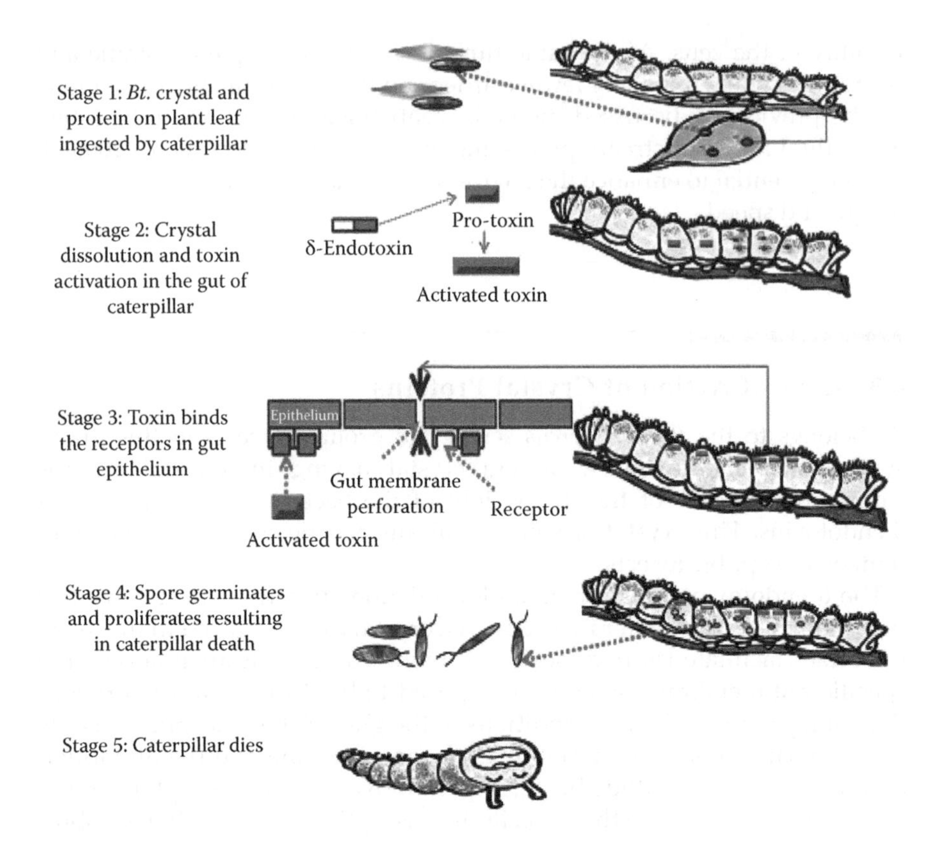
Stage 1: *Bt.* crystal and protein on plant leaf ingested by caterpillar
Stage 2: Crystal dissolution and toxin activation in the gut of caterpillar
δ-Endotoxin
Pro-toxin
Activated toxin
Stage 3: Toxin binds the receptors in gut epithelium
Epithelium
Gut membrane perforation
Receptor
Activated toxin
Stage 4: Spore germinates and proliferates resulting in caterpillar death
Stage 5: Caterpillar dies

FIGURE 6.2
Diagrammatic representation of mechanism of action of *Bacillus thuringiensis* (*Bt.*) biopesticides. (Modified from www.inchem.org/documents/ehc/ehc/ehc217.htm)

6.4 Marketing Constraints/Challenges in Biopesticide Development

The salient features of biopesticides are their eco-friendly nature and biodegradability that lower their residual effect and reduce environmental pollution associated with chemical pesticides. However, they possess a different set of challenges due to their inherent nature, such as particulate suspensions, thermal susceptibility, proneness to contamination, and pest specificity (Hall and Menn, 1999). This suggests the need of well-designed and validated formulation. In the present scenario, most of the challenges have been taken care of except the case of residual activity, which can be enhanced by various means and is an interesting point to explore. Continuous advances since the early twentieth century have contributed significantly to their widespread acceptance and use as compared to chemical pesticides.

6.5 Advantages of Biopesticides

Agricultural and forest pests have always been of a major concern to mankind. They have led to several pest outbreaks in the past and cause continuing economical losses. In order to maintain the pressing demand for food owing to the increasing population, the management of pest population has become the priority to ensure increased crop yield. Therefore, it might be envisaged that the use of chemical pesticides will be continued. It is also widely accepted that the use of agrochemicals from decades brings significant economical costs and environmental liabilities of off-target drift, environmental pollution, toxicity, resistance development, and bio-magnification (Arias-Estevez et al., 2008). Besides this, the increasing number of pesticide and fungicide-resistant strains has also become a matter of serious concern. Therefore, there is a growing need to develop alternative crop protection methods. In this perspective, the focus on controlled release of biopesticides is gaining substantial importance as they are cost-effective and results in sustainable agriculture. The advantages of using biopesticides are listed in brief in Table 6.3. The advantages and the potential role of biopesticides in sustaining environmental management can be further discussed broadly on the basis of three main components: resistant management, restricted-entry level, and residues (www.insectigen.com).

TABLE 6.3

Various Advantages and Disadvantages of Biopesticides

	Biopesticides	References
Advantages	Inherently less toxic than chemical/synthetic pesticides;	Gupta and Dixit (2010);
	Generally affect only the target pest and closely related organisms, in contrast to broad spectrum, conventional pesticides;	
	Effective in very small quantities and often decompose quickly;	Arias-Estevez et al. (2008);
	As a component of IPM programs, biopesticides can greatly decrease the use of conventional pesticides;	Chandler et al. (2011);
	Biopesticides are environment-friendly and support sustainable development of agriculture;	Leng et al. (2011);
	Less cumbersome registration regulations;	Greaves (2009);
Disadvantages	Slow; inexpensive; inconsistent efficacy; narrow host range; uncertain storage/shelf life; incompatibility with synthetic/conventional chemicals; poor grower education/awareness; natural extracts might possess uncertain compositions	Rizvi et al. (2009); Gupta and Dixit (2010); Chandler et al. (2011)

6.5.1 Resistance Management

Biopesticides function in an integrated manner and possess multiple modes of action. This reduces the probability of developing resistance in a particular insect species. Biopesticides can be an excellent component of an integrated pest management (IPM) program that incorporates different ways of pest destruction via varying mechanisms of action (Chandler et al., 2011). Biopesticides act by targeting and disrupting a particular stage of the pest's life cycle. This extends the efficacy and life span of biopesticides. Biopesticide also suppresses the mutations that may lead to resistance development.

6.5.2 Restricted-Entry Intervals

Restricted-entry intervals determine the timing of application relative to a number of factors. The majority of biopesticides has low restricted-entry levels mostly around 0–4 h with no preharvest interval (Sorg, 1999). After a biopesticide is applied, a farmer can go into the field and harvest immediately.

6.5.3 Residual Effect

Residue is a serious issue associated with pesticides. EPA has stringent regulatory guidelines for their proper application. The office of pesticide programs of the EPA has established a Biopesticide and Pollution Prevention Division (BPPD) that coordinates the registration of biopesticide products (www.epa.gov/pesticides/biopesticides). EPA generally entails less data to accomplish the registration of a new product as biopesticides (<1 year) as compared to the registration of conventional chemical pesticides that can take up to 3 years for regulatory approval.

The registration cost of a biopesticide product is usually less atrocious than for a chemical pesticide product. In the United States, the registration cost of a biopesticide can usually be measured in thousands of dollars. However, it is more than a million of dollars for the registration of chemical pesticides (Talwar et al., 2012). This also suggests the regulator's preference for biopesticides. The higher costs of chemical registration have led many high-valued crops and niche markets without effective chemical crop protection measures, leaving *Bt.* as an economical choice for many producers. This is particularly important in crops that undergo global trading and subject to international maximum residue levels as well as quarantine regulations. In the present context, the biopesticides have gained importance particularly in combination with IPM strategies. Biopesticides are a fundamental component of IPM programs and can reduce the use of conventional pesticides to a great extent while increasing the crop yields.

6.6 Biopesticides and Integrated Pest Management: A Combined Approach

According to Perkins (1982), IPM is a scientific paradigm that is of global significance. The main aim of IPM is the designing and implementation of pest management practices that could meet the needs of the farming community, governments, consumers, and stakeholders by minimizing the pest losses. IPM also protects the long-term risk of environmental pollution and risks to human health while at the same time enhancing agricultural sustainability.

The use of biopesticides in IPM provides a combined approach to pest management and agricultural sustainability in a cost-effective and eco-friendly manner. However, it must be established on a meticulous understanding of pest ecology and beneficial species as well as their interaction with the crop and the ecosystem (Altieri et al., 2009; Sutherst et al., 2011). The emerging era of insect-resistant transgenic cotton presents a genuine panorama to provide a foundation for more sustainable, economically acceptable IPM with the integration of biopesticides that offer lesser reliance on chemical-based pesticides.

6.7 Summary and Future Prospects

An extensive research on *Bt.* as bioinsecticide in general and at the molecular level has been conducted in the past and is progressing at a fast pace. Although *Bt.* has been studied for more than a century, scientists are continually exploring different avenues for its growth and development. The genetic makeup of this organism imparts fascinating properties, such as a wide range of entomotoxins active against diverse insect orders. The toxic activity of *Bt.* is highly specific and it is practically harmless to humans, animals, beneficial insects, and other organisms.

Although there is an enormous possibility to control many pests, mainly in agriculture and forestry through diverse biopesticide products, there is still some research gap in terms of biopesticide development and production technology for formulation manufacturing. Therefore, it is necessary to disseminate the knowledge of biological control programs to the end users. Nonetheless, the rise in income levels due to a growing economy coupled with increasing awareness of health-related effects of chemical pesticides has increased the demand for organic food. In view of this demand and the government's efforts to mitigate climatic change, biopesticides are going to play an important role in future pest management programs.

References

Abdel-Hameed, A. 2001. Stirred tank culture of *Bacillus thuringiensis* H-14 for production of the mosquitocidal δ-endotoxin: Mathematical modelling and scaling-up studies. *World Journal of Microbiology and Biotechnology* 17: 857–861.

Adjalle, K.D., Vu, K.D., Tyagi, R.D., Brar, S.K., Valéro, J.R., and Surampalli, R.Y. 2011. Optimization of spray drying process for *Bacillus thuringiensis* fermented wastewater and wastewater sludge. *Bioprocess and Biosystems Engineering* 34: 237–246.

Agrawal, A. and Burns, M.A. 1996. Application of membrane-based preferential transport to whole broth processing. *Biotechnology and Bioengineering* 55(4): 581–591.

Alba-Perez, A. 2001. Enhanced microbial production of natural flavours via in situ product adsorption. Ph.D. dissertation, Swiss Federal Institute of Technology Zurich (ETHZ), Zurich.

Altieri, M.A., Nicholls, C.I., and Ponti, L. 2009. Crop diversification strategies for pest regulation in IPM systems. In: *Integrated Pest Management*, eds. E.B. Radcliffe, W.D. Hutchinson, and R.E. Cancelado, pp. 116–130. Cambridge University Press, Cambridge, UK.

Arias-Estevez, M., Lopez-Periago, E., Martinez-Carballo, E., Simal-Gandara, J., Juan-Carlos, M., and Garcia-Rio, L. 2008. The mobility and degradation of pesticides in soils and the pollution of groundwater resources. *Agriculture, Ecosystems and Environment* 123(4): 247–260.

Barnabe, S., Verma, M., Tyagi, R.D., and J.R. Valero. 2005. Culture media for increasing biopesticide producing microorganism's entomotoxicity, methods of producing same, biopesticide producing microorganisms so produced and method of using same. U.S. Patent International Application PCT/CA2005/000235.

Bernhard, K. and Utz, R. 1993. Production of *Bacillus thuringiensis* insecticides for experimental and commercial uses. In: *Bacillus thuringiensis, An Environmental Biopesticide: Theory and Practice*, eds. P.F. Entwistle, J.S. Cory, M.J. Bailey, and S. Higgs, pp. 255–267. John Wiley & Sons, Chichester, UK.

Bidochka, M.J., Kamp, A., Lavender, T.M., Dekoning, J., and De Croos, J.N.A. 2001. Habitat association in two genetic groups of the insect–pathogenic fungus *Metarhizium anisopliae*: Uncovering cryptic species. *Applied Environmental Microbiology* 67: 1335–1342.

Brar, S.K., Verma, M., Tyagi, R.D., and Valero, J.R. 2006. Recent advances in downstream processing and formulations of *Bacillus thuringiensis* based biopesticides. *Process Biochemistry* 41: 323–342.

Brar, S.K., Verma, M., Tyagi, R.D., Valero, J.R., and Surampalli, R.Y. 2005. Starch industry wastewater based stable *Bacillus thuringiensis* liquid formulations. *Journal of Economic Entomology* 98(6): 1890–1898.

Brar, S.K., Verma, M., Tyagi, R.D., Valero, J.R., Surampalli, R.Y., and Banerji, S. 2004. Development of sludge based stable aqueous *Bacillus thuringiensis* formulations. *Water Science Technology* 50 (9): 229–236.

Brown, S.E., Cao, A.T., Hines, E.R., Akhurst, R.J., and East, D. 2004. A novel secreted protein toxin from the insect pathogenic bacterium *Xenorhabdus nematophila*. *Journal Biological Chemistry* 279: 14595–14601.

Buque-Taboada, E.M., Straathof, A.J.J., Heijnen, J.J., and van der Wielen, L.A.M. 2006. In situ product recovery (ISPR) by crystallization: Basic principles, design, and potential applications in whole-cell biocatalysis. *Applied Microbiology Biotechnology* 71: 1–12.

Burges, H.D. and Hurst, J.A. 1997. Ecology of *Bacillus thuringiensis* in storage moths. *Journal Invertebrate Pathology* 30: 131–139.

Carozzi, N.B., Kramer, V.C., Warren, G.W., Evola, S., and Koziel, M.G. 1991. Prediction of insecticidal activity of *Bacillus thuringiensis* strains by polymerase chain reaction product profiles. *Applied and Environmental Microbiology* 57: 3057–3061.

Chandler, D., Bailey, A.S., Tatchell, G.M., Davidson, G., Greaves, J., and Grant, W.P. 2011. The development, regulation and use of biopesticides for integrated pest management. *Philosophical Transactions of the Royal Society of Biological Sciences* 366: 1987–1998.

Christy, C. and Vermant, S. 2002. The state-of-the-art of filtration in recovery processes for biopharmaceutical production. *Desalination* 147: 1–4.

Costa, S.D., Grassano, S., and Li, J. 2008. Sweet whey based biopesticide composition. World Patent Application #2008/73843.

Crickmore, N. 2006. Beyond the spore—Past and future developments of *Bacillus thuringiensis* as a biopesticide. *Journal of Applied Microbiology* 101: 616–619.

Darnon, E., Morin, E., Belleville, M.P., and Rios, G.M. 2003. Ultrafiltration within downstream processing: Some process design considerations. *Chemical Engineering and Processing* 42: 299–309.

Divya, K. and Sankar, M. 2009. Entomopathogenic nematodes in pest management. *Indian Journal of Science and Technology* 2(7): 53–60.

Dulmage, H.T. 1970. Insecticidal activity of HD-1, a new isolate of *Bacillus thuringiensis* var. *alesti*. *Journal of Invertebrate Pathology* 15: 232–239.

Dulmage, H.T. and Rhodes, RA. 1971. Production of pathogens in artificial media. In: *Microbial Control of Insects and Mites*, eds. H.D. Burges and N.W. Hussy, pp. 507–540. Academic Press, New York.

Fang, W., Leng, B., Xiao, Y. et al. 2005. Cloning of *Beauveria bassiana* chitinase gene *Bbchit1* and its application to improve fungal strain virulence. *Applied Environmental Microbiology* 71: 363–370.

Faria, M. and Wraight, S.P. 2001. Biological control of *Bemisia tabaci* with fungi. *Crop Protection* 20: 767–778.

Fernandes, P., Prazeres, D.M.F., and Cabral, J.M.S. 2003. Membrane-assisted extractive bioconversions. *Advances in Biochemical Engineering and Biotechnology* 80: 115–148.

Fillipi, B.R., Brant, L.W., Scamehorn, J.F., and Christian, S.D. 1999. Use of micellar-enhanced ultrafiltration at low surfactant concentrations and with anionic–non ionic surfactant mixtures. *Journal of Colloid Interface Sciences* 213(1): 68–80.

Friesen, T.J., Holloway, G., Hill, G.A., and Pugsley, T.S. 2006. Effect of conditions and protectants on the survival of *Penicillium bilaiae* during storage. *Biocontrol Science Technology* 16: 89–98.

Greaves, J. 2009. Biopesticides, regulatory innovation and the regulatory state. *Public Policy and Administration* 24: 245–264.

Gulati, R., Saxena, R.K., and Gupta, R. 2000. Fermentation and downstream processing of lipase from *Aspergillus terreus*. *Process Biochemistry* 36: 149–155.

Gupta, S. and Dikshit, A.K. 2010. Biopesticides: An eco-friendly approach for pest control. *Journal of Biopesticides* 3(1): 186–188.

Hall, F.R. and Menn, J.J. 1999. *Biopesticides: Use and delivery*. Humana Press, Totowa, NJ.

Jackson, M.A. 1997. Optimizing nutritional conditions for the liquid culture production of effective fungal biological control agents. *Journal of Industrial Microbiology and Biotechnology* 19: 180–187.

Jackson, M.A. and Schisler, D.A. 2002. Selecting fungal biocontrol agents amenable to production by liquid culture fermentation. *Proceedings of 7th biocontrol working group meeting "Influence of a-biotic and biotic factors on biocontrol agents." IOBC/WPRS Bulletin*, pp. 387–391.

Jain, D.K. and Tyagi, R.D. 1992. Leaching of heavy metals from anaerobic sewage sludge by sulfur oxidizing bacteria. *Enzyme and Microbial Technology* 14: 376–383.

Jaronski, S.T. 2007. Soil ecology of the entomopathogenic ascomycetes: A critical examination of what we (think) we know. In: *Use of Entomopathogenic Fungi in Biological Pest Management*, eds. K. Maniana and S. Ekesi, 91–144. Research Sign Post, Trivandrum, India.

Kaelin, P., Morel, P., and Gadani, F. 1994. Isolation of *Bacillus thuringiensis* from stored tobacco and *Lasioderma serricorne* (F.). *Applied Environmental Microbiology* 60: 19–25.

Keller, K., Friedmann, T., and Boxman, A. 2001. The bioseparation needs for tomorrow. *Trends in Biotechnology* 19(11): 438–441.

Khuzhamshukurov, N.A., Yusupov, T.Y., Khalilov, I.M., Guzalova, A.G., Muradov, M.M., and Davranov, K.D. 2001. The insecticidal activity of *Bacillus thuringiensis* cells. *Applied Biochemistry and Microbiology* 37(6): 596–598.

Knowles, B.H. 1994. Mechanism of action of *Bacillus thuringiensis* insecticidal δ-endotoxins. *Advances Insect Physiology* 24: 273–308.

Kuo, W.S. and Chak, K.F. 1996. Identification of novel Cry-type genes from *Bacillus thuringiensis* strains on the basis of restriction fragment length polymorphism of the PCR-amplified DNA. *Applied and Environmental Microbiology* 62: 1369–1377.

Lachhab, K., Tyagi, R.D., and Valero, J.R. 2001. Production of *Bacillus thuringiensis* biopesticides using wastewater sludge as a raw material: Effect of inoculum and sludge solids concentration. *Process Biochemistry* 37(2): 197–208.

Lager, R.J., Joshi, L., Bidochka, M.J., and Roberts, D.W. 1996. Construction of an improved mycoinsecticide over expressing a toxic protease. *Proceedings of the National Academy Sciences U S A* 93: 6349–6354.

Leng, P., Zhang, Z., Pan, G., and Zhao, M. 2011. Applications and development trends in biopesticides. *African Journal of Biotechnology* 10(86): 19864–19873.

Lye, G.J. and Woodley, J.M. 1999. Application of in situ product removal techniques to biocatalytic processes. *Trends in Biotechnology* 17: 395–402.

Magan, N. 2001. Physiological approaches to improving the ecological fitness of fungal biocontrol agents. In: *Fungi as Biocontrol Agents: Progress, Problems and Potential*, eds. T.M. Butt, C. Jackson, and N. Magan, 239–252. CABI Publishing, Wallingford, UK.

Markels, J.H., Lynn, S., and Radke, C.J. 1995. Cross-flow ultra filtration of micellar surfactant solutions. *AIChE Journal* 41(9): 2058–2066.

Moazami, N. 2007. Biopesticides production. In: *Encyclopedia of Life Support Systems*, 6 (Industrial Biotechnology). EOLSS Publishers Co., Paris, France.

Mohammedi, S., Subramanian, S.B., Yan, S., Tyagi, R.D., and Valero, J.R. 2006. Molecular screening of *Bacillus thuringiensis* strains from wastewater sludge for biopesticide production. *Process Biochemistry* 41: 829–835.

Myers, J.H., Simberloff, D., Kuris, A.M., and Cary, J.R. 2000. Eradication revisited: Dealing with exotic species. *Trends in Ecology Evolution* 15: 316–320.

Perkins, J.H. 1982. *Insects, Experts and the Insecticide Crisis: The Quest for New Pest Management Strategies*. Plenum Press, New York.

Prabhakar, A. and Bishop, A.H. 2009. Effect of *Bacillus thuringiensis* naturally colonising *Brassica campestris* var. *chinensis* leaves on neonate larvae of *Pieris brassicae*. *Journal of Invertebrate Pathology* 100: 193–194.

Priest, F. 1977. Extracellular enzymes synthesis in the genus *Bacillus*. *Bacteriological Reviews* 41: 711–753.

Reid, S.J., Surgrue, J.A., and Thomson, J.A. 1986. Industrial applications of a cloned neutral protease gene in *Bacillus subtilis*. *Applied Microbiology and Biotechnology* 24: 311–318.

Rizvi, P.Q., Choudhary, R.A., and Ali, A. 2009. Recent advances in biopesticides. In: *Microbial Strategies for Crop Improvement*, eds. M.S. Khan, A. Zaidi, and J. Musarrat, 185–202. Springer Berlin, Heidelberg, Germany.

Rojas, J.V., Gutierrez, E., and De la Torre, M. 1996. Primary separation of the entomopathogenic products of *Bacillus thuringiensis*. *Biotechnological Progress* 12: 564–566.

Rowe, G.E. and Margaritis, A. 2004. Bioprocess design and economic analysis for the commercial production of environmentally friendly bio insecticides from *Bacillus thuringiensis* HD-1 *kurstaki*. *Biotechnology and Bioengineering* 86(4): 377–388.

Sachdeva, V., Tyagi, R.D., and Valero, J.R. 2000. Production of biopesticides as a novel method of wastewater sludge utilization/disposal. *Water Science and Technology* 42: 211–216.

Sanahuja, G., Banakar, R., Twyman, R.M., Capell, T., and Christou, P. 2011. *Bacillus thuringiensis*: A century of research, development and commercial applications. *Plant Biotechnology Journal* 9: 283–300.

Schugerl, K. 2000. Integrated processing of biotechnology products. *Biotechnological Advances* 18: 581–599.

Schugerl, K. and Hubbuch, J. 2005. Integrated bioprocesses. *Current Opinion in Microbiology* 8: 294–300.

Sharpe, E.S., Herman, A.I., and Toolan, S.C. 1981. Foam flotation process for separating *Bacillus thuringiensis* sporulation products. U.S. Patent 4,247,644.

Shooner, F., Bousquet, S., and Tyagi, R.D. 1996. Isolation, phenotypic, characterization and phylogenetic position of a novel, facultatively autotrophic, moderately thermophilic bacterium *Thiobacillus thermosulfatus* sp. nov. *International Journal of Systematic Bacteriology* 46(2): 409–415.

Smith, R. and Couche, G.A. 1991. The phylloplane as a source of *Bacillus thuringiensis*. *Applied and Environmental Microbiology* 57: 311–315.

Soper, R.S. and Ward, M.G. 1981. Beltsville symposia in agricultural research. *Biological Control in Crop Production* 5: 161–180.

Sorg, B.A. 1999. Multiple chemical sensitivities. *Agrichemical and Environmental News* 155: 1.

Stark, D. and von Stockar, U. 2003. In situ product removal (ISPR) in whole cell biotechnology during the last 20 years. *Advances in Biochemical Engineering/Biotechnology* 80: 149–175.

Sutherst, R.W., Constable, F., Finlay, K.J., Harrington, R., Luck, J., and Zalucki, M.P. 2011. Adapting to crop pest and pathogen risks under a changing climate. *Wiley Interdisciplinary Reviews: Climate Change* 2(2): 220–237.

Szewczyk, B., Hoyos-Carvajal, L., Paluszek, M., Skrzecz, I., and Lobo de Souza, M. 2006. Baculoviruses re-emerging biopesticides. *Biotechnology Advances* 24: 143–160.
Taborsky, V. 1992. Small-scale processing of microbial pesticides. FAO Agricultural Services Bulletin No. 96. Food and Agriculture Organization of the United Nations, Rome, Italy.
Talwar, R.S., Durenja, P., and Rathore, H.S. 2012. Biopesticides. In: *Pesticides—Evaluation of Environmental Pollution*, eds. H.S. Rathore and L.M.L. Nollet. CRC Press, Boca Raton, FL.
Tamez-Guerra, P., McGuire, M.R., Medrano-Roldan, H., and Galan-Wong, L.J. 1996. Sprayable granule formulations of *Bacillus thuringiensis*. *Journal of Economic Entomology* 89: 1424–1430.
Teera-Arunsiri, A., Suphantharika, M., and Ketunuti, U. 2003. Preparation of spray-dried wettable powder formulations of *Bacillus thuringiensis* based biopesticides. *Journal of Economic Entomology* 96: 292–299.
Tirado-Montiel, M.L., Tyagi, R.D., and Valero, J.R. 2001. Wastewater treatment sludge as a raw material for the production of *Bacillus thuringiensis* based biopesticides. *Water Research* 35(16): 3807–3816.
Tirado-Montiel, M.L., Tyagi, R.D., Valero, J.R., and Surampalli, R.Y. 2003. Production biopesticides using wastewater sludge as a raw material—Effect of process parameters. *Water Science and Technology* 48(8): 239–246.
Tzeng, Y.M., Tsun, H.Y., and Chang, Y.N. 1999. Recovery of *thuringiensin* with cetylpyridinium chloride using micellar-enhanced ultrafiltration process. *Biotechnological Progress* 15: 580–586.
Ulbricht, M. 2006. Advanced functional polymer membranes. *Polymer* 47: 2217–2262.
Uribe, D. and Khachatourians, G.G. 2004. Restriction fragment length polymorphism of mitochondrial genome of the entomopathogenic fungus *Beauveria bassiana* reveals high intraspecific variation. *Mycology Research* 108: 1070–1078.
van Regenmortel, M.H.V., Fauquet, C.M., Bishop, D.H.L. et al. 2000. Virus taxonomy. *Seventh Report of the International Committee on Taxonomy of Viruses*. Academic Press, New York.
Van Rie, J., Jansens, S., Hofte, H., Degheele, D., and Van Mellaert, H. 1990. Receptors on the brush border membrane of the insect midgut as determinants of the specificity of *Bacillus thuringiensis* delta endotoxins. *Applied Environmental Microbiology* 56: 1378–1385.
Vega, O.F.L. 1999. A review of *Bacillus thuringiensis* (*Bt*) production and use in Cuba. *Biocontrol News and Information* 20(1): 47–48.
Vega, F.E., Jackson, M.A., and McGuire, M.R. 1999. Germination of conidia and blastospores of *Paecilomyces fumosoroseus* on the cuticle of the silverleaf whitefly, *Bemisia argentifolii*. *Mycopathologia* 147: 33–35.
Vidyarthi, A.S., Tyagi, R.D., Valero, J.R., and Surampalli, R.Y. 2002. Studies on the production of *Bacillus thuringiensis* based biopesticides using wastewater sludge as a raw material. *Water Research* 36: 4850–4860.
von Stockar, U. and van der Wielen, L.A.M. 2003. Process integration challenges in biotechnology. Yesterday, today and tomorrow. *Advances in Biochemical Engineering and Biotechnology* 80: 9–15.
Vora, D. and Sethna, Y.I. 1999. Enhanced growth, sporulation and toxin production by *Bacillus thuringiensis* subsp. *kurstaki* in oil seed meal extract media containing cystine. *World Journal of Microbiology and Biotechnology* 15: 747–749.

Wraight, S.P., Jackson, M.A., and De Kock, S.L. 2001. Production, stabilization and formulation of fungal biocontrol agents. In: *Fungi as Biocontrol Agents: Progress, Problems and Potential*, eds. T.M. Butt, C. Jackson, and N. Magan, 253–288. CABI Publishing, Wallingford, UK.

Wraight, S.P. and Ramos, M.E. 2005. Synergistic interaction between *Beauveria bassiana*– and *Bacillus thuringiensis tenebrionis*–based biopesticides applied against field populations of Colorado potato beetle larvae. *Journal of Invertebrate Pathology* 90: 139–150.

Yeo, H., Pell, J.K., Alderson, P.G., Clark, S.J., and Pye, B.J. 2003. Laboratory evaluation of temperature effects on the germination and growth of entomopathogenic fungi and on their pathogenicity to two aphid species. *Pest Management Science* 59: 156–165.

Yezza, A., Tyagi, R.D., Valero, J.R., and Surampalli, R.Y. 2004. Scale-up of biopesticide production process using wastewater sludge as a raw material. *Journal of Industrial Microbiology and Biotechnology* 31: 545–552.

Yezza, A., Tyagi, R.D., Valero, J.R., and Surampalli, R.Y. 2005. Production of *Bacillus thuringiensis* based biopesticides by batch and fed-batch culture using wastewater sludge as a raw material. *Journal of Chemical Technology and Biotechnology* 80: 502–510.

Ying, S.H. and Feng, M.G. 2006. Medium components and culture conditions affect the thermotolerance of aerial conidia of the fungal biocontrol agent *Beauveria bassiana. Letters in Applied Microbiology* 43: 331–335.

Zamola, B., Valles, P., Meli, G., Miccoli, P., and Kajfez, F. 1981. Use of the centrifugal separation technique in manufacturing a bioinsecticide based on *Bacillus thuringiensis. Biotechnology and Bioengineering* 23: 1079–1086.

Zhang, X., Candas, M., Griko, N., Taussig, R., and Bulla, L., Jr. 2006. A mechanism of cell death involving an adenylyl cyclase/PKA signalling pathway is induced by the Cry1Ab toxin of *Bacillus thuringiensis. Proceedings of the National Academy of Sciences U S A* 103: 9897–9902.

Zhou, X., Chen, S., and Yu, Z. 2004. Effects of spray drying parameters on the processing of a fermentation liquor. *Biosystems Engineering* 88(2): 193–199.

Zouari, N., Dhouib, A., Ellouz, R., and Jaoua, S. 1998. Nutritional requirements of strain of *Bacillus thuringiensis* subsp. *kurstaki* and use of gruel hydrolysate, for the formulation of a new medium for delta-endotoxin production. *Applied Biochemistry and Biotechnology* 69: 1–52.

Part III

Biotechnology in Food Processing

7

Production of Fermented Foods

Sanjeev K. Soni, Raman Soni, and Chetna Janveja

CONTENTS

7.1 Introduction 221
7.2 Fermentation Technologies in Food Preparation 221
7.2.1 Submerged State Fermentation (SmF) 222
7.2.2 Solid-State Fermentation 222
7.2.3 Surface Fermentation 223
7.3 Microorganisms as the Agents of Fermentation in Foods 223
7.3.1 Bacteria 223
7.3.1.1 Lactic Acid Bacteria 224
7.3.1.2 Acetic Acid Bacteria 224
7.3.1.3 Bacteria of Alkaline Fermentations 225
7.3.2 Yeasts 225
7.3.3 Molds 226
7.4 Factors Affecting Fermentation 226
7.4.1 pH 226
7.4.2 Moisture Content 226
7.4.3 Nutrient Content 227
7.4.4 Osmotic Pressure 227
7.4.5 Antimicrobial Substances 227
7.4.6 Biological Structures 227
7.4.7 Temperature 228
7.4.8 Presence of Gases 228
7.5 Starter Cultures and Their Role 228
7.5.1 Benefits of Using Starter Cultures 229
7.5.2 Preparation of Starter Cultures 229
7.6 Indigenous Fermented Foods 230
7.6.1 Fermented Milk Products 230
7.6.1.1 Cheese 230
7.6.1.2 Yogurt 240
7.6.1.3 Cultured Buttermilk 242
7.6.1.4 Cultured Cream 242
7.6.1.5 Acidophilous Milk 242
7.6.1.6 Butter and Ghee 243
7.6.1.7 Shrikhand 243

7.6.1.8 Koumiss 244
7.6.1.9 Kefir 244
7.6.2 Fermented Cereal Products 245
7.6.2.1 Bread 245
7.6.2.2 Pasta 246
7.6.2.3 Bhatura 247
7.6.2.4 Nan 247
7.6.2.5 Jalebis 247
7.6.2.6 Ogi 248
7.6.2.7 Uji 248
7.6.2.8 Kenkey 249
7.6.2.9 Mahewu 249
7.6.2.10 Injera 249
7.6.2.11 Kisra 250
7.6.2.12 Kishk 251
7.6.2.13 Bushera 251
7.6.2.14 Togwa 251
7.6.2.15 Dolo and Pito 253
7.6.2.16 Kaffir Beer and Chibuku 253
7.6.3 Fermented Legume Products 253
7.6.3.1 Tempeh 254
7.6.3.2 Natto 255
7.6.3.3 Miso 255
7.6.3.4 Soy Sauce 256
7.6.3.5 Sufu 256
7.6.3.6 Oncom 257
7.6.3.7 Dawadawa 258
7.6.3.8 Warri/Wadiyan 258
7.6.3.9 Bhallae 259
7.6.3.10 Vada/Vadai 259
7.6.3.11 Papadam 259
7.6.4 Fermented Cereal–Legume Products 260
7.6.4.1 Idli 260
7.6.4.2 Dosa 261
7.6.4.3 Dhokla 262
7.6.4.4 Adai 262
7.6.5 Fermented Vegetable Products 263
7.6.5.1 Sauerkraut 263
7.6.5.2 Pickles 264
7.6.6 Meat and Fish Products 265
7.6.6.1 Sausages 265
7.6.6.2 Fish Paste 266
7.6.6.3 Fish Sauce 267
7.7 Nutritional and Health Benefits of Fermented Foods 267

7.8 Food Safety Aspects of Fermented Foods..268
7.8.1 Effect of Fermentation on Pathogenic Organisms..................269
7.9 Summary and Future Prospects..269
References...270

7.1 Introduction

Microorganisms are associated with all the foods that we eat and are responsible for the formation of certain foods through the process of fermentation and may also be directly used as a source of food in the form of single-cell proteins and food supplements in the form of pigments, amino acids, vitamins, organic acids, enzymes, and lipidic substances and also inhibit other toxin-producing microbes. Microbial activity in foods enhances their shelf life, makes them more attractive in terms of appearance and flavors that are more appetizing, and is easily digestible. The biological agents involved in fermentations include lactic acid bacteria (LAB) and alcohol-producing yeasts and molds, occurring in succession or in combination. In general, bacteria cause acidification and leavening, yeasts ferment the sugars to produce alcohol, and esters impart desirable flavors in the products, while molds bring about the saccharification and proteolysis of the starting materials.

Wine fermentation was probably the first fermentation known to mankind. During the last 150 years, the scientific basis of winemaking has gradually become clearer, and many new varieties of wine have evolved (Pretorius, 2000). Most of the wines are generally prepared from overripened grapes, but commercial wines are also made from several fruits other than grapes. Indigenous fermented legume and cereal products form an important part of human diet in Southeast Asia including India, the Middle East, and Africa. They are also becoming popular in the developed world due to their high nutritional value and organoleptic characteristics. Fermented dairy products also constitute a vital part of human diet in many parts of the world. Pickles from vegetables and fruits like mango, gooseberry, and lemons are prepared and are used throughout India. More than their nutritive value, they act as food adjuncts and appetizers as they add palatability to foods. Indian pickles are a very big industry now and the products are popular throughout the world (Soni and Arora, 2000).

7.2 Fermentation Technologies in Food Preparation

Fermentation is an energy-generating process in which certain microorganisms are capable of converting carbohydrates into smaller organic compounds. It involves the controlled large-scale cultivation of microorganisms for producing and recovering valuable products by the process of fermentation technology, which uses microbiology, biochemistry, and engineering in an integrated fashion. Fermentation technologies can be differentiated on the basis of state in which the microorganisms are cultivated and include submerged, surface, and solid-state culture technologies (Ghose and Sahai, 1979; Ju and Afolabi, 1999; Mitchell et al., 2006; Singhania et al., 2009). Industrial fermentations may be carried out batchwise either as fed-batch operations or as continuous cultures. Batch and fed-batch operations are quite common, continuous fermentations being relatively rare. Typically, continuous fermentations are started as batch cultures and feeding begins after the microbial population has reached a certain concentration.

7.2.1 Submerged State Fermentation (SmF)

It refers to those fermentations wherein microorganisms employed grow in submerged state within the fermentation media that is generally agitated. The common examples of useful food products or supplements obtained by this fermentation include alcoholic beverages, lactic acid, vinegar, citric acid, amino acids, vitamins, single-cell proteins, probiotics, and enzymes. Most commonly used submerged fermenters in food processing include stirred tank fermenter, bubble column fermenter, airlift fermenter, fluidized bed fermenter, and trickle bed fermenters.

7.2.2 Solid-State Fermentation

It refers to the growth of microorganisms on the surface of the solid medium that occurs under static conditions in the absence of free-flowing water. The common examples of useful food products or their supplements obtained by this fermentation include cheese, indigenous fermented foods, enzymes, citric acid, etc. Solid-state fermentation (SSF) processes exhibit several advantages over surface fermentations, including improved product characteristics, higher product yields and productivities, easier product recovery, and reduced energy requirements. The use of SSF for food processing practiced since centuries has been reviewed by Soccol and Vandenberghe (2003). Primarily, two types of solid supports are mainly used in SSF processes. First are the SSF solid substrates like starch or lignocellulose residues or agro-industrial sources such as grains and grain by-products, cassava, potato, rice, beans, and sugar beet pulp. In these cases, substrate is also used as the source of carbon and nutrients for microbial growth (Tengerdy and Szakacs, 2003). Other SSF processes use inert natural or artificial solid supports like perlite, amberlite, and polyurethane foam. In the latter processes, the support is used only as an attachment for the microorganism. From the engineering point of view, inert supports are better, because they do not change their geometric and physical characteristics due to the microbial growth, allowing a better control of heat and mass transfer (Ooijkaas et al., 2000). The common solid-state fermenters used in the processing of foods into value-added products include tray fermenter, solid-bed fermenter, rotary disk fermenter, rotary drum fermenter, agitated tank fermenter, and continuous screw fermenter.

7.2.3 Surface Fermentation

The surface fermentation also known as biofilm fermentation involves the cultivation of either a population or a community of microorganisms living attached to a surface. Biofilms can be developed on either biotic or abiotic surfaces from a single species or as a community derived from several species (Fenchel 2002). This way of growth is the prevailing lifestyle of microorganisms including bacteria, yeast, and filamentous fungi (Armstrong et al., 2001). Filamentous fungi are naturally adapted to growth on surfaces and in these conditions they show a particular physiological behavior that is different to that in submerged culture. Differential physiological behavior of most attached fungi corresponds principally to a higher production and secretion of enzymes and also to a morphological differentiation that is absent in submerged cultures (Biesebeke et al., 2002). For a long time, SSF has been considered to be advantageous due to water limitation of the system so that a higher product concentration and volumetric productivity are attained. However, for an industrial application, scaling up an SSF process is a necessity that is very difficult to achieve. Submerged fermentation is the process of choice for industrial operations related to production of fungal enzyme due to the very well-known engineering aspects such as fermentation modeling, bioreactor design, and process control.

7.3 Microorganisms as the Agents of Fermentation in Foods

Processing of foods by fermentation not only ensures increased shelf life and microbiological safety of a food but may also make some foods more digestible, and as in case of cassava, fermentation reduces toxicity of the substrate. Although much fermentation are traditionally dependent on inoculation from a previous batch, starter cultures are available for ensuring consistency of process and product quality (Caplice and Fitzgerald, 1999). The most common groups of microorganisms involved in food fermentations include bacteria, yeasts, and molds.

7.3.1 Bacteria

The most important bacteria desirable in food fermentations are the LAB, which have the ability to produce lactic acid from carbohydrates. Other important bacteria, especially in the fermentation of fruits and vegetables, are the acetic acid bacteria (Soni et al., 2012).

7.3.1.1 Lactic Acid Bacteria

The LAB represent a group of gram-positive, non-respiring, non-spore-forming, cocci or rods that are functionally related by their ability to produce lactic acid during homo- or heterofermentative metabolism. The acidification and enzymatic processes accompanying the growth of LAB impart the key flavor, texture, and preservative qualities to a variety of fermented foods. Industrial applications of the LAB rely on six key beneficial and nonpathogenic species: *Lactococcus, Lactobacillus, Leuconostoc, Pediococcus, Oenococcus,* and *Streptococcus.* The fermentation of milk with intestinal species of *Lactobacillus acidophilus* and/or *Bifidobacterium bifidum* has been applied for number of products (Driessen and Boer, 1989). Besides metabolic end products, some strains of LAB also secrete antimicrobial proteinaceous compounds, termed bacteriocins, which kill closely related bacteria. On the basis of their chemical, structural, and functional properties, bacteriocins could be classified into three main classes (Nes et al., 1996).

LAB carry out the conversion of carbohydrate, in the absence of oxygen. Some families are homofermentative, while others are heterofermentative. *Lactobacillus acidophilus, Lactobacillus bulgaricus, Lactobacillus plantarum, Lactobacillus casei, Lactobacillus pentoaceticus, Lactobacillus brevis,* and *Lactobacillus thermophilus* are examples of lactic acid–producing bacteria involved in food fermentations, but overall, *Lactobacillus plantarum,* a homofermenter, produces high acidity and plays a major role in all vegetable fermentations. Homofermenters convert sugars primarily to lactic acid, while heterofermenters produce about 50% lactic acid, 25% acetic acid and ethyl alcohol, and 25% carbon dioxide. These other compounds are important as they impart particular tastes and aromas to the final product. Organisms from the gram-positive Propionibacteriaceae family are responsible for the flavor and texture of some fermented foods, especially Swiss cheese, where they are responsible for the formation of "eyes" or holes in the cheese. These bacteria break down lactic acid into acetic and propionic acids and carbon dioxide. The LAB generally used in milk fermentation are from two groups including gram-positive cocci belonging to *Streptococcus thermophilus, Streptococcus lactis, Streptococcus lactis* subsp. *diacetylactis,* and *Streptococcus cremoris* and gram-positive asporogenous rod-shaped bacteria belonging to *Lactobacillus bulgaricus* and *Lactobacillus acidophilus.* Many strains of LAB belonging to *Lactobacillus acidophilus, Bifidobacterium* spp., and *Lactobacillus casei* have probiotic properties with a number of health benefits. Interactions among lactic acid starter and probiotic bacteria have been investigated to establish adequate combinations of strains to manufacture probiotic dairy products (Vinderola et al., 2002).

7.3.1.2 Acetic Acid Bacteria

Species of *Acetobacter,* the acetic acid producers, are the second group of bacteria of importance in food fermentations. *Acetobacter* are important in the production of vinegar from fruit juices and alcohols (Soni et al., 2012).

Acetobacter can cause undesirable oxidation of alcohol to acetic acid resulting in a vinegary off-taste in wine also if they are not properly stored/handled. The most desirable action of acetic acid bacteria is in the production of vinegar. The organisms involved in vinegar production usually grow at the top of the substrate, forming a jelly-like mass. This mass is known as "mother of vinegar." The mother is composed of both acetic acid bacteria and yeasts, which work together. The principal bacteria are *Acetobacter aceti, Acetobacter xylinum,* and *Acetobacter ascendens.* It is important to maintain an acidic environment to suppress the growth of undesirable organisms and to encourage the presence of desirable acetic acid–producing bacteria.

7.3.1.3 Bacteria of Alkaline Fermentations

A third group of bacteria, which includes the *Bacillus* species, responsible for alkaline fermentations. *Bacillus subtilis, Bacillus licheniformis,* and *Bacillus pumilus* are the dominant species, causing the hydrolysis of protein to amino acids and peptides and releasing ammonia, which increases the alkalinity and makes the substrate unsuitable for the growth of spoilage organisms (Soni et al., 2012). Alkaline fermentations are more common with protein-rich foods such as soybeans and other legumes.

7.3.2 Yeasts

The most beneficial yeasts in terms of desirable food fermentation are from the *Saccharomyces* family, especially *Saccharomyces cerevisiae.* Yeasts play an important role in the food industry as they produce enzymes that favor desirable chemical reactions such as the leavening of bread and the production of alcohol and invert sugar. Some yeasts are chromogenic and produce a variety of pigments, including green, yellow, and black. Others are capable of synthesizing essential B group vitamins. Although there is a large diversity of yeasts and yeast-like fungi, only a few (ascomycetous yeasts or members of genus *Candida*) are commonly associated with the production of fermented foods. Different strains of *Saccharomyces cerevisiae* are commonly used in fermented foods and fruit- and vegetable-based beverages. All strains of this genus ferment glucose and many ferment other plant-derived carbohydrates such as sucrose, maltose, and raffinose. In the tropics, *Saccharomyces pombe* is the dominant yeast in the production of traditional fermented beverages, especially those derived from maize and millet.

Of the various yeast genera, *Brettanomyces, Candida, Cryptococcus, Debaryomyces, Dekkera, Hanseniaspora, Kloeckera, Kluyveromyces, Metschnikowia, Pichia, Rhodotorula, Saccharomyces, Saccharomycodes, Schizosaccharomyces,* and *Zygosaccharomyces* are associated with winemaking. *Kloeckera apiculata* and *Hanseniaspora uvarum* are the predominant yeasts intimately associated on the surface of grape berries, accounting for roughly 50%–75% of the total yeast population (Fleet, 1998).

7.3.3 Molds

Molds are also important organisms, both as spoilers and preservers of foods. Certain molds (*Aspergillus* species) produce undesirable toxins and contribute to the spoilage of foods; however, others impart characteristic flavors to foods and others produce enzymes, such as amylase for bread making. Molds from the genus *Penicillium* are associated with the ripening and flavor of cheeses. Molds are aerobic and therefore require oxygen for growth.

7.4 Factors Affecting Fermentation

Microbial growth is greatly affected by the chemical and physical nature of their surroundings. For successful growth of microorganisms, it is essential not only to supply proper and balanced nutrients but also to maintain physical and chemical parameters. Many of these factors are a part of the fermentation medium composition and thus categorized as internal factors, while others include the properties of the environment and are categorized as external factors (Soni, 2007; Frazier and Westhoff, 2008). The parameters that can affect the growth and metabolic activities of microorganisms are discussed in the proceeding sections.

7.4.1 pH

Every organism has a pH range with minimum and maximum values. Microorganisms perform best at pH around 7.0 (6.5–7.5). As foods differ from each other in terms of their pH, they show variation in the microbial types associated with them. Bacteria are more strict in their requirements for pH than yeasts and molds. Molds have a pH range of 0–11, yeasts grow over pH 1.5–8.5, and LAB have the broadest pH range of 3.2–10.5 among bacteria. pH is affected by inherent acidity due to native ingredients as well as by biological acidity, due to the action of certain microorganisms. pH affects the functioning of enzymes and transport of nutrients into the cell. Microorganisms tend to change the pH of the environment to bring it to the optimal value. When an organism grows in an acidic medium, its metabolic activities result in the medium becoming less acidic and vice versa.

7.4.2 Moisture Content

Water is one of the most essential requirements for life. Thus, its availability becomes a most important factor for the fermenting organism to grow and utilize the substrate. The availability of water depends on two factors—the water content of the surrounding environment and the concentration of solutes (salts, sugars, etc.) dissolved in the water. Water requirement for the living organisms can be defined in terms of water activity (a_w), generally defined as the vapor pressure of a liquid divided by that of pure water at the same temperature. Water activity of pure water is 1.0, while that of most fresh foods is 0.99. Each organism has a range of a_w over which it can grow. Lowering of a_w below optimum increases lag phase, decreases growth rate and size of the final population.

7.4.3 Nutrient Content

Carbohydrates are the main carbon compounds consisting of mono-, di-, oligo-, and polysaccharide sugars and are most commonly used as energy source. Other carbon compounds like esters, alcohols, peptides, amino acids, and organic acids may also act as energy source. In the absence of carbohydrates, a limited number of microorganisms can obtain their energy from fats and split products of proteins. Nitrogen sources in the form of proteins can be hydrolyzed by certain microorganisms. Vitamins are required by all the organisms. Some microorganisms are able to manufacture some or all of the vitamins needed.

7.4.4 Osmotic Pressure

High substrate concentrations have been shown to inhibit microbial growth and fermentation performance as a result of high osmotic pressure and low water activity. In order to maintain cell viability and increased fermentation rate at high sugar concentrations, it would be necessary to reduce the detrimental effects of high osmotic pressure.

7.4.5 Antimicrobial Substances

These may be originally present or added purposely or developed during processing of foods or during growth of microorganisms in foods. Examples of naturally present antimicrobial substances include lactenins and anticoliform factor in fresh milk, lysozyme in egg white, and essential oils in spices. The purposely added antimicrobial substances include organic acids, SO_2, nitrite, nitrate, NaCl, sugar, etc., and the antimicrobial substances developing during the processing of foods are lipid oxidation products, developing during heating of fats and furfural, hydroxymethyl furfural, developing during heating of sugars. Examples of inhibitory substances developing during microbial activity are acids including lactic, acetic, and propionic; alcohols; peroxides; or even antibiotics.

7.4.6 Biological Structures

The natural covering of fruits and vegetables, testa of seeds, shell of nuts, eggs, and hide of animals act as barriers for the entry and growth of microorganisms, and thus, as inner part of a whole, healthy tissue of living plants and animals is either sterile or low in microbial content.

7.4.7 Temperature

All forms of life are greatly influenced by temperature. Temperature influences the rate of chemical reactions and protein structure integrity thus affecting rates of enzymatic activity. At low temperature, enzymes are not denatured; therefore, every 10°C rise in temperature results in rise of metabolic activity and growth of microorganisms. Thus, high temperature kills microorganism by denaturing enzymes, by inhibiting transport carrier molecules, or by changing membrane integrity.

7.4.8 Presence of Gases

The microorganisms that are completely dependent on atmospheric oxygen for growth are called obligate aerobes, whereas those that do not require oxygen for growth but grow well in its presence are called facultative anaerobes. Aerotolerants ignore O_2 and can grow in its presence or absence. In contrast, obligate anaerobes do not tolerate the presence of oxygen at all and ultimately die. Few microorganisms require oxygen at very low level (2%–10%) of concentration and are called microaerophiles. The latter are damaged by the normal atmospheric level of oxygen (20%).

7.5 Starter Cultures and Their Role

Fermentation starters are preparations to assist the beginning of the fermentation process of various foods and fermented drinks. A starter culture is a microbiological culture that actually performs fermentation. Many products are fermented using LAB and, in some yeasts and molds, are used as starter cultures. The microorganisms exploited in the fermentation of the products make substantial contribution to taste, flavor, texture, enhanced nutrition, digestibility, and detoxification of the end products. Back-slopping technique is also exploited in preparation of fermented products and this entails inoculation of raw material with residue from the previous batch (Holzapfel, 1997). Numerous traditional fermented foods are still prepared by this technique, in addition to utilization of starter culture technique.

Starter culture as described by Holzapfel (1997) refers to a preparation containing high numbers of live microorganisms, which may be inoculated to a food raw material to produce desirable changes. Exploitation of starters is very common and they serve to accelerate fermentation processes (Mugula et al., 2003), thus leading to improved and more predictable fermentation process, improved safety, and reduced hygienic risks (Kimaryo et al., 2000), and contribute to desirable sensory attributes (Annan et al., 2003).

7.5.1 Benefits of Using Starter Cultures

The primary aim of application of starter cultures is to improve the fermentation process. The starter cultures help (i) to drive the fermentation process, (ii) rapid pH drop of the product, (iii) aroma compound accumulation during product ripening, (iv) beneficial health effects, (v) improve the commercial and hygienic quality of the product, and (vi) overall process standardization. The microorganisms used mostly originate from the foods to which they are applied and are selected based on viability, competitiveness, adaptability to the substrate, and desired properties (Holzapfel, 1997, 2002). The competitiveness depends on the type, quality of substrate, and microbial interaction during fermentation.

Fermented products produced using starters are usually of consistent quality. The utilization of lactic acid starter cultures in amylase-rich flours has been found to provide considerable opportunity for increase in nutrient density (Nout and Motarjemi, 1997). Starter cultures of LAB and yeast have also been associated with reduction of tannins and high levels of disulfide cross-linkages in sorghum prolamin proteins (Khetarpaul and Chauhan, 1989). This significantly improves protein digestibility and quality of many cereal products (Ali et al., 2003).

7.5.2 Preparation of Starter Cultures

The preparation of starter cultures is based on the isolation of pure cultures from mixed population of traditional fermented foods, followed by taxonomic identification and metabolic characterization, which includes evaluation of the growth and competitive behavior and the adaptability to a particular substrate. Therefore, selection for suitable starter cultures should take into consideration the behavior of the strains either singly or in mixed cultures under defined conditions and in the food substrate (Holzapfel, 2002). The selected strains for starter should be able to survive and remain viable throughout the fermentation process and be capable of outcompeting and eliminating pathogenic and food spoilage microorganisms through their metabolic activities. Additionally, starter cultures should generate desirable sensory qualities, eliminate antinutritional factors, and reduce hygienic and toxicological risks in fermented foods. Nonetheless, effectiveness of these starter organisms could only be achieved through stable preparation of cultures. In the Western World, commercial starter cultures are available in freeze-dried form (Leroy and De Vuyst, 2004) and the prospect of applying freeze-dried starter cultures in fermented products is realistic. The dehydration way of preparing starter cultures holds potential and this has been demonstrated in the preparations of lafun in which a small amount of lafun flour from previous fermentation is added to the soaked chopped cassava roots (Padonou et al., 2009). Traditional drying of kivunde (Kimaryo et al., 2000) also provides possibility for the dried product to be used as a starter culture. Cabinet and drum drying of aflata (Nche et al., 1994) has also been possible and indicated potential for distribution as a dehydrated starter culture. The commercial mahewu is also available in dehydrated form, which contains viable LAB. This way of preparing and preserving fermented cereal and cassava products presents a tremendous and important foundation for development and practical application of dehydrated starters for small-scale processing of fermented products. Starter cultures clearly contribute to improved quality and microbial safety of the final product (Holzapfel, 1997; Kimaryo et al., 2000; Annan et al., 2003).

7.6 Indigenous Fermented Foods

Indigenous fermented products form an important part of human diet in Southeast Asia including India, the middle East, and Africa. They are also becoming popular in the developed world due to their high nutritional value and organoleptic characteristics. There are several varieties of fermented foods commonly used in various geographic regions of the world differing from each other on the basis of the substrates employed, the method of preparation, the nature and uses of various products (Table 7.1). The fermented foods can be divided into various categories on the basis of the substrates and product use.

7.6.1 Fermented Milk Products

Milk has been considered as one of the most natural and highly nutritive part of a daily balanced diet. Currently, the integration of advanced scientific knowledge with traditional information is gaining incredible momentum toward developing the concept of potential therapeutic foods. At present, the best-known examples of therapeutic foods are fermented milk products containing health-promoting probiotic bacteria (Nagpal et al., 2012). Fermented dairy products are popular because of their being healthy and low fat foods. Some of the common fermented milk products are discussed in the following sections.

7.6.1.1 Cheese

Cheese is a milk product in which casein is precipitated in the form of a curd that holds most of the fat and other suspended materials while allowing water and dissolved constituents to drain away. Besides milk, selected bacteria, a milk-clotting agent, and sodium chloride are used to manufacture cheese. The conversion of milk into cheese involves several interrelated operations, namely, coagulation, acidification, water removal, salt addition, and ripening.

TABLE 7.1

Some Important Indigenous Fermented Foods of the World

S. No.	Fermented Food	Country/Place	Ingredient(s)	Nature of Product	Product Use	Dominant Microorganism(s)
Fermented milk products						
1	Cultured cream	United States, Russia, Central Europe	Cow's milk cream	Sour gel/cream-containing butter-like aromatic flavor	Topping on salads, vegetables, fruits, meat preparation, and in cake filling	LAB
2	Cultured buttermilk	United States, Europe, Asia	Cow's skim milk, cream	Viscous fluid milk	Health drink	*Streptococcus lactis*, *Leuconostoc cremoris*
3	Yogurt	United States, Europe, Asia	Cow/goat/sheep milk, condensed skim milk, cream, nonfat dry milk, sucrose, fruits	Flavored curd	Nutritive food supplement	*Streptococcus thermophilus*, *Lactobacillus delbrueckii*
4	Acidophilus milk	United States, Russia	Cow's milk	Acid-fermented fluid milk	Health drink	*Lactobacillus acidophilus*
5	Kefir	Russia	Cow/goat/sheep's milk	Acid and alcoholic milk	Carbonated drink	LAB and yeasts
6	Koumiss	Russia	Mare's milk	Acid and alcoholic milk	Carbonated drink	LAB and yeasts
7	Cottage cheese	United States	Cow's milk	Natural unripened soft cheese	Protein and fat supplement, taste enhancer	LAB, *Propionibacterium roquefortii*

(*continued*)

TABLE 7.1 (continued)

Some Important Indigenous Fermented Foods of the World

S. No.	Fermented Food	Country/Place	Ingredient(s)	Nature of Product	Product Use	Dominant Microorganism(s)
Fermented cereal products						
8	Ang-Kak (Chinese red rice)	China, Taiwan, Philippines, Thailand	Rice	Solid	Coloring other foods	*Monascus purpureus*
9	Ambali	India	Millet flour	Semisolid	All-time food	*Leuconostoc mesenteroides*
10	Bhatura	India (North)	White wheat flour	Deep-fried	Breakfast bread	LAB
11	Fermented rice	India	Rice	Semisolid	Breakfast	Yeasts
12	Hopper (appa)	Sri Lanka	Rice or white wheat flour and coconut water	Semisolid	Breakfast	LAB
13	Injera	Ethiopia	Tef	Flat bread	Staple food	LAB
14	Jalebi	India and Pakistan	White wheat flour	Deep-fried pretzel	Confection food	*Lactobacillus fermentum*, *Lactobacillus buchneri*, *Streptococcus lactis*
15	Kako and koko sour water	Ghana	Pearl millet	Porridge and beverage	Staple food	LAB
16	Kenkey	Ghana	Maize	Semisolid	Breakfast, lunch, and supplement to fish stews	*Lactobacillus fermentum*, *Lactobacillus reuteri*, *Candida krusei*, *Saccharomyces cerevisiae*
17	Kirario	Kenya	Green maize + millet	Porridge	Traditional drink in many of the social and cultural festivities	LAB

18	Kisra	Sudan	Sorghum flour	Spongy bread	Staple food	*Pediococcus pentosaceus, Lactobacillus brevis, Lactobacillus fermentum, Lactobacillus reuteri, Leuconostoc amylovorus, D. hansenii, Candida intermedia*
19	Kivunde	Tanzania	Cassava	Flour for porridges		*Lactobacillus plantarum*
20	Kulcha	India (North) and Pakistan	White wheat flour	Flat bread	Staple food	LAB and yeasts
21	Lafun	Benin and Nigeria	Cassava	Flour for porridges	Staple food	Bacteria
22	Mahewu	South Africa	Maize meal	Liquid	Beverage food	LAB
23	Mawe	Benin and Togo		Sourdough for breads, beverages, porridges		*Lactobacillus fermentum, Lactobacillus reuteri, Lactobacillus brevis, Lactobacillus curvatus, Lactobacillus buchneri, Pediococcus acidilactici, Lactobacillus lactis, Lactobacillus salivarius, Candida krusei, Candida kefyr, Candida glabrata, Saccharomyces cerevisiae*
24	Nan	India (North), Pakistan, Iran, and Afghanistan	White wheat flour	Breads	Staple food	*Saccharomyces cerevisiae,* LAB

(continued)

TABLE 7.1 (continued)

Some Important Indigenous Fermented Foods of the World

S. No.	Fermented Food	Country/Place	Ingredient(s)	Nature of Product	Product Use	Dominant Microorganism(s)
25	Ogi	Nigeria	Maize, millet, or sorghum	Beverages	Breakfast and infant food	LAB, *Cephalosporium* sp., *Fusarium* sp., *Aspergillus* sp., *Saccharomyces cerevisiae*, *Candida* sp.
26	Pozol	Mexico	White maize	Porridges	Beverage or porridge	LAB, yeasts, and fungi
27	Puto	Philippines	Rice	Semisolid	Breakfast and snack food	LAB, *Saccharomyces cerevisiae*
28	Shamsy bread	Egypt	Wheat flour	Spongy bread	Staple food	—
29	Ting	Botswana, South Africa	Sorghum	Sourdough for preparation of porridges	Protein supplement with meat, soup, or vegetables for lunch or supper	*Lactobacillus fermentum*, *Pediococcus acidilactici*, *Lactobacillus plantarum*
30	Uji	Kenya	Maize, sorghum, millet	Sour porridge	Breakfast food for children	*Lactobacillus fermentum*, *Lactobacillus cellobiosus*, *Lactobacillus buchneri*
31	Bushera	Uganda	Sorghum	Beverage	Weaning beverage	*Lactobacillus plantarum*, *Lactobacillus paracasei*, *Lactobacillus brevis*, *Lactobacillus fermentum*, *Lactobacillus delbrueckii*, *Streptococcus thermophilus*
32	Obiolor	Nigeria	Sorghum, millet, maize	Beverage	Health drink	*Lactobacillus plantarum*, *Streptococcus lactis*, *Bacillus* spp.

33	Togwa	Tanzania	Sorghum, maize, millet, and cassava	Beverage	Health drink	*Lactobacillus brevis*, *Lactobacillus cellobiosus*, *Lactobacillus fermentum*, *Lactobacillus plantarum*, *Pediococcus pentosaceus*, *Candida pelliculosa*, *Candida tropicalis*, *Issatchenkia orientalis*, *Saccharomyces cerevisiae*
34	Kunun-zaki	Nigeria	Millet	Beverage	Health drink	*Lactobacillus fermentum*, *Lactobacillus leichmannii*, *Saccharomyces cerevisiae*
35	Kwete	Uganda	Maize + malted millet	Beverage	Thirst-quenching beverage during hot days	*Lactobacillus*, *Lactococcus*, yeasts
36	Pito–dolo	Ghana, Togo, Nigeria, Faso, Ivory, Mali	Sorghum	Alcoholic beverage	Energy drink	*Saccharomyces cerevisiae*, *Lactobacillus fermentum*
37	Kaffir beer	South Africa	Maize	Alcoholic beverage	Drink	*Lactobacillus* spp., yeasts
38	Chibuku	Botswana, Zimbabwe	Sorghum	Alcoholic beverage	Drink	*Lactococcus raffinolactis*, *Lactobacillus plantarum*, *Lactococcus lactis*, *Lactobacillus delbrueckii*, *Streptococcus* spp.
39	Bojalwa	Botswana	Sorghum	Alcoholic beverage	Drink	Unknown

(continued)

TABLE 7.1 (continued)

Some Important Indigenous Fermented Foods of the World

S. No.	Fermented Food	Country/Place	Ingredient(s)	Nature of Product	Product Use	Dominant Microorganism(s)
Legume products						
40	Bhallae	India	Black gram	Deep-fried patties	Snack after soaking in curd, water	*Lactobacillus fermentum* *Leuconostoc mesenteroides*, yeasts
41	Chee-fan	China	Soybean, whey curd	Solid, cheese-like	Eaten fresh	*Mucor* spp., *Aspergillus glaucus*
42	Khaman	Western India	Bengal gram	Spongy cake	Breakfast food	Yeasts
43	Kenima	Nepal, Sikkim, Darjeeling district of India	Soybeans	Solid	Snack	*Bacillus subtilis*, *Enterococcus faecium*, yeasts
44	Ketjap	Indonesia	Black soybeans	Syrup	Seasoning agent	*Aspergillus oryzae*
45	Meitauza	China, Taiwan	Soybean cake	Solid	Fried in oil or cooked with vegetables	*Actinomucor elegans*
46	Meju	Korea	Soybeans	Paste	Seasoning agent	*Aspergillus oryzae*, *Rhizopus* sp.
47	Miso	Japan, China	Soybeans/soybeans and rice	Paste	Soup base, seasoning agent	*Aspergillus oryzae*, *Torulopsis etchellsii*, *Lactobacillus* sp., *Saccharomyces rouxii*
48	Natto	Northern Japan	Soybeans	Solid	Roasted or fried in oil used as a meat substitute	*Bacillus natto*
49	Oncom	Indonesia	Peanut press cake	Solid	Roasted or fried in oil, used as a meat substitute	*Neurospora* spp.

50	Papadam	India	Black gram and spices	Circular tortilla-like wafers	Condiment	*Leuconostoc mesenteroides, Lactobacillus fermentum, Streptococcus faecalis, Saccharomyces cerevisiae*
51	Soybean milk	China, Japan	Soybeans	Liquid	Drink	LAB
52	Sufu	China, Taiwan	Soybean, whey curd	Solid	Soybean cheese, condiment	*Actinomucor elegans*
53	Soy sauce	Japan, China, Philippines	Soybeans/soybeans and wheat	Liquid	Seasoning agent for meat, fish, cereals, and vegetables	*Aspergillus oryzae, Aspergillus sojae, Lactobacillus* sp., *Saccharomyces rouxii*
54	Tempe	Indonesia and vicinity	Soybeans	Solid	Fried in oil, roasted or used as meat substitute in soup	*Rhizopus oligosporus*
55	Vadai	India	Black gram	Deep-fried patties	Snack	Yeasts and LAB
56	Warris	Northern India	Black gram and spices	Ball-like hollow, brittle	Spicy condiment, eaten with vegetables, legumes, rice	*Candida, Saccharomyces cerevisiae, Leuconostoc mesenteroides*
Cereal legume products						
57	Dhokla	Western India	Rice, Bengal gram	Spongy cake	Condiment	*Lactobacillus fermentum, Lactobacillus lactis, Lactobacillus delbrueckii*
58	Dosa	Southern India	Rice, black gram	Spongy, pancake	Staple food	*Leuconostoc* spp., *lactobacillus* spp., *Saccharomyces* spp.

(*continued*)

TABLE 7.1 (continued)

Some Important Indigenous Fermented Foods of the World

S. No.	Fermented Food	Country/Place	Ingredient(s)	Nature of Product	Product Use	Dominant Microorganism(s)
59	Hama natto	Japan	Wheat flour, whole soybeans	Raisin-like	Flavoring agent for meat soft and fish, eaten as snack	*Aspergillus oryzae*, *Streptococcus* sp., *Pediococcus* sp.
60	Idli	Southern India	Rice, black gram	Spongy	Breakfast food	*Leuconostoc mesenteroides*, *Streptococcus faecalis*
61	Kecap	Indonesia and vicinity	Wheat, soybeans	Liquid	Condiment, seasoning agent	*Aspergillus oryzae*, *Lactobacillus* sp., *Pediococcus* sp.
62	Tao-si	Philippines	Wheat flour, soybeans	Semisolid	Seasoning agent	*Micrococcus*, *Streptococcus*, *Pediococcus*
63	Taotjo	East Indies	Roasted wheat meal or glutinous rice, soybeans	Semisolid	Condiment	—
Fermented vegetable products						
64	Sauerkraut	Europe	Cabbage	Sour cabbage	Food adjunct	*Leuconostoc mesenteroides*
65	Pickles	India, Pakistan	Mango, gooseberry, lemons	Acidified fruits and vegetables	Food adjunct	*Leuconostoc* spp., *Lactobacillus brevis*
Fermented meat and fish products						
66	Sausages	Europe	Meat	Acidified and cured meat	Staple food	*Pediococcus acidilactici*, *Lactobacillus plantarum*
67	Fish paste	Europe	Large fish	Paste	Staple food	LAB
68	Fish sauce	Europe	Small fish	Sauce	Condiment	LAB

There are 510 cheese varieties based principally on their moisture content and milk species. These can be distinguished between cow, sheep, goat, and buffalo milk cheeses and hard cheese (<42% water), semihard (43%–55% water), soft (>55% water), fresh rennet, fresh acid, and fresh cheese.

7.6.1.1.1 General Steps Involved in Cheese Production

Good quality milk is taken, standardized for its casein content, pasteurized, and transferred to a cheese vat. The milk is brought to 30°C and inoculated with the required amount of starter cultures. After about half an hour, rennet extract is added, and the milk is well stirred and allowed to coagulate/set at that very temperature of milk (30°C). After coagulation of milk/formation of firm clot, the curd is cut with vertical and horizontal knives into curd cuvettes. The temperature of the contents in the cheese vat is raised slowly till 40°C by circulation of the hot water in the double-jacketed cheese vat. The cut curd pieces are teased again with hand so that they do not join together. As the acid development takes place, moisture from the curd particles is squeezed, and curd particles get heavier and heavier and start setting at the bottom of the cheese vat. The cheese whey is drained out. The coagulation becomes firm, as the curd is losing water. The cut curd pieces are collected together in the cheese vat itself and a big mass results. Several blocks are prepared and placed onto the floor of the vat. These are turned upside down twice or thrice. They can be placed one above the other to apply pressure so that more of the moisture is drained off. This is continued for 2–3 h. This process is called mellowing. The mill cuts the whole block into small pieces, which are then salted by powdered salt or dipped in brine solution. The resultant is known as green cheese. After about 4 weeks or so, the green cheese is removed from the hoofs and placed in cold storage overnight and then paraffined. The blocks of cheese are then placed in the ripening (curing) room at a specific temperature and humidity generally for 3–6 months and then are ready for use. The complete process for cheese production has been depicted in Figure 7.1.

Method of ripening of cheese depends upon the variety of cheese to be manufactured, and accordingly the conditions of ripening (relative humidity, temperature, growth of microorganisms, treatment of fresh curd, etc.) are varied. To produce hard cheese, (i) the growth and activity of microorganisms on the surface of cheese are discouraged, (ii) salt is added to the curd before pressing or the pressed cheese is salted on the surface, (iii) relative humidity of the air in the ripening (curing) room is kept fairly low to discourage surface growth but high enough to prevent excessive evaporation of moisture, and (iv) loss of moisture is minimized by coating the surface with paraffin or a plastic film. To produce soft cheese, (i) the growth and activity of microorganisms on the surface of cheese is encouraged, (ii) the cheese is salted on the outside (it determines the kind of organisms that grow there), and (iii) relative humidity is kept high to permit surface growth of the organisms. Most of the cheeses are ripened at temperatures between 5°C and 16°C.

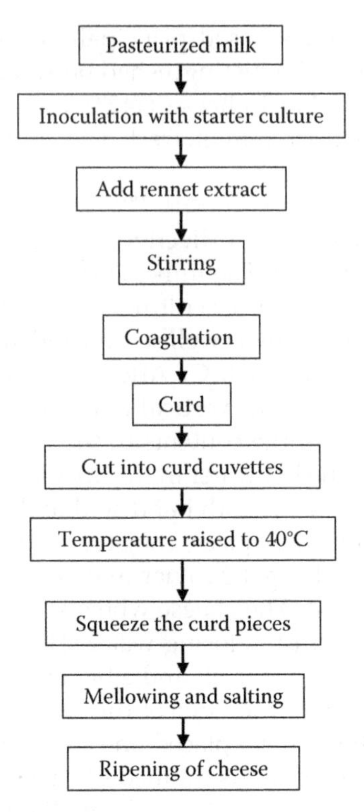

FIGURE 7.1
A flow diagram for the preparation of cheese.

7.6.1.2 Yogurt

Yogurt is the most popular fermented milk product. It is produced by fermentation of milk using thermophilic bacteria. Generally, it is made from milk by adjusting its solids to requisite level either by concentrating the milk partially and removing 15%–20% water or by supplementing nonfat dry milk/concentrated skim milk or other milk-derived ingredients including casein, sodium and calcium caseinates, and whey and whey protein concentrates (Sabban, 1986; Huang, 2000). Other permitted additives include nutritive carbohydrate sweeteners, coloring agent, stabilizers, and some fruit preparations that also impart peculiar flavors to yogurt (Tamime and Robinson, 2007). The major sweetener used for yogurt preparation is sucrose, which affects production of lactic acid and flavor by yogurt culture. The plain yogurt, however, contains no added sugar or flavor. The stabilizers, in the form of gelatin, algin and sodium alginate, and locust bean gum, are added to provide a smooth and gel-like texture to the yogurt and reduce availability of free water in the product, hence avoiding its wheying off or syneresis. The generalized flow sheet indicating the main steps involved in the production of yogurt is depicted in Figure 7.2.

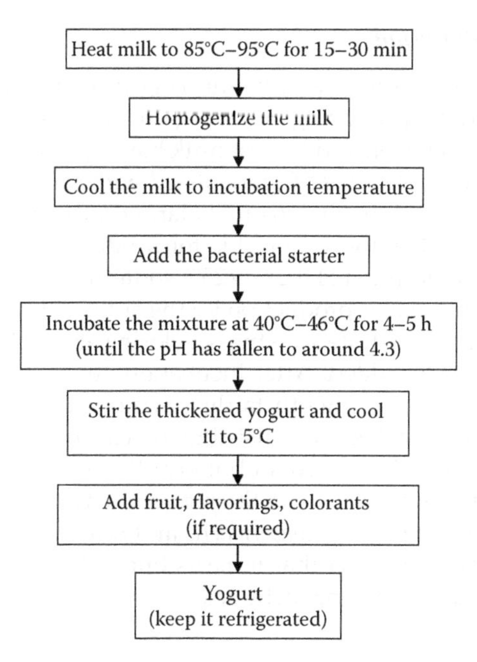

FIGURE 7.2
Flowchart for the preparation of yogurt.

The microflora involved in yogurt fermentation include the culture of *Streptococcus thermophilus* and *Lactobacillus delbrueckii* ssp. *bulgaricus* growing in the temperature range of 40°C–45°C (Guarner et al., 2005). Equal cell numbers of both of these obligatory homofermentative bacteria are desirable for flavor and texture production in the yogurt. Their relationship in yogurt is mutualistic and they grow better in mixed culture as compared to monoculture and the acid production is also more in former case. In the beginning of fermentation, streptococci grow fast resulting in the production of lactic acid, acetic acid, acetaldehyde, diacetyl, and formic acid, which are responsible for imparting typical flavor to the yogurt. The streptococcal growth lowers the pH of yogurt mix to 5.0–5.5. This change in oxidation reduction potential as well as the availability of formic acid in the medium stimulates the growth of *Lactobacillus delbrueckii* ssp. *bulgaricus*, which exhibits a strong proteolytic activity and liberates a number of amino acids from the milk proteins and further lowers the pH to 3.8–4.4. The accumulation of these amino acids further stimulates the growth of *Streptococcus thermophilus*. However, the nitrogen requirement of the latter is much lower than the amino acids present in the medium at this stage. Hence, the finished yogurt contains quite high amounts of five amino acids, which mainly include glutamic acid and proline.

7.6.1.3 Cultured Buttermilk

Cultured buttermilk is prepared by culturing pasteurized skim milk with lactic acid and aroma-producing microorganisms. The main ingredient is skimmed milk or homogenized low fat milk having fat content of ~1.7%. The milk is pasteurized, cooled to 22°C, inoculated with starter culture, and incubated at 21°C–24°C for 14–16 h to ensure balanced growth of strains that produce acid and those that are responsible for producing characteristic aroma (diacetyl flavor). The addition of 0.2%–9.25% sodium citrate to milk provides a precursor to enhance flavor production by the culture. In addition, some stabilizers, nutritive carbohydrate sweetness, coloring agents, butter flakes, whey solids, etc., may also be added. After incubation, the acid curd at pH 4.6–4.7 is agitated gently and the smooth fairly viscous buttermilk thus obtained is cooled to 5°C and packed and distributed within 24 h (Kosikowski and Mistry, 1997). The microflora used for buttermilk preparation includes homofermentative *Streptococcus lactis* and *Streptococcus cremoris* that produce lactic acid and *Streptococcus lactis* ssp. *diacetylactis* and *Leuconostoc mesenteroides* ssp. *cremoris* that produce diacetyl that imparts butter flavor and also produces some acetoin besides producing lactic acid.

7.6.1.4 Cultured Cream

This is a very popular dairy product used as topping on salads, vegetables, fruits, and meat preparation and in confectionery as cake filling, etc. It is prepared in a manner similar to that of buttermilk, the only difference being the fat content of milk, which is very high (minimum 18%). Further consistency-wise, this product is an acid gel. The cream with desired fat content is separated from the whole milk. This cream is used fresh with relatively low bacterial count for the production of cultured cream. During cream separation from milk, the bacteria tend to concentrate in cream as it constitutes the lighter phase thereby rendering it quite vulnerable to spoilage. Hence, it is pasteurized at 74°C for 30 min, homogenized, ripened using lactic and aroma-producing bacteria, and packaged for marketing. For culturing or souring the cream, the same strains are used for buttermilk production. As in buttermilk, addition of 0.15%–0.20% sodium citrate may enhance flavor production as it is metabolized by *Streptococcus lactis* ssp. *diacetylactis* and *Leuconostoc cremoris* to produce the aroma-giving compounds including diacetyl and volatile acids (Hugenholtz et al., 2000; García-Quintáns et al., 2009).

7.6.1.5 Acidophilous Milk

This is a fermented dairy product known predominantly for its therapeutic value. The milk contains a disaccharide, the lactose that is assimilated very slowly by the human body. In fact, it remains unchanged till it reaches the intestine of the body where *Lactobacillus acidophilus* bacteria are naturally present, which metabolize this disaccharide thereby producing the acidic conditions that suppress the growth of gas-forming organisms in the intestine (Berger et al., 2007). The preparation of this product involves pasteurization of whole or skim milk, its homogenization, and fermentation using *Lactobacillus acidophilus* at 37°C till the acidity reaches 0.65%. This product is however too sour and hence is not easily palatable to most of the consumers. Hence, efforts have been made to develop some novel products using some other bacteria like *Lactobacillus bifidus* in combination with *Lactobacillus acidophilus* and employing a manufacturing process similar to that of yogurt (Chandan and O'Rell, 2006).

7.6.1.6 Butter and Ghee

The preparation procedure for both the products is common up to the point of butter preparation. It involves boiling and cooling of milk followed by its inoculation with a starter culture or a portion of previous day curd. The material is allowed to ferment overnight and is churned thereafter with a wooden or metal churn, following which the butter rises to the top and is separated. The microflora involved in this fermentation mainly includes lactobacilli followed by streptococci, coliforms, and aerobic spore-forming bacteria (Krishnaswamy and Laxminarayana, 1948). *Streptococcus lactis* ssp. *diacetylactis* as in other fermented dairy products produces flavor-imparting compounds like diacetyl, acetaldehyde and acetoin, and carbon dioxide.

The butter prepared as earlier is subjected to heating at 100°C–140°C so that the water content evaporates, followed by filtration through a muslin cloth to attain a clarified product called ghee. This product is whitish in color if prepared from buffalo's milk and is yellowish in color if prepared from cow's milk due to the presence of carotenoids in the latter. As the heating at high temperature kills all the lactic as well as other microflora, this product can be stored without refrigeration and is used for cooking purposes.

7.6.1.7 Shrikhand

It is a popular lactic acid–fermented dairy product used as a dessert in western and northern parts of India, mainly in Maharashtra. The curd is made out of milk by boiling the latter, cooling, and inoculating with LAB or a portion of the previous day's curd. The curd thus obtained is drained by hanging in a muslin bag for 6–8 h. The whey is drained out and the solid mass left behind, known as "chakka," is mixed thoroughly with 45%–50% sugar (Aneja et al., 1977). To this semisolid mass, some flavoring substances like cardamom and saffron are added and it is garnished with nuts, etc. Under refrigeration, the product can be stored for over a month's time, while at room temperature, it is to be consumed within 2–3 days. The commonly used starter bacteria are strains of *Lactobacillus lactis* spp. *lactis, Lactobacillus lactis* spp. *cremoris, Lactobacillus lactis* spp. *lactis* biovar. *diacetylactis, Lactobacillus delbrueckii* spp. *bulgaricus,* and *Streptococcus thermophilus* used either singly or in combination of two cultures (Patel and Chakraborty, 1988).

7.6.1.8 Koumiss

Koumiss is fermented milk made up of mare's milk using a mixed culture of yeast. It can be prepared by first dissolving lactose in water and then adding it to the milk, mixing the yeast culture and brown sugar thoroughly, adding a little of the milk mixture to make it a thin paste, then adding that to the rest of the milk solution and stirring well. These are then stored in very strong bottles and held at 50°F–60°F. Each day each bottle is wrapped individually in several layers of cloth after shaking it gently for about 10 min to prevent the casein from coagulating. The cloth is necessary as a safety precaution, as there is a great deal of CO_2 buildup inside the bottle and it might explode. The koumiss will be ready in 3–5 days.

7.6.1.9 Kefir

The term kefir is of Turkish origin. It is a fermented milk drink made by inoculating cow, goat, or sheep milk with kefir grain. Kefir grains are a combination of bacteria and yeasts in a matrix of proteins, lipids, and sugars, and this symbiotic matrix forms "grains" that resemble cauliflower. The generalized flow sheet indicating the main steps involved in the production of kefir is depicted in Figure 7.3.

The traditional method involves the direct addition of kefir grains (2%–10%) to milk in a loosely covered acid-proof container, which is agitated one or more times a day. It is not filled to capacity, allowing room for some expansion as the kefiran and carbon dioxide gas produced cause the liquid level

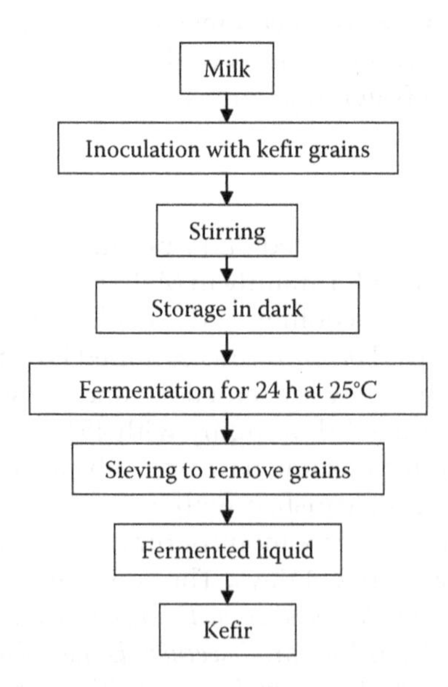

FIGURE 7.3
Flowchart for the preparation of kefir.

to rise. If the container is not light-proof, it should be stored in the dark to prevent degradation of vitamins and inhibition of the culture. After a period of fermentation lasting around 24 h, ideally at 20°C–25°C, the grains are removed from the liquid by sieving and reserved as the starter for a fresh lot of milk to be fermented to kefir. The fermented liquid that contains live microflora from the grain may now be consumed as a beverage, used in recipes, or kept aside for several days to undergo a slower secondary fermentation that further thickens and sours the liquid. The shelf life of the beverage is 2–3 days when stored under non-refrigerated conditions.

7.6.2 Fermented Cereal Products

The multiple beneficial effects of cereals can be exploited in different ways leading to the design of novel cereal foods or cereal ingredients that can target specific populations. Cereals can be used as fermentable substrates for the growth of probiotic microorganisms (Charalampopoulos et al., 2002; Kohajdová and Karovicova, 2007). Cereals contain water-soluble fiber, such as β-glucan and arabinoxylan, oligosaccharides, such as galacto- and fructooligosaccharides, and resistant starch, which have been suggested to fulfill the prebiotic concept. The development of new technologies of cereal processing that enhance their health potential and the acceptability of the food product are of primary importance (Charalampopoulos et al., 2002).

Cereal fermentations involve mainly the processing of maize, rice, sorghum, and the millets. Wheat, barley, corn, millet, and soybeans are also used in addition to rice. The important fermented products obtained from cereals are discussed in the following sections.

7.6.2.1 Bread

Bread is one of the oldest fermented foods, dating back to the Neolithic era. The first breads produced were cooked versions of a grain paste, made from ground cereal grains and water (Jaine, 1999). It is often made from a wheat-flour dough that is cultured with yeast, allowed to rise, and finally baked in an oven. Owing to its high levels of gluten that gives the dough sponginess and elasticity, wheat is the most common grain used for the preparation of bread. The bread is also made from the flour of rye, barley, maize, and oats, usually in combination with wheat flour.

The protein of the wheat flour is unique in that when the flour is mixed with water in certain proportions, the protein forms an elastic colloidal mass or dough that can hold gas, forming a spongy structure when baked. The bread dough is usually leavened by bread yeasts that ferment the sugars in the dough and produce mainly carbon dioxide and alcohol. However, other actively gas-forming microorganisms such as wild yeasts, coliform bacteria, saccharolytic *Clostridium* spp., and heterofermentative LAB can also be used instead of bread yeasts for leavening.

The sponge consisting about 65% of the total flour plus a portion of the total dough water, yeast, and ammonium salts to stimulate the yeasts and other salts including $KBrO_3$, KIO_3, CaO_2, and $(NH_4)_2S_2O_8$ to improve the dough characteristics is prepared and discharged into a trough where it undergoes a fermentation for 4.5 h in a controlled environment. From a starting temperature of about 25°C, the final temperature increases by ~6°C due to the exothermic reactions brought about by yeast activity. At the end of sponge formation, the sponge is transferred into a dough mixer and mixed until the dough has been transformed from a sticky, wet-appearing mixture into a smooth, cohesive dough, characterized by a glossy sheen. As *Saccharomyces cerevisiae* evolves carbon dioxide, the gas diffuses into previously formed gas vesicles and is retained in the dough due to the unique nature of gluten. The mixed dough is placed in troughs and allowed to rest for 20–30 min during which the dough recovers from mechanical stress, relaxes, and is better able to undergo the remaining processing stages. In the next stage, the dough is divided by cutting into pieces of desired weight by a machine. The machine volumetrically divides the dough into pieces and discharges them into cylinders from where they are fed into the bread pans. Pans containing the dough pieces are placed in fermentation units called proof boxes for fermentation prior to baking. The environment in these units is typically maintained at 35°C–43°C and a relative humidity of 80%–95%. The dough pieces expand in the pans to a desired volume, a process usually requiring ~60 min. The proofed loaves are then placed in an oven for baking. Gas within the dough expands; steam and alcohol vapors also contribute to this expansion. Enzymes are active until the bread reaches about 75°C. At this temperature, the starch gelatinizes and the dough structure is set. With the bread surface temperature reaching 130°C to 140°C, sugars and soluble proteins react chemically to produce an attractive crust color. The center of the loaf does not exceed 100°C temperature. Remaining stages in the bread-making process include cooling of the baked bread, slicing, wrapping, and distribution to stores for sale to the consumer.

7.6.2.2 Pasta

Pasta is a wheat-derived (type of noodle) staple food of traditional Italian cuisine. It is made from an unleavened dough of a durum wheat flour mixed with water and formed into sheets or various shapes, then cooked. Pastas may be divided into two broad categories: dried (*pasta secca*) and fresh (*pasta fresca*). The raw material for pasta products is generally durum wheat. The use of durum semolina leads to higher rheological properties of the dough, as well as greater color and cooking quality of the product (Troccoli et al., 2000). Non-durum wheat ingredients can also be used to produce specifically blended pasta (Fuad and Prabhasankar, 2010; De Noni and Pagani, 2010).

Pasta products generally do not require any microbial bioprocess to achieve the final product. However, several potential biotechnological approaches have been made recently to study LAB as a new driver of innovation

(Capozzi et al., 2012). Several studies have focused on LAB isolation and characterization in durum wheat and experimental applications of LAB in pasta making (Corsetti et al., 2007; di Cagno et al., 2005; Ricciardi et al., 2005; Valerio et al., 2009; De Angelis et al., 2010; Russo et al., 2010; Capozzi et al., 2011). Concerning microbial dynamics, Russo et al. (2010) studied the dominant LAB during the pasta-manufacturing process. The majority of strains belonged to *Pediococcus pentosaceus* and *Enterococcus faecium* species. During the technological steps, an increase of LAB population was observed from kneading to extrusion, followed by a continuous decrease, probably in reason of the raising temperature encountered in extrusion and preheating.

7.6.2.3 Bhatura

This is a leavened, deep-fried bread made from white wheat flour, consumed as a breakfast food and snacks with masala chana in Northern India. Dough is prepared by mixing white wheat flour with water and kneading for 20 min at room temperature. The dough is supplemented with 1% common salt and inoculated with a portion of ripe batter from the previous batch and kept in the open to undergo fermentation for 8–14 h. The dough is then made into balls, which are then flattened by hand using vegetable oil to 12–18 cm in diameter before deep-frying in oil.

7.6.2.4 Nan

This is leavened flat bread with central pouch. Nan is generally consumed as a staple food by the people in India, Pakistan, Afghanistan, and Iran. Its dough is raised by using white wheat flour, salt, inoculum from the previous batch, and water (100:2:10:50). It is kneaded by hand at room temperature and allowed to ferment in the open for 6–8 h. Sometimes the dough is prepared by sponge dough method where yogurt is added as inoculum along with baker's yeast and some chemical leaveners. The flattened dough is then transferred onto a circular pad of cotton cloth and plastered onto the inner wall of a clay oven called tandoor. Before baking, some spices like caraway seeds (*Carum carvi*), coriander (*Coriandrum sativum*), and kalaunji (*Buchanania lanzan*) are sprinkled on it. The finished product is obtained by baking at 120°C–150°C for 10–15 min.

7.6.2.5 Jalebis

These are pretzel-like, syrup-filled confectionary food prepared from fermented wheat-flour dough by deep-frying. These are prepared by mixing together fine wheat flour (1 kg), curd (200 g), and water (650 mL) to form a thick batter and is fermented overnight. The fermented batter is squeezed through cloth with a hole of about 4 mm diameter and deposited as continuous spirals into heated oil or ghee (160°C–180°C). In about a minute, these spirals become light brown and are removed from the frying pan. The excess fat is drained away and the jalebi is immediately immersed into the sugar syrup having consistency of honey for 2–3 min and then removed. Often rose water or any other flavoring agent is added to the syrup (Batra, 1981). Mixed fermentation is carried out by LAB, *Lactobacillus fermentum, Lactobacillus lactis,* and *Lactobacillus buchneri* with *Streptococcus lactis* and *Streptococcus faecalis* and lowering the pH from an initial 4 to 3. The yeast *Saccharomyces cerevisiae* is also active, produces CO_2, and causes about a 10% increase in volume during fermentation. The contents of nitrogen and total sugars have been found to decrease during fermentation (Ramakrishnan, 1977; Batra, 1981).

7.6.2.6 Ogi

Ogi is a lactic-fermented maize gruel, which is a staple food in most parts of Nigeria and other parts of West Africa and serves as a weaning food for infants. Ogi is often marketed as a wet cake wrapped in leaves or transparent polythene bags. It is diluted to a solid content of 8%–10% and boiled into a pan, or cooked and turned into a stiff gel called "agidi" or "eko" prior to consumption. It is prepared by steeping maize in warm water for 1–2 days, after which wet milling is done and filtrate is allowed to ferment naturally. The souring of ogi is determined by individual taste and the preferred pH of the product is 3.5–3.7. The predominant microorganisms involved in fermentation of ogi are mixed population of LAB and yeasts. Among LAB, *Lactobacillus fermentum, Lactobacillus plantarum, Lactobacillus brevis, Lactobacillus curvatus, Lactobacillus buchneri* and yeasts, *Candida humicola, Candida krusei,* and *Saccharomyces cerevisiae* are the most commonly isolated (Odunfa and Adeyele, 1985; Nago et al., 1998). LAB are the predominant organisms in the fermentation responsible for lactic acid production, while yeasts contribute to flavor development.

7.6.2.7 Uji

Uji is a traditional thin gruel (porridge) prepared from fermentation of maize, millet, sorghum, or cassava flours. Single or mixtures of these flours may be used in preparing uji. It is a lactic-fermented product with a pH range of 3.5–4.0. The high acidity in fermented uji confers some antimicrobial effect and increases shelf life of the product. The slurry is prepared by mixing flour and water (30% w/v) and allowed to ferment at ambient temperature near fire for 2–5 days. The fermented slurry is diluted with water, boiled, and sweetened with sugar prior to consumption. Uji is used for weaning and also widely consumed by adults or preschool and school children for breakfast (Mbugua and Njenga, 1991) in Kenya and many other parts of East Africa. According to a survey conducted by Mbugua (1977), uji is sold to factory and construction site workers by small-scale producers in many urban areas of Kenya. This clearly suggests that gruel and porridges prepared from cereals or tuber roots (cassava) act as most important sources of energy for all age groups. Uji-fermented slurry is sometimes sun-dried, packaged, and stored prior to boiling. *Lactobacillus plantarum* is mainly responsible for souring of uji and also some heterofermentative strains of *Lactobacillus fermentum*, *Lactobacillus cellobiosus*, and *Lactobacillus buchneri* have been reported during fermentation of uji (Mbugua, 1984). These strains are also responsible for the development of flavor and taste in the final product.

7.6.2.8 Kenkey

Kenkey is a maize-based bread product widely consumed in Ghana. It is taken for breakfast together with tea, sardines, or chili. Traditional preparation of kenkey entails steeping maize kernels in water for 2 days, after which the softened kernels are wet milled into grits and split into two halves. The first half is kneaded with water into dough that is allowed to ferment for 2–3 days and the other half is mixed with water and boiled to obtain aflata, which is a gelatinized mass. The fermented dough and aflata are mixed together (1:1) to form a sticky dough, which is molded into balls and wrapped in plantain leaves and boiled in water for several hours to produce ready-to-eat kenkey bread. A mixed flora of LAB (*Lactobacillus fermentum* and *Lactobacillus reuteri*) and yeasts (*Candida krusei*, *Saccharomyces cerevisiae*) are the dominant microorganisms in fermented kenkey (Jespersen et al., 1994).

7.6.2.9 Mahewu

This is a fermented maize meal commonly consumed as a staple among black South Africans. It is traditionally prepared by adding one part of maize meal to nine parts of boiling water. The suspension is cooked for 10 min, allowed to cool, and then transferred to a fermentation container. At this stage, wheat flour (about 5% of the maize meal used) is added to serve as a source of inoculum. Fermentation occurs in a warm sunny place within 24 h. Mahewu is currently produced on an industrial scale (Figure 7.4) as a dry food product that is marketed as a precooked ready-to-mix powder. *Streptococcus lactis* is the main fermenting organism in traditionally prepared mahewu, which offers some advantages over ogi in that the initial fermentation by wild fungi, etc., is eliminated by boiling both the maize meal and water for steeping. Furthermore, it is precooked and requires only mixing prior to consumption.

7.6.2.10 Injera

Injera is an Ethiopian-fermented bread made from tef (*Eragrostis tef*), which is a major contributor of the daily protein intake of Ethiopian population. Traditional preparation of injera involves mixing *tef* flour with water (ratio 1:2) and addition of a starter locally referred to as *ersho*, which is a fluid saved from previously fermented dough (16% by weight of flour) and left to ferment for 3 days at ambient temperature. The fermented dough is kneaded and thinned to a batter by addition of water equal to the original weight of flour.

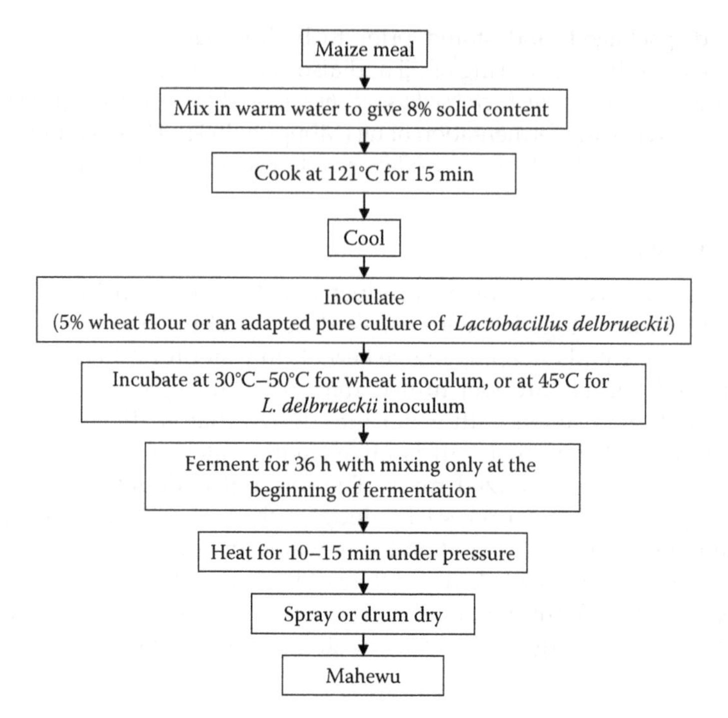

FIGURE 7.4
Flowchart for the traditional preparation of mahewu.

The thin batter is left further to ferment and rise for about 2 h prior to cooking on hot injera clay griddle. A small portion of batter is saved to serve as starter for the next batch. Yeast (*Candida guilliermondii*) is the major microorganism involved in the fermentation of injera, together with *Lactobacillus bulgaricus* and some *Lactobacillus* species (Blandino et al., 2003).

7.6.2.11 Kisra

Kisra is fermented flat sorghum bread, which constitutes a major food intake of the staple diet for the majority of Sudanese population. Traditional production of kisra entails mixing sorghum flour with water in ratio 1:2 and adding a small amount of previously fermented dough to the mixture to act as a starter. Depending on the amount of starter used, fermentation is allowed at 37°C for 6–24 h. Fermentation has been demonstrated to improve protein availability and digestibility in sorghum. An apparent increase in protein content in sorghum has been reported after 4 days of fermentation, and remarkable changes in the content of amino acids such as cysteine, methionine, and threonine have been noted (Chavan and Kadam, 1989; Abasiekong, 1991). Consequently, fermentation of kisra leads to nutritional improvement of the final product. Strains of *Pediococcus pentosaceus*, *Lactobacillus brevis*, *Lactobacillus fermentum*,

Lactobacillus reuteri, Leuconostoc amylovorus, and yeasts (*Debaryomyces hansenii* and *Candida intermedia*) are dominant in kisra dough (Mohammed et al., 1991; Hamad et al., 1992, 1997).

7.6.2.12 Kishk

This is a fermented product prepared from parboiled wheat and milk. It is consumed in Egypt and in most Arabian countries. During the preparation of kishk, wheat grains are boiled until soft, dried, milled, and sieved in order to remove the bran. Milk is separately soured in earthenware containers, concentrated, and mixed with the moistened wheat flour thus prepared, resulting in the preparation of a paste called hamma. The hamma is allowed to ferment for about 24 h, following which it is kneaded and two volumes of soured salted milk are added prior to dilution with water. Alternatively, milk is added to the hamma and fermentation is allowed to proceed for a further 24 h. The mass is thoroughly mixed, formed into balls, and dried. Kishk is a highly nutritious food, having a protein content of about 23.5%. It is highly digestible and has high biological value. Microorganisms responsible for fermentation include *Lactobacillus plantarum, Lactobacillus brevis, Lactobacillus casei, Bacillus subtilis,* and yeasts.

7.6.2.13 Bushera

Bushera is a traditional sorghum beverage, which is extensively consumed in southwest Uganda. It is prepared from germinated or non-germinated sorghum grains. It is used for weaning, as a beverage. Traditional preparation of bushera is presented in Figure 7.5.

Red or brown sorghum grain varieties are used for the production of bushera. Wood ash is added to speed up fermentation and increase the sweetness of bushera. When non-germinated sorghum is used for production of bushera, malt of millet and sorghum are added to increase sweetness and enhance flavor. In sour bushera, fermentation time is prolonged to 2–4 days as compared to 1 day for sweet bushera. Back slopping is also practiced in the production of bushera, but this has been considered to lead to fast production of acid and hence excessive sourness. Lactobacilli are predominant in the household-fermented bushera (Muyanja et al., 2003). LAB isolated during fermentation of bushera consist of *Lactobacillus fermentum, Lactobacillus paracasei, Lactobacillus brevis, Lactobacillus delbrueckii,* and *Lactobacillus plantarum,* which are dominant in the later stages of bushera fermentation. The dominance of *Lactobacillus plantarum* in the late stages of fermentation is attributed to its high acid tolerance.

7.6.2.14 Togwa

Togwa is a nonalcoholic beverage, which can be prepared from cassava, maize, sorghum, millet, or their combinations. Togwa is widely consumed in Tanzania. Traditional preparation of togwa involves mixing maize and

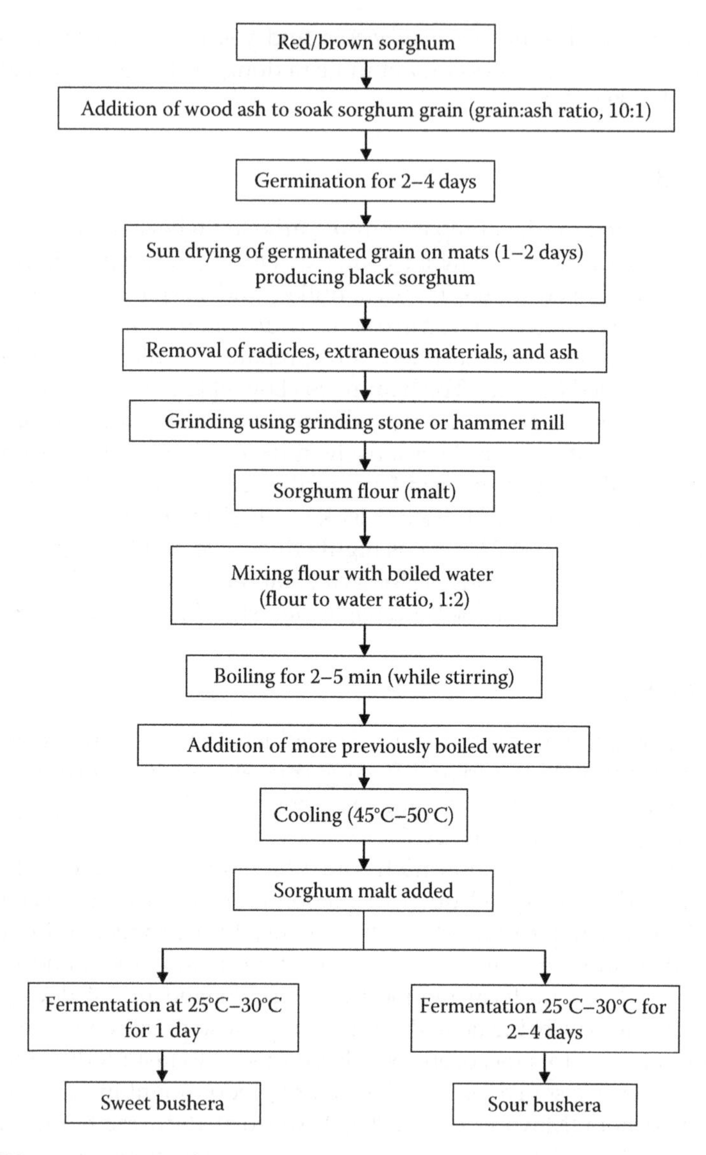

FIGURE 7.5
Flow diagram for production of sweet and sour sorghum bushera.

sorghum or sorghum and millet (ratio 1:1 w/w) into a slurry. The slurry is mixed with water (1:9 w/v) and boiled into a gruel, which is later cooled to around 30°C. The cooled gruel can be supplemented with a small amount of sorghum malt or back slopped with remains from the previous togwa and allowed to ferment for 1–3 days. LAB (*Lactobacillus brevis, Lactobacillus cellobiosus, Lactobacillus fermentum, Lactobacillus plantarum, Pediococcus pentosaceus*) and yeasts (*Candida pelliculosa, Candida tropicalis, Issatchenkia orientalis,*

Saccharomyces cerevisiae) are the predominant microorganisms associated with the fermentation of togwa (Mugula et al., 2001).

7.6.2.15 Dolo and Pito

These are popular sorghum alcoholic beverages widely consumed in West Africa. African cereal-based beverages have improved nutritive value compared to unfermented cereal grains and thereby contributing as sources of energy, proteins, minerals, and vitamins in the diet. Traditional preparation of these beers entails the use of malt and whole grain, which contributes to their high nutritional value. Bansah (1990) described pito to contain organic acids, sugars, amino acids, proteins, and vitamins after fermentation. Dolo is largely consumed in Burkina Faso, Ivory Coast, and Mali (Konlani et al., 1996). Pito is common in Ghana, Nigeria, and Togo (Demuyakor and Ohta, 1993; Sanni and Lonner, 1993). The traditional processing of these beers involves steeping of sorghum grains in water, germination to produce malt, mashing, fermentation, cooking, and alcoholic fermentation of the wort with yeast (*Saccharomyces cerevisiae*) from the previous fermentation. *Lactobacillus fermentum* has been found to be the predominant LAB involved in spontaneous acidification of dolo and pito wort, while yeasts are responsible for alcoholic fermentation (Sawadogo-Lingani et al., 2008).

7.6.2.16 Kaffir Beer and Chibuku

These are sorghum-based beers that are largely consumed by the Bantu people of South Africa, Botswana, and Zimbabwe. Kaffir beer is prepared using South African red sorghum varieties that confer it a pinkish color after fermentation. It is sour in taste and appears opaque because of high content of suspended solids. Kaffir beer is consumed in the active stage of fermentation as it has a short shelf life. The main steps involve malting, mashing, souring, boiling, and alcoholic fermentation (Hornsey, 2003). Chibuku is also packaged in a similar manner as kaffir beer, in hard 1 L tetra pack or plastic containers. However, in the traditional process, the sorghum malt is produced by steeping in water for 8–24 h, after which the grain is allowed to sprout for 5–7 days. The malted grains are sun-dried and milled into a powder. The small amount of uncooked malt is added to act as inoculum, unlike in commercial production, in which *Saccharomyces cerevisiae* is added as a starter. LAB and yeasts are thought to be the predominant microorganisms during fermentation. These African beers are a source of carbohydrates and dietary fiber, and since they contain yeasts, they are also an excellent source of B vitamins (Hamaker, 2008).

7.6.3 Fermented Legume Products

Soybeans and black grams are the principal legumes used in the preparation of a variety of fermented foods in different parts of the world. Soybeans are fermented as such, as a paste or as a liquid. Peanuts and locust bean are also fermented in some parts of the world. Black gram is generally used as the starting substrate for the production of a variety of fermented legume products in India.

7.6.3.1 Tempeh

Tempeh is a traditional soy product originally from Indonesia. It is made by natural culturing and controlled fermentation process that binds soybeans into a cake form, similar to a very firm vegetarian burger patty. It is a whole soybean product with different nutritional characteristics and textural qualities (Shurtleff and Aoyagi, 2001). It has a firm texture and an earthy flavor that becomes more pronounced as it ages (Bennett and Sammartano, 2008). Tempeh production begins with whole soybeans, which are softened by soaking and dehulled, and then partly cooked. A mild acidulent, usually vinegar, may be added to lower the pH and create a selective environment that favors the growth of the tempeh mold over competitors. A fermentation starter containing the spores of the fungus *Rhizopus oligosporus* is mixed in. The beans are spread into a thin layer and are allowed to ferment for 24–36 h at a temperature around 30°C. Under conditions of lower temperature or higher ventilation, gray or black patches of spores may form on the surface. This sporulation is normal on fully mature tempeh. A mild ammonia smell may accompany good tempeh as it ferments, but it should not be overpowering. The traditional preparation of tempeh is shown in Figure 7.6.

Tempeh production involves two distinct fermentations. The first one occurring during soaking with bacterial culture, which results in the

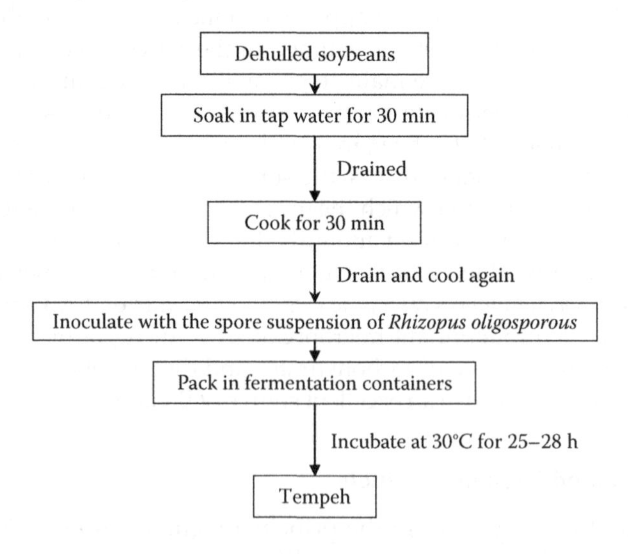

FIGURE 7.6
Flow sheet for tempeh preparation.

acidification of beans, and the second fermentation results in the overgrowth of bean cotyledons by mold mycelium. The number of species of bacteria present in the commercial tempeh is few. Among bacteria, *Klebsiella pneumoniae* being predominant followed by *Bacillus* and *Micrococcus*, fungi belonging to *Aspergillus, Mucor, Penicillium,* and *Rhizopus,* and yeasts belonging to *Trichosporon* are widely used. The soy carbohydrates in tempeh become more digestible as a result of the fermentation process.

7.6.3.2 Natto

Fermented whole soybeans are known as natto in Japan, tu-si in China, tao-si in the Philippines, and tnua-nao in Thailand. These are consumed with boiled rice or used as a seasoning agent with cooked meat, seafoods, and vegetables. There are three main types of natto prepared in Japan. Itohiki natto, referred to simply as natto, is prepared by steaming the beans for about 15 min and inoculating with *Bacillus natto,* a variant strain of *Bacillus subtilis.* Yukiwari natto, a second type of fermented whole soybean product, is made by mixing itohiki natto with salt and rice koji, the source of enzymes to hydrolyze the soybean components in fermentation, produced by *Aspergillus oryzae.* In the case of Hamdel natto, the third major type of fermented whole soybean product in Japan, the soybeans after soaking in water for about 4 h are steamed without pressure and are then inoculated after being cooled with koji. Bacteria such as *Micrococcus, Streptococcus,* and *Pediococcus* are also reported to be widely distributed on the surface and in the inner part of hama natto.

7.6.3.3 Miso

Miso is a traditional Japanese seasoning produced by fermenting rice, barley, and/or soybeans, with salt and fungus. High in protein and rich in vitamins and minerals, miso is typically salty, but its flavor and aroma depend on various factors in the ingredients and fermentation process. Miso is the fermented product made from rice and soybeans. The essential microorganisms involved in fermentation of miso include molds *Aspergillus oryzae* and *Rhizopus oligosporus* and a yeast *Saccharomyces rouxii,* but under traditional conditions, LAB such as *Pediococcus halophilus* or *Streptococcus faecalis* are also found to be present, contributing flavor through the production of acids.

Miso is very healthy and contains isoflavones, saponins, soy protein, and live enzymes. There are many varieties of miso. All are made from soybeans or cereals and a special koji. Koji are grains (mainly rice, also barley) or soybeans that are fermented with the mold *Aspergillus oryzae.* During the production of koji, these *Aspergillus* molds will produce a lot of enzymes that will later break down the proteins and carbohydrates of the substrate. The traditional preparation of miso is shown in Figure 7.7.

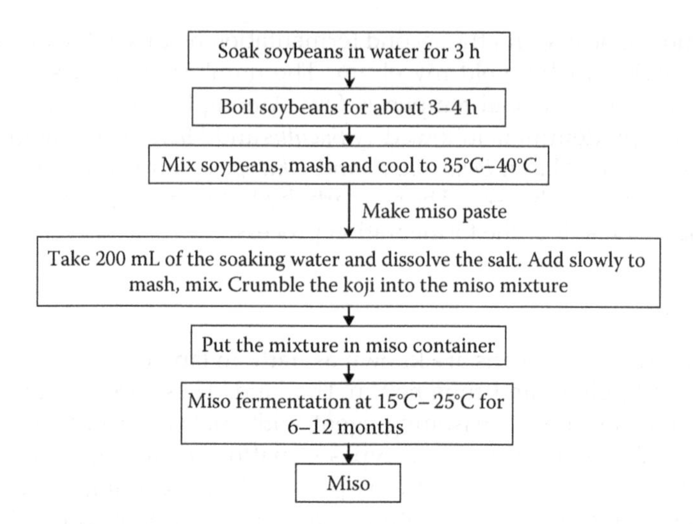

FIGURE 7.7
Flow sheet for miso preparation.

7.6.3.4 Soy Sauce

Soy sauce is a condiment produced from fermented paste of boiled soybeans, roasted grain, brine, and *Aspergillus oryzae* or *Aspergillus sojae* molds. After fermentation, the paste is pressed, producing a liquid and a cake of soy and cereal residue, which is usually reused as animal feed. Soy sauce fermentation involves two distinct basic processes. The first one includes the fermentation with microorganisms and the other chemical process involves the use of acids to promote the hydrolysis of the ingredients. The essential microorganisms in soy sauce fermentation are the molds *Aspergillus oryzae* or *Aspergillus sojae*, the homofermentative LAB *Pediococcus cerevisiae* or *Lactobacillus delbrueckii*, and the yeast *Saccharomyces rouxii*. Koji, the source of enzymes used in soy sauce, is prepared by culturing a number of mixed strains of *Aspergillus oryzae* or *Aspergillus sojae* either on steam-polished rice in Japan or a mixture of wheat bran and soybean flour in China. It is then mixed with saline water to form a mash. The traditional preparation of soy sauce is shown in Figure 7.8.

7.6.3.5 Sufu

Sufu is a traditional fermented soybean curd originating in China. It is a soft cheese-like product with a spreadable creamy consistency, and it has been consumed widely as an appetizer for centuries in China (Zhang and Shi, 1993; Han et al., 2001, 2003).

For preparation of sufu, raw soybeans are soaked in water, then ground with water, and the liquid extract is filtered off, and a milky fluid results, which is colloquially called soya milk. The protein in this extract can be

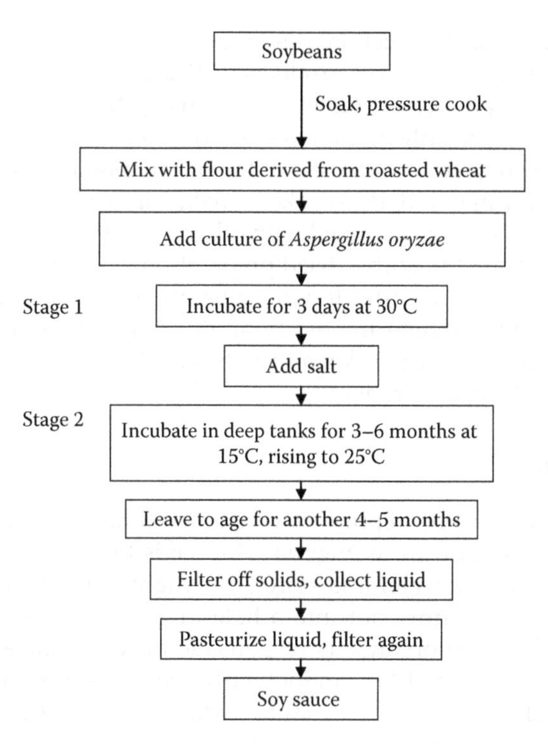

FIGURE 7.8
Flow sheet for soy sauce preparation.

precipitated with calcium/magnesium salts to give the curd called "tofu." For sufu production, this is cut into small cubes, dried for 10 min in an oven at 100°C, then inoculated with *Actinomucor elegans* and incubated until mold mycelium completely covers the tofu. The cubes are then placed in brine, which may contain rice wine, red rice, soy sauce, and various other flavorings possibly after the molded cubes have been dried and salted for 3–4 days. The steeping in brine lasts for about 3 months after which the product is ready for consumption. Protein degradation is one of the important biochemical events during sufu fermentation. Post-fermentation is the main stage responsible for the hydrolysis of protein together with the increase in the content of amino-type nitrogen and free amino acid.

7.6.3.6 Oncom

It is a fermented peanut press cake product consumed largely in Indonesia as a roasted snack, covered with water and seasoned with salt or sugar. Usually, oncom is made from the by-products of the production of other foods—soybean tailings left after making tofu, peanut press cake left after the oil has been pressed out, cassava tailings while extracting the starch, coconut press cake remaining after oil has been pressed out, or when coconut milk has been produced. Since oncom production uses by-products to make food, it increases the economic efficiency of food production. Black oncom is made by using *Rhizopus oligosporus* and red oncom is made by using *Neurospora intermedia* var. *oncomensis*, and it is the only human food produced from *Neurospora* sp.

Steinkraus (1995) gave an overview of the production that involves peanut press cake as a raw material. Press cake is soaked in water for 3–4 h and then drained, crumbled to open it, mixed with other ingredients, steamed, formed into flat cakes, inoculated with a dried preparation of the mold, covered with the banana leaves after transfer to woven bamboo trays, and fermented for 36–48 h at ambient temperature and humidity. Under these conditions, the molds grow through the cake. *Neurospora* has pink spores and these impart their color to the finished product.

7.6.3.7 Dawadawa

It is a fermented locust bean product, consumed largely in West Africa, and is also known as daddowa in Nigeria. The seeds after removing the powdery pulp are boiled in water and stored overnight. The swollen cotyledons are then boiled for 30 min, put into a hole in ground, and left to ferment for 2–3 days. The microorganisms responsible for dawadawa fermentation have not been determined but undoubtedly include spore-forming bacilli, LAB, and probably yeast. New variants of dawadawa condiment have also been produced by traditional fermentation of African yam bean and soybean seeds and its sensory evaluation has been carried out by Wokoma and Aziagba (2001). Processed substrates were incubated for 72 h at room temperature in earthen pots lined with washed fresh banana leaves.

7.6.3.8 Warri/Wadiyan

Warris, in the form of spicy, hollow, brittle, friable 5–8 cm in diameter balls, are very popular in Northern India and used as condiment in cooking with vegetables, legumes, or rice. Dehulled black gram (*Phaseolus mungo*) grains after being washed and soaked overnight in water are ground to a paste. The paste is then spiced with asafetida (*Ferula foetida*, 0.5%–1.0%), caraway (*Carum carvi*, 0.5%–1.0%), cardamom (*Elettaria cardamomum*, 1.0%), cloves (*Syzygium aromaticum*, 0.4%–0.5%), fenugreek (*Trigonella foenum-graecum*, 1.0%), ginger (*Zingiber officinale*, 8.0%), and red pepper (*Capsicum annum*, 1.0%) and molded into small balls (Batra and Millner, 1974), which are allowed to undergo fermentation and drying in the open air for 4–8 days. Several bacteria and yeasts constituting the natural flora of black grams, spices added to the dough, and the surroundings are associated with warri fermentation (Batra and Millner, 1974, 1976; Batra, 1981; Reddy et al., 1982; Sandhu and Waraich, 1984; Sandhu et al., 1986; Sandhu and Soni, 1988). The microbial population dynamics, the predominant types, their prevalence, and succession suggest the involvement of bacteria alone or in combination with lesser proportion of yeasts during the warri fermentation (Sandhu and Soni, 1989; Soni and Sandhu, 1990). Seasonal variations generally affect the development and prevalence of microorganisms during fermentations; summers were found to favor higher bacterial load (10^{10} to 10^{12}/g) than winters (10^{9} to 10^{10}/g). *Bacillus subtilis* occurs mainly during summers, whereas *Leuconostoc mesenteroides, Streptococcus faecalis,* and *Lactobacillus fermentum* are the principal bacteria associated during both the seasons. Yeasts including *Candida vartiovaarai, Kluyveromyces marxianus,* and *Candida krusei* are largely encountered during winters, while *Saccharomyces cerevisiae, Pichia membranaefaciens, Trichosporon beigelii,* and *Hansenula anomala* are commonly involved during both the seasons.

7.6.3.9 Bhallae

Bhallae are snack foods prepared from unspiced black gram paste, prepared in the same way as that of warri, but without spices, and fermented for 12–18 h. The paste is pan or deep-fried and consumed as a snack as such or after briefly soaking in spiced tamarind water or curd. The microflora involved in bhallae fermentation generally comprise yeasts (Sandhu and Waraich, 1984) and bacteria (Sandhu et al., 1986). *Lactobacillus fermentum* and *Leuconostoc mesenteroides* including *Bacillus subtilis, Streptococcus faecalis, Flavobacter,* and *Achromobacter,* while *Kluyveromyces marxianus* and *Trichosporon beigelii* are most predominant yeast species in bhallae fermentation. *Trichosporon pullulans, Hansenula anomala, D. hansenii, Pichia membranaefaciens, Saccharomyces cerevisiae,* and *Candida curvata* are other yeasts involved in the fermentation.

7.6.3.10 Vada/Vadai

This is another black gram–based snack food of India prepared from unspiced legume paste and is deep-fried in oil as bhallae, but the main difference is that the dough is comparatively finer and the incubation period is 18–24 h. Legume beans are soaked in water for 4–6 h, then ground to a fine paste, and left to ferment at ambient temperature for few hours after the addition of salt, chilies, and sometimes onions. The paste is then made into balls and deep-fried in vegetable oil. The microflora involved in fermentation include bacteria alone or in combination with yeasts (Soni, 1987). The bacteria are found in the range of 10^{8}–10^{10} per gram and belong to *Leuconostoc mesenteroides, Lactobacillus fermentum, Streptococcus faecalis,* and *Bacillus subtilis.* Yeast numbers are reported up to 10^{6} per gram and generally belong to *D. hansenii, Trichosporon beigelii, Hansenula anomala, Saccharomyces cerevisiae, Kluyveromyces marxianus, Trichosporon pullulans, Pichia membranaefaciens,* and *Candida curvata.*

7.6.3.11 Papadam

These are thin, crisp, circular tortilla-like wafers of legume, cereal, or flour of starch crops eaten as snack or with meals either roasted or deep-fried. The dough is prepared and spiced as in the case of warris, but without fenugreek and ginger, rolled into wafers of 10–20 cm in diameter and 0.5 mm thickness. These are then fermented and dried in the open air for 4–6 h. Both bacteria and yeasts have been isolated from papadam (Batra and Millner, 1974; Soni and Sandhu, 1990). Bacteria are generally in the range of 10^8–10^{10} per gram belonging mainly to three types including *Leuconostoc mesenteroides, Lactobacillus fermentum,* and *Streptococcus faecalis,* while yeasts vary up to 10^3 per gram and belong to six types. Of these, *Saccharomyces cerevisiae* was the most predominant followed by *Hansenula anomala, Trichosporon beigelii, Candida krusei, Pichia membranaefaciens,* and *D. hansenii.*

7.6.4 Fermented Cereal–Legume Products

This is the most dominant group of fermented foods next to dairy products consumed in India. However, in contrast to established fermented milk products, cereal–legume-based traditional foods have not made any significant impact on the population other than that depending heavily on cereal products as staple diet. The different cereal–legume products are discussed in the following sections.

7.6.4.1 Idli

Idli is the most popular among the cereal–legume-based foods of India. The traditional method of idli preparation consists of washing and soaking rice and black gram in the proportion of 2:1 for 2–4 h at ambient temperatures and grinding them separately. The batter is then allowed to ferment for 18–30 h by means of natural microflora or by inoculating with fermented batter from a previous batch after the addition of ~1% salt. The fermented batter is then poured with a ladle on to greased round-shaped molds and steam-cooked for 10–20 min to give a fine textural idli. Attempts were made to substitute black gram dal with other legumes such as common bean (Steinkraus et al., 1967), soybean (Ramakrishnan et al., 1976; Akolkar and Parekh, 1983; Soni and Sandhu, 1989a,b), great northern bean (Sathe and Salunkhe, 1981), and mung bean (Soni and Sandhu, 1989a,b). The flow sheet for idli production is depicted in Figure 7.9.

Both bacteria and yeasts have been reported to participate in idli fermentation. However, acid and gas production has been found to be mostly dependent upon the bacterial growth. Black gram, the leguminous component of idli batter, not only serves as an effective substrate but also provides the maximum number of microorganisms for fermentation. Black gram dal soaked in water has a higher concentration of soluble nutrients to support the growth of LAB. Bacteria identified as part of the microflora responsible for the production of good idli include *Leuconostoc mesenteroides, Lactobacillus coryniformis, Lactobacillus delbrueckii, Lactobacillus fermentum, Lactobacillus lactis, Streptococcus faecalis,* and *Pediococcus cerevisiae,* while the yeast flora generally involved in fermentation include *Torulopsis candida, Trichosporon pullulans,*

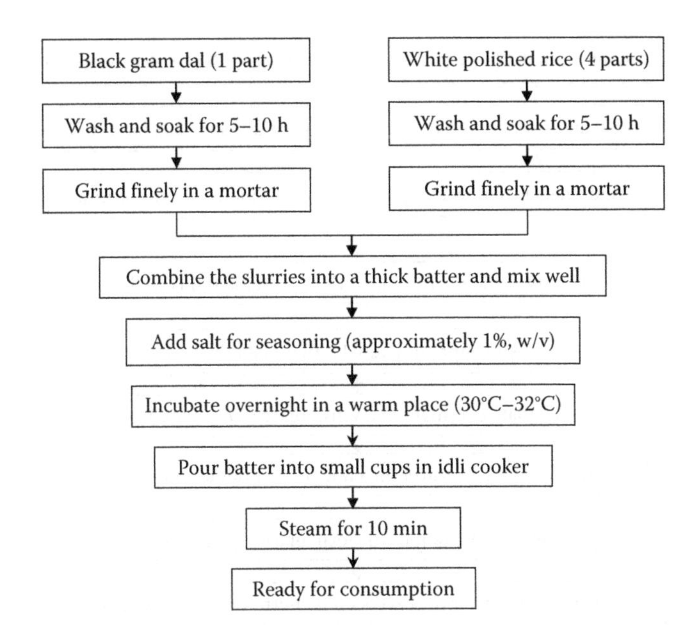
Black gram dal (1 part)
Wash and soak for 5–10 h
Grind finely in a mortar
White polished rice (4 parts)
Wash and soak for 5–10 h
Grind finely in a mortar
Combine the slurries into a thick batter and mix well
Add salt for seasoning (approximately 1%, w/v)
Incubate overnight in a warm place (30°C–32°C)
Pour batter into small cups in idli cooker
Steam for 10 min
Ready for consumption

FIGURE 7.9
Flowchart for the production of traditional Indian idli.

Candida cacaoi, Debaryomyces tamarii, Rhodotorula graminis, Candida fragicola, Candida kefyr, Wingea robertsii, Issatchenkia terricola, Hansenula anomala, Candida glabrata, Candida tropicalis, Candida sake, Candida krusei, and *Torulopsis holmii* (Sandhu and Soni, 1991).

7.6.4.2 Dosa

Dosa is a thin, crisp, fried, pancake-like staple food of southern India. It is prepared by soaking 1–3 parts of rice and 1 part of black gram dal or yellow lenticel in water for 4–6 h at ambient temperature, grinding to a fine paste by adding 2.0–2.5 parts (w/w) water, and mixing the two together to make a free running batter. The batter mixture is then allowed to ferment overnight for 12–24 h after the addition of about 1% salt, by natural microflora or by inoculating with the fermented batter of the previous batch that serves as the starter.

Leuconostoc mesenteroides is the most commonly encountered bacterium followed by *Streptococcus faecalis, Lactobacillus fermentum, Bacillus amyloliquefaciens, Lactobacillus delbrueckii, Bacillus subtilis, Pediococcus cerevisiae, Bacillus polymyxa,* and *Enterobacter* sp., while *Saccharomyces cerevisiae* is the most predominant yeast involved in the fermentation followed by *Trichosporon pullulans, Kluyveromyces marxianus, Candida kefyr, Candida krusei, Streptococcus faecalis,* and *Bacillus amyloliquefaciens* (Soni et al., 1985, 1986; Sandhu et al., 1986; Sandhu and Soni, 1988; Soni and Sandhu, 1989b, 1990). The traditional preparation of dosa is shown in Figure 7.10.

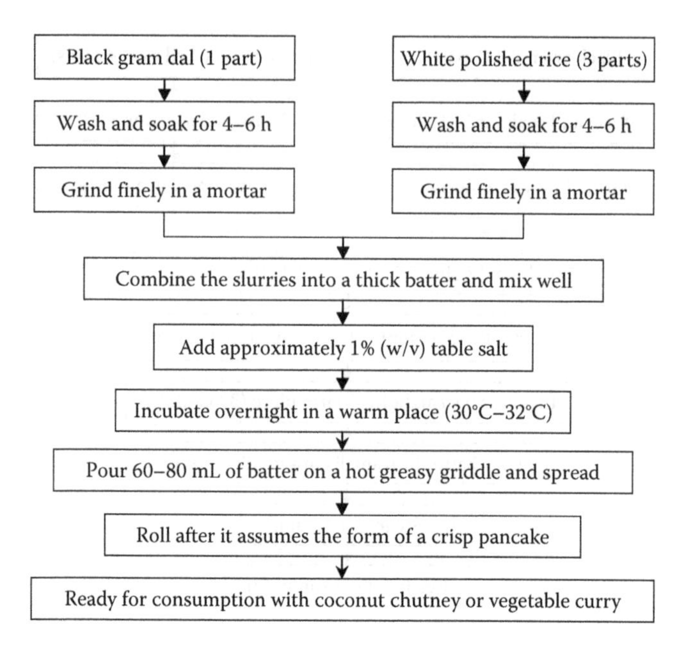

FIGURE 7.10
Flowchart for the production of traditional Indian dosa.

7.6.4.3 Dhokla

This is a fermented food commonly used in Gujarat, prepared from a batter made of coarsely ground rice and Bengal gram (*Cicer arietinum*) dal powder in a 3:1 proportion, mixed with one part of curd and adequate water, and fermented overnight at room temperature. Green chilies and table salt are added and the batter is poured on greased plates and steamed in a suitable pan for 20 min. After cooling, the dhokla is cut into squares seasoned with coriander leaves and green chilies, mustard, and asafetidas (Joshi et al., 1989). Bacteria belonging to *Lactobacillus fermentum*, *Lactobacillus lactis*, *Lactobacillus delbrueckii*, and *Leuconostoc mesenteroides* and yeast belonging to *Hansenula silvicola* from fermenting batter are generally associated with dhokla fermentation (Joshi et al., 1989). The flavor of the product could be due to the acetoin and less volatile fatty acids, acetic acid, propionic acid, isobutyric acid, and isovaleric acid that show a marked rise during fermentation (Wiseblatt and Kohn, 1960). Fermentation improves the antioxidant activities of cereal–legume-mixed batters during the preparation of traditional steam pancakes (Moktan et al., 2011).

7.6.4.4 Adai

This is made of rice and split legumes like dosa and eaten with sauce or chutney or sugar. Ten parts rice and three parts each of black gram dal, Bengal gram dal, and red gram dal are cleaned well and soaked in water for 2–3 h, coarsely ground, and left to ferment for 2–3 h in summer and overnight during winters, after the addition of some salt. On completion of this, 1.5 parts grated coconut, 1–2 parts finely chopped onions, and 0.6 parts chili powder or green chopped chilies are added with enough water to make a batter. About 40 g batter is spread on to an oily hot plate (about 10 mm thick), browned on both sides, and served hot with coconut chutney or chutney powder with oil.

7.6.5 Fermented Vegetable Products

Vegetables are highly perishable commodities and hence need to be used within a short span of time after harvesting. In order to avoid wastage of this highly nutritive commodity, simple techniques like salting, pickling, and dehydration of vegetables are in use since centuries. With the advent of modernization in the area of food processing, various processed vegetable products have become commercially available. Generally, the sequence of microbial predominance during fermentation of vegetables is salt-resistant LAB, heterofermentative *Leuconostoc* spp., and *Lactobacillus brevis* followed by homofermentative *Lactobacillus plantarum* and *Pediococcus* spp. These bacteria result in acid production and lowering of pH, thereby favoring the growth of yeasts at this stage. The fermentation with the said microflora results in conversion of sugars present in vegetables to lactic acid, acetic acid, carbon dioxide, and several other metabolites. The presence of these end products in the fermentation medium inhibits the growth of undesirable microorganisms that may deteriorate the substrate thereby ensuring a long shelf life of the fermented product.

7.6.5.1 Sauerkraut

Sauerkraut is a product obtained by lactic acid fermentation of shredded cabbage and in literal sense is the sour or acid cabbage. After removal of outer broken or dirty leaves of cabbage, the hard central core of the cabbage is shredded mechanically. This shredded cabbage is termed as "slaw," to which 2.25%–2.5% salt is added with a purpose to extract water from the cabbage by osmosis, suppressing the growth of undesirable microflora and making the conditions conducive for growth of LAB. The LAB *Leuconostoc mesenteroides*, a heterofermentative species, has been observed to initiate the lactic fermentation as its growth is much faster than other LAB especially over a wide range of temperature and salt concentration. *Leuconostoc mesenteroides* produces acid resulting in lowering pH and suppressing the growth of undesirable microorganisms. Along with acid, carbon dioxide is also produced, which replaces the air in the vessel and results in anaerobic conditions and stimulates the growth of many LAB. The growth of heterofermentative bacterial species like *Lactobacillus brevis*, *Lactobacillus plantarum*, and *Pediococcus cerevisiae* at this stage results in the production of major end products in the sauerkraut-like lactic acid, carbon dioxide, acetic acid, and

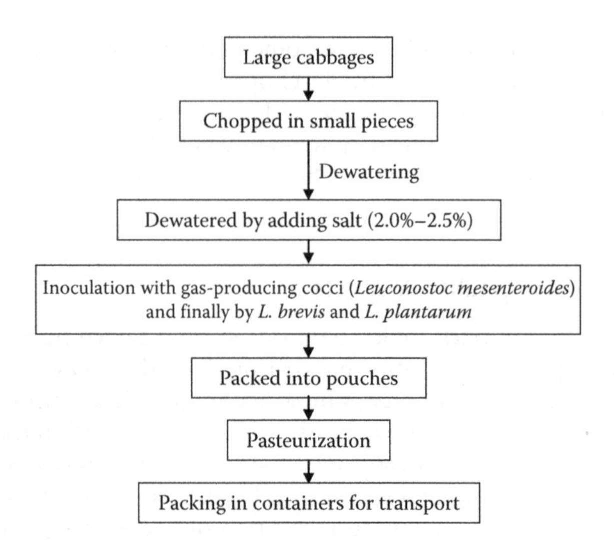

FIGURE 7.11
Flowchart for the production of traditional sauerkraut.

ethanol along with some minor products like volatile compounds. The total acidity of final product ranges between 1.7% and 2.3% and the ratio of acetic/lactic, i.e., volatile to nonvolatile acid, is about 1–4. The time required for fermentation is 1–2 months. The optimum temperature for sauerkraut fermentation is around 21°C. A variation of temperature alters the activity of the microbial process and affects the quality of the final product. Salt plays an important role in initiating the sauerkraut process and affects the quality of the final product. The addition of too much salt may inhibit the desirable bacteria, although it may contribute to the firmness of the kraut. The traditional preparation of sauerkraut is shown in Figure 7.11.

7.6.5.2 Pickles

Pickles from vegetables and fruits like mango, gooseberry, and lemons are a big Indian industry and the products are popular throughout the world. More than their nutritive value, they act as food adjuncts and appetizers as they add palatability to Indian foods. The spices in the pickle aid digestion by stimulating the flow of gastric juices. The vegetables being used for pickles are carrot, cucumber, cabbage, bitter gourd, beans, chilies, pepper, jackfruit, garlic, ginger, brinjal, and onion. Salting or curing is done by adding dry salt or brine to the raw materials, which contain plenty of water. The weak brine that is formed contains nutrients like proteins, carbohydrates, and minerals, and this helps an active fermentation by lactic cultures, which come as surface contamination from raw materials. The organisms involved are salt-resistant LAB, belonging initially to heterofermentative *Leuconostoc* spp. and *Lactobacillus brevis* that are generally replaced by homofermentative *Lactobacillus plantarum* and *Pediococcus* spp.

Coliform bacteria, particularly *Enterobacter* and *Klebsiella* spp., are also involved. As acids are produced, pH falls down, and yeasts become more active. Along with lactic acid, small quantities of acetic acid, alcohol, and sometimes hydrogen and carbon dioxide are also produced during fermentation. There is a considerable degree of change in the physical appearance of the fruit or the vegetable during the salting or curing process.

7.6.6 Meat and Fish Products

Meat and fish being highly perishable, their preservation through fermentation has been practiced since ancient times. "Fermentation" is mainly referred to as transformation of organic substances into simpler compounds by the action of microorganisms or enzymes. In case of meat, the fermentation mainly involves the action of bacteria that lowers the pH, thereby suppressing pathogenic as well as spoilage microflora (Murali et al., 1882; Sankaran et al., 1984; Murali et al., 1985). Among fermented meat products, fermented sausages are the most popular and important product. Fermented fish serves as major protein supplement worldwide. The fish preservation techniques being used since antiquity include use of salt and/or sun drying. Salt mainly brings out dehydration thereby decreasing the moisture content to such a level that spoilage does not occur. Moreover, salt has antibacterial action as well. The fish and salt are packed in layers and allowed to remain as such for long durations, during which the proteolytic enzymes present in fish are liberated, which act on muscles and membranes and release the solubilized proteins that are degraded further due to enzymatic action resulting in fermentation of fish paste if incubated for shorter periods and fish sauce if incubated for longer periods.

7.6.6.1 Sausages

The raw material used for the preparation of sausages is finally chopped meat. The meat is chilled or partially frozen followed by comminution in a meat grinder or cutter. At this stage, its curing is done by adding salt and nitrite/nitrate to develop desired pink color and taste. The curing salt when added brings the initial water activity (a_w) to about 0.95 that inhibits or suppresses the growth of pathogenic or undesired microflora and favors the growth of lactobacilli, staphylococci, and micrococci. Sodium chloride also solubilizes proteins from the microfibrillar structure of meat that improves the taste of meat products. Sodium or potassium nitrite is added to the comminuted meat for the purpose of developing required color, inhibiting autoxidative processes that lead to rancidity and favors growth of ripening microflora that mainly include lactobacilli and nonpathogenic catalase-positive cocci (Wirth, 1991). Further, spices and seasonings are blended into this mixture, which mainly include ground pepper, garlic, mace, and cardamom; some sausages may also have red pepper, mustard, etc. (Vandendriessche et al., 1980; Nes and Skjelkvale, 1982). Some of these spices

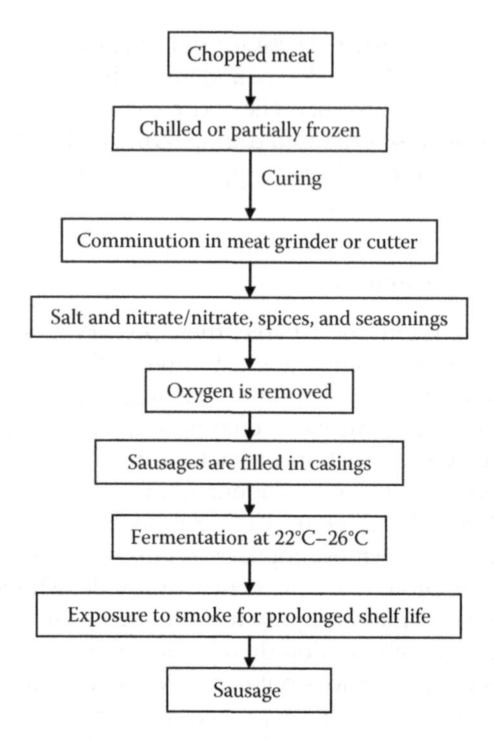

FIGURE 7.12
Flowchart for the production of traditional sausage.

contain manganese that favors the growth of LAB as the latter require for their enzymatic activities (Kandler, 1983; Zaika and Kissinger, 1984). Comminuted fatty tissue is also added to the mixture, which is then filled into casings. The casings should allow water evaporation from sausages and then allowed to ferment in ripening chamber at about 22°C–26°C. After fermentation, the product is exposed to smoke produced by controlled combustion of wood. As the smoke has antimicrobial activity, it results in the prolonged shelf life of the product.

The microflora of refrigerated fresh meat mainly includes psychrotrophic *Pseudomonas* and small numbers of LAB. Lactic acid production lowers the pH and water activity and suppresses the growth of pathogenic and spoilage bacteria. They also add to the aroma of final product due to the production of aroma compounds like autein, diacetyl, and 2,3-butanediol as a result of their lipolytic activity. The preparation of traditional sausage is shown in Figure 7.12.

7.6.6.2 Fish Paste

This is a staple food very popular in many countries all over the world. Large fish or most of other fishes that may not be suitable for fish sauce production can be used for production of fish paste. The fish is cleared and mixed with salt in a ratio of 3:1 and allowed to ferment in vats during which the proteolytic enzymes of fish start degrading the tissues. The excess liquids generated are drained off, which also results in loss of salt, thereby allowing use of resultant paste as a staple in the diet due to its lesser shelf life. The paste normally contains about 35% solids, 13% proteins, and 20%–25% salt.

7.6.6.3 Fish Sauce

The preparation of fish sauce by Romans and Greeks has been known since the ancient times. It is prepared from small fish, which are mashed/hand pressed and are put in layers with salt in the earthen pots (ratio of fish to salt is usually 3:2). The pots are kept buried in the ground for 6–18 months. Subsequently, the supernatant formed is collected and is used as first-grade sauce. In case of some fatty fish, the oil comes on the top, which is separated and sold separately because of its high nutritive value due to the presence of polyunsaturated fats in it. After collection of first-grade sauce, the residue is extracted twice or thrice to produce second- or third-grade sauce. The biochemical changes involved in fish sauce preparation mainly include conversion of insoluble fish proteins into soluble form due to the action of enzymes present in the fish and the enzymes of naturally present microorganisms. The proteins are converted into amino acids and polypeptides. The fish sauce has very peculiar aroma and flavor. The ammoniacal flavor is mainly due to the presence of ammonia and trimethylamine formed as a side product during conversion of insoluble fish proteins to soluble form. The cheesy aroma is due to the presence of low molecular weight fatty acids and the meaty aroma is probably due to the presence of a large number of volatiles in the product.

7.7 Nutritional and Health Benefits of Fermented Foods

Generally, a significant increase in the soluble fraction of a food is observed during fermentation. Fermentation results in a lower proportion of dry matter in the food and the concentrations of vitamins, minerals, and proteins appear to increase when measured on a dry-weight basis (Adams, 1990). The quantity as well as quality of the food proteins is expressed by biological value, and often the content of water-soluble vitamins is generally increased, while the antinutritional factors show a decline during fermentation (Paredes-López and Harry, 1988). Single as well as mixed culture fermentation of pearl millet flour with yeast and lactobacilli significantly increased the amount of total soluble sugars, reducing and nonreducing sugar, with a simultaneous decrease in its starch content (Khetarpaul and Chauhan, 1990). The protein efficiency ratio (PER) of wheat has been found to increase on fermentation, partly due to the increase in availability of lysine. PER value of the mixture was raised to a level comparable to casein by the fermentation process (Hesseltine and Wang, 1980). Fermentation may not increase the content of protein and amino acids unless ammonia or urea is added as a nitrogen source to the fermentation media (Reed, 1981). The relative nutritional value (RNV) of maize has also been found to increase from 65% to 81% upon germination and fermentation of the flour made of the germinated maize. It gave a further increase in RNV to 87% (Lay and Fields, 1981). During fermentation, certain microorganisms produce vitamins at a higher rate as compared to the others. The content of thiamine and riboflavin in dhokla and ambali was found to be 50% higher after fermentation. Similarly, fermented milk products in general showed an increase in folic acid content and a slight decrease in vitamin B_{12} while other B vitamins were slightly affected (Alm, 1982) as compared to raw milk.

One of the reasons for increasing interest in fermented foods is their ability to promote the functions of the human digestive system. This contribution is called probiotic effect. A fermented food product or live microbial food supplement that has beneficial effects on the host by improving intestinal microbial balance is generally understood to have probiotic effect (Fuller, 1989). The possible role of LAB in lowering cholesterol concentration and various mechanisms by which it may be possible has been discussed by Haberer et al. (1997). Studies on supplementation of infant formula with *Lactobacillus acidophilus* showed that the serum cholesterol in infants was reduced from 147 mg/mL to 119 mg/100 mL (Harrison and Peat, 1975). In an in vitro study, the ability of 23 bacterial strains of LAB isolated from various fermented milk products to bind cholesterol was investigated. Studies on antimutagenic activity of milk fermented with mixed cultures of various LAB and yeast showed that the fermented milks produced with mixed cultures of LAB had a wider range of activity against mutagens than those produced with a single strain of LAB (Tamai, 1995). Some LAB present in fermented milk products play an important role in the immune system of the host after colonization in the gut (De Simone, 1986). A study by Perdigon et al. (1995) showed that *Lactobacillus casei* could prevent enteric infections and stimulate secretory IgA in malnourished animals but also translocate bacteria, while yogurt could inhibit growth of intestinal carcinoma through increased activity of IgA, T cells, and macrophages. Lactobacilli are also involved in reducing or eliminating procarcinogens and carcinogens in the alimentary canal (Reddy et al., 1983; Mital and Garg, 1995). The enzymes, glucuronidase, azoreductase, and nitroreductase, present in the intestinal canal, are known to convert procarcinogens to carcinogens (Goldin and Gorbach, 1984).

7.8 Food Safety Aspects of Fermented Foods

It has been estimated that more than 13 million infants and children under 5 years of age die annually in the tropical regions of the world (Kenneth and Anne, 1989). After respiratory infections, diarrhea is the most common illness and has the greatest negative impact upon the growth of infants and young children. The causes of diarrhea traditionally have been ascribed to water supply and sanitation (Motarjemi et al., 1993). Foods prepared under unhygienic conditions and heavily contaminated with pathogenic organisms play a major role in child mortality through a combination of diarrheal diseases, nutrient malabsorption, and malnutrition. All food items contain microorganisms of different types and in different amounts. Dominance of a particular microorganism depends upon several factors, and sometimes microorganisms initially present in very low numbers in the food but increase afterward. In contrast to fermented meat, fish, dairy, and cereal products, fermented vegetables have not been recorded as a significant source of microbial food poisoning (Fleming and McFeeters, 1981).

7.8.1 Effect of Fermentation on Pathogenic Organisms

Although *Salmonella, Campylobacter, Shigella, Vibrio, Yersinia,* and *Escherichia* are the most common organisms associated with bacterial diarrheal diseases, other enterotoxigenic genera, including *Pseudomonas, Enterobacter, Klebsiella, Serratia, Proteus, Providencia, Aeromonas, Achromobacter,* and *Flavobacterium,* have also been reported (Nout et al., 1989a,b). In a study, a group of children fed with lactic acid–fermented gruel had a mean number of 2, 1 diarrhea episodes compared to 3, 5 for the group fed with unfermented gruel (Lorri and Svanberg, 1994). Similarly, Adams (1990) suggested that LAB are inhibitory to many other microorganisms when they are cultured together, and this is the basis of the extended shelf life and improved microbiological safety of lactic-fermented foods. *Lactobacillus* species can produce a variety of metabolites, including lactic and acetic acids that lower pH, that are inhibitory to competing bacteria, including psychrotrophic pathogen (Briedt and Fleming, 1997).

7.9 Summary and Future Prospects

Over the years, LAB have become an integral part of the human food chain through the continuous use of variety of lactic-fermented foods. Besides imparting desirable attributes, the primary role of LAB had been preservation, attributed to the production of lactic acid, hydrogen peroxide, and secondary metabolites, including bacteriocins. As the traditional methods of preparing indigenous foods are simple and the fermentations are brought about by the microflora coming from staples and the environment leading to improved digestibility and nutritional value, there is a possibility of development of various undesirable microorganisms and the production of toxic substances by certain species, which is a point of serious concern to the food microbiologists. There is much to be learned about the role of individual microorganisms, and standardization of their levels by artificially inoculating the ingredients with different microbes along with the optimization of various physicochemical factors like temperature of incubation, pH of dough batter, and supplementation of some free sugar in the fermenting ingredients may lead to further improvements in the nutritional and organoleptic quality of the end products.

Starter cultures for fermented foods are today developed mainly by design rather than by screening. The design principles are based on knowledge of bacterial metabolism and physiology as well as on the interaction with the food product. The design tools available are food-grade tools for genetic, metabolic, and protein engineering and an increased use of laboratory automation and high-throughput screening methods. The large body of new data will influence the future patterns of regulation. It can either become a promoting force for the practical use of biotechnology to make better and safer products or limit the use of starter cultures to a few strains with official approval. Successful cultures based on modern technology are expected to be launched in the areas of probiotics, bioprotection, general improvement of yield and performance for the existing culture market, and probably the introduction of cultures for fermenting other food products. A scientific basis for dramatic innovations that could transform the culture industry is currently being established.

References

Abasiekong, S.F. 1991. Protein and fat contents of crushed grains of maize and sorghum. *Journal of Applied Bacteriology* 70: 391–393.

Adams, M.R. 1990. Topical aspects of fermented foods. *Trends in Food Science and Technology* 1: 141–144.

Akolkar, P.N. and Parekh, L.J. 1983. Nutritive value of soy idli. *Journal of Food Science and Technology* 20: 1–4.

Ali, M.A.M., El Tinay, A.H., and Abdalla, A.H. 2003. Effect of fermentation on the in vitro protein digestibility of pearl millet. *Food Chemistry* 80: 51–54.

Alm, L. 1982. Effect of fermentation on B-vitamin content of milk in Sweden. *Journal of Dairy Science* 65: 353–359.

Aneja, R.P., Vyas, M.N., Nanda, I.L., and Thareja, V.K. 1977. Development of an industrial process for the manufacture of shrikhand. *Journal of Food Science and Technology* 14: 159–163.

Annan, T.N., Poll, L., Sefa-Dedeh, S., Plahar, A.W., and Jakobsen, M. 2003. Influence of starter culture combination of *Lactobacillus fermentum*, *Saccharomyces cerevisiae* and *Candida krusei* on aroma in Ghanaian maize dough fermentation European food. *Research and Technology* 216: 377–384.

Armstrong, E.N., Yan, K.G., Boy, P.C., and Wright-Burgess, J.G. 2001. The symbiotic role of marine microbes on living surfaces. *Hydrobiologia* 461: 37–40.

Bansah, D. 1990. Traditional brewing of *pito*. Process and product characteristics. M.Phil. Thesis, University of Ghana, Legon, Ghana.

Batra, L.R. 1981. Fermented cereals and gram legumes of India and vicinity. In: *Advances in Biotechnology*, eds. M. Moo Young and C.W. Robinson, pp. 547–554. Pergamon Press, Toronto, Ontario, Canada.

Batra, L.R. and Millner, P.D. 1974. Some Asian fermented foods and beverages and associated fungi. *Mycologia* 66: 942–950.

Batra, L.R. and Millner, P.D. 1976. Asian fermented foods and beverages. *Developments in Industrial Microbiology* 17: 117–128.

Bennett, B.L. and Sammartano, R. 2008. *The Complete Idiot's Guide to Vegan Cooking*. Alpha Books, Penguin Group, Inc., USA, p. 17.

Berger, B., Pridmore, R.D., Barretto, C. et al. 2007. Similarity and differences in the *Lactobacillus acidophilus* group identified by polyphasic analysis and comparative genomics. *Journal of Bacteriology* 189: 1311–1321.

Biesebeke, R., Ruijter, G., Rahardjo, Y.S.P. et al. 2002. *Aspergillus oryzae* in solid state and submerged fermentations: Progress report on a multidisciplinary project. *FEMS Yeast Research* 2: 245–248.

Blandino, A., Al-Aseeri, M.E., Pandiella, S.S., Cantero, D., and Webb, C. 2003. Cereal based fermented foods and beverages. *Food Research International* 36: 527–543.

Breidt, F. and Fleming, H.P. 1997. Using lactic acid bacteria to improve the safety of minimally processed fruits and vegetables. *Food Technology* 51: 44–46.

Caplice, E. and Fitzgerald, G.F. 1999. Food fermentations: Role of microorganisms in food production and preservation. *International Journal of Food Microbiology* 50: 131–149.

Capozzi, V., Menga, V., Digesu, A.M. et al. 2011. Biotechnological production of vitamin B2-enriched bread and pasta. *Journal of Agricultural and Food Chemistry* 59: 8013–8020.

Capozzi, V., Russo, P., Fragasso, M., DeVita, P., Fiocco, D., and Spano, G. 2012. Biotechnology and pasta-making: Lactic acid bacteria as a new driver of innovation. *Frontiers in Microbiology* 3: 94.

Chandan, R.C. and O'Rell, K.R. 2006. Ingredients for Yogurt manufacturing. In: *Manufacturing Yogurt and Fermented Milks*. ed. R.C. Chandan, vol. 11, pp. 179–193. Blackwell Publishing, Oxford, U.K.

Charalampopoulos, D., Wang, R., Pandiella, S.S., and Webb, C. 2002. Application of cereals and cereal components in functional foods: A review. *International Journal of Food Microbiology* 79: 131–141.

Chavan, J.K. and Kadam, S.S. 1989. Nutritional improvement of cereals by fermentation. *Critical Reviews in Food Science and Nutrition* 28: 349–400.

Corsetti, A., Settanni, L., Chaves Lõpez, C., Felis, G.E., Mastrangelo, M., and Suzzi, G. 2007. A taxonomic survey of lactic acid bacteria isolated from wheat (*Triticum durum*) kernels and non-conventional flours. *System Applied Microbiology* 30: 561–571.

De Angelis, M., Cassone, A., Rizzello, C.G. et al. 2010. Mechanism of degradation of immunogenic gluten epitopes from *Triticum turgidum* L. var. *durum* by sourdough lactobacilli and fungal proteases. *Applied and Environmental Microbiology* 76: 508–518.

De Noni, I. and Pagani, M.A. 2010. Cooking properties and heat damage of dried pasta as influenced by raw material characteristics and processing conditions. *Critical Reviews in Food Science and Nutrition* 50: 465–472.

De Simone, C. 1986. Microflora, youghurt and the immune system. *International Journal of Immunotherapy* 11(Suppl): 19–23.

Demuyakor, B. and Ohta, Y. 1993. Characteristics of single and mixed culture fermentation of Pito beer. *Journal of Science and Food Agriculture* 62: 401–408.

di Cagno, R., de Angelis, M., Alfonsi, G. et al. 2005. Pasta made from durum wheat semolina fermented with selected lactobacilli as a tool for a potential decrease of the gluten intolerance. *Journal of Agricultural and Food Chemistry* 53: 4393–4402.

Driessen, F. and Boer, R. 1989. Fermented milks with selected intestinal bacteria a healthy trend in new products. *Netherlands Milk and Dairy Journal* 43: 369–382.

Fenchel, T. 2002. Microbial behavior in a heterogeneous world. *Science* 296: 1068–1071.

Fleet, G.H. 1998. The microbiology of alcoholic beverages. In: *Microbiology of Fermented Foods*, ed. B.J.B. Wood, vol. 1, p. 217. Blackie Academic and Professional, Glassgow, U.K.

Fleming, H.P. and McFeeters, R.F. 1981. Use of microbial cultures: Vegetable products. *Food Technology* 35: 84–88.

Frazier, W.C. and Westthoff, D.C. 2008. *Food Microbiology*, 4th edn. Tata McGraw-Hill Companies Inc., New York.

Fuad, T. and Prabhasankar, P. 2010. Role of ingredients in pasta product quality: A review on recent developments. *Critical Reviews in Food Science and Nutrition* 50: 787–798.

Fuller, R. 1989. Probiotic in man and animals. *Journal of Applied Bacteriology* 66: 365–378.

García-Quintáns, N., Repizo, G., Martín, M., Magni, C., and López, D. 2009. Activation of the diacetyl/acetoin pathway in *Lactococcus lactis* subsp. *lactis* bv *diacetylactis* CRL264 by acidic growth. *Applied and Environmental Microbiology* 74: 1988–1996.

Ghose, T.K. and Sahai, V. 1979. Production of cellulases by *Trichoderma reesei* QM 9414 in fed-batch and continuous-flow culture with cell recycle. *Biotechnology and Bioengineering* 21: 283–296.

Goldin, B.R. and Gorbach, S.L. 1984. The effect of oral administration on *Lactobacillus* and antibiotics on intestinal bacterial activity and chemical induction of large bowel tumors. *Developments in Industrial Microbiology* 25: 139–150.

Guarner, F., Perdigon, G., Corthier, G., Salminen, S., Koletzko, B., and Morelli, L. 2005. Should yoghurt cultures be considered probiotic. *British Journal of Nutrition* 93: 783–786.

Haberer, P., Holzapfel, W.H., and Wagner, H. 1997. Moegliche rollevon milchsaeurebakterienbeider cholesterinsenkung im blutserum. *Mitteilungsblattder Bundesanstaltfuer Fleischforschung Kulmbach* 36: 202–207.

Hamad, S.H., Boecker, G., Vogel, R.F. and Hammes, W.P. 1992. Microbiological and chemical analysis of fermented sorghum dough for Kisra production. *Applied Microbiology* 37: 728–731.

Hamad, S.H., Dieng, M.C., Ehrmann, M.A., and Vogel, R.F., 1997. Characterization of the bacterial flora of Sudanese sorghum, flour and sorghum sourdough. *Journal of Applied Microbiology* 28: 764–770.

Hamaker, B.R. 2008. *Technology of Functional Cereal Products*. CRC Press LLC, Boca Raton, FL.

Han, B.Z., Beumer, R.R., Rombouts, F.M., and Nout, M.J.R. 2001. Microbiological safety and quantity of commercial sufu—A Chinese fermented soybean food. *Food Control* 12: 541–547.

Han, B.Z., Rombaouts, F.M., and Nout, M.J.R. 2003. Review: Sufu—A Chinese fermented soybean food. *International Journal of Food Science* 81: 27–34.

Harrison, V.C. and Peat, G. 1975. Serum cholesterol and bowel flora in the new born. *American Journal of Clinical Nutrition* 28: 1351–1355.

Hesseltine, C.W. and Wang, H.L. 1980. The importance of traditional fermented foods. *Bioscience* 30: 402–404.

Holzapfel, W. 1997. Use of starter culture in fermentation on a household scale. *Food Control* 8: 241–258.

Holzapfel, W. 2002. Appropriate starter culture technologies for small-scale fermentation in developing countries. *International Journal of Food Microbiology* 75: 197–212.

Hornsey, I.S. 2003. *A History of Beer and Brewing*, pp. 20–22. Royal Society of Chemistry, Cambridge, U.K.

Huang, H.T. 2000. Science and civilisation in China. *Biology and Biological Technology, Part 5, Fermentations and Food Science*, vol. 6. Cambridge University Press, Cambridge, U.K.

Hugenholtz, J., Kleerebezem, M., Starrenburg, M., Delcour, J., De Vos, W., and Hols, P. 2000. *Lactococcus lactis* as a cell factory for high-level diacetyl production. *Applied and Environmental Microbiology* 66: 4112–4114.

Jaine, T. 1999. Bread. In: *The Oxford Companion to Food*, 2nd edn., eds. A. Davidson and T. Jaine, pp. 96–98. Oxford University Press, Oxford, U.K.

Jespersen, L., Halm, M., Kpodo, K., and Jakobsen, M. 1994. Significance of yeasts and moulds occurring in maize dough fermentation for kenkey production. *International Journal of Food Microbiology* 24: 239–248.

Joshi, N., Gadhole, S.H., and Pradnya, K. 1989. Microbial and biochemical changes during dhokla fermentation with special reference to flavor compounds. *Journal of Food Science and Technology* 26: 13–15.

Ju, L.K. and Afolabi, O.A. 1999. Wastepaper hydrolysate as soluble inducing substrate for cellulase production in continuous culture of *Trichoderma reesei*. *Biotechnology Progress* 15: 91–97.

Kandler, O. 1983. Carbohydrate metabolism in lactic acid bacteria. *Antonie van Leeuwenhoek* 49: 209–224.

Kenneth, H., and Anne, R.P. 1989. Child mortality in the developing world. *Population and Development Review* 15: 680.

Khetarpaul, N. and Chauhan, B.M. 1989. Effect of fermentation by pure cultures of yeasts and lactobacilli on phytic acid and polyphenol content of pearl millet. *Journal of Food Science* 54: 780–781.

Khetarpaul, N. and Chauhan, B.M. 1990. Fermentation of pearl millet flour with yeasts and lactobacilli: In vitro digestibility and utilisation of fermented flour for weaning mixtures. *Plant Foods for Human Nutrition* 40: 167–173.

Khetarpaul, N. and Chauhan, B.M. 1991. Effect of natural fermentation on phytate and polyphenolic content and in-vitro digestibility of starch and protein of pearl millet (*Pennisetum typhoideum*). *Journal of the Science of Food and Agriculture* 55: 189–195.

Kimaryo, V.M., Massawe, G.A., Olasupo, N.A., and Holzapfel, W.H. 2000. The use of a starter culture in the fermentation of cassava for the production of "kivunde," a traditional Tanzanian food product. *International Journal of Food Microbiology* 56: 179–190.

Kohajdová, Z. and Karovicova, J. 2007. Fermentation of cereals for specific purpose. *Journal of Food and Nutrition Research* 46: 51–57.

Konlani, S., Delgenes, J.P., Moletta, R., Traoré, A., and Doh, A. 1996. Isolation and physiological characterisation of yeasts involved in sorghum beer production. *Food Biotechnology* 10: 29–40.

Kosikowski, F.V. and Mistry, V.V. 1997. *Cheese and Fermented Milk Foods*, 3rd edn., ed. F.V. Kosikowski, pp. 360–361. LLC, Westport, CT.

Krishnaswamy, M.A. and Laxminarayana, H. 1948. Microflora in butter. *Current Science* 17: 159.

Lay, M.M.G. and Fields, M.L. 1981. Nutritive value of germinated corn and corn fermented after germination. *Journal of Food Science* 46: 1069–1073.

Leroy, F. and De Vuyst, L. 2004. Lactic acid bacteria and functional starter cultures for the food fermentation industry. *Trends in Food Science and Technology* 15: 67–78.

Lorri, W. and Svanberg, U. 1994. Lower prevalence of diarrhoea in young children fed lactic acid-fermented cereal gruels. *Food and Nutrition Bulletin* 15: 57–63.

Mbugua, S.K. 1984. Isolation and characterisation of lactic acid bacteria during the traditional fermentation of uji. *East African Agriculture and Forestry Journal* 50: 36–43.

Mbugua, S.K. 1977. *The Survey on the Recipes for Uji. The Extent and Mode of Consumption in the Kenyan Urban Areas.* Department of Food Science and Technology, University of Nairobi, Nairobi, Kenya.

Mbugua, S.K. and Njenga, J. 1991. The antimicrobial activity of fermented uji. *Ecology of Food and Nutrition* 28: 191–198.

Mital, B.K. and Garg, S.K. 1995. Anticarcinogenic, hypocholesterolemic and antagonistic activities of *Lactobacillus acidophilus*. *CRC Critical Reviews in Microbiology* 21: 175–214.

Mitchell, D.A., Berovic, M., and Krieger, N. 2006. Solid state fermentation bioreactor fundamental introduction and review. In: *Solid Fermentations Bioreactors Fundamentals of Design and Operation*, eds. D.A. Mitchell, M. Berovic, and N. Kreiger, pp. 1–12. Springer, New York.

Moktan, B., Roy, A., and Sarkar, P.K. 2011. Antioxidant activities of cereal–legume mixed batters as influenced by process parameters during preparation of dhokla and idli, traditional steamed pancakes. *International Journal of Food Science and Nutrition* 62: 360–369.

Motarjemi, Y., Käferstein, F., Moy, G., and Quevedo, F. 1993. Contaminated weaning food: A major risk factor for diarrhoea and associated malnutrition. *Bulletin of the World Health Organization* 71: 79–92.

Mugula, J.K., Narvhus, J.A., and Sørhaug, T. 2003. Use of starter cultures of lactic acid bacteria and yeasts in the preparation of togwa, a Tanzanian fermented food. *International Journal of Food Microbiology* 83: 307–318.

Mugula, J.K., Nnko, S.A.M., and Sørhaug, T. 2001. Changes in quality attributes during storage of togwa, a lactic acid fermented gruel. *Journal of Food Safety* 21: 181–194.

Murali, H.S. 1984. Studies on the improvement of raw mutton quality using microorganisms, M.Sc. Thesis. University of Mysore, Mysore, India.

Murali, H.S., Leela, R.K., Sankaran, R., and Sharma, T.R. 1982. Studies on the effect of lactic culture on the improvement of raw meat colour. Presented at *23rd Annual Conference of AMI*, CFTRI, Mysore, India.

Murali, H.S., Leela, R.K., Sankaran, R., and Sharma, T.R. 1985. Effect of some lactic cultures on the natural microflora of mutton. *Chemie, Mikrobiologie Technologie der Lebensmittel* 9: 19–23.

Muyanga, C.M.B.K., Narvhus, J.A., Treimo, J., and Langsrud, T. 2003. Isolation, characterisation and identification of lactic acid bacteria from *bushera*: A Ugandan traditional fermented beverage. *International Journal of Food Microbiology* 80: 201–210.

Nago, C.M., Hounhouigan, D.J., Akissoe, N., Zanou, E., and Mestres, C. 1998. Characterization of the Beninese traditional ogi, a fermented maize slurry: Physiological and microbiological aspects. *International Journal of Food Science and Technology* 33: 307–315.

Nagpal, R., Behare, P.V., Kumar, M. et al. 2012. Milk, milk products, and disease free health: An updated overview. *Critical Reviews in Food Science and Nutrition* 52: 321–333.

Nche, P.F., Nout, M.J.R., and Rombouts, F.M. 1994. The effect of cow pea-supplementation on the quality of kenkey, a traditional Ghanaian fermented maize. *Journal of Cereal Science* 19: 191–197.

Nes, I.F., Diep, D.B., Havårstein, L.S., Brurberg, M.B., Eijsink, V., and Holo, H. 1996. Biosynthesis of bacteriocins in lactic acid bacteria. In: *Lactic Acid Bacteria: Genetics, Metabolism and Applications*, eds. G. Venema, J.H.J. Huis in't veld, and J. Hugenholtz, pp. 17–32. Kluwer Academic Publishers, Dordrecht, the Netherlands.

Nes, I.F. and Skjelkavale, R. 1982. Effect of natural spices and oleoresins on *Lactobacillus plantarum* in the fermentation of dry sausage. *Journal of Food Science* 47: 1618–1625.

Nout, M.J.R. and Motarjemi, Y. 1997. Assessment of fermentation as a household technology for improving food safety. *Food Control* 8: 221–226.

Nout, M.J.R., Rombouts, F.M., and Hautvast, G.J. 1989a. Accelerated natural lactic fermentation of infant food formulations. *Food and Nutrition Bulletin* 11: 65–73.

Nout, M.J.R., Rombouts, F.M., and Havelaar, A. 1989b. Effect of accelerated natural lactic fermentation of infant food ingredients on some pathogenic microorganisms. *International Journal of Food Microbiology* 8: 351–361.

Ooijkaas, L.P., Weber, F.J., Buitelaar, R.M., Tramper, J., and Rinzema, A. 2000. Defined media and inert supports: Their potential as solid-state fermentation production systems. *Trends in Biotechnology* 18: 356–360.

Pandonou, S.W., Nielsen, D.S., Hounhouigan, D.J., Thorsen, L., Nago, C.M., and Jakobsen, M. 2009. The microbiota of Lafun, an African traditional cassava food product. *International Journal of Food Microbiology* 133: 22–30.

Paredes-López, O. and Harry, G.I. 1988. Food biotechnology review: Traditional solid-state fermentations of plant raw materials—Application, nutritional significance and future prospects. *CRC Critical Reviews in Food Science and Nutrition* 27: 159–187.

Patel, R.S. and Chakraborty, B.K. 1988. Shrikhand: A review. *Indian Journal of Dairy Sciences* 41: 1–6.

Perdigon, G., Alvarez, S., Rachid, M., Aguero, G., and Gobbato, N. 1995. Immune system stimulation by probiotics. *Journal of Dairy Science* 78: 1597–1606.

Pretorius, I.S. 2000. Tailoring wine yeasts for the new millennium: Novel approaches to the ancient art of winemaking. *Yeast* 16: 675–729.

Ramakrishnan, C.V. 1977. The use of fermented foods in India. In: *Symposium on Indigenous Fermented Foods*. Bangkok, Thailand.

Ramakrishnan, C.V., Parekh, L.J., Akolkar, P.N., Rao, G.S., and Bhandari, S.D. 1976. Studies on soy idli fermentation. *Plant Foods for Man* 2: 1–2, 15–33.

Reddy, B.S., Ekelund, G., Bohe, M., Engle, A., and Domellöf, L. 1983. Metabolic epidemiology of colon cancer: Dietary pattern and fecal sterol concentrations of three populations. *Nutrition and Cancer* 5: 34–40.

Reddy, N.R., Pierson, M.D., Sathe, S.K., and Salunkhe, D.K. 1982. Legume based fermented foods: Their preparation and nutritive quality. *CRC Critical Reviews in Food Science and Nutrition* 17: 335–370.

Reed, G. 1981. Use of microbial cultures: yeast products. *Food Technology* 35: 89–94.
Ricciardi, A., Parente, E., Piraino, P., Paraggio, M., Romano, P. 2005. Phenotypic characterization of lactic acid bacteria from sourdoughs for altamura bread produced in Apulia (southern Italy). *International Journal of Food Microbiology* 98: 63–72.
Russo, P., Beleggia, R., Ferrer, S., Pardo, I., and Spano, G. 2010. A polyphasic approach in order to identify dominant lactic acid bacteria during pasta manufacturing. *Food Science and Technology* 43: 982–986.
Sabban, O.F. 1986. Savoir-faire oublier: Le travail du lait en Chine ancienne, vol. 21, pp. 31–67. Memoirs of the Research Institute for Humanistic Studies, Kyoto University, Shimbun, Japan.
Sandhu, D.K. and Soni, S.K. 1988. Optimization of physico-chemical parameters for Indian dosa batter fermentation. *Biotechnology Letters* 10: 227–232.
Sandhu, D.K. and Soni, S.K. 1989. Microflora associated with Indian Punjabi warri fermentation. *Journal of Food Science and Technology* 26: 21–25.
Sandhu, D.K. and Soni, S.K. 1991. Microbial prevalence and succession associated with idli fermentation. *Indian Journal of Microbiology* 31: 285–289.
Sandhu, D.K., Soni, S.K., and Vikhu, K.S. 1986. Distribution and role of yeast in Indian fermented foods. In: *Yeast Biotechnology*, eds. P. Tauro and R.K. Vashishta, pp. 142–148. Haryana Agricultural University Press, Hisar, India.
Sandhu, D.K. and Waraich, M.K. 1984. Distribution of yeasts in indigenous fermented foods with a brief review of literature. *Kawaka* 12: 73–85.
Sankaran, R., Murali, H.S., Leela, R.K., and Sharma, T.R. 1984. Enhancement of colour and texture of mutton by the use of lactic acid bacteria. *Journal of Food Science and Technology* 23: 172–175.
Sanni, A.L. and Lonner, C. 1993. Identification of yeasts isolated from Nigerian traditional alcoholic beverages. *Food Microbiology* 10: 517–523.
Sathe, S.K. and Salunkhe, D.K. 1981. Fermentation of the great northern bean (*Phaseolus vulgaris* L.) and rice blends. *Journal of Food Science* 46: 1374.
Sawadogo-Lingani, H., Diawara, B., Traore, A.S., and Jakobsen, M. 2008. Technological properties of *Lactobacillus fermentum* in the processing of dolo and pito, West African sorghum beers, for selection of starter cultures. *Journal of Applied Microbiology* 104: 873–882.
Shurtleff, W. and Aoyagi, A. 2001. *The Book of Tempeh*. Ten Speed Press, Berkeley, CA, p. 145.
Singhania, R.R., Patel, A.K., Soccol, C.R., and Pandey, A. 2009. Recent advances in solid-state fermentation. *Biochemical Engineering Journal* 44: 13–18.
Soccol, C.R. and Vandenberghe, L.P.S. 2003. Overview of applied solid-state fermentation in Brazil. *Biochemical Engineering Journal* 13: 205–218.
Soni, S.K., Sandhu, D.K., and Vilkhu, K.S. 1985. Studies on dosa: An indigenous Indian fermented food: Some biochemical changes accompanying fermentation. *Food Microbiology* 2: 175–181.
Soni, S.K., Sandhu, D.K., Vilkhu, K.S., and Kamra, N. 1986. Microbiological studies on dosa fermentation. *Food Microbiology* 3: 45–53.
Soni, S.K. 1987. Studies on some Indian fermented foods—Microbiological and chemical aspects, Ph.D. Thesis. Guru Nanak Dev University, Amritsar, India.
Soni, S.K. 2007. Microbes and ethanol production. In: *Microbes: A Source of Energy for 21st Century*, ed. S.K. Soni, pp. 301–358. New India Publishing Agency, New Delhi, India.

Soni, S.K. and Arora, J.K. 2000. Indian fermented foods: Biotechnological approaches. In: *Food Processing: Biotechnological Applications*, eds. S.S. Marwaha and J.K. Arora, pp. 143–190. Asiatech Publishers Pvt. Ltd., New Delhi, India.

Soni, S.K. and Marwaha, S.S. 2003. Cereal products: Biotechnological approaches for their production. In: *Biotechnology: Strategies in Agro-Processing*, eds. S.S. Marwaha and J.K. Arora, pp. 236–266. Asiatech Publishers Pvt. Ltd., New Delhi, India.

Soni, S.K. and Sandhu, D.K. 1989a. Fermentation of idli: Effects of changes in the raw materials and physico-chemical conditions. *Journal of Cereal Science* 10: 227–238.

Soni, S.K. and Sandhu, D.K. 1989b. Nutritive improvement of Indian dosa batters by yeast enrichment and black gram replacement. *Journal of Fermentation and Bioengineering* 68: 52–55.

Soni, S.K. and Sandhu, D.K. 1990. Indian fermented foods: Microbiological and biochemical aspects. *Indian Journal of Microbiology* 30: 135–157.

Soni, S.K., Soni, R., and Bansal, N. 2012. Genetic manipulation of food microorganisms. In: *Food Biotechnology: Principles and Practice*, eds. V.K. Joshi and R.S. Singh, pp. 157–232. I.K. International Publishing House Pvt. Ltd., New Delhi, India.

Steinkraus, K.H. 1995. *Handbook of Indigenous Fermented Foods*, 2nd edn., pp. 11–110. Marcel Dekker, New York.

Steinkraus, K.H., van Veen, A.G., and Theibeau, D.B. 1967. Studies on idli—An Indian fermented black gram rice food. *Food Technology* 21: 916–919.

Tamai, Y., Oishi, H., Nakagawa, I., Watanabe, Y., Shinmoto, H., Kuwabara, Y., Yamato, K., and Nagai, S. 1995. Antimutagenic activity of the milk fermented by mixed-cultured with various lactic acid bacteria and a yeast. *Journal of Japanese Society of Food Science and Technology [Nippon Shokuhin Kogyo Gakkaishi]* 42: 383–387.

Tamime, A.Y. and Robinson, R.K. 2007. *Yoghurt Science and Technology*. Woodhead Publishing Ltd., Cambridge, U.K.

Tengerdy, R.P. and Szakacs, G. 2003. Bioconversion of lignocellulose in solid substrate fermentation. *Biochemical Engineering Journal* 13: 169–179.

Troccoli, A., Borrelli, G.M., De Vita, P., Fares, C., and Di Fonzo, N. 2000. Mini review: Durum wheat quality: A multidisciplinary concept. *Journal of Cereal Science* 32: 99–113.

Valerio, F., Favilla, M., De Bellis, P., Sisto, A., de Candia, S., and Lavermicocca, P. 2009. Antifungal activity of strains of lactic acid bacteria isolated from a semolina ecosystem against *Penicillium roqueforti*, *Aspergillus niger* and *Endomyces fibuliger* contaminating bakery products. *Systematic and Applied Microbiology* 32: 438–448.

Vandendriessche, F., Vandekerckhove, P., and Demeyer, D. 1980. The influence of some spices on the fermentation of a Belgian dry sausage. In: *Proceedings of the 26th European Meeting of Meat Research Workers*, vol. 2, pp. 128–133. American Meat Science Association, Colorado Springs, CO.

Vinderola, C.G., Mocchiutti, P., and Reinheimer, J.A. 2002. Interactions among lactic acid starter and probiotic bacteria used for fermented dairy products. *Journal of Dairy Science* 85: 721–729.

Wirth, F. 1991. Restricting and dispensing with curing agents in meat products. *Fleisch-wirtschaft* 71: 1051–1054.

Wiseblatt, L. and Kohn, F.W. 1960. Some solatile aromatic compounds in fresh bread. *Cereal Chemistry* 37: 55–56.

Wokoma, E.C., and Aziagba, G.C. 2001. Microbiological, physical and nutritive changes occurring during the natural fermentation of African yam bean into dawa dawa. *Global Journal of Pure and Applied Sciences* 7: 219–224.

Zaika, L.L. and Kissinger, J.C. 1984. Fermentation enhanced by spices: Identification of active component. *Journal of Food Science* 49: 5–9.

Zhang, G.Y., Shi, Y.G. 1993. The history of sufu production. *Journal of Food and Fermentation Industry*. 6: 72–74.

8

Functional Foods

Kulwant S. Sandhu and Sarabjeet K. Sra

CONTENTS

8.1 Introduction 280
8.2 Definition of Functional Foods 281
8.3 Synbiotics 282
8.4 Prebiotics 282
8.4.1 Common Properties of Prebiotics 283
8.4.2 Classification of Prebiotics 284
8.4.2.1 Galacto-Oligosaccharides 285
8.4.2.2 Fructo-Oligosaccharides 285
8.4.2.3 Raffinose, Stachyose, and Soybean Oligosaccharides 286
8.4.2.4 Xylo-Oligosaccharides 286
8.4.3 Role of Prebiotics in Human Health 286
8.5 Probiotics 287
8.5.1 Common Properties of Probiotics 289
8.5.2 Criteria for Probiotic Strains 289
8.6 Lactobacillus Bacteria as Therapeutic Agents 290
8.6.1 Maintenance of Normal Intestinal Microflora 290
8.6.2 Enhancement of the Immune System 290
8.6.3 Reduction of Lactose Intolerance 290
8.6.4 Reduction of Serum Cholesterol Levels 291
8.6.5 Anticarcinogenic Activity 291
8.6.6 Protection against Ulcers 291
8.6.7 Protection against Allergies/Eczema 291
8.7 Synergistic Effect of Combination of Prebiotics and Probiotics as Synbiotic 292
8.7.1 Possible Clinical Applications of Synbiotic Health Food 293
8.7.2 Mode of Action of Synbiotics 294
8.8 Nutraceuticals 295
8.8.1 Classification of Nutraceuticals 296
8.8.1.1 Dietary Supplements 296
8.8.1.2 Functional Foods 296
8.8.1.3 Medical Foods 296
8.8.1.4 Farmaceuticals 296

8.8.2 Nutraceuticals and Human Health 297
8.8.3 Bioproduction of Nutraceuticals 298
8.8.3.1 Lipid-Based Nutraceuticals 298
8.8.3.2 Polar Lipids 298
8.8.3.3 Long-Chain Polyunsaturated Fatty Acids 299
8.8.3.4 Large-Molecule Nutraceuticals 299
8.8.3.5 Small-Molecule Nutraceuticals 300
8.8.4 Bioavailability 301
8.8.5 Safety 301
8.9 Health Benefits of Functional Foods 302
8.9.1 Omega-3 Fatty Acids 302
8.9.2 Isoflavones 305
8.9.3 Tocopherols and Tocotrienols 306
8.9.4 Flavonoids 306
8.9.5 Carotenoids 307
8.9.6 Limonoids 308
8.9.7 Organosulfur Compounds 309
8.9.8 Curcumin 309
8.9.9 Capsaicin 310
8.9.10 Isothiocyanates 310
8.9.10.1 Erucin 310
8.9.10.2 Sulforaphane 310
8.9.11 Oleuropein 311
8.9.12 Xylitol 311
8.9.13 Phytosterols and Phytostanols 311
8.9.14 Proteins and Peptides 312
8.9.14.1 Milk and Milk Products 312
8.9.14.2 Cereals 313
8.9.14.3 Legumes 313
8.9.14.4 Egg and Meat 314
8.9.14.5 Gamma-Aminobutyric Acid 314
8.9.15 Fiber 314
8.10 Summary and Future Prospects 315
References 316

8.1 Introduction

It is mainly the advances in understanding the relationship between nutrition and health that resulted in the development of the concept of functional foods. This is a practical and new approach to achieve optimal health status by promoting the state of well-being and possibly reducing the risk of diseases. According to British Nutrition Foundation, a dietary ingredient that affects its host in a targeted manner so as to exert a positive effect (to justify a health claim) can be classified as "functional ingredient." Functional ingredients are available abundantly in natural foods such as vegetables, fruits, cereals, nuts, milk, and milk-based products. Many herbs, algae, and microorganisms are also rich in functional ingredients (Cocks et al., 1995). The foods containing such ingredients are known as functional foods, having health-giving properties in addition to their nutritional value. Functional food is basically a food derived from naturally occurring raw materials that is taken as a part of daily diet and has this additional functionality. Theoretically, a functional food has to be 100% natural, i.e., free from any additives such as synthetic color or flavor or any type of preservative. The functionality can include such things as prevention and recovery from a specific disease, enhancement of immunity, control of physical and mental states, and slowing down of aging process.

Functional foods must remain foods, mostly composed of bulk ingredients (fruits, vegetables, cereals, nuts, milk, and milk-based products), and they must demonstrate their effects in amounts that can be expected for normal consumption patterns. They are not pills or capsules but part of a normal diet. Broadly, it can be said that functional foods are one step ahead of healthy natural foods in assisting the therapeutic process of the body toward substitution of medicines. All functional foods have a common denominator, i.e., they affect beneficially one or more target functions in the body and beneficial effect can be expected when they are consumed as part of normal food habit (Varshney, 2002).

Large interest has recently increased in the development of functional food products that may provide a health benefit beyond that of conventional nutrients (vitamins, minerals, etc.). Exhaustive work has been done on functional foods in the last two decades and a very vast literature is available on them (Siro et al., 2008; Sloan, 2008; Abdel, 2010). Functional foods with reference to healthy nutrition, probiotic bacteria, omega-3 fatty acid preparations, vitamins, minerals, and phytosterols have been discussed previously (Krieger-Mettbach, 2009). Foods rich in antioxidants and simultaneously characterized by a low glycemic index can reduce the risk of increased postprandial (after eating a meal) oxidative stress, which is one of the constituents for the onset of several chronic diseases (Scazzina et al., 2008).

8.2 Definition of Functional Foods

The Japanese (Foods for Specified Health Use [FOSHU]), European (Functional Food Science in Europe [FUFOSE]), and Dutch (Specific Health Promoting Food [SHF]) have given different definitions for functional foods. According to Japanese definition, functional food is any food that has a positive impact on an individual's health, physical performance, or state of mind, in addition to its nutritive value. The modified definition of FUFOSE with some additions and changes has given a clear picture of technological aspects of foods having health benefits. According to this, a functional food can be a natural food, a food to which a component has been added, or a food from which a component has been removed by technological or biotechnological means. It can also be a food where the nature of one or more components has been modified, or a food in which the bioavailability of one or more components has been modified, or any combination of these possibilities. The process elaborating the functional component was added, removed, or modified is not relevant for the definition of functional foods. A functional food might be functional for all members of a population or for particular groups of the population, which might be defined, e.g., by age or by genetic constitution (Chadwick et al., 2003). It is an elaborate definition, which includes the components covered by Japanese and Dutch definitions; thus, it is a complete definition of functional foods. In addition to synbiotics, prebiotics and probiotics, nutraceuticals, and the health benefits of functional foods have been discussed in detail in the following sections.

8.3 Synbiotics

A synbiotic is a combination of a live microbial food supplement (probiotic) and a prebiotic; specifically, the latter should be a substrate for the probiotic that acts to increase its survival/activity in the host gastrointestinal tract while also stimulating indigenous beneficial bacteria. The advantages of adding a prebiotic in a probiotic product during its manufacturing process as well as the beneficial effects were observed following synbiotic ingestion. The combination of probiotics and prebiotics as synbiotics has a synergistic effect in treating the diseases. Preliminary results have shown beneficial effects on biomarkers of diseases such as ulcerative colitis and colorectal cancer (Saulnier, 2007). The synbiotic health foods concept is being used by many European dairy drink and yogurt manufacturers in products in many countries (see Table 8.1.)

8.4 Prebiotics

The concept of prebiotics is relatively new. Prebiotics can be described as nondigestible food ingredients that beneficially affect the host by selectively stimulating the growth of one or a limited number of bacterial species already established in the colon and thus improve host health (Gibson et al., 1995). According to another definition, prebiotics are the nondigestible food components (usually

TABLE 8.1

List of Dairy-Based Synbiotic Food Products in Different Countries

Product	Manufacturer and Country
Actifit	Emmi, Switzerland
Probioplus	Migros, Switzerland
Symbalance	Tonilait, Switzerland
Proghurt	Ja Naturlich Naturprodukte, Austria
Fysiq	Mona, The Netherlands
Vifit	Sudmilch/Stassano, Belgium, Germany, U.K.
Fyos	Nutricia, Belgium
"On Guard" (liquid yogurt)	LBL Foods, Columbus, NJ
Impact	Synbiotic power food concentrate for sportsman, Australia
Orafti's synergy-1	National Cancer Institute, Rockville, MD
B-Activ	Mother Dairy, India
Nesvita	Nestle, India
Amul Prolife	Amul, India
Amul Prolite	Amul, India
Amul Sugarfree	Amul, India
Yakult	Yakult Danone India (YDI) Private Limited, India

Sources: Kumar, R. et al., *Indian Food Ind.*, 27, 57, 2008; Rediff, Probiotic drugs mart to grow more, 2007, http://www.rediff.com/money/2007/jun/05probiotic.htm (accessed July 2, 2012); Business Standard, Probiotics market needs more awareness, 2008, http://www.businessstandard.com/common/news_article.php?leftnm=5&autono=311953 (accessed July 2, 2012).

oligosaccharides), which evade digestion by mammalian enzymes in the upper regions of gastrointestinal tract, reach the colon in intact state, and are metabolized by beneficial numbers of indigenous microbiota (Gibson and Roberfroid, 1995). Prebiotics are the food and fuel for the beneficial bacteria.

8.4.1 Common Properties of Prebiotics

Properties of prebiotics reported by different research workers are as follows:

- They have low digestibility and poor absorption in the upper gastrointestinal tract (Gibson and Roberfroid, 1995; Ziemer and Gibson, 1998).
- Be a selective substrate for one or a limited number of potentially beneficial bacteria in the colon-culture protagonist (Gibson and Roberfroid, 1995; Gibson et al., 1995).
- Ability to alter the colonic microflora toward a healthier composition or selectively stimulate the growth and/or activity of intestinal bacteria associated with health and well-being (Gibson and Roberfroid, 1995).

- Help to increase the absorption of certain minerals such as calcium, magnesium, phosphorous, and iron (Coudray et al., 1997; Scholz-Ahrens et al., 2001).
- May have a favorable effect on the immune system and provide improved resistance against infection (Cummings and Macfarlane, 2002; Saavedra and Tschernia, 2002).

8.4.2 Classification of Prebiotics

Prebiotics can be classified into three categories, i.e., disaccharides, oligosaccharides, and resistant starch (Figure 8.1). Oligosaccharides are found as major components of many natural products (e.g., plant cells, milk) in either free or combined form. These are carbohydrates, having 3–10 degree of polymerization (DP). Free oligosaccharides are natural constituents of all placental mammals and can also be found in bacteria, fungi, plants, etc. The disaccharides, lactulose and galactobiose, also exhibit similar functional properties and are usually regarded as oligosaccharides.

Functional oligosaccharides have positive effects on human health, both in the prevention and in the treatment of chronic diseases. Therefore, there is great interest in health benefits of the functional oligosaccharides. The functional oligosaccharides of various origins (viruses, bacteria, plants, and fungi) have been used extensively as pharmacological supplements, food ingredients, in processed food to aid weight control, for regulation of glucose control in diabetic patients, and reducing serum lipid levels in hyperlipidemics and some other acute and chronic diseases (Qiang et al., 2009).

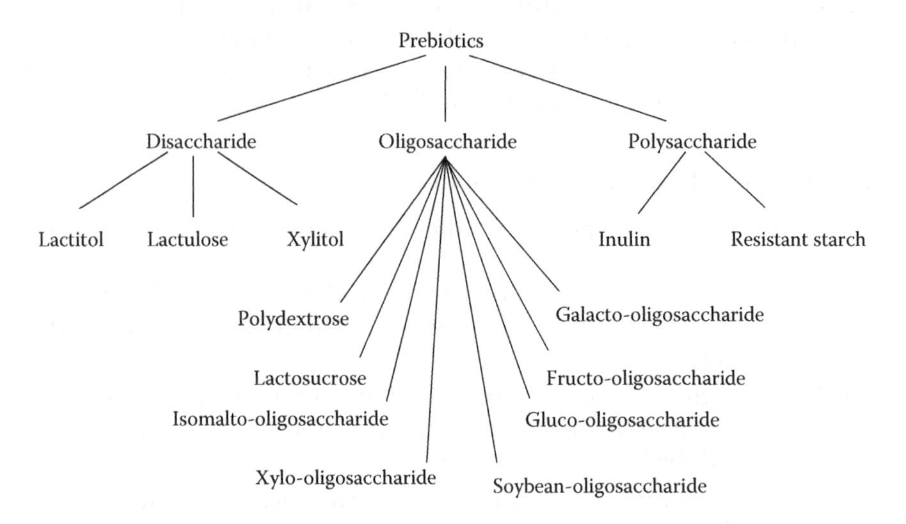

FIGURE 8.1
Classification of prebiotics. (Adapted from Magalhães, M.S. et al., *Functional Foods Forum*, University of Turku, Finland, p. 122, 2011.)

More than 20 different types of nondigestible oligosaccharides are on the world market, which are extracted from natural sources (e.g., raffinose and soybean oligosaccharide), obtained by enzymatic hydrolysis of polysaccharides (e.g., xylo-oligosaccharide [XOS] and isomalto-oligosaccharide), or produced by enzymatic transgalactosylation (e.g., galacto-oligosaccharide [GOS] and fructo-oligosaccharide [FOS]) (Sako et al., 1999). Eight different kinds of oligosaccharides have been licensed as FOSHU by the Ministry of Health and Welfare, Japan, including GOS, FOS, lactosucrose, XOS, soybean-oligosaccharide, raffinose, and isomalto-oligosaccharide (Tanaka and Matsumoto, 1998). Some oligosaccharides are described here.

8.4.2.1 Galacto-Oligosaccharides

GOSs are industrially synthesized from lactose by enzymatic transgalactosylation using β-galactosidases mainly of bacterial origin (e.g., *Bacillus circulans*). These consist of a chain of galactose molecules usually with a glucose molecule at the reducing terminus, varying in chain length (DP 3–8) and linkages. Commercially available GOSs, in the liquid and granular forms, are prepared according to the need of consumer use and cost. Various GOSs are found in human breast milk. They have certain properties that are related to their use in food such as the following: these are transparent and colorless, 40% sweeter than that of sucrose, having slightly higher viscosity than high fructose–glucose syrup, and stable at 100°C–160°C and pH 2–3.

GOSs have become the focus of a great deal of attention in the field of functional foods, owing to their known health benefits and potential to improve the quality of many foods. Because of these properties, they are currently used as low-calorie sweeteners in fermented milk products, confectioneries, breads, and beverages (Ah-Reum and Deok-Kun, 2010).

8.4.2.2 Fructo-Oligosaccharides

Chemically, FOSs are short- and medium-length chains of β-D-fructans in which fructan units are bound by a β-2,1 osidic linkage. Fructose polymers are naturally occurring in a number of vegetables and fruits as β-2,1-linked inulin or β-2,6-linked levan. FOSs are mainly found in chicory, Jerusalem artichoke, and onion. The use of FOS as novel functional ingredients in foods has been reviewed (Nitschke and Umbelino, 2002).

In view of the value of FOS for the production of functional foods containing cereal components, FOS and lysine contents of durum wheat bread made from a mixture of unripe wheat grains (a potential source of FOS) at 10%, 20%, and 30% and remilled semolina were determined (D'Egidio et al., 2001). Statistical analysis of the results indicated that FOS content was generally good at all unripe wheat grain concentrations, as were the major qualitative characteristics of the bread and bioavailability of the lysine. According to another study, cereal grains harvested at an early stage of kernel development have FOS concentration two to three times higher than those in mature grains; therefore, they may be used in functional foods (D'Egidio et al., 2005). Pasta samples made from semolina mixed with 30% immature wheat grain meal from wheat cv. Diulio or Simeto were investigated. The results showed that incorporation of immature wheat grain meal did not affect cooking quality of the pasta; however, it reduced in vitro starch digestibility by 7%–8%. The use of immature durum wheat grains (which contain FOS) has been reported in pasta, bread, and cookies (D'Egidio et al., 2008). FOS, dietary fiber (DF), and resistant starch contents and in vitro starch digestibility were assessed. Addition of 30% immature durum wheat to pasta increased FOS content by 20% without adverse effects on cooking quality. In yeast-leavened bakery products such as bread, FOS underwent hydrolysis; it was suggested that immature durum wheat would be more usefully added to chemically leavened products such as cookies.

In a randomized, double-blinded, placebo-controlled clinical study, the impact of a jelly containing short-chain fructo-oligosaccharides (sc-FOSs) and *Sideritis euboea* extract on human fecal microbiota has been studied (Mitsou et al., 2009). The daily intake of a jelly containing sc-FOS and *S. euboea* extract is well tolerated and demonstrated significant bifidogenic properties in healthy volunteers consuming their usual diets.

8.4.2.3 Raffinose, Stachyose, and Soybean Oligosaccharides

Soybean oligosaccharides (SOSs) can be readily isolated from soybean extract and consist of raffinose and stachyose as oligosaccharide as well as sucrose, glucose, and fructose. Raffinose and stachyose are not digested in human upper intestine but are readily fermented by colonic bacteria.

8.4.2.4 Xylo-Oligosaccharides

XOSs are β-1,4-linked xylose oligomers with DP of 2 to around 7 and industrially produced exclusively in Japan from xylan by partial enzymatic hydrolysis using endoxylanase (EC 3.2.1.8). Xylan, similar to hemicelluloses, usually constitutes plant cell walls in conjunction with cellulose and pectin. XOSs are fermented by a limited number of colonic bacteria such as *Bifidobacterium* spp., *Lactobacillus* spp., *Bacteroides* spp., and *Peptostreptococcus* spp. (Tanaka and Sako, 2003).

8.4.3 Role of Prebiotics in Human Health

The benefits of prebiotics have been elaborated by different authors (Ziemer and Gibson, 1998; Niness, 1999; Macfarlane et al., 2006). They have been summarized here and are shown schematically in Figure 8.2:

- Aid in normal gastrointestinal functions
- Reduce digestive illness
- Boost immunity

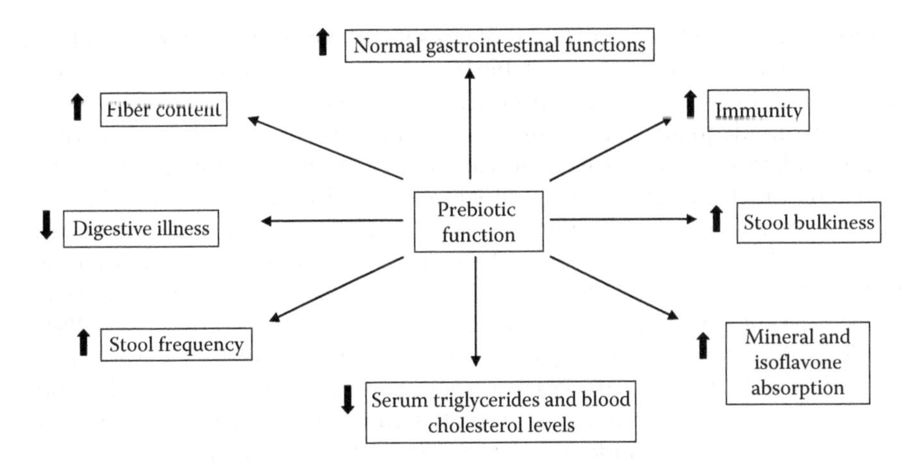

FIGURE 8.2
A schematic presentation of functions of prebiotics.

- Increase fiber content
- Prevent allergenic colitis, traveler's diarrhea, and necrotizing enterocolitis
- Increase biomass of probiotics and bulking of stool
- Increase stool frequency in constipated patients
- Decrease serum triglycerides and blood cholesterol levels in hypercholesterolemic patients
- Enhance mineral and isoflavone absorption

8.5 Probiotics

Probiotics are friendly or good bacteria described as "living drug," which are good for life. They are now considered as an army of microorganisms running inside the human body and fighting against a variety of digestive disorders. Probiotics are nonpathogenic organisms (yeast or bacteria, especially lactic acid bacteria) in foods that can exert a positive influence on the host's health (Marteau et al., 2001). Probiotics are foods containing live microorganisms that actively enhance health of consumers by improving the balance of microflora in the gut, when ingested live in sufficient numbers (Fuller, 1992). A more comprehensive definition was given by Salminen (1996), as "Probiotic is a live microbial culture or cultured dairy product, which beneficially influences the health and nutrition of the host." The Food and Agricultural Organization/World Health Organization (FAO/WHO, 2002) defined probiotics as, "Live microorganisms which when administered in adequate amounts confer a health benefit to the host."

Probiotics may play a beneficial role in several medical conditions, including diarrhea, gastroenteritis, irritable bowel syndrome, inflammatory bowel disease (Crohn's disease and ulcerative colitis), cancer, depressed immune function, inadequate lactose digestion, infant allergies, failure to thrive, hyperlipidemia, hepatic diseases, *Helicobacter pylori* infections, and genitourinary tract infections, all of which are found to improve with the use of probiotics (Brown and Valiere, 2004). The probiotics bacteria most commonly studied include members of the genera *Lactobacillus, Bifidobacterium* (*Bifidobacterium bifidum, Bifidobacterium adolescentis, Bifidobacterium longum, Bifidobacterium infantis*), and *Streptococci* (e.g., *Streptococcus salivarius* spp. *thermophilus, Streptococcus lactis*) (Gibson and Roberfroid, 1995). *Saccharomyces boulardii* (McFarland et al., 1994), *Escherichia coli* (Kruis et al., 1997), and *Enterococcus* strains are used as probiotics in nonfood formats. Many probiotic strains have been identified, studied, and commercialized. Probiotic-containing products are common in Japan and Europe (Lee et al., 1999; Sanders and Huis in't Veld, 1999). A list of microorganisms currently used in probiotic products around the world is given in Table 8.2.

TABLE 8.2

Commonly Used Microorganisms as Probiotics

Lactobacilli	Bifidobacteria	Other Lactic Acid Bacteria	Non-Lactic Acid Bacteria
L. acidophilus	*B. animalis*	*Enterococcus faecium*	*Bacillus cereus*
L. casei	*B. breve*	*Enterococcus faecalis*	*Escherichia coli*
L. johnsonii	*B. infantis*	*Lactococcus lactis*	*Bacillus subtilis*
L. reuteri	*B. longum*	*Streptococcus thermophilus*	*Saccharomyces boulardii*
L. salivarius	*B. adolescentis*	*Streptococcus diacetylactis*	*Saccharomyces cerevisiae*
L. paracasei	*B. lactis*	*Streptococcus intermedius*	*Clostridium butyricum*
L. fermentum	*B. bifidum*		*Propionibacterium freudeneichii*
L. plantarum	*B. thermophilum*		
L. crispatus			
L. gasseri			
L. rhamnosus			
L. delbrueckii spp. *bulgaricus*			
L. brevis			
L. cellobiosus			
L. curvatus			

Sources: Collins, M.D. and Gibson, G.R., *Am. J. Clin. Nutr.*, 69, 1052S, 1999; Salminen, S. and Ouwehend, A.C., *Encyclopedia Dairy Sci.*, 4, 2315, 2003; Brown, A.C. and Valiere, A., *Nutr. Clin. Care*, 7, 56, 2004.

8.5.1 Common Properties of Probiotics

- Should be nonpathogenic, nontoxic, and nonallergic (Holzapfel et al., 1998).
- Should be capable of surviving and metabolizing in upper gastrointestinal tract secretion in the gut environment, e.g., resistant to low pH, organic acids, bile juice, saliva, and gastric acid (Marteau et al., 1993; Gismondo et al., 1999).
- Should be of human origin, genetically stable, and capable of remaining viable for long periods in field condition (Holzapfel et al., 1998).
- Should have the ability to modulate immune response and provide resistance to disease through improved immunity or by the production of antimicrobial substance in the gut (Kaila et al., 1992; Salminen et al., 1996; Huis in't Veld et al., 1998; Buts et al., 1999).
- Should have a good adhesion to human intestinal tract (colonization) and influence on gut mucosal permeability (Bernet et al., 1994; Ouwehend et al., 2001).
- Should be antagonistic against carcinogenic (*Lactobacillus acidophilus, bifidobacterium, Lactobacillus plantarum,* and *Lactobacillus rhamnosus*) and pathogenic organisms (Filho-Lima et al., 2000; Ray et al., 2001; Shah, 2009).
- Should possess clinically proven health benefits, e.g., gastrointestinal disorders, persistent diarrhea, *Clostridium difficile* colitis, antibiotic-associated diarrhea, acute infantile gastroenteritis, and inflammatory bowel diseases (Vanderhoof et al., 1999; Marteau et al., 2001; Gill and Guarner, 2004; Bai and Quyang, 2006).
- Should have technological properties for commercial viability and stability of desired characteristics during processing, storage, and transportation. Sensory properties should be compatible with food use (Huis in't Veld and Shortt, 1996).

8.5.2 Criteria for Probiotic Strains

It should have an established safety record. By law, only safe bacteria (generally regarded as safe [GRAS]) are accepted on the market. The health effects should be documented. The bacteria must produce "on measurable beneficial health effects" and the scientific evidence must be for a specific strain. The bacteria must be alive when reaching the human gut in order to provide the health benefits. The number of live bacteria (colony-forming units [CFU]) is guaranteed until the end of shelf life. It means that there should be sufficient dose of friendly bacteria not only at the time of manufacture but also at the time of consumption (Kumar et al., 2008).

8.6 *Lactobacillus* Bacteria as Therapeutic Agents

In humans, *Lactobacilli* (e.g., *L. acidophilus, Lactobacillus casei, Lactobacillus delbrueckii*) are commonly used as probiotic cultures, either as single species or in mixed culture with other bacteria. *Lactobacilli* are most widely used in functional foods and play an important role as therapeutic agents. Specific information pertaining to *Lactobacillus* has been summarized here under the following heads.

8.6.1 Maintenance of Normal Intestinal Microflora

A particular strain of bacteria used in yogurt, *Lactobacillus* strain GG, aids in the treatment and prevention of antibiotic-associated diarrhea, traveler's diarrhea, and acute diarrhea in children. In adults, this particular strain of *Lactobacillus* has been shown to stimulate bowel function by altering the microflora and suppressing fermentation in the intestine. Yogurt with *Lactobacillus gasseri* may be beneficial for older adults with "atopic gastritis," a condition that predisposes to intestinal infections and constipation (Salminen, 1996).

8.6.2 Enhancement of the Immune System

The use of probiotics in immunocompromised patients appears promising. Children with HIV infections have episodes of diarrhea and frequently experience malabsorption associated with possible bacterial overgrowth. Administration of *L. plantarum* 299v (can be given safely to immunocompromised hosts) may have a positive effect on immune response and has the potential to improve the growth and development. Clinical testing has focused mostly on immune cell levels and not on actual incidence of disease (Parvez et al., 2006). An immunomodulator comprising the lactic acid bacterium *Lactobacillus reuteri,* which produces reuterin and has dehydratase activity, has also been reported (Hidetoshi et al., 2006).

8.6.3 Reduction of Lactose Intolerance

It is well established that the intake of yogurt enhances lactose digestion in individuals with low intestinal levels of lactase, the enzyme necessary to digest lactose (milk's sugar). The beneficial effect appears to be a consequence of the lactic acid bacteria in fermented milk increasing lactase activity in the small intestine (Aim, 1982; Gilliland and Kim, 1984).

8.6.4 Reduction of Serum Cholesterol Levels

Different strains of lactic acid bacteria in cultured dairy foods appear to have different effects on blood cholesterol levels. In a study of older adults, intake of about 1 cup of yogurt with live cultures per day for 1 year prevented an increase in total and low-density lipoprotein (LDL) cholesterol levels in blood (Pearce, 1996; Anderson and Gilliland, 1999). Elderly hypertensive patients who consumed fermented milk with a starter containing *Lactobacillus helveticus* and *Saccharomyces cerevisiae* experienced reductions in systolic and diastolic blood pressures (Hata et al., 1996). Characteristics of antihypertensive peptides and the role of lactic acid bacteria (*L. helveticus, Lactococcus lactis,* and *L. delbrueckii* spp. *bulgaricus*) in functional foods with antihypertensive properties have been reported (Flambard, 2002). *L. plantarum* and *Lactobacillus paracasei* strains isolated from the Italian Castelmagno PDO cheese also showed an in vitro cholesterol-lowering activity (Belviso et al., 2009).

8.6.5 Anticarcinogenic Activity

Consumption of *L. casei* might delay the recurrence of urinary bladder tumors (Aso and Akazan, 1992; Aso et al., 1995). The possibility that lactic acid bacteria in cultured dairy foods may protect against certain cancers such as colorectal cancer and possibly breast cancer has been investigated. Studies indicate that specific bacterial strains of *Lactobacillus* reduce the growth of cancer cells and the activity of fecal carcinogenic enzymes implicated in the development of colon cancer. Intake of yogurt has been shown to reduce colorectal tumors (Wollowski et al., 1999).

8.6.6 Protection against Ulcers

Probiotic bacteria may inhibit the gastric colonization and activity of *H. pylori,* which is associated with gastritis, peptic ulcers, and gastric cancer (Macfarlane and Cummings, 2002). An inhibition of *H. pylori* infection was also shown in humans consuming *Lactobacillus johnsonii* (Michetti et al., 1999).

8.6.7 Protection against Allergies/Eczema

Treatment with *Lactobacillus* GG for 4 weeks has been found to alleviate intestinal inflammation in infants with atopic eczema/dermatitis syndrome and cow's milk allergy (Viljanen et al., 2005).

Therapeutic properties of different strains of *Lactobacillus* alone and in combination with other bacteria (Parvez et al., 2006) have been given in Table 8.3.

TABLE 8.3

Various Special Therapeutic or Prophylactic Properties of Specific Strains of Bacteria

Microflora	Associated Actions
Bifidobacteria spp., *En. faecium*, *Lactobacillus* strains	Reduced incidence of neonatal necrotizing enterocolitis. Decreased duration of acute diarrhea from gastroenteritis. Administration of multiple organisms, predominantly *Lactobacillus* strains shown to be effective in ameliorating pouchitis. Lactose digestion improved. Decreased diarrhea and symptoms of intolerance in lactose-intolerant individuals, children with diarrhea, and individuals with short-bowel syndrome. Microbial interference therapy—the use of nonpathogenic bacteria to eliminate pathogens and as an adjunct to antibiotics. Improved mucosal immune function, mucin secretion, and prevention of disease
L. acidophilus	Significant decrease in diarrhea in patients receiving pelvic irradiation, decreased polyps, adenomas, and colon cancer in experimental animals. Prevented urogenital infection with subsequent exposure to three pathogens *E. coli, Klebsiella pneumoniae, Ps. aeruginosa*
L. plantarum	Lowered serum cholesterol levels, reduced incidence of diarrhea in day care centers when administered to only half of the children. Especially effective in reducing inflammation in inflammatory bowel, e.g., enterocolitis in rats, small bowel bacterial overgrowth in children, pouchitis. Reduced pain and constipation of irritable bowel syndrome. Reduced bloating, flatulence, and pain in irritable bowel syndrome in controlled trial
L. reuteri	Positive effect on immunity in HIV+ children. Shortened the duration of acute gastroenteritis and acute diarrhea
L. rhamnosus, *L. salivarius*, *Bacteroides* spp.	Enhanced cellular immunity in healthy adults in controlled trial. Suppressed and eradicated *H. pylori* in tissue cultures and animal models by lactic acid secretion. Suppressed chronic colitis, gastritis, arthritis (increased bacterial urease activity in chronic juvenile arthritis)
Sa. boulardii (yeast)	Reduced recurrence of *C. difficile* diarrhea, effects on *C. difficile* and *Klebsiella oxytoca* resulted in decreased risk and/or shortened duration of antibiotic-associated diarrhea. Shortened the duration of acute gastroenteritis. Decreased only functional diarrhea, but not any other symptoms of irritable bowel syndrome

Source: Parvez, S. et al., *J. Appl. Microbiol.*, 100, 1171, 2006.

8.7 Synergistic Effect of Combination of Prebiotics and Probiotics as Synbiotic

As is obvious from the foregoing discussion, the prebiotics and probiotics both have a number of beneficial health effects. For improving the effectiveness, researchers have combined a specific prebiotic with a specific probiotic to form a unique synbiotic compound that creates an opportunity to provide targeted health benefits and a huge boost in immunity. Considerable attention has been given in recent years to the concept of combining prebiotics and probiotics as synbiotics, in order to produce health-enhancing functional food ingredients (Gibson and Roberfroid, 1995). A few instances have already been reported earlier under the heading of synbiotics. The best synbiotic combinations currently available include *Bifidobacteria* and FOSs and *Lactobacillus* with inulin.

A synbiotic functional food product derived from cooked pulses by fermentation has been patented (Soldatov and Soldatov, 2011). There was an increased number of probiotic bacteria reaching the colon in a viable form; stimulation in the colon growth and implantation of exogenous and endogenous beneficial bacteria; anti-inflammatory, antimutagenic, and anticarcinogenic activities; production of bioactive compounds; and increased shelf life of the food product. The concept of synbiotics in functional foods, with particular reference to the results of a study investigating the health impacts of a product combining lactitol with lactic acid bacteria, has been discussed by Williamson (2009). Nutritional compositions of a preparation containing *N*-acetyl-lactosamine and/or an oligosaccharide containing *N*-acetyl-lactosamine and a probiotic *Lactobacillus* sp. have been patented along with its use for the prevention and treatment of pathogenic infection of the gastrointestinal and upper respiratory tracts (Sprenger, 2009).

A synbiotic combination could improve the survival of the bacteria cussing the upper part of the gastrointestinal tract, thereby enhancing their effects in the large bowel (Roberfroid, 2000). The main reason for using a synbiotic is that a true probiotic without its prebiotic food does not survive well in the digestive system. To enhance viability, not only on the shelf but also in the colon, the product must follow for greater attachment and growth rate of the healthy bacteria in order to minimize the growth of harmful bacteria. Synbiotic foods supply the beneficial bacteria to multiply, encouraging the body's own defense mechanisms to manage disease. A synbiotic will also suppress the development of putrefactive process in the stomach and intestines, thus preventing the occurrence of food allergies and number of diseases such as ulcerous colitis, constipation, diarrhea, cancer, and gastrointestinal infections.

8.7.1 Possible Clinical Applications of Synbiotic Health Food

There is a long list of clinical applications of synbiotics. The main possible applications of synbiotic health food are discussed here.

- Synbiotics are helpful for the prevention and control of diarrhea. They control antibiotic-associated diarrhea, traveler's diarrhea, necrotizing enterocolitis, and infectious diarrhea (including *C. difficile* diarrhea, rotavirus diarrhea, and HIV/AIDS diarrhea). They prevent the growth of *H. pylori* (responsible for gastritis, duodenal and gastric ulcers, and some malignancies) (Rolfe, 2000; Marteau et al., 2001).

- Synbiotics prevent gastroenteritis (inflammation of the mucous membrane of the intestines), atopic dermatitis, atopic eczema, asthma, allergic rhinitis, cow's milk allergy in infants, dental caries, alcohol-induced liver damage, and various types of cancer. They are reported to prevent constipation, abdominal pain, flatulence, and bloating (Nobaek et al., 2000).
- They are helpful in the management of diseases like Crohn's disease, ulcerative colitis, pouchitis, hypertension, illness-related weight loss, and obesity. Synbiotics improve the immune system and lactose digestion, delay the process of aging, reduce chances of hepatic encephalopathy, and normalize the liver and kidney functions. They help to increase the absorption of certain minerals and prevent osteoporosis (Goldin, 1998; Niness, 1999; Brown and Valiere, 2004).

8.7.2 Mode of Action of Synbiotics

The potential implications of synbiotics and their mode of action have been discussed (Kumar et al., 2008) as follows:

- Antimicrobial qualities may result more so from the probiotic components than the prebiotic. Beneficial bacteria reinforce the intestinal walls by crowding out pathogenic organisms, thereby helping to prevent their attachment to where they can cause disease. Probiotic also stimulates antigen-specific and nonspecific immune responses.
- Anticarcinogenic qualities are not well understood, but what is speculated is that the sugars that are fermented by the bacteria are converted into substances that may inhibit the growth of carcinogenic cells. Prebiotic sugars help to increase calcium and magnesium in the colon, which in turn assist in controlling the rate of cell turnover and the formation of insoluble bile or fatty acids salts, which in the soluble form have damaging effects. Probiotic bacteria have the ability to bind to some carcinogens and inactivate them. This may directly inhibit the growth of some tumors, as well as any bacteria that may convert matters into carcinogens.
- Antiallergenic aspects may be helpful in such cases as food allergies since the probiotic bacteria help to reinforce the barrier function of the intestinal wall, thereby possibly preventing the absorption of some antigens.
- Antidiarrheal aspects are successful through the crowding out of pathogens known to cause diarrhea. Since they also strengthen the intestinal wall, this too would help to prevent diarrhea.
- Osteoporosis prevention may occur as a result of an improvement in mineral absorption and balance. This ability comes as a result of oligosaccharides (3–6 sugar chains) that bind such minerals as calcium

and magnesium in the small intestine but releasing them in the large intestine where they are better absorbed. Fatty acids are also formed from the fermentation process and will then assist with the absorption of these minerals.

- They aid in reducing of serum fats and blood sugars in some individuals; although not well understood, prebiotics appear to lower triglyceride levels, as well as total cholesterol and LDL cholesterol levels. As for its ability to normalize serum glucose levels, prebiotics may help by delaying gastric emptying and/or shorten the transit time through the small intestine. In addition, a substance released through fermentation (propionate) appears to make better use of the glucose molecule conversion.
- Regulation of immune system occurs through increased immunoglobulin A (IgA) since probiotic bacteria are known to have the ability to increase the levels of circulating IgA. In addition, they enhance nonspecific immune mechanism, such as increasing phagocyte activity.
- They are helpful in treating liver-related brain dysfunction. Using synbiotics and fermentable fiber has a beneficial effect not only on the liver but also on the brain dysfunction that affects many who have liver disease. This dysfunction results in changes in behavior, intelligence, consciousness, and neuromuscular function.

8.8 Nutraceuticals

The term nutraceutical is a hybrid or contraction of nutrition and pharmaceutical. Nutraceutical is a product isolated or purified from a biological material that is generally sold in medicinal form and is not usually associated with food. A nutraceutical is demonstrated to have a physiological benefit or provide protection against chronic disease. Nutraceutical is defined as any substance that may be considered a food or part of a food and provides medical or health benefits, including the prevention and treatment of disease. Such products may range from isolated nutrients, dietary supplements, genetically engineered designer foods, herbal products, and processed foods such as cereals, soups, and beverages. In general, nutraceuticals are considered as components of traditional and exotic foods that have the potential to augment human health. The substance may be part of intact food source, or a part of processed food or may be a fortified or enriched substance in the food or may be provided as dietary supplement.

Nutraceuticals are natural compounds used to supplement the diet as a means to increase the uptake of important nutrients in the diet of humans and animals. Nutraceutical compounds are becoming an important commodity in the United States and globally as market demand increased for both purified and supplemented forms. In the 1990s, nutraceuticals from natural resources were most widely explored in plants, marine organisms, and microorganisms, in particular actinomycetes and fungi like *Sa. cerevisiae* and *Gibberella fujikuroi* (Giordano et al., 1999). Nutraceuticals can also be obtained from other eukaryotic systems such as plants, algae, and fungi due to the development of biopharmaceuticals primarily from animal tissue culture. Many nutraceuticals are extracted and purified from plants (e.g., antioxidants) and animals (e.g., fish oil rich in essential fatty acids), which are an important component of the total nutraceutical market.

8.8.1 Classification of Nutraceuticals

Nutraceuticals are classified into four categories as dietary supplements, functional foods, medical foods, and pharmaceuticals (Chaturvedi et al., 2011).

8.8.1.1 Dietary Supplements

They contain dietary ingredients that may include vitamins, minerals, herbs or other botanicals, amino acids, and substances such as enzymes, organ tissues, glandulars, and metabolites and are intended to be taken in tablet, capsule, powder, soft gel, gel cap, or liquid form. It should not be represented for use as a conventional food or as a sole item of a meal or the diet and has to be clearly labeled as being a dietary supplement.

8.8.1.2 Functional Foods

As defined earlier, they give a specific medical or physiological benefit, other than a purely nutritional effect.

8.8.1.3 Medical Foods

Medical foods are formulated in such a way that they can be consumed or administered internally under the supervision of a physician and are intended for the specific dietary management of a disease or condition for which distinctive nutritional requirements, on the basis of recognized scientific principles, are established by medical evaluation.

8.8.1.4 Farmaceuticals

Farmaceuticals is a combination of two words: farm and pharmaceuticals. It refers to medically valuable compounds produced from modified agricultural crops or animals (usually through biotechnology). The term farmaceuticals is more frequently associated, in agricultural circles, with medical applications of genetically engineered crops or animals.

8.8.2 Nutraceuticals and Human Health

In recent days, nutraceuticals are appealing to nutritionists and health-care professionals because of the growing knowledge that supports the claimed health benefits. They are demonstrated to have a physiological benefit or provide protection against chronic diseases (Kanu et al., 2007), improve health (Chen et al., 2008, 2009; Obiro et al., 2008), delay the aging process, and increase life expectancy (Morganti, 2007). Some examples of these nutraceuticals are probiotics, antioxidants, and phytochemicals. There is a long list of nutraceuticals. More than 200 nutraceutical compounds, detailing their chemical properties, biochemical activity, and dietary sources, have been studied (Wildman, 2007). Marine nutraceuticals and functional foods covering a variety of commercially available and newly developing value-added products from marine sources have been discussed by Shahidi and Barrow (2008). There are few examples of nutraceuticals with their reported medicinal value and food sources:

- Antioxidants: resveratrol from red grape products; flavonoids from citrus, tea, wine, and dark chocolate; anthocyanins found in berries
- Reducing hypercholesterolemia: soluble DF products, such as psyllium seed husk
- Cancer prevention: broccoli (sulforaphane [SF])
- Improved arterial health: soy or clover (isoflavonoids)
- Lowered risk of cardiovascular disease: α-linolenic acid (ALA) from flax or chia seeds and omega-3 fatty acids in fish oil

A traditional medicine has been prepared from green oats to improve mental health and cognitive function (Wullschleger, 2007). The feasibility of using green oat extracts as a natural ingredient in food supplements, nutraceutical foods, and functional foods to improve mental health and decrease stress levels and the onset of mood swings has been explored. In addition, many botanical and herbal extracts such as ginseng and garlic oil have been developed as nutraceuticals. Nutraceuticals are often used in nutrient premixes or nutrient systems in the food and pharmaceutical industries. The potential functions of nutraceuticals are so often related to the maintenance and improvement of health that is necessary to distinguish it from being a food ingredient and a drug. There exists a "gray" or overlapping area that has developed between nutraceuticals as dietary supplement and pharmaceuticals. If nutraceutical products are sold under the pharmaceutical banner, strict regulations are implemented to make sure that they cause no harm to the consumers, the active principle is highly concentrated, and dosage is strictly controlled. If it is marketed as dietary supplements or as a functional food, the stringent regulations do not bound them and only the claims made play an important role in deciding their usage (Soundarajan et al., 2010).

8.8.3 Bioproduction of Nutraceuticals

Bioprocessing technologies for the production of nutraceutical compounds have been described (Walker et al., 2007). From bioproduction point of view, nutraceuticals are classified into following classes.

8.8.3.1 Lipid-Based Nutraceuticals

Lipids include many broad classes of neutral and partially polar compounds with differing physiological function. Some contain important regulatory function such as sterols (e.g., cholesterol, phytosterol, and ergosterol) produced by plants as well as fungi such as *Sa. cerevisiae* and *G. fujikuroi* (Giordano et al., 1999). Some antioxidants are the class of phytosterols found not only in plant materials such as rice bran (Godber et al., 1994) but also in microorganisms like algae (Abalde et al., 1991). These antioxidants include tocopherols and tocotrienols beneficial in lowering cholesterol as well as preventing cardiovascular diseases (Lloyd et al., 2000). Tocopherols are also believed to have anticarcinogenic effects (Tarber and Packer, 1995; Dunford, 2001). Many neutral lipids in mono-, di-, and triacylglycerol forms act as functional precursors to other metabolites or act as an energy storage source. Most lipids consumed by humans and animals are derived from plants and animals due to their relative abundance in these sources. Bioproduction of lipids from microbial sources are also well known but have only recently been established at the commercial level for production, primarily for lipids with considerable bioactivity such as very-long-chain unsaturated compounds with high degrees of omega-3 fatty acids or phospholipids and sulfolipids (SLs) geared toward cancer research from algal or fungal sources (Walker et al., 2007).

8.8.3.2 Polar Lipids

Polar lipids are often associated with vital membrane structure and function depending on the lipid class (e.g., SL, phosphatidylcholine, phosphatidylethanolamine), length and position of constituent fatty acids, and degree of saturation (Csabai and Joo, 2003). Sphingolipids (e.g., sphingomyelin, a phospholipid and a ceramide) are another special class of bioactive compounds associated with cell membranes that act as lipid mediators important to intra- and extracellular signaling, regulation of cell growth, differentiation, and apoptosis. Breakdown of the abundant membrane phospholipid, sphingomyelin, by sphingomyelinases produces sphingosine and ceramide (Liu et al., 1997). Sphingosine serves as a precursor to sphingosine-1-phosphate, which has cancer-promoting properties such as cell proliferation and angiogenesis, while ceramide primarily functions physiologically as a cancer-preventive agent with properties of apoptosis and growth inhibition, thus together, placing the cell into a life-and-death balance (Ogretment and Hannun, 2004).

Sulfoquinovosyl diacylglycerol (SQDG) is characterized by its unique sulfonic acid head group, a 6-deoxy-6-sulfo-glucose (Benning, 1998). SQDG extracted from cyanobacteria has been shown to inhibit HIV-1 in cultured human lymphoblastoid T-cell lines and thus was placed as high priority for further research by the National Cancer Institute (Gustafson et al., 1989). SQDG was also found to reduce proliferation and viability of human gastric carcinoma cells SNU-1 by induction of apoptosis to direct necrosis at micromolar concentrations (Quasney et al., 2001). These results indicate potential of whole algae or algal extracts containing SLs as chemotherapeutic or preventive dietary supplements.

8.8.3.3 Long-Chain Polyunsaturated Fatty Acids

Polyunsaturated fatty acids (PUFAs) are a class of compounds that have a variety of important functions in biological systems. Studies have shown that the longer-chain fatty acids, in particular eicosapentaenoic acid (EPA; C20:5n3), arachidonic acid (ARA; C20:4n6), and docosahexaenoic acid (DHA; C22:6n3), have important roles as biosynthetic precursors, as cellular membrane components, and as protective agents against oxidative stress (Iqbal and Valivety, 1997; Holman, 1998). The relative levels of these compounds have been found to have profound effects on human health. Isolation of highly efficient oleaginous microorganisms and development of related fermentation technologies may lead to fermentation as an alternative to agricultural and animal processes for the production of PUFAs (Certik and Shimizu, 1999). Some microbial species produce high yields of certain PUFAs, which include *Mortierella alpina* (Shimizu et al., 1988; Higashiyama et al., 1998; Koike et al., 2001), *Mortierella elongata* (Bajpai et al., 1991a,b; Bajpai and Bajpai, 1993), *Pythium irregulare* (Stinson et al., 1991; O'Brien et al., 1993, Cheng et al., 1999; Hong et al., 2002), *Pythium ultimum* (Ghandi and Weete, 1991; Strednasky et al., 2000), and *Entomophthora exitalis* (Kendrick and Ratledge, 1992). *P. irregulare* is regarded as one of the most promising microbial species for possible commercial production of EPA due to its high EPA yield. The current commercial supplies of long-chain PUFAs primarily come from various oil seed plants, such as flax or borage, and from marine fish oil.

8.8.3.4 Large-Molecule Nutraceuticals

Large-molecular nutraceuticals are generally made up of larger polymers typically in the 103–106 KDa range for proteins and as high as 1010 KDa for polynucleotides. These are divided into two main classes:

1. Polysaccharide compounds

   Many polysaccharides have been identified for therapeutic activity. Poly-branched β-D-glucans isolated from the cell walls of *Sa. cerevisiae* (Muller et al., 1997) and *Agrobacterium* sp. (Kim et al., 2003) have shown potent nonspecific immune activation of macrophage immune cells. T-cells, NK-cells, and B-cells have shown high anti-AIDS activity with low side effects (Kim et al., 2003). A novel sulfated polysaccharide, calcium spirulan (Ca-SP), isolated from the cyanobacterium *Spirulina platensis* has been found to have both antiviral and anticancer properties (Hiyashi and Hayashi, 1996; Mishima and Murata, 1998). Chitosan is the deacetylated form of chitin found in abundance in arthropod exoskeletal layers, but extractable chitosan is also found in various fungi such as the industrial microorganism *Rhizopus oryzae* (Tan et al., 1996; New and Stevens, 2002; Hu et al., 2004). Chitosan is extensively marketed as a weight loss nutritional supplement that binds to lipids, lowers serum cholesterol, and improves low-density lipoprotein to high-density lipoprotein (LDL/HDL) ratios. Chitosan has also shown anticarcinogenic properties (Kiode, 1998) as well as a tissue scaffold for implants and promotes healing of wounds (Madihally and Matthew, 1999).

2. Proteins and nucleotides

   The advances in biotechnology have enabled production of many desirable compounds from animals and plants. Future production of novel proteins such as scytovirin, isolated from the cyanobacterium *Scytonema varium*, which displays anticytopathic activity against laboratory strains of HIV-1 (Bokesch et al., 2003), is expected to become increasingly important as new research developments are reported.

8.8.3.5 Small-Molecule Nutraceuticals

Many natural nutraceuticals are small molecules with molecular weight less than 600 Da and typical radius of 0.5 nm. These often include monomer compounds such as sugars, fatty acids, amino acids, and organic acids. However, they are very diverse in chemical characterization in terms of their chemical polarity, natural existence in plants/animals, and biological activities. Many important small-molecular nutraceuticals not only often found in plants but also present in high concentration within some microbial communities include various classes of antioxidant and alkaloids that stimulate the immune system by activation of T-cells, scavenge for free radicals, and inhibit tremors and analgesia (Nalik et al., 2001). Compounds of commercial interest include carotenoids that function as provitamin A, astaxanthin from red yeast *Phaffia rhodozyma* (Lim et al., 2002), canthaxanthin, and lycopene (Sabio et al., 2003).

A compound known to have anticonvulsant properties is the extracellular alkaloid compound pimprinine produced by *Streptomyces* sp. (Nalik et al., 2001). Hydroxytyrosol is a powerful antioxidant naturally present in olive oil and is synthesized in strains of *Pseudomonas aeruginosa* (Allouche et al., 2004). Cis-β-carotene is produced in *Dunaliella salina* algal culture and in fungi *Blakeslea trispora, Neurospora crassa* (Rabbani et al., 1998; Gessler et al., 2002), and *Ph. rhodozyma* when stimulated by light (Ruddat and Garber, 1983).

8.8.4 Bioavailability

Bioavailability, which can be thought of as the "absorption rate" of a supplement product, is one of the main challenges in finding effective nutraceutical products. Among unprocessed foods, not all foods are broken down and digested as effectively. Nutraceuticals with poor absorption rates result in nutrients being disposed off from the body without providing any nutritional or medicinal benefit. This aspect needs to be investigated thoroughly because the monitoring of absorption of nutraceuticals is very complicated. In the development of effective nutraceutical products, bioavailability plays an important role. The bioavailability of a substance that is in natural state will be more in comparison to manufactured product (Chaturvedi et al., 2011).

8.8.5 Safety

Unlike pharmaceutical drugs, within the United States, nutraceutical products are widely available and monitored with the same level of scrutiny as "dietary supplements." Within the oversight of the Federal Food and Drug Administration (FDA), unlike many other countries such as Canada, the use of broad-based definitions creates inconsistent credibility distinguishing the standards, function, and effectiveness between "nutraceuticals" and "dietary supplements." Within this loose regulatory oversight, legitimate companies producing nutraceuticals provide credible scientific research to substantiate their manufacturing standards, products, and consumer benefits and differentiate their products from "dietary supplements." Despite the international movement within the industry, professional organizations, academia, and health regulatory agencies to add specific legal and scientific criterion to the definition and standards for nutraceuticals, within the United States, the term is not regulated by the FDA. The FDA still uses a blanket term of "dietary supplement" for all substances without distinguishing their efficacy, manufacturing process, supporting scientific research, and increased health benefits. In 2005, the National Academies' Institute of Medicine and National Research Council created a blue-ribbon committee to create an improved framework for the FDA to evaluate dietary supplements.

Nutraceuticals hold great potential, as an alternative to substance obtained by plant. Yet sometimes they also cause harmful effect as seen with ephedrine, a widely used botanical ingredient in weight loss products. Nowadays, people are more conscious about their health and these products offer the promised health benefits. But danger is associated with some products due to lack of solid information about interaction and side effects (Chaturvedi et al., 2011). Understanding the safety of compounds listed in this novel field of nutraceuticals will enable the consumers and manufactures to use it properly for the benefit of mankind.

8.9 Health Benefits of Functional Foods

Health benefits of prebiotics, probiotics, and synbiotics have been discussed earlier in detail. Fruits and vegetables contain important functional ingredients. There are numerous bioactive chemicals present in a variety of foods that perform specific functions in human health. Many nutritive and non-nutritive antioxidants, including vitamins A, C, and E, as well as selenium, coenzyme Q10, and other phytochemicals, such as anthocyanins, are being incorporated into functional foods because of their wide range of health benefits. Antioxidants reduce the damage caused by free radicals (compounds with an extra electron). These free radicals can react with and damage unsaturated fatty acids, proteins, and DNA, which is the basis of one theory of how cancers are formed. However, the oxidative stress theory proposes that prooxidants can form when the body exceeds the antioxidant level necessary to avoid oxidation, resulting in stroke, cancers, and heart disease. Research is still underway in this area to determine the minimum and maximum levels of antioxidants that should be consumed daily (Poonia and Dabur, 2009). The important ingredients and biomolecules, their sources, and potential health benefits are presented in Table 8.4. The health benefits of functional foods other than prebiotics and probiotics have been discussed here in relation to specific ingredients and biomolecules present in foods.

8.9.1 Omega-3 Fatty Acids

Omega-3 fatty acids are essential PUFAs found naturally in the diet. The three main omega-3 fatty acids are ALA, which may provide beneficial effects against certain cancers, EPA, and DHA, which helps build infant's brain tissue. Omega-3 fatty acids are found mainly in fish and certain nuts and seeds such as flaxseed.

Humans have a limited capacity to synthesize EPA, ARA, and DHA from short-chain fatty acids. Additionally, typical American diets are extremely low in these compounds. The importance of EPA and ARA in human diet lies in that they are the biosynthetic precursors to the eicosanoid system that controls inflammatory and anti-inflammatory responses. There is increasing evidence that a variety of disorders, such as heart disease and hypertension, are related to malfunctions of the eicosanoid system caused by dietary imbalance of long-chain PUFA. DHA accumulates preferentially in the brain where it has been found to have roles both in neural impulse transmission and in protecting the brain from oxidative stress. Dietary deficiencies of DHA have been linked to bipolar disorder and schizophrenia. The diets rich in n-3 long-chain fatty acids tended to offset age-related degenerative diseases as a result of increased n-3/n-6 ratios in the blood and fatty acid profile in the brain (Connor, 2000). As a result, it has been found that dietary

TABLE 8.4

Sources and Potential Benefits of Functional Food Ingredients

Class/Components	Source	Potential Benefit	References
Carotenoids			
α-Carotene	Carrots	Neutralizes free radicals that damage cells	Burke et al. (2005)
β-Carotene	Fruits, vegetables	Neutralizes free radicals	
Lutein	Green vegetables	Contributes to the maintenance of a healthy vision	
Lycopene	Tomato products	Reduces the risk of prostate cancer	
Zeaxanthin	Eggs, citrus, corn	Contributes to the maintenance of a healthy vision	
Collagen hydrolysate			
Collagen hydrolysate	Gelatin	Improves osteoarthritis	Zaque (2008)
Dietary fiber			
Insoluble fiber	Wheat bran	Reduces the risk of breast and/or colon cancer	Slavin (2000)
β-Glucan	Oats	Reduces the risk of cardiovascular disease	
Soluble fiber	Psyllium	Reduces the risk of cardiovascular disease	
Whole grains	Cereal grains	Reduce the risk of cardiovascular disease	
Fatty acids			
Omega-3 fatty acids—DHA/EPA	Tuna, fish, marine oils	Reduce the risk of cardiovascular disease and improve mental and visual functions	Harris et al. (2006)
Conjugated linoleic acid (CLA)	Cheese, meat products	Improves body composition, decreases the risk of certain cancers	
Flavonoids			
Anthocyanidins	Fruits	Reduce the risk of certain types of cancer	Mazza and Miniati (1993)
Catechins	Tea	Reduce the risk of certain types of cancer	
Flavanones	Citrus	Reduce the risk of certain types of cancer	

(continued)

TABLE 8.4 (continued)

Sources and Potential Benefits of Functional Food Ingredients

Class/Components	Source	Potential Benefit	References
Flavones	Fruits/ vegetables	Reduce the risk of certain types of cancer	
Phenols			
Caffeic acid, ferulic acid	Fruits, vegetables, citrus	Reduce the risk of degenerative diseases, heart disease, eye disease	Scalbert and Williamson (2000)
Plant sterols			
Stanol ester	Corn, soy, wheat, fortified table spreads, stanol ester dietary supplements	Lowers blood cholesterol	Dev et al. (2011)
Polyols			
Sugar alcohols (xylitol, sorbitol, mannitol, lactitol)	Some chewing gums and other food applications	May reduce risk of dental caries (cavities)	Dev et al. (2011)
Isothiocyanates			
Sulforaphane	Cruciferous vegetables, cauliflower, broccoli, cabbage, kale, horseradish	Reduces the risk of certain types of cancer. May enhance detoxification of undesirable compounds and bolster cellular antioxidant defenses	Dev et al. (2011)
Saponins			
Saponins	Soybeans, soy foods, soy protein-containing foods	Lower LDL cholesterol; contain anticancer enzymes	
Soy protein			
Soy protein	Soybeans and soy foods	5 g/day reduces the risk of heart disease	
Phytoestrogens			
Isoflavones—daidzein, genistein, glycitein	Soybeans and soy foods	Reduce menopause symptoms	
Lignans	Flax, rye, vegetables	Protect against heart disease and some cancers, lower LDL cholesterol, total cholesterol and triglycerides	

TABLE 8.4 (continued)

Sources and Potential Benefits of Functional Food Ingredients

Class/Components	Source	Potential Benefit	References
Sulfides/thiols			
Diallyl sulfide	Onions, garlic, olives, leeks, scallions	Lowers LDL cholesterol, maintains health immune system	
Allyl methyl trisulfide, dithiolthiones	Cruciferous vegetables	Lower LDL cholesterol, maintains health immune system	
Tannins			
Proanthocyanidins	Cranberry products, cocoa, chocolate	Improve urinary tract health, reduce risk of cardiovascular disease	Kandil et al. (2002)

Source: Poonia, A. and Dabur, R.S., *Indian Food Ind.*, 28, 53, 2009.

supplementation of PUFA has considerable efficacy in the treatment of these conditions, and a considerable dietary supplement market has developed around these PUFAs (Mace and Halliwell, 2004). Beneficial health effects from the consumption of certain fish oils have been attributed to the presence of the essential PUFA, EPA, and DHA. These omega-3 fatty acids have been linked to a reduced risk of coronary heart disease (CHD), arthritis, inflammation, hypertension, psoriasis, other autoimmune disorders, and cancer (Simopoulos, 1989). PUFAs are currently marketed as dietary supplements at health food stores in the form of concentrated fish oils. Supplementation into baby foods has lately received greater interest. PUFAs are also prescribed medications for humans and pets.

Currently, the recommendation from various health organizations is to consume 0.5–2.0 g of omega-3 fatty acids per day. However, the Food and Nutrition Board might begin reviewing the data of omega-3 fatty acids, possibly leading to a formal recommendation in the near future. The FDA has approved menhaden fish oil as a GRAS ingredient, making it a prime candidate for functional foods.

8.9.2 Isoflavones

In soybean, isoflavones exist in the entire plant including seeds, leaves, stems, seedlings, and roots, whereas soybean hypocotyls have comparatively higher concentration of isoflavones. The main isoflavones found in soy are genistein, daidzein, and glycitein.

Isoflavones have four major potential beneficial effects, i.e., heart disease prevention (Anthony et al., 1998); cancer prevention, particularly with respect to breast (Peterson and Barnes, 1991; Wu et al., 1996; Messina et al., 1997; Cline and Hughes, 1998), prostate (Griffiths et al., 1998; Stephens, 1999), and colon (Messina and Bennink, 1998) cancers; bone mass density increase to prevent osteoporosis (Anderson and Carner, 1997); and reduction of postmenopausal syndromes in women (Knight et al., 1996). Consumption of tofu and other soy foods may be associated with the low incidence of breast cancer in Japanese women (Aldercreutz et al., 1991; Aldercreutz, 1998). Although the epidemiological data are inconsistent, those studies that show protective effects indicate that as little as one serving (e.g., 8 oz soy milk, 4 oz tofu) of soy per day containing approximately 30 mg of isoflavones is associated with a reduced cancer risk. Soy in the form of isolates and concentrates can be added to a range of products, including beverages (Chang, 2002).

8.9.3 Tocopherols and Tocotrienols

Oils and cereal grains are the richest source of vitamin E in human diet and provide valuable raw material for the production of tocopherol and tocotrienol concentrates. Intake of tocopherols improves the insulin action in diabetic and nondiabetic subjects (Paolisso et al., 1993). They also have protective effects against the progression of CHDs as well as act as chemoprotective agents that block or suppress mutation, promotion, and proliferation in different types of cancer, including hormonal and alimentary-canal-related cancers (Patterson et al., 1997).

8.9.4 Flavonoids

Flavonoids and their glycosides include anthocyanins, flavonols, phenolic acids and their derivatives, ellagic acid derivatives, and resveratrol. There are various sources of flavonoids including citrus fruits, onion, apples, broccoli, berries, pomegranate, and tea. Flavonoids are regarded as being antiallergenic, anti-inflammatory, antiviral, antiproliferant, and anticarcinogenic and also act in certain metabolic pathways in mammals (Middleton and Kandaswami, 1994). Flavonoids are very active toward a great number of enzyme systems in mammals. Studies have shown that the mortality from CHDs is inversely correlated with the intake of flavonoids in the diet. The antioxidant properties of flavonoids act as cancer-preventing agents and play an inhibitory role in various stages of tumor development in animals (Hollman and Katan, 1998). They also help in preventing diseases such as diabetes, inflammation, ulcers, vision deterioration, and aging in vivo.

Naringin is an active principle of grapefruit. Grapefruit juice is a traditional beverage often consumed with food for its health benefits. The antioxidant activity and antiaging effects are some proven health benefits of grapefruit juice. It has cholesterol-lowering properties (Patil et al., 2006). Nutritional compositions containing punicalagins, the predominant pomegranate tannin, have been used to enhance the immune system and reduce allergic inflammatory responses in infants and children (Jouni et al., 2010).

The nutritional compositions comprised a protein source, a fat source, a carbohydrate source, and punicalagins. Extracts from berry fruits rich in anthocyanins are mainly used for the treatment of cutaneous capillary fragility, for symptoms of venous insufficiency, and hemorrhoids. The downregulation of iNOS and COX2 by blueberry extracts can be mediated through NF-κB signaling pathway (Lau et al., 2009). The beneficial effects of blueberries may involve direct modulation of oxidative stress and/or inflammatory signaling cascades. The antiobesity action of licorice flavonoid oil is controlled by regulation of rate-limiting enzymes in fatty acid synthetic and oxidative pathways in the liver (Kamisoyama et al., 2008). A healthy and functional food with obesity-suppressing activity has also been described (Hak et al., 2008). The food was manufactured using an extract or raw juice of purple-colored potato as a major or minor ingredient, or by adding the extract or raw juice to various foods, which can be used to treat a wide range of obesity patients, including adolescents, adults, and the elderly people.

8.9.5 Carotenoids

Carotenoids are tetraterpenoids, of which there are over 600 carotenoids identified in plants, nonphotosynthetic bacteria, yeasts and molds (Krinsky and Johnson, 2005; Stahl and Sies, 2005), and some animal foods such as salmon, lobster, and egg yolk (Braun and Cohen, 2007). They are responsible for the red, orange, pink, and yellow pigment in natural foods (McGuire and Beerman, 2007). They have been considered as one of the best biological markers of fruit and vegetable intake (Semba et al., 2007).

Carotenoids have been associated with a vast range of diseases, especially degenerative diseases. Studies into carotenoids continue to demonstrate the importance of these phytochemicals. These include oxidative stress, vitamin A deficiency, asthma, chronic obstructive pulmonary disease, Alzheimer's disease, cystic fibrosis, and many more (Braun and Cohen, 2007). There is strong evidence from observational epidemiological studies that people consuming a higher level of fruit and vegetables have a reduced risk of developing certain types of cancer (Krinsky and Johnson, 2005). Further investigation suggested that carotenoids may be the chemoprotective agent due mainly to their high antioxidant activity (Krinsky and Johnson, 2005), but also through intracellular gap junction communication, interference with cell proliferation, hormonal and immune system modulation, and upregulation of phase II in the detoxification system (Rao and Agarwal, 1999; Karppi et al., 2009). Early observational epidemiological studies suggested a high-carotenoid diet reduced the risk of cardiovascular disease through antioxidant activity in particular. A high dietary intake of β-carotene has been associated with age-related macular degeneration and cataract prevention. The role of supplementation is not so clear. In an antioxidant combination, β-carotene could be effective (Braun and Cohen, 2007). Carotenoids have been shown to provide photoprotection from ultraviolet-induced free radical damage that can eventually suppress immune function (Hughes, 2001; Alves-Rodrigues and Shao, 2004; Roberts et al., 2009). Carotenoids may work better together in skin protection as they act by quenching oxygen species repairing ultraviolet light damage, modulating enzyme activity and gene expression, enhancing cell–cell communications, and suppressing cellular responses and inflammation (Stahl and Sies, 2005; Braun and Cohen, 2007). β-Carotene and other provitamin A carotenoids are required for conversion to vitamin A. β-Carotene is converted to retinoic acid, which is regulated by vitamin A status. Conversion may be enhanced by α-tocopherol and zinc (Braun and Cohen, 2007).

Lycopene is the major carotenoid pigment found in tomatoes, making up approximately 80%–90% of the total carotenoids in the common variety of tomatoes. Recent studies found that the consumption of lycopene-rich foods may reduce the risk of cancers of the prostate, breast, digestive tract, bladder, cervix, lung, and other epithelial cell types (Olson, 1986; Micozzi et al., 1990; Levy et al., 1995). They also help in reducing the susceptibility of lymphocyte DNA to oxidative damage. Lycopene had a preventive effect on atherosclerosis by protecting plasma lipids from oxidation.

8.9.6 Limonoids

Limonoids are a group of structurally related triterpene derivatives found in citrus fruits. The primary ones are limonin and nomilin. Citrus is recognized as one of the most healthful component of human diet. Epidemiological studies have shown that citrus consumption can significantly reduce an individual's risk for a variety of different cancers including cancers of the oral cavity, stomach, lung, colon, rectum, esophagus, larynx, and pancreas (Block, 1991). In a study, it was shown that a variety of limonoids including deacetylnomilin, obacunone, nomilin, and a mixture of limonoid glucosides were potent inhibitors of proliferation of estrogen receptor-negative (MDA-MB-435 cells) and estrogen receptor-positive (MCF-7 cells) human breast cancer cells (Guthrie et al., 2000). It has been reported that limonoids may have additional health-promoting properties. The limonoids were found to be primarily responsible for LDL cholesterol-lowering properties (Kuroska et al., 2000). Limonin and nomilin were found to have anti-HIV activity (Battinelli et al., 2003). The two limonoids inhibited HIV replication in a variety of different cellular systems. Both of the compounds were also found to inhibit in vitro HIV-1 protease activity.

Citrus fruits (particularly oranges, mandarins, lemons, and limes) contain significant quantities of D-limonene in the peel and smaller quantities in the pulp. High therapeutic ratio of D-limonene in the chemotherapy of rodent cancers suggests that D-limonene may be an efficacious chemotherapeutic agent for human malignancies (Crowell et al., 1994; Elson and Yu, 1994). Solvent D-limonene preparation, injected directly into the biliary system, can dissolve or disintegrate the retained cholesterol gallstones in humans (Igimi et al., 1976; Schenk et al., 1980). It also enhances the percutaneous absorption of drugs such as anti-inflammatory agents (Okabe et al., 1994).

8.9.7 Organosulfur Compounds

Organosulfur compounds are the most medicinally significant components. Garlic contains nearly four times (11–35 mg/g fresh garlic) as much sulfur-containing compounds (per g fresh weight) as onions, broccoli, cauliflower, and apricots. Organosulfur compounds have cholesterol as well as blood pressure lowering properties. In addition to this, garlic may prevent and possibly play a curative role in arteriosclerosis therapy by preventing plaque formation, increasing plaque regression, or both. Organosulfur compounds are also responsible for the antioxidant, anticancer, and antimicrobial effects (Holub et al., 2002).

Garlic and onion have been widely tested for their antidiabetic potential. Both these species have been shown to be hypoglycemic in different diabetic animal models in limited human trials (Srinivasan, 2005). The hypoglycemic potency of garlic and onion is attributed to the disulfide compounds present in them, di-(2-propenyl) disulfide and 2-propenylpropyl disulfide, respectively, which cause direct or indirect stimulation of insulin secretion by the pancreas (Kumari et al., 1995; Augusti and Sheela, 1996). In addition, they may also have an insulin-sparing action by protecting from sulfhydryl inactivation by reacting with endogenous thiol-containing molecules such as cysteine, glutathione, and serum albumins.

8.9.8 Curcumin

Curcumin has been reported to have anti-lithogenic (prevention of gallstone formation from cholesterol crystals), anti-inflammatory, and antimutagenic properties. Turmeric is a spice claimed to possess beneficial hypoglycemic effect and to improve glucose tolerance in a limited number of studies (Tank et al., 1990). Nephropathy is a common complication in chronic diabetes. High blood cholesterol is an added risk factor that determines the rate of decline of kidney function in a diabetic situation. Dietary curcumin (of turmeric) has been found to have a promising ameliorating influence on the severity of renal lesions in streptozotocin diabetic rats (Babu and Srinivasan, 1998, 1999). Hypocholesterolemic effect of this spice as well as the ability to lower lipid peroxidation under diabetic condition is implicated in the amelioration of renal lesions. The effects of turmeric powder and processed sulfur on the weight gain, body fat deposition, and lipid profile of serum and liver in Wistar rats were investigated (Kim et al., 2011). The results showed that turmeric powder along with sulfur can reduce the weight gain and body fat deposition and improve serum and liver lipid profile in rats fed with a high fat diet.

8.9.9 Capsaicin

Capsaicin is the pungent principle component of red chili that has been shown to be useful in diabetic neuropathy (The Capsaicin Study Group, 1992). In an 8-week double-blind placebo-controlled study with parallel randomized treatment conducted by 12 independent investigators involving 219 patients, topical application of 0.075% capsaicin cream was effective in pain management. Capsaicin acts as digestive stimulant and has thermogenic influence in obesity-related insulin-resistant patients. Capsaicin has been found to possess antiseptic and antioxidant properties.

8.9.10 Isothiocyanates

Isothiocyanates (ITCs) in *Brassica* spp. exhibit anticancer properties related to the induction of phase II detoxification enzymes. The variations in ITCs in baemuchae (×Brassicoraphanus), which was newly generated by an interspecific hybrid cross between radish and Chinese cabbage, have been evaluated (Sooyeon et al., 2009). Baemuchae has potential to be utilized in functional foods due to the presence of SF.

8.9.10.1 Erucin

Erucin (ER) is a dietary ITC present in cruciferous vegetables, such as rocket salads (*Eruca sativa* Mill., *Diplotaxis* sp.), that has been recently considered a promising cancer chemopreventive phytochemical. Biological activity of ER was investigated on human lung adenocarcinoma A549 cells, analyzing its effects on molecular pathways involved in apoptosis and cell cycle arrest, such as PARP-1 cleavage, p53, and p21 protein expression. ER affects the A549 cell proliferation, enhancing significantly p53 and p21 protein expression in a dose-dependent manner ($p < 0.001$). PARP-1 cleavage occurs only after exposure to high concentrations of ER (50 μM), in accordance with previous studies showing similar bioactivity of other ITCs (Melchini et al., 2009). The induction of p53, p21, and PARP-1 cleavage may participate in the antiproliferative activity of ER in human lung adenocarcinoma A549 cells. Comparison of data with those obtained with the ITC SF, structurally related to ER, underlines the strong relationship between structural analogy of ITCs and their biological activity. The ability of dietary compounds to modulate molecular mechanisms that affect cancer cell proliferation is certainly a key point of the cancer prevention potential by functional foods.

8.9.10.2 Sulforaphane

SF, a phytochemical constituent of cruciferous vegetables, has received much attention as a potential cancer chemopreventive compound. Epidemiological and dietary studies have revealed an association between high intake of cruciferous vegetables and decreased cancer risk. Recent studies suggest that Nrf2-mediated signaling, which controls the expression of many genes responsible for carcinogen detoxification and protection against oxidative stress, is regulated by SF. The Nrf2-mediated signaling pathways, particularly in relation to MT gene induction and the apoptosis-inducing effects of SF, have also been studied (Yeh and Yen, 2009).

8.9.11 Oleuropein

Health benefits of oleuropein are discussed with particular reference to cardioprotective properties of an oleuropein-containing olive leaf extract. It is suggested that the leaf extract can be used in a variety of functional foods and supplements to help lower high blood pressure and improve heart health (Frutarom, 2009).

8.9.12 Xylitol

Xylitol, a five-carbon sugar alcohol ($C_5H_{12}O_5$), has a relative sweetness equal to sucrose, a negative heat of solution, and is nonfermentable by cariogenic bacteria (Makinen, 1978). Once ingested by humans, 80%–85% of xylitol consumed is metabolized by the liver, and 20%–80% of this portion is converted to glucose depending on metabolic need for glucose. The slow conversion of xylose to glucose reduces insulin stimulation. Due to these properties, it has found many uses in the food industry, especially in confectionaries, gums, oral hygiene products, and diabetic foods (Drapcho, 2007). Xylitol is found naturally at low levels in fruits and vegetables, e.g., plums contain 1% xylitol by weight, while strawberries, raspberries, eggplant, spinach, and pumpkin contain 0.1%–0.3% xylitol.

8.9.13 Phytosterols and Phytostanols

Phytosterols are plant-based components currently being evaluated for their health benefits. These naturally occurring plant sterols have chemical structure similar to that of cholesterol, but are not synthesized by the human body. The most abundant phytosterols are sitosterol and campesterol. Phytostanols are the saturated forms of phytosterols and are found in less abundance. The most common phytostanols are sitostanol and campestanol. Incorporation of these plant sterols into foods began when a process was identified that allowed these products to be esterified and solubilized into foods.

Both phytosterols and phytostanols might lower blood cholesterol levels. This beneficial effect is thought to be related to the effects of phytosterols on absorption of dietary and biliary cholesterol in the gut. The consumption of various plant sterol-enriched beverages is effective in lowering plasma cholesterol; the lipid-lowering potential of plant sterol in a soymilk format has been investigated (Rideout et al., 2009). As compared to 1% dairy milk, consumption of low- and moderate-fat plant sterol-enriched soy beverages represents an effective dietary strategy to reduce circulating lipid concentration in normal to hypercholesterolemic individuals by reducing intestinal cholesterol absorption. The efficacy of single versus multiple doses of plant sterols on circulating lipid level and cholesterol trafficking has been investigated (AbuMweis et al., 2009). It was concluded that in order to provide an optimal cholesterol-lowering impact, plant sterols should be consumed as smaller doses given more often, rather than one large dose. Phytosterols could be incorporated in the diet not only to lower the cardiovascular disease risk but also to potentially prevent cancer development. The potential effects and mechanisms of action of phytosterols on different forms of cancer have also been reviewed (Woyengo et al., 2009).

8.9.14 Proteins and Peptides

Detailed information on bioactive proteins and peptides as functional foods and nutraceuticals has been compiled (Mine et al., 2010). The antioxidative and anti-inflammatory properties of proteins and peptides have been highlighted. Development of antihypertensive food products and mechanism for antihypertensive activity has been updated. It has been reported that food protein-derived bioactive peptides act as inhibitors for calmodulin, a protein that plays important roles in maintaining physiological functions of cells and body organs. The peptide-based immunotherapies have been discussed for food allergy treatment.

8.9.14.1 Milk and Milk Products

The health-promoting proteins and peptides in colostrums and whey of bovine milk have been described. The best characterized bioactive bovine whey proteins include immunoglobulins (lgs), lactoferrin, lactoperoxidase, and growth factors. Lactoferrin's protective effect ranges from direct antimicrobial activities against a large panel of microorganisms including bacteria, viruses, fungi, and parasites to anti-inflammatory and anticancer activities. The development of lactotripeptides and their use as antihypertensive food ingredients, with particular reference to successful isolation of isoleucine–proline–proline (IPP) tripeptide from milk and use of IPP in functional foods aimed at blood pressure control, have been discussed (Heyden, 2009). The antihypertensive peptides originate from fermented milk products and enzymatic hydrolysis of food proteins such as milk casein, whey proteins, and fish meat. The antihypertensive peptides have angiotensin I-converting enzyme inhibitory activities. They may be considered to be an ideal food derived from natural functional ingredients to help blood pressure within normal range. A meta-analysis of five randomized controlled clinical trials and the blood-pressure-lowering effects of *L. helveticus*-fermented milk products containing casein-derived bioactive tripeptides (IPP and valine–proline–proline [VPP]) has been examined (Jauhiainen et al., 2009). The meta-analysis confirmed that *L. helveticus*-fermented milk containing IPP and VPP reduces blood pressure in hypertensive subjects. Functional foods containing bioactive milk peptides can thus be useful as part of a strategy for dietary treatment and prevention of hypertension.

8.9.14.2 Cereals

Wheat proteins have various functions. Glutelins and gliadins account for 40%–60% of total protein. Although the amount of albumin is not high, it contains amylase inhibitor that inhibits amylase activity and retards carbohydrate digestion, preventing the increase in postprandial hyperglycemia. However, amylase is resistant to proteolysis by digestive enzymes and may be the cause of allergies such as baker's asthma. On the other hand, gliadins sometimes become a cause of wheat-dependent exercise-induced anaphylaxis and celiac disease. The epitopes in gliadins for wheat-dependent exercise-induced anaphylaxis and celiac diseases comprise glutamine-rich tandem repeat motifs, the cleavage in the modification of the motifs is effective to reduce their allergenicity. Some peptides such as Ile-Ala-Pro from wheat ab-gliadin and Ile-Val-Tyr from wheat germ prevent hypertension by inhibiting ACE. Rice bran protein concentrates and isolates have multifactorial functionality including their medicinal value. The residual proteins being of high molecular weight and with considerable water-binding capacity may be used as hydrocolloids, which have demonstrated health benefits in improving lipid profile, controlling hyperglycemia, and lowering glycemic index of foods (Mine et al., 2010).

8.9.14.3 Legumes

Soy protein is one of the vegetable proteins examined extensively for lipid-lowering effect in humans. Soy peptides stimulate hypolipidemic events and accelerate antiobesity effects. Soy proteins have potential to address the problem of metabolic syndrome. They may have the ability to prevent some lifestyle diseases. Soybean oligopeptides were evaluated for the antioxidant activities and in vitro and in vivo antihypertensive effects. SOS was found to exhibit 1,1-diphenyl-2-picrylhydrazyl radical scavenging effect ($IC_{50} = 4.5 \pm 0.13$ mg/mL) and significantly inhibited lipid peroxidation in linoleic acid oxidation system ($IC_{50} = 1.2 \pm 0.09$ mg/mL). SOS had potent angiotensin I-converting enzyme inhibitory activity ($IC_{50} = 1.1 \pm 0.06$ mg/mL) and antihypertensive effect in spontaneously hypertensive rats at a dose of 200 mg/kg. SOS could be a natural antioxidative or antihypertensive compound in the medicine and food industries (Cai et al., 2012). The proteins and peptides from legumes have nutraceutical/pharmaceutical/therapeutic biological activities. The proteins from *Cicer arietinum* (chickpeas), *Pisum sativum* (green peas), and *Vigna unguiculata* (cowpeas) have also got special attention.

8.9.14.4 Egg and Meat

Avian eggs contain bioactive proteins and peptides. Lysozymes and antibodies are found in eggs. Eggs contain more than 60 various types of proteins. The presence of biologically active proteins, especially minor egg proteins, has raised interest in developing novel agents from eggs against chronic diseases such as cancer. The hydrolysates and peptides from fish muscles and collagen and their ACE inhibitory and antioxidant activities may lead to their applications as bioactive ingredients in functional foods, conventional foods, and nutraceuticals (Mine et al., 2010).

8.9.14.5 Gamma-Aminobutyric Acid

Gamma-aminobutyric acid (GABA) is a nonprotein amino acid found in potato. It is a significant component of the free amino acids pool in most prokaryotic and eukaryotic cells. It has many biological functions such as neurotransmission and induction of hypotensive, diuretic, and tranquilizer effects. Food-derived peptides have been reported as regulators of satiety (Matsui and Tanaka, 2010). An antiobesity agent or food that contains processed stems and leaves of *Angelica keiskei*, as well as a GABA or its derivative, was found to have superior weight-increase-suppressing and body-fat-reducing effects (Takagaki, 2006). Effectiveness of defatted rice germ enriched with GABA as a functional food with tranquilizing properties was investigated (Okada et al., 2000). The most common mental symptoms during the menopausal and presenile period such as sleeplessness, somnipathy, and depression were improved in more than 65% of the patients with such symptoms. Overall improvement was observed in 75% of the patients.

8.9.15 Fiber

Benefits of fiber include sugar reduction, low glycemic index, prebiotic effect, cholesterol excretion, and effects on mineral absorption (Cochet, 2006). Many studies suggest that daily consumption of fiber reduces the risk of chronic diseases, including cancer and cardiovascular disease. Recommended fiber intake is 25–35 g of fiber each day, although most Americans consume about half of that amount (Poonia and Dabur, 2009). There are two types of fiber—soluble and insoluble. It has been found that one dose of psyllium fiber per day helped lower total cholesterol levels by 7% and LDL cholesterol by 15%. Oat bran (β-glucan) and psyllium are the only two fibers allowed by FDA to make claims related to cholesterol reduction and heart disease. However, numerous other soluble fibers could also potentially be used in functional foods (Poonia and Dabur, 2009). A functional DF supplement for intestinal health, comprising xylitol-coated ground psyllium husk, has been described (Lee, 2005). The ground psyllium husk facilitates intestinal functions and increases fecal volume, while the xylitol coating makes consumption easier as compared to a conventional plain psyllium husk.

Nondigestible fibers, known as oligosaccharides, have been shown to benefit the host by selectively stimulating the activity of bacteria already in the colon. They also reduce the risk of some chronic diseases by lowering blood triglyceride levels, increasing HDL cholesterol, increasing stool weight and frequency, controlling blood glucose, and possibly preventing colon cancer.

Pectin displays a wide range of physiological and nutritional effects important to human nutrition and health. Pectin is a DF, because it is not digested by enzymes produced by the human body. Pectin intake reduces the risk of colorectal cancer. A significant suppression of the incidence of chemically induced colon cancer has been reported when rats were fed a 10% pectin diet at the promotion stage of carcinogenesis (Heitman et al., 1992). Dietary pectin was also found to affect the absorption and bioavailability of β-carotene (Rock and Swendseid, 1992).

DF intakes do not currently meet recommendations in many Western countries, and it is important to find new ways to increase intakes. β-Glucan provides a possible option for fiber enrichment, because of its ability to reduce blood cholesterol levels and balance blood glucose and insulin response after meals. This study looked at whether providing foods enriched with β-glucan would be a feasible strategy for improving consumer's DF intake.

The interactions between different DF preparations and other active components, including nutritional and nonnutritional, were studied under simulated biological conditions, and the potential links of these defined interactions with the health claims of increased DF intake were considered (Sun et al., 2007). Results indicated that interactions between DF and other food components occur during their movement through the human gastrointestinal tract. The inhibitory effects of soluble and insoluble fibers on lipolysis catalyzed by lipase were observed. Plant fibers and their polysaccharides significantly influence food properties such as antioxidant activity and bioavailability. Thus, not only the physiological and nutritional functions of DF, but also the beneficial synergistic effects resulting from their interactions with other food components, could be delivered to consumers through tailoring of functional food formulations. The role of fiber in health benefits has been discussed in detail under Section 8.4.

8.10 Summary and Future Prospects

Awareness of the world communities about the beneficial effects of functional foods on human health is on the upswing. Today's consumer has more and faster access to nutrition information through the media. With increased awareness of the consumer regarding the link between diet and health, health-conscious people are becoming more and more diet conscious. It is important to note that probiotics, prebiotics, and synbiotics are gaining popularity. The synbiotic combination has the ability to heal and regulate the intestinal flora, particularly after the destruction of microorganisms following antibiotic, chemotherapy, or radiation therapy. Without the beneficial microorganisms throughout the digestive system, proper digestion, absorption, and/or manufacture of nutrients cannot take place. The functional foods that are marketed with claims of heart disease reduction focus primarily on the major risk factors, i.e., cholesterol, diabetes, and hypertension. Some of the most innovative products are designed to be enriched with "protective" ingredients, believed to reduce risk. They may contain, e.g., soluble fiber (from oat and psyllium), useful for lowering both cholesterol and blood pressure, or fructans, effective in diabetes. Phytosterols and stanols lower LDL cholesterol in a dose-dependent manner. Soy protein is more hypocholesterolemic in subjects with very high initial cholesterol and recent data also indicate favorable activities in the metabolic syndrome. Fatty acid (n-3) appears to exert significant hypotriacylglycerolemic effects, possibly partly responsible for their preventive activity. Dark chocolate is gaining much attention for its multifunctional activities, useful for the prevention of both dyslipidemia and hypertension.

Further, the physiological studies of naturally occurring biomolecules need to be thoroughly investigated to find their bioavailability and their real health benefits. Nutrigenomics (effect of diet on gene expression) work to be studied for the development of personalized foods for improvement of health of specific target groups. Development of dietary patterns, functional foods, and supplements that are designed to improve genome health maintenance in humans with specific genetic backgrounds may provide an important contribution to a new optimum health strategy based on the diagnosis and individualized nutritional treatment of genome instability. There is need to explore new biomolecules for their functional properties. In India, there is vast flora and fauna that provides herbs and natural foods rich in medicinal properties. In fact, with the selection/consumption of safe and balanced foods with natural therapeutic or functional ingredients, the diseases can be alleviated from human kind.

References

Abalde, J., Fabreagas, J., and Herrero, C. 1991. β-Carotene, vitamin C and vitamin E content of the marine microalga *Dunaliella tertiolecta* cultured with different nitrogen sources. *Bioresource Technology*, 38: 121–125.

Abdel, S.A.M. 2010. Functional foods: Hopefulness to good health. *American Journal of Food Technology*, 5: 86–99.

AbuMweis, S.S., Vanstone, C.A., Lichtenstein, A.H., and Jones, P.J.H. 2009. Plant sterol consumption frequency affects plasma lipid levels and cholesterol kinetics in humans. *European Journal of Clinical Nutrition*, 63: 747–755.

Ah-Reum, P. and Deok-Kun, O. 2010. Galacto-oligosaccharide production using microbial beta-galactosidase: Current state and perspectives. *Applied Microbiology and Biotechnology*, 85: 1279–1286.

Aim, L. 1982. Effect of fermentation on lactose, glucose and galactose content in milk and suitability of fermented milk products for lactose-deficient individuals. *Journal of Dairy Science*, 65: 346–352.

Aldercreutz, H. 1998. Epidemiology of phytoestrogens. *Baillieres Clinical Endocrinology and Metabolism*, 12: 605–623.

Aldercreutz, H., Honjo, H., Higashi, A., Fotsis, T., Hamalainen, E., Hasegawa, T., and Okadia, H. 1991. Urinary excretion of lignans and isoflavonoid phytoestrogens in Japanese men and women consuming a traditional Japanese diet. *American Journal of Clinical Nutrition*, 54: 1093–1100.

Allouche, N., Damak, M., Ellouz, R., and Sayadi, S. 2004. Use of whole cells of *Pseudomonas aeruginosa* for synthesis of the antioxidant hydroxytyrosol via conversion of tyrosol. *Applied and Environmental Microbiology*, 70: 2105–2109.

Alves-Rodrigues, A. and Shao, A. 2004. The science behind lutein. *Toxicology Letters*, 150: 57–83.

Anderson, J.J.B. and Carner, S.C. 1997. The effects of phytoestrogens on bone. *Nutrition Research*, 17: 1617–1632.

Anderson, J.W. and Gilliland, S.E. 1999. Effect of fermented milk (yoghurt) containing *Lactobacillus acidophilus* L1 on serum cholesterol hypercholesterolemic humans. *Journal of the American College of Nutrition*, 18: 43–50.

Anthony, M.S., Clarkson, T.B., and Williams, J.K. 1998. Effects of soy isoflavones on athero-sclerosis: Potential mechanisms. *American Journal of Clinical Nutrition*, 68: 1390S–1393S.

Aso, Y. and Akazan, H. 1992. Prophylactic effect of a *Lactobacillus casei* preparation on the recurrence of superficial bladder cancer. *Urologia Internationalis*, 49: 125–129.

Aso, Y., Akazan, H., Kotake, T., Tsujamoto, T., and Imai, K. 1995. Preventive effect of a *Lactobacillus casei* preparation on the recurrent of superficial bladder cancer in a double-blind trial. *European Urology*, 27: 104–109.

Augusti, K.T. and Sheela, C.G. 1996. Antiperoxide effect of S allyl cysteine sulfoxide, an insulin secretagogue in diabetic rats. *Experientia*, 52: 115–119.

Babu, P.S. and Srinivasan, K. 1998. Amelioration of renal lesions associated with diabetes by dietary curcumin in experimental rats. *Molecular and Cellular Biochemistry*, 181: 87–96.

Babu, P.S. and Srinivasan, K. 1999. Renal lesions in streptozotocin induced diabetic rats maintained in onion and capsaicin containing diets. *Journal of Nutritional Biochemistry*, 10: 477–483.

Bai, A.P. and Quyang, Q. 2006. Probiotics and inflammatory bowel diseases. *Postgraduate Medical Journal*, 82: 376–382.

Bajpai, P. and Bajpai, P.K. 1993. Eicosapentaenoic acid (EPA) production from microorganisms: A review. *Journal of Biotechnology*, 30: 161–183.

Bajpai, P., Bajpai, P.K., and Ward, O.P. 1991a. Effects of ageing *Mortierella mycelium* on production of arachidonic and eicosapentaenoic acids. *Journal of the American Oil Chemists Society*, 68: 775–780.

Bajpai, P., Bajpai, P.K., and Ward, O.P. 1991b. Eicosapentaenoic acid (EPA) formation: Comparative studies with *Mortierella* strains and production by *Mortierella elongate. Mycological Research*, 95: 1294–1298.

Battinelli, L., Mengoni, F., Lichtner, M., Mazzanti, G., Saija, A., Mastroianna, C.M., and Vullo, V. 2003. Effect of Limonin and nomilin on HIV-1 replication on infected human mononuclear cells. *Planta Medica*, 69: 910–913.

Belviso, S., Giordano, M., Dolci, P., and Zeppa, G. 2009. In vitro cholesterol-lowering activity of *Lactobacillus plantarum* and *Lactobacillus paracasei* strains isolated from the Italian Castelmagno PDO cheese. *Dairy Science and Technology*, 89: 169–176.

Benning, C. 1998. Biosynthesis and function of the sulfolipid sulfoquinovosyl diacylglycerol. *Annual Review of Plant Physiology and Plant Molecular Biology*, 49: 53–75.

Bernet, M.F., Brassart, D., Neeser, J.R., and Servin, A.L. 1994. *Lactobacillus acidophilus* LA-1 binds to human intestinal cell lines and inhibits cell attachments and cell invasion by enterovirulent bacteria. *Gut*, 35: 483–489.

Block, G. 1991. Vitamin C and cancer prevention: The epidemiologic evidence. *American Journal of Clinical Nutrition*, 53: 270s–282s.

Bokesch, H.R., O'Keefe, B.R., McKee, T.C. et al. 2003. A potent novel anti-HIV protein from the cultured cyanobacterium *Scytonema varium*. *Biochemistry*, 42: 2578–2584.

Braun, L. and Cohen, M. 2007. *Herbs and Natural Supplements—An Evidence based Guide* (2nd edn.), Elsevier, Sydney, Australia.

Brown, A.C. and Valiere, A. 2004. Probiotics and medical nutrition therapy. *Nutrition in Clinical Care*, 7: 56–68.

Burke, J.D., Curran-Celentano, J., and Wenzel, A.J. 2005. Diet and serum carotenoid concentrations affect macular pigment optical density in adults 45 years and older. *Journal of Nutrition*, 135: 1208–1214.

Business Standard. 2008. Probiotics market needs more awareness. http://www.businessstandard.com/common/news_article.php?leftnm=5&autono=311953 (accessed July 2, 2012).

Buts, J.P., Keyer, N.D., Marandi, S. et al. 1999. *Saccharomyces boulardii* upgrades cellular adaptation after proximal enterectomy in rats. *Gut*, 45: 89–96.

Cai, M., Gu, R., Li, C. et al. 2012. Pilot-scale production of soybean oligopeptides and antioxidant and antihypertensive effects in vitro and *in vivo*. *Journal of Food Science and Technology*, Published online April 25, 2012 (DOI: 10.1007/s13197–012–0701–4).

Certik, M. and Shimizu, S. 1999. Biosynthesis and regulation of microbial polyunsaturated fatty acid production. *Journal of Bioscience and Bioengineering*, 87: 1–14.

Chadwick, R., Henson, S., Moseley, B. et al. 2003. *Functional Foods*. Springer, New York.

Chang, S.K.C. 2002. Isoflavones from soybeans and soy foods. In: *Functional Foods: Biochemical and Processing Aspects*, eds. J. Shi, G. Mazza, and M.L. Maguer, pp. 39–69. CRC Press, Boca Raton, FL.

Chaturvedi, S., Sharma, P.K., Garg, V.K., and Bansal, M. 2011. Role of nutraceuticals in health promotion. *International Journal of PharmTech Research*, 3: 442–448.

Chen, Z.Y., Jiao, R., and Ma, K.Y. 2008. Cholesterol-lowering nutraceuticals and functional foods. *Journal of Agricultural and Food Chemistry*, 56: 8761–8773.

Chen, Z.Y., Peng, C., Jiao, R., Wong, Y.M., Yang, N., and Huang, Y. 2009. Antihypertensive nutraceuticals and functional foods. *Journal of Agricultural and Food Chemistry*, 57: 4485–4499.

Cheng, M.H., Walker, T.H., Hulbert, G.J., and Raman, D.R. 1999. Fungal production of eicosapentaenoic and arachidonic acids from industrial waste streams and crude soybean lipid. *Bioresource Technology*, 67: 101–110.

Cline, J.M., and Hughes, C.L. Jr. 1998. Phytochemicals for the prevention of breast and endometrial cancer. *Cancer Treatment Research*, 94: 107–134.

Cochet, D. 2006. Fighting the flab. *Ingredients, Health and Nutrition*, 9: 19–20.

Cocks, S., Wrigley, S.K., Chicarelli-Robinson, M.I., and Smith, R.M. 1995. High-performance liquid chromatography comparison of supercritical-fluid extraction and solvent extraction of microbial fermentation products. *Journal of Chromatography*, 697: 115–122.

Collins, M.D. and Gibson, G.R. 1999. Probiotics, prebiotics and synbiotics: Approaches for modulating the microbial ecology of gut. *American Journal of Clinical Nutrition*, 69: 1052S–1057S.

Connor, W.E. 2000. Importance of n-3 fatty acids in health and disease. *American Journal of Clinical Nutrition*, 71: 171S–175S.

Coudray, C., Betanger, J., Castiglia-Delavaud, C., Remesy, C., Vermorel, M., and Rayssignuir, Y. 1997. Effect of soluble or partly on absorption and balance of calcium, magnesium, iron and zinc in healthy young men. *European Journal of Clinical Nutrition*, 51: 375–380.

Crowell, P.L., Elson, C.E., Bailey, H.H., Elegbede, A., Haag, D.J., and Gould, M.N. 1994. Human metabolism of experimental cancer therapeutic agent *d*-limonene. *Cancer Chemotherapy and Pharmacology*, 35: 31–37.

Csabai, P. and Joo, F. 2003. Reactivity of the individual lipid classes in homogenous catalytic hydrogenation of model and biomembranes detected by MALDI-TOF mass spectrometry. *Catalysis Communications*, 4: 275–280.

Cummings, J.H. and Macfarlane, G.T. 2002. Gastrointestinal effects of prebiotics. *British Journal of Nutrition*, 87: S145–S151.

D'Egidio, M.G., Novaro, P., Nardi, S., Cecchini, C., and Colucci, F. 2001. Durum wheat bread enriched with immature wheat grain. I. Functional properties and quality evaluation by image analysis. *Tecnica Molitoria*, 52: 1201–1207.

D'Egidio, M.G., Pagani, M.A., and Casiraghi, M.C. 2008. Fructans in durum wheat: An opportunity for functional foods. *Tecnica Molitoria International*, 59: 156–162.

D'Egidio, M.G., Zardi, M., Pagani, M.A., Casiraghi, M.C., and Cecchini, C. 2005. Pasta products enriched with immature wheat grain: Technological and nutritional properties. *Tecnica Molitoria International*, 56: 89–97.

Dev, R., Kumar, S., Singh, J., and Chauhan, B. 2011. Potential role of nutraceuticals in present scenario: A review. *Journal of Applied and Pharmaceutical Sciences*, 1: 26–28.

Drapcho, C.M. 2007. Microbial modeling as basis for bioreactor design for nutraceutical production. In: *Functional Food Ingredients and Nutraceuticals*, ed. J. Shi, pp. 237–265. CRC Press, Boca Raton, FL.

Dunford, N.T. 2001. Health benefits and processing of lipid based nutritionals. *Food Technology*, 55: 38–43.

Elson, C.E. and Yu, S.G. 1994. The chemoprevention of cancer by mevalonate-derived constituents of fruit and vegetables. *Journal of Nutrition*, 124: 607–614.

Filho-Lima, J.V.M., Vieira, E.C., and Nicoli, J.R. 2000. Antagonistic effect of *Lactobacillus acidophilus, Saccharomyces boulardii* and *Escherichia coli* combination against experimental infections with *Shigella flexneri* and *Salmonella enteritidis* spp. *Typhimurium* in gnotobiotic mice. *Journal of Applied Microbiology*, 88: 365–370.

Flambard, B. 2002. Role of bacterial cell wall proteinase in antihypertension. *Sciences Des Aliments*, 22: 209–222.

Frutarom. 2009. It's a matter of the heart. *Nutraceuticals-Now*, (Winter): 10–11.

Fuller, R. 1992. *Probiotics: The Scientific Basis*, p. 398. Chapman and Hall, London, U.K.

Gessler, N.N., Sokolow, A.V., Bykhovshy, V.Y., and Belozerskaya, T.A. 2002. Superoxide dismutase and catalase activities in carotenoid—Synthesizing fungi *Blakeslea trispora* and *Neurospora crassa* fungi in oxidative stress. *Applied Biochemistry and Microbiology*, 38: 205–209.

Ghandi, S.R. and Weete, J.D. 1991. Production of polyunsaturated fatty acids arachidonic acid and eicosapentaenoic acid by the fungus *Pythium ultimum*. *Journal of General Microbiology*, 137: 1825–1830.

Gibson, G.R., Beatty, E.R., Wang, X., and Cummings, J.H. 1995. Selective stimulation of bifidobacteria in the human colon by oligofructose and inulin. *Gastroenterology*, 108: 975–982.

Gibson, G.R. and Roberfroid, M.B. 1995. Dietary modulation of the human colonic microbiota: Introducing the concept of prebiotics. *Journal of Nutrition*, 125: 1401–1412.

Gill, H.S. and Guarner, F. 2004. Probiotics and human health: A clinical perspective. *Postgraduate Medical Journal*, 80: 516–526.

Gilliland, S.E. and Kim, H.S. 1984. Effect of viable starter culture bacteria in yoghurt on lactose utilization in humans. *Journal of Dairy Science*, 67: 1–6.

Giordano, W., Avalos, J., Fernandez-Martin, R., Cerda-Olmedo, E., and Domenech, C.E. 1999. Lovastatin inhibits the production of gibberellins but not sterols or carotenoid biosynthesis in *Gibberella fujikuroi*. *Microbiology*, 145: 2997–3002.

Gismondo, M.R., Drago, L., and Lombardi, A. 1999. Review of probiotics available to modify gastrointestinal flora. *International Journal of Antimicrobial Agents*, 12: 287–292.

Godber, J.S., Shin, T.S., Saska, M., and Wells, J.H. 1994. Rice bran: As a viable source of high value chemicals. *Louisiana Agriculture*, 37: 13–17.

Goldin, B.R. 1998. Health benefits of probiotics. *British Journal of Nutrition*, 80: 203S–207S.

Griffiths, K., Denis, L., Turkes, A., and Morton, M.S. 1998. Phytoestrogens and diseases of the prostate gland. *Baillieres Clinical Endocrinology and Metabolism*, 12: 625–647.

Gustafson, K.R., Cardelina, J.H., Fuller, R.W. et al. 1989. AIDS-antiviral sulfolipids from cyanobacteria (blue-green algae). *Journal of the National Cancer Institute*, 81: 1254–1258.

Guthrie, N., Hasegawa, S., Manners, G., and Carroll, K.K. 2000. *Citrus Limonoids Functional Chemicals in Agriculture and Foods*, pp. 164–174. American Chemical Society, Washington, DC.

Hak, T.-L., Youn, S.-K., Sung, H.-K., Yun, H.-R., and Min, H.-L. 2008. Healthy and functional food for obesity patient using purple-colored potato. U.S. Patent 20080171103.

Harris, W.S., Assaad, B., and Poston, W.C. 2006. Tissue omega-6/omega-3 fatty acid ratio and risk for coronary artery disease: A review. *American Journal of Cardiology*, 98: 19i–26i.

Hata, Y., Yamamoto, M., Ohni, M., Nakajima, K., Nakampura, Y., and Takano, T. 1996. A placebo-controlled study the effect of sour milk on blood pressure in hypertensive subjects. *American Journal of Clinical Nutrition*, 64: 767–771.

Heitman, D.W., Hardman, W.E., and Cameron, I.L. 1992. Dietary supplementation with pectin and guar gum on 1,2-dimethylhydrazine-induced colon carcinogenesis in rats. *Carcinogenesis*, 13: 815–818.
Heyden, L. 2009. A story from the heart: Past and present of lactotripeptide IPP. *Alimentaria*, 407: 68–71.
Hidetoshi, M., Taketo, S., Kurumi, H. et al. 2006. Immunomodulator and food modulating immunity. Japanese Patent PN: JP2006257077.
Higashiyama, K., Yaguchi, T., Akimoto, K., Fujikawaa, S., and Shimizu, S. 1998. Enhancement of arachidonic acid production by *Mortierella alpina* 1S-4. *Journal of the American Oil Chemists Society*, 75: 1501–1505.
Hiyashi, T. and Hayashi, K. 1996. Calcium spirulan, an inhibitor of enveloped virus replication from Blue-Green Alga *Spirulina platensis*. *Journal of Natural Products*, 59: 83–87.
Hollman, P.C. and Katan, M.B. 1998. Bioavailability and health effects of dietary flavonols in man. *Archives of Toxicology Supplement*, 20: 237–248.
Holman, R.T. 1998. The slow discovery of the importance of omega-3 essential fatty acids in human health. *Journal of Nutrition*, 128: 427S–433S.
Holub, B.J., Arnott, K., Davis, J.P., Nagpurkar, A., and Peschell, J. 2002. Organosulfur compounds from garlic. In *Functional Foods: Biochemical and Processing Aspects*, eds. J. Shi, G. Mazza, and M.L. Maguer, pp. 213–238. CRC Press, Boca Raton, FL.
Holzapfel, W.H., Haberer, P., Snel, J., Schillinger, U., and Huis in't Veld, J.H.J. 1998. Overview of gut flora and probiotics. *International Journal of Food Microbiology*, 41: 85–101.
Hong, H., Datla, N., Reed, D.W., Covello, P.S., MacKenzie, S.L., and Qiu, X. 2002. High level production of gamma-linolenic acid in *Brassica juncea* using a Delta 6 desaturase from *Pythium irregular*. *Plant Physiology*, 129: 354–362.
Hu, K.J., Hu, J.L., Ho, K.P., and Yeung, K.W. 2004. Screening of fungi for chitosan producers and copper adsorption capacity of fungal chitosan and chitosanaceous materials. *Carbohydrate Polymers*, 58: 45–52.
Hughes, D.A. 2001. Dietary carotenoids and human immune function. *Nutrition*, 17: 823–827.
Huis in't Veld, J.H., Jos-Bosschaert, M.A.R., and Shortt, C. 1998. Health aspects of probiotics. *Food Science and Technology Today*, 12: 46–50.
Huis in't Veld, J. and Shortt, C. 1996. Selection criteria for probiotic microorganisms. In: *Gut Flora and Health—Past, Present and Future*, eds. R.A. Leeds and I.R. Rowland, *International Congress and Symposium Series no* 219: pp. 27–36. The Royal Society of Medicine Press Ltd., London, U.K.
Igimi, H., Hisatsugu, T., and Nishimura, M. 1976. The use of d-limonene preparation as a dissolving agent of gallstones. *The American Journal of Digestive Diseases*, 21: 926–939.
Iqbal, G. and Valivety, R. 1997. Polyunsaturated fatty acids, part 1: Occurrence, biological activities of applications. *Trends in Biotechnology*, 15: 401–409.
Jauhiainen, T., Korpela, R., Vapaatalo, H., Turpeinen, A.M., and Kautiainen, H. 2009. Bioactive milk peptides and blood pressure. *Agro Food Industry Hi-tech*, 20: 26–28.
Joint FAO/WHO working group report on drafting guidelines for the evaluation of probiotics in food. Ontario, Canada, April 30 and May 1, 2002.
Jouni, Z., Rai, D., and Rangavajla, N. 2010. Nutritional compositions containing punicalagins. U.S. Patent 20100004334.

Kaila, M., Isolauri, E., Soppi, E., Virtanen, E., Laine, S., and Arvilommi, H. 1992. Enhancement of the circulating antibody secreting cell response in human diarrhoea by a human *Lactobacillus* strain. *Pediatric Research*, 32: 141–144.

Kamisoyama, H., Honda, K., Tominaga, Y., Yokota, S., and Hasegawa, S. 2008. Investigation of the anti-obesity action of licorice flavonoid oil in diet-induced obese rats. *Bioscience Biotechnology and Biochemistry*, 72: 3225–3231.

Kandil, F.E., Smith, M.A., Rogers, R.B. et al. 2002. Composition of a chemopreventive proanthocyanidin-rich fraction from cranberry fruits responsible for the inhibition of 12–0-tetradecanoyl phorbol-13-acetate (TPA)-induced ornithine decarboxylase (ODC) activity. *Journal of Agricultural and Food Chemistry*, 50: 1063–1069.

Kanu, P.J., Kerui, Z., Kanu, J.B., Huiming, Z., Haifeng, Q., and Kexue, Z. 2007. Biologically active components and nutraceuticals in sesame and related products: A review and prospect. *Trends in Food Science and Technology*, 18: 599–608.

Karppi, J., Kurl, S., Nurmi, T., Rissanen, T.H., Pukkala, E., and Nyyssonen, K. 2009. Serum lycopene and the risk of cancer: The Kuopio Ischaemic Heart Disease risk factor (KIHD) study. *Annals of Epidemiology*, 19: 512–518.

Kendrick, A. and Ratledge, C. 1992. Lipid formation in the oleaginous mold *Entomophthora-exitalis* grown in continuous culture: Effects of growth, rate, temperature and dissolved oxygen tension or polyunsaturated fatty acids. *Applied Microbiology and Biotechnology*, 37: 18–22.

Kim, J., Mandal, P.K., Choi, K., Pyun, C., Hong, G., and Lee, C. 2011. Beneficial dietary effect of turmeric and sulphur on weight gain, fat deposition and lipid profile of serum and liver in rats. *Journal of Food Science and Technology*, Published online October 22, 2011 (DOI: 10.1007/s13197–011–0569–8).

Kim, M., Ryu, K., Choi, W., Rhee, Y., and Lee, I. 2003. Enhanced production of 1→3 β-D-glucan by a mutant strain of *Agrobacterium species. Biochemical Engineering Journal*, 16: 163–168.

Kiode, S.S. 1998. Chitin-chitosan: Properties, benefits and risks. *Nutrition Research*, 18: 1091–1101.

Knight, D.C., Wall, P.L., and Eden, J.A. 1996. A review of phytoestrogens and their effects in relation to menopausal symptoms. *Australian Journal of Nutrition and Diet*, 53: 5–11.

Koike, Y., Cai, H.J., Higashiyama, K., Fujikawa, S., and Park, E.Y. 2001. Effect of consumed carbon to nitrogen ratio on mycelial morphology and arachidonic acid production in cultures of *Mortierella alpine. Journal of Bioscience and Bioengineering*, 91: 382–389.

Krieger-Mettbach, B. 2009. Functional foods. *Fleischerei*, 60: 30.

Krinsky, N.I. and Johnson, E.J. 2005. Carotenoid actions and their relation to health and disease. *Molecular Aspects of Medicine*, 26: 459–516.

Kruis, W., Schutz, E., Fric, P., Fixa, B., Judmaier, G., and Stolte, M. 1997. Double blind comparison of an oral *Escherichia coli* preparation and mesalazine in maintaining remission of ulcerative colitis. *Alimentary Pharmacology and Therapeutics*, 11: 853–858.

Kumar, R., Rani, B., Maity, T.K., and Misra, A.K. 2008. Synbiotics: Nature's ultimate power food for human health. *Indian Food Industry*, 27: 57–64.

Kumari, K., Mathew, B.C., and Augusti, K.T. 1995. Anti-diabetic and hypolipidemic effects of S-methyl cysteine sulfoxide isolated from Allium cepa. *Indian Journal of Biochemistry and Biophysics*, 32: 49–54.

Kuroska, E.M., Hasegawa, S., and Manners, G. 2000. *Citrus Limonoids Functional Chemicals in Agriculture and Foods*, pp. 175–184. American Chemical Society, Washington, DC.

Lau, F.C., Joseph, J.A., McDonald, J.E., and Kalt, W. 2009. Attenuation of iNOS and COX2 by blueberry polyphenols is mediated through the suppression of NF-kappaB activation. *Journal of Functional Foods*, 1: 274–283.

Lee, J.K. 2005. Xylitol coated functional dietary fiber supplement. U.S. Patent 20050249858.

Lee, Y.K., Nomoto, K., Salminen, S., and Gorbach, S.L. 1999. *Handbook of Probiotics*, John Wiley & Sons Inc., New York.

Levy, J., Bisin, E., Feldman, B. et al. 1995. Lycopene is a more potent inhibitor of human cancer cell proliferation then either α-carotene or β-carotene. *Nutrition and Cancer-An International Journal*, 24: 257–266.

Lim, G., Lee, S., Lee, E., Haam, S., and Kim, W. 2002. Separation of astaxanthin from red yeast *Phaffia rhodozyma* by supercritical carbon dioxide extraction. *Biochemical Engineering Journal*, 11: 181–187.

Liu, B., Obeid, L.M., and Hannun, Y.A. 1997. Sphingomyelinases in cell regulation. *Seminars in Cell and Developmental Biology*, 8: 311–322.

Lloyd, B.J., Sibernmorgen, T.J., and Beers, K.W. 2000. Effect of commercial processing on antioxidants in rice bran. *Cereal Chemistry*, 77: 551–555.

Mace, K. and Halliwell, B. 2004. Influence of diet on ageing and longevity. In *Bioprocesses and Biotechnology for Functional Foods and Nutraceuticals*, eds. J.B. German and J.R. Neeser, Chapter 16. CRC Press, Boca Raton, FL.

Macfarlane, G.T. and Cummings, J.H. 2002. Probiotics, infection and immunity. *Current Opinion in Infectious Diseases*, 15: 501–506.

Macfarlane, S., Macfarlane, G.T., and Cummings, J.H. 2006. Prebiotics in the gastrointestinal tract: A review. *Alimentary Pharmacology and Therapeutics*, 24: 701–714.

Madihally, S.V. and Matthew, H.W.T. 1999. Porous chitosan scaffolds for tissue engineering. *Biomaterials*, 20: 1133–1142.

Magalhães, M.S., Salminen, S., Anna, P., Marchelli, R., Ferreira, C.L., and Tommola, J. 2011. *Functional Foods Forum*, p. 122. University of Turku, Finland.

Makinen, K.K. 1978. *Biochemical Principles of the Use of Xylitol in Medicines and Nutrition with Special Consideration of Dental Aspects*, Birkhauser Verlag, Basel, Switzerland.

Marteau, P.R., De-Vrese, M., Cellier, C.J., and Schrezenmeir, J. 2001. Protection from gastrointestinal diseases with the use of probiotics. *American Journal of Clinical Nutrition*, 73: 430S–436S.

Marteau, P., Pochart, P., Bouhmik, Y., and Desjeux, J.F. 1993. Method for microorganisms in human: The case of *Bifidobacteria*. *Dairy Science Abstract*, 55: 1205.

Matsui, T. and Tanaka, M. 2010. Antihypertensive peptides and their underlying mechanisms. In: *Bioactive Proteins and Peptides as Functional Foods and Nutraceuticals*, eds. Y. Mine, E. Li-Chan, and B. Jiang, pp. 43–54. Blackwell Publishing Ltd., Ames, IA.

Mazza, G. and Miniati, E. 1993. *Anthocyanins in Fruits, Vegetables, and Grains*, CRC Press, Boca Raton, FL.

McFarland, L.V., Surawicz, C.M., Greenberg, R.N. et al. 1994. A randomized placebo-controlled trial of *Saccharomyces boulardii* in combination with standard antibiotics for *Clostridium difficile* disease. *The Journal of the American Medical Association*, 271: 1913–1918.

McGuire, M. and Beerman, K.A. 2007. *Nutritional Sciences: From Fundamentals to Foods*, Thomson Wadsworth, Belmont, CA.

Melchini, A., Costa, C., Traka, M. et al. 2009. Erucin, a new promising cancer chemopreventive agent from rocket salads, shows anti-proliferative activity on human lung carcinoma A549 cells. *Food and Chemical Toxicology*, 47: 1430–1436.

Messina, M., Barnes, S., and Setchell, K.D. 1997. Phytoestrogens and breast cancer. *Lancet*, 350: 971–972.

Messina, M. and Bennink, M. 1998. Soy foods, isoflavones and risk of colonic cancer: A review of the in vitro and in vivo data. *Baillieres Clinical Endocrinology and Metabolism*, 12: 707–728.

Michetti, P., Dorta, P.H., Wiesel, D. et al. 1999. Effect of whey-based culture supernatant of *Lactobacillus acidophilus* (*johnsonii*) La 1 on *Helicobacter pylori* infection in humans. *Digestion*, 60: 203–209.

Micozzi, M.S., Beecher, G.R., Taylor, P.R., and Khachik, F. 1990. Carotenoid analyses of selected raw and cooked foods associated with a lower risk for cancer. *Journal of the National Cancer Institute*, 82: 282–288.

Middleton, E. and Kandaswami, C. 1994. The impact of plant flavonoids on mammalian biology: Implications for immunity, inflammation and cancer. In: *The Flavanoids: Advances in Research since 1986*, ed. J.B. Harborne, pp. 619–652. Chapman & Hall, London, U.K.

Mine, Y., Li-Chan, E., and Jiang, B. 2010. *Bioactive Proteins and Peptides as Functional Foods and Nutraceuticals*, Blackwell Publishing Ltd., Ames, IA.

Mishima, T. and Murata, J. 1998. Inhibition of tumor invasion and metastasis by calcium spirulan (Ca-SP) a novel sulfated polysaccharides derived from a blue green algae, *Spirulina platensis*. *Clinical and Experimental Metastasis*, 16: 541–550.

Mitsou, E.K., Turunen, K., Anapliotis, P., Zisi, D., Spiliotis, V., and Kyriacou, A. 2009. Impact of a jelly containing short-chain fructo-oligosaccharides and *Sideritis euboea* extract on human faecal microbiota. *International Journal of Food Microbiology*, 135: 112–117.

Morganti, P. 2007. Functional foods for a better wellness. *NutraCos*, 6: 28–30.

Muller, A., Ensley, H., Pretus, H. et al. 1997. The application of various protic acids in the extraction of (1→3)-beta-D-glucan from *Saccharomyces cerevisiae*. *Carbohydrate Research*, 299: 203–208.

Nalik, S.R., Harindran, J., and Varde, A.B. 2001. Pimprinine an extracellular alkaloid produced by *Streptomyces* CDRIL-312: Fermentation, isolation and pharmacological activity. *Journal of Biotechnology*, 88: 1–10.

New, N. and Stevens, W.F. 2002. Production of fungal chitosan by solid substrate fermentation followed by enzymatic extractions. *Biotechnology Letters*, 24: 131–134.

Niness, K.R. 1999. Inulin and Oligofructose: What are they? *Journal of Nutrition*, 129: 1402S–1406S.

Nitschke, M. and Umbelino, D.C. 2002. Fructooligosaccharides: New functional ingredients. *Boletim da Sociedade Brasileira de Ciência e Tecnologia*, 36: 27–34.

Nobaek, S., Johansson, M.L., Molin, G., Ahme, S., and Jeppsson, B. 2000. Alteration of intestinal microflora is associated with reduction in abdominal bloating and pain in patients with irritable bowel syndrome. *American Journal of Gastroenterology*, 95: 1231–1238.

Obiro, W.C., Tao-Zhang., and Bo-Jiang. 2008. The nutraceutical role of the *Phaseolus vulgaris* alpha-amylase inhibitor. *British Journal of Nutrition*, 100: 1–12.

O'Brien, D.J., Kurantz, M.J., and Kwoczak, R. 1993. Production of eicosapentaenoic acid by the filamentous fungus *Pythium irregular. Applied Microbiology and Biotechnology*, 40: 211–214.

Ogretment, B. and Hannun, Y.A. 2004. Bioactive sphingolipids in cancer pathogenesis and treatment. *Nature Reviews Cancer*, 4: 604–616.

Okada, T., Sugishita, T., Murakami, T. et al. 2000. Effect of the defatted rice germ enriched with GABA for sleeplessness, depression, autonomic disorder by oral administration. *Journal of the Japanese Society for Food Science and Technology*, 47: 596–603.

Okabe, H., Suzuki, E., Sayito, T.,Takayama, K., and Nagai, T. 1994. Development of novel transdermal system containing *d*-limonene and ethanol as absorption enhancers. *Journal of Controlled Release*, 32: 243–247.

Olson, J. 1986. Carotenoid, vitamin A and cancer. *Journal of Nutrition*, 116: 1127–1130.

Ouwehend, A.C., Tolkko, S., and Salminen, S. 2001. The effect of digestive enzymes on the adhesion of probiotic bacteria *in-vitro. Journal of Food Science*, 66: 856–859.

Paolisso, G., D'Amore, A., Giugliano, D., Ceriello, A., Varricchio, M., and D'Onofrio, E. 1993. Pharmacologic doses of vitamin E improve insulin action in healthy subjects and non-insulin-dependent diabetic patients. *American Journal of Clinical Nutrition*, 57: 650–656.

Parvez, S., Malik, K.A., Kang, S.A., and Kim, H.Y. 2006. Probiotics and their fermented food products are beneficial for health. *Journal of Applied Microbiology*, 100: 1171–1185.

Patil, B.S., Brodbelt, J.S., Miller, E.G., and Turner, N.D. 2006. Potential health benefits of citrus: An overview. In: *Potential Health Benefits of Citrus*, pp. 1–16. American Chemical Society, Washington, DC.

Patterson, R.E., White, E., Kristal, A.R., Neuhouser, M.L., and Potter, J.D. 1997. Vitamin supplements and cancer risk: The epidemiologic evidence. *Cancer Causes and Control*, 8: 786–802.

Pearce, J. 1996. Effect of milk and fermented dairy products on the blood cholesterol content and profile of mammals in relation to coronary heart disease. *International Dairy Journal*, 6: 661–672.

Peterson, G. and Barnes, S. 1991. Genistein inhibition of the growth of human breast cancer cells: Independence from estrogen receptors and multi-drug resistance gene. *Biochemical and Biophysical Research Communications*, 179: 661–667.

Poonia, A. and Dabur, R.S. 2009. Functional foods—New opportunities and challenges in production. *Indian Food Industry*, 28: 53–61.

Qiang, X., YongLie, C., and Qian-Bing, W. 2009. Health benefit application of functional oligosaccharides. *Carbohydrate Polymers*, 77: 435–441.

Quasney, M.E., Carter, L.C., Oxford, C., Watkins, S.M., Gershwin, M.E., and German, J.B. 2001. Inhibition of proliferation and induction of apoptosis in SNU-1 human gastric cells by the plant sulfolipid sulfoquinovosyldiacylglycerol. *Journal of Nutritional Biochemistry*, 12: 310–315.

Rabbani, S., Beyer, P., Johannes, V.L., Hugueney, P., and Kleinig, H. 1998. Induced β-carotene synthesis driven by triacylglycerol deposition in the unicellular alga *Dunaliella bardawil. Plant Physiology*, 116: 1239–1248.

Rao, A.V. and Agarwal, S. 1999. Role of lycopene as antioxidant carotenoid in the prevention of chronic diseases: A review. *Nutrition Research*, 19: 305–323.

Ray, B., Miller, K.W., and Jain, K.M. 2001. Bacteriocins of lactic acid Bacteria: Current prospective. *International Journal of Microbiology*, 41: 01–21.

Rediff. 2007. Probiotic drugs mart to grow more. http://www.rediff.com/money/2007/jun/05probiotic.htm (accessed July 2, 2012).
Rideout, T.C., Chan, Y.M., Harding, S.V., and Jones, P.J.H. 2009. Low and moderate-fat plant sterol fortified soymilk in modulation of plasma lipids and cholesterol kinetics in subjects with normal to high cholesterol concentrations: Report on two randomized crossover studies. *Lipids in Health and Disease*, 8: 45.
Roberfroid, M.B. 2000. Prebiotics and probiotics: Are the functional foods? *American Journal of Clinical Nutrition*, 71: 1682S–1687S.
Roberts, R.L., Green, J., and Lewis, B. 2009. Lutein and zeaxanthin in eye and skin health. *Clinics in Dermatology*, 27: 195–201.
Rock, C.L. and Swendseid, M.E. 1992. Plasma beta-carotene response in humans after meals supplemented with dietary pectin. *American Journal of Clinical Nutrition*, 55: 96–99.
Rolfe, R.D. 2000. The role of probiotic cultures in the control of gastrointestinal health. *Journal of Nutrition* 130: 396S–402S.
Ruddat, M. and Garber, E.D. 1983. Biochemistry, physiology, and genetics of carotenogenesis in fungi. In: *Secondary Metabolism and Differentiation in Fungi*, eds. J.W. Bennett and A. Ciegler, pp. 95–152. CRC Press, Boca Raton, FL.
Saavedra, J.M. and Tschernia, A. 2002. Human studies with probiotics and prebiotics: Clinical implications. *British Journal of Nutrition*, 87: S241–S246.
Sabio, E., Lozano, M., Montero de Espinosa, V. et al. 2003. Lycopene and B-carotene extraction from tomato processing waste using supercritical CO_2. *Industrial and Engineering Chemistry Research*, 42: 6641–6646.
Sako, T., Matsumoto, K., and Tanaka, R. 1999. Recent progress on research applications of non-digestible. *International Dairy Journal*, 9: 69–80.
Salminen, S. 1996. Functional dairy foods with *Lactobacillus* strain GG. *Nutrition Reviews*, 54: S99–S101.
Salminen, S., Isolauri, E., and Salminen, E. 1996. Clinical use of probiotics for stabilizing the gut mucosal barrier: Successful strain and future challenges. *AntonieVanLeeuwenhoek*, 70: 347–358.
Salminen, S. and Ouwehend, A.C. 2003. Probiotics: Application in dairy products. *Encyclopedia of Dairy Sciences*, 4: 2315–2322.
Sanders, M.E. and Huis in't Veld, J. 1999. Bringing a probiotic containing functional food to the market: Microbiological product, regulatory and labeling issues. *AntonieVanLeeuwenhoek*, 76: 293–315.
Saulnier, D.M.A. 2007. Synbiotics: Making the most of probiotics and prebiotics by their combinations? *Food Science and Technology Bulletin: Functional Foods*, 4: 9–19.
Scalbert, A. and Williamson, G. 2000. Dietary intake and bioavailability of polyphenols. *Journal of Nutrition* 130: 2073S–2085S.
Scazzina, F., Rio, D.D., Serventi, L., Carini, E., and Vittadini, E. 2008. Development of nutritionally enhanced tortillas. *Food Biophysics*, 3: 235–240.
Schenk, J., Dobronte, Z., Koch, H., and Stolte, M. 1980. Studies on tissue compatibility of d-limonene as a dissolving agent of cholesterol gallstones. *Zeitschrift Fur Gastroenterologie*, 18: 389–394.
Scholz-Ahrens, K.E., Schaafsma, G., Van Den Heuvel, E.G.H.M., and Schrezenmeir, J. 2001. Effect of prebiotics on mineral metabolism. *American Journal of Clinical Nutrition*, 73: 459S–464S.
Semba, R.D., Lauretani, F., and Ferrucci, L. 2007. Carotenoids as protection against sarcopenia in older adults. *Archives of Biochemistry and Biophysics*, 458: 141–145.

Shah, R.K. 2009. Functional fermented dairy products with synbiotics—A new horizon for Indian dairy industry & consumers. *Indian Dairyman*, 61: 33–41.

Shahidi, F. and Barrow, C. 2008. *Marine Nutraceuticals and Functional Foods*, CRC Press, Boca Raton, FL.

Shimizu, S., Kawashima, H., Shinmen, Y., Akimoto, K., and Yamada, H. 1988. Production of eicosapentaenoic acid by *Mortierella fungi*. *Journal of the American Oil Chemists Society*, 65: 1455–1459.

Simopoulos, A.P. 1989. Summary of the NATO advanced research workshop on dietary w-3 and w-6 fatty acids: Biological effects and nutritional essentiality. *Journal of Nutrition*, 119: 521–528.

Siro, I., Kapolna, E., Kapolna, B., and Lugasi, A. 2008. Functional food. Product development, marketing and consumer acceptance: A review. *Appetite*, 51: 456–467.

Slavin, J.L. 2000. Mechanisms for the impact of whole grain foods on cancer risk. *Journal of the American College of Nutrition*, 19: 300S–307S.

Sloan, A.E. 2008. Top 10 functional food trends. *Food Technology*, 62: 25–43.

Soldatov, A. and Soldatov, I. 2011. Synbiotic food products comprising pulses and methods for manufacturing the same. Patent PN: CA 2672477 A1.

Sooyeon, L., Joonhee, L., and Jong, K.-K. 2009. Analysis of isothiocyanates in newly generated vegetables, baemuchae (x*Brassicoraphanus*) as affected by growth. *International Journal of Food Science and Technology*, 44: 1401–1407.

Soundarajan, J.J., Kagliwal, L., and Singhal, R.S. 2010. Safety considerations of drug-nutraceutical interaction. *Indian Food Industry*, 29: 29–43.

Sprenger, N. 2009. Synbiotic mixture. Patent PN: EP 2100523 A1.

Srinivasan, K. 2005. Plant foods in the management of diabetes mellitus: Spices as potential antidiabetic agents. *International Journal of Food Sciences and Nutrition*, 56: 399–414.

Stahl, W. and Sies, H., 2005. Bioactivity and protective effects of natural carotenoids. *Biochimica et Biophysica Acta*, 1740: 101–107.

Stephens, F.C. 1999. The rising incidence of breast cancer in women and prostate cancer in men. Dietary influences: A possible preventive role for nature's sex hormone modifiers—The phytoestrogens: A review. *Oncology Reports*, 6: 865–870.

Stinson, E.E., Kwoczak, R., and Kurantz, M. 1991. Effect of cultural conditions or production of eicosapentaenoic acid by *Pythium irregular*. *Journal of Industrial Microbiology*, 8: 171–178.

Strednasky, M., Conti, E., and Salaris, A. 2000. Production of polyunsaturated fatty acids by *Pythium ultimum* in solid-state cultivation. *Enzyme and Microbial Technology*, 26: 304–307.

Sun, W.D., Melton, L.D., and Skinner, M.A. 2007. Dietary fibers, functional foods and human well-being. *Asia Pacific Journal of Clinical Nutrition*, 16: S76.

Takagaki, K. 2006. Anti-obesity agent and food containing the anti-obesity agent. Japanese Patent PN: JP 2006306840 A.

Tan, S.C., Tan, T.K., Wong, S.M., and Khor, E., 1996. The chitosan yield of zygomycetes at their optimum harvesting time. *Carbohydrate Polymers*, 30: 239–242.

Tanaka, R. and Matsumoto, K. 1998. Recent progress on prebiotics in Japan, including galacto-oligosaccharides. *Bulletin of the International Dairy Federation*, 336: 21–27.

Tanaka, R. and Sako, T. 2003. Prebiotics: Types. In: *Encyclopedia of Dairy Science*, eds. H. Roginski, J.W. Fuquay, and P.F. Fox, pp. 2256–2268. Academic Press, London, U.K.

Tank, R., Sharma, R., Sharma, T., and Dixit, V.P. 1990. Anti-diabetic activity of *Curcuma longa* in Alloxan induced diabetic rats. *Indian Drugs*, 27: 587–589.

Tarber, M.G. and Packer, L. 1995. Vitamin E beyond antioxidant function. *American Journal of Clinical Nutrition*, 62: 501–509.

The Capsaicin Study Group 1992. Effect of treatment with capsaicin on daily activities of patients with painful diabetic neuropathy. *Diabetes Care*, 15: 159–165.

Vanderhoof, J.A., Whitney, D.B., Antonsson, D.L., Hanner, T.L., Lupo, J.V., and Young, R.J. 1999. Lactobacillus GG in the prevention of antibiotic-associated diarrhoea in children. *Journal of Pediatrics*, 135: 564–568.

Varshney, S.C. 2002. Role of functional foods in diet. *Indian Food Industry*, 21: 41–43.

Viljanen, M., Kuitunen, M., Haahtela, T., Juntunen-Backman, K., Korpela, R., and Savilahti, E. 2005. Probiotic effects on faecal inflammatory markers and on faecal IgA in food allergic atopic eczema/dermatitis syndrome infants. *Pediatric Allergy and Immunology*, 16: 65–71.

Walker, T.H., Drapcho, C.M., and Chen, F. 2007. Bioprocessing technology for production of nutraceutical compounds. In: *Functional Food Ingredients and Nutraceuticals*, ed. J. Shi, pp. 211–236. CRC Press, Boca Raton, FL.

Wildman, R.E.C. 2007. *Handbook of Nutraceuticals and Functional Foods*, CRC Press, Boca Raton, FL.

Williamson, A. 2009. Synbiotics: Tomorrow's nutritional buzzword? *Asia Pacific Food Industry* (April): 36–37.

Wollowski, I., Ji, S.T., Bakalinsky, A.T., Neudecker, C., and Pool-Zobel, B.L. 1999. Bacteria used for the production of yoghurt inactivate carcinogens and prevent DNA damage in the colon of rats. *Journal of Nutrition*, 129: 77–82.

Woyengo, T.A., Ramprasath, V.R., and Jones, P.J.H. 2009. Anticancer effects of phytosterols. *European Journal of Clinical Nutrition*, 63: 813–820.

Wu, A.H., Horn-Ross, P.L., Nomura, A.M. et al. 1996. Tofu and risk of breast cancer in Asian—Americans. *Cancer Epidemiology Biomarkers and Prevention*, 5: 901–906.

Wullschleger, C. 2007. Green oat extract: Mental health and cognitive function. *Food Engineering and Ingredients*, 32: 30–33.

Yeh, C.T. and Yen, G.C. 2009. Chemopreventive functions of sulforaphane: A potent inducer of antioxidant enzymes and apoptosis. *Journal of Functional Foods*, 1: 23–32.

Zaque, V. 2008. A new view concerning the effects of collagen hydrolysate intake on skin properties. *Archives of Dermatological Research*, 300: 479–483.

Ziemer, C.J. and Gibson, G.R. 1998. An overview of probiotics, prebiotics and synbiotics in the functional food concepts: Perspective and future strategies. *International Dairy Journal*, 8: 473–479.

9

Enzymes in Food Processing

Ivana G. Sandri, Luciani T. Piemolini-Barreto, and Roselei C. Fontana

CONTENTS

9.1 Introduction ... 329
9.2 Advantages of Microbial Enzymes ... 330
9.3 Mechanism of Enzyme Action ... 330
9.4 Enzyme Engineering ... 332
9.5 Production and Purification of Enzymes ... 333
9.6 Enzyme Immobilization Techniques ... 335
9.7 Applications of Enzymes in Food Industries ... 337
9.7.1 Meat Processing ... 337
9.7.2 Enzymes in Milk and Cheese Industries ... 338
9.7.3 Enzymes in Baking Industry ... 339
9.7.4 Enzymes in Fruit Juice Industry ... 341
9.7.4.1 Case Study: Effect of Pectinases in Polyphenol Extraction ... 342
9.7.5 Enzymes in Wine Production ... 342
9.8 Enzyme Biosensors ... 345
9.9 Summary and Future Prospects ... 347
References ... 347

9.1 Introduction

For centuries, enzymes have been used in a wide variety of applications in the food industries. They can change and improve the functional, nutritional, and sensory properties of ingredients and products. They are obtained from animal tissues, plants, bacteria, and fungi including yeast. Plant-derived commercial enzymes include α-amylase (bread baking), β-amylase (production of high-malt syrups), proteolytic enzyme papain, bromelain and ficin (meat tenderizing), and some other special enzymes like lipoxygenase from soybean (bread dough improvement). Animal-derived enzymes include proteinases like chymosin (coagulation of milk in cheese making), rennin (cheese production), and lipase (cheese flavors). However, most enzymes are produced by microorganisms, including α-amylases (bread making), amyloglucosidases (corn syrup production), cellulases (fruit liquefaction in juice production), glucose oxidases (baking applications), lipases and esterases (flavor enhancement in cheese products; modification of fat function by interesterification; synthesis of flavor esters), pectinases (clarification of fruit juices by depectinization), and protease (milk coagulation for cheese making; improvement of bread dough).

9.2 Advantages of Microbial Enzymes

The bulk of enzymes, in terms of both quantity and variety, are derived from microorganisms. Microbial enzymes have two advantages over animal and plant enzymes. First, they are economical and can be produced on a large scale within limited space and time frame. The amount of enzyme produced depends on the scale of production, the microbial strain, and growth conditions. It can be easily extracted and purified. Second, there are technical advantages in producing enzymes by using microorganisms as they are capable of producing a wide variety of enzymes; grow in a wide range of environmental conditions; show genetic flexibility, i.e., they can be genetically manipulated to increase enzyme yield; and have short generation times. Table 9.1 summarizes examples of sources of important microbial enzymes (Benejam et al., 2009; Bhalla and Chatanta, 2000; Couri et al., 2008; Jurado et al., 2002; Law, 2002; Oort, 2010a; Sandri et al., 2011).

9.3 Mechanism of Enzyme Action

The basic mechanism by which enzymes catalyze reactions begins with the substrate (or substrates) binding to the active site on the enzyme. The active site is the specific region of the enzyme that combines with the substrate. The specificity of enzymes is determined by the complementary shape, charge, and hydrophilic/hydrophobic characteristics of the substrates and their 3D organization (Figure 9.1).

Enzymes can act in several ways, whereby each enzyme lowers the energy needed for the reaction to occur or to proceed (Oort, 2010a). Briefly, these mechanisms act by

1. Lowering the activation energy by creating an environment in which the transition state is stabilized. This can be achieved by binding and thus stabilizing the transition-state conformation of the substrate/product molecules.

TABLE 9.1
Examples of Microbial Enzymes Used in Food Processing

Enzyme	EC Number	Source	Industrial Use
α-Amylase	3.2.1.1	*Aspergillus* spp. *Bacillus* spp. *Microbacterium imperiale*	Bread making
α-Acetolactate	4.1.1.5	*Bacillus subtilis*	Reduction of wine-maturation time
β-Galactosidase	3.2.1.23	*Aspergillus* spp. *Kluyveromyces* spp.	Milk processing
Amyloglucosidase	3.2.1.3	*A. niger* *Rhizopus* spp.	Corn syrup production
Cellulase	3.2.1.4	*A. niger* *Trichoderma* spp.	Fruit liquefaction in juice production
Chymosin	3.4.4.3	*Aspergillus awamori* *Kluyveromyces lactis*	Coagulation of milk for cheese making
Glucose oxidase	1.1.3.4	*A. niger* *Penicillium chrysogenum*	Oxygen removal from food packaging
Inulinase	3.2.1.7	*Aspergillus* spp. *Kluyveromyces* *Paenibacillus* spp. *Penicillium* spp.	Production of fructose Production of oligosaccharides
Lipase	3.1.1.3	*Aspergillus* spp. *Candida* spp. *Rhizomucor miehei* *Penicillium roqueforti* *Rhizopus* spp. *B. subtilis*	Flavor enhancement in cheese products Fat function modification by interesterification Synthesis of flavor esters
Pectinase (polygalacturonase)	3.2.1.15	*Aspergillus* spp. *Penicillium funiculosum*	Clarification of fruit juices and wine by depectinization
Protease	3.4.21.24	*Aspergillus* spp. *R. miehei* *Cryphomectria parasítica* *Penicillium citrinum* *Rhizopus niveus* *Bacillus* spp.	Milk coagulation for cheese making Hydrolyzate production for soups and savory foods Bread dough improvement

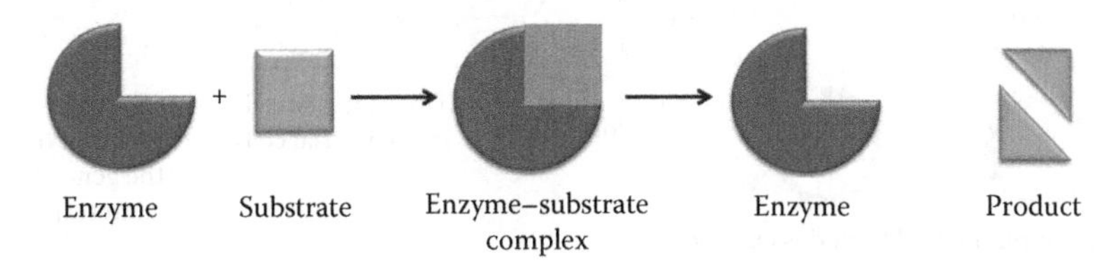

FIGURE 9.1
Mechanism of enzyme action.

2. Lowering the energy of the transition state but without distorting the substrate by creating an environment with the opposite charge distribution to that of the transition state.
3. Providing an alternative pathway—Temporarily reacting with the substrate to form an intermediate enzyme–substrate complex, which would be impossible in the absence of the enzyme.
4. Effect of decreasing entropy—Reducing the reaction entropy change by bringing substrates together in the correct orientation to react.

9.4 Enzyme Engineering

Enzyme engineering has enabled the increasing competitiveness of enzymatic technologies in food industry, contributing to the selection of new enzymes, mainly through recombinant DNA technologies (which allow the modification of kinetic properties and stability), and the development of new solutions in terms of technology reactors and enzyme immobilization techniques and design of the reaction medium. The basis of recombinant DNA technology used for enzyme engineering is summarized in Figure 9.2.

Enzyme engineering studies the forces and interactions within the 3D structure of an enzyme, which can be changed modifying its amino acid sequence by manipulating the gene (Lehmann and Wyss, 2001). Understanding the protein structures is essential to determine the sites for site-directed mutation, mutagenesis, and recombination, where the environmental adaptation is reproduced in vitro in a much hastened timescale, improving the properties of an enzyme to make more efficient industrial processes (Wong, 2002). The alterations are performed based on the increasing knowledge on the

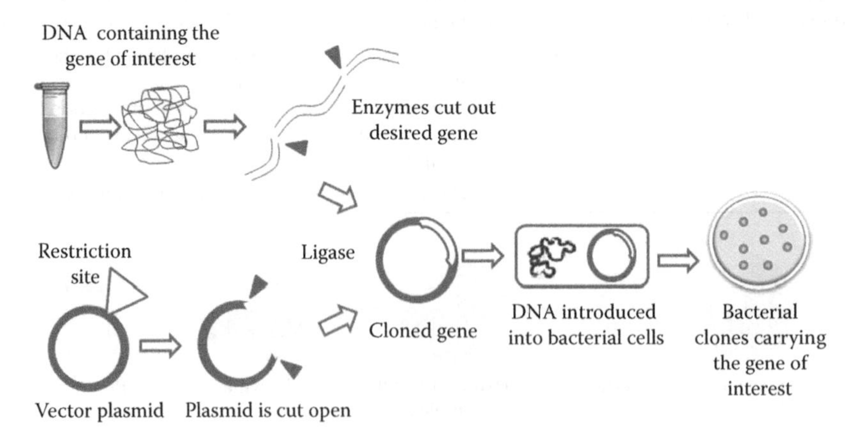

FIGURE 9.2
Schematic representation of the recombinant DNA technology.

enzyme's structure and functions. Computer-assisted design provides data on amino acid and protein sequences. Adequate processing of the data allows the formulation of general rules predicting the effect of mutations on enzyme properties (Vieceli et al., 2006).

To improve enzyme stability against high temperatures was one of the first experiments of protein engineering by understanding the chemistry of protein unfolding and hydrophobic forces. Since thermal stability is determined by a series of short- and long-range interactions, this can be improved by several amino acid substitutions in a single mutant. However, improvements not only were restricted to thermostability but also addressed other characteristics such as increased catalytic efficiency and altered substrate specificity and selectivity of the enzyme.

9.5 Production and Purification of Enzymes

The majority of industrially used food enzymes are produced using microorganisms. While choosing the microbial strain for the production of enzyme, several aspects are considered. The microorganism should have a Generally Recognized as Safe (GRAS) status. The organism should be able to produce high amount of the desired enzyme in a reasonable life time frame. The optimization microbial of a fermentation process includes media composition, microbial cultivation, and process conditions. Fermentations are most often run in the fed-batch mode, although batch and continuous processes are also used (Dodge, 2010). Some enzymes are best produced while the culture is actively growing, and faster growth results in faster enzyme production. This is called growth-associated enzyme production. Metabolic enzymes are examples of proteins that are typically growth associated. Many hydrolytic enzymes are typically produced as secondary metabolites. Such enzymes are produced during slow specific growth rates or even under nongrowth conditions. This is called nongrowth-associated production.

The industrial fermentation processes are based on submerged and solid-state cultivation. Surface fermentations typically employ moist nonsubmerged particulate or nonparticulate substrates (grains, vegetables, agro-industrial residues, fresh cheese), preferably sterile and formed into high surface area matrices. In some cases, a mix of nutrients is also added to the solid substrate and microbial seed is grown in submerged fermentation process before inoculating the matrix.

The major challenges of the production design are oxygen supply, removal of gases and heat of respiration from the substrate matrix, homogeneous conditions, and contamination control. The final stages of production consist of downstream processing steps aimed at recovering the maximum amount of enzymes produced during fermentation. The choice of specific operations is dictated not just by the nature of the fermentation broth, the biochemical and biophysical properties of the enzyme, and the refining needs of the product but also for practical considerations. This constrains the source of raw materials and reduces flexibility to change materials once a product has been approved for food use.

If the specifications of the enzyme product require lower concentration of the nonenzyme components in the crude concentrate, some form of purification is required (Linke and Berger, 2011). The separation of permeable solids can be accomplished by centrifugation and filtration. Impurities, such as biopolymers, can often be reduced by careful selection of the conditions used during cell separation: pH, salt content, and flocculant type. If the soluble impurities can be induced to form a solid phase, it can simply be removed by filtration. Each protein-purification step usually results in some degree of product loss. Therefore, an ideal protein purification strategy is one in which the highest level of purification is reached in the fewest steps. The selection steps to be used are dependent on the size, charge, solubility, and other properties of the target enzyme. Ideally, precipitation results both in concentration and in purification. The chromatography is another choice for making purified enzyme. This is usually the most expensive option and is only economical for a small number of enzymes that are used in very low doses (Dodge, 2010; Panesar et al., 2010a). The main steps used in enzyme purification are shown in Figure 9.3.

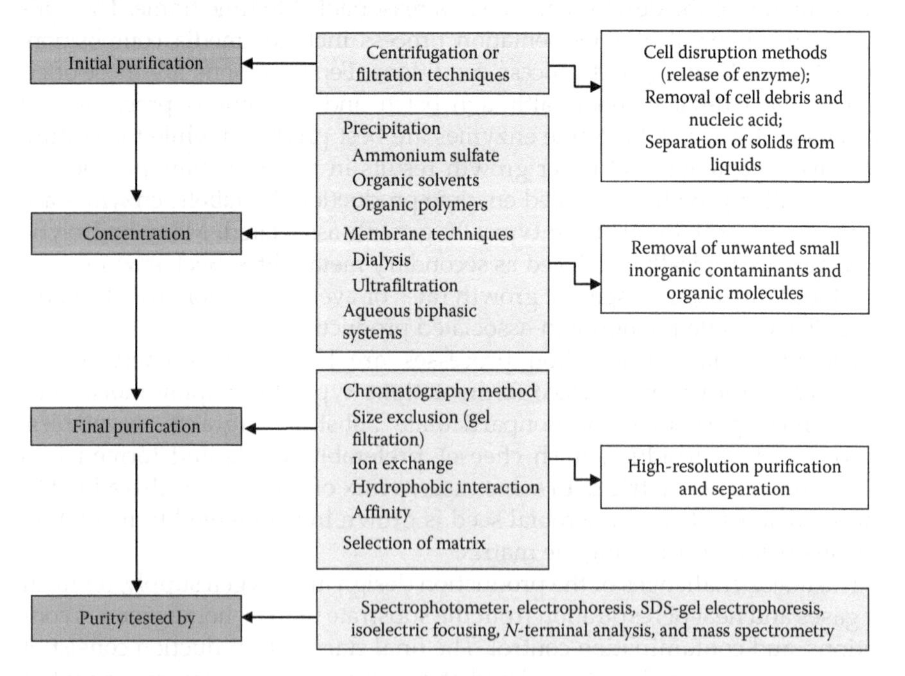

FIGURE 9.3
Main steps used in enzyme purification.

The main problem usually observed during the downstream process is the irreversible loss of the 3D structure of an enzyme, resulting in a loss of catalytic activity. Increasing the number of separation operations for improved refining means is to accept the concomitant reduction of activity with each step. Considering these problems, there is an obvious demand for efficient alternatives for enzyme purification with sufficient selectivity, maximal recovery, adequate enrichment, and high preservation of enzyme activity besides acceptable costs and minimal time expense.

9.6 Enzyme Immobilization Techniques

Immobilization has several economic and technical advantages: reduction of enzyme costs, more efficient reactor performances, less downstream processing, products are easily separable from biocatalysts, and in some cases, enzyme properties are favorably altered by immobilization (Danial et al., 2010). This technology is in principle applicable to all soluble substrates. Solid substrates, however, have to be treated in batch or semicontinuous processes. The use of a relatively expensive catalyst as an enzyme requires, in many instances, its recovery and reuse to make an economically feasible process. Moreover, the use of an immobilized enzyme permits to greatly simplify the design of the reactor and control of the reaction. Thus, immobilization is usually a requirement to the use of an enzyme as an industrial biocatalyst.

The techniques usually employed to immobilize enzymes on solid supports are mainly based on chemical and physical mechanisms. Chemical immobilization mainly includes enzyme attachment to the matrix by covalent bonds and cross-linking between enzyme and matrix. Physical mechanisms involve (1) the adsorption of enzyme molecules onto a porous support or on a matrix containing ion exchange residues or (2) the entrapment within an insoluble gel matrix (Brena and Batista-Viera, 2006; Choi, 2004; Sheldon, 2007). The different techniques for immobilization of enzyme are diagrammatically presented in Figure 9.4.

Immobilization can be performed by several methods, namely, entrapment/microencapsulation, binding to a solid carrier, and cross-linking of enzyme aggregates, resulting in carrier-free macromolecules (Brena and Batista-Viera, 2006). The simpler method of immobilization, based on a weak binding force, such as the adsorption of protein on the surface of insoluble supports provides a small perturbation of the native structure of the enzyme but promotes the leak of the adsorbed protein from the support during use. This can occur especially when there is a mild change in temperature, pH, or ionic strength, or simply because a substrate is present.

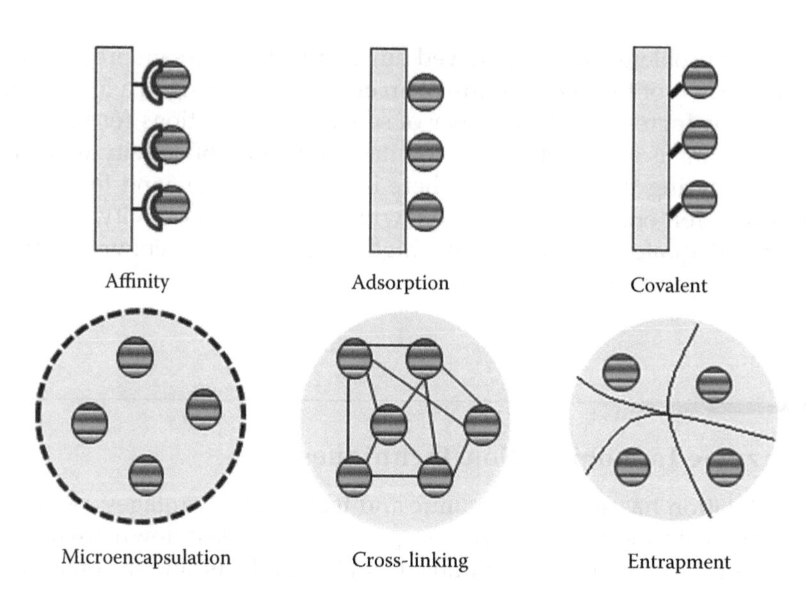

FIGURE 9.4
Schematic representations of the basic methods of enzyme immobilization.

Aiming to improve the performance of biocatalysts for food processing, several works are being developed in order to evaluate different methods and supports for enzyme immobilization. In Table 9.2, some examples of enzymes immobilized to improve the performance of relevant enzymes in food processing are given (Anjani et al., 2007; Chopra et al., 2010; Kailasapathy and Lam, 2005; Murty et al., 2002; Nunes et al., 2011; Pal and Khanum, 2011; Panesar et al., 2010b; Rajagopalan and Krishnan, 2008; Tumturk et al., 2008).

The integration of enzymes and food processing is a well-established approach; however, many researches are continuously being made to make enzymatic applications still more effective and diversified (Blanco et al., 2007). These efforts are directed for the design of new/improved biocatalysts that are more stable (temperature and pH), less dependent on metal ions, and less susceptible to inhibitory agents and changes in environmental conditions while maintaining the targeted activity or evolving novel activities. Hence, it can be foreseen that efforts will be toward the development of immobilized biocatalyst that can be used in different reactor configurations and that comply with the economic requirements for large-scale application. All these strategies either isolated or preferably suitably integrated have been put into practice for food applications to improve existing processes or to implement new ones with the latter often combined with the output of new products resulting from novel enzymatic activities.

TABLE 9.2

Examples of Immobilized Enzymes with Potential Applications in Food Processing

Immobilization Method	Enzymes	Support	Application
Entrapment	Flavourzyme (fungal protease and peptidase)	Calcium alginate	Cheese ripening
Entrapment	α-Amylase	Calcium alginate	Hydrolysis of starch
Covalent	Xylanases	Sodium alginate	Beverage industries (bakery goods, coffee, starch, plant oil, and juice manufacture)
Adsorption entrapment covalent	β-Galactosidases	Chitosan calcium alginate silica, gelatin	Hydrolysis of lactose in milk synthesis of galacto-oligosaccharides
Adsorption	Glucose isomerase	Duolite A7 resin	Production of fructose syrup
Entrapment	Invertases	Sodium alginate, chitosan, gelatin	Hydrolysis of sucrose for production of inverted sugar syrup
Adsorption	Lipases	Celite, cellulose, silica gel	Hydrolysis of olive oil

9.7 Applications of Enzymes in Food Industries

Some examples of the most important applications of enzymes in the food industry are briefly discussed in this section.

9.7.1 Meat Processing

The cooked meat is considered tender if it can be chewed easily and has the desired texture. Tenderization is influenced by many factors and involves several biochemical reactions led by endogenous enzymes. Examples are endogenous peptidases, especially cathepsins. To increase the tenderization effect, addition of proteolytic enzymes is one of the most popular methods.

Proteolytic enzymes from plant sources have received special attention as being active over a wide range of temperatures and pH. Papain extracted from papaya latex is the most commonly used proteolytic enzyme in meat tenderizers (Melendo et al., 1996). In addition to papain, bromelain extracted from pineapple stem and ficin extracted from fig latex are used in meat tenderization (Katsaros et al., 2009). Interesting new proteases from plant origin are cucumis and ginger extracts. Powdered cucumis extract from the Kachri fruit as well as ginger proteases has been successfully used to tenderize various types of meat. Ginger extract has proven to be especially effective in increasing collagen solubilization (Naveena and Mendiratta, 2004).

Due to the biochemical consistency and structure of meat, it can be challenging to evenly distribute the tenderizing protease into meat pieces. Possible methods are, e.g., spraying, injecting, dipping, and marinating. Dipping of meat pieces in solution containing proteolytic enzymes or marinating the meat in enzyme solution has been widely used. The problem with this method is the poor penetration of the enzyme into the meat pieces, resulting in overtenderized surface and mushy texture whereas the interior remains are unaffected. Injection of proteolytic enzyme solution directly into meat pieces has been shown to be a more effective way of tenderization than marinating in enzyme-containing solution (Fogle et al., 1982).

Different enzymes or enzyme mixtures are suitable for different meat cuts. For fine-tuned applications, the substrate specificity and activity profiles of applied enzyme preparations should be known to adjust the enzyme levels and processing conditions to suit each other. For example, transglutaminase (TGase; protein-glutamine γ-glutamyltransferase) is an enzyme with the ability to improve the functional characteristics of protein such as texture, flavor, and shelf life (Ahhmed et al. 2007). TGase initially attracted interest because of its capacity to reconstitute small pieces of meat into a steak (Romero de Ávila et al., 2010). Lipases (triacylglycerol lipase) can be used for flavor development in sausage production (Miklos et al., 2011), whereas phytase (myoinositol hexaphosphate phosphohydrolase) has been used to reduce production costs of meat foods as well as to reduce environmental contamination (Lim and Lee, 2009; Selle et al., 2009).

9.7.2 Enzymes in Milk and Cheese Industries

Milk is a wholesome natural food widely used as an ingredient in many formulated foods. Milk contains proteins, specifically caseins that maintain its liquid form. Proteases are used to hydrolyze caseins during cheese production, specifically kappa casein, which stabilizes micelle formation preventing coagulation. Use of rennet cheese production is the oldest application of enzyme in the milk industry. Rennet and rennin are general terms for any enzyme used to coagulate milk. Sources of rennet are animals such as calves, sheep, and goats and microorganisms. However, enzyme preparations derived from microorganisms have been used for this purpose (Chazarra et al., 2007; El-Bendary et al., 2007; Llorente et al., 2004; Low et al., 2006; Naz et al., 2009; Shieh et al., 2009; Yu and Chou, 2005).

The production of different types of cheese uses proteolytic enzymes, of which rennin is the most widely used (De Lima et al., 2008). Rennin acts on κ-casein in milk, specifically hydrolyzing the Phe–Met bonds and releasing a strongly hydrophilic C-terminal fragment, susceptible to complexation with calcium ions and allows the coprecipitation of all fractions of milk casein (Ageitos et al., 2006).

Cow milk also contains whey proteins such as lactalbumin and lactoglobulin (Jovanovic et al., 2007). The denaturing of these whey proteins by using proteases results in a creamier yogurt product. Denaturation of whey proteins is also essential for cheese production.

Lipases are extensively used in the dairy industry for the hydrolysis of milk fat. The dairy industry uses lipases to modify the length of fatty acid chains to enhance the flavor of various cheeses. Current applications also include the acceleration of cheese ripening and the lipolysis of butter, fat, and cream (Sharma et al., 2001). Lipases and peptidases contribute to the taste and aroma of cheese by producing not only peptides, amino acids, and fatty acids but also deaminases, decarboxylases, aminases, and transaminases, which produce different amino acids, amines, keto acids, aldehydes, and ammonia.

Enzymes can also be used to hydrolyze lactose in milk and its by-products. Lactose (milk sugar) is a disaccharide sugar composed of glucose and galactose and a fermentable substrate. It can be fermented outside of the body to produce cheese, yogurt, and acidified milk. It can be fermented within the large intestine of human beings with low expression of lactase enzyme in the intestinal mucosa to ferment the disaccharide to its absorbable simple hexose sugars (Solomons, 2002). Hydrolysis of lactose to glucose and galactose by β-galactosidase is an important process in the food industry, due to the potentially beneficial effects on absorption of foods containing lactose (Jurado et al., 2002). In addition, this enzyme catalyzes the formation of galacto-oligosaccharides, which are prebiotic additives for the so-called healthy foods (Grosová et al., 2008). β-Galactosidase is one of the relatively few enzymes that have been used in large-scale processes in both free and immobilized forms. Hydrolysis of whey lactose to glucose and galactose by immobilized β-galactosidase comes as a tool for the commercial use of this feedstock (Mariotti et al., 2008).

Several problems in nutritional intolerance to lactose (Lomer et al., 2007; Playne et al., 2003), organoleptic elaboration of milk with sweetened flavor (Jelen and Tossavainen, 2003), or improvement in the technological and sensorial characteristics of foods containing hydrolyzed lactose from milk or whey, such as increased solubility (avoidance of lactose crystallization and the grainy aspect of ice-cream and condensed or powdered products), can be solved by its hydrolysis by β-galactosidase (Panesar et al., 2006).

9.7.3 Enzymes in Baking Industry

The use of enzymes to produce bakery products increased considerably thereby completely or partially replacing chemical additives in recent years for meeting the demand of consumers for healthier products. They are used in baking in order to improve the rheological properties of dough, influencing the quality of the final product, such as volume, color, flavor, aroma, structure, tenderness, and shelf life.

Starch and its derivatives are essential for human nutrition and are subjected to various industrial applications. The application of enzyme allows a selective degradation of starch under moderate conditions. The enzymes used in the processing of starch are commonly known as amylases. The enzymes used in the processing of starch are commonly known as amylases, which exist in various types, allowing the use of many alternative paths according to the source of starch, to improve their conversion (Couri et al., 2008).

Wheat and wheat flour are deficient in certain enzymes that facilitate their use in dough preparation. For dough making, fungal α-amylase is often added to flour. The enzyme partially hydrolyzes starch into glucose, maltose, and dextrin. The action of α-amylase is possible because during the grinding of wheat, the structure of the starch changes, thus making the polymer susceptible to enzymatic attack and also because of the ability of many amylases to hydrolyze raw starch (Couri et al., 2008; Gupta et al., 2003).

Introduction of α-amylase into the dough increases loaf volume, reduces the aging rate (retrogradation) of starch, and extends the period of rising in the oven thereby increasing the maximum height of the mass (Gupta et al., 2003). This enzyme also helps in the formation of reducing sugars, which in the presence of heat promotes the browning of the bread crust (Maillard reaction). In cake, cookies, and cracker production, α-amylases are used to decrease viscosity (Benejam et al., 2009). Today, many enzyme preparations such as peptidases, lipases, xylanases, pullulanases, pentosanases, cellulases, and oxidases (glucose oxidase, lipoxygenase) are used in the baking industry for different purposes (Caballero et al., 2007).

In bread making, lipoxygenases act by oxidizing the carotenoids present in the flour, thus promoting their bleaching. These enzymes also increase the strength of the flour and improve their rheological properties and increase the stability and viscosity of the gluten network, contributing to the bread volume (Junqueira et al., 2003).

Glucose oxidase can be used to promote the oxidation of the protein matrix and also as a conditioner for the preparation of wheat flour dough (Bonet et al., 2007). It catalyzes the oxidation of β-D-glucose to β-D-gluconolactone and hydrogen peroxide. Hydrogen peroxide promotes the oxidation of SH groups of two cysteine residues for SS cross-links in the gluten network. The cross-linked proteins in wheat flour result in flour fortification and functional improvements in the technological properties of baked products, such as improved volume and better bread crust.

Xylanases are widely used in baking because of their positive effect on the properties of the dough during mixing and fermentation and the quality of bread (Butt et al., 2008). In general, xylanases with specificity for water-extractable arabinoxylans (WEAX) reduce the dough viscosity by reducing the size of these molecules. Also xylanases due to their specificity for water-unextractable arabinoxylans (WUAX) depolymerize the fraction leading to a redistribution of water arabinoxylose (AX) for the phases of gluten and starch and resulting in higher viscosity and smoothness of the dough, thereby representing a positive role in dough rheology and gas retention (Oort, 2010b).

Lipases can produce mono- and diglycerides from lipids, thereby improving the specific volume and crumb softness and slowing bread aging (Purhagen et al., 2011). In addition, lipase enzymes provide increased strength and elasticity to the dough. Lipases also act on the gluten network (Moayedallaie et al., 2010). This can be explained in part by the generation of free fatty acids, which can be oxidized by endogenous lipoxygenases, leading to a higher oxidation potential, which in turn can positively affect the formation of the gluten network. Furthermore, the addition of lipase to the dough helps in increasing the elasticity, i.e., the ability to retain air and aroma.

Proteases are used to reduce mixing time; to improve the dough machinability; to improve the gas retention due to better extensibility; to improve the grain and crumb texture; and to improve the color, flavor, and the water absorption. The increasing usage of proteases is related to the fact that the flour available on the market is getting richer in proteins (Oort, 2010b).

9.7.4 Enzymes in Fruit Juice Industry

In the fruit juice industry, pectic enzymes are used to promote the degradation of pectin. In several processes, it is common to utilize different enzymes (Kashyap et al., 2001). The commercial pectolytic enzyme preparations basically contain mixtures of activities, namely, main and secondary activities. The main activities of pectolytic enzymes are polygalacturonases, pectinesterases, and pectin lyases. The major secondary activities are cellulases, hemicellulases, proteases, and beta-glucanase, and these are contained in commercially available pectinase products (depending on the production strain), but these are usually not standardized.

The synergistic effect of combining cellulase and pectinase is crucial in the enzyme treatment of many fruit pulps to increase yield (Demir et al., 2001). In case of berries, e.g., color is an important quality parameter and the addition of enzyme preparations containing cellulases (Bhat, 2000) can lead to extraction of color. In clear bright juices (apple, pear, and grape), enzymes increase juice production during pressing and filtration and promote the removal of suspended matter. In case of cloudy juices, polygalacturonase quickly degrades the pectic acid formed thus preventing precipitation (Sandri et al., 2011).

During fruit processing, pectinases are used after the raw material is cut so as to macerate the pulp until partial or total liquefaction, thus reducing processing time and improving extraction of the fruit components (Kaur et al., 2004). During the extraction of juice, pectin and other polysaccharides, together with phenolic compounds and proteins, are dispersed in the solution to increase viscosity and turbidity. In nonclarified juices, pectinases can be used at different stages, allowing a better extraction of sugars and soluble solids, resulting in higher yields. After extraction, pectinases are added to clarify the juice and reduce its viscosity to destabilize flocculating substances, thus resulting in their coagulation and precipitation and consequent clarification. During the enzymatic treatment, there is an increase in the size of insoluble particles due to the reduction of electrostatic repulsion between the colloidal particles and grouping them for removal during filtration.

The complete depectinization by using pectinase enzymes provides a good clarification and filtration of the juice. Depectinization of juices after pressing is needed to obtain a juice with low viscosity. In the production of juice concentrates, depectinization is required to prevent gelation during storage or concentration (Mojsov et al., 2011). When the maceration process is performed without adding pectinolytic enzymes, a highly viscous pectin gel is formed, which hinders the pressing and lowers the yield. When the pulp does not receive any enzymatic treatment, the colloidal substances are suspended leading to juice turbidity.

The extent and duration of enzymatic treatments depend on the composition of juices. As an example, see "Case Study Effect of Pectinases in Polyphenol Extraction" in Section 9.7.4.1.

9.7.4.1 Case Study: Effect of Pectinases in Polyphenol Extraction

In this case study, effectiveness of pectinase was evaluated for the extraction of polyphenol extraction in juices of strawberry guava, mulberry, and raspberry. A crude enzymatic extract obtained with *Aspergillus niger* LB23 in a solid-state fermentation process was compared with commercial preparations. The commercial enzymatic preparations are produced by *A. niger*, *Aspergillus aculeatus*, and *Trichoderma reesei*. The experimental enzymatic extracts as well as the commercial preparations were diluted to 10 units (U) of total pectinases per mL. Then, to each 10 mL of juice, 1 mL of the enzyme was added. With pectinases' total activity at 1 U/mL of fruit juice, a reaction was conducted at 30°C and 50°C for 60 min. Figure 9.5 shows the percentage increase in polyphenol extraction by using pectinases as compared to control (i.e., without pectinase). In both cases, the highest percentage increase in polyphenol was recorded at a temperature of 50°C. The crude enzymatic preparation obtained from *A. niger* LB23 reached similar results of polyphenol extraction as achieved with commercial products. The efficiency of the pectinolytic enzyme depends on the temperature, substrate type, and nature of the substrate and enzymatic complex used in the preparation.

9.7.5 Enzymes in Wine Production

Wine is an alcoholic beverage made from the fermentation of fresh grapes or fresh grape juice. Fermentation processes for winemaking involved different enzymatic activities present in the substrate (wine) and originated from the microbial agent used (yeast).

Pectinases naturally occur in grapes, but grape enzymes are poorly active under the pH conditions and SO_2 levels associated with winemaking

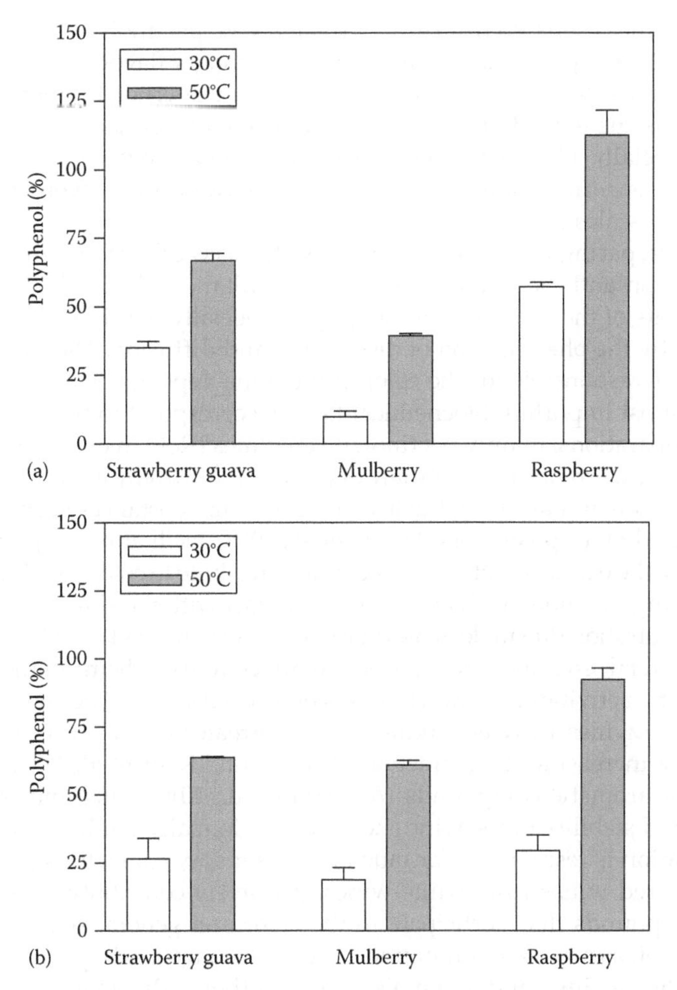

FIGURE 9.5
Polyphenol (%) in juices of strawberry guava, mulberry, and raspberry, by using pectinases during 60 min, with different enzymatic treatments: (a) commercial preparation and (b) experimental enzymatic extract.

practice (Guérin et al., 2009). The enzymes from exogenous sources are viewed as processing aids and these can reinforce or replace the enzymatic activity of grape (Bruchmann and Fauveau, 2010). Basically, pectinolytic enzymes can be used during the vinification process to (1) enhance must extraction by degrading structural polysaccharides that interfere with the extraction process, (2) increase the volume of free-run juice and reduce the pressing time during grape crushing, (3) settle many suspended particles in must before or after fermentation, and (4) ensure clarification and filtration of wine.

White wines prepared from grapes treated with pectinolytic enzymes tend to mature more quickly and usually have a stronger flavor of the fruit compared to untreated wines. In red wine, a matter of crucial importance and attention is the extraction and diffusion of fine components of the grape's skin, especially phenolic compounds and aromas that contribute to the color and sensorial characteristics of the wine (Ribéreau-Gayon et al., 2003; Romero-Cascales et al., 2008).

When preparing red wine by following the classical process, the alcoholic fermentation and maceration take place simultaneously. In this phase, the constituents of the solid part of the grape, especially the skin, are carried to the must by the phenomenon of dissolution and diffusion. The extraction of substances responsible for the color of the wine depends on several factors, and the most important biochemical factors correspond to the use of enzymatic preparations, mainly pectinolytic enzymes (Roldán et al., 2006).

Pectinases degrade pectic substances of the wine, thereby increasing the rate of clarification and avoiding the loss of important substances such as anthocyanins and other precursors (Iranzo et al., 1998), facilitating grape pressing, increasing the extraction of substances that contribute to color and flavor of the wine during the time in which it stays in contact with the skins, and improving wine filtration due to decreased viscosity (Fernández et al., 2000).

Commercial enzyme preparations with cellulosic, hemicellulosic, and pectinolytic activities are known as maceration enzymes. The combined use of these enzymes in winemaking ensures greater efficiency in skin maceration by increasing the extraction of pigments (Kaur et al., 2004) and the release of aromatic compounds (Armada et al., 2010), thus improving the quality and stability of the wine (Revilla and Gonzalez-SanJose, 2003).

Maceration is responsible for much of the sensory characteristics that differentiate red wines from white wines, decisively contributing to the phenolic compounds that participate in the color and general structure of the wine as well as aromatic compounds, nitrogen compounds, polysaccharides (especially pectins), and minerals, among others (Ribéreau-Gayon et al., 2003). The most common red wine making procedure is characterized by the solid parts of grapes (skins, seeds, and sometimes parts of the stalk) remaining in contact with the must, and the length of this contact period depends on the characteristics and the quality desired for the wine.

Enzymes can either have an increasing or decreasing effect on wine color (Bautista-Ortin et al., 2005) and anthocyanin content (Revilla and Gonzalez-SanJose, 2003). The color of red wine is not only due to the anthocyanins extracted from grape skins during crushing, pressing, and fermentation but also due to various products formed during vinification and aging. In particular, tannins (i.e., proanthocyanidins, which are oligomers and polymers of flavan-3-ols such as catechins) play a significant role in the taste of wines (Cheynier et al., 2006) and contribute to the color stability of red wines as they react with anthocyanins to form derived pigments such as tannin–anthocyanin and anthocyanin–tannin adducts (Salas et al., 2003).

9.8 Enzyme Biosensors

A biosensor can be defined as a device incorporating a biological sensing element connected to a transducer to convert an observed response into a measurable signal, whose magnitude is proportional to the concentration. Basically, in enzymatic biosensors, enzymes are immobilized in a potentiometric, amperometric, optometric, calorimetric, or piezoelectric transduction (Shantilatha et al., 2003). The principle of detection by a biosensor is based on the specific interaction between the analyte of interest and the recognition element. As a result of this specific interaction, changes are produced in one or several physicochemical properties (pH, electron transference, heat transference, change of potential or mass, variation of optical). This system transforms the response of the recognition element into an electronic signal indicative of the presence of the analyte under study or proportional to its concentration in the sample. The action of the biochemical transducer over the applied system (enzyme-catalyzed reaction) results in the change of a physical property or in the beginning of a process (electron flux originated by a redox reaction), which is sensed and converted into an electrical signal by the physical transducer (electrode under constant potential) (Guilbault et al., 2004; Rogers, 2006; Velasco-Garcia and Mottram, 2003).

Biosensors currently represent powerful tools for analysis with numerous applications in the food industry, mainly in biotechnological instruments. The most important characteristics of these devices to be competitive with other technologies in the food industry are their specificity, high sensitivity, short response time, capacity to be incorporated into integrated systems, the facility to automate them, capacity to work in real time, versatility, low production cost, possibility of regeneration, and simplicity involved in constructing the devices (Mello et al., 2010). The major advantages of biosensors over other techniques are their ability to operate in natural samples with little or no pretreatment as well as their ability for online performance in order to facilitate the quality control of products during processing, which is of great importance in the food industry (Apetrei et al., 2011; Ghasemi-Varnamkhasti et al., 2012). The traditional methods to identify food contaminants include physicochemical, serological, and biological; however, many of these require large quantities of prepared samples, long analysis time, and lack sufficient sensitivity and selectivity.

The use of biosensors based on immobilized enzymes solves many problems such as enzyme loss (especially if expensive), maintenance of enzyme stability, and shelf life of the biosensor. It also reduces the time of enzymatic response and offers disposable devices that can be easily used in stationary or flow systems. Among the commercially available enzymes, the most often used biosensors are oxidoreductases, especially glucose oxidase, horseradish peroxidase, and alkaline phosphatase, because they are very stable when catalyzing oxide reduction reactions (Mello and Kubota, 2002).

In many cases, multi-enzymatic chains are employed, where the enzyme that generally recognizes the analyte does not act directly on it but rather interact with some product derived from it.

Industrial applications for biosensors include monitoring fermentation broths or food processing procedures by detecting concentrations of glucose and other fermentative end products. Biosensors can also be used to monitor ethanol formation, e.g., during apple storage in a controlled atmosphere or the development of rot in tubercles like potatoes, or any other application where ethanol accumulation can be associated with loss of quality. Likewise, research has been conducted to analyze the content of some organic acids and sugars as indicators of fruit and vegetable maturity. The high sensitivity of enzymatic biosensors allows the detection of microorganisms such as *Escherichia coli, Salmonella* sp., and *Staphylococcus aureus* and pesticides and herbicides (Killard and Smyth, 2000).

The potential uses of biosensors in food are numerous and each application has its own requirements in terms of concentration of the analyte to be measured, required output precision, the necessary volume of the sample, time required for the analysis, time required to prepare the biosensor or to reuse it, and cleanliness requirements of the system. To evaluate safety, quality, and process control in food, different biosensors have been developed, which are described in Table 9.3 (Ghasemi-Varnamkhasti et al., 2012; Huet et al., 2010; Mello and Kubota, 2002; Moretto et al., 1998; Palmisano et al., 2000; Tsai et al., 2003).

Food safety involves ensuring the production and marketing of safe food, thus ensuring consumer's health. The quantity and types of food additives incorporated into food products are regulated by the legislation of each country, and their detection and quantification are important to

TABLE 9.3

Applications of Enzymatic Biosensors in Food Industry

Applications	Sample Source	Component Analyzed
Food safety	Residues of pesticides	Parathion, propoxur, carbaryl, paraoxon
	Fertilizers	Nitrate, phosphate, nitrite
	Heavy metals	Cadmium, copper, nickel, zinc, mercury
	Antibiotic residues	Aminoglycosides, sulfonamides, quinolones, β-lactams, tetracyclines
Food quality	Grape juice, wine, milk, yogurt	Glucose
	Beer, wine	Ethanol, glycerol
	Milk	Lactose
	Egg yolk, flour, soya	Lecithin
	Freshness indices	Inosine, amines, octapine
Process control	Sugars	Glucose, lactose, fructose
	Alcohols	Ethanol
	Amino acids	Lysine

prevent fraud and malpractice by manufacturers as well as allergies and other adverse effects to certain population groups. Based on this need, biosensors are used to detect xenobiotic substances, i.e., substances external to the food product, such as additives and pesticides and components of the food itself like toxins of various sources. The advent of multisensor systems such as electronic and bioelectronic tongues (or array of biosensors) has shown a bright future for the quality enhancement in food industries, and many researchers have worked on the application of such systems instead of traditional systems (Apetrei et al., 2011; Del Valle, 2010; Escuder-Gilabert and Peris, 2010).

9.9 Summary and Future Prospects

The food industry is benefiting from different applications of enzymes including major advances in the development of enzymatic biosensors with different transduction systems that can be applied in such areas as food safety, quality, and process control, being focused mainly on determining composition, contamination of primary materials, and processed foods. In the area of food safety, enzymatic biosensors allow to identify the presence of highly toxic organic contaminants and anti-nutritional elements that affect the food chain, either accidentally or intentionally. Demands for sensitivity, specificity, speed, and accuracy of analytical measurements have generated considerable interest in developing biosensors as diagnostic tools in the food industry.

References

Ageitos, J.M., Vallejo, J.A., Paza, M., and Villa, T.G. 2006. Fluorescein thiocarbamoyl-kappa-casein assay for the specific testing of milk-clotting proteases. *Journal of Dairy Science*, 89: 3770–3777.

Ahhmed, A.M., Kawahara, S., Ohta, K., Nakade, K., Soeda, T., and Muguruma, M. 2007. Differentiation in improvements of gel strength in chicken and beef sausages induced by transglutaminase. *Meat Science*, 76: 455–462.

Anjani, K., Kailasapathy, K., and Phillips, M. 2007. Microencapsulation of enzymes for potential application in acceleration of cheese ripening. *International Dairy Journal*, 17: 79–86.

Apetrei, C., Alessio, P., Constantino, C.J.L. et al. 2011. Biomimetic biosensor based on lipidic layers containing tyrosinase and lutetium bisphthalocyanine for the detection of antioxidants. *Biosensors and Bioelectronics*, 26: 2513–2519.

Armada, L., Fernández, E., and Falqué, E. 2010. Influence of several enzymatic treatments on aromatic composition of white wines. *LWT-Food Science Technology*, 43: 1517–1525.

Bautista-Ortin, A.B., Martinez-Cutillas, A., Ros-Garcia, J.M., López-Roca, J.M., and Gómez-Plaza, E. 2005. Improving colour extraction and stability in red wines: The use of maceration enzymes and enological tannins. *International Journal of Food Science and Technology*, 40: 867–878.

Benejam, W., Steffolani, M.E., and León, A.E. 2009. Use of enzyme to improve the technological quality of a panettone like baked product. *International Journal of Food Science and Technology*, 44: 2431–2437.

Bhalla, T.C. and Chatanta, D.K. 2000. Application of enzymes in food processing. In: *Food Processing: Biotechnological Applications*, eds. S.S. Marwaha and J.K. Arora, pp. 123–142, Asiatech Publishers Inc., New Delhi, India.

Bhat, M.K. 2000. Research review paper cellulases and related enzymes in biotechnology. *Biotechnology Advances*, 18: 355–383.

Blanco, R.M., Terreros, P., Munoz, N., and Serra, E. 2007. Ethanol improves lipase immobilization on a hydrophobic support. *Journal of Molecular Catalysis B: Enzymtic*, 47: 13–20.

Bonet, A., Rosell, C.M., Pérez-Munuera, I., and Hernando, I. 2007. Rebuilding gluten network of damaged wheat by means of glucose oxidase treatment. *Journal of the Science of Food Agriculture*, 87: 1301–1307.

Brena, B.M. and Batista-Viera, F. 2006. Immobilization of enzymes and cells. In: *Immobilization of Enzymes on Electrodes*, ed. J.M. Guisan, pp. 15–30, Humana Press Inc., Totowa, NJ.

Bruchmann, A. and Fauveau, C. 2010. Enzymes in potable alcohol and wine production. In: *Enzymes in Food Technology*, eds. R.J. Whitehurst and M. van Oort, pp. 195–207, Blackwell Publishing Ltd, Oxford, U.K.

Butt, M.S., Tahir-Nadeem, M., Ahmad, Z., and Sultan, M.T. 2008. Xylanases and their applications in baking industry. *Food Technology and Biotechnology*, 46: 22–31.

Caballero, P.A., Gómez, M., and Rosell, C.M. 2007. Improvement of dough rheology, bread quality and bread shelf-life by enzymes combination. *Journal Food Engineering*, 81: 42–53.

Chazarra, S., Sidrach, L., López-Molina, D., and Rodríguez-López, J.N. 2007. Characterization of the milk-clotting properties of extracts from artichoke (*Cynara scolymus*, L.) flowers. *International Dairy Journal*, 17: 1393–1400.

Cheynier, V., Dueñas-Paton, M., Salas, E. et al. 2006. Structure and properties of wine pigments and tannins. *American Journal of Enology and Viticulture*, 57: 298–305.

Choi, M.M.F. 2004. Progress in enzyme-based biosensors using optical transducers. *Microchimica Acta*, 148: 107–132.

Chopra, H.K., Panesar, P.S., Panesar, B., Marwaha, S.S., Dua, A., and Kennedy, J.F. 2010. Immobilized enzymes in food processing. In: *Enzymes in Food Processing: Fundamentals and Potential Applications*, eds. P.S. Panesar, S.S. Marwaha, and H.K. Chopra, pp. 261–270, I.K. International Publishing House Pvt. Ltd., New Delhi, India.

Couri, S., Park, Y., Pastore, G., and Domingos, A. 2008. Enzimas na produção de alimentos e bebidas. In: *Enzimas em Biotecnologia: Produção, Aplicação e Mercado*, eds. E.P.S. Bon, M.A. Ferrara, and M.L. Corvo, pp. 153–177, Interciência Ltda, Rio de Janeiro, Brazil.

Danial, E.N., Elnashar, M.M., and Awad, G.E.A. 2010. Immobilized inulinase on grafted alginate beads prepared by the one-step and the two-steps methods. *Industrial and Engineering Chemistry Research*, 49: 3120–3125.

De Lima, C.J.B., Cortezi, M., Lovaglio, R.B., Ribeiro, E.J., Contiero, J., and De Araújo, E.H. 2008. Production of rennet in submerged fermentation with the filamentous fungus *Mucor miehei* NRRL 3420. *World Applied Sciences Journal*, 4: 578–585.

Del Valle, M. 2010. Electronic tongues employing electrochemical sensors. *Electroanalysis*, 22: 1539–1555.

Demir, N., Acar, J., Sarioglu, K., and Mutlu, M. 2001. The use of commercial pectinase in fruit juice industry, part III: Optimization of enzymatic liquefaction of carrot pulp by using immobilized commercial pectinase. *Journal of Food Engineering*, 47: 275–280.

Dodge, T. 2010. Enzymes in food technology. In: *Production of Industrial Enzymes*, eds. R.J. Whitehurst and M. van Oort, pp. 44–56, Blackwell Publishing Ltd, Oxford, U.K.

El-Bendary, M.A., Moharam, M.E., and Ali, T.H. 2007. Purification and characterization of milk clotting enzyme produced by *Bacillus sphaericus*. *Journal of Applied Sciences Research*, 3: 695–699.

Escuder-Gilabert, L. and Peris, M. 2010. Review: Highlights in recent applications of electronic tongues in food analysis. *Analytica Chimica Acta*, 665: 15–25.

Fernández, M., Úbeda, J.F., and Briones, A.I. 2000. Typing of non *Saccharomyces* yeasts with enzymatic activities of interest in wine-making. *International Journal of Food Microbiology*, 59: 29–36.

Fogle, D.R., Plimpton, R.F., Ockerman, H.W., Jarenback, L., and Persson, T. 1982. Tenderization of beef: Effect of enzyme, enzyme level, and cooking method. *Journal of Food Science*, 47: 1113–1118.

Ghasemi-Varnamkhasti, M., Rodríguez-Méndez, M.L., Mohtasebi, S.S. et al. 2012. Monitoring the aging of beers using a bioelectronic tongue. *Food Control*, 25: 216–224.

Grosová, Z., Rosenberg, M., and Rebroš, M. 2008. Perspectives and applications of immobilized β-galactosidase in food industry-a review. *Czech Journal of Food Science*, 26: 1–14.

Guérin, L., Sutter, D.-H., Demois, A., Chereau, M., and Trandafir, G. 2009. Determination of activity profiles of the main commercial enzyme preparation used in winemaking. *American Journal of Enology and Viticulture*, 60: 322–331.

Guilbault, G.G., Pravda, M., Kreuzer, M., and O'Sullivan, C.K. 2004. Biosensors-42 years and counting. *Analytical Letters*, 37: 1481–1496.

Gupta, R., Gigras, P., Mohapatra, H., Goswami, V.K., and Chauman, B. 2003. Microbial α-amylase: A biotechnological perspective. *Process Biochemistry*, 38: 1599–1616.

Huet, A.-C., Fodey, T., Haughey, S.A., Weigel, S., Elliott, C., and Delahaut, P. 2010. Advances in biosensor-based analysis for antimicrobial residues in foods. *Trends in Analytical Chemistry*, 29: 1281–1294.

Iranzo, J.F.U., Brionez-Pérez, A.I., and Izquierdo-Canãs, P.M. 1998. Study of the oenological characteristics and enzymatics activities of wine yeasts. *Food Microbiology*, 15: 399–406.

Jelen, P. and Tossavainen, O. 2003. Low lactose and lactose-free milk and dairy products—Prospects, technologies and applications. *Australian Journal of Dairy Technology*, 58: 161–165.

Jovanovic, S., Barac, M., Macej, O., Vucic, T., and Lacnjevac, C. 2007. SDS-PAGE analysis of soluble proteins in reconstituted milk exposed to different heat treatments. *Sensors*, 7: 371–383.

Junqueira, R.M., Rocha, F., Moreira, M.A., and Castro, I.A. 2003. Effect of proofing time and wheat flour strength on bleaching, sensory characteristics, and volume of French breads with added soybean lipoxygenase. *Cereal Chemistry*, 84: 443–449.

Jurado, E., Camacho, F., Luzón, G., and Vicaria, J.M. 2002. A new kinetic model proposed for enzymatic hydrolysis of lactose by a β-galactosidase from *Kluyveromyces fragilis*. *Enzyme and Microbial Technology*, 31: 300–309.

Kailasapathy, K. and Lam, S.H. 2005. Application of encapsulated enzymes to accelerate cheese ripening. *International Dairy Journal*, 15: 929–939.

Kashyap, D.R., Vohra, P.K., and Tewari, R. 2001. Application of pectinases in the commercial sector: A review. *Bioresource Technology*, 77: 215–227.

Katsaros, G.I., Katapodis, P., and Taoukis, P.S. 2009. High hydrostatic pressure inactivation kinetics of the plant proteases ficin and papain. *Journal Food Engineering*, 91: 42–48.

Kaur, G., Kumar, S., and Satyanarayana, T. 2004. Production, characterization and application of a thermostable polygalacturonase of a thermophilic mould *Sporotrichum thermophile* Apinis. *Bioresource Technology*, 94: 239–243.

Killard, A.J. and Smyth, M.R. 2000. Separation-free electrochemical immunosensor strategies. *Analytical Letters*, 33: 1451–1465.

Lehmann, M. and Wyss, M. 2001. Engineering proteins for thermostability: The use of sequence alignments versus rational design and directed evolution. *Current Opinion in Biotechnology*, 12: 371–375.

Law, B.A. 2002. The nature of enzymes and their action in foods. In: *Enzymes in Food Technology*, eds. R.J. Whitehurst and B.A. Law, pp. 1–18, Academic Press, Boca Raton, FL.

Lim, S.-J. and Lee, K.-J. 2009. Partial replacement of fish meal by cottonseed meal and soybean meal with iron and phytase supplementation for parrot fish *Oplegnathus fasciatus*. *Aquaculture*, 290: 283–289.

Linke, D. and Berger, R.G. 2011. Foaming of proteins: New prospects for enzyme purification processes. *Journal of Biotechnology*, 152: 125–131.

Llorente, B.E., Brutti, C.B., and Caffini, N.O. 2004. Purification and characterization of a milk-clotting aspartic proteinase from globe artichoke (*Cynara scolymus* L.). *Journal of Agricultural and Food Chemistry*, 52: 8182–8189.

Lomer, M.C.E., Parkes, G.C., and Sanderson, J.D. 2007. Review article: Lactose intolerance in clinical practice-myths and realities. *Alimentary Pharmacology and Therapeutics*, 27: 93–103.

Low, Y.H., Agboola, S., Zhao, J., and Lim, M.Y. 2006. Clotting and proteolytic properties of plant coagulants in regular and ultrafiltered bovine skim milk. *International Dairy Journal*, 16: 335–343.

Mariotti, M.P., Yamanaka, H., Araujo, A.R., and Trevisan, H.C. 2008. Hydrolysis of whey lactose by immobilized β-Galactosidase. *Brazilian Archives of Biology and Technology*, 51: 1233–1240.

Melendo, J.A., Beltrán, J.A., Jaime, I., Sancho, R., and Roncalés, P. 1996. Limited proteolysis of myofibrillar proteins by bromelain decreases toughness of coarse dry sausage. *Food Chemistry*, 57: 429–433.

Mello, L.D., Ferreira, D.C.M., and Kubota, L.Y. 2010. Enzymes as analytical tools in food processing. In: *Enzymes in Food Processing: Fundamentals and Potential Applications*, eds. P.S. Panesar, S.S. Marwaha, and H.K. Chopra, pp. 303–325, I.K. International Publishing House Pvt. Ltd., New Delhi, India.

Mello, L.D. and Kubota, L.T. 2002. Review of the use of biosensors as analytical tools in the food and drink industries. *Food Chemistry*, 77: 237–456.

Miklos, R., Xu, X., and Lametsch, R. 2011. Application of pork fat diacylglycerols in meat emulsions. *Meat Science*, 87: 202–205.

Moayedallaie, S., Mirzaei, M., and Paterson, J. 2010. Bread improvers: Comparison of a range of lipases with a traditional emulsifier. *Food Chemistry*, 122: 495–499.

Mojsov, K., Ziberoski, J., Bozinovic, Z., and Petreska, M. 2011. Comparison of effects of three commercial pectolytic enzyme preparations in white winemaking. *Applied Technologies and Innovations*, 4: 34–38.

Moretto, L.M., Ugo, P., Zanata, M., Guerriero, P., and Martin, C.R. 1998. Nitrate biosensor based on the ultrathin-film composite membrane concept. *Analytical Chemistry*, 70: 2163–2166.

Murty, V.R., Bhat, J., and Muniswaran, P.K.A. 2002. Hydrolysis of oils by using immobilized lipase enzyme: A review. *Biotechnology and Bioprocess Engineering*, 7: 57–66.

Naveena, B.M. and Mendiratta, S.K. 2004. The tenderization of buffalo meat using ginger extract. *Journal of Muscle Foods*, 15: 235–244.

Naz, S., Masud, T., and Nawaz, M.A. 2009. Characterization of milk coagulating properties from the extract of *Withania coagulans*. *International Journal of Dairy Technology*, 62: 315–320.

Nunes, P.A., Pires-Cabral P., and Ferreira-Dias, S. 2011. Production of olive oil enriched with medium chain fatty acids catalysed by commercial immobilized lipases. *Food Chemistry*, 127: 993–998.

van Oort, M. 2010a. Enzymes in food technology-introduction. In: *Enzymes in Food Technology*, eds. R.J. Whitehurst and M. van Oort, pp. 1–16, Blackwell Publishing Ltd., Oxford, U.K.

van Oort, M. 2010b. Enzymes in bread making. In: *Enzymes in Food Technology*, eds. R.J. Whitehurst and M. van Oort, pp. 103–35, Blackwell Publishing Ltd., Oxford, U.K.

Pal, A. and Khanum, F. 2011. Covalent immobilization of xylanase on glutaraldehyde activated alginate beads using response surface methodology: Characterization of immobilized enzyme. *Process Biochemistry*, 46: 1315–1322.

Palmisano, F., Rizzi, R., Centonze, D., and Zambonin, P.G. 2000. Simultaneous monitoring of glucose and lactate by an interference and cross-talk free dual electrode amperometric biosensor based on electropolymerized thin films. *Biosensors and Bioelectronics*, 15: 531–539.

Panesar, P.S., Chopra, H.K., and Marwaha, S.S. 2010a. Fundamentals of enzymes. In: *Enzymes in Food Processing: Fundamentals and Potential Applications*, eds. P.S. Panesar, S.S. Marwaha, and H.K. Chopra, pp. 1–43, I.K. International Publishing House Pvt. Ltd., New Delhi, India.

Panesar, P.S., Kumari, S., and Panesar, R. 2010b. Potential applications of immobilized β-galactosidase in food processing industries. *Enzyme Research*, 10: 1–16.

Panesar, P.S., Panesar, R., Singh, R.S., Kennedy, J.F., and Kumar, H. 2006. Microbial production, immobilization and applications of β-D-galactosidase. *Journal of Chemical Technology and Biotechnology*, 81: 530–543.

Playne, M.J., Bennett, L.E., and Smithers, G.W. 2003. Functional dairy foods and ingredients. *Australian Journal of Dairy Technology*, 58: 242–264.

Purhagen, J.K., Sjöö, M.E., and Eliasson, A.-C. 2011. Starch affecting anti-staling agents and their function in freestanding and pan-baked bread. *Food Hydrocolloids*, 25: 1656–1666.

Rajagopalan, G. and Krishnan, C. 2008. Immobilization of malto-oligosaccharide forming α-amylase from *Bacillus subtilis* KCC103: Properties and application in starch hydrolysis. *Journal of Chemical Technology and Biotechnology*, 83: 1511–1517.

Revilla, I. and Gonzalez-SanJose, M.L. 2003. Compositional changes during the storage of red, wines treated with pectolytic enzymes: Low molecular-weight phenols and flavan-3-ol derivative levels. *Food Chemistry*, 80: 205–214.

Ribéreau-Gayon, P., Dubourdieu, D., Donèche, B., and Lonvaud, A. 2003. *Tratado de Enologia*. Microbiologia del vino Vinificaciones, Buenos Aires, Argentina.

Rogers, K.R. 2006. Recent advances in biosensor techniques for environmental monitoring. *Analytica Chimica Acta*, 568: 222–231.

Roldán, A., Palacios, V., Peñatez, X., Benitez, T., and Pérez, L. 2006. Use of *Trichoderma* enzymatic extracts on vinification of Palomino fino grapes in the Sherry region. *Journal of Food Engineering*, 75: 375–382.

Romero de Ávila, M.D., Ordónez, J.A., De la Hoz, L., Herrero, A.M., and Cambero, M.I. 2010. Microbial transglutaminase for cold-set binding of unsalted/salted pork models and restructured dry ham. *Meat Science*, 84: 747–754.

Romero-Cascales, I., Fernández-Fernández, J., Roz-García, J.M., Lópezz-Roca, J.M., and Gómez-Plaza, E. 2008. Characterisation of the main enzymatic activities present in six commercial macerating enzymes and their effects on extracting colour during winemaking of Monastrell grapes. *International Journal of Food Science and Technology*, 43: 1295–1305.

Salas, E., Fulcrand, H., Meudec, E., and Cheynier, V. 2003. Reactions of anthocyanins and tannins in model solutions. *Journal of Agricultural and Food Chemistry*, 51: 7951–7961.

Sandri, I.G., Fontana, R.C., Barfknecht, D.M., and Silveira, M.M. 2011. Clarification of fruit juices by fungal pectinases. *LWT-Food Science Technology*, 44: 2217–2222.

Selle, P.H., Cowieson, A.J., and Ravindran, V. 2009. Consequences of calcium interactions with phytate and phytase for poultry and pigs. *Livestock Science*, 124: 126–141.

Shantilatha, P., Varma, S., and Mitra, C.K. 2003. Designing a simple biosensor. In: *Advances in Biosensors: Perspectives in Biosensors*, eds. B.D. Malhotra and A.P.F. Turner, pp. 3–10, Elsevier Science B.V., Amsterdam, the Netherlands.

Sharma, R., Chisti, Y., and Banerjee, U.C. 2001. Production, purification, characterization and applications of lipases. *Biotechnology Advances*, 19: 627–662.

Sheldon, R.A. 2007. Enzyme immobilization: The quest for optimum performance. *Advanced Synthesis and Catalysis*, 349: 1289–1307.

Shieh, C.-J., Thi, L.-A.P., and Shih, I.-L. 2009. Milk-clotting enzymes produced by culture of *Bacillus subtilis* natto. *Biochemical Engineering Journal*, 43: 85–91.

Solomons, N.W. 2002. Fermentation, fermented foods and lactose intolerance. *European Journal of Clinical Nutrition*, 56: 50–55.

Tsai, H.-C., Doong, R.-A., Chiang, H.-C., and Chen, K.-T. 2003. Sol-gel derived urease-based optical biosensor for the rapid determination of heavy metals. *Analytica Chimica Acta*, 481: 75–84.

Tumturk, H., Demirel, G., Altinok, H., Aksoy, S., and Hasirci, N. 2008. Immobilization of glucose isomerase in surface-modified alginate gel beads. *Journal of Food Biochemistry*, 32: 234–246.

Velasco-Garcia, M.N. and Mottram, T. 2003. Biosensor technology addressing agricultural problems. *Biosystems Engineering*, 84: 1–12.

Vieceli, J., Müllegger, J., and Tehrani, A. 2006. Computer-assisted design of industrial enzymes: The resurgence of rational design and in silico mutagenesis. *Industrial Biotechnology*, 2: 303–308.

Wong, D.W.S. 2002. Recent advances in enzyme development. In: *Handbook of Food Enzymology*, eds. J.R. Whitaker, A.G.J. Voragen, and D.W.S. Wong, pp. 379–387, Marcel Dekker, New York.

Yu, P.-J. and Chou, C.-C. 2005. Factors affecting the growth and production of milk-clotting enzyme by *Amylomyces rouxii* in rice liquid medium. *Food Technology and Biotechnology*, 43: 283–288.

10

Production of Polysaccharides

Gordon A. Morris and Stephen E. Harding

CONTENTS

10.1 Introduction 355
10.2 Xanthan 363
10.3 Gellan 366
10.4 Cyanobacterial Polysaccharides 368
10.5 Polysaccharides from Lactic Acid Bacteria 369
10.6 Hyaluronic Acid 370
10.7 Other Important Bacterial Polysaccharides 372
  10.7.1 Xylinan 372
  10.7.2 Dextrans 373
  10.7.3 Pullulan 374
  10.7.4 Scleroglucan and Schizophyllan 375
10.8 Polysaccharide Vaccines 375
10.9 Summary and Future Prospects 377
Acknowledgment 377
References 377

10.1 Introduction

Over the last 20 years, there has been an expanding interest in the production of polysaccharides produced extracellularly by microorganisms for food, pharmaceutical, and medical applications including vaccines (Jones, 2005; Freitas et al., 2011). The wide range of their usefulness derives from a great diversity of structural, conformational, and functional properties (Morris and Harding, 2009; Pereira et al., 2009; Freitas et al., 2011) even though they are built up from very similar building blocks: pyranose (six-membered) or furanose (five-membered) carbohydrate ring structures (Table 10.1).

Polysaccharides made by microorganisms are secreted from the cell to form a layer over the surface of the organism (Sutherland, 1999), often of substantial depth in comparison with the cell dimensions. Because of their

TABLE 10.1

Commercial Bacterial Exopolysaccharides: Structures, Properties, Applications, and Market

Polysaccharide	Structure	Charge
Xanthan	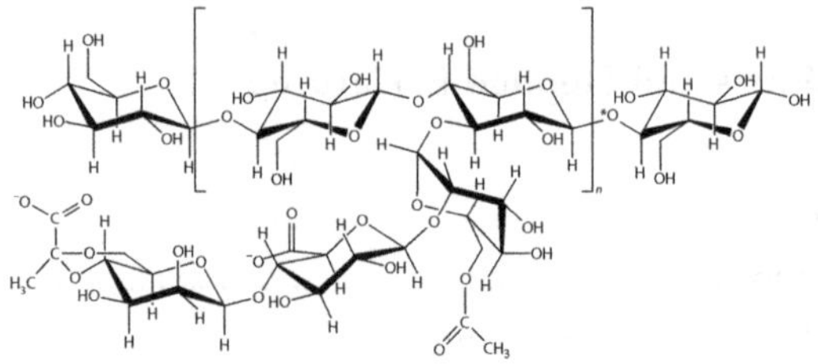	Negative
Gellan	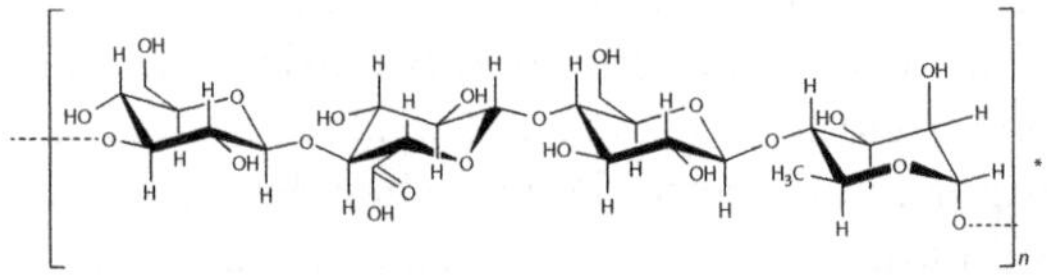	Negative
Cyanobacterial polysaccharides	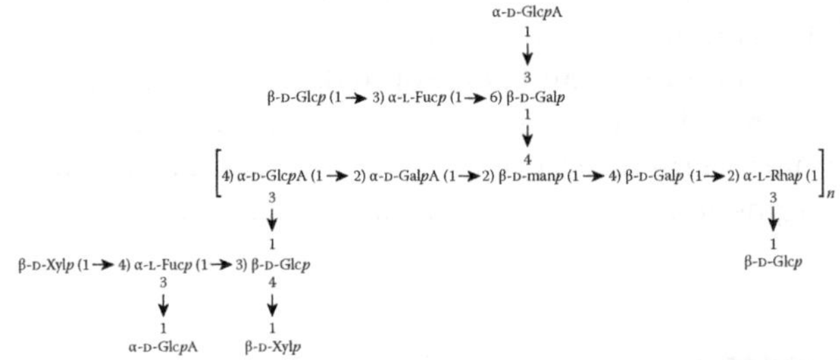 α-D-GlcpA 1 ↓ 3 β-D-Glcp (1→ 3) α-L-Fucp (1→ 6) β-D-Galp 1 ↓ 4 [4) α-D-GlcpA (1→ 2) α-D-GalpA (1→2) β-D-manp (1→ 4) β-D-Galp (1→ 2) α-L-Rhap (1]$_n$ 3 ↓ 1 (on α-D-GlcpA): β-D-Xylp (1→ 4) α-L-Fucp (1→ 3) β-D-Glcp 3 ↓ 1 (on α-L-Fucp): α-D-GlcpA 4 ↓ 1 (on β-D-Glcp): β-D-Xylp 3 ↓ 1 (on α-L-Rhap): β-D-Glcp	Negative
Polysaccharides from lactic acid bacteria	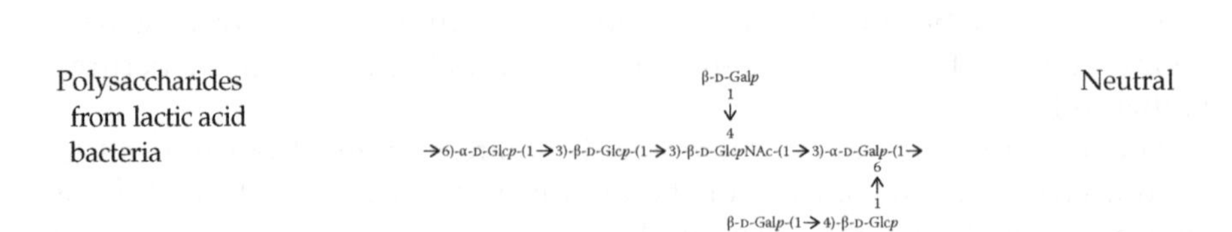	Neutral

Molecular Weight (g/mol)	Properties	Applications	Market (Metric Tons)	Price (US$)/kg	References
$(2–50) \times 10^6$	Hydrocolloid—high viscosity yield at low shear rates even at low concentrations; stability over wide temperature, pH, and salt concentration ranges	Foods Petroleum industry Pharmaceuticals Cosmetics and personal care products Agriculture	96,000	3–5	Dea et al. (1977); Dhami et al. (1995); Morris et al. (1977, 2001); Sworn (2009)
5.0×10^5	Hydrocolloid—stability over wide pH range Gelling capacity Thermoreversible gels	Foods Pet food Pharmaceuticals Research: agar substitute and gel electrophoresis	NA	55–66	Milas et al. (1990); Sutherland (1999); Valli and Clark (2010)
8×10^4 to 2×10^6	Shear-thinning behavior	Emulsifiers Viscosifiers Medicines Bioflocculants Heavy metal removal	NA	NA	Bar-or and Shilo (1987); Choi et al. (1998); De Philippis et al. (2001, 2007); Falchini et al. (1996); Li et al. (2001a,b); Mazor et al. (1996); Moreno et al. (2000); Morris et al. (2001); Pereira et al. (2009); Vincente-García et al. (2004)
3×10^4 to 3×10^6	GRAS	Viscosifiers Prebiotic	NA	NA	Laws et al. (2001, 2008); Leivers (2011); Xiao-Mei et al. (2004)

(continued)

TABLE 10.1 (continued)

Commercial Bacterial Exopolysaccharides: Structures, Properties, Applications, and Market

Polysaccharide	Structure	Charge
Xylinan	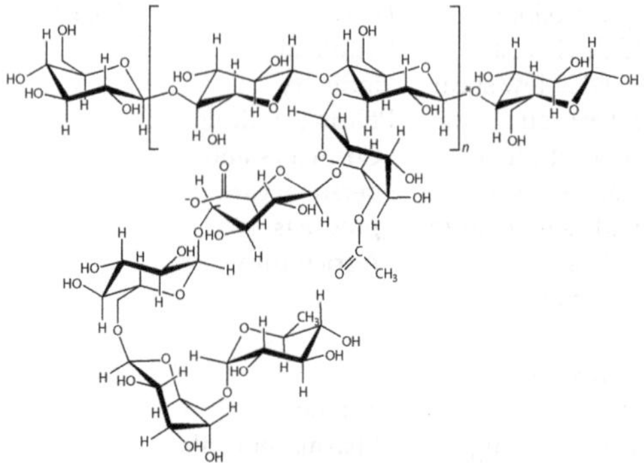	Negative
Pullulan	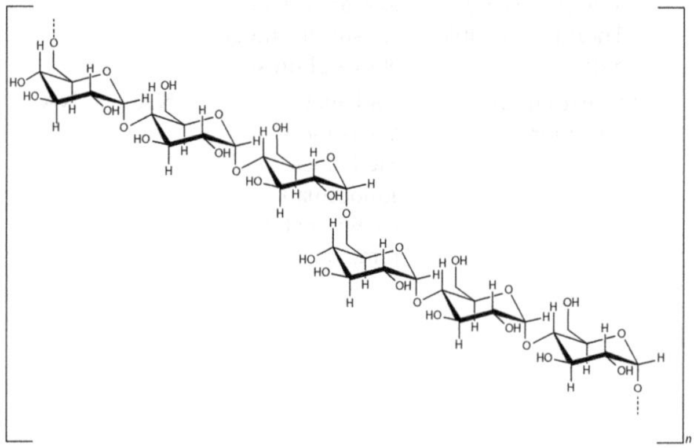	Neutral
Dextran	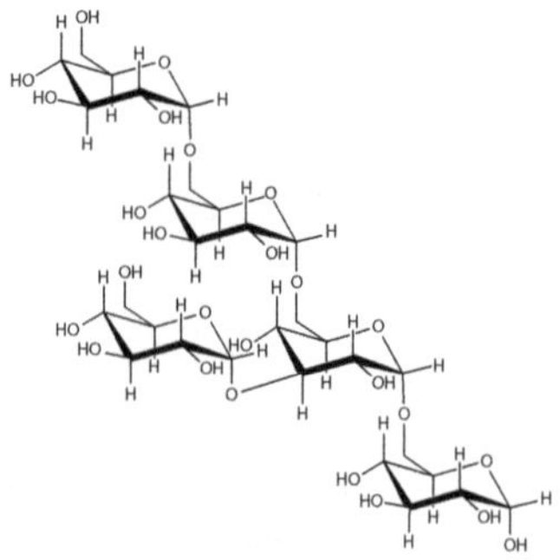	Neutral

Molecular Weight (g/mol)	Properties	Applications	Market (Metric Tons)	Price (US$)/ kg	References
~2.5×10^6	Synergistic gels with konjac glucomannan and galactomannans	Food industry, viscosifier, and gelling agent It is also a component in Nata, a sweet confectionery popular in Japan and the Philippines.	NA	NA	Berth et al. (1996a,b); Griffin et al. (1997a,b); Harding et al. (1996); Morris et al. (1989); Ridout et al. (1998a,b)
10^3–10^6	Nontoxic Odorless Tasteless	Starch replacement (not digested by mammalian amylases) Denture adhesive	2,000	9–15	Israilides et al. (1999); Kawahara et al. (1984); Oku et al. (1979); Singh et al. (2008)
10^6 – 10^9	Nonionic Good stability Newtonian fluid behavior	Foods Pharmaceutical industry: blood volume expander Chromatographic media	2,000	3–5	Jiang et al. (2004); Singh et al. (2008)

(*continued*)

TABLE 10.1 (continued)

Commercial Bacterial Exopolysaccharides: Structures, Properties, Applications, and Market

Polysaccharide	Structure	Charge
Scleroglucan and schizophyllan	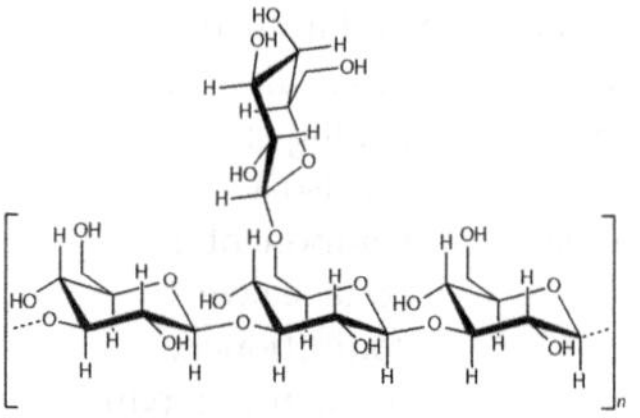	Neutral
Hyaluronic acid	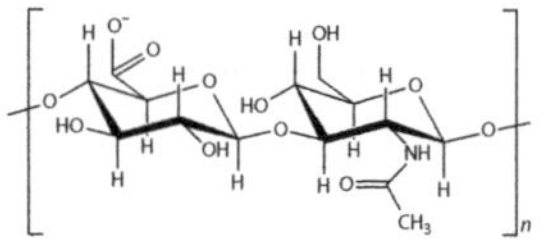	Negative
Polysaccharide vaccines	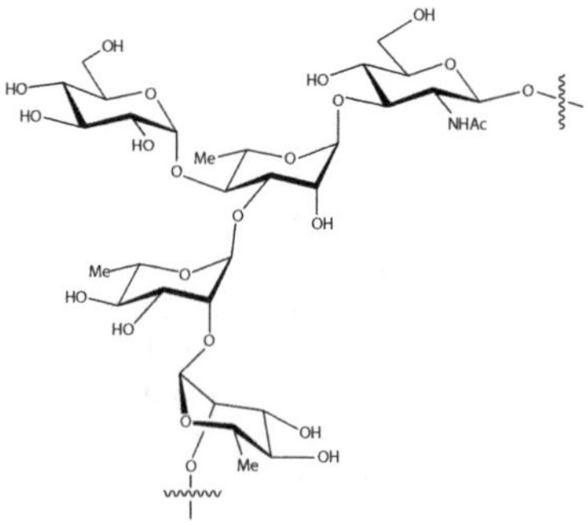	Neutral/ negative

Source: Adapted from Freitas, F. et al., *Trends Biotechnol.*, 29, 388, 2011.

Molecular Weight (g/mol)	Properties	Applications	Market (Metric Tons)	Price (US$)/kg	References
5×10^5	Hydrogen-bond-stabilized triple helix Stimulate an immune response against tumor cells	Skin and hair care Cosmetic industry	NA	NA	Biver et al. (1986); Kitamura et al. (1994); Kurachi et al. (1990); Yanaki et al. (1980)
$1–3 \times 10^6$	Biological activity Highly hydrophilic Biocompatible Non-Newtonian solution even at low concentrations	Medicine Solid culture media	NA	100,000	Armstrong and Johns (1997); Chong et al. (2005); Cleland and Wang (1970); Esposito et al. (2005); Hokputsa et al. (2003); Kogan et al. (2007)
1×10^5 to 1×10^7	Immunogenic	Medicine	—	Variable	Jones (2005); Lindenberg (1990); Phalipon et al. (2006); Robbins et al. (2011)

position, they are characterized as exopolysaccharides, to distinguish them from any polysaccharides that might be found within the cell. Their functions are thought to be mainly protective, either as a general physical barrier preventing the access of harmful substances or more specific as a way of binding and neutralizing bacteriophages. In appropriate environments, they may prevent dehydration.

They may also prevent phagocytosis by other microorganisms or the cells of the immune system. The capsular polysaccharides are often highly immunogenic and may have evolved their unusual diversity as a way of avoiding antibody responses: advantage of this feature can be taken in the development of vaccines (Jones, 2005). They also have a role in the adhesion and penetration of the host. Since in the case of plants this will involve interaction with polysaccharide structures in the cell walls, there are clearly possibilities of specific interactions.

Some secreted polysaccharides can be involved in pathogenicity. *Pseudomonas aeruginosa*, commonly found in respiratory tract infections, produces alginate that contributes to blockage in the respiratory tract and leads to further infection, while similar blockages of phloem in plants have been described (Griffin et al., 1996a). However, the secreted polysaccharides themselves present virtually no known toxicity problems and many are harvestable at low cost in large quantities, making them attractive for biotechnological use.

Microbes are known to produce nearly all the major plant polysaccharides such as glucans, alginate-like materials, and even cellulose—as well as the complex bacteria-specific materials. Genetic manipulation of bacteria has been studied for longer and is in general much easier than that for higher organisms so that they are an obvious target for both manipulation of biosynthetic pathways and the expression of heterologous genes to produce especially desirable enzymes. Polysaccharides are not of course under the direct control of genetic material in the way that enzymes are and must be approached indirectly by manipulating the biosynthetic or degradation pathways by way of the enzymes responsible.

This chapter is focused on the production, properties, and biomodifications of a selected range of important microbial polysaccharides, as it is impossible to cover each biopolymer comprehensively. The focus of this chapter is primarily on the commercial importance on xanthan, but it also considers a "xanthan-like" exopolysaccharide—gellan, and the increasingly important group of cyanobacterial polysaccharides, some of which are also "xanthan-like"—as well as lactic acid bacteria polysaccharides, dextrans, xylinan, pullulan, scleroglucan, schizophyllan, and microbial hyaluronic acid (HA) as representative examples. The contents of this chapter also cover the current and future application of exopolysaccharides as vaccines against serious disease such as meningitis in infants.

10.2 Xanthan

Xanthan is produced by *Xanthomonas campestris*—the bacterium responsible for cabbage blight (Leach et al., 1957; Lilly et al., 1958)—and is grown largely on glucose, which it converts with high efficiency (80%) to the xanthan gum. Typically, for a fermentation product, the raw material costs are small, but the major cost involves the cost of recovering the gum from the culture medium. The current worldwide production of xanthan gum is around 90,000 ton a year (Freitas et al., 2011). It has a β(1→4)-linked glucan main chain with alternating residues substituted on the third position with a trisaccharide chain containing two mannose and one glucuronic acid residue (Figure 10.1) and thus a charged polymer. Some of the mannose residues may also carry acetyl groups (Morris et al., 1977; Donot et al., 2012). It is useful because it forms relatively rigid rodlike structures in solution at ambient temperatures, though they convert to the random configuration on heating (Morris et al., 1977). These rods are able to align themselves—rather like agarose and the κ- and ι-carrageenans, with the unsubstituted regions of galactomannans, such as guar and locust bean gum, to produce fairly rigid mixed gels with applications in food manufacture (Dea et al., 1977; Morris et al., 1977; Ross-Murphy et al., 1983; Dea, 1989; Goycoolea et al., 1994, 1995a,b; Morris and Foster, 1994; Fitzsimons et al., 2007; Sworn, 2009).

The primary structure (as worked out by sequential degradation and methylation analysis) consists of a repeating unit of β-D-glucose, and in this respect, it resembles cellulose, with the important difference that C3 of every alternate Glc*p* residue is substituted by a negatively charged trisaccharide involving mannose and acetylated mannose with some of the terminal β-D-Man*p* residues substituted at positions C4 and C6 by pyruvate ($CH_3 \cdot CO \cdot COO^-$). Thus, xanthan (except under highly acidic conditions) is a highly charged polyanion. Figure 10.1 shows the structure of a (pyruvalated) alternate repeat section of xanthan.

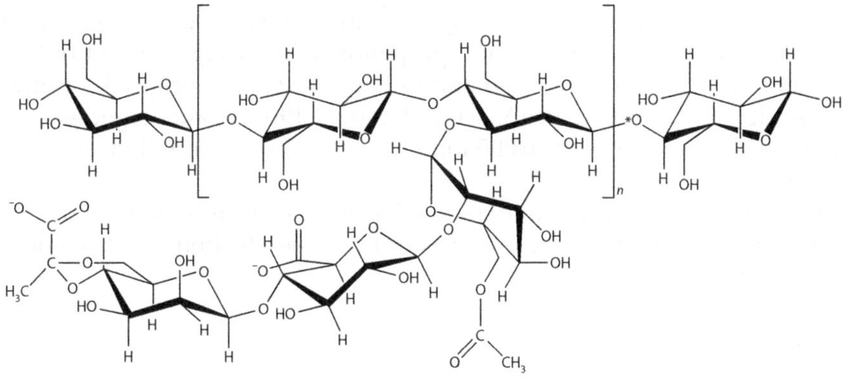

FIGURE 10.1
Xanthan repeat unit showing a trisaccharide side chain with pyruvalated end mannose unit.

X-ray fiber diffraction studies of xanthan (Millane et al., 1989) have shown the presence of some helical structure; therefore, its extended rigid rod conformation is not only due to charge repulsion effects. There is still uncertainty as to whether this helix is a double helix (Holzwarth and Prestridge, 1977; Paradossi and Brant, 1982; Sato et al., 1984; Yevlampieva et al., 1999) or two side-by-side single helices (Morris et al., 1977; Milas and Rinaudo, 1979; Norton et al., 1984). Using a combination of electron microscopy and light scattering, a mass per unit length, M_L, of (1950 ± 250) g/mol·nm was estimated (Stokke et al., 1989; Kitamura et al., 1991). From x-ray fiber diffraction studies, a corresponding value of 950 g/mol·nm was obtained, and a duplex or double-helical structure was inferred for the native structure. High temperatures will reversibly melt this helix to give a more random structure (Kitamura et al., 1991), and as this loss of order occurs independently from the twofold reduction in molecular weight, this has been used as evidence for the side-by-side helix model (Norton et al., 1984; Kitagawa et al., 1985). A recent paper (Lad et al., 2013) demonstrated the importance of the careful interpretation of x-ray patterns in probing xanthan fine structure.

The primary laboratory and commercial fermentation medium has been known for over 50 years for *X. campestris* growth, and xanthan production is a phosphate-buffered (pH ~ 7) broth containing D-glucose (30 g/L) (or sucrose, starch, hydrolyzed starch) NH_4Cl, $MgSO_4$, trace salts with 5 g/L casein (or soybean) hydrolysate. Xanthan is produced through aerobic fermentation at 28°C temperature and the production is further stimulated by the presence of pyruvic, succinic, or other organic acids. The xanthan produced in this way is very similar to the naturally produced xanthan by the microbe living on a cabbage. In the commercial process, the oxygen uptake from the broth is controlled to a rate of 1 mmol/L·min. Under these fermentation conditions, the bacterium is an extremely efficient enzyme mini-factory converting >70% of the substrate (D-glucose or related) to polymeric xanthan. After the completion of the fermentation process, the bacterium is separated by centrifugation and the xanthan thus produced is precipitated with methanol or 2-propanol at 50% weight concentration. The xanthan slurry is then dried and milled for use. The original commercial producer of xanthan was Kelco Ltd. (now CPKelco) and, together with other producers, produces 90,000 ton of xanthan gum annually worldwide (Freitas et al., 2011).

The biosynthetic process undertaken by the bacterium (Sutherland, 1989, 1999) confirmed the processes proposed for the production of other microbial polysaccharides:

1. Substrate uptake
2. Substrate metabolism
3. Polymerization
4. Modification and extrusion

These molecules are extremely large, for example, the weight-average molecular weight, M_w, of xanthan has been found by ultracentrifuge techniques to be as high as 6 million g/mol (Dhami et al., 1995; Morris et al., 2001), and stiff, with a molecular persistence length, L_p, of ~150 nm. Xanthan gives very viscous solutions (intrinsic viscosities, $[\eta]$, of several thousand mL/g) and indeed is one of the largest of the aqueous soluble polysaccharides. The very high viscosity at low concentrations (e.g., at 0.5%, a viscosity of ~1000 cP has been observed at room temperature) makes it ideal as a thickening and suspending agent.

One of the most important commercial properties of xanthan gum is its ability to produce a large increase in the viscosity of a liquid by adding a very small quantity of gum ~0.5%–1% (Sworn, 2009). The viscosity of xanthan gum solutions decreases at higher shear rates (shear thinning or pseudoplasticity). There has been considerable interest in improving the weak gelation characteristics of xanthan by inclusion of galactomannans (locust bean gum and guar gum) or konjac glucomannan into mixtures (Dea et al., 1977; Morris et al., 1977; Ross-Murphy et al., 1983; Dea, 1989; Morris and Foster, 1994; Goycoolea et al., 1995a,b; Fitzsimons et al., 2007; Sworn, 2009). Deacetylating the xanthan side chains seems to enhance the synergistic interactions (Shatwell et al., 1991a,b). Gel-like behavior of xanthan can be also be improved by the addition of calcium ions (Ross-Murphy et al., 1983; Mohammed et al., 2007) and/or by repeated freeze–thaw cycles (Giannouli and Morris, 2003).

The key sites for modification are the first and terminal mannose residues on the trisaccharide side chains (particularly extent of acetylation of the first and pyruvalation of the latter) and the helical backbone by forming noncovalent interactions with galactomannans: these interactions appear to be also affected by the substitutions in the side chains (Shatwell et al., 1991a,b). With respect to the trisaccharide side chains, there are two approaches to modification: changing the physiological conditions of fermentation and the use of different pathovars or strains of *X. campestris*. The pathovars *X. campestris pv. phaseoli* and *X. campestris pv. oryzae* yield virtually acetyl-free and pyruvate-free xanthan gums, respectively.

The other approach (Vanderslice et al., 1989) has been to clone and sequence clusters of xanthan biosynthetic genes and to isolate in sufficient quantity these genetic mutants deficient or defective in one or more of these biosynthetic enzymes to give "polytetramer" (lacking in the terminal mannosyl or pyruvalated mannose group) and "polytrimer" (lacking in the addition of the adjacent glucuronic acid residue). Although yields of 50% polytetramer have been reported, the attempts to produce polytrimer or other xanthan variants have not been encouraging. Furthermore, the polytetramer variant of xanthan does not form synergistic gels with locust bean gum or konjac glucomannan (Foster and Morris, 1994).

Xanthan gum has been approved as food grade by the U.S. Food and Drug Administration for nearly 30 years. This makes it attractive not only as a food product (Sanderson, 1982; Bylaite et al., 2005; Sworn, 2009; Freitas et al.,

2011; Palaniraj and Jayaraman, 2011) but also for use in packaging material in contact with food and also for use in pharmaceutical and products having biomedical applications. Its uses chiefly derive from its solubility in hot and cold water conditions and its very high thickening and suspending potential that in turn derives from the very high viscosity of its suspensions. Despite the high viscosity, xanthan suspensions exhibit high shear thinning, which means that they also flow easily (i.e., good pourability). For food applications, besides high viscosity and thickening and suspending ability, high acid stability of xanthan suspensions adds to their value. This makes it highly popular in sauces, syrups and toppings, and salad dressings. In drinks, the addition of xanthan together with carboxymethylcellulose adds "body" and acts as a thickener to the liquid and assists in uniform distribution of fruit pulp, etc. It is also used to add body to dairy products. The high freeze–thaw stability of xanthan suspensions makes them particularly attractive for the frozen food industry. The high suspending and stability properties are also taken advantage of by the animal feed industry for transporting liquid feeds with added vitamins and other supplements that would otherwise sediment out with transport or storage time. It has also been suggested as an additive to fruit drinks to reduce tooth decay (West et al., 2004). For pharmaceuticals, xanthan has recently been added to the list of hydrophilic matrix carriers, along with chitosan, cellulose ethers, modified starches, and scleroglucan (Melia, 1991). Tablets containing 5% xanthan gum were shown under low-shear conditions to give successful controlled release of acetaminophen into stomach fluid, and tablets containing 20% xanthan successfully carried a high loading (50%) of the drug theophylline. The high suspension stability of xanthan is used in pharmaceutical cream formulations and in barium sulfate preparations. This high cream stability is taken advantage of also by the cosmetic industry, including toothpaste technology facilitating the suspension of ingredients (high viscosity) and the easy brushing onto and off the teeth (high shear thinning). Uniform pigment dispersal along with other ingredients and long-time stability makes xanthan a good base for shampoos.

Just to reinforce the diversity of application, in textiles, the suspension stabilizing property of xanthans makes them ideal for producing sharp prints from dyes with a minimum risk of running for application to textiles and (in conjunction with guar) carpets. Furthermore, xanthan has been reported to exhibit antitumor activity (Takeuchi et al., 2009).

10.3 Gellan

Inspired by the example of xanthan, other bacterial polysaccharides with useful properties have been developed. In the case of gellan, this was the result of a systematic search for a polysaccharide with specific properties

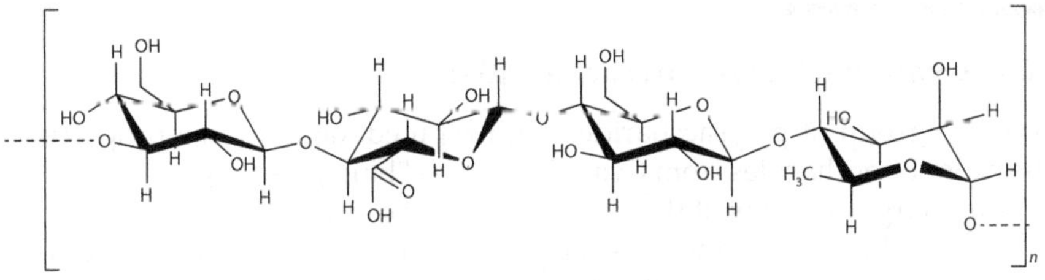

FIGURE 10.2
Repeat unit of gellan in deacetylated form.

followed by the identification of the organism (Valli and Clark, 2009). Gellan is obtained from cultures of *Sphingomonas paucimobilis* (formerly known as *Sphingomonas elodea, Auromonas elodea,* or *Pseudomonas elodea*). It has a linear structure with a repeat unit of a tetrasaccharide (Figure 10.2), each with one carboxyl group and, in the native state, one acetyl group (O'Neill et al., 1983). Due to the presence of the carboxyl group, it is therefore sensitive to calcium levels but has some rheological properties similar to those of xanthan charge density and adopts a double-helical conformation in solution (Milas et al., 1990). The double helix molecular weight is reported to be ~5 million with an intrinsic viscosity of ~3,500 mL/g (Milas et al., 1990). Gels produced on untreated gellans are weak and rubbery although deacylation produces hard, brittle gels. Gellan gelation is promoted by the presence of cations (Grasdalen and Smidsrød, 1987; Sutherland, 1999; Morris et al., 2012), and divalent cations (e.g., Ca^{2+} or Mg^{2+}) are more efficient than monovalent cations (e.g., K^+ or Na^+).

Gellan is usually grown on relatively simple growth media containing (Bajaj et al., 2007) the following:

1. A carbon source (e.g., glucose, fructose, maltose, sucrose, or mannitol) at concentrations in the range of 2%–4%
2. A nitrogen source (e.g., yeast extract, ammonium, or potassium nitrate)
3. Other components (e.g., amino acids and nucleotide phosphate sugars)

It is then isolated from the media by ethanol or 2-propanol precipitation from the culture medium after heating at ~90°C to kill the bacteria and may be partially deacetylated by alkali treatment. Deproteinization is an important step in the recovery process, and recent research has shown that deproteinization with alkaline protease is the most suitable method in terms of gellan recovery and deproteinization efficiency (Wang et al., 2007). It was intended to be a competitor to xanthan, though before permission for food use was obtained, it was also promoted as an agar substitute, particularly for use in growth media. It is one of many film-forming polysaccharides being considered for use in implants for insulin in the treatment of diabetes (Li et al., 2001a,b).

10.4 Cyanobacterial Polysaccharides

Another group of polysaccharides with reputed xanthan-like properties is the exopolysaccharides from Cyanobacteria—"blue-green algae"—known to produce large amounts of these substances (Li et al., 2001).

Like xanthan, the cyanobacterial polysaccharides are relatively complex containing on average 6–10 different saccharide residues (De Philippis et al., 2001; Pereira et al., 2009), for example, xylose, arabinose, fucose, galactose, rhamnose, mannose, glucose, and uronic acids (De Philippis and Vincenzini, 1998; Li et al., 2001a,b; Pereira et al., 2009). Charged groups such as sulfate and pyruvate together with hydrophobic groups, for example, acetates, have also been reported (De Philippis et al., 2001). All the cyanobacterial polysaccharides studied so far are branched structures (Pereira et al., 2009).

Due to their diversity of bacteria studied, cyanobacterial exopolysaccharides have been grown under a wide variety of conditions (Li et al., 2001a,b). Nitrogen seems to have a negative effect on polysaccharide production (De Philippis et al., 1998; Moreno et al., 1998), whereas temperature, salinity, illumination, and exposure to UV radiation can have either positive or negative effects on exopolysaccharide production in cyanobacteria (Li et al., 2001a,b) and is worth noting that optimal exopolysaccharide production does not necessarily coincide with those for cell growth (Moreno et al., 1998, 2000; Li et al., 2001a,b). The daily production of cyanobacterial polysaccharides from different sources is up to 2 g/L (De Philippis et al., 1998), which is small in comparison to xanthan production that can be up to 10 g/L (De Philippis and Vincenzini, 1998).

Cyanobacterial exopolysaccharides have been shown (Figure 10.3) to exhibit xanthan-like shear-thinning properties (Cesaro et al., 1990; Navarini et al., 1990, 1992; De Philippis and Vincenzini, 1998; Moreno et al., 1998; Morris et al., 2001; Vincente-García et al., 2004). Polysaccharide molecular weights have been measured in the range of 8×10^4 to 2×10^6 g/mol (De Philippis and Vincenzini, 1998; Moreno et al., 2000; Morris et al., 2001; Vincente-García et al., 2004; Pereira et al., 2009), and "gross" macromolecular conformations have been estimated to be a random coil (Cesaro et al., 1990; Navarini et al., 1992) for the exopolysaccharide from *Cyanospira capsulata* (CC-EPS), a rigid rod-type polysaccharide (Morris et al., 2001) from *Aphanothece halophytica* GR02 (AH-EPS), or intermediate between random coil and rigid rod for the polysaccharide from *Anabaena* sp. ATCC 33047 (Moreno et al., 2000).

These exopolysaccharides have potential biotechnological importance as conditioners of soils for the improvement of their water-holding capacity, emulsifiers, viscosifiers, medicines, bioflocculants, and heavy metal removal (Bar-or and Shilo, 1987; Falchini et al., 1996; Mazor et al., 1996; Choi et al., 1998; De Philippis and Vincenzini, 1998; Burja et al., 2001; Li et al., 2001a,b;

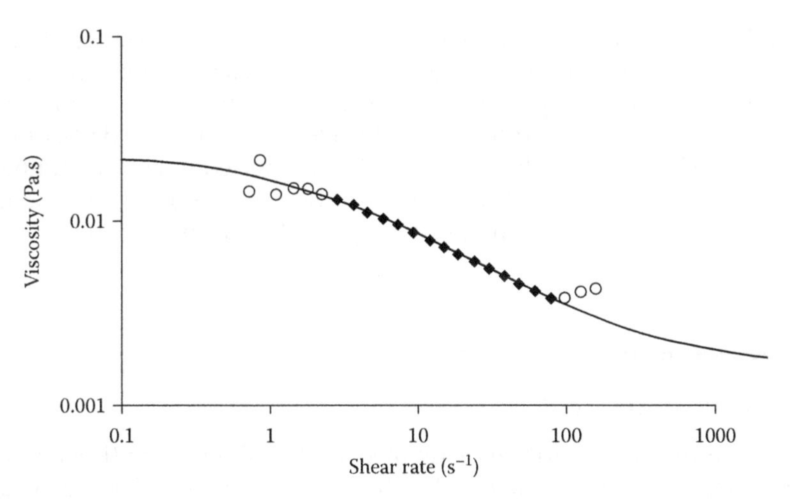

FIGURE 10.3
Viscosity and fitting for the shear-thinning cyanobacterial exopolysaccharide from *A. halophytica GR02* (AH-EPS) at 0.6 g/L measured at 25°C. Circles: points not used for fitting. Square: points used for fitting. Line: fitted curve. (Adapted from Morris, G.A., Hydrodynamic investigation of polysaccharides and their interactions with casein, PhD dissertation, University of Nottingham, Nottingham, U.K., 2001.)

De Philippis et al., 2007; Pereira et al., 2009; Singh et al., 2011). A pectin-like polysaccharide has also been described, which also shows affinity to metal cations (Plude et al., 1991).

10.5 Polysaccharides from Lactic Acid Bacteria

Exopolysaccharides are a complex family of polysaccharides, which can be divided into two subgroups: homopolysaccharides and heteropolysaccharides (Gänzle and Schwab, 2005). Homopolysaccharides can be further divided into glucans and fructans (Leivers, 2011). Some examples of homopolysaccharides are dextran, mutan, and levan from *Leuconostoc* and *Streptococcus* spp. (Ruas-Madiedo et al., 2002; Gänzle and Schwab, 2005; Leivers, 2011).

The heteropolysaccharides are formed by repeating units that most often contain a combination of D-glucose, D-galactose, L-rhamnose, and *N*-acetylmannosamine (ManNAc) (Leivers, 2011). Noncarbohydrate substituents such as phosphate may also be present (Ruas-Madiedo et al., 2002).

Lactic acid bacterial polysaccharides are "Generally Regarded as Safe" (GRAS) and are therefore permitted for use in the food industry. It is a common misconception that the lactic acid bacterial polysaccharides increase the viscosity of yogurt, for example; however, this is not the case as it is more likely that the extracellular polysaccharide (EPS) will interrupt the casein network (Ruas-Madiedo et al., 2002) and therefore has a negative effect on viscosity. Lactic acid bacterial EPSs are also associated with a number of spoilage issues in dairy products (Bouman, et al., 1982; Somers et al., 2001) and alcoholic beverages (Williamson, 1959; Duenas et al., 1995; de Nadra and de Saad, 1995). However, they have also been reported to have a number of health benefits including antitumor (Kitazawa et al., 1991), immunomodulatory (Zubillaga et al., 2001), and potential prebiotic effects of fructans.

10.6 Hyaluronic Acid

The HA industry is worth an estimated US $1 billion/year (Kogan et al., 2007), and the medical-grade HA sells at US $40,000–60,000/kg (Chong et al., 2005). Potentially the most interesting of the glycosaminoglycans, this biopolymer is found in the joints, skin (Chong et al., 2005), vitreous humor of the eyes (Meyer and Palmer, 1934), and umbilical cord (Weissmann and Meyer, 1954) of vertebrates. It is also produced by *Staphylococcus* and some *Streptococci* (Chong et al., 2005; Kim et al., 2006; Kogan et al., 2007). The HA layer may enable virulent strains of *Streptococci* to attack immune systems of unidentified higher organisms (Chong et al., 2005; Yamada and Kawasaki, 2005).

HA is a linear polymer consisting of alternating β-(1→4) *N*-acetyl-D-glucosamine and β-(1→3) D-glucuronic acid (Figure 10.4) (Weissmann and Meyer, 1954) resulting in stiff chains due to strong interchain hydrogen bonding (Benincasa et al., 2002; Chong et al., 2005) and water bridges between acetamido and carboxylate groups (Benincasa et al., 2002). HA therefore adopts a conformation between that of a random coil and rigid rod in dilute solution (Cleland and Wang, 1970; Hokputsa et al., 2003).

HA is commercially extracted from rooster combs or produced by *Streptococcus zooepidemicus* using microbial fermentation technique (Kim et al., 2006).

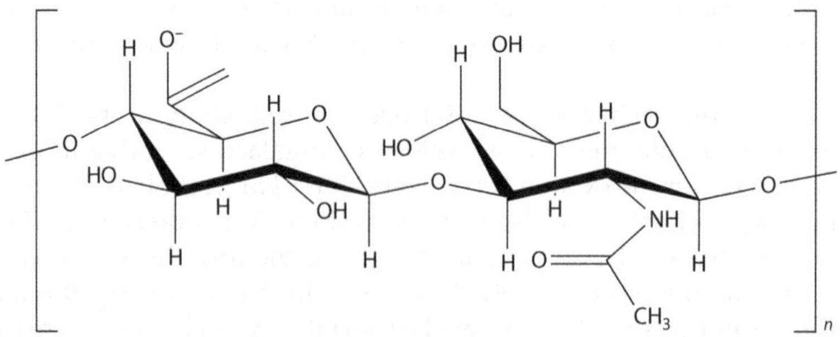

FIGURE 10.4
Repeat unit of HA.

The fermentation medium typically includes yeast extract, glutamine, mineral salts (sodium chloride, potassium hydrogen phosphate, and magnesium sulfate), and glucose (Armstrong and Johns, 1997; Kim et al., 2006). HA quality, in terms of yield and molecular weight, has been studied under various conditions such as temperature, pH, agitation speed, and aeration rate (Armstrong and Johns, 1997; Chong et al., 2005; Kim et al., 2006). Typical growth media contains glucose, peptone, and yeast extract (Schiraldi et al., 2010). Nonoptimal growth conditions favor the production of high molecular weight HA in larger amounts (Armstrong and Johns, 1997; Chong et al., 2005; Kim et al., 2006). Due to the increased viscosity of the growth medium, the production of HA at concentrations greater than ~5–10 g/L becomes impractical (Yamada and Kawasaki, 2005; Schiraldi et al., 2010).

Due to its high molecular weight of ~1 to 3×10^6 g/mol (Cleland and Wang, 1970; Armstrong and Johns, 1997; Hokputsa et al., 2003) and semiflexible coil conformation (Cleland and Wang, 1970; Hokputsa et al., 2003), HA behaves as a non-Newtonian solution even at low concentrations (Chong et al., 2005). The critical coil overlap concentration, c*, has been reported to be molecular weight dependent where the product of $c^{*} \cdot M_w \approx 2{,}800{,}000\ g^2/L \cdot mol$. Therefore, HA solutions enter the semi-dilute regime at concentrations of approximately 1.5 g/L and the concentrated regime (independent of molecular weight) at >15 g/L (Chong et al., 2005).

HA molecules of high molecular weight have also been shown to have mucoadhesive and anti-inflammatory effects (Chong et al., 2005), and patients require less frequent injections of higher molecular weight material for arthritis treatment by viscosupplementation, which is important in patient acceptance of a treatment (Chong et al., 2005).

Chemically modified HAs are available commercially, examples of which include HA benzyl esters (Fidia farmaceutici, Abano Terme, Italy) and divinyl sulfone cross-linked HA (Genzyme, Cambridge, MA). Internally cross-linked (or autocross-linked) HA molecules have also been prepared (Esposito et al., 2005). HA nanofibers for wound healing are also now commercially available. Modifications in terms of molecular weight and molecular weight distributions have also been achieved by means of changes in the growth medium.

Genetic modification of the non-HA-producing bacteria is an important development in bacterial HA production. The introduction of the *hasA* gene into *Escherichia coli* or *Bacillus subtilis* results in the production of HA (Weigel et al., 1997; Yamada and Kawasaki, 2005), and although there is no improvement in quality (in terms of molecular weight and yield), any possible contaminations by streptococcal exo- and endotoxins are eliminated (Yamada and Kawasaki, 2005; Kogan et al., 2007), and HA produced by *B. subtilis* has received GRAS status and is likely to be released in the market in the near future (Kogan et al., 2007).

HA is used extensively in the medical, cosmetic, and food industries, and its applications have been reviewed recently (Kogan et al., 2007). The most important industrial applications include viscosurgery, viscosupplementation, viscoaugmentation, viscoseparation (Chong et al., 2005; Kogan et al., 2007), and drug delivery using HA-microspheres for controlled drug delivery (Pereswetoff-Morath, 1998; Esposito et al., 2005). In the United States, approximately 1.3 million cosmetic procedures utilized HA-based products costing a total of approximately $730 million (American Society for Aesthetic Plastic Surgery, Cosmetic Surgery National Data Bank, 2006 Statistics, http://www.surgery.org/sites/default/files/Stats2010_1.pdf).

Although not strictly speaking HA, the "HA-like" exopolysaccharide from *Vibrio diabolicus*, which can be found in deep-sea hydrothermal vents, shows great potential in bone healing (Zanchetta et al., 2003a,b; Poli et al., 2009).

10.7 Other Important Bacterial Polysaccharides

10.7.1 Xylinan

Xylinan (or acetan) is a complex anionic exopolysaccharide extracted during the commercial production of bacterial cellulose by *Acetobacter xylinum* as a water-soluble by-product (Ishida et al., 2002). The structural repeat unit is similar to that of xanthan and consists of cellulosic backbone (…→4) β-D-Glc*p*(1→…) where alternate glucopyranose residues are substituted at the 3-position with a pentasaccharide side chain (Morris et al., 1989):

α-L-Rha*p*-(1→6)-β-D-Glc*p*-(1→6)-α-D-Glc*p*-(1→4)-β-D

-GlcA*p*-(1→2)-α-D-Man*p*-(1→3)

The branched glucopyranose residues together with the mannopyranose residue are partially O-acetylated at the sixth position (Ojinnaka et al., 1996; Sutherland, 1997). Analysis from x-ray diffraction (Morris et al., 1989) and atomic force microscopy (Kirby et al., 1995) shows that xylinan adopts a helical structure in both the solid and solution phases. Xylinan chains upon heating undergo thermoreversible helix (ordered state)—coil (disordered state) transition (Ojinnaka et al., 1996). These molecules are of average-weight M_w, i.e., ~2.5 million g/mol (Berth et al., 1996a,b; Harding et al., 1996). Furthermore, the double-helical structure results in a stiff macromolecular chain with a molecular persistence length L_p of ~100 nm and a mass per unit length M_L of ~2500 g/mol·nm (Berth et al., 1996a,b) resulting in viscous solutions, [η], ~ several thousand mL/g (Berth et al., 1996a). The Mark–Houwink *a* parameter of 0.90 was found, which is consistent with a rigid conformation (Berth et al., 1996a). Intermolecular binding has been demonstrated between xylinan and the industrially important konjac glucomannan and galactomannan resulting in the formation of synergistic gels (or increased viscosity at lower concentrations) (Ridout et al., 1998a,b). Chemical modification of xylinan has been through deacetylation of the 6-linked O-acetyl groups (Christensen and Smidsrød, 1996; Ridout et al., 1998a). It has been shown that acetylation does not prevent helix formation (Ojinnaka et al., 1996), but it does reduce the synergistic interaction with glucomannans or galactomannans (Ridout et al., 1998a). Xylinan synthesis involves several genes including *ace*A, *ace*B, and *ace*C (Griffin et al., 1996a,b; 1997a,b; Edwards et al., 1999), and it has been shown that chemical mutagenesis of the native bacterium results in a strain that produces a polysaccharide where the side chain is a tetrasaccharide (van Kranenburg et al., 1999). Furthermore, this novel xylinan variant has higher viscosity. A further novel polysaccharide with a trisaccharide side chain was designed by deactivating the *ace*P gene (Edwards et al., 1999). Xylinan is used in the food industry as a viscosifier and gelling agent (van Kranenburg et al., 1999). It is also a component in Nata, a sweet confectionery popular in Japan and the Philippines (Griffin et al., 1997a,b; van Kranenburg et al., 1999).

10.7.2 Dextrans

Dextrans are mostly $\alpha(1\rightarrow6)$D-glucopyranosyl polymers with molecular weights up to ~1 × 106 g/mol more or less branched through 1→2, 1→3, or 1→4 links. In most cases, the length of the side chains is short, and the branched residues vary between 5% and 33%. The major commercial dextran is about 95% 1→6 linked and 5% 1→3 linked and is made by selected strains of *Leuconostoc* (Figure 10.5).

Dextran is produced commercially by the fermentation of sucrose-rich media. Molecular weight and yield depend on the process variables such as temperature, pH, and sucrose concentration (Sarwat et al., 2008). The dextran is then purified from the medium by ethanol precipitation. Acid hydrolysis is often used to control molecular weight, though fungal dextranases can also be used. The product with an average molecular weight of about 60,000 g/mol is used in medicine as a blood extender, while fractions of defined molecular weights (e.g., the Pharmacia "T" series, e.g., T500 Dextran, which stands for dextran of weight-average molecular weight 500,000 g/mol) are familiar in laboratories and, to some extent, like the pullulan "P" series, serve as polysaccharide standards in molecular weight calibrations. "Blue dextran" is a well-known marker for the void volume for gel filtration studies. Dextrans are also used as part of incompatible phase separation systems, usually with polyethylene glycol. Their solubility in both aqueous and nonaqueous media renders them good membrane formers using electrospinning techniques (Jiang et al., 2004).

Dextrans have found very wide applications in laboratory work because these are particularly free from positive interactions with proteins. The interactions can be almost entirely characterized as co-exclusion. This has found application, similar to pullulans, in calibrating gel filtration media. However, cross-linked dextran gels show different effects. For example, they swell and shrink in a way related to the osmotic pressure of the solvent system and can be used to make miniature osmometers (Ogston and Silpananta, 1970).

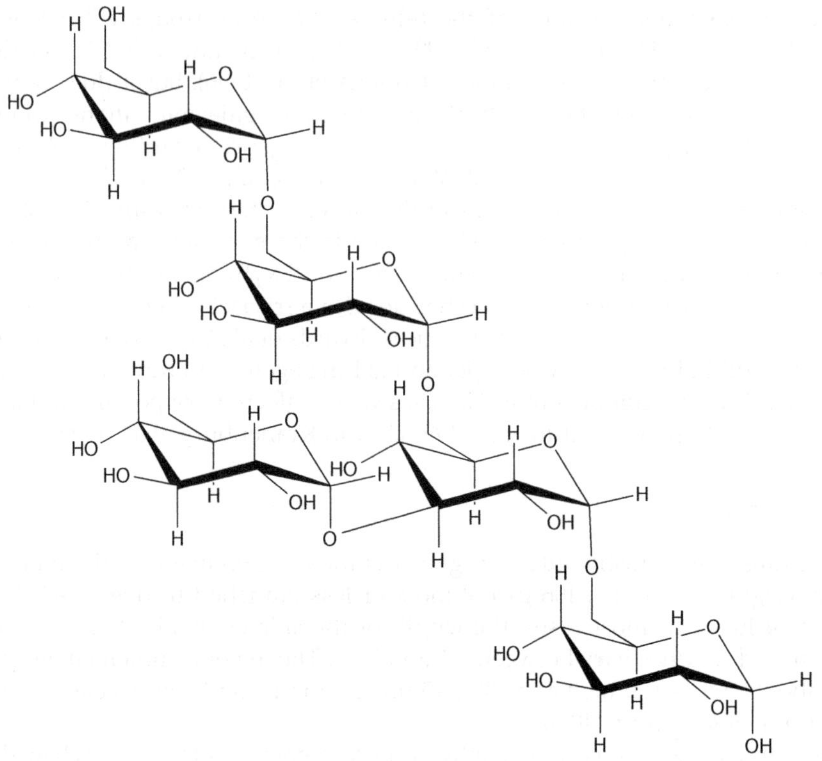

FIGURE 10.5
Repeat unit of dextran showing an $\alpha(1\rightarrow3)$ branch.

10.7.3 Pullulan

Pullulan is an α-glucan made up from maltotriose units linked by $\alpha(1\rightarrow6)$ bonds. It is obtained from *Aureobasidium pullulans* and is hydrolyzed by pullulanase to yield maltotriose. It is not attacked by digestive enzymes of the human gut and is used to form films. Production is now substantial and has found particular application in formulating snack foods in Japan based on cod roe, powdered cheese, and as a packaging film for ham. It is water soluble, with molecular weights in the range 5,000–900,000 g/mol, with straight unbranched chains, and behaves as a very flexible molecule, with properties of those of a "random coil" (with a molecular persistence length, L_p, of approximately 2 nm) according to a combination of sedimentation coefficient and intrinsic viscosity measurements (Kawahara et al., 1984). It has been proposed as a "standard polysaccharide" in the sense that it is so near to random coil behavior, and readily obtainable in very reproducible form, that it could be used for comparative tests with other polysaccharides.

10.7.4 Scleroglucan and Schizophyllan

The study of microbial polysaccharides would not be complete without at least a brief consideration of two fungal polysaccharides, which are attracting increasing commercial interest: the weak-gelling scleroglucan and schizophyllan systems. They are both large, neutral polysaccharides of weight-average molecular weights of approximately 500,000 g/mol. They both appear from x-ray diffraction to exist as hydrogen-bond-stabilized triple helices, stabilized by hydrogen bonds with resultant extra-rigid rod-like properties in solution: they have virtually the largest persistence lengths known for polysaccharides—~150 nm for schizophyllan (Yanaki et al., 1980) and 200 nm for scleroglucan (Biver et al., 1986). The existence of the triple helix for scleroglucan has been further supported by electron microscopic/light scattering measurements of the mass per unit length ($M_L = 2100 \pm 200$) g/mol·nm (Kitamura et al., 1994) along similar lines to that which supported the duplex model for xanthan as discussed earlier. The Mark–Houwink viscosity "a" parameter of 1.7 for schizophyllans of $M_w < 500{,}000$ g/mol is again among the highest known for a polysaccharide and is on the rigid rod limit (Yanaki et al., 1980), and these molecules have been used to flocculate small colloidal particles (Ferretti et al., 2003). For chains of $M_w > 500{,}000$ g/mol, the "a" parameter falls to ~1.2 and corresponds to slightly more flexibility as the polymer length increases.

Chemically also they are very similar, with a backbone of repeating β(1→3)-linked glucose residues:

$$\ldots\!\rightarrow\!3)\ \beta\text{-}\textsc{d}\text{-Glc}p(1\!\rightarrow\ldots$$

Both on scleroglucan and schizophyllan, every third residue has a β(1→6)-linked D-glucose branch, which protrudes from the triple helix. It has been demonstrated that using electron microscopy, certain denaturation–renaturation treatments cause the formation of interesting ring structures or "macrocycles" (Stokke et al., 1993). In common with other branched glucans, they appear to stimulate an immune response against tumor cells (Kurachi et al., 1990) and, particularly scleroglucan, have been considered for use in cosmetics (as part of skin and hair products) for application in pesticides (to assist binding to foliage), and along with xanthan and other polysaccharides, advantage can be taken of its high capacity to bind water and high heat stability in oil well drilling fluids.

10.8 Polysaccharide Vaccines

No survey of microbial polysaccharides would be complete without reference to the use of polysaccharides in the production of vaccines against serious disease. Certain types of pathogenic bacteria such as *Streptococcus pneumoniae,*

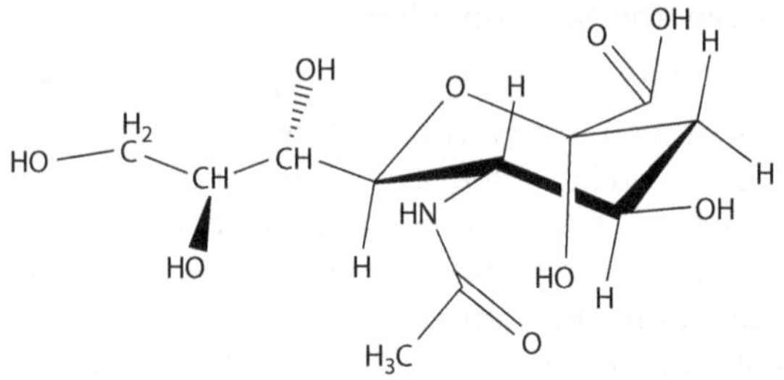

FIGURE 10.6
N-acetyl neuraminic acid.

Neisseria meningitidis ("meningococcus"), and *Haemophilus influenzae* type B, besides producing harmful or dangerous toxins, also produce high molecular weight capsular polysaccharides that are harmless. They do help the said bacteria to establish an infection and help hide cell surface components from immune recognition and complement activation (Abdelhameed et al., 2012; Harding et al., 2012). Purified extracts of polysaccharides may themselves be immunogenic and can be used at least in principle to produce immunity against the organism which is producing them. As a result, vaccines against some of these organisms are now available.

The polysaccharides themselves consist of repeat sequences of saccharide residue. One residue that appears frequently is *N*-acetyl neuraminic acid, a type of "sialic acid" (Figure 10.6). This residue is common on the surface of membrane glycoproteins and also on mucosal surfaces. Thus, considerable care is required against possible autoimmunity problems. A more serious problem with polysaccharide vaccines is that their effects are not generally long lasting: The repeat sequences produce a T-cell-independent IgM rather than IgG response in infants with little immunological memory effect and are ineffective for infants <2 years old. To counter this, recently a lot of attention has been paid to the development of *conjugate* vaccines where the polysaccharides or repeat sequences from them are covalently attached via a linker to an appropriate protein carrier usually based on a bacterial toxoid from diphtheria or tetanus.

Such conjugates have been successful in enhancing IgG antibody production and stimulating T-cell-dependent immunity against diseases such as meningitis. Two recent articles from the National Institute of Biological Standards in London provide an excellent description of current developments in this area (Suker et al., 2004; Jones, 2005).

In generating these vaccines, important issues have to be addressed with regard to the reproducibility of preparations relating to not only antigenicity and chemical purity but also physical properties including molecular weight and molecular weight distribution limits, and in this regard, the use of techniques, for example, SEC-MALLs, analytical ultracentrifugation, viscometry for physical characterization (Abdelhameed et al., 2012; Harding et al., 2012), and NMR and GC-MS for chemical characterizations, will become increasingly important. This highlights the increasing difference in characterization requirements for a microbial polysaccharide for food and biomedical/pharmaceutical application, examples of which are xanthan and vaccines.

10.9 Summary and Future Prospects

Polysaccharides from plant are still the most important source of hydrocolloids currently on the market (Freitas et al., 2011), mainly due to the high cost of carbon sources used for bacterial fermentation. The production of microbial polysaccharides is gaining importance as virtually they have no known toxicity problem; also the relative simplicity of bacterial genetics compared to those of higher organisms is making them the best source for genetic manipulation with improved production capabilities. Genetic manipulations can make the microorganisms potentially useful for the production of value-added tailor-made polysaccharides having a specific functionality. This suggests that the future market for bacterial polysaccharides is not in competition with plant polysaccharides but in niche applications where traditional polysaccharides are unable to compete.

Finally, the virtual lack of toxicity is providing hope in the development of effective conjugate vaccines against serious disease, provided that issues concerning autoimmunity and product reproducibility are properly taken care of (Jones, 2005; Abdelhameed et al., 2012; Harding et al., 2012).

Acknowledgment

The authors thank Prof. Edwin Morris, Emeritus Professor at the University of Nottingham, for his help.

References

Abdelhameed, A.S., Morris, G.A., Adams, G.G. et al. 2012. An asymmetric and slightly dimerized structure for the tetanus toxoid protein used in glycoconjugate vaccines. *Carbohydrate Polymers*, 90: 1831–1835.

Armstrong, D.C. and Johns, M.R. 1997. Culture conditions affect the molecular weight properties of hyaluronic acid produced by *Streptococcus zooepidemicus*. *Applied Environmental Microbiology*, 63: 2759–2764.

Bajaj, I.B., Survase, S.A., Saudagar, P.S., and Singhal, R.S. 2007. Gellan gum: Fermentative production, downstream processing and applications. *Food Technology and Biotechnology*, 45: 341–354.

Bar-or, Y. and Shilo, M. 1987. Characterization of macromolecular flocculants produced by *Phormidium sp. StrainJ-1* and by *Anabaenopsis circularis PCC6720. Applied Environmental Microbiology*, 53: 2226–2230.

Benincasa, M.-A., Cartoni, G., and Fratte, C.D. 2002. Flow field fractionation and characterization of ionic and neutral polysaccharides of vegetable and microbial origin. *Journal of Chromatography A*, 967: 219–234.

Berth, G., Dautzenburg, H., Christensen, B.E., Rother, G., and Smidsrød, O. 1996a. Physicochemical studies on xylinan (acetan). I. Characterization by gel permeation chromatography on Sepharose Cl-2B coupled to static light scattering and viscometry. *Biopolymers*, 39: 709–719.

Berth, G., Dautzenburg, H., Christensen, B.E., and Smidsrød, O. 1996b. Physicochemical studies on xylinan (acetan). II. Characterization by static light scattering. *Biopolymers*, 39: 7219–7728.

Biver, C., Lesec, J., Allain, C., Salome, L., and Lecourtier, J. 1986. Rheological behavior and low-temperature sol-gel transition of scleroglucan solutions. *Polymer Communications*, 27: 351–353.

Bouman, S., Lund, D.B., Driessen, F.M., and Schmidt, D.G. 1982. Growth of thermoresistant streptococci and deposition of milk constituents on plates of heat exchangers during long operating times. *Journal of Food Protection*, 45: 806–812.

Burja, A.M., Banaigs, B., Abou-Mansour, E., Burgess, J.G., and Wright, P.C. 2001. Marine cyanobacteria-a prolific source of natural products. *Tetrahedron*, 57: 9347–9377.

Bylaite, E., Adler-Nissen, J., and Meyer, A.S. 2005. Effect of xanthan on flavor release from thickened viscous food model systems. *Journal of Agricultural and Food Chemistry*, 53: 3577–3583.

Cesaro, A., Liut, G., Bertocchi, C., Navarini, L., and Urbani, R. 1990. Physicochemical properties of the exocellular polysaccharide from *Cyanospira capsulata. International Journal of Biological Macromolecules*, 12: 79–84.

Choi, W.C., Yoo, S., Oh, I., and Park, S.H. 1998. Characterization of an extracellular flocculating substance produced by a planktonic cyanobacterium, *Anabaena* sp. *Biotechnology Letters*, 20: 643–646.

Chong, B.F., Blank, L.M., Mclaughlin, R., and Nielsen, L.K. 2005. Microbial hyaluronic acid production. *Applied Environmental Microbiology*, 66: 341–351.

Christensen, B.E. and Smidsrød, O. 1996. Dependence of the content of unsubstituted (cellulosic) regions in prehydrolyzed xanthans on the rate of hydrolysis by *Trichoderma reesei* endoglucanase. *International Journal of Biological Macromolecules*, 18: 93–99.

Cleland, R.L. and Wang, J.L. 1970. Ionic polysaccharides. III. Dilute solution properties of hyaluronic acid fractions. *Biopolymers*, 9: 799–810.

Dea, I.C.M. 1989. Industrial polysaccharides. *Pure and Applied Chemistry*, 61: 1315–1322.

Dea, I.C.M., Morris, E.R., Rees, D.A., Welsh, E.J., Barnes, H.A., and Price, J. 1977. Associations of like and unlike polysaccharides: Mechanism and specificity in galactomannans interacting with bacterial polysaccharides and related systems. *Carbohydrate Research*, 57: 249–272.

De Philippis, R., Paperi, R., and Sili, C. 2007. Heavy metal sorption by released polysaccharides and whole cultures of two exopolysaccharides-producing cyanobacteria. *Biodegradation*, 18: 181–187.

De Philippis, R., Sili, C., Paperi, R., and Vincenzini, M. 2001. Exopolysaccharide producing cyanobacterial and their possible exploitation: A review. *Journal of Applied Phycology*, 13: 293–299.

De Philippis, R. and Vincenzini, M. 1998. Exocellular polysaccharides from cyanobacteria and their possible applications. *FEMS Microbiology Reviews*, 22: 151–175.

Dhami, R., Harding, S.E., Jones, T., Hughes, T., Mitchell, J.R., and To, K.-M. 1995. Physicochemical studies on a commercial food-grade xanthan. I. Characterization by sedimentation velocity, sedimentation equilibrium and viscometry. *Carbohydrate Polymers*, 27: 93–99.

Donot, F., Fontana, A., Baccou, J.C., and Schorr-Galindo, S. 2012. Microbial exopolysaccharides: Main examples of synthesis, excretion, genetics and extraction. *Carbohydrate Polymers*, 87: 951–962.

Duenas, M., Irastorza, A., Fernandez, K., and Bilbao, A. 1995. Heterofermentative *Lactobacilli* causing ropiness in basque country ciders. *Journal of Food Protection*, 58: 76–80.

Edwards, K.J., Jay, A.J., Colquhoun, I.J., Morris, V.J., Glasson, M.J., and Griffin, A.M. 1999. Generation of a novel polysaccharide by inactivation of the aceP gene from the acetan biosynthetic pathway in *Acetobacter xylinum*. *Microbiology*, 145: 1499–1506.

Esposito, E., Menegatti, E., and Cortesi, R. 2005. Hyaluronan-based microspheres as tools for drug delivery: A comparative study. *International Journal of Pharmaceutics*, 288: 35–49.

Falchini, L., Sparvoli, E., and Tomaselli, L. 1996. Effect of *Nostoc* (Cyanobacteria) inoculation on the structure and stability of clay soils. *Biology and Fertility of Soils*, 23: 346–352.

Ferretti, R., Stoll, S., Zhang, J., and Buffle, J. 2003. Flocculation of haematite particles by a comparatively large rigid polysaccharide: Schizophyllan. *Journal of Colloid and Interface Science*, 266: 328–338.

Fitzsimons, S.M., Mulvihill, D.M., and Morris, E.R. 2007. Denaturation and aggregation processes in thermal gelation of whey proteins resolved by differential scanning calorimetry. *Food Hydrocolloids*, 21: 638–644.

Foster, T.J. and Morris, E.R. 1994. Xanthan polytetramer–Conformational stability as a barrier to synergistic interaction. In: *Gums and Stabilisers for the Food Industry* 7, eds. P.A. Williams and G.O. Phillips, pp. 281–289. Oxford University press, Oxford, U.K.

Freitas, F., Alves, V.D., and Reis, M.A.M. 2011. Advances in bacterial exopolysaccharides: From production to biotechnological applications. *Trends in Biotechnology*, 29: 388–398.

Gänzle, M.G. and Schwab, C. 2005. Exopolysaccharide production by intestinal lactobacilli. In: *Probiotics and Prebiotics: Scientific Aspects*, ed. G.W. Tannock, pp. 83–96. Horizon Scientific Press, Wymondham, U.K.

Giannouli, P. and Morris, E.R. 2003. Cryogelation of xanthan. *Food Hydrocolloids*, 17: 495–501.

Goycoolea, F.M., Morris, E.R., and Gidley, M.J. 1995a. Screening for synergistic interactions in dilute polysaccharide solutions. *Carbohydrate Polymers*, 28: 351–358.

Goycoolea, F.M., Richardson, R.K., Morris, E.R., and Gidley, M.J. 1995b. Stoichiometry and conformation of xanthan in synergistic gelation with locust bean gum or konjac glucomannan: Evidence for heterotypic binding. *Macromolecules*, 28: 8308–8320.

Grasdalen, H. and Smidsrød, O. 1987. Gelation of gellan gum. *Carbohydrate Polymers*, 7: 371–393.

Griffin, A.M., Edwards, K.J., Gasson, M.J., and Morris, V.J. 1996a. Identification of the structural genes involved in bacterial exopolysaccharide production. In: *Biotechnology and Genetic Engineering Reviews*, Vol. 13, ed. M.P. Tombs, pp. 1–18. Intercept Ltd., Andover, MA.

Griffin, A.M., Edwards, K.J., Morris, V.J., and Gasson, M.J. 1997a. Genetic analysis of acetan biosynthesis in *Acetobacter xylinum*: DNA sequence analysis of the aceM gene encoding an UDP-glucose dehydrogenase. *Biotechnology Letters*, 19: 469–474.

Griffin, A.M., Morris, V.J., and Gasson, M.J. 1996b. Identification, cloning and sequencing the aceA gene involved in acetan biosynthesis in *Acetobacter xylinum. FEMS Microbiology Letters*, 137: 115–121.

Griffin, A.M., Poelwijk, E.S., Morris, V.J., and Gasson, M.J. 1997b. Cloning of the aceF gene encoding the phosphomannose isomerase and GDP-mannose pyrophosphorylase activities involved in acetan biosynthesis in *Acetobacter xylinum. FEMS Microbiology Letters*, 154: 389–396.

Harding, S.E., Abdelhameed, A.S., Morris, G.A. et al. 2012. Solution properties of capsular polysaccharides from *Streptococcus pneumoniae. Carbohydrate Polymers*, 90: 237–242.

Harding, S.E., Berth, G., Hartmann, J., Jumel, K., Cölfen, H., and Christensen, B.E. 1996. Physicochemical studies on xylinan (acetan). III. Hydrodynamic characterization by analytical ultracentrifugation and dynamic light scattering. *Biopolymers*, 39: 729–736.

Hokputsa, S., Jumel, K., Alexander, C., and Harding, S.E. 2003. Hydrodynamic characterisation of chemically degraded hyaluronic acid. *Carbohydrate Polymers*, 52: 111–117.

Holzwarth G. and Prestridge E.B. 1977. Multistranded helix in xanthan polysaccharide. *Science*, 19: 757–759.

Ishida, T., Sugano, Y., Nakai, T., and Shoda, M. 2002. Effects of acetan on production of bacterial cellulose by *Acetobacter xylinum. Bioscience, Biotechnology, and Biochemistry*, 66: 1677–1681.

Israilides, C., Smith, A., Scanlon, B., and Barnett, C. 1999. Pullulan from agro-industrial waste. In: *Biotechnology and Genetic Engineering Reviews*, Vol. 16, ed. S. E. Harding, pp. 309–324, Intercept Ltd., Andover, MA.

Jiang, H., Fang, D., Hsiao, B.S., Chu, B., and Chen, W. 2004. Optimization and characterization of dextran membranes prepared by electrospinning. *Biomacromolecules*, 5: 326–323.

Jones, C. 2005. Vaccines based on the cell surface carbohydrates of pathogenic bacteria. *Anais da Academia Brasileira De Ciências*, 7: 293–324.

Kawahara K., Ohta, K., Miyamoto, H., and Nakamura, S. 1984. Preparation and solution properties of pullulan fractions as standard samples for water-soluble polymers. *Carbohydrate Polymers*, 4: 335–356.

Kim, S-J., Park, S-Y., and Kim, C.-W. 2006. A novel approach to the production of hyaluronic acid by *Streptococcus zooepidemicus. Journal of Microbiology and Biotechnology*, 16: 1849–1855.

Kirby, A.R., Gunning, A.P., Morris, V.J., and Ridout, M.J. 1995. Observation of the helical structure of the bacterial polysaccharide acetan by atomic force microscopy. *Biophysical Journal*, 68: 360–363.

Kitagawa, H., Sato, T., Norisuye, T., and Fujita, H. 1985. Optical rotation behavior of xanthan in mixtures of water and cadoxen. *Carbohydrate Polymers*, 5: 407–422.

Kitamura, S., Hori, T., and Kurita, K. et al. 1994. An antitumor, branched (1→3)-beta-D-glucan from a water extract of fruiting bodies of *Cryptoporus volvatus*. *Carbohydrate Research*, 263: 111–121.

Kitamura, S., Takeo, K., Kuge, T., and Stokke, B.T. 1991. Thermally induced conformational transitions of double-stranded xanthan in aqueous salt solutions. *Biopolymers*, 31: 1243–1255.

Kitazawa, H., Toba, T., Itoh, T., Kumano, N., Adachi, S., and Yamaguchi, T. 1991. Antitumoral activity of slime-forming encapsulated *Lactococcus lactis* subsp. *cremoris* isolated from Scandinavian ropy sour milk, "viili". *Animal Science and Technology (Japan)*, 62: 277–283.

Kogan, G., Šoltés, L., Dtern, R., and Gemeiner, P. 2007. Hyaluronic acid: A natural polymer with a broad range of biomedical and industrial applications. *Biotechnology Letters*, 29: 17–25.

van Kranenburg, R., Boels, I.C., Kleerebezem, M., and de Vos, W.M. 1999. Genetics and engineering of microbial exopolysaccharides for food: Approaches for the production of existing and novel polysaccharides. *Current Opinion in Biotechnology*, 10: 498–504.

Kurachi, K., Ohno, N., and Yadomae, T. 1990. Preparation and antitumor-activity of hydroxyethylated derivatives of 6-branched (1→3)-β-D-glucan obtained from the culture filtrate of *Sclerotinia sclerotiorum* ifo-9395. *Chemical & Pharmaceutical Bulletin Tokyo*, 38: 2527–2531.

Lad, M., Todd, T., Morris, G.A., MacNaughtan, W., Sworn, G., and Foster, T.J. 2013. On the origin of sharp peaks in the x-ray diffraction patterns of xanthan powders. *Food Chemistry*, 139: 1146–1151.

Laws, A., Chadha, M., Chacon-Romero, M., Marshall, V., and Maqsood, M. 2008. Determination of the structure and molecular weights of the exopolysaccharide produced by *Lactobacillus acidophilus* 5e2 when grown on different carbon feeds. *Carbohydrate Research*, 343: 301–307.

Laws, A., Gu Y., and Marshall, V. 2001. Biosynthesis, characterisation, and design of bacterial exopolysaccharides from lactic acid bacteria. *Biotechnology Advances*, 19: 597–625.

Leach, J.G., Lilly, V.G., Wilson, H.A., and Purvis, M.R. Jr. . 1957. Bacterial polysaccharides: The nature and function of the exudate produced by *Xanthomonas phaseoli*. *Phytopathology*, 47: 13–120.

Leivers, S. 2011. Characterisation of bacterial exopolysaccharides. PhD dissertation, University of Huddersfield, Huddersfield, U.K.

Li, P., Harding, S.E., and Liu, Z. 2001a. Cyanobacterial exopolysaccharides: Their nature and potential biotechnological applications. In: *Biotechnology and Genetic Engineering Reviews*, Vol. 18, ed. S.E. Harding, pp. 375–404, Intercept Ltd., Andover, MA.

Li, J., Kamath, K., and Dwivedi, C. 2001b. Gellan film as an implant for insulin delivery. *Journal of Biomaterials Applications*, 15: 321–343.

Lilly, V.G., Wilson, H.A., and Leach, J.G. 1958. Bacterial polysaccharides. II. Laboratory-scale production of polysaccharides by species of *Xanthomonas*. *Applied Microbiology*, 6: 105–108.

Mazor, G., Kidron, G.J., Vonshak, A., and Abeliovich, A. 1996. The role of cyanobacterial exopolysaccharides in structuring desert microbial crusts. *FEMS Microbiology Ecology*, 21: 121–130.

Melia, C.D. 1991. Hydrophilic matrix sustained release systems based on polysaccharide carriers. *Critical Review in Therapeutic Drug*, 8: 395–421.

Meyer, K. and Palmer, J.W. 1934. The polysaccharide of the vitreous humor. *The Journal of Biological Chemistry*, 107: 629–634.

Milas, M. and Rinaudo, M. 1979. Conformational investigation on the bacterial polysaccharide xanthan. *Carbohydrate Research*, 76: 189–196.

Milas, M., Shi, X., and Rinaudo, M. 1990. On the physicochemical properties of gellan gum. *Biopolymers*, 30: 451–464.

Millane, R.P., Narasaiah, T.V., and Arnott, S. 1989. On the molecular structures of xanthan and genetically engineered xanthan variants with truncated side chains. In: *Biomedical and Biotechnological Advances in Industrial Polysaccharides*, eds. V. Crescenzi, I.C.M. Dea, S. Polaletti, S.S. Stivala, and I.W. Sutherland, pp. 469–478, Gordon and Breach, New York.

Mohammed, Z.H., Haque, A., Richardson, R.K., and Morris, E.R. 2007. Promotion and inhibition of xanthan 'weak-gel' rheology by calcium ions. *Carbohydrate Polymers*, 70: 38–45.

Moreno, J., Vargas, M.A. Olivares, H., Rivas, J., and Guerrero, M.G. 1998. Exopolysaccharide production by the cyanobacterium *Anabaena sp*. ATCC 33047 in batch and continuous culture. *Journal of Biotechnology*, 60: 175–182.

Morris, G.A. 2001. Hydrodynamic investigation of polysaccharides and their interactions with casein, PhD dissertation, University of Nottingham, Nottingham, U.K.

Morris, V.J., Brownsey, G.J., Cairns, P., Chilvers, G.R., and Miles, M.J. 1989. Molecular origins of acetan solution properties. *International Journal of Biological Macromolecules*, 11: 326–328.

Morris, E.R. and Foster, T.J. 1994. Role of conformation in synergistic interactions of xanthan. *Carbohydrate Polymers*, 23: 133–135.

Morris, G.A. and Harding, S.E. 2009. Polysaccharides, Microbial. In: *Encyclopedia of Microbiology*, 3rd ed., ed. M. Schaechter, pp. 482–494. Elsevier, Amsterdam, the Netherlands.

Morris, G.A., Li, P., Puaud, M., Liu, Z., Mitchell, J.R., and Harding S.E. 2001. Hydrodynamic characterisation of the exopolysaccharides from the halophilic cyanobacteria *Aphanothece halophytica* GR02: A comparison with xanthan. *Carbohydrate Polymers*, 44: 261–268.

Morris, E.R., Nishinari, K., and Rinaudo, M. 2012. Gelation of gellan—A review. *Food Hydrocolloids*, 28: 373–411.

Morris, E.R., Rees, D.A., Young, G., Walkinshaw, M.D., and Darke, A. 1977. Order–disorder transition for a bacterial polysaccharide in solution: A role for polysaccharide conformation in recognition between *Xanthomonas* pathogen and its plant host. *Journal of Molecular Biology*, 110: 1–16.

de Nadra, M.C.M. and de Saad, A.M.S. 1995. Polysaccharide production by *Pediococcus pentosaceus* from wine. *International Journal of Food Microbiology*, 27: 101–106.

Navarini, L., Bertocchi, C., Cesaro, A., Lapasin, R., and Crescenzi, V. 1990. Rheology of aqueous solutions of an extracellular polysaccharide from *Cyanospira capsulata*. *Carbohydrate Polymers*, 122: 169–187.

Navarini, L., Cesaro, L., and Ross-Murphy, S.B. 1992. Viscoelastic properties of aqueous solutions of an exocellular polysaccharide from cyanobacteria. *Carbohydrate Polymers*, 18: 265–272.

Norton, I.T., Goodall, D.M., Frangou, S.A., Morris, E.R., and Rees, D.A. 1984. Mechanism and dynamics of conformational ordering in xanthan polysaccharide. *Journal of Molecular Biology*, 175: 371–394.

Ogston, A.G. and Silpananta, J. 1970. Thermodynamics of interaction between sephadex and penetrating solute. *Biochemical Journal*, 116: 171–175.
Ojinnaka, C., Jay, A.J., Colquhoun, I.J., Brownsey, G.J., Morris, E.R., and Morris, V.J. 1996. Structure and conformation of acetan polysaccharide. *International Journal of Biological Macromolecules*, 19: 149–156.
Oku, T., Yamada, K., and Hosoya, N. 1979. Effects of pullulan and cellulose on the gastrointestinal tract of rats. *Nutrition and Food Science*, 32: 235–241.
O'Neill, M.A., Selvendran, R.R., and Morris, V.J. 1983. Structure of the acidic extracellular gelling polysaccharide produced by *Pseudomonas elodea*. *Carbohydrate Research*, 124: 123–133.
Palaniraj, A. and Jayaraman, V. 2011. Production, recovery and applications of xanthan gum by *Xanthomonas campestris*. *Journal of Food Engineering*, 106: 1–12.
Paradossi, G. and Brant, D.A. 1982. Light-scattering study of a series of xanthan fractions in aqueous-solution. *Macromolecules*, 15: 874–879.
Pereira, S., Zille, A., Micheletti, E., Moradas-Ferreira, P., De Philippis, R., and Tamagnini, P. 2009. Complexity of cyanobacterial exopolysaccharides: Composition, structures, inducing factors and putative genes involved in their biosynthesis and assembly. *FEMS Microbiology Reviews*, 33: 917–941.
Pereswetoff-Morath, L. 1998. Microspheres as nasal drug delivery systems. *Advanced Drug Delivery Reviews*, 29: 185–194.
Phalipon, A., Costachel, C., Grandjean, C., Thuizat, A., Guerreiro, C., and Tanguy, M. 2006. Characterization of functional oligosaccharide mimics of the *Shigella flexneri* serotype 2a O-antigen: Implications for the development of a chemically defined glycoconjugate vaccine. *Journal of Immunology*, 176: 1686–1694.
Plude, J.L., Parker, D.L., Schommer, O.J. et al. 1991. Chemical characterization of the polysaccharide from the slime layer of the cyanobacterium *Microcystis flos-aquae* C3–40. *Applied and Environmental Microbiology*, 57: 1696–1700.
Poli, A., Anzelmo, G., and Nicolaus, B. 2009. Bacterial exopolysaccharides from extreme marine habitats: Production, characterization and biological activities. *Marine Drugs*, 8: 1779–1802.
Ridout, M.J., Brownsey, G.J., and Morris, V.J. 1998a. Synergistic interactions of acetan with Carob or Konjac mannan. *Macromolecules*, 31: 2539–2544.
Ridout, M.J., Cairns, C., Brownsey, G.J., and Morris, V.J. 1998b. Evidence for the intermolecular binding between deacetylated acetan and the glucomannan Konjac mannan. *Carbohydrate Research*, 309: 375–379.
Robbins, J.B., Schneerson, R., Xie, G., Hanson, L.Å., and Miller, M.A. 2011. Capsular polysaccharide vaccine for Group B *Neisseria meningitidis, Escherichia coli* K1, and *Pasteurella haemolytica* A2. *Proceedings of the National Academy of Sciences*, 108: 17871–17875.
Ross-Murphy, S.B., Morris, V.J., and Morris, E.R. 1983. Molecular viscoelasticity of xanthan polysaccharide. *Faraday Symposia of the Chemical Society*, 18: 115–129.
Ruas-Madiedo, P., Hugenholtz, J., and Zoon, P. 2002. An overview of the functionality of exopolysaccharides produced by lactic acid bacteria. *International Dairy Journal*, 12: 163–171.
Sanderson, G.R. 1982. The interactions of xanthan gum in food systems. *Progress in Food and Nutrition Science*, 6: 77–87.
Sarwat, F., Qader, S.A.U., Aman, A., and Ahmed, N. 2008. Production and characterization of a unique dextran from an indigenous *Leuconostoc mesenteroides* CMG713. *International Journal of Biological Sciences*, 4: 379–386.

Sato, T., Norisuye, T., and Fujita, H. 1984. Double-stranded helix of xanthan in dilute-solution—Evidence from light-scattering. *Polymer Journal*, 16: 341–350.

Schiraldi, C., la Gatta, A., and de Rosa, M. 2010. Biotechnological production and application of hyaluronan. In: *Biopolymers*, ed. M. Elnashar, pp. 387–412, Sciyo, Rijeka, Croatia.

Shatwell, K.P., Sutherland, I.W., Ross-Murphy, S B., and Dea, I.C.M. 1991a. Influence of the acetyl substituent on the interaction of xanthan with plant polysaccharides—I. Xanthan—Locust bean gum systems. *Carbohydrate Polymers*, 14: 29–51.

Shatwell, K.P., Sutherland, I.W., Ross-Murphy, S.B., and Dea, I.C.M. 1991b. Influence of the acetyl substituent on the interaction of xanthan with plant polysaccharides—II. Xanthan—Guar gum systems. *Carbohydrate Polymers*, 14: 115–130.

Singh, R.S., Saini, G.K., and Kennedy, J.F. 2008. Pullulan: Microbial sources, production and applications. *Carbohydrate Polymers*, 73: 515–531.

Singh, R.K., Tiwari, S.P., Rai, A.K., and Mohapatra, T.M. 2011. Cyanobacteria: An emerging source for drug discovery. *The Journal of Antibiotics*, 64: 401–412.

Somers, E.B., Johnson, M.E., and Wong, A.C. 2001. Biofilm formation and contamination of cheese by nonstarter lactic acid bacteria in the dairy environment. *Journal of Dairy Science*, 84: 1926–1936.

Stokke, B.T., Elgsaeter, A., and Kitamura, S. 1993. Macrocyclization of polysaccharides visualized by electron microscopy. *International Journal of Biological Macromolecules*, 15: 63–68.

Stokke, B.T., Smidsrød, O., and Elgsaeter, A. 1989. Electron microscopy of native xanthan and xanthan exposed to low ionic strength. *Biopolymers*, 28: 617–637.

Suker, J., Corbel, M.J., Jones, C., Feavers, I.M., and Bolgiano, B. 2004. Standardization and control of meningococcal C conjugate vaccines. *Expert Review of Vaccines*, 3: 89–96.

Sutherland, I.W. 1989. Microbial polysaccharides—Biotechnological products of current and future potential. In: *Biomedical and Biotechnological Advances in Industrial Polysaccharides*, eds. V. Crescenzi, I.C.M. Dea, S. Paoletti, S.S. Stivala, and I.W. Sutherland, pp. 123–132, Gordon and Breach, New York.

Sutherland, I.W. 1997. Microbial exopolysaccharide structural subtleties and their consequences. *Pure and Applied Chemistry*, 69: 1911–9117.

Sutherland, I.W. 1999. Microbial polysaccharide products. In: *Biotechnology and Genetic Engineering Reviews*, Vol. 16, ed. S.E. Harding, pp. 217–229, Intercept Ltd., Andover, MA.

Sworn, G. 2009. Xanthan gum. In: *Food Stabilisers, Thickeners and Gelling Agents*, ed. A.Imeson, pp. 325–342, Wiley-Blackwell, Oxford, U.K.

Takeuchi, A., Kamiryou, Y., Yamada, H. et al. 2009. Oral administration of xanthan gum enhances antitumor activity through Toll-like receptor 4. *International Immunopharmacology*, 9: 1562–1567.

Valli, R. and Clark, R. 2009. Gellan gum, In: *Food Stabilisers, Thickeners and Gelling Agents*, ed. A. Imeson, pp. 325–342, Wiley-Blackwell, Oxford, U.K.

Vanderslice, R.W., Doherty, D.H., Capage, M.A. et al. 1989. Genetic engineering of polysaccharide structure in *Xanthomonas campestris*. In: *Biomedical and Biotechnological Advances in Industrial Polysaccharides*, eds. V. Crescenzi, I.C.M. Dea, S. Paoletti, S.S. Stivala, and I.W. Sutherland, pp. 145–156, Gordon and Breach, New York.

Vincente-García, V., Ríos-Leal, E., Colderón-Domínguez, G., Cañizares-Villaneuva, R.O., and Olvera-Ramírez, R. 2004. Detection, isolation and characterization of the exopolysaccharide produced by a strain of *Phormidium* 94a isolated from an arid zone in Mexico. *Biotechnology and Bioengineering*, 85: 306–310.

Wang, X., Yuan, Y., Wang, K., Zhang, D., Yang, Z., and Xu, P. 2007. Deproteinization of gellan gum produced by *Sphingomonas paucimobilis* ATCC 31461. *Journal of Biotechnology*, 128: 403–407.

Weigel, P.H., Hascall, V.C., and Tammi, M. 1997. Hyaluronan synthases. *The Journal of Biological Chemistry*, 272: 13997–14000.

Weissmann, B. and Meyer, K. 1954. The structure of hyalobiuronic acid and hyaluronic acid from the umbilical cord. *Journal of the American Chemical Society*, 76:1753–1757.

West, N.X., Hughes, J.A., Parker, D. et al. 2004. Modification of soft drinks with xanthan gum to minimize erosion: A study *in situ*. *British Dental Journal*, 196: 478–481.

Williamson, D.H. 1959. Studies on *lactobacilli* causing ropiness in beer. *Journal of Applied Bacteriology*, 22: 392–402.

Xiao-Mei, G.U., Hou-Ming, W.U., and Gui-Rong, M.A. 2004. Isolation, purification and structural elucidation of EPS-I, an extracellular polysaccharide from *Enterococcus durans*. *Chemical Journal of Chinese Universities*, 25:1288–1290.

Yamada, T. and Kawasaki, T. 2005. Microbial synthesis of hyaluronan and chitin: New approaches. *Journal of Bioscience and Bioengineering*, 99: 521–528.

Yanaki, T., Norisuye, T., and Fujita, H. 1980. Triple helix of *Schizophyllum commune* polysaccharide in dilute solution 3. Hydrodynamic properties in water. *Macromolecules*, 13: 1462–1466.

Yevlampieva N.P., Pavlov G.M., and Rjumtsev E.I. 1999. Flow birefringence of xanthan and other polysaccharide solutions. *International Journal of Biological Macromolecules*, 26: 295–301.

Zanchetta, P., Lagarde, N., and Guezennec, J. 2003a. A new bone-healing material: A hyaluronic acid-like bacterial exopolysaccharide. *Calcified Tissue International*, 72: 74–79.

Zanchetta, P., Lagarde, N., and Guezennec, J. 2003b. Systemic effects on bone healing of a new hyaluronic acid-like bacterial exopolysaccharide. *Calcified Tissue International*, 72: 232–236.

Zubillaga, M., Weill, R., Postaire, E., Goldman, C., Caro, R., and Boccio, J. 2001. Effect of probiotics and functional foods and their use in different diseases. *Nutrition Research*, 21: 569–579.

11

Production of Sweeteners

Pedro Fernandes and Joaquim M.S. Cabral

CONTENTS

11.1 Introduction 387
11.2 Types of Sweeteners 389
11.2.1 Saccharides 389
11.2.1.1 Sugars 389
11.2.1.2 Oligosaccharides 391
11.2.2 Nonsaccharides 391
11.2.2.1 Sugar Alcohols 391
11.2.2.2 Natural High-Intensity Sweeteners 393
11.2.2.3 Synthetic High-Intensity Sweeteners 395
11.3 Sweeteners from Starch 397
11.3.1 Corn/Glucose Syrups and Maltose Syrups 399
11.3.2 Production of High-Fructose Corn Syrups 402
11.4 Invert Sugar 404
11.5 Fructose Syrup from Inulin (Inulin Syrup) 406
11.6 Summary and Future Prospects 409
Acknowledgment 410
References 410

11.1 Introduction

Developments in the production of sweeteners have been closely related to some relevant events in the fields of biotechnology and chemistry (Ferrier, 2006). The consumption of sweeteners also keeps on increasing alongside with the level of development of countries, as a result of changes in dietary patterns (Ruprecht, 2005). Sweetening agents find application in the food industry for the production of a large number of food products (viz., soft drinks, juices, pastry, chewing gums). They are also used in the pharmaceutical industry in the formulation of both liquid and solid preparations, in order to mask the taste of drugs, give consistency to liquid preparations, and coat pills (Priya et al., 2011). The ideal sweetener has to comply simultaneously with a set of requirements: (1) be highly effective in minimal concentrations; (2) be stable at

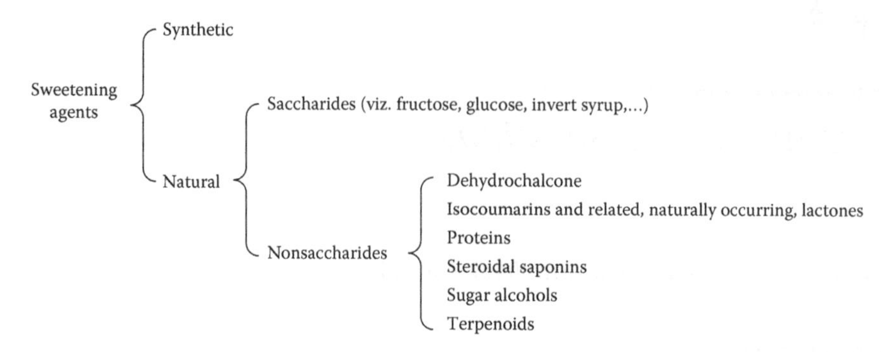

FIGURE 11.1
A schematic division of sweeteners. (From Priya, K. et al., *J. Pharm. Res.*, 4, 2034, 2011.)

a wide range of temperatures to which the formulations are exposed; (3) prolonged use must not convey deleterious effect on health; (4) present a low, if any, calorific value; and (5) production should be easy, reproducible, and have a low cost (Priya et al., 2011). A single sweetener hardly meets all these requirements. The alternative relies on making available a wide array of sweeteners, so that the requirements for the different roles may be fulfilled. The currently available sweeteners are either natural or synthetic, have a saccharide or nonsaccharide nature, and some display high-intensity qualities (e.g., requiring minimal amounts to produce the required sweetening role). An overview of the schematic division of these sweeteners is depicted in Figure 11.1.

Currently, the most relevant sweeteners when sales volumes are considered are sucrose and high-fructose corn syrup (HFCS), the latter a starch derivative (O'Donnell, 2005; Ruprecht, 2005). Sucrose and HFCS are both considered nutritive sweeteners, a group of compounds that besides providing pleasurable sensations in tasting also provide energy. This group is composed of two carbohydrates: sugars and sugar alcohols. As opposed to this group, there are some compounds that are used in very small amounts and impart an intense sweet taste while conveying virtually no energy. They are thus termed nonnutritive sweeteners or intense sweeteners (Shwide-Slavin et al., 2012). Since nonnutritive sweeteners and sugar alcohols can replace sugar sweeteners, they are also known as alternative sweeteners, sugar replacers, sugar substitutes, or macronutrient substitutes (American Dietetic Association, 2004).

Sugar sweeteners are typically mono- and disaccharides, although the term sugar is typically evocative of sucrose. This disaccharide, composed of glucose and fructose, was the first pure sugar and clearly the main player in the sugar market up to the early 1970s, where its leading role was jeopardized, particularly in the United States, with the upsurge of HFCS. Actually, the market share of HFCS increased abruptly between 1972 and 1985 as a result of technological developments in the starch processing industry, as well as of the sudden increase in the world market price for sucrose (Haley and Dohlman, 2009). These features allowed producers to place the market price of HFCS slightly lower than that of sucrose, although such gap has been gradually reducing (Haley and Dohlman, 2009; Mosier and Ladisch, 2011). Despite the competition placed by HFCS and other sweeteners, sucrose retains its leading position in the world market. Also irrespective of the advent of more powerful sweeteners, sucrose is still a reference, or "gold standard" for sweeteners, when taste delivery, mouthfeel, bulking properties, and other functional characteristics are addressed (Godshall, 2007a; O'Donnell, 2005).

11.2 Types of Sweeteners

11.2.1 Saccharides

11.2.1.1 Sugars

Examples of sugars are fructose, glucose, lactose, maltose, sucrose, corn sweeteners, HFCS, honey, invert sugar, maple syrup, and concentrated fruit juice. These more traditional sweeteners meet the specifications for Generally Recognized As Safe (GRAS) status, despite some concern existing about increasing sweetener intakes as related to optimal nutrition and health. The sweetness of relevant sugars is given in Figure 11.2. Based on the Atwater

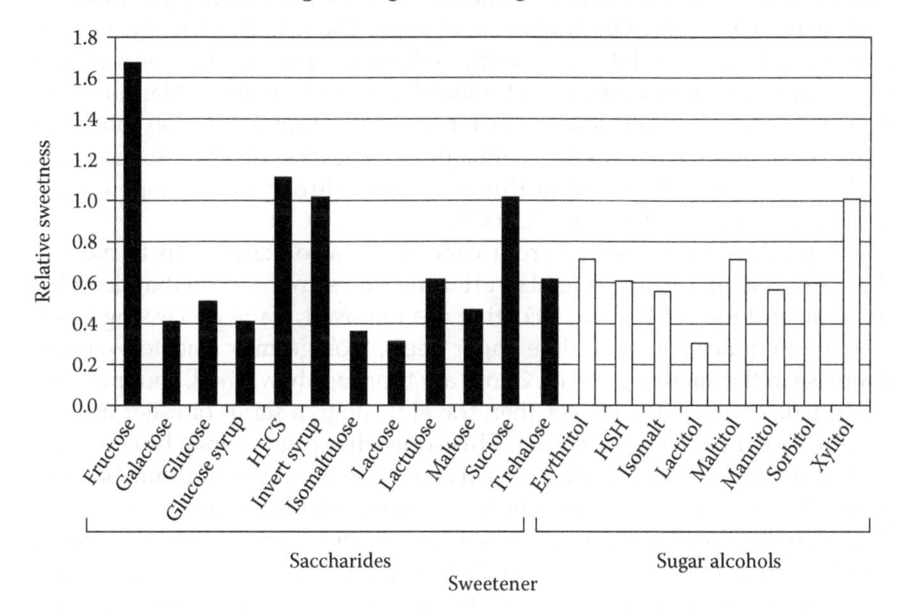

FIGURE 11.2
Relative sweetness of relevant saccharides (black bars) and sugar alcohols (white bars) in relation to sucrose. The sweetness of sucrose is set as 1.0. (From American Dietetic Association., *J. Am. Diet. Assoc.*, 104, 255, 2004; Godshall, M.A., *Sugar J.*, 69, 12, 2007a; Chattopadhyay, S. et al., *J. Food Sci. Technol.*, DOI 10.1007/s13197-011-0571-1, 2011.)

system, which basically relies on the determination of heats of combustion, sugars have a caloric value of 16.7 kJ/g (O'Donnell, 2005). Sugars and high-intensity sweeteners are often blended in order to provide a product that exhibits the taste profile of sugar while allowing for a significant reduction in calories (Rahtjen and Schwarz, 2007).

Sucrose is by far the most widely disseminated sweetener. Sucrose is obtained mostly from beet or cane sugar, although other sources for sucrose production are also used: date, palm, maple, and sorghum syrups (Belitz et al., 2009).

Beet processing starts with harvesting, usually within mid-August to mid-September. The beets are thoroughly washed and separated from unwanted material, so that they can be conveyed to the extraction process. Beets are sliced into thin chips, about 2–3 mm thick and 4–7 mm wide, and leached with slightly acidic water (pH around 5.7), containing Ca^{2+} in order to stabilize skeletal substances from the slices. This is performed at temperatures within 70°C–80°C, which favors extraction and eliminates unwanted microorganisms (Asadi, 2007; Belitz et al., 2009). The resulting raw sugar-rich juice is purified by adding lime and carbonation gas. This allows the removal of all insoluble substances and of some non-sugar-soluble substances and results in a thermostable juice with low hardness and easy filtration quality. Ion exchangers have been also introduced in the purification process in order to further soften the juice, enhance the removal of non-sugar material, and decrease the risk of fouling in later stages. The resulting thin syrup with around 17% (w/w) solids is concentrated, preferably by rapid evaporation under vacuum, to lower the risk of colored product formation (Maillard reaction and caramelization), leading to a thick juice, about 70% (w/w) solids and a sucrose content of 65% (w/w), with a purity in excess of 90%. Molasses and white sugar are obtained from this thick juice through multistage crystallization (Asadi, 2007; Belitz et al., 2009).

The production of sucrose from cane sugar also starts with harvesting (hand- or machine-cut), preferably after the canes are fired, so that dead leaf material and part of the waxy coating are removed. Cane processing has to start immediately, since, unlike sugar beets, storage may lead to excessive decrease in the sucrose content. Canes are thoroughly washed, chopped, and shredded. The shredded canes then travel through a series of roller mills so that the sugar juice is poured out. This squeezing process may be combined with extraction by mixing the pressed cane with hot water or dilute hot cane juice. The repeated squeezing allows a sucrose yield exceeding 95% (w/w). The syrup is limed for neutralization and clarification. Further processing emulates that of beet sugar (Asadi, 2007; Belitz et al., 2009). Corn sweeteners and derivatives are obtained from the hydrolysis of starch. This can eventually be coupled with glucose isomerization, to yield HFCS. Invert sugar syrup is obtained through hydrolysis of sucrose, whereas fructose-rich syrups may be obtained from further processing of HFCS or of invert sugar syrup or from the hydrolysis of inulin (Fernandes, 2010).

11.2.1.2 Oligosaccharides

Examples of this class of sweeteners are fructo-, galacto-, gentio-, palatinose-, and xylo-oligosaccharides. Two different approaches are used for the production of oligosaccharides. Either they are synthesized de novo, using glycosidase and glycosyltransferase activities, and synthesized by chemical glycosylation or they are obtained through biological, chemical, or physical methods (Patel and Goyal, 2011; Torres et al., 2010). These sweeteners have a low caloric value since they are low digestible or indigestible but fermentable; their caloric value is lower than that of sugars, within 4.2–8.4 kJ/g, and they have a low cariogenicity nature (Nakakuki, 2002; Otaka, 2006; Roberfroid, 1999). The relative sweetness is somewhat low, typically 50% or less of sucrose. They are mostly used as prebiotics, as they have a positive impact on the intestinal microbiota. They can selectively stimulate the growth or activity of beneficial bacteria in the colon (e.g., *Lactobacillus* spp. and *Bifidobacterium* spp.) while inhibiting the growth of harmful bacteria (e.g., *Enterobacterium* and *Enterococcus*). Furthermore, prebiotics are likely to favor the absorption of minerals (Godshall, 2007a). Prebiotic nature is not exclusively of oligosaccharides; it is also displayed by other low-calorie sweeteners such as lactulose, lactitol, and tagatose (Chattopadhyay et al., 2011; Godshall, 2007a).

11.2.2 Nonsaccharides

11.2.2.1 Sugar Alcohols

Erythritol, hydrogenated starch hydrolysates (including hydrogenated glucose, maltitol, and sorbitol syrups), isomaltitol, lactitol, maltitol and maltitol syrups, mannitol, sorbitol, tagatol, D-trehalose, and xylitol are sugar alcohols that are considered either GRAS or approved food additives. These polyols are listed on food labels when a nutrient claim is made, such as sugar-free (Godshall, 2007a). Unlike sugars, sugar alcohols are typically noncariogenic, since they are not converted to acids by oral bacteria that produce cavities, thereby providing a functional food benefit (Patra et al., 2009). Another potential benefit of sugar alcohols, particularly for people with diabetes, is their low glycemic response (which measures the impact of a food in blood sugar), as compared to some sugars (Figure 11.3) (American Dietetic Association, 2004; Patra et al., 2009).

The glycemic response to the intake of a carbohydrate (or other nutrient) depends on the amount, rate of digestion, absorption, and metabolism of the ingested carbohydrate (Aller et al., 2011). Thus, and in order to quantify the effect of carbohydrate content of a food on the postprandial blood glucose concentration, the glycemic index (GI) was introduced (Aller et al., 2011; Belitz et al., 2009). Since they are incompletely absorbed in the small intestine and despite being carbohydrates, sugar alcohols provide an amount of energy that averages 8.4 kJ/g, which is about half of that from typical sugars (American Dietetic Association, 2004). Also as a consequence of incomplete absorption, sugar alcohols, such as mannitol and sorbitol, may cause flatulence or have a laxative effect

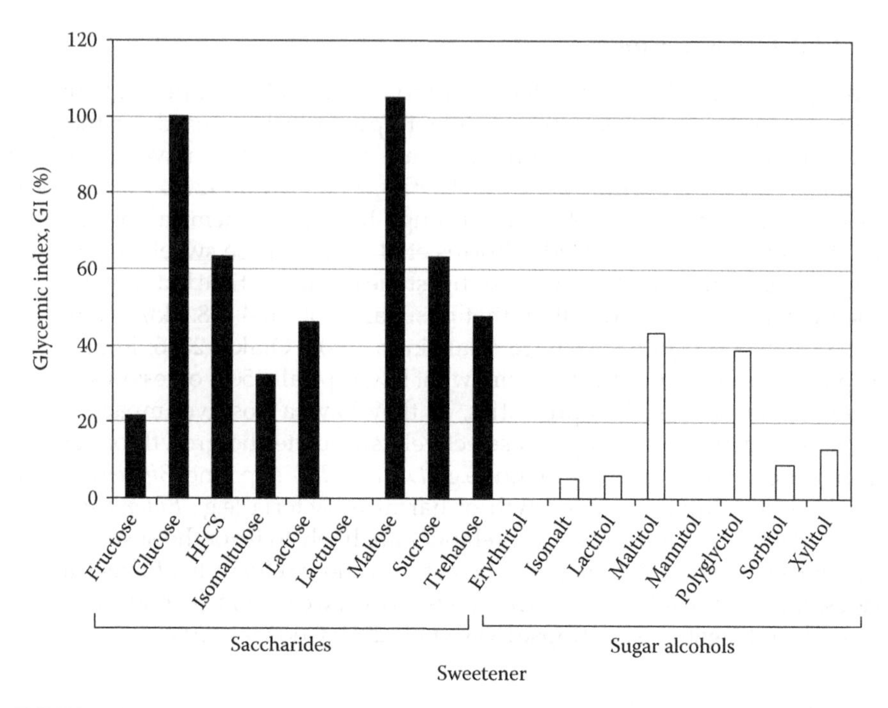

FIGURE 11.3
GI of some relevant saccharides (black bars) and sugar alcohols (white bars). The reference value is the increase in blood sugar after the intake of 50 g of glucose, which corresponds to a GI of 100%. Lactulose, erythritol, and mannitol display a null GI. (From Godshall, M.A., *Sugar J.*, 69, 12, 2007a; Belitz, H.-D. et al., 2009. *Food Chemistry*, 4th edn., Springer-Verlag, Berlin, Germany; Aller, E.E.J.G. et al., *Nutrients*, 3, 341, 2011.)

when consumed in excess (Godshall, 2007a). The sweetness of sugar alcohols varies within 30%–100% of sucrose (Figure 11.2); hence, they are often mixed with high-intensity sweeteners for a sweeter flavor (Gelov, 2011). Within sugar alcohols, the most produced are sorbitol, mannitol, and xylitol. Representative of their relevance is the estimated worldwide production of sorbitol of 700,000 ton/year (Climent et al., 2011). Industrial, large-scale production of sugar alcohols currently relies on the catalytic hydrogenation of the carbonyl group of the corresponding aldose (viz., glucose) or ketose (viz., fructose). Although the reaction is apparently simple, it can be complicated by several factors. Thus, in aqueous solution, the sugar molecules coexist as acyclic aldehydes and ketoses and as cyclic furanoses and pyranoses. Reaction kinetics is thus complex and this pattern is further enhanced since side reaction also occurs (viz., hydrolysis, isomerization, oxidative dehydrogenation). Furthermore, since heterogeneous catalysts are involved, mass transfer limitations often take place, and catalyst deactivation occurs. Hydrogenation is carried out at relatively high temperatures and pressures, typically within 80°C–250°C and within 4–15 MPa (Hirth et al., 2011), usually in slurry reactors, where catalyst with an average size under 100 μm contacts a gas–liquid mixture, but continuous operation with trickle bed or monolith reactors has been implemented (Salmi et al., 2010). Ni-based catalysts are the preferred catalysts, but these are prone to deactivation due to nickel leaching (Climent et al., 2011; Salmi et al., 2010); hence, other catalysts have also been tested. Ruthenium catalysts are the most promising (Climent et al., 2011; Hirth et al., 2011; Salmi et al., 2010). The use of biologic agents to produce sugar alcohols as an alternative to chemical methods presents a safer and greener approach, which in addition favors product specificity. Both enzyme- and fermentation-based processes have been studied for sugar alcohol production, with the former being significantly hampered due to the costs related to preparation of enzymes and cofactor regeneration. The latter approach is thus clearly favored for the production of sugar alcohols from crude feedstocks (Akinterinwa et al., 2008; Godshall, 2007a). In particular, erythritol is produced at commercial scale by fermentation from glucose, using either *Aureobasidium* sp. or *Moniliella pollinis* (Godshall, 2007a; Sasman et al., 2007; Taniguchi, 2004). Recently, a fermentative process anchored on the osmotolerant *Pseudozyma tsukubaensis* KN75 was shown to have potential for commercial application (Jeya et al., 2009). Biotechnological production of xylitol, mostly anchored in yeast cells (viz., *Candida* spp., *Pichia* spp.), or in engineered strains (viz., *Saccharomyces* spp., *Escherichia coli,* lactic acid bacteria), has also been extensively scrutinized, but to date, commercial production still relies on chemical hydrogenation (Akinterinwa et al., 2008; Chen et al., 2010; Godshall, 2007a; Monedero et al., 2010). Intensive research efforts are also being made toward the development of a competitive bioprocess for mannitol production, mostly anchored in lactic acid bacteria and in selected yeasts, as well as in engineered *E. coli* and *Bacillus megaterium* (Akinterinwa et al., 2008; Song and Vieille, 2009). The biotechnological production of sorbitol has mostly relied on *Zymomonas mobilis* (Akinterinwa et al., 2008; Erzinger and Vitolo, 2006; Liua et al., 2010). In addition, the use of engineered lactic acid bacteria has been also successfully tapped (Monedero et al., 2010).

11.2.2.2 Natural High-Intensity Sweeteners

These are compounds isolated from extracts of plants and fruits and are being commercialized (or about to be so) as either sweeteners or flavor enhancers (Priya et al., 2011). Like synthetic high-intensity sweeteners, natural high-intensity sweeteners carry out their role with minimal amounts, conveying a much sweeter taste than sucrose (Figure 11.4) while providing a low (if any) caloric input (Priya et al., 2011).

Some of the best known are terpenoid steviosides (e.g., extracts from *Stevia rebaudiana*). Among them are glycosides, such as stevioside and rebaudioside A–E, which have a noncariogenic nature (Brahmachari et al., 2011; Priya et al., 2011) and glucuronide saponins from licorice root, such as glycyrrhizin and apioglycyrrhizin (Kitagawa, 2002). Ammoniated glycyrrhizin, which is 100-fold sweeter than sucrose, is already commercially available (Priya et al., 2011); the same applies to mogroside V, a triterpene glycoside from *Siraitia grosvenorii,* also known as Luohan Guo (Chaturvedula and Prakash, 2011; He and

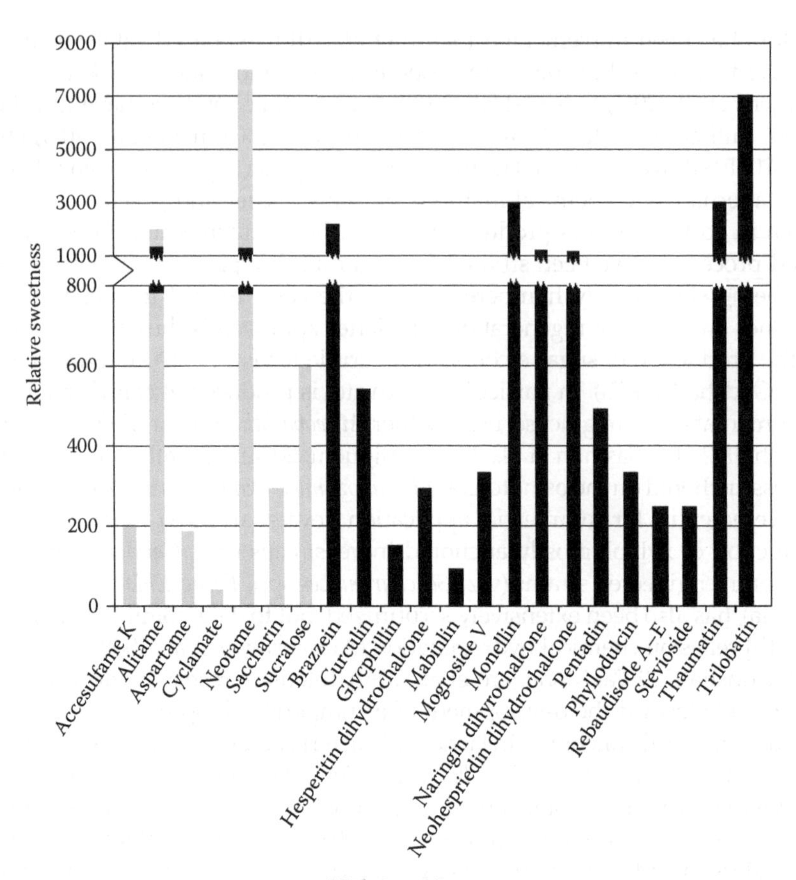

FIGURE 11.4
Relative sweetness of natural (black bars) and artificial (light gray bars) high-intensity sweeteners. Values are relative to sucrose sweetness, which is set as 1.0. (From Godshall, M.A., *Sugar J.*, 69, 12, 2007a; Chattopadhyay, S. et al., *J. Food Sci. Technol.*, DOI 10.1007/s13197-011-0571-1, 2011; Priya, K. et al., *J. Pharm. Res.*, 4, 2034, 2011.)

Zhou, 2000; Zhang et al., 2011). Sweet-tasting proteins from the fruits of African and Asian plants have also been extracted and identified. Among these are brazzein (from West African *Pentadiplandra brazzeana*); curculin (Malayan *Curculigo latifolia*); mabinlin (from Chinese *Capparis masaiki*); miraculin (from West African *Richadella dulcifica*), which rather than a sweetener displays the odd property of being able to convert sour stimuli to sweetness; monellin (from West African *Dioscoreophyllum cumminsii Diels*); pentadin (from West African *P. brazzeana*); and thaumatin (from West African *Thaumatococcus daniellii Benth*) (Faus and Sisniega, 2005; Koizumi et al., 2011; Priya et al., 2011). Phyllodulcin, an isocoumarin derivative, is another natural sweetener, obtained from the leaves of *Hydrangea serrata* (a plant indigenous of Japan and China but also found in North and South America and in temperate hills of India) (Priya et al.,

2011; Ujihara et al., 1995). Dihydrochalcone derivatives, such as glycyphyllin, extracted from the leaves of *Smilax glyciphylla* (Rennie, 1886), and trilobatin, extracted from leaves of *Lithocarpus litseifolius* (Nance) Rehd. (Fagaceae) (Ruilin et al., 1982), have also been identified as natural high-intensity sweeteners. Given the potential of natural dihydrochalcone molecules as sweetening agents, several derivatives have been synthesized. A typical process starts with a flavanone glycoside. This is converted to the corresponding chalcone molecule, which is then reduced to dihydrochalcone (Givaudan & Cie, 1976; Priya et al., 2011). Examples are neohesperidin dihydrochalcone (Robertson et al., 1974), naringin dihydrochalcone (Tang et al., 2011), and hesperetin dihydrochalcone glucoside (Nutrilite Products Inc. 1974). Steroidal saponins, such as osladin, polypodoside A, pterocaryosides A and B, and strogin, isolated from plants, have also been shown to display intense sweetening activity (Priya et al., 2011).

11.2.2.3 Synthetic High-Intensity Sweeteners

Synthetic high-intensity sweeteners include acesulfame-K, alitame, aspartame, cyclamate, neotame, saccharin, and sucralose. These synthetic sweeteners have no caloric value (Godshall, 2007a).

11.2.2.3.1 Acesulfame-K

The process for the synthesis of acesulfame-K starts with the reaction between diketene and sulfamic acid to produce acetoacetamide-N-sulfonic acid. The latter cyclizes to the acesulfame ring system in the presence of sulfur trioxide. Acesulfame reacts in the presence of potassium hydroxide to form acesulfame-K (Haber et al., 2006).

11.2.2.3.2 Alitame

Synthesis of alitame, L-α-aspartyl-N-(2,2,4,4-tetramethyl-3-thietanyl)-D-alaninamide, is a proprietary multistep process. Still it is known that it involves two intermediates, i.e., (S)-[(2,5-dioxo-(4-thiazolidine)] acetic acid and (R)-2-amino-N-(2,2,4,4-tetramethyl-3-thietanyl)propanamide. The resulting product is purified through crystallization of an alitame/4-methylbenzenesulfonic acid adduct, which is followed by further purification and a final recrystallization from water (Brennan and Hendrick, 1983; Lee, 2007: 397–399).

11.2.2.3.3 Aspartame

Aspartame is the common name for the dipeptide α-L-aspartyl-L-phenylalanine methyl ester (L-Asp-L-Phe methyl ester), where L-aspartic acid (L-Asp) and L-phenyl alanine (L-Phe-Ala) are joined through a methyl ester link (O'Donnell, 2005). Although this sweetener can be chemically synthesized, β-aspartame, an isomer with bitter taste, is formed as a by-product in the process and has to be fully separated from the α-isomer (Liese et al., 2006). The enzyme-based production of this sweetener relies in the use of thermolysin (EC 3.4.24.27), a neutral, thermostable zinc endopeptidase that typically hydrolyzes peptide bonds containing hydrophobic amino acid residues. Under adequate operational conditions, thermolysin catalyzes the reverse reaction resulting in the synthesis of the precursor of aspartame (Z-L-Asp-L-Phe methyl ester), starting from D,L-Phe methyl ester and L-Asp (protected). After hydrogenolysis, for removal of the protective group, the precursor is converted into aspartame (Bommarius and Riebel-Bommarius, 2004). The regiospecificity of thermolysin eliminates the need to protect the β-carboxyl group of L-aspartic acid. Nevertheless, protection of the α-amino group is still needed to prevent the synthesis of poly-L-aspartic acid (Bucke and Chaplin, 1990). Besides, thermolysin displays no or hardly any esterolytic activity; hence, the methyl ester group essential for the sweet taste of the peptide is preserved (Bommarius and Riebel-Bommarius, 2004). The synthesis of Z-L-Asp-L-Phe methyl ester is performed under thermodynamic controlled mode, since in the absence of acyl enzyme formation, the synthesis cannot be kinetically controlled (Hagen, 2006). In order to shift the equilibrium to favor synthesis, different approaches have been suggested such as addition of organic solvents, either water soluble or insoluble; precipitation of the product; or complexation of the product in an equilibrium-driven sequential reaction. A mixture of the latter two is used in the industrial production of aspartame, relying on the fact that, under excess of L-phenylalanine methyl ester, the protected aspartame forms a poorly soluble carboxylate anion that precipitates from the reaction mixture and is easily removed by filtration (Bommarius and Riebel-Bommarius, 2004; Hagen, 2006; Liese et al., 2006). The feasibility of continuous operation for the production of aspartame using immobilized thermolysin, in solid carriers, as cross-linked enzyme crystals, or in membrane reactors, and operating in either aqueous/organic two-phase systems or low water systems, has also been evidenced at laboratory scale (West, 2007). Irrespectively of the approach, the key issue for successful synthesis requires a compromise between initial reaction rates and final yields (West, 2007).

11.2.2.3.4 Cyclamate

Cyclamate is the sodium salt of cyclamic acid (a calcium salt of said acid being also commercially available for use in low-sodium diets), and its synthesis is performed through the sulfonation of cyclohexylamine (Belitz et al., 2009; Lee, 2007). While the sweetness is compared with sucrose, the sweetness from cyclamate builds to a maximum more slowly but persists longer (Hunt et al., 2012). Its sweetness is relatively low for a high-intensity sweetener, but it is synergistic with plenty of other sweeteners (Godshall, 2007a). Cyclamate had widespread use in the United States in the 1960s where it was granted GRAS status, but after extensive toxicity studies in animals, cyclamate was banned in the United States, although it still used in other countries (Godshall, 2007a; Lee, 2007).

11.2.2.3.5 Neotame

Neotame, N-(N-(3,3-dimethylbutyl)-L-aspartyl)-L-phenylalanine-1-methyl ester, is a derivative of aspartame, and its synthesis is carried out from aspartame and 3,3-dimethylbutyraldehyde through reductive alkylation followed by purification and further processing (Prakash, 2007). A chemoenzymatic method for neotame production has been developed, which relies on the regioselective hydrolysis of neotame esters, using either a lipase or an esterase (Prakash and Zhao, 2001).

11.2.2.3.6 Saccharin

Saccharin is synthesized by purely chemical processes (Chattopadhyay et al., 2011). In the more basic route, toluene reacts with chlorosulfonic acid to produce *o*- and *p*-toluene sulfonyl chloride. Treatment with ammonia leads to the corresponding toluene sulfonamides with the *o*-isomer being isolated and oxidized to *o*-sulfamoylbenzoic acid, which is cyclized to saccharin by heating. Another approach relies on the production of saccharin starting with anthranilic acid, which is diazotized through the addition of sodium nitrite and sulfuric acid. The resulting solution is added to sodium sulfite and the product formed is esterified with methanol and oxidized with chlorine to produce *o*-carbomethoxy benzenesulfonyl chloride. Amidation of the latter under excess ammonia results in ammonium saccharin, which is converted to saccharin by neutralization with sulfuric acid. This second approach with slight changes has been implemented into continuous mode of operation (Bakkal and Nabors, 2012; Chattopadhyay et al., 2011; Lee, 2007).

11.2.2.3.7 Sucralose

Sucralose is a trichlorinated sucrose derivative synthesized by several steps of selective protection and deprotection to produce a trichlorodisaccharide, which, in a final step, is deprotected to give sucralose (Chattopadhyay et al., 2011). Several approaches have been suggested to perform synthesis of sucralose in higher yields and low toxicity environment. In a recent process, sucrose is transformed into sucrose-6-acetate using an azo reagent as catalyst and acetic acid as an acylating agent, sucrose-6-acetate is converted to sucralose-6-acetate using an adequate chlorinating agent and trichloroacetonitrile as catalyst, and finally, sucralose-6-acetate is converted into sucralose under KOH/methanol (Wang et al., 2011a).

11.3 Sweeteners from Starch

Starch is the main reserve material of plants. Actually, this carbohydrate is a mixture of two polysaccharides, typically 70%–80% amylopectin and 20%–30% amylose. Amylopectin is a branched polymer composed of α-D-glucopyranosyl units mostly linked by (1→4) bonds with branches resulting from (1→6) linkages. On the other hand, amylose is mostly a linear polymer (Figure 11.5) composed of (1→4)-linked α-D-glucopyranosyl units (Shannon et al., 2009).

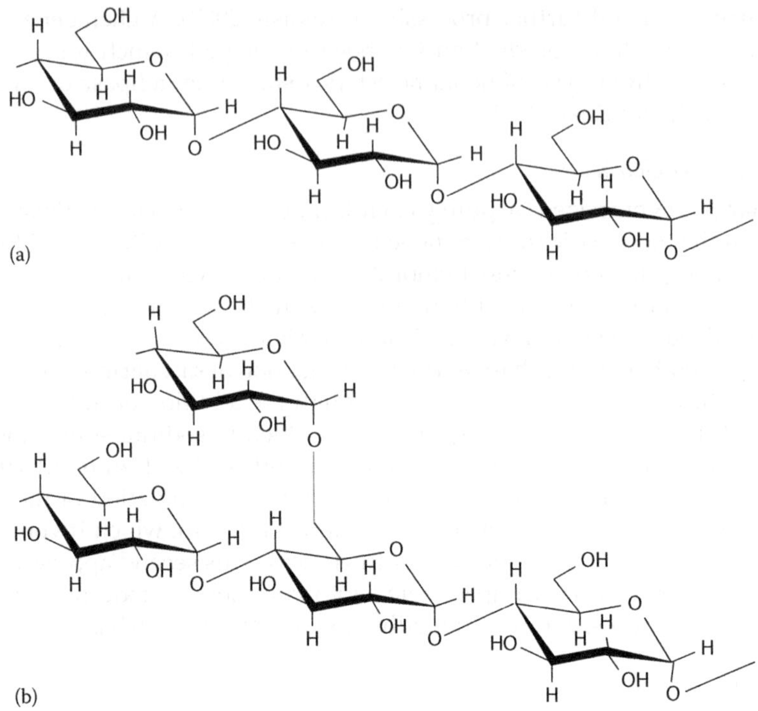

(a)

(b)

FIGURE 11.5
Structures of amylose (a) and of amylopectin (b), major components of starch.

Starch hydrolysis leads to a wide array of products, such as oligosaccharides, maltodextrins and glucose syrups, HFCS, crystalline fructose, and dextrose (Barrie et al., 2009; Belitz et al., 2009; Bisgård-Frantzen and Svendsen, 2008; Fernandes, 2010; Hobbs, 2009; Wang et al., 2011b). Starch (corn) sweeteners are often classified according to their dextrose equivalence (DE), a parameter that accounts for the total reducing sugars calculated as D-glucose on a dry weight basis. Accordingly, nonhydrolyzed starch has a DE of zero, whereas anhydrous D-glucose has a DE of 100 (Hobbs, 2009). An overall scheme for the production of starch sweeteners is given in Figure 11.6.

Starch processing for sweetener production involves three steps: (1) gelatinization, where the nanogram-sized starch granules are dissolved to form a viscous suspension; (2) liquefaction, where starch is partially hydrolyzed, concomitantly leading to a viscosity decay; and (3) saccharification, where further hydrolysis is promoted, resulting in the production of glucose and maltose by further hydrolysis (Barrie et al., 2009; Liese et al., 2006). The former can be later partially isomerized to fructose, resulting in HFCS.

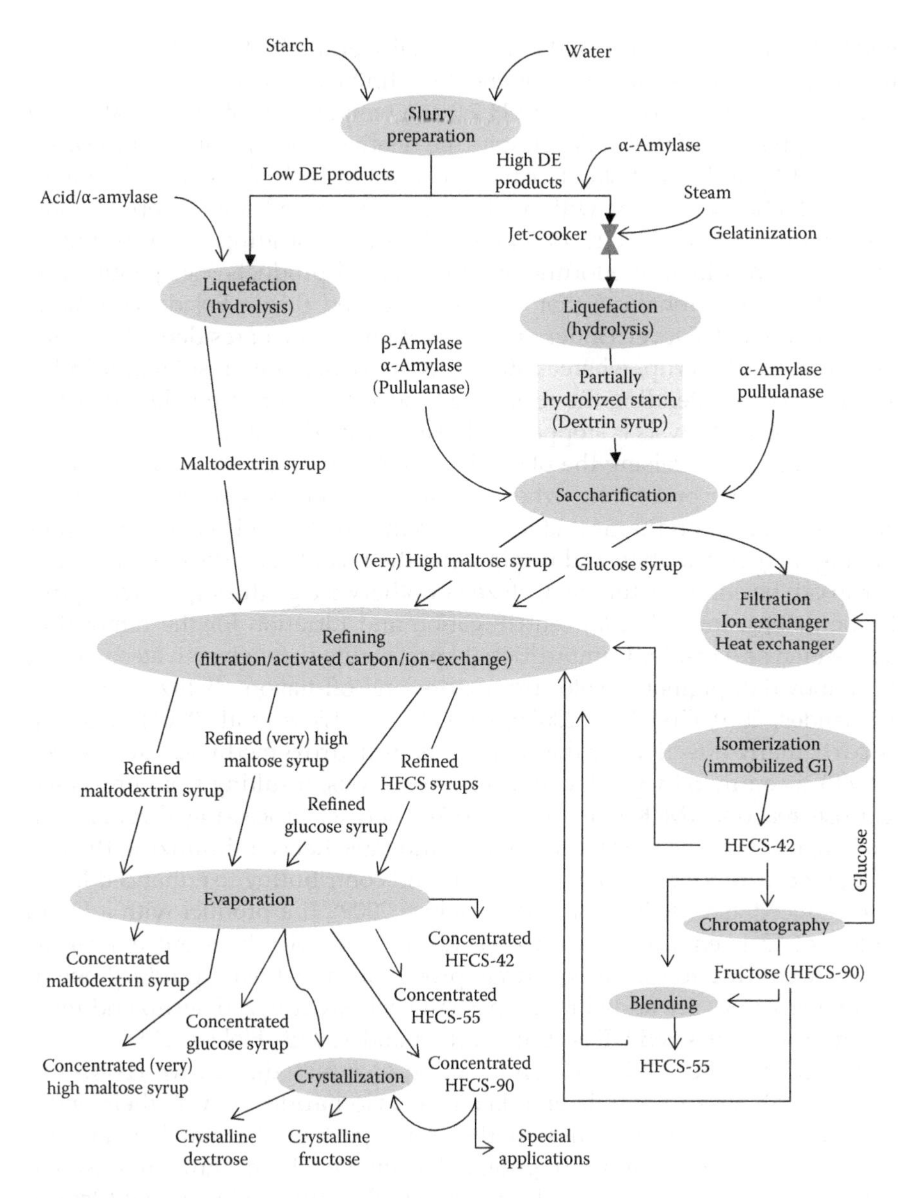

FIGURE 11.6
A schematic overview of the production of sweeteners from starch.

11.3.1 Corn/Glucose Syrups and Maltose Syrups

The processes for hydrolysis typically start with aqueous starch slurry of 30%–45% (w/w). When starch syrups (e.g., glucose or maltose syrups) are envisaged, starch saccharification is performed by either enzymatic (α- and β-amylases, amyloglucosidase, pullulanase) or acid hydrolysis (sulfuric or hydrochloric acid) or a mixed process (Belitz et al., 2009; Fernandes, 2010). In acid hydrolysis of starch, α-1,4 and α-1,6 links are cleaved. A typical continuous process is carried out at pH 2.0 at a temperature of 140°C–160°C and under a pressure of 0.54 MPa. Despite the relative random nature of the acid-catalyzed hydrolysis of starch, careful control of hydrolysis actually allows the production of syrups with a DE within 25–45 and with a reproducible carbohydrate profile. On the other hand, hydrolysis leading to syrups with a DE over 55 results in the formation of unwanted products (e.g., gentiobiose, isomaltose, and trehalose) that convey undesired flavor. Residence time is determinant in DE level. Under given conditions, 8 min of residence time may lead to a 42 DE syrup, whereas 10 min of residence time results in a 55 DE syrup. Further extending the residence time leads to unacceptable color formation. The hydrolysis is stopped by the addition of soda carbonate in a neutralizer tank, thus raising the pH to 4.5–5.0. Again, in a manner reminiscent of the beet sugar processing, when lime and carbonated gas are added, soda addition creates conditions that favor the removal of proteins and fats by promoting their flocculation and minimizing the risk of color development. The syrup can be either further hydrolyzed or otherwise go through several purification steps, involving (a) centrifugation and filtration for the removal of suspended and insoluble impurities; (b) processing through activated carbon, for removal of pigments, color precursors, and off-flavors (Belitz et al., 2009; Fernandes, 2010; Godshall, 2007b; Hobbs, 2009; Liese et al., 2006) (this step is particularly effective for the removal of 5-(hydroxymethyl)-2-furaldehyde (HMF), a common by-product of acidic hydrolysis, resulting from unwanted glucose reaction (McKibbins et al., 1962); and (c) processing through ion-exchange columns, in order to remove minerals, hence minimizing the risk of later occurrence of Maillard reactions and contributing to enhanced flavor and reduced ash (Belitz et al., 2009; Hobbs, 2009). If a product with a DE in excess of 55 is envisaged, a mixed acid–enzyme hydrolysis process can be used. According to the carbohydrate profile of the targeted product, one or several enzymes are added to the partially acid-hydrolyzed syrup and incubated until the desired DE is attained (Fernandes, 2010; Hobbs, 2009).

α-Amylases (EC 3.2.1.1) are endo-acting enzymes that cleave (1→4)-α-D-glucosidic bonds of starch or related polysaccharides. Given their endo-mechanism, glucose residues in the inner part of the starch chain are attacked; hence, the chain is fragmented in smaller chains. Enzymes used in starch processing are of fungal (viz., *Aspergillus* spp.) and bacterial (*Bacillus* spp.) sources. The carbohydrate profile of a starch suspension processed with α-amylase comprehends mostly maltose, maltotriose, maltotetraose, and maltopentaose. Apart from these, glucose, maltohexaose, and maltoheptaose are also produced. Most of the α-amylases used in starch processing have pH optima at about 6.5 and are usually stabilized by Ca^{2+} ions. To favor process integration in starch processing, efforts have been made for engineering bacterial α-amylases, aiming at enhanced thermostability and activity in low pH media and independency of metal ions (Fernandes, 2010; Hobbs, 2009).

β-Amylases (EC 3.2.1.2) are exo-acting enzymes, which cleave (1→4)-α-D-glucosidic bonds. However, β-amylases bind glucose residues at the nonreducing ends of the starch chains. These enzymes are from plant sources, such as sweet potatoes, soybeans, barley, and wheat, or from bacterial sources, such as *Bacillus* spp. These enzymes produce β-maltose and a high molecular weight β-dextrin, which are formed when the enzyme reaches an α-(1→6) branch linkage, which it cannot break and are accordingly used in the production of high-maltose syrups (van der Maarel, 2009).

Glucoamylases (amyloglucosidases or saccharifying amylases) (EC 3.2.1.3) are obtained from fungal sources, such as *Aspergillus* spp. and *Rhizopus* spp., and from yeasts. Glucoamylases catalyze the hydrolysis of both (1→4) and (1→6) α-D-glucosidic bonds in starch, although the rate of hydrolysis of the former vastly exceeds (around 600-fold) that of the latter. An exo-acting enzyme, glucoamylase, leads to the release of β-D-glucose from the nonreducing ends of starch and related poly- and oligosaccharides (Robyt, 2009).

Pullulanases (EC 3.2.1.41) are enzymes obtained from bacterial sources, which hydrolyze the α-D-α-(1→6)-glucosidic linkages of pullulan, an essentially linear polysaccharide composed mostly of repeating maltotriose units. Pullulanases also cleave α-(1→6) linkages of amylopectin and glycogen. Since pullulanases specifically hydrolyze branch linkages to produce linear chains, they are called debranching enzymes. Pullulanases are used to complement the action of glucoamylases in the production of glucose syrups with high DE, as well as in the production of extremely-high-maltose syrups, with up to 90% levels (Belitz et al., 2009; Fernandes, 2010).

Purely enzymatic processes for starch saccharification process performed in batch mode have gained relevance, when either maltose or glucose syrups are aimed at. A typical methodology for the production of such syrups starts with the preparation of a starch slurry of 30%–40% (w/w). After adjusting the pH to 6.0 and adding calcium ions and a thermostable α-amylase, the slurry is pumped into a jet-cooker, where steam at 105°C is injected to gelatinize the starch, in a procedure that lasts for 5 min. The gelatinized starch is then transferred to stirred tanks, cooled down to around 95°C, and incubated for further 1–2 h, to allow for hydrolysis until the required DE is reached (Godshall, 2007b; Wang et al., 2011b).

The syrup can then be processed as previously referred prior to concentration by evaporation to the intended level of dry solids, or proceed to the saccharification step, where, after pH adjustments, glucoamylase, pullulanase, α-amylase, or β-amylase is added to further hydrolyze the liquefied starch into maltodextrins, maltose, or glucose syrups. The process is mostly performed within 55°C–70°C and for periods of 24–90 h, again depending on the enzymes used and on the intended DE. When high-glucose syrups are targeted, glucoamylase and pullulanase are used for saccharification (Belitz et al., 2009; Fernandes, 2010; Hobbs, 2009; Liese et al., 2006) (Table 11.1).

When glucose syrups are aimed at, the combined use of glucoamylase and pullulanase over liquefied starch leads to a syrup with high DE (97)

TABLE 11.1

Examples of Maltose Syrups

Category of Syrup	DE	Maltose Level% (w/w)
High-conversion syrup	≈62	40
High-maltose syrup	48–52	≈50
Very-high-maltose syrup	50–60	70–80
Ultrahigh-maltose syrups	≈60	>80 (82–88)

Sources: Hobbs, L., Sweeteners from starch: Production, properties and uses, In: *Starch: Chemistry and Technology*, 3rd edn., eds. J.N. BeMiller and R.L. Whistler, pp. 797–832, Academic Press, New York, 2009; Fernandes, P., Enzymes in sugar industries, In: *Enzymes in Food Processing: Fundamentals and Potential Applications*, eds. P.S. Panesar, S.S. Marwaha, and H.K. Chopra, pp. 165–197, I.K. International Publishing House Pvt. Ltd, New Delhi, India, 2010.

and dextrose titer (96%), which can be used as raw material for the production of HFCS, or otherwise undergoes purification, as described for other corn-derived syrups, and concentration by evaporation and crystallization to yield monohydrated crystalline dextrose. The 96% dextrose liquid may also be further refined by adsorption chromatography to yield a 99% dextrose titer, prior to bleaching, demineralization, and concentration by evaporation (Hobbs, 2009; van der Maarel, 2009).

11.3.2 Production of High-Fructose Corn Syrups

After the activated carbon and ion-exchange steps, the pH of the glucose-rich syrup is adjusted to a value within 7.5–8.0, magnesium or cobalt ions are added, and the syrup is passed through a heat exchanger to set the temperature within 50°C–60°C and is deaerated. The syrup is then fed to a packed-bed reactor operated in continuous mode with residence times as low as 20 min, where glucose is isomerized to fructose by immobilized glucose isomerase. The low residence time is required in order to minimize the by-product formation; in order to carry out the intended bioconversion in a cost-effective manner, a high concentration of immobilized enzyme is needed (Fernandes, 2010; Liese et al., 2006). Glucose isomerase (D-xylose ketol isomerase; EC 5.3.1.5) displays higher affinity to D-xylose than to D-glucose. The latter is, however, enough for this enzyme to be naturally used for the isomerization of D-glucose to D-fructose (Bhosale et al., 1996). Main sources of the enzyme are *Streptomyces* spp., *Bacillus coagulans*, and *Actinoplanes missouriensis*. Given the intracellular nature of most glucose isomerase enzymes, when aiming for large-scale applications, the preferred form of immobilization involves cross-linking of partially disrupted cells, although several methods have been developed for enzyme recovery and purification

TABLE 11.2

Characteristics of HFCS Syrups

	Fructose (%)	Glucose (%)	Oligosaccharides (%)	Relative Sweetness[a]
HFCS-42	42	52	6	1.0
HFCS-55	55	41	4	>1.0
HFCS-90	90	9	1	1.2

Sources: Godshall, M.A., *Sugar J.*, 69, 12, 2007a; Belitz, H.-D. et al., 2009. *Food Chemistry*, 4th edn., Springer-Verlag, Berlin, Germany; Robyt, J.F., Enzymes and their action on starch, In: *Starch: Chemistry and Technology*, 3rd edn., eds. J.N. BeMiller and R.L. Whistler, pp. 238–292, Academic Press, New York, 2009.

[a] Relative sweetness as related to sucrose. The sweetness of sucrose is set as 1.0.

(Fernandes, 2010; Godshall, 2007b; Liese et al., 2006). Glucose isomerization is slightly endothermic and is reversible, with an equilibrium conversion of about 50% at 55°C. These features limit fructose yield, which is typically of 42% at the output of the enzymatic process. Still operational half-lives of individual packed-bed reactors of about 200 h are typical (Belitz et al., 2009; Fernandes, 2010; Liese et al., 2006). Thus, plant operation relies on the use of several reactors operating in parallel and containing enzymes of different ages. The fructose level of the output of each of the reactors can be controlled by changing the flow rate and temperature. Once the intended conversion is achieved, the pH is lowered to 4–5 (to minimize the risk of Maillard reactions), and syrup is pumped through activated carbon and ion-exchange resins and then concentrated by evaporation to a dry solid titer of 70%–80% (Liese et al., 2006). The syrup is thus termed HFCS-42 (Table 11.2) and quite often provides a suitable replacement for sucrose solutions.

For some applications, a flavor sweeter than that conveyed by HFCS-42 is required, and in those cases, the level of fructose in the corn syrup must be increased. Despite some alternatives, the industrial-scale approach to cope with this requirement relies on adsorption chromatography, where the different affinity of glucose and fructose to ion-exchange resins is taken advantage of (Hobbs, 2009). A typical implementation anchors on the use of polystyrene cation-exchange resins in the calcium form. Fructose molecules form a complex with the calcium ions and are hence selectively retained. Glucose and remaining oligosaccharides are eluted with hot water. Zeolites in calcium form have been introduced in order to improve the productivity of the process. The use of anion-exchange resins in bisulfate form has also been evaluated, but the lower stability of said resins hampers their application at production scale (Hobbs, 2009). The technology of simulated moving bed, based on the SAREX process, which was developed by UOP, has been implemented for the separation of fructose from glucose and remaining (oligo)saccharides. This process, which has a recovery yield in excess of 90%, starts with the demineralization of HFCS-42 syrup, which is then pumped into the separator. The final product is a fructose-rich syrup, with a purity of at least 90%. The process is carried out at temperatures that overcome diffusion limitation while providing an operating pressure that assures liquid-phase operation (Hobbs, 2009). The raffinate, in case of D-glucose-rich fraction, is blended with the syrup that comes from the saccharification step. The fructose-rich fraction is either blended to HFCS-42 to yield a syrup with 55% fructose, hence termed HFCS-55; isolated as a separate product stream corresponding to a rich (90%) fructose syrup (HFCS-90); or crystallized to 99% (at least) pure fructose (Fernandes, 2010; Hobbs, 2009).

11.4 Invert Sugar

Invert sugar is obtained from the hydrolysis of sucrose and as such is composed of equal parts (50:50) of glucose and fructose. The sweetening power of invert syrup exceeds that of sucrose, since fructose is sweeter that sucrose, while glucose is marginally less sweet than sucrose, and conveys a pleasant flavor to food products. Besides their role as sweetener, liquid invert syrup is commonly used in food products, from baking goods to hold moisture. Moreover, liquid invert syrup contributes to enhance the shelf life of food products (Godshall, 2007b).

Sucrose inversion can be carried through strong acid resins, acid hydrolysis, or enzyme (invertase)-catalyzed hydrolysis. The former method is hardly used, since several impurities result from the process, although the use of zeolites has shown some interesting results (Moreau et al., 2000). A typical acid hydrolysis process is carried out by heating the sucrose solution to a temperature of 90°C–100°C and adding a mineral acid (viz., sulfuric acid, hydrochloric acid) so that the pH drops to 1.0–2.0. The solution is held in a stirred reactor during a residence time within 45–75 min, which allows full sucrose inversion. The pH of the invert sugar is then adjusted to 6.0. The resulting solution is clarified. The temperature is lowered to 75°C. Precipitates are filtered off and the resulting syrup is further purified by processing through activated carbon and ion-exchange resins. After this, the syrup is concentrated by evaporation to the desired concentration (Granguillhome et al., 2005).

To overcome by-products, such as HMF formation, enzymatic hydrolysis is often used instead of relatively harsh acid hydrolysis. The selective invertase (EC 3.2.1.26) promotes the hydrolysis of terminal nonreducing β-D-fructofuranoside residues in β-D-fructofuranosides (Purich and Allison, 2000). Several microorganisms can be used to produce invertase, but yeast strains, such as *Saccharomyces cerevisiae* or *Saccharomyces carlsbergensis*, are the preferred sources. Invertase can be of intra- or extracellular nature. The optimal pH and temperature ranges between 3.5°C and 5.5°C and 25°C and 90°C, most commonly around 4.5°C and 55°C, respectively. Moreover, invertase has been

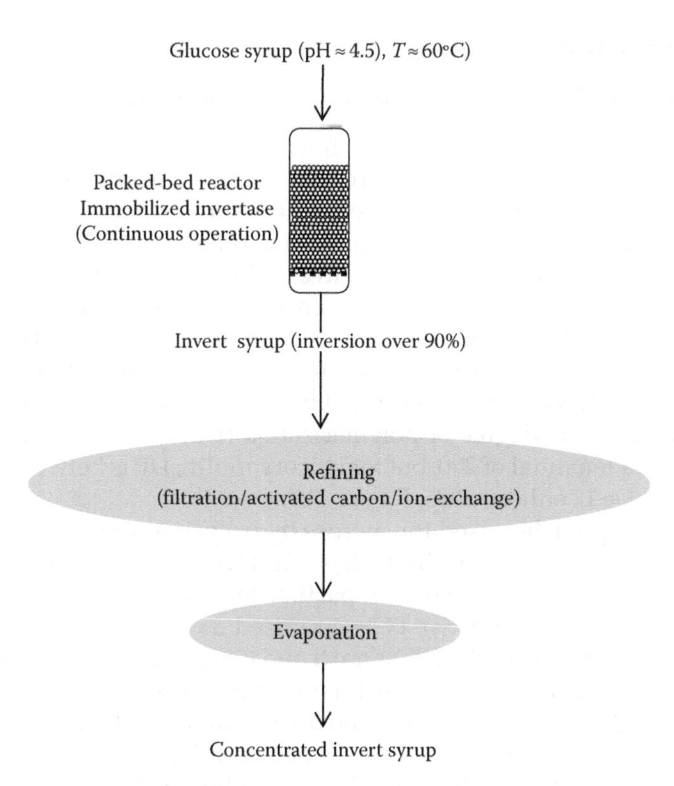

FIGURE 11.7
A schematic overview of the production of invert syrup from glucose.

proved more effective than H^+ as catalyst for sucrose hydrolysis (Fernandes, 2010; Whitaker, 2004). Enzymatic hydrolysis is preferably carried out in continuous operation, using invertase immobilized on a solid support, such as agglomerated diethylaminoethyl cellulose mixed with polystyrene (DEAE cellulose), calcium alginate, Eupergit, and chitosan beads, which is packed in a fixed bed column (Buchholz et al., 2005; Fernandes, 2010; Heikkilä et al., 1995; Lazcano et al., 1993; Serna-Saldivar and Rito-Palomares, 2008). The temperature and pH of the process depend on the enzyme source, but values around 4.5 and within 45°C–60°C are common (Figure 11.7). Residence time varies depending on the biocatalysts, substrate concentration, and inversion level. Some examples refer to 120–220 min for inversion levels above 75% for 60%–75% (w/w) sucrose syrups (Lazcano et al., 1993) or 10 min, for 95% conversion of 50%–55% (w/w) sucrose syrups (Serna-Saldivar and Rito-Palomares, 2008).

The invert sugar syrup can be further processed by filtration, activated carbon, and ion-exchange columns for purification prior to storage, or can be pumped to a chromatographic setup for the recovery and purification of fructose- and glucose-rich fractions, in a process similar to the one described for HFCS (Fernandes, 2010; Heikkilä et al., 1995).

11.5 Fructose Syrup from Inulin (Inulin Syrup)

Inulin syrup refers to the fructose-rich (at least 80%) hydrolysate of inulin (European Commission—Agriculture Directorate-General, 2004). Inulin is a key carbohydrate reserve material found in the roots and tubers of plants, among which burdock, chicory, dahlia, dandelion, garlic, Jerusalem artichoke, leek, or onion are common sources. Chicory roots are, however, the raw material for inulin extraction when industrial-scale applications are considered (Franck and De Leenheer, 2005; Kaur and Gupta, 2002). Inulin is a polydisperse carbohydrate material consisting mostly, if not solely, of (2→1) fructosyl–fructose links, eventually (but not necessarily) starting by a glucose moiety. The degree of polymerization (DP) varies according to the source, with a maximal of 200, but for chicory inulin, DP is between 2 and 70 (Franck and De Leenheer, 2005; Figure 11.8).

The three approaches used for sucrose hydrolysis aiming at invert sugar production also apply for inulin hydrolysis. Hydrolysis can either be total, leading to fructose-rich syrup, or partial, leading to the production of fructooligosaccharides (FOSs). FOSs are used as prebiotics and are commercially available from BENEO-Orafti and Cosucra, among others (Ricca et al., 2007; Singh and Singh, 2010). Enzymes involved in inulin hydrolysis are accordingly termed inulinases and are isolated mostly from *Aspergillus* spp., *Kluyveromyces* spp., *Streptomyces* spp., and *Bacillus stearothermophilus* (Basso et al., 2010; Fernandes, 2010; Singh and Singh, 2010). Based on their reaction mechanism toward the substrate, inulinases can be divided in

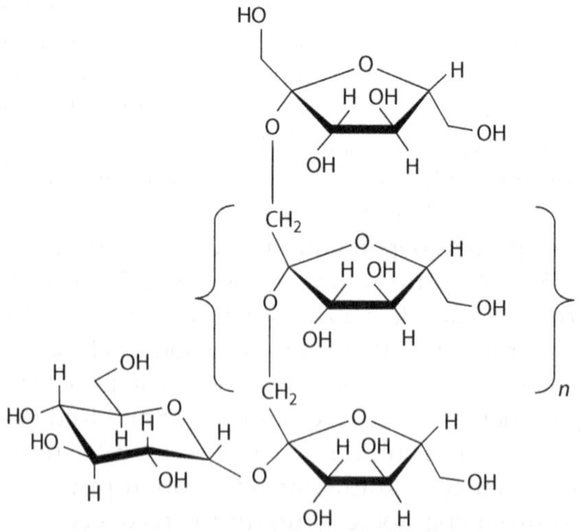

FIGURE 11.8
Structure of inulin; $n = 2–70$.

endo- and exoinulinases. Endoinulinase (E.C.3.2.1.7) promotes the endohydrolysis of (2→1)-β-D-fructosidic linkages in inulin, thereby randomly hydrolyzing internal linkages of the polysaccharide, typically yielding FOS of relatively low DP. Exoinulinases (EC 3.2.1.80) promote the hydrolysis of the terminal nonreducing (2→1)- and (2→6)-linked β-D-fructofuranose residues in inulin and derived fructans, thereby successively releasing fructose molecules (Fernandes, 2010; Singh and Singh, 2010). Ion-exchange resins and zeolite catalysts have been used for full inulin hydrolysis (Abasaeed and Lee, 1995; Yamazaki and Matsumoto, 1986). In a typical ion-exchange resin process, 5%–10% warm aqueous solution of inulin is pumped through a column containing a strong acid cation-exchange resin. This results in an effluent with a pH within 2–3, which is hydrolyzed by increasing the temperature to a range within 70°C–100°C. A residence time of about 15 min allows the complete hydrolysis. The syrup is then cooled to room temperature, filtered, processed successively through activated carbon and through weak anion-exchanger resin, thus increasing the pH of the liquid to 6.5–7.0, and finally concentrated to 40%–70% solids. The sweet fructose syrup containing oligofructans can be used as a truly "healthy" sweetener, particularly suitable for elderly people and diabetics (Yamazaki and Matsumoto, 1986). The pulp obtained after the juice extraction is rich in protein and can be used as feed (Abasaeed and Lee, 1995; Yamazaki and Matsumoto, 1986). Full enzymatic inulin hydrolysis, which is used for commercial production of inulin syrup, is typically carried out using Fructozyme L, an inulinase preparation from Novozymes, which displays both endo- and exo-activity (FOSs are obtained using endoinulinase) (Partida et al., 1998; Smits and De Leenheer, 1999). Typical commercial operation used for the production of Raftisweet (a fructose syrup from BENEO-Orafti) is carried out at 60°C and pH 4.5, allowing a product with a fructose titer over 85%, with a low titer of oligosaccharide (under 2%) and a no "burned" aftertaste, typical of acid hydrolysis processes (Franck and De Leenheer, 2005). The fructose syrup is then filtered, demineralized, and decolorized according to standard techniques and is further concentrated by evaporation to yield a syrup with a typical concentration of 75% dry solids. The incubation period in the batch process depends upon the titer of the inulin solution and on enzyme dosage. In a process aiming at improving this classical procedure, when a 15% inulin solution and 2.4 units of Fructozyme L/g inulin (dry substance) were used, an incubation period of 20 h led to a syrup with 95% fructose. Increasing the enzyme dosage to 20 units/g inulin (dry substance) decreases the incubation period to 2.5 h (Leenheer and Boot, 1998).

The production of high-fructose syrups using immobilized inulinases has also been performed since the late 1970s (Ricca et al., 2007; Singh and Gill, 2006). Some recent examples have been given in Table 11.3.

A wide array of supports such as glass beads, ion-exchange resins, natural and synthetic hydrogels, grafted hydrogels, or methacrylic polymers have been

TABLE 11.3

Some Recent Examples on the Use of Immobilized Inulinases for the Production of High-Fructose Syrups from Inulin

Support	Enzyme	Comments	Reference
Chitin, casein, and calcium alginate; anion exchange and affinity matrices[a]	Extracellular thermostable inulinase from *Aspergillus fumigatus*	The Michaelis constant values of the different immobilized inulinases varied from 0.2 to 0.5 mM (1.3 for the free enzyme). Immobilization improved the thermal stability of the enzyme. Operational half-lives within 22–45 days, under operation at 60°C and 2.5% inulin were reported.	Gill et al. (2006)
Sepabeads	Fructozyme L	Computational methods allowed a better understanding of enzyme–substrate interaction and contributed to rationalize the experimental strategies for enzyme immobilization. Operation with a fluidized-bed reactor proved feasible.	Ricca et al. (2010)
Amberlite IRC 50 functionalized with PEI and GA	Fructozyme L	Immobilization enhanced (sixfold) the half-life of the biocatalyst, when compared to the free form. The immobilized biocatalyst was effectively reused in 18 batch runs for the hydrolysis of a 5% inulin solution at 40°C.	Catana et al. (2007)
Porous xerogel	Fructozyme L	Immobilization enhanced thermal stability. The immobilized biocatalyst was effectively reused in 22 batch runs for the hydrolysis of a 5% inulin solution at 50°C. Xerogel biocatalyst doped with magnetic particles displayed lower operational stability.	Santa et al. (2011)
Calcium alginate modified with polyimines and cross-linked with GA	Extracellular inulinase from *Penicillium chrysogenum* P36	Immobilization enhanced by 5°C the optimum temperature for enzyme activity and improved thermal stability. The immobilized biocatalyst retained over 95% of its activity after 20 reuses, under 50°C and 2% inulin solution.	Elnashar et al. (2009)
Duolite A568	Exoinulinase from *K. marxianus* YS-1	The packed bed was run continuously for 75 days, under 50°C, pH 5.5, and 5% inulin, with a half-life of 72 days. Volumetric productivity, 44.5 g reducing sugars/L/h; fructose yield, 53.3 g/L	Singh et al. (2008)

GA, glataraldehyde; PEI, polyethylenimine.

[a] ConA-linked amino-activated silica beads, QAE-Sephadex, DEAE-Sephacel, Dowex, and Amberlite.

used for inulinase immobilization (Ricca et al., 2007, 2010; Singh and Gill, 2006). Particular care has been given to the use of relatively cheaper carriers, as opposite to commercially available expensive carriers such as agarose, so that an economically feasible process may result (Elnashar et al., 2009; Ricca et al., 2007; Singh and Gill, 2006). Immobilization has been shown to enhance the thermal stability of inulinases. This enables prolonged operation at temperatures of 50°C–60°C, which reduces the risk of contamination and enhances inulin solubilization (Elnashar et al., 2009; Ricca et al., 2007). In order to develop attractive process proposals, immobilized inulinases have been mostly used in stirred batch reactors and in continuous packed-bed reactors (Ricca et al., 2007; Singh and Gill, 2006, Table 11.3). Less common is the use of fluidized-bed reactors and shell-tube membrane reactors (Ricca et al., 2007, 2010). In all cases, reactor operation is typically carried out at temperatures within 40°C–60°C, pH within 4.5–5.5, and substrate concentrations within 2.5%–10%. Under selected conditions, more than 15 consecutive batch runs with conversion yields exceeding 80% have been reported (Ricca et al., 2007; Singh and Gill, 2006, Table 11.3). Packed-bed reactors filled with inulinases immobilized in different ion-exchange resins have displayed half-lives within 30–70 days under continuous operation. Conversion yields in excess of 90% after 14 days of operation, with residence times of 1.25 h, have been reported (Ricca et al., 2007; Singh et al., 2008). A more rational approach for immobilization of inulinase has been developed, which relied on computational analysis of three-dimensional endo- and exoinulinase models. Such analysis provided the basis for the understanding of enzyme–substrate interactions; it also allowed to highlight structural features that are likely to influence the efficiency of enzymes also after immobilization (Basso et al., 2010). This is likely to lead to the development of a cost-effective biocatalyst and concomitantly to a commercial continuous process for the production of fructose syrups from inulin hydrolysis.

11.6 Summary and Future Prospects

The use of sweeteners both as final products and in formulation in food and pharmaceutical sectors is well established. Nevertheless, dedicated research is being carried out toward more sustainable production methodologies and more efficient products/formulations. The fulfillment of such endeavors requires the commitment of the different sectors involved in the production of sweetener from the research toward the screening and isolation of new natural compounds, or the identification of pathways to synthesize sweeteners, to the technology to make their production at commercial scale economically feasible. These need to be coupled with the further studies on the impact of sweetener consumption in public health.

The role of biotechnology in the production of sweeteners has been consistently expanding. Its application abridges far different areas, from the improvement of sweetener-producing crops to the introduction of biocatalysts in the design of processes for the large-scale production of sweeteners. This pattern not only endured but has also been gaining relevance since enzyme-based processes are gradually replacing chemical processes. Such trend can be imparted to the higher selectivity of biocatalysts and to their ability to operate under mild condition, hence saving resources. Within such scope, much is due to the identification of more efficient enzymes, displaying enhanced thermal, pH, and operational stability. This is the outcome of dedicated screening or of genetic manipulation as well as of the identification of further enzymatic activities. The development of immobilization strategies, enabling the reuse of biocatalysts and widening the range of modes of operation, has also contributed to the implementation of enzyme-based processes at commercial scale, but there is still a considerable gap between laboratory-/bench-scale results and implementation at a large scale of said processes. More rational approaches for enzyme immobilization, such as combining further insight on protein structure and physical and chemical properties of support matrices, are being developed. These efforts are foreseen to contribute for the development of competitive biocatalysts in the near future, thus allowing for further dissemination of enzyme-based, cost-effective processes for sweetener production.

Acknowledgment

The author, Pedro Fernandes, acknowledges Fundação para a Ciência e Tecnologia (FCT) for the contract under Programme Ciência 2007.

References

Abasaeed, A.E. and Lee, Y.Y. 1995. Kinetics of inulin hydrolysis by zeolite LZ-M-8. *Hungarian Journal of Industrial Chemistry*, 24: 149–153.

Akinterinwa, O., Khankal, R., and P.C. Cirino. 2008. Metabolic engineering for bioproduction of sugar alcohols. *Current Opinion in Biotechnology*, 19: 461–467.

Aller, E.E.J.G., Abete, I., Astrup, A., Martinez, J.A., and van Baak, M.A. 2011. Starches, sugars and obesity. *Nutrients*, 3: 341–369.

American Dietetic Association. 2004. Position of the American Dietetic Association: Use of nutritive and nonnutritive sweeteners. *Journal of the American Dietetic Association*, 104: 255–275.

Asadi, M. 2007. *Beet-Sugar Handbook*. John Wiley & Sons, Inc, Hoboken, NJ.

Bakkal, A.I. and Nabors, L.B. 2012. Saccharin. In: *Alternative Sweeteners*, 4th edn., ed. L. B. Nabors, pp. 151–158. CRC Press, Boca Raton, FL.
Barrie, F.N., Vikso-Nielsen, A., Olsen, H.S., and Pedersen, S. 2009. Process for hydrolysis of starch. Patent Application US20090142817.
Basso, A., Spizzo, P., Ferrario, V. et al. 2010. Endo- and exo-inulinases: Enzyme-substrate interaction and rational immobilization. *Biotechnology Progress*, 26: 397–405.
Belitz, H.-D., Grosch, W., and Schieberle, P. 2009. *Food Chemistry*. 4th edn., Springer-Verlag, Berlin, Germany.
Bhosale, S.H., Rao, M.B., and Deshpande, V.V. 1996. Molecular and industrial aspects of glucose isomerase. *Microbiological Reviews*, 60: 280–300.
Bisgård-Frantzen, H. and Svendsen, A. 2008. Starch debranching enzymes. Patent US7816113.
Bommarius, A.S. and Riebel-Bommarius, B.R. 2004. *Biocatalysis Fundamentals and Applications*. 1st edn., Wiley-VCH Verlag GmbH & Co. KGaA, Weinheim, Germany.
Brahmachari, G., Mandal, L.C., Roy, R., Mondal, S., and Brahmachari, A.K. 2011. Stevioside and related compounds—Molecules of pharmaceutical promise: A critical overview. *Archiv der Pharmazie—Chemistry in Life Sciences*, 1: 5–19.
Brennan, T.M. and Hendrick M.E. 1983. Branched amides of L-aspartyl-D-amino acid dipeptides. Patent US4411925.
Buchholz, K., Kasche, V., and Bornscheuer, U.T. 2005. *Biocatalysts and Enzyme Technology*. Wiley-VCH Verlag GmbH & Co. KGaA, Weinheim, Germany.
Bucke, C. and Chaplin, M.F. 1990. *Enzyme Technology*. Cambridge University Press, Cambridge, U.K.
Catana, R., Eloy, M., Rocha, J.R., Ferreira, B.S., Cabral, J.M.S., and Fernandes, P. 2007. Stability evaluation of an immobilized enzyme system for inulin hydrolysis. *Food Chemistry*, 101: 260–266.
Chattopadhyay, S., Raychaudhuri, U., and Chakraborty, R. 2011. Artificial sweeteners—A review. *Journal of Food Science and Technology*, DOI 10.1007/s13197-011-0571-1.
Chaturvedula, V.S.P. and Prakash, I. 2011. Kaempferol glycosides from *Siraitia grosvenorii*. *Journal of Chemical and Pharmaceutical Research*, 3: 799–804.
Chen, X., Jiang, Z.-H., Chen, S., and Qin. W. 2010. Microbial and bioconversion production of D-xylitol and its detection and application. *International Journal of Biological Sciences*, 6: 834–844.
Climent, M.J., Corma, A., and Iborra, S. 2011. Converting carbohydrates to bulk chemicals and fine chemicals over heterogeneous catalysts. *Green Chemistry*, 13: 520–540.
Elnashar, M.M.M., Danial, E.N., and Awad, G.E.A. 2009. Novel carrier of grafted alginate for covalent immobilization of inulinase. *Industrial and Engineering Chemistry Research*, 48: 9781–9785.
Erzinger, G.S. and Vitolo, M. 2006. *Zymomonas mobilis* as catalyst for the biotechnological production of sorbitol and gluconic acid. *Applied Biochemistry and Biotechnology*, 131: 787–794.
European Commission—Agriculture Directorate-General. 2004. The common organisation of the market in sugar. http://ec.europa.eu/agriculture/markets/sugar/reports/descri_en.pdf (accessed December 28, 2011).
Faus, I. and Sisniega, H. 2005. Sweet-tasting proteins. In: *Polysaccharides and Polyamides in the Food Industry—Properties, Production, and Patents*, eds. A. Steinbüchel and S.K. Rhee, pp. 687–706. Wiley-VCH Verlag GmbH & Co. KGaA, Weinheim, Germany.

Fernandes, P. 2010. Enzymes in sugar industries. In: *Enzymes in Food Processing: Fundamentals and Potential Applications*, eds. P.S. Panesar, S.S. Marwaha, and H.K. Chopra, pp. 165–197. I.K. International Publishing House Pvt. Ltd, New Delhi, India.

Ferrier, R.J. 2006. An historical overview. In: *The Organic Chemistry of Sugars*, eds. D.E. Levy and P. Fügedi, pp. 3–24. Taylor & Francis Group, Boca Raton, FL.

Franck, A. and De Leenheer. L. 2005. Inulin. In: *Polysaccharides and Polyamides in the Food Industry. Properties, Production, and Patents*, eds. A. Steinbüchel and S.K. Rhee, pp. 281–322. Wiley-VCH Verlag GmbH & Co. KGaA, Weinheim, Germany.

Gelov, T. 2011. No-calorie sweetener compositions. Patent Application US20110027446.

Gill, P.K., Manhas, R.K., and Singh, P. 2006. Hydrolysis of inulin by immobilized thermostable extracellular exoinulinase from *Aspergillus fumigatus*. *Journal of Food Engineering*, 76: 369–375.

Givaudan & Cie. 1976. Sweetening compositions. Patent GB1428945.

Godshall, M.A. 2007a. The expanding world of nutritive and non-nutritive sweeteners. *Sugar Journal*, 69: 12–20.

Godshall, M.A. 2007b. Sugar and other sweeteners. In: *Kent and Riegel's Chemistry and Biotechnology*, 11th edn., ed. J.A. Kent, pp. 1657–1693, Springer, New York.

Granguillhome, E.C.R., Barrañon, J.A.C., and Garza, J.J.G. 2005. Process for the production of invert liquid sugar. Patent US6916381.

Haber, B., Lipinski, G.-W.R., and Rathjen, S. 2006. Acefulfame K. In: *Sweeteners and Sugar Alternatives in Food Technology*, ed. H.L. Mitchell, pp. 63–85, Blackwell Publishing Ltd, Oxford, U.K.

Hagen, J. 2006. *Industrial Catalysis: A Practical Approach*, 2nd edn., Wiley-VCH Verlag GmbH & Co. KGaA, Weinheim, Germany.

Haley, S. and Dohlman, E. 2009. Sugar and Sweeteners Outlook/SSS-256/October 5, 2009. Economic Research Service, USDA. http://www.ers.usda.gov/publications/sss/2009/SSS256.pdf (accessed December 10, 2011).

He, W. and Zhou, J.H. 2000. Process for extracting triterpene glycosides from botanical sources. Patent US6124442.

Heikkilä, H., Hyöky, G., Niittymäki, P., Viljava, T., and Myöhänen, T. 1995. Process for producing glucose and fructose from sucrose. European Patent EP0553126B1.

Hirth, T., Schweppe, R., Graf, J., Busch, R., and Pursch, M. 2011. Process for the production of sugar alcohols. Patent US7968704.

Hobbs, L. 2009. Sweeteners from starch: Production, properties and uses. In: *Starch: Chemistry and Technology*, 3rd edn., eds. J.N. BeMiller and R.L. Whistler, pp. 797–832, Academic Press, New York.

Hunt, F., Bopp, B.A., and Price, P. 2012. Cyclamate. In: *Alternative Sweeteners*, 4th edn., ed. L. B. Nabors, pp. 93–116, CRC Press, Boca Raton, FL.

Jeya, M., Lee, K.-M., Tiwari, M.K. et al. 2009. Isolation of a novel high erythritol-producing *Pseudozyma tsukubaensis* and scale-up of erythritol fermentation to industrial level. *Applied Microbiology and Biotechnology*, 83: 225–231.

Kaur, N. and Gupta, A.K. 2002. Applications of inulin and oligofructose in health and nutrition. *Journal of Biosciences*, 27: 703–714.

Kitagawa, I. 2002. Licorice root: A natural sweetener and an important ingredient in Chinese medicine. *Pure and Applied Chemistry*, 74: 1189–1198.

Koizumi, A., Tsuchiya, A., Nakajima, K.-I. et al. 2011. Human sweet taste receptor mediates acid-induced sweetness of miraculin. *Proceedings of the National Academy of Sciences*, 108: 16819–16824.

Lazcano, R.R., Perez, A.C., and Baele, N.F. 1993. Method and apparatus for the production of glucose-fructose syrups from sucrose using a recombinant yeast strain. Patent US5270177.

Lee, T.D. 2007. Sweeteners. In: *Kirk-Othmer Encyclopedia of Chemical Technology*, Vol. 24, 5th edn., pp. 224–252, John Wiley & Sons, Hoboken, NJ.

Leenheer, L. and Boot, K. 1998. Process for producing a fructose syrup rich in fructose. European Patent Application EP0822262.

Liese, A., Seelbach, K., Buchholz, A., and Haberland, J. 2006. Processes. In: *Industrial Biotransformations*, 2nd edn., eds. A. Liese, K. Seelbach, and C. Wandrey, pp. 147–513, Wiley-VCH Verlag GmbH & Co. KGaA, Weinheim, Germany.

Liua, C., Donga, H., Zhonga, J., Ryu, D.D.Y., and Bao, J. 2010. Sorbitol production using recombinant *Zymomonas mobilis* strain. *Journal of Biotechnology*, 148: 105–112.

van der Maarel, M.J.E.C. 2009. Starch-processing enzymes. In: *Enzymes in Food Technology*, 2nd edn., eds. R.J. Whitehurst and M. van Oort, pp. 320–331, Wiley-Blackwell, Singapore.

McKibbins, S.W., Harris, J.F., Seaman, J.F., and Neill, W.K. 1962. Kinetics of the acid catalyzed conversion of glucose to 5-hydroxymethyl-2-furadehyde and levulinic acid. *Forest Products Journal*, January: 17–22.

Monedero, V., Pérez-Martínez, G., and Yebra, M.J. 2010. Perspectives of engineering lactic acid bacteria for biotechnological polyol production. *Applied Microbiology and Biotechnology*, 86: 1003–1015.

Moreau, C., Durand, R., Aliès, F., Cotillon, M., Frutz, T., and Théoleyre, M.-A. 2000. Hydrolysis of sucrose in the presence of H-form zeolites. *Industrial Crops and Products*, 11: 237–242.

Mosier, N.S. and Ladisch, M.R. 2011. *Modern Biotechnology: Connecting Innovations in Microbiology and Biochemistry to Engineering Fundamentals*. John Wiley & Sons, Inc., Hoboken, NJ.

Nakakuki, T. 2002. Present status and future of functional oligosaccharide development in Japan. *Pure and Applied Chemistry*, 74: 1245–1251.

Nutrilite Products Inc. 1974. Salts of dihydrochalcone derivatives and their use as sweeteners. Patent GB1347202.

O'Donnell, K. 2005. Carbohydrate and intense sweeteners. In: *Chemistry and Technology of Soft Drinks and Fruit Juices*, ed. P.R. Ashurst, pp. 68–89, Blackwell Publishing Ltd., Oxford, U.K.

Otaka, K. 2006. Functional oligosaccharide and its new aspect as immune modulation. *Journal of Biological Macromolecules*, 6: 3–9.

Partida, V.Z., Lopez, A.C., and Gomez, A.J.M. 1998. Method of producing fructose syrup from agave plants. Patent US5846333.

Patel, S. and Goyal, A. 2011. Functional oligosaccharides: Production, properties and applications. *World Journal of Microbiology and Biotechnology*, 27: 1119–1128.

Patra, F., Tomar, S.K., and Arora, S. 2009. Technological and functional applications of low-calorie sweeteners from lactic acid bacteria. *Journal of Food Science*, 74: R16–R23.

Prakash, I. 2007. Synthesis of N-[N-(3,3-dimethylbutyl)-L-α-aspartyl]-L-phenylalanine 1-methyl ester using 3,3-dimethylbutyraldehyde precursors. Patent US7288670.

Prakash, I. and Zhao, R.Y. 2001. Chemoenzymatic synthesis of neotame. Patent US 6627431.

Priya, K., Gupta, V.R.M., and Srikanth, K. 2011. Natural sweeteners: A complete review. *Journal of Pharmacy Research*, 4: 2034–2039.

Purich, D.L. and Allison, R.D. 2000. *Handbook of Biochemical Kinetics*. Academic Press, New York.

Rahtjen, S. and Schwarz, S. 2007. Mixture of fructose-containing sweeteners with ternary of quaternary high-intensity sweetener blends. European Patent EP1764004.

Rennie, E.H. 1886. LXXXV—Glycyphyllin, the sweet principle of *Smilax glycyphylla*. *Journal of the Chemical Society, Transactions*, 49: 857–865.

Ricca, E., Calabrò, V., Curcio, S., Basso, A., Gardossi, L., and Iorio, G. 2010. Fructose production by inulinase covalently immobilized on Sepabeads in batch and fluidized bed bioreactor. *International Journal of Molecular Sciences*, 11: 1180–1189.

Ricca, E., Calabrò, V., Curcio, S., and Iorio, G. 2007. The state of the art in the production of fructose from inulin enzymatic hydrolysis. *Critical Reviews in Biotechnology*, 27: 129–145.

Roberfroid, M.B. 1999. Caloric value of inulin and oligofructose. *The Journal of Nutrition*, 129: 1436–1437.

Robertson, G.H., Clark, J.P., and Lundin, R. 1974. Dihydrochalcone sweeteners: Preparation of neohesperidin dihydrochalcone. *Industrial & Engineering Chemistry Product Research and Development* 13: 125–129.

Robyt, J.F. 2009. Enzymes and their action on starch. In: *Starch: Chemistry and Technology*, 3rd edn., eds. J.N. BeMiller and R.L. Whistler, pp. 238–292, Academic Press, New York.

Ruilin, N., Tanaka, T., Zhou, J., and Tanaka, O. 1982. Phlorizin and trilobatin, sweet dihydrochalcone-glucosides from leaves of *Lithocarpus-litseifolius* (Nance) Rehd. (Fagaceae). *Agricultural and Biological Chemistry*, 46: 1933–1934.

Ruprecht, W. 2005. The historical development of the consumption of sweeteners—A learning approach. *Journal of Evolutionary Economics*, 15: 247–272.

Salmi, T., Murzin, D., Mäki-Arvela, P., Wärnå, J., Eränen, K., Kumar, N., and Mikkola, J.-P. 2010. Catalytic engineering in the processing of biomass into chemicals. In: *Novel Concepts in Catalysis and Chemical Reactors*, eds. A. Cybulski, J.A. Moulijn, and A. Stankiewicz, pp. 163–188, Wiley-VCH Verlag GmbH & Co. KGaA, Weinheim, Germany.

Santa, G.L., Bernardino, S.M., Magalhães, S. et al. 2011. From inulin to fructose syrups using sol–gel immobilized inulinase. *Applied Biochemistry and Biotechnology*, 165: 1–12.

Sasman, T., Head, W.A., and Cameron, C.A. 2007. Process for producing erythritol. Patent Application US20070037266.

Serna-Saldivar, S.R. and Rito-Palomares, M.A. 2008. Production of invert syrup from sugarcane juice using immobilized invertase. Patent US7435564.

Shannon, J.C., Garwood, D.L., and Boyer, C.D. 2009. Genetics and physiology of starch development. In *Starch: Chemistry and Technology*, 3rd edn., eds. J. BeMiller and R.L. Whistler, pp. 24–72, Academic Press, New York.

Shwide-Slavin, C., Swift, C., and Ross, T. 2012. Nonnutritive sweeteners: Where are we today? *Diabetes Spectrum*, 25: 104–110.

Singh, R.S., Dhaliwal, R., and Puri, M. 2008. Development of a stable continuous flow immobilized enzyme reactor for the hydrolysis of inulin. *Journal of Industrial Microbiology and Biotechnology*, 35: 777–782.

Singh, P. and Gill, P.K. 2006. Production of inulinases: Recent advances. *Food Technology and Biotechnology*, 44: 151–162.

Singh R.S. and Singh, R.P. 2010. Production of fructooligosaccharides from inulin by endoinulinases and their prebiotic potential. *Food Technology and Biotechnology*, 48: 435–450.

Smits, G. and De Leenheer, L. 1999. Process for the manufacture of chicory inulin, hydrolysates and derivatives of inulin, and improved chicory inulin products, hydrolysates and derivatives. WIPO Patent Application WO/1999/037686.

Song, S.H. and Vieille, C. 2009. Recent advances in the biological production of mannitol. *Applied Microbiology and Biotechnology*, 84: 55–62.

Tang, D.M., Zhu, C.F., Zhong, S.A., and Zhou, M.D. 2011. Extraction of naringin from pomelo peels as dihydrochalcone's precursor. *Journal of Separation Science*, 34: 113–117.

Taniguchi, H. 2004. Carbohydrate research and industry in Japan. *Starch/Stärke*, 56: 1–5.

Torres, D.P.M, Gonçalves, M.P.F., Teixeira, J.A., and Rodrigues, L.R. 2010. Galacto-oligosaccharides: Production, properties, applications, and significance as prebiotics. *Comprehensive Reviews in Food Science and Food Safety*, 9: 438–454.

Ujihara, M., Shinozaki, M., and Kato, M. 1995. Accumulation of phyllodulcin in sweet-leaf plants of *Hydrangea serrata* and its neutrality in the defence against a specialist leaf mining herbivore. *Researches on Population Ecology*, 37: 249–257.

Wang, F., He, H., Yang, X., Yu, Y., and Fan, Z. 2011a. Method of sucralose synthesis yield. Patent US7884203.

Wang, L., Underwood, J.M., Peters, B., Gregory, K.L., and Lester, K. 2011b. Starch hydrolysis. Patent Application US20110091938.

West, S. 2007. Production of flavours, flavour enhancers and other protein-based speciality products. In: *Novel Enzyme Technology for Food Applications*, ed. R. Rastall, pp. 183–204, Woodhead Publishing Limited, Abington, MA.

Whitaker, J.R. 2004. Factors affecting enzyme activity in foods. In: *Proteins in Food Processing*, ed. R.Y. Yada, pp. 270–291, CRC Press, Boca Raton, FL.

Yamazaki, H. and Matsumoto, K. 1986. Production of fructose syrup. Patent US4613377.

Zhang, M., Yang, H., Zhang, H., Wang, Y., and Hu, P. 2011. Development of a process for separation of mogroside V from *Siraitia grosvenorii* by macroporous resins. *Molecules*, 16: 7288–7301.

12

Production of Biocolors

Nuthathai Sutthiwong, Yanis Caro, Philippe Laurent, Mireille Fouillaud, Alain Valla, and Laurent Dufossé

CONTENTS

12.1 Introduction 418
12.2 *Monascus* Pigments, an Old Story for Asians 421
 12.2.1 *Monascus* Pigment 421
 12.2.2 Fungal Metabolites 421
 12.2.3 Methods of Production 422
  12.2.3.1 Submerged Fermentations 422
  12.2.3.2 Solid-State Fermentations 422
 12.2.4 Methods to Control Mycotoxin Production 423
 12.2.5 *Monascus*-Like Pigments from Nontoxigenic *Penicillium* Species 424
12.3 Microbial Anthraquinones 425
 12.3.1 Arpink Red from *P. oxalicum* 425
 12.3.2 Other Microbial Anthraquinones 426
12.4 Riboflavin, a Yellow Food Colorant but Also the Vitamin B_2 429
12.5 Fluorescent Pink from the Red Microalga *Porphyridium*, Phycobiliproteins 429
12.6 Phycocyanin, the Marine Blue from *Porphyridium* 430
12.7 Current Carotenoid Production Using Microorganisms 432
 12.7.1 β-Carotene 432
  12.7.1.1 β-Carotene from *B. trispora* 432
  12.7.1.2 β-Carotene from *Phycomyces blakesleeanus* 434
  12.7.1.3 β-Carotene from *Mucor circinelloides* 434
 12.7.2 Lycopene 435
  12.7.2.1 Lycopene from *B. trispora* 435
  12.7.2.2 Lycopene from *Fusarium sporotrichioides* 436
 12.7.3 Astaxanthin 437
  12.7.3.1 Astaxanthin from *X. dendrorhous*, Formerly *Phaffia rhodozyma* 437
  12.7.3.2 Astaxanthin from *Agrobacterium aurantiacum* and Other Bacteria 438

12.7.4 Zeaxanthin....................438
12.7.5 Canthaxanthin....................439
12.7.6 Torulene and Torularhodin....................440
12.8 Summary and Future Prospects....................440
Acknowledgment....................442
References....................442

12.1 Introduction

Nature is rich in colors (minerals, plants, microalgae, etc.) and pigment-producing microorganisms (fungi, yeasts, bacteria) are quite common (Figure 12.1). Currently, the vast majority of the natural food colorants permitted in the European Union and the United States are derived by extraction of the pigments from raw materials obtained from the flowering plants of the kingdom Plantae. The production of many existing natural colorants of plant origin has a disadvantage of dependence on the supply of raw materials, which are influenced by agro-climatic conditions—in addition, their chemical profile may vary from batch to batch. Moreover, many of the pigments derived from the contemporary sources are sensitive to heat, light, and oxygen, and some may even change their color in response to pH changes as in case of anthocyanins. Until recently, problem of color loss and stability in products could be easily tackled by using synthetic pigments, such as azo dyes, originally derived from coal tar. This view has changed over the last 5 years as

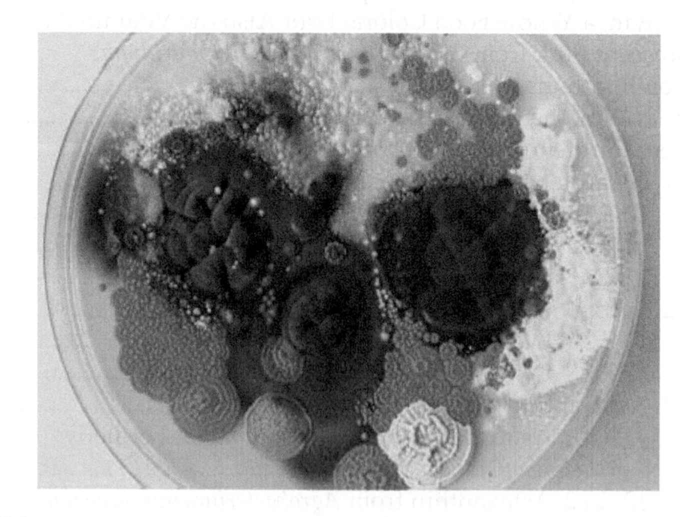

FIGURE 12.1
Spontaneous development of pigmented microorganisms at the surface of nutritive agar Petri dish.

concerns about possible adverse health effects of synthetic colors have grown (Southampton study, McCann et al., 2007).

Among the molecules produced by microorganisms are carotenoids, melanins, flavins, phenazines, quinones, bacteriochlorophylls, and more specifically monascins, violacein, or indigo (Figure 12.2) (Dufossé, 2004; Kerr, 2000; Plonka and Grabacka, 2006). The success of any pigment produced by fermentation depends upon its acceptability in the market, regulatory approval, and the size of the capital investment required to bring

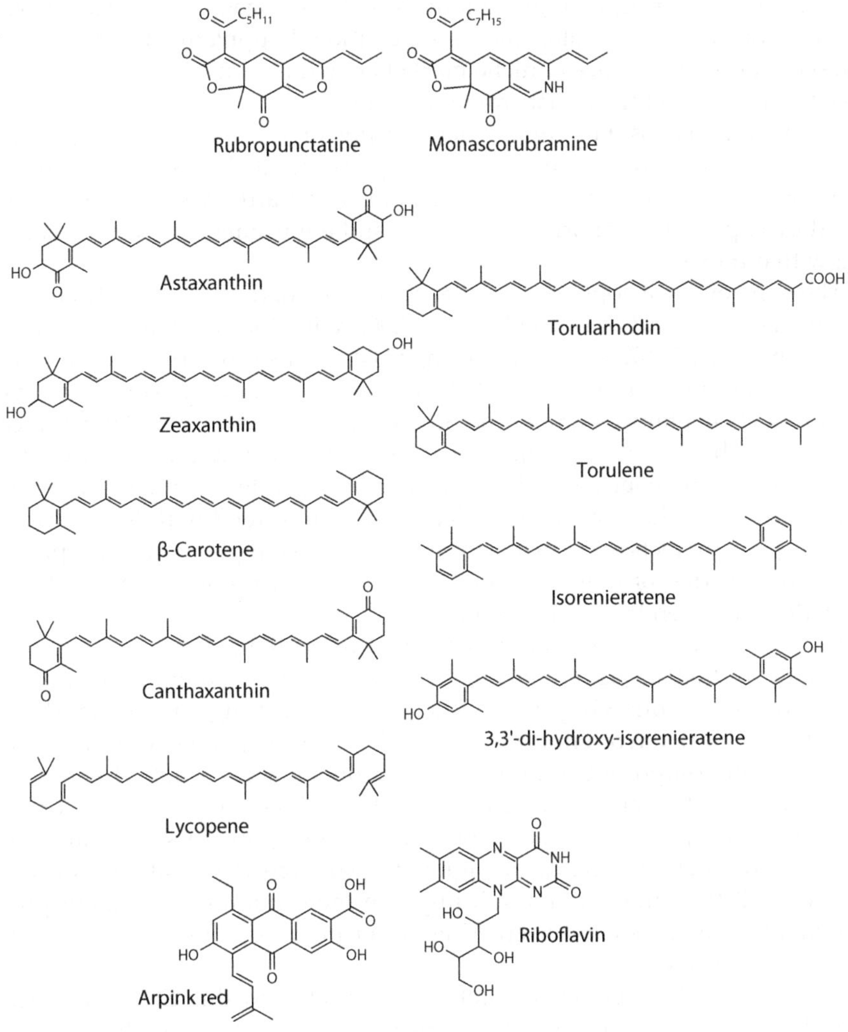

FIGURE 12.2
Some microbial food-grade pigments.

the product to market. A few years ago, some expressed doubts about the successful commercialization of fermentation-derived food-grade or cosmetic-grade pigments because of the high capital investment requirements for fermentation facilities and the extensive and lengthy toxicity studies required by regulatory agencies. Public perception of biotechnology-derived products also had to be taken into account. Nowadays, some fermentative food-grade pigments are on the market: *Monascus* pigments, astaxanthin from *Xanthophyllomyces dendrorhous*, arpink red from *Penicillium oxalicum*, riboflavin from *Ashbya gossypii*, and β-carotene from *Blakeslea trispora*. The successful marketing of pigments derived from algae or extracted from plants, both as a food color and as a nutritional supplement, reflects the presence and importance of niche markets in which consumers are willing to pay a premium for "all natural ingredients."

Colors can serve as the primary identification of food and are also a protective measure against the consumption of spoiled food. Colors of foods create physiological and psychological expectations and attitudes that are developed by experience, tradition, education, and environment: "We inevitably eat with our eyes."

The controversial topic of "synthetic dyes in food" has been discussed for many years and was amplified in 2007 with the Southampton study (Mc Cann et al., 2007; Schab and Trinh, 2004) and its transcription in a legal frame (i.e., the use of warning labels in Europe about a hyperactivity link for products containing any of the Southampton colors is mandatory since July 2010). The scrutiny and negative assessment of synthetic food dyes by the modern consumer have given rise to a strong interest in natural coloring alternatives. Some companies decided to "color food with food," using mainly plant extracts or pigments from plants, for example, red from paprika, beetroots, berries, or tomato; yellow from saffron or marigold; orange from annatto; and green from leafy vegetables.

Penetration of the fermentation-derived ingredients into the food and cosmetic industries is increasing year after year. Examples could be taken from the following fields: thickening or gelling agents (xanthan, curdlan, gellan), flavor enhancers (yeast hydrolysate, monosodium glutamate), flavor compounds (gamma-decalactone, diacetyl, methyl-ketones), acidulants (lactic acid, citric acid), etc. Efforts have been made in order to reduce the production costs of fermentation pigments compared to those of synthetic pigments or pigments extracted from natural sources (Dufossé, 2006). Innovations will improve the economy of pigment production by isolating new or creating better microorganisms, by improving the processes.

This chapter focuses on research works related to this field published over the past 10 years by private companies or academic laboratories, with an emphasis on pigments for food use. As recently described by our group, there is "a long way from the Petri dish to the market place" and thus to the product on store shelves.

12.2 *Monascus* Pigments, an Old Story for Asians

12.2.1 *Monascus* Pigment

Monascus is cultivated on solid medium in Asian countries to produce a red colorant named "Anka" used as a food ingredient. In a Chinese medical book on herbs published in the first century, this term "ang-kak" or "red mold rice" was first mentioned. Red mold rice has been used as a food colorant or spice in cooking. In 1884, a purple mold was isolated on potato and linseed cakes and was named it *Monascus ruber.* This ascomycete was so named as it has only one polyspored ascus. Then in 1895, another strain was isolated from the red mold rice obtained from the market in Java, Indonesia. This fungus was named *Monascus purpureus.* Then several other species were isolated around the world. *Monascus* is often encountered in oriental foods, especially in Southern China, Japan, and Southeastern Asia. Currently, more than 50 patents have been issued in Japan, the United States, France, and Germany, concerning the use of *Monascus* pigments for food. Annual consumption of *Monascus* pigments in Japan moved from 100 tons in 1981 to 600 tons at the end of the 1990s and was valued at $1.5 million. New food applications, like the coloration of processed meats (sausage, hams), marine products like fish paste, surimi, and tomato ketchup were described (Blanc et al., 1994).

12.2.2 Fungal Metabolites

The main metabolites produced by *Monascus* are polyketides, which are formed by the condensation of one acetylcoA with one or more malonylcoA with a simultaneous decarboxylation as in the case of lipidic synthesis. The metabolites mainly consist of the pigments, monacolins, and, under certain conditions, of a mycotoxin (Juzlova et al., 1996).

Monascus pigments are a group of fungal metabolites called azaphilones, which have similar molecular structures as well as similar chemical properties. Two molecular structures of the *Monascus* pigments are shown on Figure 12.2. Ankaflavin and monascine are yellow pigments, rubropunctatine and monascorubrine are orange, and rubropunctamine and monascorubramine are purple. The same color exists in two molecular structures differing in the length of the aliphatic chain. These pigments are produced mainly in the cell-bound state.

They have low water solubility, are sensitive to heat, are unstable in the pH range of 2–10, and fade with light. A number of methods have been patented in order to make water-soluble pigments. The principle is the substitution of the replaceable oxygen in monascorubrine or rubropunctatine by nitrogen of the amino group of various compounds such as amino acids, peptides, and proteins, changing the color from orange to purple. *Monascus* pigments can be reduced and oxidized and can react with other products, especially amino acids, to form various derivative products sometimes called the complexed pigments. Glutamyl-monascorubrine and glutamyl-rubropunctatine were isolated from the broth of a submerged culture.

Stability of the pigments is affected by acidity, temperature, light, oxygen, water activity, and time. It was shown that these pigments added to sausages or canned pâté remained stable for 3 months' storage at 4°C, while their stability ranged from 92% to 98%. Thus, the main patents have focused on the solubilization, the stability, and the extraction in solution of pigments. The pigments can easily react with amino group-containing compounds in the medium such as proteins, amino acids, and nucleic acids, to form water-soluble pigments.

A series of hypocholesteremic agents have been isolated from *Monascus* and named monacolin J, K, and L. These polyketides were first isolated from cultures of *Penicillium citrinum* and they can inhibit specifically the enzyme controlling the rate of cholesterol biosynthesis. They are currently used in China in traditional and modern medicine.

Antibacterial properties of *Monascus* were first mentioned in 1977. The so-called monascidin A was effective against *Bacillus, Streptococcus,* and *Pseudomonas.* It was shown that this molecule was citrinin and its production by various *Monascus* species was studied using different culture media and conditions.

12.2.3 Methods of Production

12.2.3.1 Submerged Fermentations

Considerable contradiction exists in the published works as to the best carbon source for red pigment production in liquid cultures. Traditionally cultured on breads and rice, *Monascus* grows on every amylaceous substrate. *Monascus* grows quite well on starch, dextrines, glucose, maltose, and fructose. High production of pigments was achieved using glucose and maltose. The nitrogen source seems to have more importance than the carbon source and ammonium, and peptones as nitrogen sources gave superior growth and pigment concentrations compared to nitrate. The best results were obtained using glucose and histidine. The carbon/nitrogen ratio was also shown to be important: at a value close to 50 g/g, growth would then be favored, while in the region of 7–9 g/g, pigmentation would be favored.

12.2.3.2 Solid-State Fermentations

The classical Chinese method consists of inoculating steamed rice grains spread on big trays with a strain of *Monascus anka* and incubating in an aerated and temperature-controlled room for 20 days (Babitha et al., 2007). In these types of cultures, moisture content, oxygen, and carbon dioxide levels in the gas environment, as well as cereal medium composition, are the most important parameters to control.

Moisture content is a very important parameter. Red pigments were produced in plastic bags containing rice grains. It was observed that pigmentation occurred only at a relatively low initial moisture level (26%–32%). Initial substrate moisture content regulated pigmentation as it was found that glucoamylase activity increased along with a rise in the initial substrate moisture content. Therefore, at high moisture content, as high enzyme activity was produced, glucose was rapidly liberated in high amounts (120 g/L) with ethanol formation, which inhibited pigmentation.

Thus it was confirmed that solid culture was superior to liquid culture for red pigment production by *M. purpureus*. This result has been attributed to the derepression of pigment synthesis in solid systems due to the diffusion of intracellular pigments into the surrounding solid matrix. In submerged fermentation, the pigments normally remain in the mycelium due to their low solubility in the usually acidic medium.

Levels of oxygen and carbon dioxide in the gas environment influence pigment production significantly while affecting growth to a lesser extent in solid-state culture. With *M. purpureus* on rice, maximum pigment yields were observed at 0.5×10^5 Pa of oxygen partial pressure in closed pressure vessels. However, high carbon dioxide partial pressures progressively inhibited pigment production, with complete inhibition at 10^5 Pa. In a closed aeration system with a packed-bed fermentor, oxygen partial pressures ranging from 0.05 to 0.5×10^5 Pa at constant carbon dioxide partial pressures of 0.02×10^5 Pa gave high pigment yields with a maximum at 0.5×10^5 Pa of oxygen, whereas lower carbon dioxide partial pressures at constant oxygen partial pressures of 0.21×10^5 Pa gave higher pigment yields. Maximum oxygen uptake and carbon dioxide production rates were observed at 70–90 and 60–80 h, respectively, depending on the gas environment. Respiratory quotients were close to 1.0 except at 0.05×10^5 Pa of oxygen and 0.02×10^5 Pa of carbon dioxide partial pressures.

When studying various cereal media, it was shown that the best results were obtained using "mantou" meal (yeast-fermented wheat meal).

12.2.4 Methods to Control Mycotoxin Production

In order to chemically identify the so-called monascidin A discussed by some Chinese scientists in their papers as a component suitable for the preservation of food, this compound was isolated and chemical investigations using mass spectrometry (MS) and nuclear magnetic resonance spectroscopy (NMR) were undertaken. Monascidin A was characterized as citrinin (Xu et al., 2003). Thus, in order to avoid the production of this toxin, various strains were screened in order to see if all were toxinogenic, and it was shown that in the species of *Monascus* available in public collections, nontoxigenic strains were obtainable.

Another way to avoid the production of citrinin can be through controlling the biosynthesis of the metabolite. To control the biosynthesis, the metabolic pathway has to be investigated. The metabolic pathway is the same for citrinin and the pigment: the polyketide pathway in which condensation of acetates and malonates occurred. In the case of the pigment, there is at the end of the pathway an esterification of a fatty acid on the chromophore to obtain the colored molecules. Several modifications of the culture conditions are possible in order to increase the pigment production or reduce the citrinin one: addition of fatty acids, change of the nitrogen source. Adding fatty acids to the medium was effective in favoring the synthesis of pigment, but the citrinin production remained unchanged.

The final modification of the culture conditions was the replacement of glutamic acid by other amino acids. *M. ruber* was cultivated in a liquid medium containing glucose and various amino acids, and histidine was found to be the most effective nitrogen source regarding citrinin production inhibition. When the pathway of histidine assimilation was investigated, it was shown that during its catabolism, one molecule of hydrogen peroxide was produced per molecule of consumed histidine and it is known that peroxidases can destroy citrinin in the presence of hydrogen peroxide. So the production of citrinin can be avoided by control of the medium especially by the selection of a suitable amino acid, usually histidine.

Despite the enormous economic potential of *Monascus* pigment, it does not lead to a commercial exploitation in the Western world, mainly because of ignorance and also because of reluctance to change from food public agencies. Indeed, these agencies do not approve *Monascus* pigments for use in the food industry, although they do appear to be nontoxic if correctly used. Thus, even though species of *Monascus* have been consumed in the Far East for many years, this does not help the pigment to gain approval in the European Union or the United States.

12.2.5 *Monascus*-Like Pigments from Nontoxigenic *Penicillium* Species

A screening for novel producers of *Monascus*-like pigments was conducted among ascomycetous fungi belonging to *Penicillium* species that are not reported to produce citrinin or any other known mycotoxins (Mapari et al., 2005, 2008). Monascorubrin, xanthomonasin A, and threonine derivatives of rubropunctatin were identified in the extract of *Penicillium aculeatum* IBT 14263, and monascorubrin was identified in the extract of *Penicillium pinophilum* IBT 13104. None of the tested *Penicillium* extracts showed the presence of citrinin. Thus, the present study brought out two novel promising sources of yellow, orange, and purple-red *Monascus*-like food pigments in the species of Penicillia that do not produce citrinin and opened the door to look for several more new promising sources of natural food colorants in the species of Penicillia.

12.3 Microbial Anthraquinones

12.3.1 Arpink Red from *P. oxalicum*

Many patents from Ascolor s.r.o. (Czech Republic) relate to a new fungus strain having the properties to produce a red colorant that can be applied in the food and cosmetic industries (WO9950434, CZ285721, EP1070136, US6340586) (Sardaryan et al., 2004). The strain *P. oxalicum* var. Armeniaca CCM 8242, obtained from soil, produces a chromophore of the anthraquinone type (Figure 12.2). Some strains of the same species are effective as biological control agents, for example, reduction of the incidence of *Fusarium* wilt of tomato under glasshouse and field conditions. Others have been described for the production of milk-clotting enzyme.

The cultivation of the fungus in liquid broth requires carbohydrates (such as sucrose, molasses), nitrogen (corn extract, yeast autolysate, or extract), zinc sulfate, and magnesium sulfate. The optimum conditions for performing the microbiological synthesis are pH value 5.6–6.2 and temperature 27°C–29°C. On the second day of incubation, a red colorant is released into the broth, increasing up to 1.5–2.0 g/L of broth after 3–4 days of incubation. After biosynthesis of the red colorant is completed, the liquid from the broth is filtered or centrifuged for being separated from the biomass. The liquid is then acidified to pH 3.0–2.5 to precipitate the colorant. The precipitate is dissolved in ethyl alcohol and filtered. Following the removal of alcohol, the colorant in the crystalline form is obtained, i.e., dark red powder.

The colorant gives a raspberry-red color in aqueous solution, stable at pH over 3.5. Neutral solutions are stable even after 30 min of boiling and color shade does not change in relation with pH.

Many toxicological data are available on this red pigment: acute oral toxicity in mice, 90-day subchronical toxicological study, acute dermal irritation/corrosion, acute eye irritation/corrosion, antitumor effectiveness, micronucleus test in mice, AMES test (*Salmonella typhimurium* reverse mutation assay), estimation of antibiotic activity, and results of estimation of five mycotoxins. A new patent on arpink red was filed in 2001 with claims of anticancer effects of the anthraquinone derivatives and applications within the food and pharmaceutical fields.

After evaluating all the materials provided by the company Ascolor Biotech s.r.o., the Codex Alimentarius Commission (Rotterdam meeting, March 11–15, 2002) made the following statement: "there will not be any objections to use the red colouring matter Arpink Red" in meat products in the amount up to 100 mg/kg, meat and meat product analogues in the amount up to 100 mg/kg, nonalcoholic drinks in the amount up to 100 mg/kg, alcoholic drinks in the amount up to 200 mg/kg, milk products in the amount up to 150 mg/kg, ice creams in the amount up to 150 mg/kg, and confectionery in the amount up to 300 mg/kg.

12.3.2 Other Microbial Anthraquinones

Anthraquinoid molecules are derivatives of "9,10-anthraquinone" (which is also called 9,10-anthracenedione or 9,10-dioxoanthracene, i.e., an aromatic organic compound with formula $C_{14}H_8O_2$ and whose ketone groups are on the central ring B). In general, for each anthraquinoid molecule, there are eight possible hydrogens that can be substituted. Anthraquinoid pigments are a class of naturally occurring pigments that are mainly distributed in nature, especially from insects or tinctorial plants (e.g., *Rubia, Galium, Rheum*). Many anthraquinones are colored and they provide the most important red dyes and lakes used in artistic paintings. They are relatively stable and light-fast and they give a bright color. These properties led to their use for selected applications at present, such as in textile dyeing, printing applications, and cosmetic formulation (e.g., alizarin pigment from madder), and for some of them in food manufacturing (e.g., red carminic acid from cochineal as coloring agent for beverages or processed meat). Industrially, microbial anthraquinone pigments were first isolated from cultures of *P. oxalicum var. Armeniaca* as mentioned earlier (case of arpink red). However, the extraction, isolation, and characterization of other microbial anthraquinones have more recently been reported in the literature from some filamentous fungi with different shades such as red, bronze, maroon, and yellow (Figure 12.3).

For example, the anthraquinone pigment emodin (orange) has been isolated and identified from cultures of both *P. citrinum* and *Penicillium islandicum* (Duran et al., 2002; Frisvad, 1989; Mapari et al., 2009). The red pigment produced by strain of *Isaria farinosa* was recently elucidated as a chromophore of the anthraquinone type (Velmurugan et al., 2010). Similarly, the red pigment produced by *Paecilomyces sinclairii*, which was beforehand discovered but uncharacterized (Cho et al., 2002), is certainly of identical chemical nature according to Velmurugan et al. (2010), i.e., an amino group linked to an anthraquinone structure. Some strains of *Eurotium* could produce other microbial anthraquinones such as physcion (yellow), flavoglaucin (yellow), auroglaucin (orange), and erythroglaucin (red). From both strains of *Aspergillus*, i.e., *A. sulphureus* and *A. westerdijkiae*, viopurpurin (purple), which is an anthraquinone pigment, has been isolated and identified. In the same way, *Aspergillus glaucus* has been identified as a possible source of emodin (orange) and physcion (yellow) pigments. Strains of *Emericella purpurea* could synthesize both azaphilone and anthraquinone pigments such as epurpurins A–C (yellow) (Hideyuki et al., 1996). Some red anthraquinone pigments have also been isolated from cultures of *Fusarium oxysporum* and *Fusarium moniliforme*; moreover, the dyeing potential of the red pigment produced by *F. oxysporum* was assessed for woolen materials by Nagia and El-Mohamedy (2007). *Curvularia lunata* was also known to produce different anthraquinone pigments like chrysophanol (red), helminthosporin (maroon), and cynodontin (bronze). The catenarin (red) pigment has been isolated and identified from cultures of different *Drechslera*, i.e., from *D. teres, D. graminea, D. tritici-repentis, D. phlei*, and *D. dictyoides*. From *Drechslera avenae*,

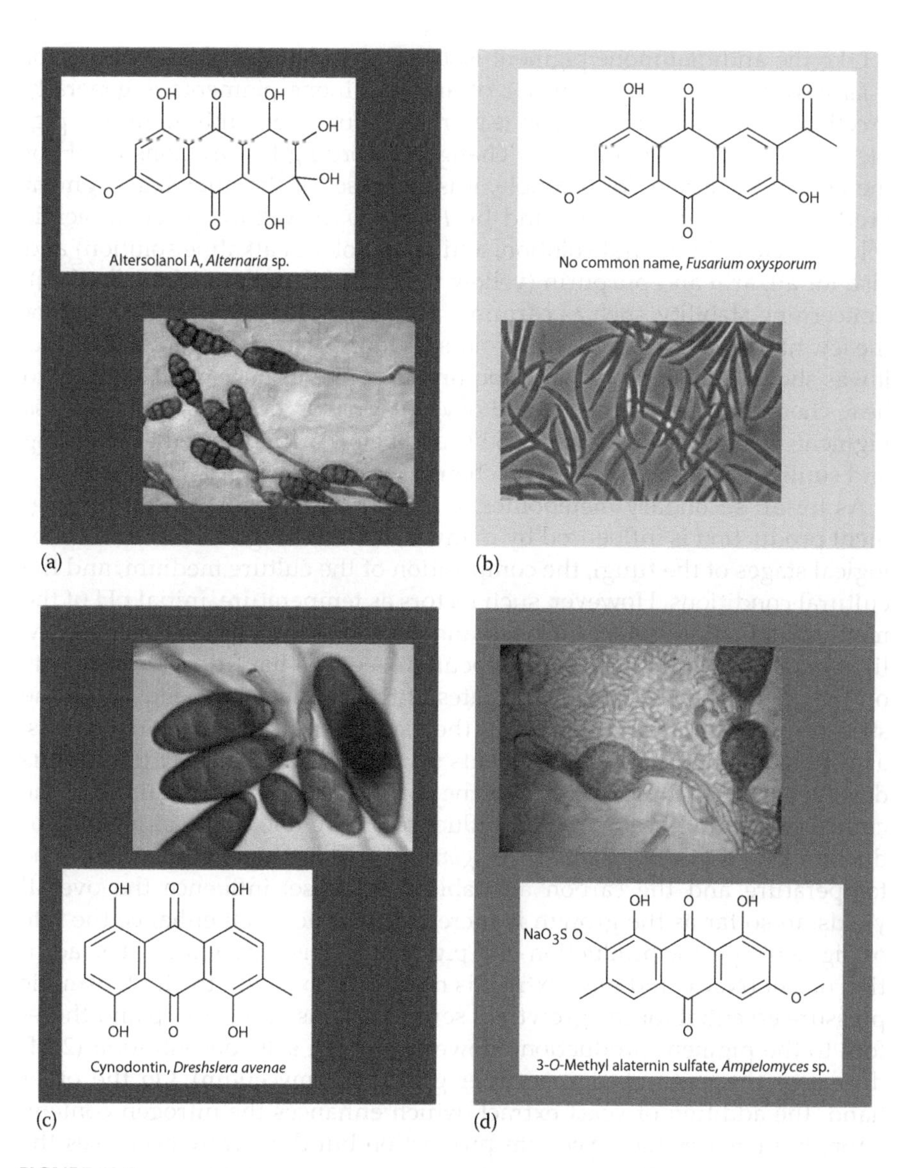

FIGURE 12.3
Some anthraquinones from fungal origin: (a) maroon, (b) bronze, (c) yellow, (d) red-orange.

more particularly, both anthraquinone pigments have been isolated and identified: cynodontin (bronze) and helminthosporin (maroon) (Duran et al., 2002; Engström et al., 1993). Other anthraquinone pigments like averythrin have been isolated and identified from a culture of *Herpotrichia rhodosticta*, which is a member of the genus *Pyrenochaeta* (Van Eljk and Roeijmans, 1984). Some strains of *Dermocybe* sp. were also well known to produce physcion (yellow) and dermocybin pigments.

Like the anthraquinone pigment isolated and identified from cultures of *I. farinosa*, the solubility in water of anthraquinone pigments is generally excellent, with some exceptions (e.g., rhein). For some anthraquinone pigments, the color of the molecule changed according to variations in pH of the aqueous solution; for example, it is the case for the microbial pigment produced both by *I. farinosa* and by *Pa. sinclairii* (yellow to red in acidic solution, violet in neutral solution, and pale violet in alkaline solution) and also for alizarin and purpurin (yellow to red [acid]; red to violet [alkaline]). Concerning stability, such as carminic acid from cochineal, which is one of the few natural and water-soluble colorants that resist degradation with time, it was shown that microbial anthraquinone pigments have good stability to heat, chemical oxidation, light, and oxygen. Most of them, like the *I. farinosa* pigment, were relatively stable at 60°C and below and subjected to steaming and sunlight exposure showed no change of color.

As for all secondary metabolites, it is frequently reported that the pigment production is influenced by many factors. Among them, the morphological stages of the fungi, the composition of the culture medium, and the cultural conditions. However, such factors as temperature, initial pH of the medium, agitation speed, nitrogen and carbon concentrations, osmolarity, light exposition, and age of the inoculum seem to have different impacts on the yields and the production rates of the pigments, depending on the strains involved. As an example, in the glaucus group, the nutritive needs, as well as the nature of the pigments produced, can change if the fungus develops the cleistothecial form or the conidial stage. The initial pH of the culture seems to have not much influence on the rate of the pigment production by *Eurotium cristatum* (*As. glaucus* group) (Anke et al., 1980). The temperature and the carbon availability (glucose) influence the overall yields, in so far as the growth is increased, but does not enhance the rate of pigment-specific production (mg pigment/g dried mycelium). The addition of salt (NaCl) and malt extract is necessary to obtain the high osmotic pressure essential for the growth of several strains of the group and therefore to the pigment production. However, a high salt concentration (2 M) decreases the yield of pigment (per g of dried mycelium). On the other hand, the addition of yeast extract, which enhances the nitrogen content, strongly increases the mycelium production but drastically decreases the pigment synthesis. The negative effect of a high N content on the pigment production has been noticed in several cases. Moreover, it was inferred from the results that for all the species of the *As. glaucus* group studied, the mycelium contains considerably larger amounts of pigments than the culture filtrates. The level of pigment production by *Penicillium* sp. rises with agitation speed up to 200 rpm and is reduced thereafter (Gunasekaran and Poorniammal, 2008). The inoculum age for an optimal production of pigment was about 4 days. Some experiments on the suitable aeration rates for pigment production by *Pa. sinclairii* showed that the maximum yield was achieved with 3.5 vvm, whereas the maximum biomass concentration was obtained with an aeration rate of 1.5 vvm (Cho et al., 2002). It seems that this parameter had complex effects, due to the relation between the mycelium morphology (highly branched or highly vacuolated) and the quality of oxygen transfers. The optimization of pigment production in filamentous fungi could therefore be based on some common concepts but requires specific adaptations depending on the producing strains.

12.4 Riboflavin, a Yellow Food Colorant but Also the Vitamin B_2

Riboflavin (vitamin B_2) has a variety of applications as a yellow food colorant. Its use is permitted in most countries. Applications include dressings, sherbet, beverages, instant desserts, ice creams, tablets, and other products. Riboflavin has a special affinity for cereal-based products, but its use in these applications is somewhat limited due to its slight odor and naturally bitter taste. There are numerous microorganisms that produce riboflavin fermentatively. Riboflavin fermentation (Burgess et al., 2009) could be classified into three categories: weak overproducers (100 mg/L or less, e.g., *Clostridium acetobutylicum*), moderate overproducers (up to 600 mg/L, e.g., yeasts such as *Candida guilliermondii* or *Debaryomyces subglobosus*), and strong overproducers (over 1 g/L, e.g., the fungi *Eremothecium ashbyii* and *A. gossypii*) (Kapralek, 1962).

12.5 Fluorescent Pink from the Red Microalga *Porphyridium*, Phycobiliproteins

The red microalga genus *Porphyridium* is a source of biochemicals possessing nutritional and therapeutical values. These biochemicals include polysaccharides (having anti-inflammatory and antiviral properties), long-chain polyunsaturated fatty acids, carotenoids such as zeaxanthin, and fluorescent phycobiliproteins (Bermejo Roman et al., 2002).

The phycobiliproteins are accessory photosynthetic pigments, aggregated in the cell as phycobilisomes, which are attached to the thylakoid membrane of the chloroplast. The red phycobiliproteins, phycoerythrin, and the blue phycobiliprotein, phycocyanin, are soluble in water and can serve as natural colorants in food, cosmetics, and pharmaceuticals. Chemically, the phycobiliproteins are built up of chromophores—the bilins—which are open-chain tetrapyrroles, covalently linked via thioether bonds to an apoprotein.

The microalgae are cultured in bioreactors under solar or artificial light in the presence of carbon dioxide and salts. The bioreactors could be closed systems made of polyethylene sleeves rather than open pools. Optimal conditions for pigment production are low to medium light intensity and medium temperatures (20°C–30°C) (Kathiresan et al., 2007).

Pigment extraction is achieved by cell breakage, extraction into water or buffered solution, and centrifugation to separate out the filtrate. The filtrate may then be partly purified and sterilized by microfiltration and spray dried or lyophilized.

Porphyridium sp. is the source of a fluorescent pink color. The main *Porphyridium* sp. phycobiliproteins are β-phycoerythrin and β-phycoerythrin. Maximum absorbance of a 1% solution of B-phycoerythrin in a 1 cm cuvette is at 545 nm, and the fluorescence emission peak is at 575 nm. Batch culture of *Porphyridium* sp. outdoors yields approximately 200 mg of colorant/L of culture after 3 days; the phycoerythrin level in the colorant is about 15%. A higher concentration of phycoerythrin, up to 30%, can be achieved under optimal algal culture conditions. The pinkish-red color can be used to color confections, gelatin deserts, and dairy products. The quantity of color required for 1 kg of food varies from 50 to 100 mg/kg (Yaron and Arad, 1993).

The color is stable at 60°C for 30 min and has a long shelf life at pH 6–7. As an ingredient in dry food preparations stored under low humidity conditions, it is very stable. A number of patents have been granted for use of the red color from *Porphyridium* in foods.

In addition to its coloring properties, red phycoerythrin possesses a yellow fluorescence. Opportunities for exploiting this property for special effects in food are under study. A range of foods that fluoresce under natural and UV light were prepared and tested. These include transparent lollipops made from sugar solutions, dry sugar-drop candies for cake decoration (that fluoresce under UV light), and soft drinks and alcoholic beverages that fluoresce at pH 5–6. Fluorescent color has also been added to alcoholic beverages containing up to 30% alcohol, but the shelf life for such products is short (Yaron and Arad, 1993).

This red color has not yet been approved for use in food or cosmetics. However, studies on rats fed with the dried biomass have not shown any adverse growth or histological effects. Future efforts should thus be devoted toward obtaining official approval of the color in foodstuffs for human consumption.

12.6 Phycocyanin, the Marine Blue from *Porphyridium*

A source of blue color is the red microalga *Porphyridium aerugineum*. This species is different from other red microalgae in that it lacks red phycoerythrin and its phycocyanin is C-phycocyanin rather than the R-phycocyanin that accompanies phycoerythrin found in many red algae and in other *Porphyridium* species. However, the biochemicals produced by *Po. aerugineum* are similar to those of other red microalgae, for example, sulfated polysaccharides, carotenoids, and lipids. Alternative source of C-phycocyanin is *Spirulina platensis*.

The algal extract of *Po. aerugineum* is blue, with maximum absorbance at a wavelength of 620 nm and a red fluorescence with maximum emission at 642 nm. The main phycobiliprotein, C-phycocyanin, is the same type of phycocyanin found in most Cyanobacteria. The chromophores are composed of phycocyanobilins, conjugated to an apoprotein via thioether bonds.

Po. aerugineum is a unicellular alga, cultured under artificial or solar light in a fresh water medium supplied with CO_2, in outdoor bioreactors. Algal growth was optimized for yield and for the properties of the blue color produced. The parameters that require close monitoring are light intensity and temperature, in order to avoid stress conditions. Stress conditions result in decreased color yield and solubility and increased biosynthesis of the polysaccharides (which encapsulate the cells and are excreted into the medium). Production of the color involves centrifugal separation of the biomass, cell breakage, and extraction. Use of a salt solution rather than water as an extraction medium increases stability of the color during extraction (Yaron and Arad, 1993).

Methods for partial exclusion of the polysaccharide from the color extract in order to enhance resolubilization of the dried color were developed. These processes included either microfiltration or coprecipitation of the polysaccharide with an added positively charged polysaccharide. Microfiltration was also used to sterilize the solution containing the produced color prior to drying. Drying was performed either by lyophilization or by spray drying. After 4 days, the yield of color reached 100 mg of product/L of batch cultured for 4 days and contained 60% phycocyanin.

The blue color reached phycocyanin levels of up to 60% of the dry matter without any further separation steps. The quantity required for coloring food was 140–180 mg of color/kg of blue food or drink. The polysaccharides accompanying the product stabilize the color and contribute added value by virtue of their functional nutritional properties. If the polysaccharides are separated out, antioxidants can be added to stabilize the color (Moreira et al., 2012). The shade of the blue color produced from *Po. aerugineum* does not change with pH. The color is stable under light but sensitive to heat. Within a pH range from 4 to 5, the blue color produced from *Po. aerugineum* is stable at 60°C for 40 min (this is not typical of blue colors from Cyanobacteria). This property is important for food uses, since many food items are acidic, particularly drinks and confections. The blue color was added to clear Pepsi® (without heat application) and to Bacardi Breezer®, and these beverages did not lose their color for at least 1 month at room temperature.

The color was very stable in dry preparations. Sugar flowers for cake decoration maintained their color for years of storage. Foods prepared with the color include gelatin and ice cream. The color was mixed with other colorants to obtain a range of shades and hues.

The blue color from *Po. aerugineum* has not been cleared for food use by the authorities, and it is not yet produced commercially. Toxicological studies carried out with other species of red microalgae have not revealed any adverse effects. Efforts should now be devoted to carrying out the required studies and procedures that will allow the use of the blue color as a substitute for synthetic colors (Eriksen, 2008).

12.7 Current Carotenoid Production Using Microorganisms

Commercial processes are already in operation or under development for the production of carotenoids by molds, yeasts, and bacteria. The production of β-carotene by microorganisms, as well as by chemical synthesis or from plant extracts, is well developed (Table 12.1), and several other carotenoids, notably lycopene, astaxanthin, zeaxanthin, and canthaxanthin, are also of interest.

12.7.1 β-Carotene

12.7.1.1 β-Carotene from B. trispora

The source organism, *B. trispora,* is a commensal mold associated with tropical plants. The fungus exists in (+) and (–) mating types (Breitenbach et al., 2012), of which the (+) type synthesizes trisporic acid, a metabolite of β-carotene. On mating the two types in a specific ratio, the (–) is stimulated by trisporic acid to synthesize large amounts of β-carotene. The mold has

TABLE 12.1

Commercial β-Carotene and β-Carotene-Containing Preparations from Various Sources

Trademark	Company	Origin
AL CARC 9004	Diana Naturals	Carrot
Altratene	Allied Industrial Corp.	Chemical synthesis
Betanat	Vitatene (Spain, DSM group)	*B. trispora*
Betatene	Cognis Nutrition & Health	*Dunaliella salina*
CaroPure	DSM	Chemical synthesis
CaroPure	DSM	*B. trispora*
Caroxan	Pot au Pin	Carrot
Lucarotin	BASF	Chemical synthesis
Mixed carotenoids	Global Palm Products	Palm oil *(Elaeis guineensis)*
Vitan	Vita-Market (Ukraine)	*B. trispora*

Data collected at Food Ingredients Europe, Paris, 2011.

shown no pathogenicity or toxicity in many experiments such as (1) standard pathogenicity tests in mice, (2) analyses of extracts of several fermentation mashes for fungal toxins, and (3) enzyme immunoassays of the final product, the β-carotene crystals, for four mycotoxins.

The production process proceeds essentially in two stages. Glucose and corn steep liquor can be used as carbon and nitrogen sources. Whey, a by-product of cheese manufacture, has also been considered, with strains adapted to metabolize lactose. In the initial fermentation process, seed cultures are produced from the original strain cultures and subsequently used in an aerobic submerged batch fermentation to produce a biomass rich in β-carotene (Papaioannou and Liakopoulou-Kyriakides, 2010). In the second stage, the recovery process, the biomass is isolated and transformed into a form suitable for isolating the β-carotene, which is extracted with ethyl acetate, suitably purified and concentrated, and the β-carotene crystallized. The final product is either used as crystalline β-carotene (purity >96%) or formulated as a 30% suspension of micronized crystals in vegetable oil. The production process is subject to Good Manufacturing Practices (GMP) procedures and adequate control of hygiene and raw materials. The biomass and the final crystalline product comply with an adequate chemical and microbiological specification, and the final crystalline product also complies with the Joint FAO/WHO Expert Committee on Food Additives (JECFA) and EU specifications as set out in Directive 95/45/EC for coloring materials in food.

The first β-carotene product from *B. trispora* was launched in 1995 at the Food Ingredients Europe business meeting in London. Following the optimization of the fermentation process, many aspects had to be addressed before the product could be marketed:

- *The microorganism.* A fungus isolated from a natural environment, not genetically modified; yield improvement achieved by classical genetics
- *Guidelines for labeling.* Natural β-carotene; natural β-carotene from *B. trispora;* fermentative, natural β-carotene; natural β-carotene from a fermentative source
- *Lobbying from other β-carotene producers (nature-identical, mixed carotenes from palm oil, β-carotene from the microalgae Dunaliella).* The EU Health and Consumer Protection Directorate General was asked to give an opinion on the safety of β-carotene from a dried biomass source, obtained from a fermentation process with *B. trispora,* for use as a coloring matter for foodstuffs
- *Safety of the fermentation-produced β-carotene.* High-pressure liquid chromatography (HPLC) analysis, stability tests, and microbiological tests have shown that the β-carotene obtained by co-fermentation of *B. trispora* complies with the EC specification for E 160 aii, also including the proportions of *cis* and *trans* isomers, and is free of mycotoxins or other toxic metabolites. Tests in vitro for gene mutations and chromosomal

TABLE 12.2

Isomers Described in "β-Carotene" from Various Sources

	Carotenoids (%)			
Source	**All *Trans*-β-Carotene**	***cis*-β-Carotene**	**α-Carotene**	**Other**
Fungus (*B. trispora*)	94	3.5	0	2.5
Chemical synthesis	98	2	0	0
Algae (*Du. salina*)	67.4	32.6	0	0
Palm oil	34	27	30	9

aberrations with the β-carotene produced by the manufacturer in the EU showed it to be free of genotoxic activity. In a 28-day feeding study in rats with the β-carotene manufactured in the EU, no adverse findings were noted at a dose of 5% in the diet, the highest dose level used. In conclusion, evaluation of the source organism and the production process yielded no grounds to suppose that the final crystalline product, β-carotene, differs from the chemically synthesized β-carotene used as a food colorant. The final crystalline fermentation product has been shown to comply with the specification for β-carotene E 160 aii listed in Directive 95/45/EC. The Committee considers that "β-carotene produced by co-fermentation of *Blakeslea trispora* is equivalent to the chemically synthesized material used as food colorant and is therefore acceptable for use as a colouring agent for foodstuffs" (Table 12.2).

Today there are other industrial productions of β-carotene from *B. trispora* in Russia and Ukraine and in León (Spain). The process has been developed to yield up to 30 mg of β-carotene/g dry mass or about 3 g/L. *B. trispora* is now also used for the production of lycopene (ACNFP, 2004).

12.7.1.2 β-Carotene from Phycomyces blakesleeanus

Another mold, *Ph. blakesleeanus*, is a potential source for various chemicals including β-carotene. The carotene content of the wild type grown under standard conditions is modest, about 0.05 mg/g dry mass, but some mutants accumulate up to 10 mg/g. As for *B. trispora*, sexual stimulation of carotene biosynthesis is essential and can increase yields up to 35 mg/g. The most productive strains of *Phycomyces* achieve their full carotenogenic potential on solid substrates or in liquid media without agitation. *B. trispora* is more appropriate for production in usual fermentors.

12.7.1.3 β-Carotene from Mucor circinelloides

Mu. circinelloides wild type is yellow because it accumulates β-carotene as the main carotenoid. The basic features of carotenoid biosynthesis, including photoinduction, are similar in *Phycomyces* and *Mucor*. *Mu. circinelloides* responds to blue light by activating biosynthesis. Wild-type strains grown in darkness contain minimal amounts of β-carotene because of the low levels of transcription of the structural genes for carotenogenesis (Almeida and Cerda-Olmedo, 2008). When exposed to a light pulse, the level of transcription of these genes increases strongly, leading to high pigment concentrations. *Mu. circinelloides* is a dimorphic fungus that grows either as yeast cells or in a mycelium form, and research is now focused on yeastlike mutants that could be useful in a biotechnological production.

12.7.2 Lycopene

Lycopene is an intermediate in the biosynthesis of all dicyclic carotenoids, including β-carotene. In principle, therefore, blocking the cyclization reaction and the cyclase enzyme by mutation or inhibition will lead to the accumulation of lycopene. This strategy is employed for the commercial production of lycopene.

12.7.2.1 Lycopene from B. trispora

A process for lycopene production by *B. trispora* is now established, with the aim of marketing this product in Europe for use as a nutritional food ingredient and dietary supplement or a food color. Lycopene is an intermediate in the β-carotene biosynthetic pathway and microbial strains that accumulate lycopene are easy to obtain by mutagenesis, molecular biology, or use of inhibitors. In the commercial process, imidazole or pyridine is added to the culture broth to inhibit the enzyme lycopene cyclase. The product, predominantly (all-*trans*)-lycopene, is formulated into a 20% or 5% suspension in sunflower oil with α-tocopherol at 1% of the lycopene level. Also available is an α-tocopherol-containing 10% and 20% lycopene cold water dispersible product. Lycopene oil suspension is intended for use as a food ingredient and in dietary supplements. The proposed level of use for lycopene in food supplements is 20 mg/day (ACNFP, 2004).

Approval for the use of lycopene from *B. trispora* was sought under regulation (EC) No 258/97 of the European Parliament and the Council concerning novel foods and novel food ingredients. The European Food Safety Authority was also asked to evaluate this product for use as a food color. It was stated in the application that it is likely that lycopene from *B. trispora* in food supplements would simply replace those supplements containing lycopene from other sources that are already being marketed so that overall consumption levels would not increase. However, incorporation of lycopene into foods would result in additional intake.

The dossier first satisfied the United Kingdom Advisory Committee on Novel Foods and Processes (ACNFP) in 2004 (ACNFP, 2004), but the Panel from the European Food Safety Authority, in opinions adopted on April 21 and October 5, 2005, was unable to conclude whether the proposed use levels of lycopene from *B. trispora* would be safe. In 2012, the placing on the market of lycopene from *B. trispora* as a novel food ingredient under Regulation (EC) No 258/97 of the European Parliament is fully authorized.

Existing authorizations and evaluations on lycopene from various sources are quite numerous. Lycopene, extracted from tomatoes, is authorized as food coloring agent within the EU (E160d) (Directive 94/36/EC) and the United States (CDR 21 73.295). This restricts the amount that can be added to foods. Lycopene was evaluated by the Scientific Committee on Food in 1975 when it was unable to allocate an Acceptable Daily Intake (ADI) but felt able to accept the use of lycopene prepared from natural foods by physical processes, without further investigations, as a coloring matter in food, provided that the amount consumed does not differ significantly from the amount consumed through the relevant foodstuffs. This opinion was reiterated by SCF in 1989. When JECFA evaluated lycopene from natural sources in 1977, they postponed a decision because of lack of data. In 1999, the SCF evaluated synthetic lycopene, but the available data were not sufficient to allow for an acceptance. The SCF concluded: "The Committee is not able to allocate an ADI and considers its use in food is unacceptable at present." Synthetic lycopene is currently used as food ingredient but is not approved for coloring matters within the EU, and it is considered Generally Recognized as Safe (GRAS) in the United States (GRAS notice No GRN 000119). In Australia and New Zealand, lycopene is permitted for use as a food color in processed foods in accordance with GMP under Schedule 3 of Standard 1.3.1 in the Food Standards Code. In Japan, tomato color, defined as "a substance composed mainly of lycopene obtained from tomato fruits," is permitted for use as a food additive under the Food Sanitation Law.

To summarize, the lycopene from *B. trispora* is considered by the EFSA Panel to be nutritionally equivalent to natural dietary lycopene, but further safety trials are necessary. While the toxicity data on lycopene from *B. trispora* and on lycopene from tomatoes do not give indications for concern, nevertheless, these data are limited and do not allow an ADI to be established. The main concern is that the proposed use levels of lycopene from *B. trispora* as a food ingredient may result in a substantial increase in the daily intake of lycopene compared to the intakes solely from natural dietary sources (Vitatene™, personal communication).

12.7.2.2 Lycopene from Fusarium sporotrichioides

The fungus *F. sporotrichioides* has been genetically modified to manufacture lycopene from the cheap corn fiber material, the "leftovers" of making ethanol. Corn fiber is abundant (the U.S. ethanol industry generates 4 million tons annually) and costs about 5 cents a pound. Distiller's dry grains with solubles could also be used as a substrate. Using a novel, general method for the sequential, directional cloning of multiple DNA sequences, the isoprenoid pathway of the fungus was redirected toward the synthesis of carotenoids. Strong promoter and terminator sequences from the fungus were added to carotenoid biosynthetic genes from the bacterium *Erwinia uredovora*, and the chimeric genes were assembled, introduced in the fungus, and expressed at levels comparable to those observed for endogenous biosynthetic genes. Cultures in laboratory flasks produced 0.5 mg lycopene/g dry mass within 6 days, and this is predicted to increase in the next few years.

12.7.3 Astaxanthin

Astaxanthin (3,3′-dihydroxy-β,β-carotene-4,4′-dione) is widely distributed in nature and is the principal pigment in crustaceans and salmonid fish. The carotenoid imparts distinctive orange-red coloration to the animals and contributes to consumer appeal in the market place. Since animals cannot synthesize carotenoids, the pigments must be supplemented in the feeds of farmed species. Salmon and trout farming is now a huge business and feeding studies have shown that astaxanthin is very effective as a flesh pigmenter. There are also reports of beneficial actions of astaxanthin for human health, so its use in supplements is of interest (Seabra and Pedrosa, 2010).

12.7.3.1 Astaxanthin from X. dendrorhous, *Formerly* Phaffia rhodozyma

Among the few astaxanthin-producing microorganisms, *X. dendrorhous* is one of the best candidates for commercial production of (3*R*, 3′*R*)-astaxanthin. Many academic laboratories and several companies have developed processes suitable for industrial production (Rodriguez-Saiz et al., 2010).

Several reports have shown that constituents in the medium, among other environmental factors, affect astaxanthin production in this yeast. The effects of different nutrients on *X. dendrorhous* have generally been studied in media containing complex sources of nutrients such as peptone, malt, and yeast extracts. By-products from agriculture were also tested, such as molasses, enzymatic wood hydrolysates, corn wet-milling coproducts, bagasse or raw sugarcane juice, date juice, and grape juice. Although such media are often convenient because they contain all nutrients, they suffer from the disadvantage of being undefined and sometimes variable in composition; this may mask important nutritional effects. For this reason, several studies have yielded results that are difficult to interpret in detail because of inadequate characterization of growth-limiting factors in the media. Thus, in order to elucidate the nature of nutritional effects as far as possible, chemically defined or synthetic media were used by some authors. In one study, 11 strains were assayed for their ability to utilize 99 different compounds as single carbon source. In a second study, carotenoid biosynthesis was increased at low ammonium or phosphate levels and stimulated by citrate. Factorial design and response surface methodology could be used to optimize the astaxanthin production. The optimal conditions stimulating the highest astaxanthin were 19.7°C temperature, 11.25 g/L carbon concentration, 6.0 pH, 5% inoculum, and 0.5 g/L nitrogen concentration. Under these conditions, the astaxanthin content was 8.1 mg/L. Fermentation strategy also has an impact on growth and carotenoid production of *X. dendrorhous*, as shown with fed-batch (e.g., limiting substrate is fed without diluting the culture) or pH-stat (i.e., a system in which the feed is provided depending on the pH) cultures. The highest biomass obtained was 17.4 g/L.

A major drawback in the use of *X. dendrorhous* is that disruption of the cell wall of yeast biomass is required before addition to animal diet, to allow intestinal absorption of the pigment. Several chemical, physical, autolytic, and enzymatic methods for cell wall disruption have been described, including a two-stage batch fermentation technique. The first stage was for "red yeast" cultivation. The second stage was the mixed fermentation of the yeast and *Bacillus circulans*, a bacterium with a high cell-wall-lytic activity.

Another starting point in optimization experiments is the generation of mutants, but metabolic engineering of the astaxanthin biosynthetic pathway is now attractive. It should be possible to manage carbon fluxes within the cell and resolve competition between enzymes such as phytoene desaturase and lycopene cyclase.

The case of *X. dendrorhous* (*Pha. rhodozyma*) is peculiar as hundreds of scientific papers and patents deal with astaxanthin production by this yeast, but the process has not been economically efficient up to now (Schmidt et al., 2011). New patents are filed almost each year, with improvement in astaxanthin yield, for example, 3 mg/g dry matter in a U.S. Patent.

12.7.3.2 Astaxanthin from Agrobacterium aurantiacum *and Other Bacteria*

Compared to the huge research effort devoted to *X. dendrorhous*, astaxanthin production by *Ag. aurantiacum* has been investigated to a lesser extent. The first description of astaxanthin biosynthesis in this bacterium was published in 1994. Astaxanthin is 1 of 10 carotenoids present. The biosynthetic pathway, the influence of growth conditions on carotenoid production, and the occurrence of astaxanthin glucoside were described in two subsequent papers, but commercial processes have not yet been developed (Yokoyama et al., 1995).

Numerous screenings have been conducted with the aim of characterizing new biological sources of astaxanthin, and positive targets were isolated such as *Paracoccus carotinifaciens* or *Halobacterium salinarum*. The latter is particularly interesting because (1) the extreme NaCl concentrations (about 20%) used in the growth medium prevent contamination with other organisms so no particular care has to be taken with sterilization; (2) NaCl concentrations under 15% induce bacterial lysis, so that no special cell breakage technique is necessary; and (3) pigments may be extracted directly with sunflower oil instead of organic solvents. This eliminates possible toxicity problems due to trace amounts of acetone or hexane and facilitates pigment assimilation by animals.

12.7.4 Zeaxanthin

Zeaxanthin (-β,β-carotene-3,3′-diol) can be used, for example, as an additive in feeds for poultry to intensify the yellow color of the skin or to accentuate the color of the yolk of their eggs (Breithaupt, 2007). It is also suitable for use as a colorant, for example, in the cosmetics and food industries, and as a health supplement that may help to prevent age-related macular degeneration.

In the mid-1960s, several marine bacteria were isolated that produced zeaxanthin. Cultures of a *Flavobacterium* sp. (ATCC 21588, classified under the accepted taxonomic standards of that time) in a defined nutrient medium containing glucose or sucrose, as carbon source, were able to produce up to 190 mg of zeaxanthin/L, with a concentration of 16 mg/g dried cellular mass. One species, *Flavobacterium multivorum* (ATCC 55238), is currently under investigation in many studies (Bhosale et al., 2003).

Recently, a zeaxanthin-producing "*Flavobacterium*" was reclassified as a new *Paracoccus* species, *Par. zeaxanthinifaciens,* and earlier findings that IPP biosynthesis occurs exclusively *via* the mevalonate pathway were confirmed. A second strain, isolated in a mat from an island of French Polynesia, is peculiar as it also produces exopolysaccharides.

Sphingobacterium multivorum, the new name for *Fl. multivorum,* was recently shown to utilize the alternative deoxyxylulose phosphate (DXP) pathway. A strain was constructed for overproduction of zeaxanthin in industrial quantities.

As more bacteria are examined, the distribution of the mevalonate and DXP pathways will become better defined, thus facilitating the metabolic engineering of microorganisms with improved production of commercially important isoprenoid compounds including carotenoids.

12.7.5 Canthaxanthin

Canthaxanthin (-β,β-carotene-4,4′-dione) has been used in aquafeed for many years in order to impart the desired flesh color in farmed salmonids. A *Bradyrhizobium* sp. strain was described as a canthaxanthin producer and the carotenoid gene cluster was fully sequenced. Interest in canthaxanthin is decreasing since the discovery of extreme overdosage, i.e., the deposition of minute crystals in the eye, a fact leading to adverse media attention in the past, and some pressure to limit its use in aquafeeds (Chandi and Gill, 2011).

A second bacterium under scrutiny for canthaxanthin production is *Haloferax alexandrinus,* which belongs to the extremely halophilic Archaea, chemo-organotrophic organisms that satisfy some of their energy requirements with light. Members of the family *Halobacteriaceae* are characterized by red-colored cells, the color in most cases being due to the presence of C_{50}-carotenoids (especially bacterioruberins) as the major carotenoids. Some species have been reported to produce C_{40}-carotenoids and ketocarotenoids as minor carotenoids. Recently, the biotechnological potential of these members of the Archaea has increased because of their unique features, which facilitate many industrial procedures. For example, no sterilization is required, because of the extremely high NaCl concentration used in the growth medium (contamination by other organisms is avoided). In addition, no cell-disrupting devices are required, as cells lyse spontaneously in fresh water.

A 1 L scale cultivation of the cells in flask cultures (6 days) under nonaseptic conditions produced 3 g dry mass, containing 6 mg total carotenoid and 2 mg canthaxanthin. Further experiments in a batch fermentor also demonstrated increases in the biomass concentration and carotenoid production.

A third example is *Gordonia jacobea* (CECT 5282), a Gram-positive, catalase-negative G+C 61% bacterium, which was isolated in routine air sampling during screening for microorganisms that produce pink colonies. Analysis of the carotenoid extracts by HPLC-MS revealed that the main pigment in the isolates is canthaxanthin. The low carotenoid content (0.2 mg/g dry mass) in the isolate does not support an industrial application, but after several rounds of mutations, a hyperpigmented mutant (MV-26) was isolated, which accumulated six times more canthaxanthin than the wild-type strain. Apart from their high pigment production, the advantages of mutants of this species from the industrial point of view are (1) the optimal temperature for growth and carotenogenesis 30°C, which is usual in fermentors; (2) the use of glucose, an inexpensive carbon source, for optimal growth and pigmentation; and (3) the fact that >90% of the total pigments can be extracted directly with ethanol, a nontoxic solvent allowed for human and animal feed. Many other culture media were tested, giving canthaxanthin from 1 to 13.4 mg/L.

12.7.6 Torulene and Torularhodin

Yeasts in the genus *Rhodotorula* synthesize carotenoids, mainly torulene (3′,4′-didehydro-β,ψ-carotene) and torularhodin (3′,4′-didehydro-β,ψ-caroten-16-oic acid) accompanied by very small amounts of β-carotene. Most of the research has focused on the species *Rhodotorula glutinis*, but some papers deal with other species such as *R. gracilis*, *R. rubra*, and *R. graminis* (Frengova and Beshkova, 2009).

Feed supplemented with a *Rhodotorula* cell mass has been found to be safe and nontoxic in animals. Its use in the nutrition of laying hens has also been documented. As the β-carotene content in wild strains of *R. glutinis* is low, efforts have been made to increase it through strain improvement, mutation, medium optimization, and manipulation of culture conditions (temperature, pH, aeration, C/N ratio). These studies mainly resulted in an increased yield of torulene and torularhodin, which are of minor interest, though some did succeed in increasing the β-carotene content up to about 70 mg/L.

12.8 Summary and Future Prospects

There are several advantages of microorganisms for the study of biosynthesis and function of pigments. Bacteria and fungi offer a tremendous resource in that they produce hundreds to thousands of various pigmented molecules.

It is likely that many more natural pigments will be isolated from this biomass (Dufossé et al., 2005).

Synthetic pigments traditionally used by food or cosmetic processors continue to be utilized with success; however, with the increasing consumer preference for natural food additives, natural colorants from plants is now a big business and most of the research efforts within the scientific field of colorants are conducted on natural ones. Regarding bacteria, yeast, or fungi (Table 12.3), despite common belief about the high production cost of fermentation pigments, two initiatives started in Europe these last years: β-carotene from the filamentous

TABLE 12.3

Microbial Production of Pigments (Already in Use as Natural Colorants or with High Potential in This Field)

Molecule	Color	Microorganism	Status[a]
Ankaflavin	Yellow	*Monascus* sp. (fungus)	IP
Anthraquinone	Red	*P. oxalicum* (fungus)	IP
Astaxanthin	Pink-red	*X. dendrorhous* (yeast), formerly *Pha. rhodozyma*	DS
Astaxanthin	Pink-red	*Ag. aurantiacum* (bacteria)	RP
Astaxanthin	Pink-red	*Par. carotinifaciens* (bacteria)	RP
Canthaxanthin	Dark red	*Bradyrhizobium* sp. (bacteria)	RP
Canthaxanthin	Dark red	*H. alexandrinus* (bacteria)	RP
Canthaxanthin	Dark red	*G. jacobea* (bacteria)	DS
Lycopene	Red	*B. trispora* (fungus)	DS
Lycopene	Red	*F. sporotrichioides* (fungus)	RP
Melanin	Black	*Saccharomyces neoformans* var. *nigricans* (yeast)	RP
Monascorubramin	Red	*Monascus* sp. (fungus)	IP
Naphthoquinone	Deep bloodred	*Cordyceps unilateralis* (fungus)	RP
Riboflavin	Yellow	*A. gossypii* (fungus)	IP
Rubrolone	Red	*Streptomyces echinoruber* (bacteria)	DS
Rubropunctatin	Orange	*Monascus* sp. (fungus)	IP
Torularhodin	Orange-red	*Rhodotorula* sp. (yeast)	DS
Zeaxanthin	Yellow	*Flavobacterium* sp. (bacteria)	DS
Zeaxanthin	Yellow	*Par. zeaxanthinifaciens* (bacteria)	RP
Zeaxanthin	Yellow	*S. multivorum* (bacteria)	RP
β-carotene	Yellow-orange	*B. trispora* (fungus)	IP
β-carotene	Yellow-orange	*F. sporotrichioides* (fungus)	RP
β-carotene	Yellow-orange	*Mu. circinelloides* (fungus)	DS
β-carotene	Yellow-orange	*Neurospora crassa* (fungus)	RP
β-carotene	Yellow-orange	*Ph. blakesleeanus* (fungus)	RP
Unknown	Red	*Penicillium purpurogenum* (fungus)	DS
Unknown	Red	*Pa. sinclairii* (fungus)	RP

[a] IP, Industrial production; DS, development stage; RP, research project.

fungi, *B. trispora* (produced by Gist-Brocades now DSM; approved in 2000 by the EU Scientific Committee on Food Safety) and arpink red from *P. oxalicum* (manufactured by Ascolor). These companies invested a lot of money as any combination of new source and/or new pigment drives a lot of experimental work, process optimization, toxicological studies, regulatory issues, and tremendous paper work. Another development under process is the production of lycopene using *B. trispora* by Vitatene, a subsidiary of Spanish penicillin firm Antibioticos (now a subsidiary of DSM, 2011). Exploration of fungal biodiversity is still going on, with special interest in water-soluble pigments. The case of *X. dendrorhous* (*Pha. rhodozyma*) is very peculiar as hundreds of scientific papers and patents deal with astaxanthin production using this yeast, and the process has not been economically efficient up to now. Microorganisms could either be used for the biosynthesis of "niche" pigments not found in plants, such as aryl carotenoids. Carotenoids play an exceptional role in the fast-growing "over-the-counter medicine" and "nutraceutical" sector. Among carotenoids under investigation for coloring or for biological properties, a small number are available from natural extracts or chemical synthesis. The list is rather short compared to the long list of 700 entries in the *Carotenoid Handbook* (Britton et al., 2004). With imagination, biotechnology could be a solution for providing additional pigments including interesting aryl carotenoids. Isorenieratene (Φ,Φ-carotene) and its monohydroxy and dihydroxy derivatives can be produced by bacteria, i.e., *Brevibacterium linens*, *Streptomyces mediolani*, or *Mycobacterium aurum*.

Research projects mixing molecular biology and pigments were investigated all over the world and it seems that current productions are not effective in terms of final yield. Nowadays, combinatorial genetic engineering is being addressed, based on an increasing number of known carotenogenic gene sequences. By combining genes, some authors were able to obtain more efficient biosynthesis, or new carotenoids, never described in nature, such as multi-hydroxylated ones, which could be very efficient as antioxidants.

Acknowledgment

Nuthathai Sutthiwong is grateful to the Royal Thai Government for the research grant obtained for preparing her Ph.D. at the University of Reunion Island, France.

References

Almeida, E.R.A. and Cerda-Olmedo, E. 2008. Gene expression in the regulation of carotene biosynthesis in *Phycomyces*. *Current Genetics*, 53: 129–137.

Anke, H., Kolthoum, I., Zähner, H., and Laatsch, H. 1980. The anthraquinones of the *Aspergillus glaucus* group. I. Occurrence, isolation, identification and antimicrobial activity. *Archives of Microbiology*, 126: 223–230.

Advisory Committee on Novel Foods and Processes (ACNFP), 2004. United Kingdom, advisory committee on novel foods and processes: Annual report 2003. pp. viii and 60.

Babitha, S., Soccol, C.R., and Pandey, A. 2007. Effect of stress on growth, pigment production and morphology of *Monascus* sp. in solid cultures. *Journal of Basic Microbiology*, 47: 118–126.

Bermejo Roman, R., Alvarez-Pez, J.M., Acien Fernandez, F.G., and Molina Grima, E. 2002. Recovery of pure β-phycoerythrin from the microalga *Porphyridium cruentum*. *Journal of Biotechnology*, 93: 73–85.

Bhosale, P., Ermakov, I.V., Ermakova, M.R., Gellermann, W., and Bernstein, P.S. 2003. Resonant Raman quantification of zeaxanthin production from *Flavobacterium multivorum*. *Biotechnology Letters*, 25: 1007–1011.

Blanc, P.J., Loret, M.O., Santerre, A.L., Pareilleux, A., Prome, D., Prome, J.C., Laussac, J.P., and Goma, G. 1994. Pigments of *Monascus*. *Journal of Food Science*, 59: 862–865.

Breitenbach, J., Fraser, P.D., and Sandmann, G. 2012. Carotenoid synthesis and phytoene synthase activity during mating of *Blakeslea trispora*. *Phytochemistry*, 76: 40–45.

Breithaupt, D.E. 2007. Modern application of xanthophylls in animal feeding—A review. *Trends in Food Science and Technology*, 18: 501–506.

Britton, G., Liaaen-Jensen, S., and Pfander, H. 2004. *Carotenoids Handbook*. Birkhäuser Publication, Basel, Switzerland.

Burgess, C.M., Smid, E.J., and van Sinderen, D. 2009. Bacterial vitamin B2, B11 and B12 overproduction: An overview. *International Journal of Food Microbiology*, 133: 1–7.

Chandi, G.K. and Gill, B.S. 2011. Production and characterization of microbial carotenoids as an alternative to synthetic colors: A review. *International Journal of Food Properties*, 14: 503–513.

Cho, Y.J., Park, J.P., Hwang, H.J., Kim, S.W., and Choi, J.W. 2002. Effect of carbon source and aeration rate on broth rheology and fungal morphology during red pigment production by *Paecilomyces sinclairii* in a batch bioreactor. *Journal of Biotechnology*, 95: 13–23.

Dufossé, L. 2004. *Pigments in Food, more than Colours*. Université de Bretagne Occidentale Publication, Quimper, France.

Dufossé, L. 2006. Microbial production of food grade pigments. *Food Technology and Biotechnology*, 44: 313–321.

Dufossé, L., Galaup, P., Yaron, A. et al. 2005. Microorganisms and microalgae as sources of pigments for food use: A scientific oddity or an industrial reality? *Trends in Food Science and Technology*, 16: 389–406.

Duran, N., Maria, F.S.T., Roseli, D.C., and Elisa, E. 2002. Ecological-friendly pigments from fungi. *Critical Reviews in Food Science and Nutrition*, 42: 53–66.

Engström, K., Brishammar, S., Svensson, C., Bengtsson, M., and Andersson, R. 1993. Anthraquinones from some *Drechslera* species and *Bipolaris sorokiniana*. *Mycological Research*, 97: 381–384.

Eriksen, N.T. 2008. Production of phycocyanin—A pigment with applications in biology, biotechnology, foods and medicine. *Applied Microbiology and Biotechnology*, 80: 1–14.

Frengova, G.I. and Beshkova, D.M. 2009. Carotenoids from *Rhodotorula* and *Phaffia*: Yeasts of biotechnological importance. *Journal of Industrial Microbiology and Biotechnology*, 36: 163–180.

Frisvad, J.C. 1989. The connection between the Penicillia and Aspergilli and mycotoxins with special emphasis on misidentified isolates. *Archives on Environmental Contamination and Toxicology*, 18: 452–467.

Gunasekaran, S. and Poorniammal, R. 2008. Optimization of fermentation conditions for red pigment production from *Penicillium* sp. under submerged cultivation. *African Journal of Biotechnology*, 7: 1894–1898.

Hideyuki, T., Koohei, N., and Ken-ichi, K. 1996. Isolation and structures of dicyanide derivatives, epurpurins A to C, from *Emericella purpurea. Chemical and Pharmaceutical Bulletin (Tokyo)*, 44: 2227–2230.

Juzlova, P., Martinkova, L., and Kren, V. 1996. Secondary metabolites of the fungus *Monascus*: A review. *Journal of Industrial Microbiology*, 16: 163–170.

Kapralek, F. 1962. The physiology of riboflavin production by *Eremothecium ashbyi. Journal of General Microbiology*, 29: 403–419.

Kathiresan, S., Sarada, R., Hattacharya, S., and Ravishankar, G.A. 2007. Culture media optimization for growth and phycoerythrin production from *Porphyridium purpureum. Biotechnology and Bioengineering*, 96: 456–463.

Kerr, J.R. 2000. Phenazine pigments: Antibiotics and virulence factors. *The Infectious Disease Review*, 2: 184–194.

Mapari, S.A.S., Hansen M.E., Meyer A.S., and Thrane U. 2008. Computerized screening for novel producers of *Monascus*-like food pigments in *Penicillium* species. *Journal of Agricultural and Food Chemistry*, 56: 9981–9989.

Mapari, S.A.S., Meyer, A.S., Thrane, U., and Frisvad, J.C. 2009. Identification of potentially safe promising fungal cell factories for the production of polyketide natural food colorants using chemotaxonomic rationale. *Microbial Cell Factories*, 8: 24–28.

Mapari, S.A.S., Nielsen, K.F., Larsen, T.O., Frisvad, J.C., Meyer, A.S., and Thrane, U. 2005. Exploring fungal biodiversity for the production of water-soluble pigments as potential natural food colorants. *Current Opinion in Biotechnology*, 16: 231–238.

Mc Cann, D., Barrett, A., Cooper, A. et al. 2007. Food additives and hyperactive behaviour in 3-year-old and 8/9-year-old children in the community: A randomised, double-blinded, placebo-controlled trial. *The Lancet*, 370: 1560–1567.

Moreira, I.D., Passos, T.S., Chiapinni, C. et al. 2012. Colour evaluation of a phycobiliprotein-rich extract obtained from *Nostoc* PCC9205 in acidic solutions and yogurt. *Journal of the Science of Food and Agriculture*, 92: 598–605.

Nagia, F.A. and El-Mohamedy, R.S.R. 2007. Dyeing of wool with natural anthraquinone dyes from *Fusarium oxysporum. Dyes and Pigments*, 75: 550–555.

Papaioannou, E.H. and Liakopoulou-Kyriakides, M. 2010. Substrate contribution on carotenoids production in *Blakeslea trispora* cultivations. *Food and Bioproducts Processing*, 88: 305–311.

Plonka, P.M. and Grabacka, M. 2006. Melanin synthesis in microorganisms—Biotechnological and medical aspects. *Acta Biochimimica Polonica*, 53: 429–443.

Rodriguez-Saiz, M., de la Fuente, J.L., and Barredo, J.L. 2010. *Xanthophyllomyces dendrorhous* for the industrial production of astaxanthin. *Applied Microbiology and Biotechnology*, 88: 645–658.

Sardaryan, E., Zihlova, H., Strnad, R., and Cermakova, Z. 2004. Arpink Red—Meet a new natural red food colourant of microbial origin. In: *Pigments in Food, more than Colours*, ed. L. Dufossé, pp. 207–208, Université de Bretagne Occidentale (Publication), Quimper, France.

Schab, D.W. and Trinh, N-H.T. 2004. Do artificial food colors promote hyperactivity in children with hyperactive syndromes? A meta-analysis of double-blind placebo-controlled trials. *Journal of Developmental and Behavioral Pediatrics*, 25: 423–434.

Schmidt, I., Schewe, H., Gassel, S. et al. 2011. Biotechnological production of astaxanthin with *Phaffia rhodozyma/Xanthophyllomyces dendrorhous. Applied Microbiology and Biotechnology*, 89: 555–571.
Seabra, L.M.J. and Pedrosa, L.F.C. 2010 Astaxanthin: Structural and functional aspects. *Brazilian Journal of Nutrition*, 23: 1041–1050.
Van Eljk, G.W. and Roeijmans, H.J. 1984. Averythrin-6-monomethyl ether, an anthraquinone pigment from *Herpotrichia rhodosticta. Experimental Mycology*, 8: 266–268.
Velmurugan, P., Lee, Y.H., Nanthakumar, K., Kamala-Kannan, S., Dufossé, L., Mapari, S.A.S., and Oh, B-T. 2010. Water-soluble red pigments from *Isaria farinosa* and structural characterization of the main colored component. *Journal of Basic Microbiology*, 50: 1–10.
Xu, G.R., Chen, Y., Yu, H.L., Cameleyre, X., and Blanc, P.J. 2003. HPLC fluorescence method for determination of citrinin in *Monascus* cultures. *Archiv Für Lebensmittelhygiene*, 54: 82–84.
Yaron, A. and Arad, S. 1993. Phycobiliproteins-blue and red natural pigments-for use in food and cosmetics. In: *Food Flavors, Ingredients and Composition, Developments in Food Science*, ed. G. Charalambous, pp. 835–838, Elsevier, London, U.K.
Yokoyama, A., Adachi, K., and Shizuri, Y. 1995. New carotenoid glucosides, astaxanthin glucoside and adonixanthin glucoside, isolated from the astaxanthin-producing marine bacterium, *Agrobacterium aurantiacum. Journal of Natural Products*, 58: 1929–1933.

13

Production of Bioflavors

Jyothi Ramamohan, Harish K. Chopra, and Shubhneet Kaur

CONTENTS

13.1 Introduction 448
13.2 Structural Classification of Flavors 449
  13.2.1 Diacetyl 450
  13.2.2 Pyrazines 450
  13.2.3 Lactones 450
  13.2.4 Terpenes 451
  13.2.5 Esters 456
  13.2.6 Carboxylic Acids 457
13.3 Biotechnological Routes for the Production of Bioflavors 459
  13.3.1 Plant Tissue Culture Method 460
  13.3.2 Enzymatic Method 460
  13.3.3 Microbial Method 463
    13.3.3.1 De Novo Synthesis of Bioflavors 463
    13.3.3.2 Biotransformation Method 466
13.4 Safety and Regulatory Aspects of Natural Flavors 470
  13.4.1 Flavor and Extract Manufacturers Association, USA 470
  13.4.2 Latest Regulation Concerning GMO Labeling 471
  13.4.3 Commission Regulation (EC) No. 50/2000 of January 10, 2000 471
  13.4.4 Food Sanitation Law in Japan 471
    13.4.4.1 Labeling System of Foods Produced by Recombinant DNA Technique 472
  13.4.5 Joint FAO/WHO Food Standards 472
13.5 Summary and Future Prospects 473
13.A Appendix: List of Traditional Food Preparation Processes 473
References 474

13.1 Introduction

Flavor is the sensory impact of a food or other substances, which is determined mainly by the chemical senses of taste and smell. The primary role of flavor in food is to make the food palatable. Many food items can be made consumable with the addition of flavors, which increases the functional and economic value of the food. The application of flavor technology depends on the identification of the sensory active compounds responsible for natural flavors. The production of characteristic flavors in foods through chemical or biochemical reactions is very complex, and understanding these concepts is a challenging job before the food chemist and technologist. In recent years, the interest in the structure–activity relationship of food flavor has increased manyfold due to the commercial importance of these compounds.

Consumers today have increased awareness of food and food products. Their demand for "natural" food flavoring agents with the perceived safety and good health associated with food products is driving flavor houses to bring nature back to their workstations, thereby opening up renewed challenges and opportunities in food research and development.

Microorganisms such as yeast, bacteria, and fungi have been adding flavor to our gourmet palettes by their routine metabolic pathways to ripen cheese; to ferment beer, whiskey, and exotic wines; as well as to enhance the whiff of our daily baked bread, cocoa, coffee, and tea. The term natural flavor or natural flavoring means the essential oil, oleoresin, essence or extractive, protein hydrolysate, distillate, or any product of roasting, heating, or enzymolysis, which contains the flavoring constituents derived from a spice, fruit or fruit juice, vegetable or vegetable juice, edible yeast, herb, bark, bud, root, leaf or similar plant material, meat, seafood, poultry, eggs, dairy products, or fermentation products thereof, whose significant function in food is flavoring rather than nutritional. Natural flavors include the natural essence or extractives obtained from plants (Code of Federal Regulation 21CFR). The flavors in most of the natural products, especially in fruits, vegetables, spices, etc., are either product or by-product of various metabolic pathways (Kumar and Panesar, 2009). The general scheme representing the main pathways involved in flavor synthesis in fruits and vegetables is shown in Figure 13.1.

Food technologists and scientists are now coaxing microbial flora to ferment and produce an array of flavor compounds and make these accessible to the food/flavor industry. The global market for flavors and fragrances has grown up to $22 billion in 2012; therefore, the flavor companies are directing their attention toward flavor compounds of biological origin in view of the safety and environmental concerns. To be labeled "natural," in most countries, the flavor compounds must be products of biotransformation, sourced either from plants or from their parts, microorganisms (bacteria, yeast, fungi), products of natural catalysts (enzymes), natural precursors, or physical extraction from natural sources. To ensure the "natural" word on the label

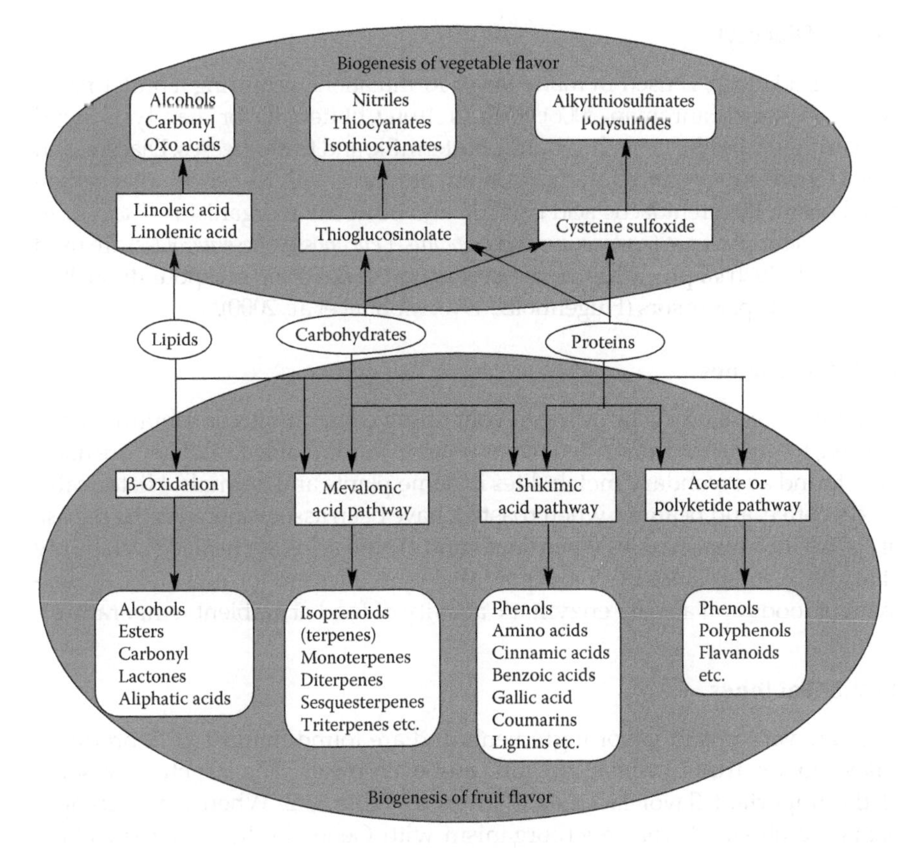

FIGURE 13.1
Biogenesis of fruit and vegetable flavors.

of end products, food and related flavoring companies have to source the natural flavors for applications in food products, as per the growing demand from discerning customers.

In this chapter, the production of "bioflavors" by different biotechnological approaches such as plant tissue culture, enzyme catalysis, and fermentation has been discussed with the emphasis on the production of economically important bioflavor compounds.

13.2 Structural Classification of Flavors

Based on their chemical structures, flavoring compounds can be classified into many different groups like alcohols, aldehydes, ketones, lactones, and pyrazines.

13.2.1 Diacetyl

Diacetyl, a ketone, is used in many flavor formulations, requiring a butter flavor note (Longo and Sanroman, 2006). With its strong, buttery flavor, diacetyl is found in many dairy products, such as butter, buttermilk, and fresh cheeses. Diacetyl and acetaldehyde are produced in insignificant amounts with respect to total carbon metabolism. Different lactic acid bacteria and other microorganisms (*Lactococcus lactis, Lactobacillus* sp., *Streptococcus thermophilus, Leuconostoc mesenteroides*) are used in many industrial processes for the production of flavoring compounds such as diacetyl or its precursors (Hugenholtz, 1993; Monnet et al., 2000).

13.2.2 Pyrazines

Pyrazines are heterocyclic nitrogen-containing compounds contributing to the roasted, toasted, fried, or grilled flavor notes to foods (Table 13.1). They are naturally found as secondary metabolites in some plants and animals. Historically, cocoa, coffee, and peanuts, when roasted, have been associated with the release of "pyrazine"-impact notes when sugar and amino acids are heated (Delahunty et al., 1993). It has since been accepted that pyrazines are formed on heat treatment of foods and also by enzymatic activity of food at ambient temperature.

13.2.3 Lactones

They are very potent flavor ingredients and are found in most milk products, baked goods, fruits, vegetables, nuts, and even meats. The structure of some of the important flavor lactones is given in Table 13.2. When flavor components are obtained from microorganism with Generally Recognized as Safe (GRAS) status through fermentation processes, they have the advantage of being labeled "natural." In addition to ketones, lactones are the primary components responsible for the cooked flavor associated with baked goods made with butter. Lactones in butter are the major source of flavor in confectionary and high-quality candies where they provide the unique, pleasurable flavors and mouthfeel associated with these products. A significant role is especially played by γ- and δ-lactones, since these substances have fruity flavors such as apricot, mango, nectarine, and peach, as well as butter and coconut flavors (Do et al., 1969; Crouzet et al., 1990). The coconut-like flavor attribute is imparted by lactones with a chain length of 8–9 carbon atoms, while fruity, peach-like flavor is imparted by lactones of chain length 10–11 and 12 carbon atoms, which usually impart an oily, fruity note, whereas δ-lactones usually smell softer and creamier. δ-Lactones occur naturally in dairy products, but there is no natural 5-decanolide or 5-dodecanolide available by other routes than by expensive and uneconomical isolation from natural products such as butter. But with the developments in the extraction of flavors from microbial sources, an invention describing the production of natural 5-decanolide and natural 5-dodecanolide, through the microbial hydrogenation of Massoi bark oil fractions, using selected strains of yeast and fungus has been patented (Laat et al., 1992).

TABLE 13.1

Structure and Flavor of Common Pyrazines

Pyrazine	Structure	Sensorial Description
2,5-Dimethylpyrazine	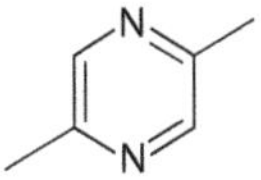	Nutty, fatty
2-Ethyl-3,5-dimethylpyrazine	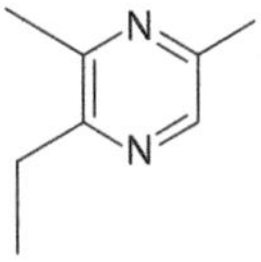	Cocoa, chocolate, nutty (burned almond, filbert-hazelnut) notes
2-Methoxy-3-isopropylpyrazine	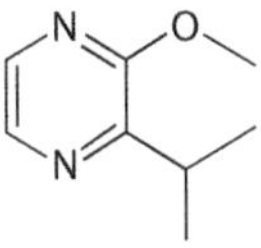	Soil, vegetable, earthy, bell pepper, pea, potato
2-Methoxy-3-isobutylpyrazine	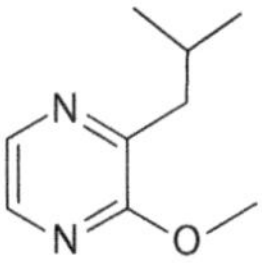	Peppery
2-Methoxy-3-methylpyrazine	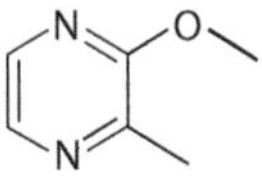	Vegetable, popcorn, potato
2-Methyl-6-ethoxypyrazine	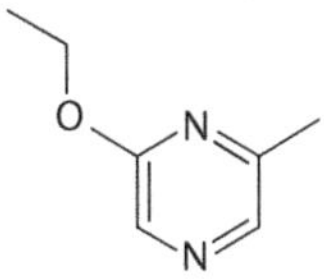	Pineapple
2,3,5-Trimethylpyrazine	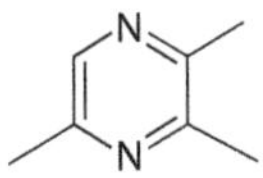	Nutty, baked potato, roasted peanut, cocoa, burned notes
2,3,5,6-Tetramethylpyrazine	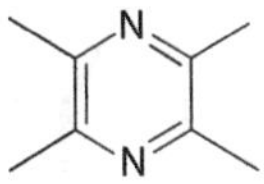	Weak, nutty, musty, chocolate odor; chocolate taste

13.2.4 Terpenes

More than 30,000 members are included in this class of natural products named terpenes, which predominantly originate from plants. They constitute an important role in the fragrance and flavor industry as they are present in many essential oils (Longo and Sanroman, 2006). The building blocks of terpenes are isopentenyl diphosphate and dimethyl allyl diphosphate. Linalool, nerol, geraniol, and citronellol (Table 13.3) are important components in a flavorist's palette among terpenes. Nerol, citronellol, and geraniol were found

TABLE 13.2

Structure, Source, and Taste of Common δ-Lactones and γ-Lactones

Lactone	Structure	Source	Taste
δ-Decalactone (5-pentapentanolide)	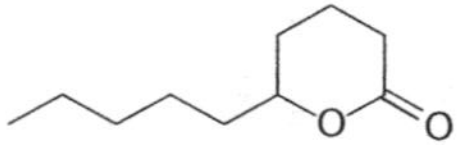	Apricot, blue cheese, burley tobacco, butter, coconut, mango, peach, raspberry, rum, strawberry, tea	Creamy, coconut, lactonic, sweet and dairy-like with milky nuances
δ-Dodecalactone (5-heptylpentanolide)	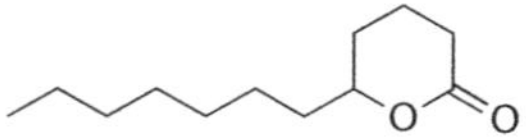	Butter, coconut, cream, milk	Sweet, creamy, lactonic, waxy, milky, coconut
Glucono-δ-lactone	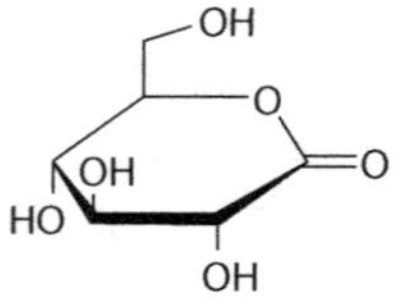	Honey, fruit juice, wine	Tangy taste to foods
δ-Hexalactone (5-methylpentanolide)	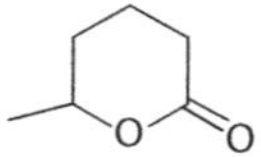	Butter, coconut, raspberry, strawberry, tea	Coconut, cream, chocolate
Jasmine lactone (Z-deca-7-en-5-olide)	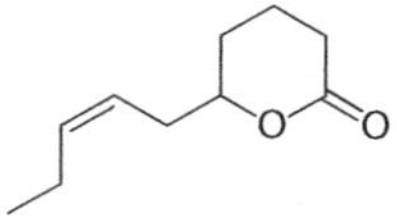	Gardenia flowers, jasmine, Osmanthus, tea, tuberose	Powerful, oily fatty fruity, peach–apricot
Massoia lactone (5-pentylpent-2-en-5-olide)	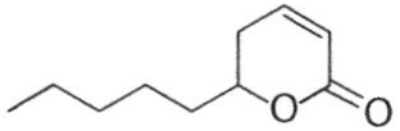	Tobacco, massoia oil, Osmanthus, sugar, molasses, tuberose, wine	Creamy, coconut, milky, green, and slightly fruity

δ-Nonalactone (5-butylpentanolide)	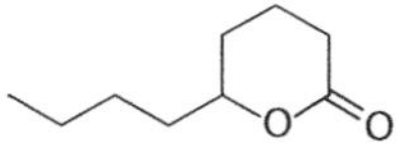 Butter, cooked beef, cooked pork, milk, whiskey, wine	Sweet, oily–fatty, coconut, fruity, dairy–creamy
δ-Octalactone (5-propylplentanolide)	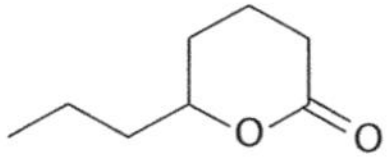 Apricot, blue cheese, burley tobacco, butter, cheddar cheese, coconut, cream, pineapple, raspberry, strawberry, yogurt	Sweet, creamy, fatty
6-Pentyl-alpha-pyrone (deca-2,4-dien-5-olide)	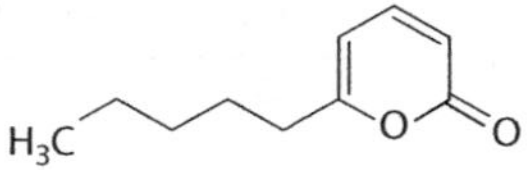 Cooked beef, mushroom, osmanthus, peach	Coconut, creamy, sweet. brown with a fatty waxy
δ-Undecalactone (5-hexylpentanolide)	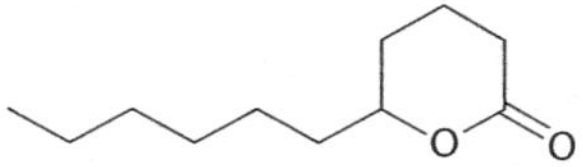 Butter, coconut, cream, milk	Creamy, peachy, milky, waxy
γ-Decalactone (4-hexylbutanolide)	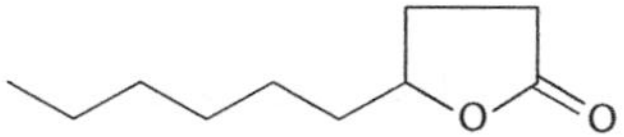 Apricot, blue cheese, butter, coconut milk (fresh), guava, mango, oriental tobacco, peach, plum, strawberry, tea, wine	Coconut, buttery sweet
γ-Dodecalactone (4-octylbutanolide)	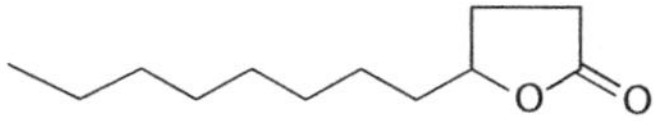 Apricot, beer, blue cheese, cheddar cheese, cooked pork, milk products, peach, pineapple, rum, strawberry	Fatty, fruity, peach odor, milky–peach
γ-Hexalactone (4-ethylbutanolide)	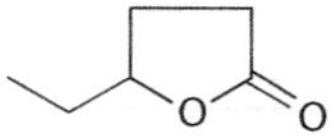 Apricot, Burley tobacco, cocoa, grape, grape brandy, mango, peach, raspberry, strawberry, wheat bread	Sweet, creamy, vanilla-like with green lactonic powdery nuances
γ-Heptalactone (4-propylbutanolide)	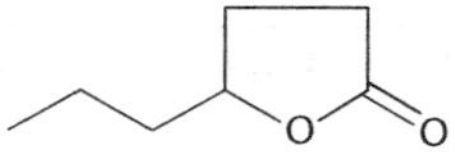 Beer, licorice, mango, roasted filbert, papaya, peach, strawberry, tea, wine	Sweet, lactonic, creamy, coconut, coumarin

(continued)

TABLE 13.2 (continued)

Structure, Source, and Taste of Common δ-Lactones and γ-Lactones

Lactone	Structure	Source	Taste
γ-Nonalactone (4-pentylbutanolide)	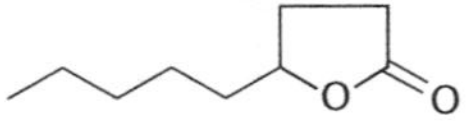	Asparagus, beer, cooked pork, licorice, mushroom, peach, roasted barley, roasted filbert, tamarind, tea, Virginia tobacco, wheat bread, whiskey, wine	Coconut
γ-Octalactone (4-butylbutanolide)	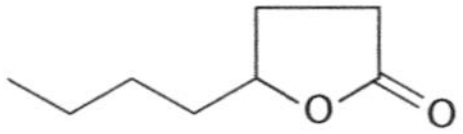	Apricot, blue cheese, butter, cooked chicken, cooked pork, licorice, milk, raspberry, roasted barley, roasted filbert (hazelnut), roasted peanut, roasted pecan	Sweet creamy, with coconut nuances
γ-Undecalactone (4-heptylbutanolide)	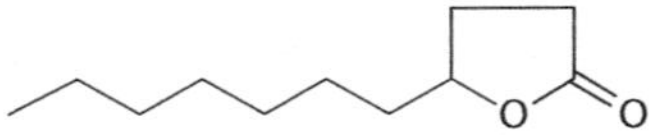	Apple, apricot, butter, cooked pork, cooked rice, passion fruit, peach	Coconut, creamy, vanilla, nutty, peach
γ-Valerolactone (4-methylbutanolide)	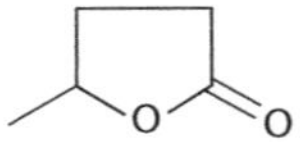	Cocoa, coffee, honey, peach, virginia tobacco, wheat bread	Sweet, hay-like, coumarinic

Source: www.leffingwell.com/glacton.htm

TABLE 13.3

Structure and Flavor Notes of Common Terpenes

Terpene	Structure	Flavor Notes
Caryophyllene	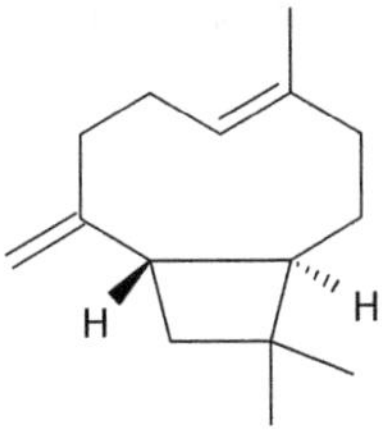 H H	Spicy, oily, citrus
Citronellol	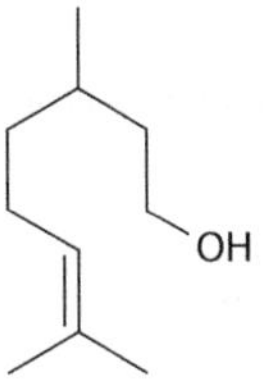 OH	Fresh, roselike, bitter taste
Farnesene	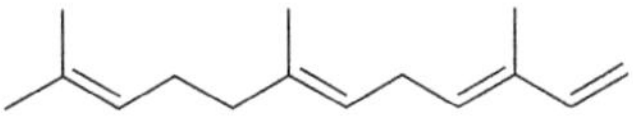	Green apple odor
Geraniol	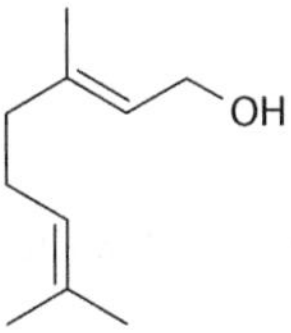 OH	Sweet, roselike, fruity
Humulene	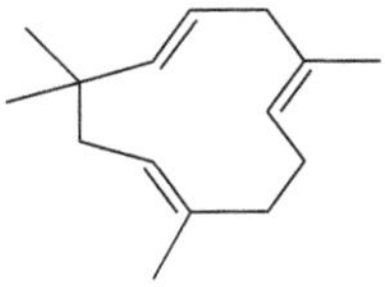	Herbal, spicy
Limonene	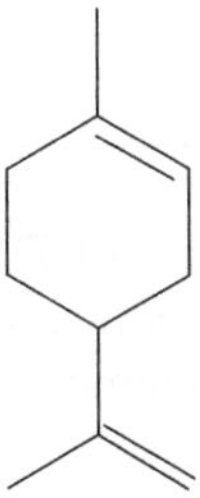	Citrus

(continued)

TABLE 13.3 (continued)

Structure and Flavor Notes of Common Terpenes

Terpene	Structure	Flavor Notes
Linalool	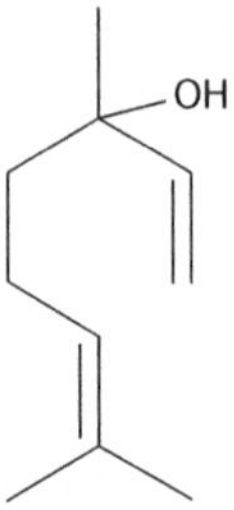 OH	Sweet, fresh, citrus, floral
Myrcene	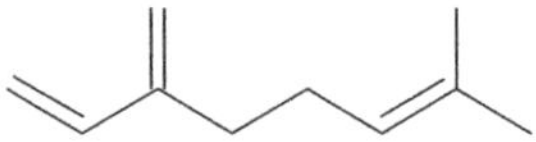	Green hop aroma
Nerol	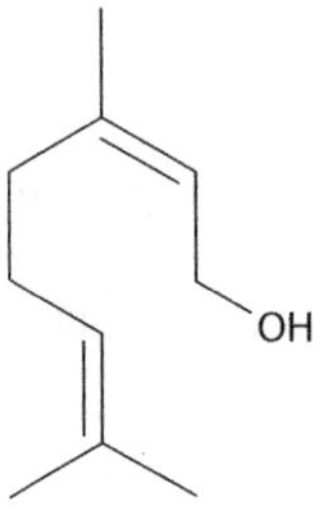 OH	Sweet, floral, roselike, slightly bitter
α-Terpineol	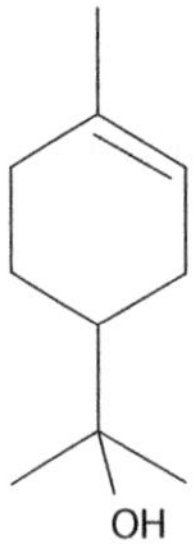 OH	Sweet, floral, fruity when diluted

to be produced by *Trametes odorata* (Collins and Halim, 1977). Citrus products obtain their characteristic flavor from terpenes. Limonene accounts for 90% of citrus oils. Citral is a prominent flavor component of lemon oil.

13.2.5 Esters

Fermented foods such as ripened cheese, lager beer, and old wine owe their aroma and typical taste to aromatic esters being generated in them, as they mature. Every fruit is recognized by typical esters present in them. Some common esters are listed in Table 13.4.

Esters are commonly used flavoring agents appreciated for providing the fruity aromas. They are employed in fruit-flavored products (i.e., beverages,

TABLE 13.4

Some Common Esters

Ester	Source
Amyl acetate	Banana
Amyl butyrates	Apricot
Ethyl butanoate	Peach
Ethyl butyrate	Pineapple
Ethyl cinnamate	Oil of cinnamon
Ethyl methanoate	Rum
Isobutyl formate	Raspberries
Methyl anthranilate	Grapes
Methyl butyrate	Apple
Methyl phenyl ethanoate	Honey
Octyl ethanoate	Orange
Pentyl ethanoate	Pear

candies, jellies, and jams), baked goods, wines, and dairy products (i.e., cultured butter, sour cream, yogurt, and cheese). Acetate esters, such as ethyl acetate, hexyl acetate, isoamyl acetate, and 2-phenylethyl acetate, are recognized as important flavor compounds in wine and other grape-derived alcoholic beverage (Longo and Sanroman, 2006).

Ethyl butyrate and ethyl caproate impart a fruity attribute to the cheddar cheese. The whiff of a good beer is quite likely to be due to the esterification of short-chain fatty acids and by yeast present during beer fermentation. Esters of short-chain fatty acids are aroma-impact compounds found in fermented dairy products. These esters are responsible for fruity flavors that can be regarded either as a defect or as an attribute by the consumer (Liu et al., 2004).

13.2.6 Carboxylic Acids

Due to the sour taste and intense smell, carboxylic acids hold prominent place in flavoring industries. These are mainly linked to sour and pungent sensations (Table 13.5). The lower carbon number carboxylic acids are mainly concerned with microbial fermentation and fruit handling objects, whereas higher acids have been observed to be linked with oxidative rancidity (Kalua et al., 2007).

Carboxylic acids have been reported to be the most abundant volatiles in roasted almond nuts by using supercritical fluid extraction method (Lasekan and Abbas, 2010). These are also found to be abundant in the oils stored in oxygen-rich environments (Kalua et al., 2007). Common acids include acetic, propionic, butyric, isobutyric, valeric, isovaleric, and lactic acids. 2-Methyl butyric acid constitutes the main aromatic compound of cranberry, whereas methyl pentanoic acid gives sweet, green, and sharp strawberry character to foods (Singh and Bhari, 2012). The sweet taste of hard cheese varieties has been increased about 1.5 units due to propionic acid fermentation (Frohlich-Wyder

TABLE 13.5

Structure and Taste of Common Carboxylic Acids

Carboxylic Acid	Structure	Taste
Acetic acid	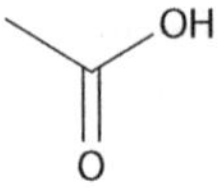 OH, O	Distinctive sour taste
Butyric acid	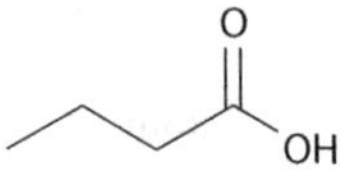 O, OH	Penetrating, diffusive sour, reminiscent of rancid butter
Caproic acid	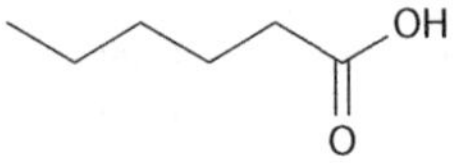 OH, O	Soapy, cheesy, fatty
Caprylic acid	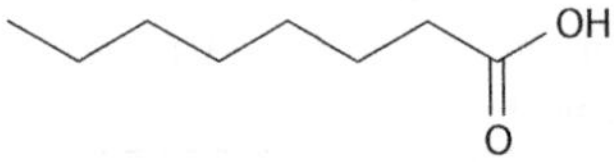 OH, O	Awful rancid taste
Citric acid	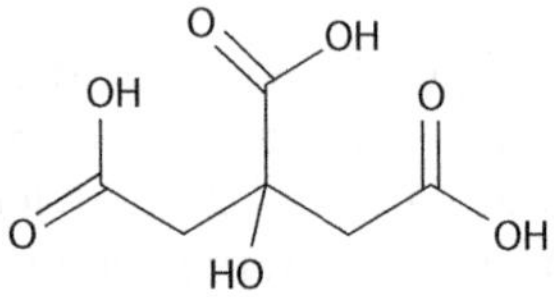 O, OH, OH, O, O, OH, HO	Sour
Isobutyric acid	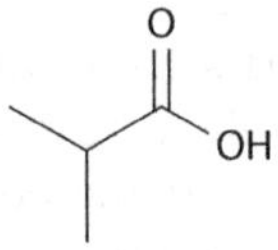 O, OH	Powerful, diffusive sour, in dilution pleasant and fruity
Isovaleric acid	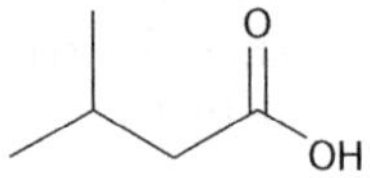 O, OH	Diffusive, acid–acrid, in dilution cheesy, unpleasant
2-Methylbutyric acid	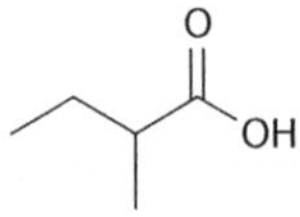 O, OH	Pungent, acrid, in dilution pleasant and fruity–sour
4-Methyl octanoic acid	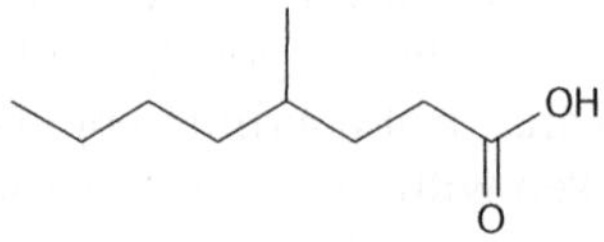 OH, O	Fatty, waxy, creamy
2,3-Dihydroxy-3-methyl pentanoic acid	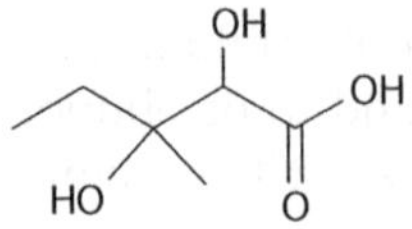 OH, OH, HO, O	Sweet, green, and sharp strawberry

TABLE 13.5 (continued)

Structure and Taste of Common Carboxylic Acids

Carboxylic Acid	Structure	Taste
Propionic acid	O, OH	Pungent, sour
Succinic acid	O, HO, OH, O	Bitter and acidic

and Bachmann, 2004). *Lactobacillus plantarum* gives highly acidified products, containing mostly lactic acid, whereas a typical nutty and sweet taste to a fermented product by the production of propionic acid. The acetic acid has been attributed by *Propionibacterium* (Hugenholtz and Kleerebezem, 1999).

13.3 Biotechnological Routes for the Production of Bioflavors

The versatile nature of flavors can be extracted by different methods, that is, direct extraction from nature, chemical transformations, and biotechnological transformations (Figure 13.2). Flavors can be directly extracted from their natural sources (plants/vegetables), such methods are not only time consuming but also result in the production of flavors with a very low yield. Furthermore, these

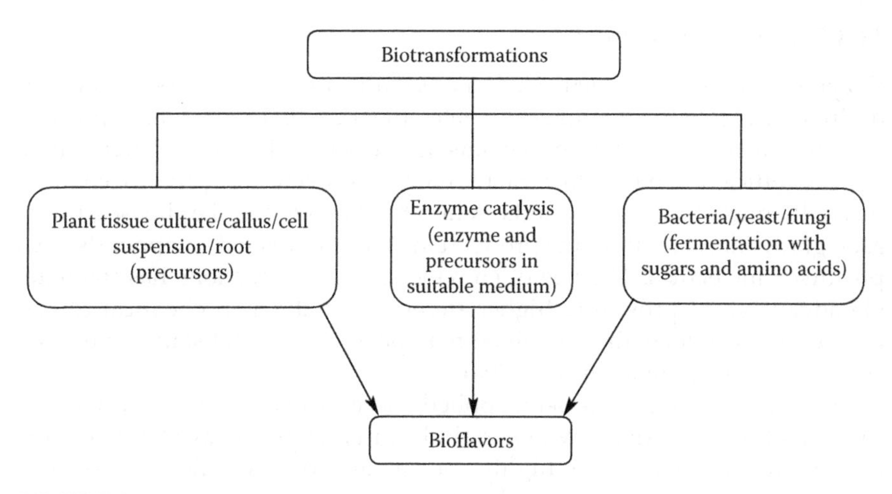

FIGURE 13.2
Various routes to production of bioflavors.

methods are strongly influenced by climatic conditions (Bicas et al., 2010). Under such circumstances, biotechnological methods can be seen as promising tools for the production of flavoring compounds. Different biotechnological methods used for the bioproduction of flavors have been discussed as follows.

13.3.1 Plant Tissue Culture Method

Plant cell culture provides a sustainable, renewable, and scalable source of plant products. However, there are a number of challenges and opportunities for the biosynthesis of flavors via plant tissue culture. Plants produce low quantities of flavors and extraction of these compounds is a quite expensive and time-consuming process. The chemical and genetic regulation of secondary metabolism of plants is to be explored fully before making any commercial production profitable as most of the flavor compounds of interest are secondary metabolites of plants (Smetanska, 2008). The biosynthesis of rosmarinic acid by cell cultures of *Coleus blumei* is one of the most prominent secondary compounds (Ulbrich et al., 1985).

Plant parts like roots of onion and garlic or their culture can be used for the production of flavors. The name and chemical structures of some of the important flavors derived from plant sources have been given in Table 13.6. A patented process for the production of onion flavor using root organ cultures of *Allium*, after successfully growing them in bioreactors, to produce onion flavor has been established (Prince et al., 1993). The use of viable aerial roots of vanilla plant for the production of natural vanillin has also been reported (Westcott et al., 1994; Rao and Ravishankar, 2000). Citrus cells have been used for the transformation of valencene into nootkatone, a constituent of grape flavor. Cocoa flavor can be produced from the cultures of *Theobroma cacao* (Singh and Bhari, 2012).

13.3.2 Enzymatic Method

Enzyme catalysis has been used successfully in recent years to produce desired flavoring compounds. Of late, the use of isolated and purified enzymes instead of microorganisms is attracting increased attention to improve efficiency and yield. Suitable enzymes specifically produce enantiomerically pure flavor compounds. A number of enzymes (i.e., lipases, proteases, glucosidase) catalyze the production of aroma-related compounds from precursor molecules. The use of enzyme-catalyzed reactions has the notable advantage of providing higher stereoselectivity than chemical routes. Besides, the products thus obtained may possess the legal status of natural substances (Demyttenaere et al., 2001).

Enzyme technology has been applied successfully to promote ester synthesis with inexpensive raw material like fatty acids and alcohol using the enzyme lipase. Lipases are highly specific and the esterification between carboxylic acids and alcohols depends on alcohol or acid chain length (Kumar et al., 2005). Researchers are exploring the feasibility of using

TABLE 13.6

Some Important Plant-Derived Flavors

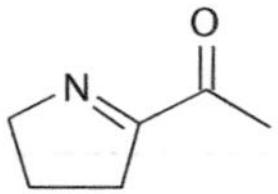

2-Acetyl-1-pyrroline (basmati rice)

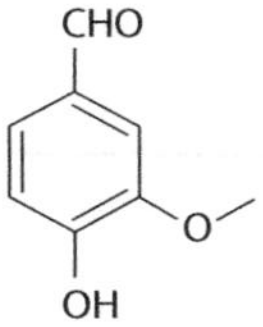

Vanillin (flat-leaved vanilla)

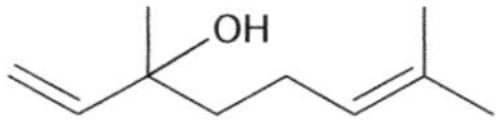

Linalool (mint, cinnamon, scented herbs, rosewood)

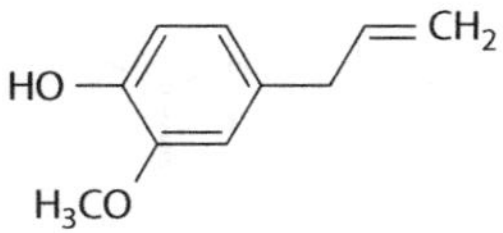

Eugenol (clove, basil, cinnamon, nutmeg, bay leaf)

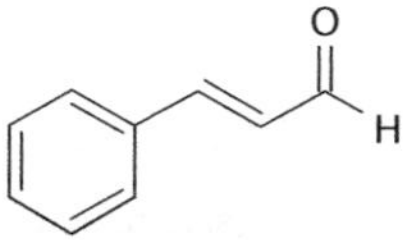

Cinnamaldehyde (cinnamon)

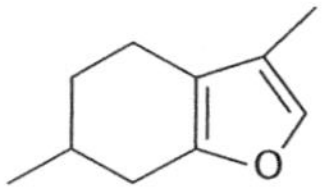

Menthofuran (peppermint)

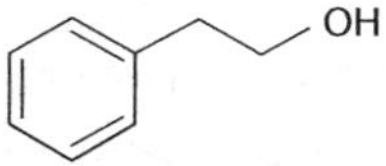

Phenethyl alcohol (rose, carnation, hyacinth, Aleppo pine, ylang-ylang, geranium, neroli, and champaca)

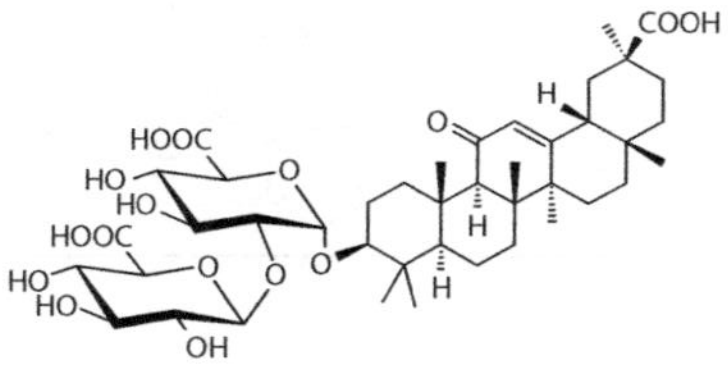

Glycyrrhizin (licorice)

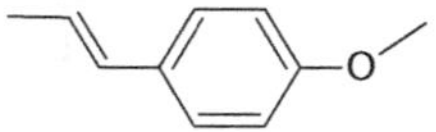

Anethol (anise, fennel, licorice, star anise)

(*continued*)

TABLE 13.6 (continued)

Some Important Plant-Derived Flavors

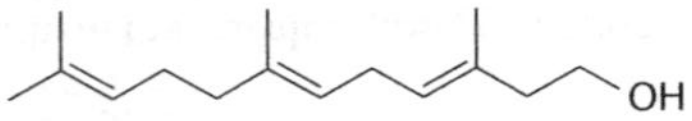 **Farnesol** (lemon grass, rose, musk, bulsum, tolu)	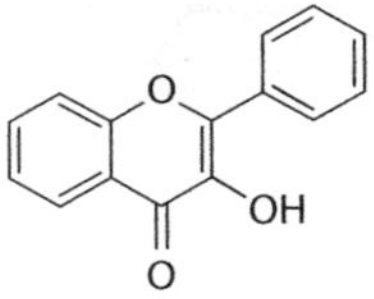 **Flavonol** (variety of fruits and vegetables)	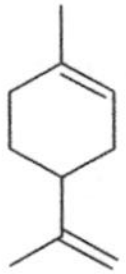 **Limonene** (citrus fruits)
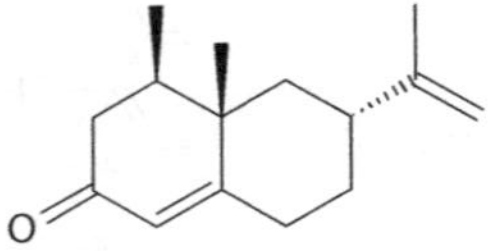 **Nootkatone** (grapefruit)	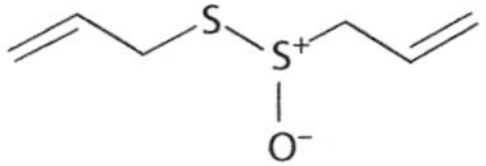 **Allicin** (garlic)	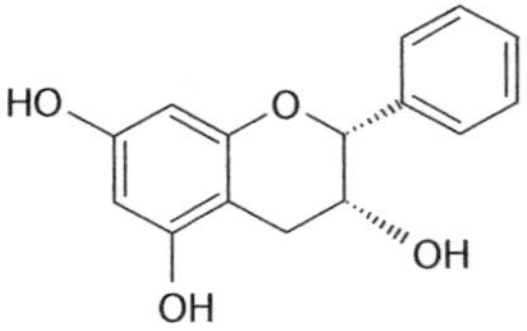 **Proanthocyanidin** (apples, maritime pine bark, cinnamon, aronia fruit, cocoa beans, grape seed, grape skin)

supercritical carbon dioxide (SC-CO_2) to increase the efficiency of several lipase enzymes to produce flavors. The high cost of SC-CO_2 extraction systems is the main drawback in commercializing production via this route. The isolation of the flavors has also been achieved by selective solvent extraction, membrane separation, solid phase adsorption, or distillation. Many researchers have also been focused on the production of esters in aqueous media to overcome the problems associated with organic, solvent-free, and supercritical systems. The lipases from *Staphylococcus epidermidis* have been used to catalyze the synthesis of many flavors in aqueous media (Longo and Sanroman, 2006).

Terpenols in grapes are mostly found in glycosidically bound forms that are odorless. Glucosidases are also used to enhance the aroma of some wines by releasing glycosidically bound volatile terpenes and flavor precursors. The addition of exogenous enzymes during or after the fermentation has been found to be the most effective way to improve the hydrolysis of the aroma precursor compounds and achieve an increase in wine flavor (Aryan et al., 1987).

Proteases also play an important role in flavor production. These enzymes allow the hydrolysis of proteins of fruits and vegetables resulting in the production of flavors due to the formation of pyrazines and alcohols (Singh and Bhari, 2012). It has also been reported that the protease treatment of by-product obtained from crayfish processing results in the production of flavors with an increase in the concentration of benzaldehyde and pyrazines (Longo and Sanroman, 2006).

Despite using the free enzymes, immobilized enzymes and enzyme engineering have also been used for the production of bioflavors. These approaches can play an important role in enhancing the production of desired flavor in a cost-effective manner. The oxidation of vanillylamine to the flavor component of vanillin using the immobilized enzyme amine oxidase from *Aspergillus niger* and monoamine oxidase from *Escherichia coli,* along with the pathway toward the formation of vanillin, has been described (Yoshida et al., 1997).

13.3.3 Microbial Method

The bioflavors can be produced using microorganisms by two ways, that is, de novo synthesis, which involves the production of bioflavors by metabolizing the microbial cells, and biotransformation, which is a single reaction catalyzed by enzymes derived from microbial sources.

13.3.3.1 De Novo Synthesis of Bioflavors

Microorganisms have been extensively used to produce "natural flavors." Yeasts, bacteria, and fungi are the main sources for "natural" aroma compounds and other flavor ingredients. The manipulation of biosynthetic and metabolic pathways of the microorganisms produces the desired metabolites

TABLE 13.7

Some Aromatic Compounds Produced by Microorganisms

Microorganisms	Aromatic Compounds	Flavor Notes
Corynebacterium sp.	Pyrazines	Roasted
B. subtilis, B. cereus, S. lactis	2-Acetyl-1-pyrroline	Nutty
Saccharomyces sp.	Acetyl pyrazine, 2-acetyl-pyridine	Bread
B. subtilis sp.	Vanillin	Creamy, sweet
Pseudomonads	Acetoin, methyl butanol	Malty, milky
Propionibacterium acidi	Acetic acid, propionic acid	Sharp, acidic
Lactobacillus sp., *S. lactis, Lactobacillus citrovorum*	Acetoin, diacetyl	Buttery
L. rhamnosus		
Penicillium sp.	Methyl ketones, 1-octene-3-ol	Blue cheese
	Phenyl-ethanol	Mushroom, rosy
Ceratocystis moniliformis	Ethyl acetate, propyl acetate	Fruity-pineapple
Ceratocystis fimbriata	Isoamyl acetate, isobutyl acetate	Banana
Geotrichum sp.	Ethyl and higher alcohol esters	Fruity, melon
Kluyveromyces sp.	Phenyl-ethyl alcohol and esters	Fruity rose
K. lactis	Terpenes	Fruity, peach
Sporobolomyces	γ-Decalactones	Pear, peach
S. odorus	γ-Decalactone	Creamy, coconut
T. viride	6-Pentyl-alphapyrone	Coconut
Trichoderma sp.	Sesquiterpenes, 6-pentyl-a-pyrone	Coconut, anise
Septoria sp.	Cinnamate derivatives	Cinnamon
Ischnoderma benzoinum	4-Methoxy benzaldehyde	Sweet, anisic notes
C. moniliformis	Citronellol, geraniol	Floral, roselike
Pseudomonas putida	L. menthol	Mint

Source: With permission from Scharpf, I.G., Seitz, E.W., Morris, A.J., and Farbood, M.I. Generation of flavor and odor compounds through fermentation processes, In *Biogeneration of Aromas*, Parliament, T.H. and Croteau, R., Eds., 323. Copyright 1986 American Chemical Society.

that are further converted into natural organic flavors like alcohols, esters, aldehydes, ketones, lactones, and terpeneoids (Scharpf et al., 1986). Some of the important flavor compounds having great industrial value produced by microorganisms are listed in Table 13.7.

The biotechnological process for vanillin production by using different microorganisms has been investigated by many researchers (Muller et al., 1998; Barghini et al., 2007; Converti et al., 2010). *Schizosaccharomyces pombe* (also known as fission yeast) and *Saccharomyces cerevisiae* (baker's yeast) have been observed to be good sources of natural vanillin (Hansen et al., 2009). The ability of various bacteria from different genera including *Pseudomonas fluorescens* to metabolize ferulic acid as carbon source leading to vanillin production has also been described (Andreoni et al., 1995).

Some of the important flavors isolated from microorganisms are as follows:

13.3.3.1.1 *Natural Butter Flavor: Diacetyl*

The production of dairy flavor compounds, such as butyric acid, lactic acid, and diacetyl in mixed cultures of lactic acid bacteria growing in starch-based media, has been reported by Escamilla et al. (2000). In dairy fermentation, the citric acid present in milk is metabolized by the lactic acid bacteria. The ability to utilize citric acid is found only in some *Leuconostoc* species. *L. lactis* has been reengineered to increase the efficiency of sugar to diacetyl production (Hugenholtz et al., 2000). *Lactobacillus rhamnosus* is a heterolactic acid bacterium, which can be used to produce flavor compounds like diacetyl and acetoin. Growth on medium containing glucose + citrate demonstrated simultaneous utilization of carbon sources. The key control points that drive the formation of acetoin and diacetyl have been deduced on studying enzyme kinetics in *L. lactis* (Hoefnagel et al., 2002).

13.3.3.1.2 *Pyrazines*

The immobilized culture of *L. lactis* subsp. *lactis* biovar. *diacetylactis* FC1 has been used for the formation of tetramethylpyrazine (Lee et al., 1996). Biogeneration of natural pyrazine derivatives has been tried using microbial fermentation and enzyme technology (Rizzi, 1988). Tetramethylpyrazine has been reported to be produced by *Corynebacterium glutamicum* and *Bacillus subtilis,* and 2-methoxy 3-alkyl pyrazine by *Pseudomonas perolens.* An aerobic fed-batch culture of *L. lactis* subsp. *lactis* biovar. *diacetylactis* FC1 (*L. diacetylactis*) has been optimized for the production of tetramethylpyrazines (Kim et al., 1994). Tetramethylpyrazine production from glucose by a newly isolated *Bacillus* mutant has also been reported (Xiao et al., 2006). Similarly, submerged culture method has been used for the optimization of pyrazine production by newly isolated *Bacillus* sp. (Nopmaneerat et al., 2011). Commercial *Bacillus natto* was cultured in the liquid media containing various amino acids, and the effect of amino acids L-threonine and L-serine on the formation of 2,5-dimethyl pyrazine, tri-methyl pyrazine, and tetra-methyl pyrazine has been examined (Sugawara et al., 1990). The organism was selected after screening cultures of alkaline-fermented soybeans, thua-nao and natto from Thailand and Japan (Sugawara et al., 1998).

13.3.3.1.3 *Lactones and Ketones*

Ricinoleic acid and castor oil are used by *Sporobolomyces odorus* or *Rhodotorula glutinis* in an efficient process for the production of γ-decalactone (Cheetham et al., 1993). The production of γ-decalactone by *Geotrichum* sp. in broth in the presence of enzymatically hydrolyzed castor oil has also been reported (Neto et al., 2004). *Tyromyces sambuceus* and *Cladosporium suaveolens* produced coconut-flavored γ-decalactone and δ-dodecalactone from ricinoleic acid and linolenic acid, respectively (Kapfer et al., 1989; Allegrone et al., 1991).

The formation of methyl ethyl ketones in *Agaricus bisporus, A. niger,* and *Trichoderma viride* has been reviewed by many researchers (Janssens et al., 1992). An alternative approach based on co-immobilization of precursors for bioflavor generation by microbial cells within beads made of food-grade gel matrix has also been used (Kogan and Freeman, 1994). The bioflavor produced in this process has been accumulated within the beads, and the flavor-retaining bead may then be employed as a food additive. Furthermore, the potential inherent in bioflavor generation by co-immobilization of filamentous fungi with an emulsion of oily precursor was demonstrated by γ-decalactone production from castor oil.

13.3.3.1.4 Terpenes

Many filamentous fungi are capable of producing terpenes by de novo synthesis method. *Ceratocystis* sp. and *T. viride* have been observed as good microbial sources for the production of fruity and coconut-flavored compounds, respectively. The fruity and floral flavors of terpenes like citronellol, linalool, and geraniol have been reported to be produced by *Kluyveromyces lactis*. Similarly, γ-decalactone (peach-smelling compound) and 4-hydroxy-cis-6-dodecenoic acid γ-lactone have been produced by *Sporobolomyces odorus*, whereas *Sporidiobolus salmonicolor* and *Yarrowia lipolytica* produce γ-decalactone from castor oil through partial β-oxidation (Vandamme, 2003).

13.3.3.1.5 Esters

Yeasts ferment grape juice to a fine wine, metabolizing the ingredients to fine esters, ethyl acetate being prominent among them. Ethyl acetate production through fermentation by *Candida utilis* has been reported (Armstrong et al., 1989). The production of esters by *Aspergillus* sp., *Candida rugosa*, and *Rhizopus arrhizus* has been investigated in organic media (Langrand et al., 1990). *Geotrichum klebahnii* yeast is also capable of producing a wide range of ethyl esters with pleasant fruity flavor (Vandamme, 2003). *Kluyveromyces marxianus* produce alcohol esters and aldehydes when grown on cassava bagasse (Mederios et al., 2000), whereas *Pichia anomala* are good producers of acetate esters (Rojas et al., 2011). Among bacterial sources, acetic acid bacteria form esters of ethyl acetate and *Clostridium pasteurianum* produced apple-odor-resembling esters (Lanza et al., 1976).

13.3.3.2 Biotransformation Method

Microbial cells can be used for the transformation of substrates into flavor compounds through fermentation. The commercialization of this approach for bioflavor production is limited due to high cost and low yield. The most studied precursor *R*-(+)-limonene has been widely used in biotransformation processes for the production of bioflavors. Different types of microorganisms like *Cladosporium* sp., *Penicillium digitatum*, and *Fusarium oxysporum* are capable of degrading this molecule for the production of flavoring compounds.

13.3.3.2.1 Biotransformation of Terpenes

The biotransformation of (*R*)-(+)- and (*S*)-(−)-limonene by fungi has been reported in which more than 60 fungal cultures were screened for their ability to bio-convert the substrate using solid-phase microextraction as the monitoring technique. It was observed that (+)- and (−)-limonene was converted to alpha-terpineol (main metabolite) by *P. digitatum*. The bioconversion of (*R*)-(−)- and (*S*)-(−)-limonene by *Corynespora cassiicola* yielded (1S, 2S, 4R)- and (1R, 2R, 4S)-limonene-1,2-diol, respectively. The bioconversions using liquid cultures were also monitored by solid-phase microextraction as a function of time (Demyttenaere et al., 2001). Furthermore, a bioprocess for the production of high concentrations of *R*-(+)-α-terpineol from *R*-(+)-limonene has been reported by Bicas et al. (2010).

The biotransformation of geraniol, nerol, and citral by *A. niger* was studied by Demyttenaere et al. (2000). A comparison has been made between submerged liquid, sporulated surface cultures, and spore suspensions. This bioconversion has also been carried out with surface cultures of *Penicillium* sp. The main bioconversion products obtained from geraniol and nerol by liquid cultures of *A. niger* were linalool and α-terpineol. The formation of terpenes, using fungi of the ascomycetes and basidiomycetes species, has been patented (Schindeler and Bruns, 1980). The nucleic acid sequence of a monoterpene synthase from sweet basil, a key enzyme for the production of geraniol, has been determined in order to allow the production of recombinant geraniol synthase (Pichersky et al., 2005). *F. oxysporum*, *Aspergillus* sp., and *Penicillium* sp. grown in cassava wastewater have also been employed for the production of −*R*(+)-α-terpineol from −*R*(+)-limonene by biotransformation method (Maróstica and Pastore, 2007).

13.3.3.2.2 Biotransformation of Esters

The biosynthesis of esters in fermented dairy products has been studied by Liu et al. (2004). The biosynthesis of flavor-active esters in dairy systems proceeds through two enzymatic mechanisms—esterification and alcoholysis. Esterification is the formation of esters from alcohols and carboxylic acids, whereas alcoholysis is the production of esters from alcohols and acylglycerols or from alcohols and fatty acyl-CoAs derived from the metabolism of fatty acids, amino acids, and/or carbohydrates. Alcoholysis is essentially a transferase reaction, in which fatty acyl groups from acylglycerols and acyl-CoA derivatives are directly transferred to alcohols, and is the major mechanism of ester biosynthesis by dairy lactic acid bacteria and yeasts (Liu et al., 2004).

13.3.3.2.3 Biotransformation of Diacetyls

Streptococcus diacetilactis synthesize diacetyl and acetylmethylcarbinol, with intermediate synthesis of α-acetolactic acid, from citric acid or pyruvic acid as shown in Figure 13.3 (Seitz et al., 1963).

13.3.3.2.4 Biotransformation of Vanillin

Vanillin, the main aromatic compound, has been produced by microbial biotransformation. As per the International Dairy Foods Association (www.idfa.org/news), vanilla continues to be America's flavor of choice in ice cream and novelties, in both supermarket and foodservice sales. This flavor is the most versatile, mixing well with toppings, drinks, and bakery desserts.

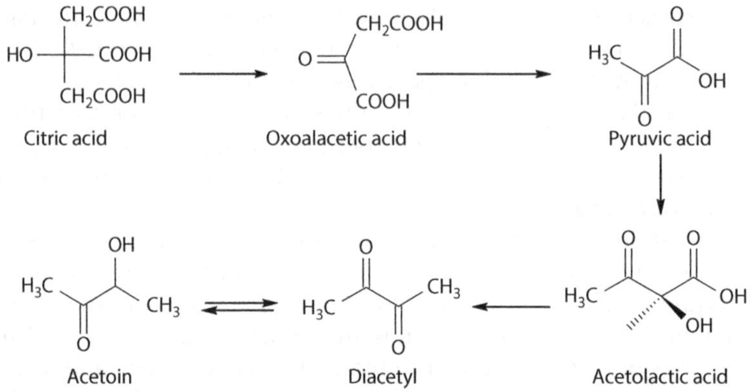

FIGURE 13.3
Biosynthesis of diacetyl from citric acid.

In 2010, 1800 MT, of cured vanilla pods valued $370 million, were imported by the United States (Data Source: Department of Commerce, U.S. Census Bureau). The vanilla plant does not produce the vanilla flavor directly, but flavor precursors are produced in the vanilla pods, which, on curing, produce the vanillin for use as vanilla flavor. Vanillin is bound to a glucoside (glucovanillin—the precursor) in the green pod and is odorless; during the fermentation/curing process, vanillin is slowly released by the inherent β-glucosidase present in the pod. Simultaneously, the polyphenol oxidase and peroxidase enzymes present in the pod initiate the browning of the pod, to impart the characteristic-rich dark brown color (Rao and Ravishankar, 2000). The potential of two *Rhodococcus* strains for vanillin production from ferulic acid and eugenol has been investigated (Plaggenborg et al., 2006). The metabolic pathway from eugenol via ferulic acid to vanillin (Figure 13.4) has been characterized at the enzymatic and molecular levels in *Pseudomonas* strains (Walton et al., 2000).

The agro-residues can also be employed for the production of flavoring compounds by biotransformation method. The lignin-containing agriculture waste can be used for enzymatic extraction of ferulic acid. This acid can be further converted to vanillin by *Pycnoporus cinnabarinus*. Similarly, the genetically engineered *E. coli* has been used for the production of vanillin from ferulic acid derived from wheat bran and corn cob hydrolysate (Bicas et al., 2010).

Natural Vanillin Ex. Turmeric and Ex. Ferulic Acid have been commercially sold by Advanced Biotech, Paterson, NJ, and labeled "Natural" in the United States and Natural Flavoring Substance in Europe as per EEC Guidelines (www.adv-bio.com). Rhovanil® Natural is a vanillin naturally obtained by fermentation from natural ferulic acid. Rhovanil® Natural has the advantage that it meets the requirements of "natural status" as defined by both European and American regulations (www.rhodia.com). "AMCAN" natural vanillin is obtained from ferulic acid fermentation and is labeled "Natural Flavor" according to EEC Regulation 1134/2008 (www.amcan.fr).

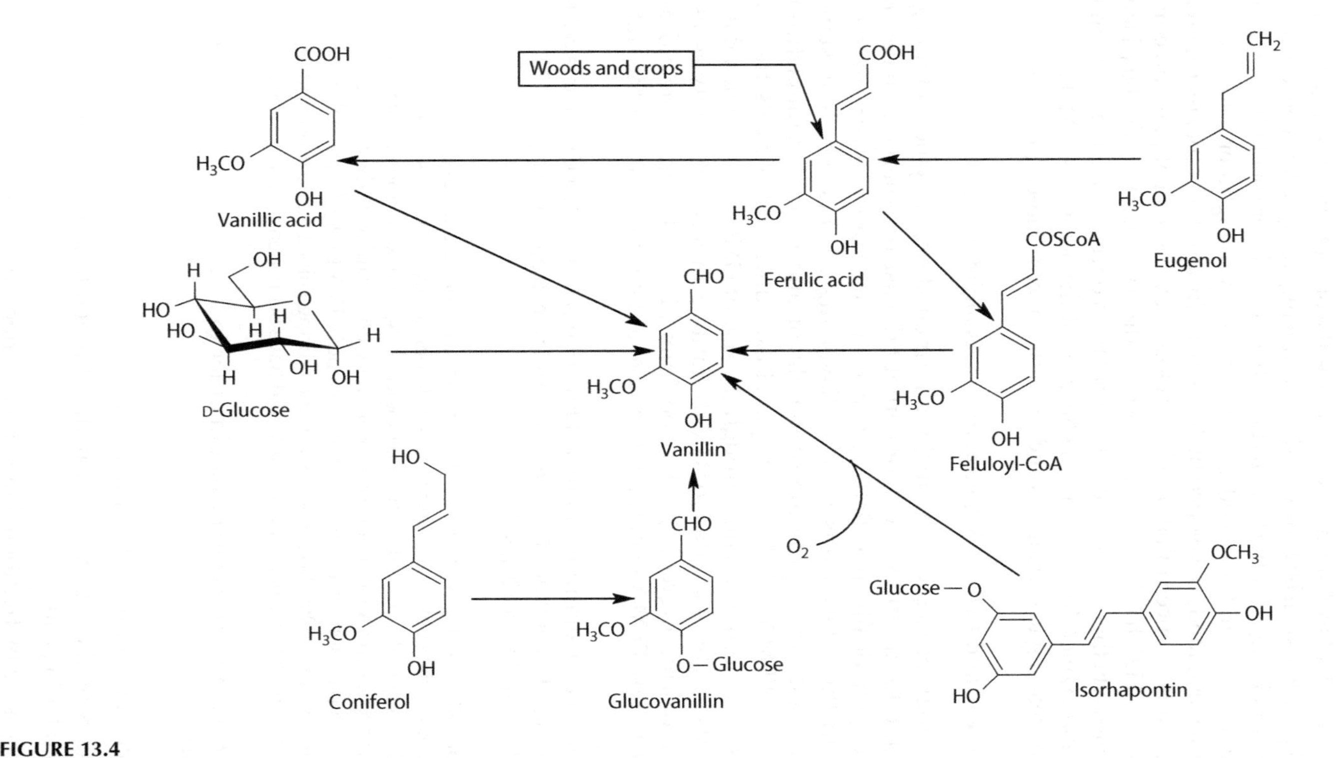

FIGURE 13.4
Production of vanillin by biotransformation method. (From Feron, G. and Wache, Y., Microbial biotechnology of food flavor production, In Shetty, K., Paliyath, G., Pometto, A. and Levin, R.E., Eds., Food Biotechnology, 2nd Edn., CRC Press, Taylor and Francis Group, Boca Raton, FL, 2005, p. 439.)

13.4 Safety and Regulatory Aspects of Natural Flavors

It is very important to determine whether the flavors obtained through biotechnological processes can be considered as natural regarding safety concerns. The U.S. Food and Drug Administration (FDA) has no basis for concluding that bioengineered foods differ from other foods in any meaningful or uniform way, or that as a class, foods developed by the new techniques present any different or greater safety concern than foods developed by traditional plant breeding methods. The FDA has issued guidance for industry, whereby voluntary labeling indicates whether foods have or have not been developed using biotechnology and bioengineering. The FDA believes that the acronyms GMO and GM may be misleading to consumers and prefer the word "bio-engineering."

13.4.1 Flavor and Extract Manufacturers Association, USA

Biotechnology processes may include fermentation, enzyme technology, and cell and tissue culture with or without the use of genetic engineering. The use of genetically modified organisms (GMOs) presents certain issues relevant to the safety assessment of flavor ingredients. The long-standing safety assessment program of the Flavor and Extract Manufacturers Association (FEMA), USA, expert panel examines and reviews the production of flavor ingredients through the use of GMOs so that such flavor ingredients can be evaluated for GRAS status. Similarly, the safety assessment of flavoring ingredients derived from plant cell and tissue culture is conducted by FEMA expert panel and evaluates the same for the status of GRAS.

The European Committee in their Directive (EC 1334/2008), Article 3.2 (c) published on December 16, 2008 and effective from January 20, 2011 has defined 'Natural flavoring substances' as

> "Natural flavoring substance" shall mean a flavoring substance obtained by appropriate physical, enzymatic or microbiological processes from material of vegetable, animal or microbiological origin either in the raw state or after processing for human consumption by one or more of the traditional food preparation processes listed in Appendix 13.A.1 (*Official Journal of the European Union Directive*, 2000). Natural flavoring substances correspond to substances that are naturally present and have been identified in nature.

The three main criteria to be met in order to label a substance as "natural flavor" are

1. Occurrence of the substance in nature
2. Origin of the substance in nature (mineral, animal, vegetable or microbiological)
3. Processing by traditional food preparation processes

It should be pointed out that products obtained from fermentation and extracted from the fermentation broth may be referred to as natural according to this regulation.

Since 1997, Community legislation has made labeling of genetically modified (GM) food mandatory for

- Products that consist of GMO or contain GMO
- Products derived from GMO, but no longer containing GMO if there is still DNA or protein resulting from the genetic modification present

13.4.2 Latest Regulation Concerning GMO Labeling

Regulation (EC) 1830/2003 of the European Parliament and of the Council of September 22, 2003, concerning the traceability and labeling of GMOs and the traceability of food and feed products produced from GMOs, and amending Directive 2001/18/EC were published in the *Official Journal of the European Union.*

13.4.3 Commission Regulation (EC) No. 50/2000 of January 10, 2000

The Commission Regulation (EC) No. 50/2000 of January 10, 2000, recommends the labeling of foodstuffs and food ingredients containing additives and flavorings that have been genetically modified or have been produced from GMOs.

Australia and New Zealand have defined their food standards under the Australia New Zealand Food Standards Code Standard 1.3.1—food additives and flavorings are defined under Section. 11. Permitted flavoring substances for the purposes of this standard are flavoring substances obtained by physical, microbiological, enzymatic, or chemical processes from material of vegetable or animal origin either in its raw state or after processing by traditional preparation process including drying, roasting, and fermentation (Australia New Zealand Food Standards Code).

13.4.4 Food Sanitation Law in Japan

Under this law, the term "natural flavoring agent" means food additives intended for use for flavoring food, which are the substances obtained from animals or plants, or mixtures thereof. Under the Food Sanitation Law, a list of plant or animal sources of natural flavorings is released for guidance, toward regulation of food additives.

The safety assessment of foods and food additives produced by recombinant DNA techniques (hereafter GM foods) is mandatory under the Food Sanitation Law. The Ministry of Health, Labour, and Welfare (MHLW) receives application, and the Food Safety Commission evaluates the safety of GM foods in terms of human health.

13.4.4.1 Labeling System of Foods Produced by Recombinant DNA Technique

In April 2001, the labeling for GM foods became mandatory in Japan. The labeling system is run by two ministries: the Ministry of Agriculture, Forestry and Fisheries (MAFF) and the MHLW. The MAFF intends to provide information to consumers in choosing food under the Japanese Agricultural Standards Law. The MHLW intends to publicize the fact that the product has undergone the safety assessment. The labeling is part of the legal safety assessment system under the Food Sanitation Law. Commercial quantities of flavoring components that can carry the "Natural" label are now available from different sources (Table 13.8).

13.4.5 Joint FAO/WHO Food Standards

According to the joint report of FAO/WHO (2004) on food labeling, "consideration may be given to the exemption from labeling of specific categories (e.g., highly processed food ingredients, processing aids, food additives, flavours) of food and food ingredients obtained through certain techniques of genetic modification/genetic engineering." Additionally, it has been stated that "no disclosure is required where the ingredient comprises less than 2% of the total weight of the product and has been used for the purposes of flavouring" (FAO/WHO, 2004).

TABLE 13.8

Natural-Labeled Flavor/Aroma Ingredients

Products (Natural)	Company/Country	Declaration
Natural Vanillin—Naturallin	AMCAN, www.amcan.fr	Natural-Europe
Natural Flavor and Fragrance Ingredients	Advanced Biotech (USA), www.adv-bio.com	Natural-USA Natural Flavoring Substance-EEC
Natural Chemicals and Isolates	Fleurchem (USA), www.fleurchem.com	Natural-USA
Pyrazine Mixture Natural Vanillin Ex. Clove and Ex. Turmeric	Natural Advantage (USA), www.natural-advantage.net	US 21 CFR101.22
Natural Flavor Molecules	Robertet (France), www.robertet.com	Natural-Europe
Natural Vanillin Ex Ferulic Acid		
Rhovanil Natural™ (Vanillin)	Rhodia® (France), www.rhodia.com	Natural-USA Natural-Europe
SYMVANIL	Symrise, www.symrise.com	Natural-USA Europe

13.5 Summary and Future Prospects

The research in food processing to generate "bioflavors" is being stimulated by the market demand from consumers who seek "natural" labels in foods/flavors. The biotechnological routes for flavor production are becoming more and more important, not only because of consumer preference for natural products, but also because bioflavoring may yield better economic benefits compared to plant extracts and synthetic flavors. Recent developments in biochemistry, recombinant technology, molecular biology, and microbiology have improved the design and technology of fermentation to improve the economic viability of preparing bioflavors. Furthermore, biotransformations represent an additional tool in the production of bioflavors catalyzing a large number of stereo- and regioselective chemical manipulations that are not easily achieved by the classical synthetic procedures. Understanding and manipulating the complex metabolic engineering using genetic techniques can also boost the efficiency of existing de novo synthesis and precursor conversion pathways for the large-scale and new flavor production in an economical and time-effective manner.

13.A Appendix: List of Traditional Food Preparation Processes

Chopping	Coating
Heating, cooking, baking, frying (up to 240°C at atmospheric pressure) and pressure cooking (up to 120°C)	Cooling
Cutting	Distillation/rectification
Drying	Emulsification
Evaporation	Extraction, incl. solvent extraction in accordance with Directive 88/344/EEC
Fermentation	Filtration
Grinding	—
Infusion	Maceration
Microbiological processes	Mixing
Peeling	Percolation
Pressing	Refrigeration/freezing
Roasting/grilling	Squeezing
Steeping	—

Source: *Official Journal of the European Union* EC 1334/2008.

References

Allegrone, G., Barbeni, M., Cardillo, R. et al. 1991. On the steric course of the microbial generation of (Z6)-gamma-dodecenolactone from (10 R,S) 10 hydroxyoctadeca-(E8,Z12)-dienoic acid. *Biotechnology Letters* 13: 765–768.

Andreoni, V., Bemasconi, S., and Bestetti, G. 1995. Biotransformation of ferulic acid and related compounds by mutant strains of *Pseudomonas fluorescens. Applied Microbiology and Biotechnology* 42: 830–835.

Armstrong, D.W., Gillies, B., and Yamazaki, H. 1989. Natural flavor produced by biotechnological processing. In: *Flavor Chemistry: Trends and Development*, G. Charalambous, Ed., pp. 105–120. American Chemical Society, Washington, DC.

Aryan, A.P., Wilson, B., Strauss, C.R., and Williams, P.J. 1987. The properties of glycosidases of *Vitis vinifera* and a comparison of their β-glucosidase activity with that of exogenous enzymes. *American Journal of Enology and Viticulture* 38: 182–188.

Australia New Zealand Food Standards Code Standard 1.3.1-Food Additives. www.comlaw.gov.au/details/F2011C00892 (accessed February 16, 2012).

Barghini, B.P., Di Gioia, D., Fava, F., and Ruzzi, M. 2007. Vanillin production using metabolically engineered *Escherichia coli* under non-growing conditions. *Microbial Cell Factories* 6: 13–23.

Bicas, J.L., Fontanilleb, P., Pastorea, G.M., and Larrocheb, C. 2010. A bioprocess for the production of high concentrations of *R*-(+)-α-terpineol from *R*-(+)-limonene. *Process Biochemistry* 45: 481–486.

Bicas, J.L., Silva, J.C., Dionisio, A.P., and Pastore, G.M. 2010. Biotechnological production of bioflavor and functional sugars. *Ciencia e Tecnologia de Alimentos* 30: 7–18.

Cheetham, P.S.J., de Rooij, J.F.M., and Maume, K.A. 1993. Method of producing gamma-hydroxydecanoic acid or its lactone by feeding a ricinoleic acid source to sp. odorus or rh. glutinis. United States Patent 5219742.

Code of Federal Regulation 21CFR Part 101.22(a)(3), www.accessdata.fda.gov (accessed February 16, 2012).

Collins, R.P. and Halim A.F. 1977. Essential oil composition of *Ceratocystis virescens. Mycologia* 69: 1129–1136.

Converti, A., Aliakbarian, B., and Domnsm A.F. 1977. Essential oil composition of acid microbial production of biovanillin. *Brazilian Journal of Microbiology* 41: 519–530.

Crouzet, J., Etievant, P., and Bayonove, C. 1990. Stone fruit: Apricot, plum, peach, cherry. In: *Food Flavor Part C, The Flavor of Fruits*, I.D. Morton and A.J. Macleod, Eds., pp. 43–91. Elsevier, NY.

Data Source: Department of Commerce, U.S. Census Bureau, Foreign Trade Statistics. www.fas.usda.gov/gats (accessed February 16, 2012).

Delahunty, C.M., Conner, J.M., Piggoot, J.R., and Paterson, A. 1993. Perception of heterocyclic nitrogen compounds in mature whisky. *Journal of the Institute of Brewing* 99: 479–482.

Demyttenaere, J.C., del Carmen Herrera, M., and De Kimpe, N. 2000. Biotransformation of geraniol, nerol and citral by sporulated surface cultures of *Aspergillus niger* and *Penicillium* sp. *Phytochemistry* 55: 363–373.

Demyttenaere, J.C., Van Belleghem, K., and De Kimpe, N. 2001. Biotransformation of (R)-(+)- and (S)-(−)-limonene by fungi and the use of solid phase micro extraction for screening. *Phytochemistry* 5: 199–208.

Do, J.Y., Salunkhe, D.K., and Olson, L.E. 1969. Isolation, identification and comparison of the volatiles of peach fruit as related to harvest maturity and artificial ripening. *Journal of Food Science* 34: 618–621.

Escamilla, M.L., Valdes, S., Soriano, S.E., and Tomasini, A. 2000. Effect of some nutritional and environmental parameters on the production of diacetyl and on starch consumption by *Pediococcus pentosaceus* and *Lactobacillus acidophilus* in submerged cultures. *Journal of Applied Microbiology* 88: 142–153.

FAO/WHO. 2004. Joint FAO/WHO Food Standards Programme. Codex Alimentarius Commission Twenty-Seventh Session, June 28–July 3, 2004, Rome. Report of the Thirty-Second Session of the Codex Committee on Food Labeling, 10–14, May 2004, Montrt FAO/WHO Fo.

Feron, G. and Wache, Y. 2005. Microbial biotechnology of food flavor production. In: *Food Biotechnology*, 2nd edn., K. Shetty, G. Paliyath, A. Pometto, and R.E. Levin, Eds., p. 439. CRC Press, Taylor and Francis Group, Boca Raton, FL.

Frohlich-Wyder, M.T. and Bachmann, H.P. 2004. Cheese with propionic acid fermentation. In: *Cheese Chemistry, Physics and Microbiology*, 3rd edn., P.F. Fox, P.L.H. McSweeney, T.M. Coagan, and Guinee, T.P., Eds., pp. 141–156. T.P. Elsevier Academic Press, London, U.K.

Hansen, E.H., Moller, B.L., Kock, G.R. et al. 2009. De novo biosynthesis of vanillin in fission yeast (*Schizosaccharomyces pombe*) and baker's yeast (*Saccharomyces cerevisiae*). *Applied and Environmental Microbiology* 75: 2764–2774.

Hoefnagel, M.H., Starrenburg, M.J., Martens, D.E. et al. 2002. Metabolic engineering of lactic acid bacteria, the combined approach: Kinetic modeling, metabolic control and experimental analysis. *Microbiology* 148: 1003–1013.

Hugenholtz, J. 1993. Citrate metabolism in lactic acid bacteria. *FEMS Microbiology Reviews* 12: 165–178.

Hugenholtz, J. and Kleerebezem, M. 1999. Metabolic engineering of lactic acid bacteria: overview of the approaches and results of pathway rerouting involved in food fermentations. *Current Opinion in Biotechnology* 10: 492–497.

Hugenholtz, J., Kleerebezem, M., Starrenburg, M., Delcour, J., deVos, W., and Hols, P. 2000. *Lactococcus lactis* as a cell factory for high-level diacetyl production. *Applied and Environmental Microbiology* 66: 4112–4114.

International Dairy Foods Association, Washington, DC. www.idfa.org (accessed February 16, 2012).

Janssens, L., DePootera, H.L., Schampa, N.M., and Vandammeb, E.J. 1992. Production of flavor by micro-organisms. *Process Biochemistry* 27: 195–215.

Japan: Food Sanitation Law, www.mhlw.go.jp/english/topics/foodsafety/foodadditives (accessed February 16, 2012).

Kalua, C.M., Allen, M.S., Bedgood, D.R., Jr., Bishop, A.G., Prenzler, P.D., and Robards, K. 2007. Olive oil volatile compounds, flavor development and quality: A critical review. *Food Chemistry* 100: 273–286.

Kapfer, G.F., Berger, R.G., and Draweti, F. 1989. Production of 4-decanolide by semicontinuous fermentation of *Tyromyces sambuceus*. *Biotechnology Letters* 11: 561–566.

Kim, K.S., Lee, H.J., Shon, D.H., and Chung, D.K. 1994. Optimum conditions for the production of tetramethylpyrazine flavor compound by aerobic fed-batch culture of *Lactococcus lactis* subsp. *lactis* biovar. *Diacetylactis* FC1. *Journal of Microbiology and Biotechnology* 4: 327–332.

Kogan, N. and Freeman, A. 1994. Development of macrocapsules containing bioflavors generated in situ by immobilized cells. *Process Biochemistry* 29: 671–677.

Kumar, R., Jayant, M., and Giridhar, M. 2005. Effect of chain length of the acid on the enzymatic synthesis of flavors in supercritical carbon dioxide. *Journal of Biochemical Engineering* 23: 199–202.

Kumar, H. and Panesar, P.S. 2009. *Food Chemistry*, pp. 421–429. Alpha Science International, Harrow, U.K.

Laat, M. De., Wilhelmes, T.A., Schaft, V., and Peter, H. 1992. Natural delta-lactones and process of the production thereof. U.S. Patent 5128261.

Langrand, G., Rondot, N., Triantaphylides, C., and Baratti, J. 1990. Short chain flavor esters synthesis by microbial lipases. *Biotechnology Letters* 12: 581–586.

Lanza, E., Ko, K.H., and Palmer, J.K. 1976. Aroma production by cultures of *Ceratocystis moniliformis*. *Journal of Agriculture and Food Chemistry* 20: 1247–1250.

Lasekan, O. and Abbas, K. 2010. Analysis of volatile flavor compounds and acrylamide in roasted Malaysian tropical almond (*Terminalia catappa*) nuts using supercritical fluid extraction. *Food and Chemical Toxicology* 48: 2212–2216.

Lee, J.-E., Woo, G., and Lee, H.J. 1996. Tetramethylpyrazine production by immobilized culture of *Lactococcus lactis* subsp. *lactis* biovar. *diacetilactis* FCI. *Journal of Microbiology and Biotechnology* 6: 137–141.

Liu, S.Q., Holland, R., and Crow, V.L. 2004. Esters and their biosynthesis in fermented dairy products: A review. *International Dairy Journal* 14: 923–945.

Longo, M.A. and Sanroman, M.A. 2006. Production of food aroma compounds. *Food Technology and Biotechnology* 44: 335–353.

Maróstica, M.R., Jr. and Pastore, G.M. 2007. Production of *R*-(+)-α-terpineol by the biotransformation of limonene from orange essential oil, using cassava waste water as medium. *Food Chemistry* 101: 345–350.

Monnet, C., Aymes, F., and Corrieu, G. 2000. Diacetyl and alpha-acetolactate overproduction by *Lactococcus lactis* subsp. lactis biovar diacetylactis mutants that are deficient in alpha-acetolactate decarboxylase and have a low lactate dehydrogenase activity. *Applied and Environmental Microbiology* 66: 5518–5520.

Muller, B., Münch, T., Muheim, A., and Wetli, M. 1998. A process for the production of vanillin. European Patent 0885968.

Neto, R.S., Pastore, R.M., and Macedo, G.A. 2004. Biocatalysis and biotransformation producing γ-decalactone. *Journal of Food Science* 69: 677–680.

Nopmaneerat, L., Adiluchayakorn, P., Sutasanawichanna, P., Pruksasri, S., and Leejeerajumnean, A. 2011. Optimization of pyrazine production from newly isolated *Bacillus* sp. by submerged culture. Paper presented at *12th ASEAN Food Conference*, Bangkok, Thailand.

Official Journal of the European Union Directive (EC 1334/2008), Article 3 c) Commission Regulation (EC) no. 50/2000 of January 10, 2000.

Pichersky, E., Iijima, Y., Lewinsohn, E., Gang, D.R., and Simon, J. 2005. Cloning and sequencing of geraniol synthase from sweet basil and use in the production of geraniol and metabolites. International Patent PCT WO 2005060553.

Plaggenborg, R., Overhage, J., Loos, A. et al. 2006. Potential of *Rhodococcus* strains for biotechnological vanillin production from ferulic acid and eugenol. *Applied Microbiology and Biotechnology* 72: 745–755.

Prince, C., Schuler, M.L., and Ithaca, N.Y. 1993. Flavor compounds from *Allium* root cultures. U.S. Patent No. 5244794.

Rao, S.R. and Ravishankar, G.A. 2000. Vanilla flavor: Production by conventional and biotechnological routes. *Journal of the Science of Food and Agriculture* 80: 289–304.

Rizzi, G.P. 1988. The biogenesis of food related pyrazines. *Food Reviews International* 4: 375–400.

Rojas, V., Gill, J.V., Pinaga, F., and Manzanares, P. 2011. Studies on acetate ester production by non-*Saccharomyces* wine yeasts. *International Journal of Food Microbiology* 70: 283–2289.

Scharpf, I.G., Seitz, E.W., Morris, A.J., and Farbood, M.I. 1986. Generation of flavor and odor compounds through fermentation processes. In: *Biogeneration of Aromas*, T.H. Parliament and R. Croteau, Eds., pp. 323–346. American Chemical Society, Washington, DC.

Schindeler, J. and Bruns, K. 1980. Process for producing monoterpenes containing aroma by fermentation. German Patent 2840143.

Seitz, E.W., Sandine, W.E., Ellike, P.R., and Day, E.A. 1963. Studies on diacetyl biosynthesis by *Streptococcus diacetilactis*. *Canadian Journal of Microbiology* 9: 431–441.

Singh, R.S. and Bhari, R. 2012. Microbial flavors: Current status and future prospects. In: *Food Biotechnology: Principles and Practices*, V.K. Joshi and R.S. Singh, Eds., pp. 691–738. I.K. International Publishing House Pvt. Ltd., New Delhi, India.

Smetanska, I. 2008. Production of secondary metabolites using plant cell cultures. *Advances in Biochemical Engineering and Biotechnology* 111: 187–228.

Sugawara, E., Ito, T., Younekura, Y., Sakurai, Y., and Odagiri, S. 1990. Effect of amino acids on microbiological pyrazines formation by *B. natto* in chemically defined liquid medium. *Nippon Shokuhin Kogyo Gakkaishi* 37: 520–523 (Japanese).

Sugawara, E., Suzuki, T., and Yoshida, Y. 1998. The comparison of pyrazine compounds in non-salted fermented soybean products. *Food Science and Technology International Tokyo* 4: 85–88.

Ulbrich, B., Weisner, W., and Arens, H. 1985. Large scale production of rosmarinic acid from plant cell cultures of *Coleus blumei*. In: *Primary and Secondary Metabolism of Plant Cell Cultures*, K.H. Neumann and E. Reinhard, Eds., pp. 293–303. Springer-Verlag, Berlin, Germany.

Vandamme, E.J. 2003. Bioflavors and fragrances via fungi and their enzymes. *Fungal Diversity* 13: 153–166.

Walton, N.J., Narbad, A., Faulds, C.B., and Williamson, G. 2000. Novel approaches to the biosynthesis of vanillin. *Current Opinion in Biotechnology* 11: 490–496.

Westcott, R.J., Cheetham, P.S.J., and Barraclough, A.J. 1994. Use of organized viable plant aerial roots for the production of natural vanillin. *Photochemistry* 35: 135–138.

Xiao, Z.J., Xie, N.Z., Liu, P.H., Hua, D.L., and Xu, P. 2006. Tetramethylpyrazine production from glucose by a newly isolated *Bacillus* mutant. *Applied Microbiology and Biotechnology* 73: 512–518.

Yoshida, A., Takenaka, Y., Tamaki, H., Frshida, A., Adachi, O., and Kumagai, H. 1997. Vanillin formation by microbial amine oxidase from vanillylamine. *Journal of Fermentation and Bioengineering* 84: 603–605.

14

Genetically Modified Foods

Konstantin Skryabin and Victor Tutelyan

CONTENTS

14.1 Introduction....480
14.2 General Concepts of Genetic Manipulation....480
14.2.1 Genetic Modification and GM Plants of the First, Second, and Future Generations....480
14.2.2 Plant Transformation Vectors and Procedures....481
14.2.3 Selection and Regeneration of Plant Transformants....483
14.2.4 Genomic Analysis of GM Plants....484
14.2.5 Control of Gene Expression....484
14.3 GM Crops in the World....485
14.3.1 Types of GM Crops....486
14.3.2 GM Crops in Russia....488
14.4 Benefits of GM Foods....488
14.5 Safety Concerns of GM Foods....489
14.5.1 General Principles....489
14.5.2 Recombinant DNA Safety....492
14.5.3 World Practices in Safety Assessment of GMOs....493
14.5.4 GMO Safety Assessment System in the Russian Federation....494
14.5.5 Labeling Issues of GM Foods....496
14.5.6 Methods for Identification and Quantification of GMOs....497
14.6 Global Status of Commercialized GM Crops....498
14.6.1 GM Soybeans....499
14.6.2 GM Maize....499
14.6.3 GM Potato....500
14.6.4 GM Cotton....501
14.7 Summary and Future Prospects....502
References....503

14.1 Introduction

Biology entered the twenty-first century and marked the beginning of a tremendous quality shift, based on accumulation and integration of leading theories and practices. Mapping on the human genome, the development of such lines of researches as genomics, proteomics, transcriptomics, and metabolomics allows precise defining of mechanisms in the future that are fundamental to health and longevity, illness and death, and clinical medicine. Generally speaking, development in all fields of biological science is conditioned by progress of genetic technologies and their wide practical use.

The term "biotechnology" means the totality of industrial methods that use living organisms and biological processes. Biotechnological methods used in wine-making, bread making, beer brewing, and cheese making based on using microorganisms are known for a long time. Modern biotechnology is based on gene (cell and genetic) engineering, which allows to receive valuable biologically active substances (antibiotics, hormones, enzymes, immunomodulators, synthetic vaccines, amino acids, etc.) and food proteins and to create new plants and breeds. The main advantage of the application of new approaches is decreased dependence on natural resources and use of economically sound and ecological methods. The creation of genetically modified (GM) plants allows to speed up the process of selection of cultural varieties with the properties that cannot be raised when using conventional methods.

14.2 General Concepts of Genetic Manipulation

14.2.1 Genetic Modification and GM Plants of the First, Second, and Future Generations

Genetic modifications make agricultural crops of the first generation resistant to pesticides, pests, and diseases. They also decrease pre/postharvest losses and improve the quality of the produce. One may assume that plants of the second generation (2010–2020) will be resistant to pathogens, influence of climate factors, soil salination, etc. They will have extended shelf life, high nutritional value, enhanced taste properties, and ability to produce immune preparations and medicines and will be free of allergens. For cultures of the third generation (2020 and later), besides the properties listed earlier, there will be alteration in the time of blossom and fruitification; change of size, shape, and number of fruits; increase in photosynthesis efficiency; and production of nutrients with the enhanced assimilation level.

All these achievements became possible due to tremendous progress in genetic engineering, plant transformation and regeneration techniques, and techniques for the analysis of GM plants. Genetic engineering has considerable theoretical and methodological potential to perform transfer of genes from one organism to another. In particular, gene technology methods comprise synthesis of genes in vitro, extraction of separate genes or inheritable structures from cells, targeted rebuilding of the separated structures, copying and replication of the isolated (or synthesized) genes or genetic structures, and combination of different genomes in one cell (Glazko and Glazko, 2001; Chrispeels and Sadava, 2002; Vorobiev, 2004).

14.2.2 Plant Transformation Vectors and Procedures

The technology of obtaining GM plants is a rather lengthy and multistep process: the main step includes isolation of target gene, construction of specialized plant transformation vectors, transfer of foreign DNA into plant cells, selection of transformants and verification of transformation events by molecular and biological methods, and regeneration of the whole plant from the transformed cells (Van den Eedea et al., 2004; Querci et al., 2007, 2010).

Efficient and controlled expression of target gene in plants often requires redesigning and modification of its coding sequence toward optimized expression in plant cells, taking into account plant codon frequencies and absence of fortuitous transcription termination and processing sites. Typical plant transformation vector contains an expression cassette with a target gene placed under the control of plant-specific promoter and terminator sequences. Popular plant promoters, for instance, are 35S promoter of cauliflower mosaic virus, *Agrobacterium tumefaciens nos*-promoter (dicotyledons), and promoters of maize alcohol dehydrogenase gene and rice actin gene (monocotyledons) (Finnegan and McElroy, 1994; Van den Eedea et al., 2004). Once appropriate recombinant plant transformation constructs have been obtained, the next step involves their introduction into plant genome. For this purpose, a range of different transformation methods are available, including *Agrobacterium,* ballistic, protoplast gene transfer methods, as well as several alternative approaches (Sanford, 1990; Kapila et al., 1997; Van den Eedea et al., 2004), as shown in Figure 14.1. Efficient delivery of vector DNA into plant protoplast may be performed with the help of electroporation, microinjections, wrapping DNA into liposomes, and fusion of bacterial spheroplasts and by chemically induced endocytosis (Kapila et al., 1997). Alternative methods of genetic transformation, such as transformation of organelles (mitochondria and chloroplasts) and use of mobile genetic elements for gene transfer into plants, did not receive widespread application (Kapila et al., 1997; Van den Eedea et al., 2004).

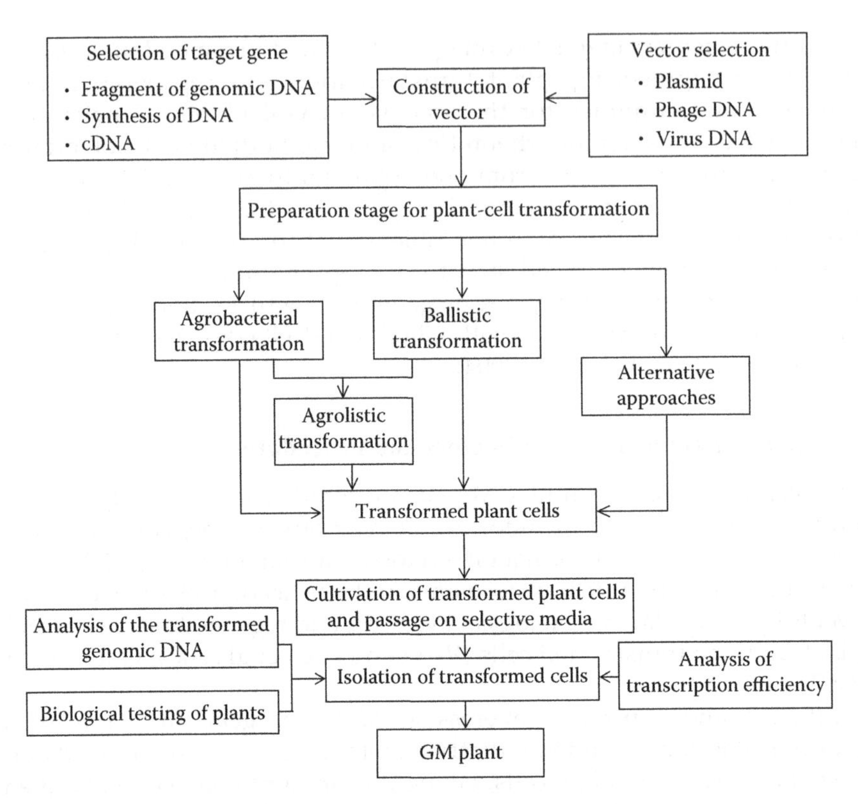

FIGURE 14.1
Different techniques used in the development of GM plants.

At present, mainly agrobacterial and ballistic methods of plant genome modification are used to produce GM crops (Newell et al., 1990). The agrobacterial method involves the use of *Agrobacterium* spp. (*A. tumefaciens*, *Agrobacterium rhizogenes*), which have the ability to cause cancerous growth (so-called crown galls) in many dicotyledonous plants. Nevertheless, efficiency of agrobacterial transformation is quite low as only 1 out of 10,000 plant cells becomes a carrier of recombinant DNA (Singer and Berg, 1991; Van den Eedea et al., 2004).

Ballistic method of transformation of plant genome, also known as microbombardment method of particle acceleration, biolistics (word derived from biology and ballistics), lies in using microbombardment of intact plant cells with gold or tungsten particles, which play the role of the carrier of recombinant DNA (Russian Federation National Methodical Recommendations, 2006). Microparticles may consist of any chemically inert metal with high molecular weight (gold, tungsten, palladium, rhodium, platinum, indium, etc.), so as not to form metal–organic complexes with DNA but possess high kinetic energy for efficient penetration in cell wall (Sanford, 1990; Van den Eedea et al., 2004).

A combined method of transformation has been developed and applied successfully recently, which got the name agrolistic. It is based on the introduction of a foreign DNA with ballistic method that includes T-DNA vector with target and marker gene along with agrobacterial *vir*-genes (Kapila et al., 1997).

14.2.3 Selection and Regeneration of Plant Transformants

The key to obtaining whole fructifying GM plant from transformed cell is the plant's capacity for totipotency, i.e., the ability to multiply, differentiate, and form mature plants from any isolated plant cell. The process of transformation is performed under aseptic conditions, so as to exclude contamination with bacteria, which may lead to false-positive results when testing with polymerase chain reaction (PCR; Russian Federation National Standard, 2003; Russian Federation National Methodical Recommendations, 2006).

Regenerants are cultivated in several passages for a period up to several months in the presence of high concentrations of selective agents. To monitor transformation and regeneration process, various marker and reporter genes are used. Frequently used antibiotic and herbicide resistance marker genes as per reported by Miki and McHugh (2004) are as follows:

- *nptII*: gene encoding neomycin phosphotransferase providing resistance to aminoglycoside antibiotics (kanamycin and others)
- *hptI*: gene encoding hygromycin phosphotransferase (hygromycin antibiotic resistance)
- *bar*: gene encoding phosphinothricin acetyltransferase (bialaphos herbicide resistance)
- *gox*: gene encoding glyphosate oxidase for glyphosate herbicide resistance

While obtaining GM plants resistant to pesticides, gene of resistance can take a role of both target and marker gene. As opposed to marker genes, the reporter ones practically do not affect the metabolism of the modified plants. In most of the cases, β-glucuronidase (GUS), green fluorescent protein, Luciferase, chloramphenicol acetyltransferase, and other reporter genes are used. Substitution of marker genes with reporter genes during the selection of GM plants is very desirable, as negative influence on human health and environment is practically excluded when using reporter genes (EFSA, 2007).

The presence of marker genes, especially antibiotic-resistant genes, is one of the main arguments against using GM products. That is why a range of methodological approaches has been developed, which allow to eliminate the marker genes in mature regenerated GM plants. Selective elimination of marker genes becomes possible by the use of certain transformation systems, which can be distinguished by a system of co-transformation, site-specific recombination, intragenome redistribution of genes by mobile elements, use of specific promoters of marker genes, and substitution of the original gene with the corresponding modified gene (GMO Compass, 2006).

At the same time, it is necessary to take into account that when the plants are exposed to strong stress by growing them on a medium containing antibiotics (kanamycin, hygromycin, or cefotaxime), DNA is exposed to hypermethylation (Schmitt et al., 1997). It is known that methylation of regulatory–promoter encoded regions of the highest eukaryote genes, including plants, turns off their expression, i.e., the probability of "silencing" of foreign gene expression to a considerable degree. Thus, the selection of GM plants on a medium with moderate concentrations of kanamycin decreases stress load and creates conditions for their normal growth and development.

14.2.4 Genomic Analysis of GM Plants

The presence of target gene, evaluation of its copy number in the host genome, and transgene expression can be evaluated by various techniques. PCR analysis is routinely used for the verification of the presence of transgene and site of integration. Southern, Northern, and Western blot analyses are used to estimate insert copy number and expression levels.

The final step of laboratory testing of GM plants includes biological research oriented to confirm stable phenotypic display of target trait. Necessity of such kind of tests to the great degree is caused by many evidences of genetic instability, which shows that the level of transgene expression as a whole is unpredictable and varies to the great extent even in transformed identical DNA constructs of the clones, received in conjunction in one experiment. It is observed that the expression level depends on many factors, in particular on the number of transgene copies and sites of their insertion. Besides, during insertion into genome, the DNA construction can be considerably rebuilt (duplications, inversions, etc.). The two important factors responsible for the significant variations in transgene expression among plant transformants are epigenetic chromosomal position effects and inappropriate promoter–enhancer interactions, which significantly increase the screening efforts aimed at the selection of lines with desired expression patterns (Singer et al., 2012).

14.2.5 Control of Gene Expression

Transgenes are generally the chimeric assembly of cloned gene from diverse organisms that is introduced into the plant genome by genetic engineering technique. In order to get stable expression and inheritance of transgenes in GM lines of the plants, the following points need to be taken into account:

- Single-copy target gene insertions are preferable, since they are less likely to be the subject of gene silencing through mechanisms of RNA interference as compared to multicopy inserts with potential rearrangements, inverted target gene copies, etc.
- Degree and length of homology of a foreign gene with recipient genome must be minimal.
- Integration of transgene replica is preferred into unmethylated region of plant genome.

Thus, the control of the transgene expression level is absolutely necessary for the next generations (Kapila et al., 1997). Many gene functions are lethal if mis-expressed, and in such cases, it may be desirable to have a tight control over gene expression. The control of gene expression for commercially valuable traits is also essential to preserve the value of desired traits, particularly in those crops that are inbred or where seeds can be saved (Beachy, 2005).

14.3 GM Crops in the World

By using the different techniques of genetic manipulation in plants (as discussed earlier), more than 50 species of GM plants are obtained and brought to the field tests in the world. By the middle of 2012, more than 196 transformation events for 25 crops are likely to be registered and approved for the production of food and feed, including more than 30 lines of selective stackers (James, 2011). In accordance with the 2012 database of the Food and Drug Administration (FDA), 88 lines of GM plants were registered and approved for field-scale production of food and feed in the United States, and 41 lines of GM cultures were registered in the European Union.

The global expansion of acreage under GM crops connected with the high efficiency and perspectives of agro-biotechnologies: within the period from 1996 to 2011, the area of cultivation of GM cultures increased by 96 times and is equal to 160 million hectare. At present, 29 countries, including eight countries of the EU, cultivate GM crops. Among them, the top five countries (United States, Brazil, Argentina, India, and Canada) occupied more than 10 million hectare areas (Figure 14.2). Brazil is emerging as a global leader in the cultivation of GM crops with increasing its hectarage of biotech crops more than any other country in the world in three consecutive years.

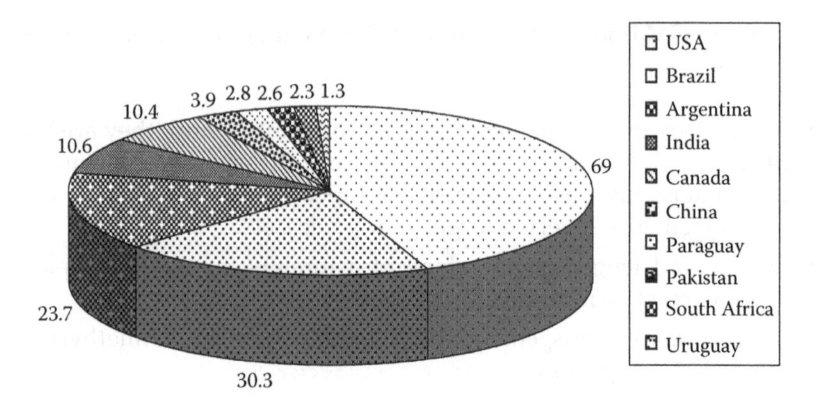

FIGURE 14.2
Country-wise area of GM crops in million hectares. (From James, C., Global status of commercialized biotech/GM crops, ISAAA brief no. 43, ISAAA, Ithaca, NY.)

The countries having <0.1 million hectare area under GM crops include Columbia, Chile, Honduras, Portugal, Czech Republic, Poland, Egypt, Slovakia, Romania, Sweden, Costa Rica, and Germany having maize as popular growing GM crop (James, 2011).

14.3.1 Types of GM Crops

The GM crops with resistance to herbicide traits in 2011 occupied 59% of the total GM acreage (93.9 million hectares). The plants with combined traits—"selection stackers"—containing two or more transformation events occupied the second place in 2011 (26%, 42.2 million hectares), whereas the cultivation of Bt-crops, resistant to pesticides, is at third place (15%, 23.9 million hectares) (James, 2011).

Plants can be made tolerant to nonselective herbicides by the insertion of the genes coding protein tolerant to herbicides (glyphosate, chlorsulfuron, and imidazolinone herbicides) or by the insertion of genes that determine the intensive metabolism of herbicides in plants (ammonium glufosinate, sodium 2,2-dichloropropionate).

Currently, glyphosate is the most widely used herbicide in the world, and therefore it seems reasonable to create a significant number of GM plants resistant to it. Glyphosate (*N*-(phosphomethyl) glycine, monoisopropylamine salt) belongs to nonselective herbicides, whose mechanism of action is based on the blocking of the synthesis of some essential aromatic amino acids such as the metabolism of shikimic acid. The key step is the synthesis of 5-enolpyruvylshikimate-3-phosphate from phosphoenolpyruvate and shikimate-3-phosphate, which is catalyzed by 5-enolpyruvylshikimate-3-phosphate synthase (EPSPS) enzyme. This enzyme is the target for glyphosate to block its mechanism. In order to obtain cultures resistant to glyphosate, the *EPSPS* gene, which codes the synthesis of EPSPS, is modified in such a way that the resulting expressed enzyme becomes tolerant to the effect of glyphosate. *Agrobacterium* sp. strain SP4 DNA or the own cloned gene of the plant, undergoing this transformation (Fichet and Brants, 2001), is used as the *EPSPS* gene source. The shikimic acid metabolic pathway that is typical for plants, algae, bacteria, fungi, and protozoa is not present in other life forms, including insects, fish, birds, humans, and other mammals (Giesy et al., 2000; Williams et al., 2000; Potrykus and Ammann, 2010).

Phosphinothricin (a herbicide), whose mechanism of action is based on the inhibition of the glutamine synthetase enzyme of plant cells, is an active substance of the ammonium glufosinate. In plants, glutamine synthetase converts ammonia into glutamine. Blocking of glutamine synthetase by glufosinate results in a rapid depletion of reserves of the plant glutamine, which leads to the accumulation of ammonia in photosynthetic tissues and, thus, to plant poisoning (Cremer, 2001). It is accepted to designate the resistance gene to glufosinate as *pat* or *bar* depending on whether it was isolated from *Streptomyces viridochromogenes* or *Streptomyces hygroscopicus*, which produce a powerful tripeptide antibiotic—bialaphos. L-Phosphinothricin is an active component of bialaphos. The *pat/bar* gene encodes the synthesis of phosphinothricin acetyltransferase enzyme, which acetylates the free NH_2 group of phosphinothricin. Plants with this inserted gene have acquired the ability to produce phosphinothricin acetyltransferase and resist glufosinate (Cremer, 2001).

The insecticidal effect of δ-endotoxin, formed in the process of sporulation of Gram-positive soil bacteria *Bacillus thuringiensis*, is based on a specific binding with insect intestinal epithelium cell receptors; this binding results in the disturbance of the osmotic balance, consequently swelling and lysis of cells. δ-Endotoxins are synthesized in the bacterial cell as protoxins and are processed in the insect's gut, turning from protoxin into active protein, toxic for scale-winged insects (Lepidoptera), two-winged insects (Diptera), and beetles (Coleoptera). The δ-endotoxins are harmless for warm-blooded animals (Saraswathy and Kumar, 2004). All described genes of δ-endotoxins are combined into five groups based on the homology in amino acid sequences and the insecticidal activity of proteins encoded by them (from *cry1* to *cry5*). A method has been developed in order to obtain a durable expression of the *cry* genes, the principle of which was based on the use of synthetic analogues of the *cry* genes, in which optimized ones for plants replaced the natural codons. Thus, to obtain the resistance in potato plants to Colorado beetle, the partially modified gene of δ-endotoxin *Cry111A* from *B. thuringiensis* subsp. *tenebrionis* is usually applied, and for obtaining maize resistant to pests, the partially modified δ-endotoxin gene, *Cry1A* from *B. thuringiensis* subsp. *kurstaki*, is used (Saraswathy and Kumar, 2004).

If there is no need to insert a gene into the plant, the only motive is to block or weaken the expression of a gene already present; the mechanism of RNA interference can be used for selective silencing of the target genes in plants.

This approach was applied to obtain GM tomato plants with improved fruit quality. Insertion of inverted copy of the polygalacturonase (PG) gene (involved in pectin destruction) triggered posttranscriptional gene silencing (PTGS) of the endogenous gene, which results in the reduction of PG content. The application of PTGS strategy resulted in the delay of the ripening process and allowed to obtain tomato plants with new properties of fruits, suitable for long-term storage (Van Eck et al., 2006).

14.3.2 GM Crops in Russia

The analysis of the situation, which has been developed in the agriculture of Russia during the last few years, indicates a drop in the production of all types of agricultural products by more than 1.5 times. With the preserved total volumes of natural and labor resources, more than 30 million hectares of highly productive agrocenosis have been withdrawn from the cultivation. In the Russian Federation, the efficiency of agro-ecosystems is very low, e.g., soybean yields (10.9 c/hectare) in Russia is one-third of the U.S. index (29.58 c/hectare) and the EU index (27.75 c/hectare). Application of modern biotechnological techniques is a new approach for creating highly productive agricultural systems and ensuring a significant increase in crop yields and livestock productivity.

In the Russian Federation, 19 lines of GMOs were approved for the population nutrition and 15 lines of GMOs were registered as feeds. The total contribution of Russia to the global biotechnology market according to the expert estimations is ~0.2%. Presently, Russia is not represented on the market of GM crops, because GM cultures are not cultivated in Russia (Skryabin, 2010).

The practical use of the new ways of transforming genome of living organisms led to the need for strict regulation of the safety assessment process of the GMOs that were intended to be used for food. When developing the GMO safety assessment approaches, the probability of the products, which may pose a potential threat to human health, was considered as these products might differ from the traditional ones by food value and toxicological and allergenic properties. The reason for these changes lies not only in the properties of the new protein and its metabolites but also in the possible recombinant DNA-based synthesis of unknown components. The lack of information on both chemical structure and the actual presence of such components in GMOs creates difficulties for identification.

14.4 Benefits of GM Foods

Everybody knows that due to the rapid rise in population and consumer awareness, there is an ever-increasing pressure on the food supply chain to meet the quantitative as well as qualitative requirements of the food worldwide. A Conference Rio+20 at Rio de Janeiro (Brazil, June 20–22, 2012) brought together governments, international institutions, NGOs, and other major groups, where countries adopted Agenda 21. Participants of Rio+20 discussed the measures that can reduce poverty while promoting decent jobs, clean energy, and a more sustainable and fair use of resources.

GM technology can help mitigate negative effects in different fields of modern life and has enormous potential for the future. The GM plants are already contributing to sustainability to providing food, feed, and fiber security, improvement in nutritional value, better quality food crops, conserving biodiversity, contributing to the alleviation of poverty and hunger, reducing agriculture's environmental footprint, and helping mitigate climate change and reducing greenhouse gases (Table 14.1).

14.5 Safety Concerns of GM Foods

14.5.1 General Principles

When considering the problem of safety of GM foods, and comparing them to the traditionally derived foods, one should take into account that the latter may also present danger to human health. This is the reason for the number of requirements during production, storage, and cooking of food. In addition, it is required to take care that the sanitary and hygienic safety of food products in the unbalanced diet may promote the development of diseases, caused by the violation of the nutritional status (atherosclerosis, diabetes, obesity, hypertension, gout, etc.). Potentially dangerous factors associated with food can be deficiencies of nutrients, microorganisms, and chemicals, which occur in food naturally (e.g., saponins in potatoes), or introduced into the food chain intentionally (e.g., food additives, residues of agrochemicals), or may be accidental (e.g., environmental pollutants). Thus, if it is proved that a GMO is as safe as its traditional analogue, it is possible to consider this GMO harmless and safe for human health when used under normal conditions.

The World Health Organization (WHO) and the UN Food and Agriculture Organization (FAO) approve the generally accepted concept of safety assessment of genetically modified organisms (GMOs) of plant origin. This approach is based on the principle of "substantial equivalence," or "composite equivalence"—the equivalence of the chemical composition of a GMO and its traditional analogue in the content of key nutrients, allergens, anti-alimentary, and toxic substances, characteristic for this type of product or determined by the properties of the transferred genes. The principle of "substantial equivalence" serves the base for establishment of safety class. If the GM product is chemically equal to the traditional analogue, the GMO

TABLE 14.1

Benefits of GM Foods

	Benefits	Examples	Crops	Countries
Food, feed, and fiber security	Higher productivity of GM food, feed, and fiber	2010: Additionally 44.1 MT food, feed, and fiber were produced by GM crops	Soybean—13.1 MT Maize—28.3 MT Cotton—2.1 MT Canola—0.65 MT	Globally
	Higher economic gains at the farmer level	2010: Additionally US$ 14 billion	—	Globally
	Improvement in nutritional value and better quality food crops	2013/2014: Golden rice that can help to reduce vitamin A deficiency	Golden rice that contains enhanced level of beta-carotene	Field trials and regulatory compliance experiments, Philippines 2013/2014
		2010: GM potato with modification of starch content	Potato "Amflora"	Sweden, Czech Republic, and Germany
	Increase in shelf life	1994: First commercially available GM tomato with "deactivated" gene. Plant was no longer able to produce an enzyme. Polygalacturonase involved in fruit softening. GM tomatoes could be left to ripen on the vine and still have a long shelf life, thus allowing them to develop their full flavor.	GM tomatoes "FlavrSavr"	U.S. market. No longer on the market nowadays
Contributing to the alleviation of poverty and hunger	Increasing income	2011: Significant input to the income of about 15 million poor farmers	GM cotton, GM maize, and GM rice	China, India, Pakistan, South Africa, Burkina Faso, Myanmar

TABLE 14.1 (continued)

Benefits of GM Foods

	Benefits	Examples	Crops	Countries
Conserving biodiversity	Land saving and preclude deforestation	2010: Additionally 44.1 MT food, feed, and fiber were produced by GM crops; that is why 14 million hectares of arable lands have been reserved.	—	Globally
Reducing agriculture's environmental footprint	Reduction in pesticide use	2010: Reduction of 43.2 million kg of active ingredients (equivalent to a saving of 11.1% pesticides)	—	Globally
	Saving on fossil fuels	2010: Saving of 1.7 billion kg of CO_2 (equivalent to reducing of 0.8 million cars on the roads)	—	Globally
	Conserving soil and moisture (due to no till technology)	2010: Soil carbon seizure equivalent to 17.6 billion kg CO_2 or removing 8 million cars off the road	—	Globally
	Increasing efficiency of water usage	2013: First drought-tolerant GM maize hybrids will be commercialized.	GM maize	United States (perspective)
		2017: First drought-tolerant GM tropical maize will be commercialized.		Sub-Saharan Africa, Latin America, and Asia (perspective)

(continued)

TABLE 14.1 (continued)

Benefits of GM Foods

	Benefits	Examples	Crops	Countries
Abiotic stresses: Climate changes	Helping mitigate climate change and reducing greenhouse gases	2010: "Speeding the breeding" strategy Programs to develop varieties and hybrids that are well adapted to more rapid changes in climatic conditions	—	Globally
Biotic stresses	Disease resistance	2011: Virus-resistant papaya	GM papaya	China
	Economic benefits with reduction in use of fungicides	2014–2015: Potato "Fortuna" Resistant to fungus *Phytophthora infestans*	Late blight potato	EU countries

is considered to be completely harmless and does not need any further study (the first class of safety), but if there are differences with the traditional analogue, i.e., the presence of new components or the absence of any components, further research should focus on the identified differences (the second class of safety). If the GM product does not fully conform to the traditional analogue, the safety assessment should be extended (the third class of safety). The principle of equivalence is only a starting point to study the safety of GMOs and the list of necessary researches can be extended regardless of the degree of equivalence of the product under investigation.

Currently most of the GMOs, produced on large/industrial scale, belong to the second class of safety, because they differ from traditional ones only by the presence of one to two proteins in the component composition. These proteins being responsible for the manifestation of a given trait, and therefore, further research is aimed at studying the properties of the new protein(s) and its/their effect on the organism.

14.5.2 Recombinant DNA Safety

When discussing the possibility of influence of recombinant DNA on the organism, it should be borne in mind that recombinant DNA and natural DNA are identical, since the nucleotide sequence rearranges due to the genetic modification, although its chemical structure does not change at all.

The use of recombinant DNA does not make any changes to the food chain, this being because in nature there are numerous variations of the nucleotide sequences in DNA. Thus, being based on the chemical equivalence and the identity of the metabolism of the modified and natural DNA, the possibility of toxic and allergenic effect of recombinant DNA may be completely excluded (Russian Federal Laws, 1996–2007; FAO, 2004; WHO, 2005).

14.5.3 World Practices in Safety Assessment of GMOs

At national levels, there are systems for evaluating the quality and safety of GMOs in the United States, Canada, the EU, the Russian Federation, Mexico, Argentina, Brazil, Chile and other countries of Latin America, Egypt, Saudi Arabia, China, Japan, India, Thailand, Philippines, Indonesia, Australia, New Zealand, and South Africa. In spite of the differences between the American and the European approaches, the main stages of the safety assessment of GMOs of plant origin in the United States and the EU are identical as a whole and include (Chassy, 2002) the following:

- The anamnestic safety analysis (the analysis of the information about the parent organism and donor organisms of the inserted genes: taxonomic characterization, the reproduction and dissemination manner description; data on the toxic, allergenic, or other adverse properties).
- The analysis of the genetic structure, methods of genetic modification, localization, and the expression level of inserted genes.
- The comparison of the chemical compositions of a GMO and its traditional analogue (composition equivalence)
- Toxicological and allergological studies of protein(s), responsible for the manifestation of task trait.
- The study of nutritional value of GMOs in young growing animals—chickens (broilers), lambs, etc. The deviation from the normal body weight gain can be regarded as an integral indicator of adverse effects on the organism as specified and unspecified effects of genetic modification may cause the change in the balance of macro- and micronutrients and consequently leads to the changes in the nutritional value of the product.

The fundamental difference between the approaches under consideration "presumed safety" in the United States, the object of the research is considered to be safe until the contrary is proved and "presumed risk"—in the EU, the object of the research is considered dangerous until the contrary is proved. Thus, the American approach adheres to the principle of composite equivalence, while the European approach adheres to the precautionary principle, which is aimed at revealing the potential negative effects of genetic modification.

14.5.4 GMO Safety Assessment System in the Russian Federation

The development of the GMOs safety assessment system in the Russian Federation has started in 1995–1996. This system not only accumulates the whole of domestic and foreign experience but also includes the latest scientific approaches, based on the achievements of the modern fundamental science: genome and proteome analysis, detection of the DNA damage and mutagenic activity, and identification of products of free radical modification of DNA and other sensitive biomarkers.

The safety assessment of GMOs is performed at the stage of the state registration. New food products, derived from GMOs of plant origin and manufactured in the Russian Federation, as well as the food products derived from GMOs of plant origin imported into the territory of the Russian Federation for the first time, are subject to the state registration. The requirements for the safety assessment are set forth in Methodical Guidelines 2.3.2.2306-07 "Medical and biological evaluation of safety of genetically modified organisms of plant origin" (Medical and Biological Safety Assessment of Plant GMO's: Methodological Guidance, 2007).

The medical and biological assessment of safety of the foods, obtained from the GMOs of plant origin, includes the following:

- Expert analysis and assessment of the data submitted by the applicant
- Expert analysis of detection methods, identification, and qualitative determination of GMOs in food products
- Medical and genetic assessments
- Assessment of functional and technological properties
- Medical and biological researches

The information analysis on the object of registration provided by the applicant determines the list and extent of necessary research; however, the fulfillment of each of the enumerated blocks of research is obligatory.

Starting from the moment of formation, the Russian system of GMO safety assessment included obligation of conducting of the chronic toxicological experiment with the duration of not less than 180 days (Tutelyan, 2007; Tyshko et al., 2011)—the European Union adopted this approach in 2004, though in the EU, the duration of the studies is only 90 days.

Toxicological studies are carried out on laboratory animals (Wistar rats, initial age is ~30–40 days), in the ratio of which the GMO under study is included (experimental group) and its traditional analogue (control group) in the maximally possible quantity, which does not disrupt the balance of the major nutrients. Dynamic monitoring of the integral (appearance, weight, etc.), hematological, biochemical, and morphological parameters is being carried during the experiment.

The use of indicators (systemic biomarkers) that reflect the level of adaptation to the environment and that are highly sensitive to foreign influence is a distinctive feature of the Russian system of safety assessment of GMOs. Special attention is paid to the systems that provide the protection of the organism from the effects of toxic compounds of exogenous and endogenous origin: the system of metabolic enzymes of xenobiotics, the system of regulation of apoptosis, and the system of the antioxidant protection. Thus, the data obtained at each stage of the experiment are of both independent and aggregate significance. The generalization and analysis of the results allow in tracing the state of the animals of both control and experimental groups at different ontogeny stages that greatly increases the diagnostic value of the research.

Along with the general toxicological studies of GMOs of plant origin risk assessment system adopted in the Russian Federation also includes the study of specific types of toxicity in the experiments in vivo (genotoxicity, immunotoxicity, allergenicity, reproductive toxicity) (Tutelyan, 2007). The study of the genotoxic effect of GMOs includes the assessment of genetic material at different levels of the structure (DNA molecules—chromosomes), studies of the immunotoxic effect includes the evaluation of immunomodulatory and sensitizing properties of GMOs in the experiment on mice of oppositely reacting lines, and the study of the allergenic action proposes assessment of the severity of the active anaphylactic shock and the intensity of the humoral immune response to the systemic anaphylaxis model with rats.

In accordance with the current research practice, the study of the GMO reproductive toxicity refers to the optional studies conducted in case of need proven. A serious constraint for the application of the reproductive toxicity assessment methods is their duration and complexity, although high informational content is an important argument in favor of conducting such studies for the health of future generations as well. Due to this, the Research Institute for Nutrition of the Russian Academy of Medical Sciences has developed an algorithm of reproductive toxicity assessment of GMOs of plant origin. The comprehensive set of methods includes (a) the study of the reproductive function by the fertile ability, hormonal status, and the level of gametogenesis in the gonads of males and females; (b) the study of the prenatal development of the offspring by the preimplantation and postimplantation mortality, biometrical fetus characteristics, and the state of the internal organs and skeletal system of the fetus; and (c) the study of the postnatal development of the offspring from the dynamics of biometrical characteristics to the parameters of physical development and the survival rate from 0 to 5th and from the 6th to 25th days of life. The present algorithm has been used in the studies of reproductive toxicity of the transgenic maize LibertyLink® (Tyshko et al., 2011) and will be further applied at the stage of the state registration of new GMOs.

The decision on state registration of GM plants in the Russian Federation is based on the whole set of the expert evaluations of materials submitted by the applicant, the results of the comprehensive medical and biological evaluation

TABLE 14.2

Russian Agencies Involved in Safety Assessment of GM Foods

Agency	Functions
Ministry of Healthcare of the Russian Federation	Designation of responsible agency for registration, medical and biological assessment of GM foods of plant origin
Federal Service for Consumer Rights Protection and Public Welfare (Rospotrebnadzor)	State registration, Federal Register, and monitoring for approved GM foods of plant origin
	Development of methodical recommendations for assessment and monitoring of GM food of plant origin
Institute of Nutrition of the Academy of Russian Medical Sciences	Medical and biological assessment of GM foods of plant origin. Development of methodical recommendations for assessment and monitoring of GM food of plant origin

of safety (results of a complex medical and biological assessments of safety), medical and genetic, and technological assessments, and the expert assessment of methods of GMO identification, as well. During the period from 1999 to 2012, a complete cycle of biomedical research has been performed for 21 lines GMOs of plant origin in Russia. The scientific base on GMO risk assessment was generated, which included the analysis of the research results, received in the framework of the procedure of GMOs registration in the Russian Federation, and the data of national and world scientific literature, devoted to the issue of safety of biotechnological products, both at the stage of registration and at the stage of post-registration monitoring.

In accordance with the international practice, the Russian Ministry for healthcare is responsible for state registration of GM foods (Table 14.2).

14.5.5 Labeling Issues of GM Foods

When treating the use of biotechnology in food production as an extremely promising avenue to solve the food supply problem, one should take into consideration a potential possibility of unintentional or intentional production of GMO by uncontrolled genetic engineering processes that can adversely affect human health in the world. Logically, legislation in the most countries imposes registration, post-registration surveillance, and labeling of the GMO-derived food. It should be stressed that labeling of a GMO food product has nothing to do with its safety: it only informs the consumer on the use of genetic engineering technology during the production of the food products (Federal Law of the Russian Federation, 2007).

The United States employs different procedures when assessing food/feed safety, and once the safety of GM line and all products derived from it is confirmed, the product is registered and all food products derived from GMO if this variety (line, event) is placed on the food market under deregulated status. FDA considers that extra labeling on the mode of production can misguide the consumers on real safety on the food product (National Research Council, 2000).

In the European Union, the control of production and life cycle of GMO-derived food imposes post-registration surveillance and obligatory labeling. The threshold for labeling of foodstuff as GMO-derived product is 0.9%. If a food product contains less than 0.9% GMO per individual ingredient (or per whole product if it consists of a single ingredient), labeling is not required, and the presence of GMO in the foodstuff is considered as technologically unavoidable admixture (Directive EU, 2001; Regulation EU, 2003).

In Russia, in order to harmonize the Russian legislation with international practices, the government issues the Federal Law "On Consumer Rights Protection" and amendments to the Part 2 of the "Civil Code of the Russian Federation," where a 0.9% threshold for GMO was set as an unintentional or technically unavoidable (adventitious) level (Customs Union Commission, 2010).

14.5.6 Methods for Identification and Quantification of GMOs

The success of GMOs during their production, processing, and testing is the essential criterion for their regulatory approval. So reliable and effective methods should be developed, which can easily but accurately differentiate between GMOs and non-GMOs. The two most common methods that are generally employed for the detection of GMOs are DNA-based methods and immunoassays as shown in Table 14.3. The PCR is a very reliable and sensitive method to detect DNA sequences resulting from genetic modification. The amplification or nonamplification of inserted gene clearly shows the presence or absence of genetic modification in the food products. Although the detection of foreign DNA sequences can be carried out by PCR, however, immunoassays are potentially useful in measuring the levels of proteins expressed by DNA sequences inserted by genetic modification (Duijn et al., 1999; Ahmed, 2002).

TABLE 14.3

Methods for Identification and Quantification of GMOs

Methods	Target
Protein detection methods	
Enzyme-linked immunosorbent assay	Novel proteins
DNA detection methods	
PCR methods for screening	Genetic elements: CaMV 35S promoter, NOS terminator
Construct-specific qualitative PCR methods	Specific DNA construct
Event-specific, PCR methods	For unequivocal identification of a specific GM event, the PCR product to be amplified must span the junction between the construct and the plant genomic DNA
Microarray detection of multiple (nonauthorized) GMOs	Recombinant DNA

14.6 Global Status of Commercialized GM Crops

The key point to ensure the independence of the country from external sources of supply is the choice of strategic crops and maintaining the highest possible reserve production capacity, minimizing dependence on climatic conditions, pathogens, and pests. World practice has yet identified a number of strategic crops, firstly, grains (wheat, rice, maize), soybeans, canola, sugar beets, and cotton. These cultures are considered as the main targets for genetic modifications. The global area of biotech crops in 2011 has been shown in Table 14.4 (National Research Council, 2010). GM wheat that is resistant to herbicide has successfully passed the test on food safety but, however, not used commercially in connection with the fact that this would lead to a sharp and dramatic redistribution of the world wheat market.

To date, commercialized GM crops are input-oriented, i.e., the incorporated traits replace or utilize inputs applied to the crops. Herbicide-resistant varieties provide crops with resistance to specific herbicides, insect-resistant traits allow the plants to produce proteins toxic to targeted insects, thus reducing the need for insecticides; virus resistance protects crops against specific viruses while minimizing sprayings and crop losses. But the current developments in genetic modification of plants have the capability to protect crop yields, improve water and soil quality, and improve feed grain safety (National Research Council, 2010). Future innovations in this field may increase efficiency in the use of water, sunlight, and fertilizer. GM traits can directly improve the nutritional qualities of the foods produced as well. Examples include crops with high vitamin or protein levels, fruits with delayed ripening, and oilseeds with lower saturated fat.

Making use of this powerful tool will be important for providing nutritious food produced in a sustainable way to the world's growing population, which is expected to grow to more than 9 billion people by 2050.

TABLE 14.4

Global Area of Biotech (GM) Crops, 2011

Crop	2011 (Millions of Hectares)	% (of Global GM Production)
Soybean	75.4	47
Maize	51.0	32
Cotton	24.7	15
Canola	8.2	5
Sugar beet	0.5	<1
Papaya	<0.1	<1
Other	<0.1	<1

Source: James, C., Global status of commercialized biotech/GM crops, ISAAA brief no. 43, ISAAA, Ithaca, NY.

14.6.1 GM Soybeans

Soybean is the dominant crop of the GM crops in the world. The purpose of genetic modification of soy significantly increases the productivity of this important food crop. Therefore, soybean varieties were developed that are resistant to pests. Biotech companies (Monsanto in collaboration with BASF) prepared multiple generations of high-yielding soybeans with the aim to boost the intrinsic yield potential of soybean. These yield traits are expected to be stacked with Genuity® Roundup Ready 2 Yield®, dicamba tolerance, and other traits on a highly effective weed management platform. Today, the most common is the world's soybean line 40-3-2, developed by Monsanto, USA, which is resistant to glyphosate pesticide. According to the USDA (2011), biotech soybean in the United States was on high adoption level—94% of total planting of soybean in the country and 75% or 75.4 million hectares of the global 100 million hectares soybean planting (FAO STAT, 2009, 2011). Eleven countries reported growing RR® soybean in 2011 (James, 2011).

As a result of circulation monitoring of food products having GM counterpart, conducted in Russia, the percentage of available GM soy is ranging from 20% to 40% depending on the region. It is shown that soybean line 40-3-2 reached 99% of all GM soybeans in the Russian market.

Further developments in this area are associated with the development of soybean, which will possess improved nutritional value and taste. Vistive® Gold Soybeans combine breeding and biotechnology to produce soybean oil with monounsaturated fat levels similar to olive oil and the low saturated fat content of canola oil. Vistive Gold has 60% less saturated fat than commodity soy. This would allow the food industry to cost-effectively eliminate trans-fats and significantly lower saturated fat content in food products in accordance with consumers' recent dietary requirements.

14.6.2 GM Maize

Corn is one of the most important crop grown in many countries and is widely used in human diets and animal feeds. The high content of thiamine (vitamin B_1) is the major constituent of maize required for normal functioning of many human organs, including brain. Thiamine content in corn far exceeds any other plant product: 100 g of corn contains about 150 mg of vitamin. Corn is also rich in vitamin H (biotin), iron, vitamins C and K, and other nutrients that make corn both ideal grain source and a starting compound for the development of functional foods and biologically active supplements.

Currently, there are more than a dozen varieties of GM maize in the world. Biotech maize is grown on 51.0 million hectares (James, 2011), which consists almost one-third, 32%, of the global maize hectarage of 159 million in 2011 (FAO STAT, 2009). It is noteworthy that in 2011, 16 countries grew GM maize, including the United States (33.9 million hectares; 88% adoption level), Brazil (9.1 million), Argentina (3.9 million), South Africa (1.9 million), and Canada (1.3 million). Most varieties of GM maize are resistant to stem moth, an insect that feeds on the stalk of corn. Several varieties of GM maize are resistant to various pesticides, and today the deployment stacked traits of *Bt* genes and herbicide tolerance become increasingly important. These stacked traits of maize are most prevalent in the United States (about 85%).

In the Russian Federation, 10 lines of GM maize passed a state registration and allowed for sale to the public and use in the food industry. However, monitoring for circulation of food products (derived from corn) showed that GM maize imported into the Russian Federation for the use in the food industry and its products consists less than 1% of the maize imported for this purpose. GM maize imported into Russia belongs to the lines MON 810, resistant to stem moth; MON 863, resistant to beetle *Diabrotica*; and NK 603, resistant to glyphosate.

Further developments in the field of production of GM maize varieties are related to changes in the structure of starch, improved technological parameters of maize, increased levels of lysine and tryptophan in the proteins of corn, and modifications of corn oil.

14.6.3 GM Potato

Russia today is one of the major potato-growing countries in the world and the total potato yield has reached 32.6 million tons (MT) (ROSSTAT, 2011). Potatoes contain almost all needed minerals and trace elements (magnesium, calcium, phosphorus, sodium, and potassium) in appropriate proportion. The vitamin C concentration in potato tubers has been recorded about 20 mg per 100 g of product. Potatoes contain a lot of fiber, which is important for regulation of the intestine. High-quality vegetable protein of potato in conjunction with animal protein curd, eggs, and cheese is the best substitute for meat.

In Russia, about 95% of potatoes are produced by private groups without following the proper production methods and treatments. Further, 40%–80% of the harvest could be lost due to pest attack. According to expert assessments, the actual annual losses caused by the Colorado potato beetle have amounted to more than 4 million tons in recent years in Russia.

Several varieties of *Bt* potato resistant to Colorado beetle developed by Monsanto are available worldwide. Two varieties Superior Newleaf and Russet Burbank Newleaf were approved in Russia for food in 2000, feed in 2003, and biosafety in 2002. IR and IR+VR potato varieties are very popular in Canada, Japan, Mexico, Philippines, and the United States.

During the last 15 years, using this approach, a number of new potato varieties, potato lines resistant to PVX virus, phytophthora, and Colorado beetle, and two biotech (*Bt*) potato varieties, namely, Elizaveta Plus and Lugovskoy Plus (Russian selection), were developed in Russia. The advantages of Russian *Bt* potatoes are reliability, cost-effectiveness, simplicity of cultivation, environmental benefits, and reduction of risks to the health of farmers, which is mainly due to no use of harmful insecticides. In 2005 and 2006, these varieties have received state approval for use in food and sale for use by the consumer (Skryabin, 2010) and have also entered into the State Register for protected breeding achievements in 2009.

Further studies have been conducted for the development of potato with improved nutritional value, e.g., a starch potato Amflora is developed only for industrial use and not for food applications. Traditionally, potato starch is used in paper manufacturing and paper coating. But the starch produced from conventional potatoes consists of two different components, amylopectin and amylose, though the latter one in many cases is not desirable. Biotechnology made it possible to suppress the gene responsible for synthesizing amylose developing a potato Amflora producing pure amylopectin—over 98% (GMO Compass, 2010). Amflora starch can be used specifically for industrial applications: for production of glossy paper and for spray with good adhesive properties. This potato (event EH92-527-1) was approved by the European Commission for commercial application in Europe.

14.6.4 GM Cotton

Cotton is known as nature's unique plant for food, feed, and fiber for clothing over the centuries. Now cotton production includes traditional cottonseed oil use and some new trends for home furnishings, industrial use, etc. Until the 1940s, cottonseed oil was the major vegetable oil produced in the United States, and now it takes the third place (annual production of 1 billion pounds) after soybean and corn oil. It represents about 5%–6% of the total America's fat and oil supply. Having a 2:1 ratio of polyunsaturated to saturated fatty acids, cottonseed oil is among the most unsaturated oils. Its fatty acid profile generally consists of 70% unsaturated fatty acids including 18% monounsaturated (oleic) and 52% polyunsaturated (linoleic) and 26% saturated, which primarily includes palmitic and stearic acid (National Cottonseed Products Association, 2002). Natural antioxidants (tocopherols) make this oil a product of interest due to its stability and long shelf life. Based on the levels of oleic, palmitic, and stearic acids, its technological properties are very profitable because of stability in the process of frying without the need for additional processing or the formation of trans-fatty acids.

GM cotton has occupied 24.7 million hectares in 2011 (James, 2011), in comparison with a global hectarage of 36 million hectares (FAO STAT, 2011) of cotton, so the biotech cotton grown in 2011 consists about 68% nowadays. The most popular in the world were insect-resistant *Bt* cotton with 17.9 million hectares (2011) and stacked cotton with 40% of all biotech cotton hectarage (2010). It is remarkable that from the beginning of cotton plantation in the United States in 1996 (insect resistant for bollworm family or lepidopteran pests), now nine countries including India, the United States, China, Pakistan, Australia, Argentine, Brazil, Burkina Faso, Colombia, Costa Rica, Mexico, Myanmar, and South Africa are cultivating GM cotton. India is the largest cotton-growing country in the world, and biotech cotton cultivation has occupied 10.6 million hectares, resulting in 88% of all produced cotton in this country. Notably, traditionally India was a producer of short, medium, and medium-long staple cotton, but as a result of hybrid technology in the 1970s and adoption of *Bt* technology in the last decade, a total cotton production has shifted from almost no long staple cotton in 1947 to 77% long staple crop in 2010–2011 (James, 2011). In accordance with ISAAA (James, 2011), roughly 67% of the cotton produced is consumed directly as food and feed with the remaining 33% used as fiber in the textile sector in India. In 2009–2010, cotton oil contributed 1.08 million tons to the total production of 7.88 million tons of edible oil from all Indian sources, which is equivalent to 13.7% (of total edible oil production). The cotton planting boom in 2011 was mainly in response to the peak prices for cotton link: US$ 4.51 per kilo compared with US$ 1.30 per kilo 2 years ago (James, 2011). In the last 10 years, India has transformed from a net importer of cotton to a net exporter of cotton. During these years, import in terms of market value has declined substantially from Rs. 2,029 crore (2001–2002) to Rs. 1,196 crore (2009–2010). Simultaneously, value of cotton export from India has increased manifold, i.e., Rs. 44 crore in 2001–2002 to Rs. 10,270 crore in 2009–2010.

The global status of other noncommercialized crops includes herbicide-tolerant biotech canola having 8.2 million hectares, which is 26% of global 31 million hectares (FAO STAT, 2009); herbicide-tolerant sugar beet (0.5 million hectares in Canada), herbicide-tolerant alfalfa (0.2 million hectares in the United States); 5,000 hectares of virus-resistant papaya (China); and sweet pepper and tomato (China) and virus-resistant papaya and squash (United States) with a total of less than 1,000 hectares (James, 2011).

14.7 Summary and Future Prospects

With the global population rising quickly toward an expected 9 billion in 2050, food, feed, and fiber demands are rising fast. It is obvious that when managed properly, GM crops can intensify agriculture sustainably. It is very difficult to predict exactly the road map of GM technology, but it could involve continued development of fast breeding technology; crops resistant to a wide range of pathogens, pests, and herbicides; novel genes related to biotic and abiotic stress; and fruits and vegetable products with longer shelf life. Further, new challenges are put forward in the development of food crops with better nutritional content, free from toxic or allergic substances, extra vitamins, and enhanced flavors, with fats and oils more suitable for human diet and for industrial purposes.

In response to new challenges, innovative crop technologies, such as plants as bioreactors for biopharmaceuticals; production of drugs and vaccines in plants; and introduction of new genetic systems, i.e., modifying photosynthesis or enabling crops to fix nitrogen, will also be continuously developed.

In summary, GM food is an excellent product of innovative technology that promises opportunity, but not a threat. GM technology is a very powerful tool and has a considerable potential for rapidly increasing the food production and the food sustainability.

References

Ahmed, F.E. 2002. Detection of genetically modified organisms in food. *Trends in Biotechnology* 20: 215–223.

Beachy, R.N. 2005. Controlling traits in transgenic plants: Tools that enhance value and reduce environmental release. In: *Agricultural Biotechnology: Beyond Food and Energy to Health and the Environment*, ed. A. Eaglesham, pp. 23–30. National Agriculture Biotechnology Council, Ithaca, NY.

Chassy, B.M. 2002. Food safety evaluation of crops produced through biotechnology. *Journal of the American College of Nutrition* 21(3): 166S–173S.

Chrispeels, M.J. and Sadava, D.E. 2002. *Plants, Genes, and Crop Biotechnology*, 2nd edn. Jones and Bartlett, Sudbury, MA.

Cremer, J. 2001. The performance of liberty in GM liberty link crops (maize, osr and sugar beet) as a broad spectrum herbicide in Europe (experience from 10 years testing). In: *Novel Approaches to Weed Control Using New Classes of Herbicides and Transgenic Plants Resistant to Herbicide*, eds. K.G. Skryabin and Y. Spiridonov, pp. 89–96. Nauka, Moscow, Russia.

Customs Union Commission (CUC). 2010. Uniform sanitary and epidemiological and hygienic requirements for products subject to sanitary and epidemiological supervision (control). Approved by Customs Union Commission No. 299, dated May 28, 2010.

Directive EU. 2001. Deliberate release into the environment of genetically modified organisms (2001/18) and regulation on genetically modified food and feed (1829/2003).

Duijn, V.J., Biert, V.R., Marcelis, B.H., Peppelman, H., and Hessing, M. 1999. Detection methods for genetically modified crops. *Food Control* 10: 375–378.

EFSA. 2007. Statement of the scientific panel on genetically modified organisms on the safe use of the *npt*II antibiotic resistance marker gene in genetically modified plants. http://www.efsa.europa.eu/fr/efsajournal/doc/742.pdf (accessed on March 25, 2013).

FAO. 2004. The state of food and agriculture 2003–2004: Agricultural biotechnology: Meeting the needs of the poor? Viale delle Terme di Caracalla, 00100 Rome, Italy. http://www.fao.org/docrep/006/Y5160E/Y5160E00.htm (accessed on March 25, 2013).

FAO STAT. 2009, 2011. http://faostat.fao.org/site/567/DesktopDefault.aspx?PageID=567#ancor

Federal Law of the Russian Federation. 2007. No 234-FZ (On amendments of the law of the Russian Federation for protection of consumer rights) and part two of the civil code of the Russian Federation.

Fichet, Y. and Brants, I. 2001. Glyphosate-tolerant sugar beet, an overview. In: *Novel Approaches to Weed Control Using New Classes of Herbicides and Transgenic Plants Resistant to Herbicide*, eds. K.G. Skryabin and Y. Spiridonov, pp. 68–75. Nauka, Moscow, Russia.

Finnegan, H. and McElroy, D. 1994. Transgene inactivation: Plants fight back! *Biotechnology* 12: 883–888.

Giesy, J.P., Dobson, S., and Solomon, K.R. 2000. Ecotoxicological risk assessment for Roundup® herbicide. *Reviews of Environmental Contamination and Toxicology* 167: 35–120.

Glazko, V.I. and Glazko, G.V. 2001. *Russian-English-Ukrainian Dictionary with definitions on Applied Genetics, DNA Technology and Bioinformatics*. Nora Print, Kyiv, Ukraine.

GMO Compass. 2006. Alternatives to antibiotic resistance markers. http://www.gmocompass.org/eng/safety/human_health/129.alternatives_antibiotic_resistance_marker_genes.html (accessed on March 25, 2013).

GMO Compass. 2010. EU: The first harvest of the *Amflora* potato. http://www.gmo-compass.org/eng/news/534.docu.html (accessed on March 25, 2013).

James, C. 2011. Global status of commercialized biotech/GM crops. ISAAA brief no.43. ISAAA, Ithaca, NY.

Kapila, J., De Rycke, R., Van Montagu, M., and Angenon, G. 1997. An *Agrobacterium*-mediated transient gene expression system for intact leaves. *Plant Science* 122: 101–108.

Medical and Biological Safety Assessment of Plant GMO's: Methodological Guidance. 2007. Medical and biological assessment of food derived from GMO's (MU 2.3.2.2306–07).

Miki, B. and McHugh, S. 2004. Selectable marker genes in transgenic plants: Applications, alternatives and biosafety. *Journal of Biotechnology* 107: 193–232.

National Cottonseed Products Association (NCPA). 2002. Cottonseed oil. http://www.cottonseed.com/publications/csobro.asp (accessed on March 25, 2013).

National Research Council, 2000. *Genetically Modified Pest-Protected Plants: Science and Regulation*. National Academy Press, Washington, DC.

National Research Council. 2010. *The Impact of Genetically Engineered Crops on Farm Sustainability in the United States*. National Academies Press, Washington, DC.

Newell, C.A., Rozman, R., Hinchee, M.A. et al. 1990. Agrobacterium-mediated transformation of *Solanum tuberosum* L. cv. "Russet Burbank." *Plant Cell Reports* 10: 30–34.

Potrykus, I. and Ammann, K. 2010. Transgenic plants for food Security in the context of development. *New Biotechnology* 27: 445–717.

Querci, M.C., Paoletti, C., and Van den Eedea, G. 2007. From sampling to quantification: Developments and harmonization of procedures for GMO testing in the European Union. In: *Collection of Biosafety Reviews*. ed. W. Craig, pp. 8–41. International Centre for Genetic Engineering and Biotechnology (ICGEB), Trieste, Italy.

Querci, M., Van Den Bulcke, M., Zel, J., Van den Eedea, G., and Broll, H. 2010. New approaches in GMO detection. *Analytical and Bioanalytical Chemistry* 396(6): 1991–2002.

ROSSTAT. 2011. Russian Federal State Statistics Service. http://www.finmarket.ru/z/nws/news.asp?id=3110712 (accessed on March 25, 2013).

Russian Federal Laws. 1996, amended in 2000 and 2010. On state regulation of genetic engineering activity; 1999. On sanitary and epidemiological well-being of the population; 2000. On the quality and safety of food product; 2007. On amendments to the federal law "On protection of consumer rights"; 2002. On environmental protection.
Russian Federation National Methodical Recommendations (Guidelines). 2006. For GM plants (soybean RR40-3-2, maize MON810 NK603, GA21, T25) quantitative identification in raw material and food with application of test kits.
Russian Federation National Standard. 2003. Raw material and food. Methodology of GM Plant Identification. PCR.
Sanford, J.C. 1990. Biolistic plant transformation. *Physiologia Plantarum* 79: 206–209.
Saraswathy, N. and Kumar, A.P. 2004. Protein engineering of δ-endotoxins of *Bacillus thuringiensis*. *Electronic Journal of Biotechnology* 7: 180–190.
Schmitt, F., Oakeley, E.J., and Jost, J.P. 1997. Antibiotics induce genome-wide hypermethylation in cultured *Nicotiana tabacum* plants. *Journal of Biological Chemistry* 272: 1534–1540.
Singer, M. and Berg, P. 1991. *Genes & Genomes*. University Science Books, Mill Valley, CA.
Singer, D.S., Liu, Z., and Cox, K.D. 2012. Minimizing the unpredictability of transgene expression in plants: The role of genetic insulators. *Plant Cell Reports* 31: 13–25.
Skryabin, K. 2010. Do Russia and Eastern Europe need GM plants? *New Biotechnology* 27(5): 593–595.
Tutelyan, V. 2007. *Genetically-Modified Food: Safety Assessment and Control*. Russian Academy of Medical Sciences, Moscow, Russia.
Tyshko, N.V., Zhminchenko, V.M., Pashorina, V.A. et al. 2011. Assessment of the impact of GMOs of plant origin on rat progeny development in three generations. *Problems of Nutrition* 80: 14–28.
USDA. 2011. Adoption of genetically engineered crops in the US. http://www.ers.usda.gov/Data/BiotechCrops/ (accessed on July 1, 2011).
Van den Eedea, G., Aarts, H., Buhk, H.J. et al. 2004. The relevance of gene transfer to the safety of food and feed derived from genetically modified (GM) plants. *Food and Chemical Toxicology* 42: 1127–1156.
Van Eck, J., Kirk, D.D., and Walmsley, A.M. 2006. Tomato (*Lycopersicum esculentum*). *Methods in Molecular Biology* 343: 459–473.
Vorobiev, A. 2004. Biotechnology and genetic engineering are priority trends of scientific and technical progress. *Vestnik Rossyskoy Akademii Meditsinskikh Nauk* 24: 8–11.
WHO (World Health Organization). 2005. Modern food biotechnology, human health and development: An evidence-based study. WHO, Geneva, Switzerland.
Williams, G.M., Kroes, R., and Munro, I.C. 2000. Safety evaluation and risk assessment of the herbicide roundup and its active ingredient, glyphosate, for humans. *Regulatory Toxicology and Pharmacology* 3: 117–165.

Part IV

Biotechnological Management of Crop Residues and By-Products of Agro-Industries

15

Production of Mushrooms

Pardeep K. Khanna and Shivani Sharma

CONTENTS

15.1 Introduction 510
15.2 Mushroom Production, Trade, and Consumption 511
  15.2.1 Global Scenario 511
  15.2.2 Indian Scenario 515
15.3 Mushroom Types 518
  15.3.1 Edible Mushrooms 518
  15.3.2 Specialty Mushrooms 518
  15.3.3 Medicinal Mushrooms 518
  15.3.4 Mycorrhizal Mushrooms 518
15.4 Button Mushroom Production 519
  15.4.1 Substrate Preparation 519
    15.4.1.1 Compost and Composting 519
    15.4.1.2 Compost Formulations 521
    15.4.1.3 Methods of Compost Preparation 521
15.5 Crop Production 527
  15.5.1 Growing Houses and Systems 527
  15.5.2 Spawn and Spawning 528
  15.5.3 Casing Soil and Casing 529
  15.5.4 Casing Materials 529
  15.5.5 Treatment of Casing Material 529
  15.5.6 Environmental Controls 529
15.6 Production Portfolio of Important Cultivated Types 530
  15.6.1 *Pleurotus* spp. 530
  15.6.2 *Lentinus edodes* 531
  15.6.3 *Auricularia polytricha* 532
  15.6.4 *Agrocybe aegerita* 532
  15.6.5 *Flammulina velutipes* or Enokitake 532
  15.6.6 *Calocybe indica* 532
  15.6.7 *Volvariella volvacea* 533
  15.6.8 *Ganoderma lucidum* 533
15.7 Nutritional Value 533
  15.7.1 Protein 534
  15.7.2 Fats/Lipids 535

15.7.3 Carbohydrates and Fibers 537
15.7.4 Minerals and Trace Elements 537
15.7.5 Vitamins 538
15.8 Medicinal Value 539
15.9 Growth Potential and Opportunities 544
15.9.1 Availability of Basic Raw Material 544
15.9.2 Suitable Technology 544
15.9.3 Suitable Environment and Diversification of Species 545
15.9.4 Low-Cost Labor 545
15.9.5 Domestic Consumption and Affordability 545
15.9.6 Export Potential 545
15.9.7 Source of Food Protein 546
15.9.8 Minimum Land Use/Intensive Use of Indoor Space 546
15.9.9 Recycling Waste Residues 547
15.9.10 Self-Employment 547
15.10 Summary and Future Prospects 547
References 548

15.1 Introduction

Mushrooms are produce of high quality and high economic value and are regarded as a valuable health food in modern society. These are a group of fungi that produce large fleshy fruiting bodies that come up from the branching mycelia infiltrating the soil or dead litter or in the wood of living or dead trees during monsoon season after thunder and rains. These macrofungi have nearly been always around, with a very long and interesting history. The early civilization of Greeks, Egyptians, Romans, Chinese, and Mexicans appreciated mushrooms both as delicacy and as therapeutic agents and used them often in religious ceremonies. Greeks regarded them as "strength food" for warriors while Romans considered them as "food of gods" and the Chinese prized mushrooms as "elixir of life." The mushrooms were worshipped (mycolatry) and consumed (mycophagy) as hallucinogens by the Mexican Indians. The Aztecs of South America referred a group of mushrooms as "teonanacatl" (god's flesh) and worshipped these as being divine. Currently, the number of recognized mushroom species has been reported to be about 14,000 (Hawksworth, 2001). Of these, about 7,000 species are considered to possess varying degrees of edibility and almost 3,000 species from 31 genera are regarded as prime edible mushrooms (Tewari, 2005). To date, only 200 of them are experimentally grown, 100 of them economically, about 20 cultivated commercially of which 10 have reached an industrial-scale production in many countries. Furthermore, about 2,000 are medicinal mushrooms with varieties of health benefits. The number of poisonous mushrooms is reported to be relatively small (approx. 1%), but estimates suggest that about 10% may have poisonous attributes while 30 species are considered to be lethal. Currently, there are still about 126,000 unknown species that are yet to be discovered (Chang, 2007).

The early history of mushroom cultivation is shrouded in antiquity. The historical record of intentionally cultivated types was given by Chang (1993). The Chinese were the first to artificially cultivate the tropical and subtropical mushrooms about 1,000 years ago, *Auricularia polytricha* in AD 600. *Flammulina velutipes* in AD 800 and *Lentinula edodes* in AD 1000, but authentic records are available for *Agaricus bisporus* whose cultivation was started in France around 1650, which spread later on to other European countries. The standard mushroom house came into existence in the United States around 1910. The first producer of pure culture virgin spawn was the American Spawn Company of St. Paul Minnesota headed by Louis F. Lambert, who advertised his brick spawn as "Lambert pure culture spawn" (Singh, 2011). Presently, mushroom cultivation has spread over 100 countries of the world and has assumed the proportion of an industrial venture and it forms an important agenda of research and development. Production of mushrooms worldwide has been steadily increasing, mainly due to contributions from developing countries such as China, India, Poland, Hungry, and Vietnam. Although major share of world mushroom production is of *A. bisporus*, the East Asian mushroom industry is diverse and produce different types of mushrooms particularly *Lentinus edodes, Pleurotus* spp., *A. polytricha,* and *Volvariella* spp. at a commercial scale. Mushroom cultivation has a special relevance to India as these represent unique exploitation of microbial technology for the bioconversion of huge quantities of available agricultural, industrial, and forestry wastes into a nutritious food. Mushrooms, which are a good protein source, may be an additional dietary supplement to make up for protein deficiency of the Indian population in addition to their satiety factor.

The cultivation of mushroom dominated by *A. bisporus* is now a well-organized industrial venture in India. Its introduction was made about five decades back, and now there are other varieties also under cultivation that include *Pleurotus* spp., *Volvariella* spp., *Calocybe indica,* and *L. edodes* that fruit under temperate, subtropical, and tropical climate.

15.2 Mushroom Production, Trade, and Consumption

15.2.1 Global Scenario

The global mushroom production as per FAO statistics (2009) has registered a steady increase from 3.89 to 6.53 million metric tons over a period of 10 years (1999–2009), thereby showing an increase of 67.7% in a

TABLE 15.1
World Production of Mushrooms

Country	MT (1999)	% Share	MT (2009)	% Share
China	2,183,006	56.04	4,680,726	71.62
United States	387,550	9.95	369,257	5.65
Netherlands	250,000	6.42	235,000	3.60
Poland	100,483	2.58	176,569	2.70
Spain	93,600	2.40	136,000	2.08
France	151,889	3.90	117,934	1.80
Italy	61,623	1.58	105,000	1.61
Canada	69,280	1.78	77,017	1.18
Japan	70,511	1.81	64,143	0.98
Indonesia	24,000	0.62	63,000	0.96
Ireland	64,800	1.66	57,747	0.88
Germany	60,000	1.54	52,000	0.80
United Kingdom	104,700	2.69	45,000	0.69
Australia	37,568	0.96	43,416	0.66
Belgium	46,000	1.18	42,208	0.65
India	14,000	0.36	38,930	0.60
South Korea	19,774	0.51	28,000	0.43
Iran	13,000	0.33	26,708	0.41
Hungry	15,901	0.41	21,950	0.34
Vietnam	14,000	0.36	20,091	0.31
Lithusania	2,500	0.06	14,056	0.22
South Africa	7,617	0.20	11,446	0.18
Israel	1,000	0.03	9,500	0.15
New Zealand	8,150	0.21	8,635	0.13
Switzerland	7,100	0.18	8,450	0.13
Denmark	8,300	0.21	7,890	0.12
Romania	4,000	0.10	7,317	0.11
Belarus	4,200	0.11	7,035	0.11
Ukraine	3,000	0.08	6,600	0.10
Thailand	9,455	0.24	6,420	0.10
All others	58,284	1.50	47,497	0.73
	3,895,291		6,535,542	

Source: FAOSTAT, 2011, World mushrooms and truffles: Production 1961–2009, United Nations.

decade (Table 15.1). The three major mushroom-producing regions of the world during this period have been Asia, United States, and EU accounting for nearly 94% of the total world production. The increased world production of mushrooms during the last decade has mainly been due to the major contributions by China, which has more than doubled its production during the last decade (FAOSTAT, 2009).

Asia's mushroom industry is more diverse, because the focus is more on specialty mushrooms. China has been the leading producer of mushrooms in the world with its share of more than 70% of the total mushroom production. Indonesia, Japan, India, Taiwan, and South Korea are the other countries in Asia contributing significantly to the pool of Asian mushrooms. In East Asian countries, specialty mushrooms are far more popular than *A. bisporus* (Chang, 2007). The Indian and Indonesian market, however, focus more on white button mushrooms. In China, 92 species were domesticated in 2002, and 50% of these were commercially cultivated (Mao, 2004).

In the United States and EU, mushroom production is overwhelmingly focused and nearly 100% dominated by *A. bisporus* (Gaze, 2005). In the United States, this variety has accounted for about 98% of total mushroom production for the last 10 years.

The production of mushrooms, primarily of *A. bisporus* in countries of EU, is mainly by the Netherlands, Poland, Spain, France, Italy, Ireland, Germany, United Kingdom, and Belgium. EU contributed about 15% to the world's total mushroom production in 2009. The percentage share of these countries in EU production chart is given in Figure 15.1.

The mushroom export and import has continuously increased in the last four decades. Global export of canned mushrooms amounted to 458,137 MT in 2008, and China accounted for 87% of total export volume of

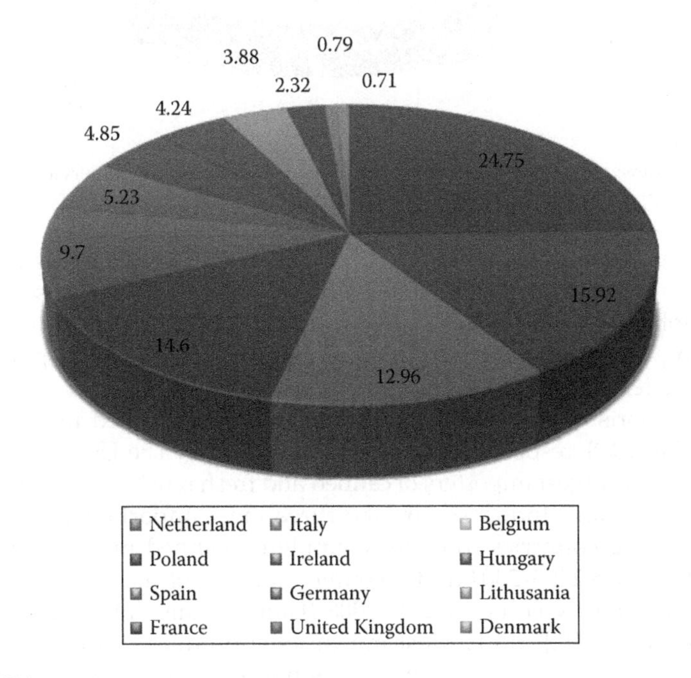

FIGURE 15.1
European Union production of mushrooms, 2008. (From Global Trade Information Service, Inc. [GTIS], World Trade Atlas Database, GTIS, Columbia, SC, 2008.)

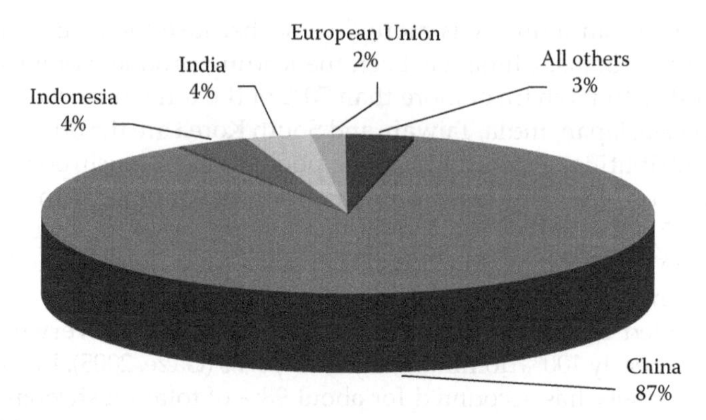

FIGURE 15.2
Canned mushrooms: global exports, 2008. (From Global Trade Information Service, Inc. [GTIS], World Trade Atlas Database, GTIS, Columbia, SC, 2008.)

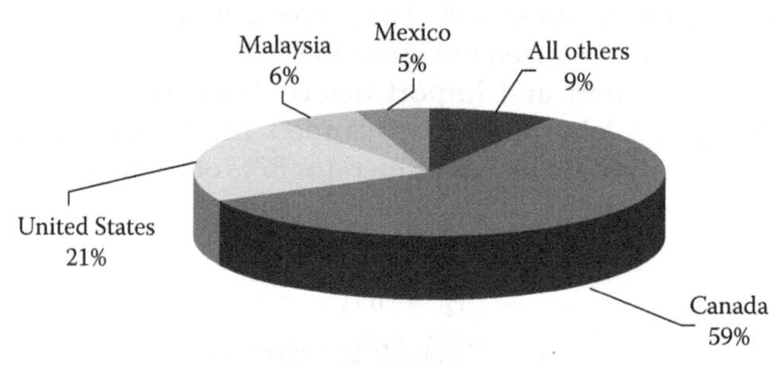

FIGURE 15.3
Fresh mushrooms: global exports, 2008. (From Global Trade Information Service, Inc. [GTIS], World Trade Atlas Database, GTIS, Columbia, SC, 2008.)

canned mushrooms (Figure 15.2) with Indonesia and India being the other major global exporter (GTIS, 2008). Canada and United States were the largest global exporter of fresh mushrooms in 2008 (Figure 15.3), together accounting for nearly 80% of the total exports, with most exports from both countries shipped to each other.

Global imports of canned and fresh mushrooms amounted to 292,267 and 90,879 MT in 2008, respectively (Figures 15.4 and 15.5). The United States and Russia were the largest importers of canned and fresh mushrooms (GTIS, 2008).

China, India, and Indonesia are the three most important global mushroom exporting countries from Asia supplying canned mushrooms to the U.S. market since 2003/2004 and accounted for 86% of the total U.S. canned mushroom import volume in 2007/2008 (United States International Trade Commission [USITC], 2010). Most mushrooms produced in EU are traded within member countries, and the EU is not a global exporter to non-EU countries. Among EU countries, the Netherlands has been the largest

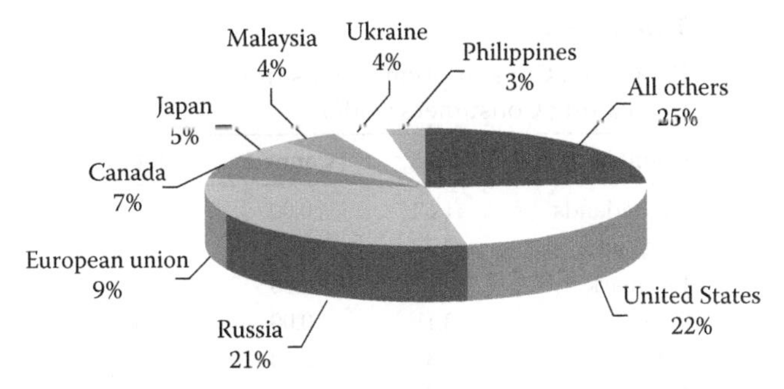

FIGURE 15.4
Canned mushrooms: global imports, 2008. (From Global Trade Information Service, Inc. [GTIS], World Trade Atlas Database, GTIS, Columbia, SC, 2008.)

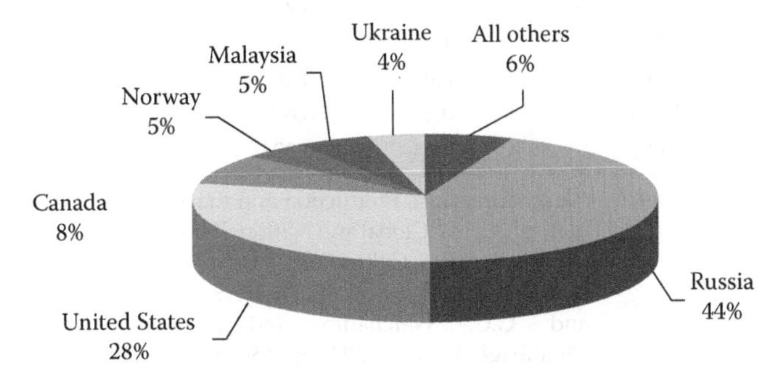

FIGURE 15.5
Fresh mushrooms: global imports, 2008. (From Global Trade Information Service, Inc. [GTIS], World Trade Atlas Database, GTIS, Columbia, SC, 2008.)

consumer, Poland the largest exporter, and United Kingdom the largest importer, whereas France and Spain were the largest producers as well as consumers of mushrooms.

China is the largest producer and consumer of mushrooms in the world followed by the United States and the Netherlands (Table 15.2). About 95% mushrooms produced in China are consumed locally, and only 5% of its total domestic production is exported and half of it is to Asian countries. The highest per capita consumption of mushrooms in 2007 was in the Netherlands (11.62 kg) followed by Belgium (4.46 kg) and Denmark (3.88 kg). Per capita consumption of mushrooms in India has increased over the period and has been estimated to be about 40 g.

15.2.2 Indian Scenario

India is not a major producer of mushrooms, but it has made its presence felt in the international trade during the last one and a half decade.

TABLE 15.2

Per Capita Consumption of Mushrooms (kg) by Leading Consumers (2007)

Country	Fresh	Canned	Total
Netherlands	11.62	0.00	11.62
Belgium	3.77	0.69	4.46
Denmark	3.22	0.66	3.88
Spain	3.11	0.00	3.11
U.K.	2.81	0.20	3.01
France	2.58	0.14	2.72
Germany	1.27	1.20	2.47
Canada	1.71	0.59	2.30
Italy	1.56	0.06	1.62
United States	1.27	0.22	1.49
China	1.16	0.00	1.16
Japan	0.60	0.26	0.86
Poland	0.35	0.00	0.35
India	0.04	0.00	0.04

Source: Wakchaure, G.C., Production and marketing of mushrooms: Global and National Scenario, In: *Mushrooms: Cultivation, Marketing and Consumption*, M. Singh, B. Vijay, S. Kamal, and G.C. Warchaure, (eds.), DMR, Chambaghat, Solan, 2011, pp. 15–20.

The production pattern of mushrooms is slightly less skewed toward *A. bisporus*, as sizeable quantities of a few other varieties are also cultivated. The production system for the most common white button mushroom consists of seasonal growing as well as hi-tech environmentally controlled cultivation. The mushroom production has registered more than 20-fold increase during the last two decades from nearly 5,000 tons in 1990 to about 113,000 tons in 2010 of which *A. bisporus* accounts for 85%–90% followed by other tropical/subtropical varieties, i.e., oyster, milky, and paddy straw mushroom, which contributes to 10%–15% of the total production. The principal white button mushroom growing states are Punjab, Haryana, Himachal Pradesh, Uttarakhand, Uttar Pradesh, Maharashtra, Andhra Pradesh, and Tamil Nadu (Table 15.3). The concentrated areas of production for button mushroom are the North Western temperate regions for seasonal growing in winter months. During the last decade, several medium- to large-sized integrated mushroom units have also come up in different states of the country. However, 30%–40% of the production is still being done under natural indoor climate conditions during the winter months.

TABLE 15.3

Mushroom Production by Tons in India by State (2010)

S. No.	State	Button	Oyster	Milky	Other Mushroom	Total Production
1.	Punjab	58,000	2,000	0	0	60,000
2.	Uttarakhand	8,000	0	0	0	8,000
3.	Haryana	7,175	0	3	0	7,178
4.	Uttar Pradesh	7,000	0	0	0	7,000
5.	Tamil Nadu	4,000	2,000	500	0	6,500
6.	Himachal Pradesh	5,864	110	17	2	5,993
7.	Orissa	36	810	0	5,000	5,846
8.	Andhra Pradesh	2,992	15	15	0	3,022
9.	Maharashtra	2,725	200	50	0	2,975
10.	Kerala	0	500	300	0	800
11.	Jammu and Kashmir	565	15	0	0	580
12.	Goa	500	20	0	0	520
13.	Bihar	400	80	0	0	480
14.	Nagaland	0	75	0	250	325
15.	Jharkhand	200	20	0	0	220
16.	Assam	20	100	5	0	125
17.	Rajasthan	100	10	0	10	120
18.	Tripura	0	100	0	0	100
19.	West Bengal	50	50	0	0	100
20.	Manipur	0	10	0	50	60
21.	Mizoram	0	50	0	0	50
22.	Chhattisgarh	0	50	0	0	50
23.	Meghalaya	25	2	0	0	27
24.	Arunachal Pradesh	20	5	0	1	26
25.	Karnataka	0	15	10	0	25
26.	Madhya Pradesh	10	5	0	0	15
27.	Gujarat	0	5	0	0	5
28.	Sikkim	1	2	0	0	3
Union Territories						
1.	Andaman and Nicobar Islands	0	100	0	0	100
2.	Chandigarh	0	0	0	0	0
3.	Dadra and Nagar Haveli	0	0	0	0	0
4.	Daman and Diu	0	0	0	0	0
5.	Delhi	3,000	50	20	0	3,070
6.	Lakshadweep	0	0	0	0	0
7.	Puducherry	0	0	0	0	0
	Total	1,00,683	6,399	920	5,313	1,13,315

Source: RMCU, Directorate of Mushroom Research, Solan, India, 2010.

15.3 Mushroom Types

Mushrooms belong to the macrofungi that can appear either below ground (hypogeous) or above ground (epigeous). Mushrooms in general can be placed under the following four categories.

15.3.1 Edible Mushrooms

Edibility may be defined by criteria that include absence of poisonous effects on humans and desirable taste and aroma. Edible mushrooms are consumed by humans for their nutritional and occasionally medicinal value. These fungal species are either harvested wild or cultivated. Some of the examples of commercially cultivated edible mushrooms include *A. bisporus*, *Pleurotus* species, *Volvariella volvacea*, *C. indica*, and *Hypsizygus tessulatus*.

15.3.2 Specialty Mushrooms

These mushrooms represent an array of species with diverse flavors, textures, and colors that have a documented food and medicinal value. Specialty mushrooms differ significantly from common button mushroom in appearance and the substrate requirement for their cultivation. Some examples of specialty mushroom include *L. edodes* (shiitake), *A. polytricha* (black ear mushroom), *Pleurotus* spp. (oyster), *F. velutipes* (enokitake), *Tremella fuciformis* (snow fungus), *Hericium erinaceus* (pimpom), *Stropharia rugosoannulata* (wine cap), and *Pholiota squarrosa* (shaggy scaly cap).

15.3.3 Medicinal Mushrooms

These mushrooms include those varieties that are used for their medicinal value or nutraceutical potential. Approximately 300 mushroom species are known to have these properties, and another 1,800 species have been identified with perspective medicinal properties. *Ganoderma lucidum* (reishi), *L. edodes* (shiitake), *H. erinaceus* (lion's mane), *Grifola frondosa* (maitake), *A. auricula* (black ear), and *Schizophyllum commune* (split gill) are few examples of the cultivated mushroom species that have been analyzed for medicinal and nutraceutical values. Several mushroom extracts, polysaccharide–Krestin (PSK), polysaccharide–polypeptide (PSP), active hexose correlated compound (AHCC), lentinan, and schizophyllan, are considered nutraceuticals, and some countries regulate them as pharmaceuticals.

15.3.4 Mycorrhizal Mushrooms

These fungi live in the soil in symbiotic association with the roots of vascular plants of different tree species in woodlands and in forest ecosystems. Many of the mycorrhizal species of mushrooms are edible. Around 5,000 species of fungi are known to form ectomycorrhizae with about 2,000 woody hosts, and these belong to Hymenomycetes in Basidiomycotina, and some are Gasteromycetes and possibly Ascomycetes (Trappe, 1962). These predominantly come under the genera *Amanita, Boletus, Tricholoma, Russula, Laccaria, Lactarius, Cantharellus, Cortinarius*, etc. These mushroom species cannot be cultivated indoor like other cultivable mushrooms. Outdoor/field cultivation technology has been developed for the cultivation of *Boletus granulatus* Linn. ex Fr., *B. edulis* Schaeff. ex Fr., *Tricholoma matsutake* (Ito et Inai) Sing., *Tuber melanosporum* Vitt. (Perigord black truffle), and *T. magnatum* Pico ex Vitt. (Rambelli, 1983).

15.4 Button Mushroom Production

White button mushroom (*A. bisporus*) is the most widely grown mushroom all over the world. Seasonal cultivators are producing this mushroom using long method of composting (LMC) and harvest about 10–15 kg of white button mushroom/100 kg compost. Environmentally controlled units are cultivating this mushroom round the year in climate-controlled cropping rooms. The yield of 20–25 kg/100 kg compost is harvested by these units. Many units because of strict environmental audit in the western countries have shifted to complete indoor composting method for the preparation of substrates, and a higher yield of 25–30 kg/100 kg compost is obtained under controlled growing conditions. The cultivation process of this mushroom includes preparation of the substrate (compost) and the crop production. Under this section, all these phases of button mushroom production have been described.

15.4.1 Substrate Preparation

The substrate preparation has undergone various innovations/improvements depending upon environmental protection laws in many developed countries. Still in the developing nations of Asia, the substrate is prepared using primitive LMC.

15.4.1.1 Compost and Composting

Cultivation of button mushroom requires the production of composted substrate produced as a result of aerobic thermogenesis by a succession of microflora. In the developed countries of Europe and America, mushroom farming is a hi-tech industry, and compost substrate is prepared either by using short methods of compost production or by indoor composting. Chinese and Indian growers excepting a few export oriented units are using methods considered rather primitive and are of low technology. Traditionally, in India, compost is prepared using base material consisting of cereal straw and/or horse/chicken manure mixed with various supplements/nutrients. Majority of the seasonal growers in India do not have compost pasteurization facility and still use the LMC. The materials used in compost preparation and their specific role have already been outlined (Hayes, 1977). The classification and description of the functional role of each category of these nutrients for composting is given as follows.

15.4.1.1.1 Base Materials

Different agricultural by-products like cereal straw (wheat, paddy, oat, barley), maize stalks, and sugarcane bagasse are being used to provide a reservoir of cellulose, hemicellulose, and lignin. These materials also provide proper physical texture and aeration during the composting process for the buildup of a succession of microflora essential for fermentation and subsequent colonization of the compost by the mushroom mycelium.

15.4.1.1.2 Supplements for Activating Fermentation

The following materials are employed for this purpose:

1. Animal manures: In India, these include horse and chicken manure used as a source of nitrogen to the compounding mixture. The N-content varies from 1% to 5% and, therefore, must be estimated in each lot. These manures also provide little carbohydrates that are released slowly during composting. In addition to nutrients, manures contribute greatly to the final bulk density of composts. Use of chicken manure is a common practice in short method of composting (SMC), however, due to difficulty in the availability of good quality horse manure. The use of chicken manure in LMC should be discouraged or, if to be used, it should be steam sterilized before incorporation.
2. Carbohydrate-rich nutrients: These include molasses, potato wastes, apple and grape pomace, and malt sprouts. These are added for the establishment of initial bacterial flora in the compost as well as minor source of nitrogen (N content 1.5%).
3. Nitrogenous fertilizers: These include organic/inorganic synthetic fertilizers like urea, ammonium sulfate, and calcium ammonium nitrate. Their N-content vary between 25% and 46% and used to maintain a correct C/N balance in the compost for the rapid growth of thermophilic microflora, as nitrogen is released quickly during fermentation raising the temperature of composting pile.

15.4.1.1.3 Concentrate Meals

In this group, both nitrogen and carbohydrates are available relatively slowly. These include wheat or rice bran, dried brewers' grains, seed meals of cotton, soya, mustard, and castor, under Indian conditions.

The seed meals have also been extensively used on Indian farms for post-composting supplementations to improve yield as their N-content vary from 3% to 12%.

15.4.1.1.4 Supplements for Rectifying Mineral Deficiencies

These include fertilizers like potash, superphosphate, and trace metal mixtures. Gypsum is also included in this group, which serves to precipitate suspended colloidal materials and obviate greasiness.

The choice of materials within each category is largely determined by cost factor, local availability, and its effect on the ecological succession of microflora.

15.4.1.2 Compost Formulations

Based upon the quantity and degree of availability of these nutrients, numerous LMC formulations based on cereal straw have been developed and standardized (Table 15.4). The use of chicken manure/horse manure in these formulations is recommended where pasteurization facilities are available.

A mixture containing 1.4%–1.5% N of the dry matter is the target at the beginning of the composting, which provides compost with C/N of 16–18:1 and N content between 1.75% and 2.0% at the completion of the process. Laborde (1991) has indicated that formulations for white button mushroom compost should be so designed that the compost mixtures have the different minerals on dry weight basis such as N (1.7%–1.8%), P_2O_5 (1.2%–1.5%), K_2O (2.0%–2.3%), CaO (1.5%–3%), and MgO (0.4%–0.5%).

15.4.1.3 Methods of Compost Preparation

Two types of technologies used in India for compost preparation are LMC and SMC. The SMC introduced around the world by Sinden and Hauser (1950, 1953) is being used with some modification by large/commercial and industrial units. But marginal and small-scale seasonal growers use LMC. Chicken manure has almost replaced horse manure as one of N-source in SMC.

15.4.1.3.1 Long Method of Composting

This is completely an outdoor process, and compost is prepared in 18–24 days in a single outdoor phase. The long duration of this method of composting is not only time consuming but also results in considerable loss of dry matter, and the compost is prone to attack by many disease-causing organisms and pests. The yield is low on such composts. It is regarded as a primitive method. The detail of one of most prevalent method of LMC procedures recommended by Punjab Agricultural

TABLE 15.4
Some Compost Formulations Used in India

Long Method		Short Method	
DMR, Solan			
Wheat and Paddy (1:1)	300 kg	Wheat straw	300 kg
CAN	9 kg	Wheat bran	15 kg
Urea	25 kg	Chicken manure	125 kg
Wheat bran	20 kg	BHC (10%)	125 g
Gypsum	20 kg	Gypsum	20 kg
PAU, Ludhiana			
Wheat and Paddy (1:1)	300 kg	Wheat and Paddy (1:1)	300 kg
CAN	9 kg	Chicken manure	60 kg
Urea	3 kg	CAN	6 kg
Superphosphate	3 kg	Superphosphate	3 kg
Muriate of Potash	3 kg	Muriate of Potash	3 kg
Wheat bran	15 kg	Wheat bran	15 kg
Gypsum	30 kg	Gypsum	30 kg
BHC (5%)	250 g	Lindane dust or	250 g
Furadan 3G	150 g	BHC (5%)	250 g
IIHR, Bangalore			
Paddy straw	150 kg	Paddy straw	300 kg
Maize stalks	150 kg	Chicken manure	150 kg
Ammonium sulfate	9 kg	Wheat bran	12.5 kg
Superphosphate	9 kg	Gypsum	10 kg
Urea	4 kg		
Rice bran	50 kg		
Cotton seed meal	15 kg		
Gypsum	12 kg		
Calcium carbonate	10 kg		
Mushroom Research Laboratory, Solan			
Wheat straw	1000 kg	Wheat straw	1000 kg
CAN	30 kg	Chicken manure	400 kg
Superphosphate	25 kg	Brewer's manure	72 kg
Urea	12 kg	Urea	14.5 kg
Sulfate of potash	10 kg	Gypsum	30 kg
Wheat bran	100 kg		
Molasses	16.6 l		
Gypsum	100 kg		
RRL, Srinagar			
Wheat straw or paddy straw	300 kg		
Molasses	12 kg		
Urea	4.5 kg		
Wheat bran	50 kg		
Muriate of potash	2 kg		
Cotton seed meal	15 kg		
Gypsum	15 kg		

University (PAU), Ludhiana, Punjab, India, for the seasonal growers is given as follows (Garcha and Khanna, 2003):

1. Requirements: Composting floor that is cemented or brick lined, preferably covered provision to store raw materials, wooden or iron molds (three in no.) of required height and width, provision of clean water and ingredients as per the formulation of PAU, India.

   Wheat straw is spread over the platform. Wetting is done for 2 days so that the straw attains a moisture level of around 75%. The fertilizers and wheat bran are mixed separately, moistened, and kept covered with a gunny bag sheet or polythene sheet for 1 day. Subsequently, this mixture is broadcast on the wetted straw mixed with forks and made into a heap (0 day). The wooden/iron boards are arranged in the form of a rectangular block and the entire mixed material is filled into it. Normally a pile of 5 ft × 5 ft ×5 ft (l × b × h) is usually achieved starting with 300 kg of base material. If the quantity is more, the boards are moved lengthwise to accommodate it so that the width and height of the pile remain the same. Too compact a pile is not desirable as anaerobic conditions would set in, the temperature of the pile will rise beyond 75°C, and secondary fermentation will lead to production of undesirable substances. Leaching of the water from the pile is avoided by carefully utilizing the water during wetting or pouring it back on top of the pile.
2. Turning schedule: Seven turnings are given to the composting pile by following the schedule of 4, 8, 12, 15, 18, 21, and 24 days along with the mixing of remaining ingredients during these turnings. The objective of the turning is to have equal amount of aeration in each portion of the pile so that uniform decomposition of the straw in the preparation of selective substrate is achieved. During each turning, one foot portions of the pile from the four sides and from the top is removed, kept separately, and moisture level is adjusted with clean water. The rest of the pile is dismantled. A new stack is then made using the wooden/iron molds where side and top portion of the pile material is placed at the center while bottom and core portion comes on the top and the sides. In this way, pile of 0 day is turned inside out on the first turning. This procedure of turning is repeated at each turning, and some additives or supplements are added as mentioned in the following:

Day 4 (first turning)	Molasses are added
Day 8 (second turning)	Plain turning is given and moisture level in the side and top portion is adjusted, if required
Day 12 (third turning)	At this turning, required quantity of gypsum is added
Day 15 (fourth turning)	Plain turning with no addition of any supplement is given
Day 18 (fifth turning)	Turning given and required quantity of Furadan 3G added
Day 21 (sixth turning)	No addition of ingredients during this turning
Day 24 (seventh turning)	The pile is dismantled, required quantity of γ-BHC is added

The compost is checked for the smell of ammonia and can be kept for 1–2 days more if the smell is detected. This substrate is now ready for spawning and filling. The whole composting process is accomplished in 26 days (−2, 0, 4, 8, 12, 15, 18, 21, and 24 days) and about 35% dry matter is lost. If paddy straw is to be used along with wheat straw (50:50, w/w), its addition to the composting pile is recommended on the eighth day of turning (Kaur and Khanna, 2001), but care should be taken to keep the total moisture content of the pile between 62% and 65%.

The compost prepared using the LMC should be dark brown in color, free from ammonia smell, has around 70% moisture, pliable (non-greasy and non-sticky), pH between 7.0 and 8.0, and should be free from insects and nematodes.

Many studies to improve the long method of compositing have been carried out, and based on these studies, a number of compost formulations are today in use in different parts of the country with the sole objective of achieving a co-relation between compost quantity and yield of mushrooms.

15.4.1.3.2 Short Method of Composting

This method of composting was introduced in early 1950s, and since then, numerous modifications and advances have been made to improve the productivity of the compost. For composting, cemented covered or uncovered compositing yards pasteurization in the form of bulk pasteurization room/tunnel or peak heating rooms are the infrastructural requirements. Compost prepared by SMC takes about 14–18 days, and the whole process is accomplished in two phases.

Phase I: This phase is an outdoor process (8–10 days) that helps to build high temperature, increased ammonia production that in turn increases the degradability of cellulose, a cell wall polysaccharide of plants, and starts with the wetting of wheat straw and chicken manure. These materials are mixed and a stack is made. After 2 days, the stack is broken; water adjusted by carefully avoiding leaching and a smaller stack is made. On 0 day, the stack is break opened, total quantity of urea is added, and a high aerobic stack of 5 ft × 5 ft length is made. Three turnings are given each after 2 days' duration with the help of turners and stacker, and gypsum is added. The ingredients are allowed to decompose outdoor under uncontrolled conditions. High temperature (70°C–80°C) is built up in the core region that results in the charring of ingredients that gives distinct brown color to the compost. The temperature around the core region ranges between 50°C and 60°C, which harbors huge population of thermophilic microflora that serves as the inoculum for the whole pile during turnings to bring about decomposition of straw ingredients.

Phase II: It is an indoor (6–8 days) biological process, which has twin objectives of pasteurization and conditioning of the compost at specific temperature for defined duration and is accomplished in a bulk chamber/tunnel or in a peak heat room.

During pasteurization, compost temperature is raised to a certain limit for a definite period of time. Compost is pasteurized between 57°C and 62°C for 6–8 h. Activity of compost is a very important factor during phase II. If it is active (green compost), its temperature rises to 57°C due to self-generation of heat or steam through boiler is supplied.

During the conditioning of the compost, ammonia generated through the activity of thermophilic microflora is converted into microbial protein, which is utilized by the mushroom mycelium during its growth at 25°C at which these microbes become inactive. Excess of ammonia is released into the atmosphere. An optimum temperature range at which these microflora grow and ammonia generation takes place is close to 45°C. The whole of compost mass is, therefore, brought to this temperature range (40°C–50°C) and conditioned for 3–5 days or till ammonia level is below 10 ppm under continuous supply of fresh air. Pasteurization and conditioning during phase II of SMC thus make the compost suitable and selective for *A. bisporus*.

The phase II of composting is accomplished either in trays by peak heating them or in a bulk chamber by maintaining carefully a very systematic and controlled regime of temperature and time, aeration and ventilation, and ammonia level. All these parameters are monitored and controlled by using sensors and probes inside the area. It has been estimated that during phase II, about 25%–30% of the compost weight is lost. Therefore, to produce 20 tons of compost from the tunnel (36 ft × 9 ft × 12 ft) or peak heating room (24 ft × 16 ft × 8 ft) to accommodate 250 standard trays, roughly 28 tons of compost should be filled for which one may start with 12 tons of raw material and for which a composting yard of 100 ft × 40 ft is large enough (Vijay and Gupta, 1995).

15.4.1.3.3 Recent Trends in Composting

A number of modifications/innovations/advances have been introduced in composting process that involve outdoor phase I, as it has been increasingly subjected to environmental audit. One of these innovations has led to the concept of indoor composting involving single phase of two phase temperature regimes, where use of phase II bulk pasteurization is also extended to phase I (formerly outdoor) as well. The concept of indoor composting has been followed with considerable success in countries like France and Italy (INRA method) and also in England, Holland, Australia, and Belgium (Anglo-Dutch method).

In the INRA indoor method (Laborde, 1991), phase I is carried out indoors at around 80°C for 2–3 days and then phase II at 50°C for 5–7 days except for a short peak heat at 58°C. Re-inoculation with pasteurized compost or with a thermophilic fungus *Scytalidium thermophilum* becomes necessary. Alternately, phase I compost is prepared in bunkers where top and side layers lose sufficient heat to prevent entire composting mass to reach very high temperature, and this cooler compost is mixed with hotter compost as an inoculant when compost is removed from the bunker (Miller, 1997). The INRA method has been named variously as environmentally controlled composting (ECC) (Gulliver et al., 1991), rapid indoor composting, or aerated rapid composting or express preparation of substrate (Laborde et al., 1993).

The Anglo-Dutch method is known as low-temperature composting. It is a single-stage process where the compost is kept in the tunnel at 45°C–50°C (47°C) for about a week so that abundant thermophilic flora develops. There is a short pasteurization phase carried out at about 60°C for 4–6 h (Gerrits and Van Griensen, 1990; Nair and Price, 1991). This process takes 6–9 days and permits excellent control, substantial saving of raw materials, and good productive compost and also increases the biomass of thermophilic microorganisms (Miller, 1997). The earlier two methods are prevalent in most parts of the world giving almost equal yields. The Directorate of Mushroom Research, Solan, India, has developed another method combining these two methods where composting is carried out in aerated bunkers.

For preparing compost by indoor composting (aerated bunkers), the ingredients are thoroughly wetted and mixed, spread over the composting yard

TABLE 15.5

Process of Composting and Their Attributes

Attributes	Long Method	Short Method	Indoor Method
Days required for compost preparation	28–30	16–20	10–12
Selectivity	Partial	Complete	Complete
Average yield (kg/100 kg compost)	10–15	20–25	25–30
Effect on environment	Polluting	Less polluting	Nonpolluting
Average compost production/ton of straw	1.75–2.0 tons	2.0–2.5 tons	3.0–3.5 tons
Average final N% in compost	1.75–2.0	2.0–2.2	2.2–2.5
Pre-wetting area for 20 ton compost output	—	60 ft × 40 ft	60 ft × 40 ft
Infrastructure required for 20 ton compost output	Outdoor composting yard (60 ft × 40 ft)	Covered composting yard (60 ft × 40 ft) +1 tunnel (36 ft × 9 ft × 12 ft)	Two phase I bunkers (45 ft × 10 ft × 10 ft + one phase II tunnel (36 ft × 9 ft × 12 ft)
Man days required for 20 ton compost output	30–35	20–25	15–20
Power requirement for 20 ton compost output	Nil	700–900 kW	800–1,000 kW
Compost handling equipments required (large farm) (>500 TPA)	Nil	Turner, filling line, hopper, regulator, Bobcats	Filling line, hopper regulator, Bobcat

Source: Vijay, B, Methods of compost preparation for white button mushroom. In: *Mushrooms: Cultivation, Marketing and Consumption*, M. Singh, B. Vijay, S. Kamal, and G.C. Warchaure (eds.), DMR, Chambaghat, Solan, 2011, pp. 49–68.

(8–10 ft height) to increase the bulk density of ingredients. The mixed ingredients are made into a high aerobic heap and left for 48 h. The material is then transferred to phase I bunker where a temperature of 70°C–75°C is obtained through the introduction of air. After 3 days of partial fermentation in phase I bunker, the entire compost mass is transferred to another bunker for another 3–4 days under the same set of conditions. Once the phase I is over, compost is transferred to phase II tunnel for the usual phase II operation as described for SMC that is to be completed in 6–7 days. It is desirable not to allow buildup of temperature beyond 75°C. The three methods of compost preparation described earlier produce a selective substrate for *A. bisporus*. The comparison of the process attributes of these three methods is given in Table 15.5.

15.5 Crop Production

The crop production includes growing system, spawning and spawn run, casing and case run, cropping, harvesting, and management.

15.5.1 Growing Houses and Systems

The low-cost rural cropping rooms for seasonal growing of this variety are generally made with thatched roof and a false polythene ceiling. The entry to the growing shed is on one end and the exhaust vents are provided on the opposite end. The walls of the shed are also prepared out of bamboos, paddy, or sarkanda (dried stems of a weed plant), and these are also covered with high-density polythene sheet covering from the outside. The polythene sheet is generally rolled to provide natural exchange of air in the growing rooms during the cropping period of this mushroom. Another type of seasonal growing rooms consists of simple cropping rooms having cemented flooring, brick walls, and provision for cross ventilation of air.

The semi-controlled growing houses are generally cold stores that have been converted to mushroom growing houses. Since such cold stores are already insulated and provided with cooling units, the only constraint for mushroom growing is provision of fresh air during cropping period. The cold stores are thus modified by providing forced aeration using blower and ducting system. The fully controlled mushroom houses are constructed to provide mechanically controlled temperature, humidity, and fresh air supply. There are three common systems used for the cultivation of button mushroom in India. These are wooden trays or box system, the shelf system, and the bag system.

The wooden trays are generally empty fruit boxes, which are slightly modified and used as growing container for button mushroom. Some small-scale seasonal growers, however, continue to use the trays systems of cultivation by making use of empty fruit boxes (apple/tomato crates), which are available in bulk during the growing season of button mushroom. Moreover, these fruit crates can be tiered one above the other. The filling depth in these wooden crates is generally kept at 15–20 cm.

The shelf system, which is very popular among the medium- and large-scale seasonal growers, is fabricated either out of bamboos (low-cost) or angle iron frames. The shelf system consists of shelves running the entire length of the cropping room with four to five vertical tiers having a gap of about 50 cm between two tiers. The width of the shelves is approximately 1.25 m, and a gap of 0.75 m is kept between adjacent shelves to allow free movement, watering, and picking of mushrooms. For making the bed, polythene sheet (150 gauge thick) is spread over the bamboo/iron frame, and the compost is spread over this sheet. The compost depth is kept between 15 and 20 cm.

The third type of growing system followed in India is bag system. The bag system has become more popular within the environmentally controlled unit, and some large seasonal growers as the bags are cheaper (compared to trays), easily disposable, and do not pose threat of contamination to the subsequent crop. Moreover, the chances of aerial contamination are also minimized, as the bags are kept closed during the spawn run period. A compost depth of 30–37.5 cm has been recommended using 60 cm × 45 cm bags of 150 gauge thickness (Dhar et al., 1985; Upadhyay and Vijay, 1988). The bags are kept on shelves made out of bamboos/iron after spawning.

15.5.2 Spawn and Spawning

Spawning of the prepared compost is carried out to provide maximum number of loci for the spread of mycelium. For grain spawn, four methods of spawning have frequently been used (Shandilya et al., 1974; Jain and Singh, 1982; Khanna and Kapoor, 2007). These are spot spawning, planting of lumps of grain spawn about 5 cm below the surface of compost about 20–25 cm apart; surface spawning, spawn is spread at the top surface of the compost and slightly mixed into the upper layer of the compost; layered spawning, the compost is successively filled with about 5 cm depth and spawned in 2–3 layers; and thorough spawning, where the spawn is mixed in the compost and then filled in the container. The optimum dose of spawn has been worked out by Patil and Shinde (1983), who observed that irrespective of the method of spawning, a spawn rate of 0.5%–0.75% fresh weight of compost is optimum for button mushroom. Bisht and Singh (1986) proposed "super spawning" or "active mycelium" spawning that involved spawning of mushroom compost using colonized compost free from contaminants as inoculum for fresh compost.

During spawn run, the temperature in the growing room is maintained at 23°C–24°C with 80%–85% humidity, and high CO_2 concentration is allowed to build up by keeping the fresh air ventilation closed. Under such conditions, it takes about 12–14 days for the spawn to colonize the compost.

Spawn run period, however, depends upon the quality and moisture content of compost, strain used, and the environmental conditions (in case of seasonal growing).

15.5.3 Casing Soil and Casing

The practice of covering uniformly the spawn colonized compost surface with a suitable soil mixture is called casing, which helps to retain moisture and support the growing pinheads and help them to become button at a quick rate of growth.

15.5.4 Casing Materials

Casing follows the spawn run phase and is essential to induce fruiting in button mushroom. A variety of casing materials are used the world over. The texture, bulk density and water holding capacity, pH, and electrical conductivity are the major physical and chemical factors that influence the yield.

Peat available in abundance in west is the widely used material for casing due to its unique properties to adsorb and release water quickly, enough porosity for air exchange, and its availability. Consequently, few materials like farm yard manure, biogas plant slurry, spent compost, and loamy-sand soil are used in solo or in different combinations as casing material in different regions of India.

15.5.5 Treatment of Casing Material

The casing materials may harbor many pests and pathogens. Before these are used by the growers, they are treated to selectively kill the harmful flora without affecting useful ones. The industrial growers having the facility of bulk chamber pasteurize the casing mixture using steam in the bulk chamber at 62°C–65°C for 6 h. Alternatively, the seasonal growers also use chemical disinfection by making use of 4% solution of formaldehyde, which has been found quite effective (Khanna and Kapoor, 2007).

Casing soil in button mushroom is a nutritionally weak medium compared to compost. Reduction in temperature after 1 week of casing by 8°C–10°C, increased ventilation, and reduced humidity provide the necessary stress for the button mushroom mycelium to induce fruiting.

15.5.6 Environmental Controls

Environmental conditions as maintained during spawn run, i.e., air temperature (23 ± 1°C), humidity (90%), and high CO_2 (1,500 ppm), are maintained in the growing room during the first week of casing after which the temperature is lowered to 15°C–17°C with fresh ventilation (30%) for 1–2 h, which reduces CO_2 level (300–1,000 ppm) and humidity to 85%. The optimal environmental parameters during various stages of cropping are given in Table 15.6.

TABLE 15.6

Optimum Environmental Parameters during Various Stages of Cropping

Parameters	Spawn Run	Case Run	Pinning	Cropping
Air temperature (°C)	23 ± 1	23 ± 1	15–17	15–17
Compost temperature (°C)	24–25	24–25	16–18	16–18
CO_2 concentration (ppm)	About 15,000	10–15,000	300–1,000	300–800
Relative humidity	90%–95%	90%–95%	About 85%	About 85%
Air changes	Nil	Nil	Fresh air (30%)	Four changes/h
	No fresh air (100% recirculation)		Recirculation (70%)	

The pin heads of button mushroom start to appear after about 1 week, and at this time, fresh air circulation in the room is further increased (four air changes/h) to allow the pin heads to develop into full buttons in the next 4–5 days. The crop appears in flushes, and generally about five flushes are harvested in a cropping period of 45–50 days. The controlled units due to mechanized controls over temperature, humidity, and CO_2 prefer to have shorter cropping period of about 30–35 days. The yields of up to 20%–25% of compost from SMC under controlled growing conditions and 10%–15% from LMC under seasonal cultivation can be obtained.

15.6 Production Portfolio of Important Cultivated Types

The production portfolio of important cultivated types of mushrooms (Figure 15.6) is discussed below.

15.6.1 *Pleurotus* spp.

Pleurotus is the most widely eaten mushroom and rank third in the global production contributing about 14% of the total annual production world over (Shukla and Jaitly, 2011). *Pleurotus* species are suitable for cultivation under both temperate and subtropical climates. These include *P. florida*, *P. pulmonarius*, *P. columbinus*, *P. sapidus*, *P. eryngii*, *P. fossulatus*, *P. djamor-cornucopiae*, *P. cornucopiae*, *P. citrinopileatus*, *P. euosmus*, *P. djamor*, *P. flabellatus*, and *P. cystidiosus*. These species are cultivated in many countries in Europe and Asia, including Italy, Germany, the Netherlands, Belgium, China, Japan, Taiwan, and India.

The substrate used for cultivation of this group of mushrooms can be wood logs and trunks, cereal straws, different kinds of plant waste materials, etc. The cultivation of *Pleurotus* can be carried out on unfermented pasteurized/unpasteurized substrates in polyethylene bags with a high

FIGURE 15.6
Some of the important cultivated mushrooms.

biological efficiency. The fruiting bodies of *Pleurotus* can be consumed fresh or sundried or can be stored as dehydrated sundried product.

15.6.2 *Lentinus edodes*

It is an edible mushroom native to East Asia that is cultivated and consumed in many Asian countries. It is considered a delicacy as well as medicinal mushroom. Japan is the world leader in the production of this mushroom, which is known by the name shiitake. It accounts for about 25% of the total world production of mushrooms. The substrate formula for *L. edodes* production involves sawdust or wheat straw supplemented with wheat bran, gypsum, and lime. In commercial-scale cultivation, polypropylene bags filled with substrate are used. It has tremendous scope for cultivation in India, both for domestic consumption as delicacy and for medicinal use.

15.6.3 *Auricularia polytricha*

It is also known as black ear mushroom. This is the fourth most popular mushroom variety in the world and has been reported to be the first cultivated mushroom in China around AD 600. It is worldwide in its distribution in temperate and subtropical regions. Extensive cultivation takes place in China, Taiwan, Thailand, and the Philippines. *Auricularia* mushroom can be grown both on wood logs and on supplemented sawdust and cereal straw in plastic bags. Traditionally, *Auricularia* cultivation is carried out on artificial logs. *Auricularia* has been used traditionally in China for stomach ailment.

15.6.4 *Agrocybe aegerita*

It is also known as black poplar mushroom. It is worldwide in distribution in temperate regions in nature from spring to autumn. It is cultivated and sold in Japan, Korea, Australia, and China. The substrates used for the cultivation of this mushroom include barley straw, wheat straw, rice straw, orange peels, etc.

Fruit bodies can be sun dried or stored in the refrigerator for 7–10 days. On an average, 300 g of fresh fruit bodies are harvested from 500 g dry wheat straw, giving 60% biological efficiency.

15.6.5 *Flammulina velutipes* or Enokitake

It is also called winter mushroom, velvet foot, or velvet stem. This mushroom can be found growing wild in China, North America, Australia, and Japan both on living and dead wood of broad leaf trees such as aspens, willows, and elms. The temperature for fruiting of these mushrooms ranges between 10°C and 14°C, and China, Japan, Korea, and Taiwan are the leading producers. This mushroom is available fresh or canned and used traditionally for soups, salads, and other dishes. The mushroom can be refrigerated for about 1 week. The cultivated types are grown in a high-carbon dioxide environment to produce long, thin stems that are white in color.

15.6.6 *Calocybe indica*

This mushroom is known as milky mushroom because of its attractive white color. This is an indigenous tropical variety of India, which fruits at around 30°C under high relative humidity. This is the third most important cultivated type in South India grown particularly in the states of Tamil Nadu and Andhra Pradesh. Its cultivation has also picked up in North India. The mushroom grows on cereal straws (wheat straw and paddy straw) and produces biological efficiency of more than 100%. The mushroom has an excellent keeping quality even at room temperature because of its fibrous texture.

15.6.7 *Volvariella volvacea*

This edible mushroom known as straw/paddy straw or Chinese mushroom is cultivated throughout East and Southeast Asia and used extensively in Asian cuisines. It can be consumed fresh as well as dry. It is a high-temperature requiring mushroom (25°C–40°C) that is grown in tropical and subtropical regions of Asia, mainly in Southeast Asia, Madagascar, Africa, and India. Only three species of the straw mushroom, *V. volvacea, V. esculanta,* and *V. diplasia,* are under artificial cultivation. In India, its cultivation is mainly restricted to Orissa. China and Taiwan are the leading producers of straw mushroom, where biological efficiency of 30%–35% is obtained on cotton waste compost. The traditional method of its cultivation is on paddy straw under both indoor and outdoor cultivation conditions where biological efficiency of 15%–20% is obtained.

Sophisticated indoor technology is recommended for an industrial-scale production of this mushroom, but low-cost technology appropriate for rural area development is normally practiced at the community levels for both indoor and outdoor cultivation. The cropping cycle of this mushroom is very short.

15.6.8 *Ganoderma lucidum*

It is also known as Lingzhi or Reishi mushroom. This is a tropical variety that fruits in a temperature range of 30°C–35°C in highly humid conditions. The most commonly treasured species are *G. lucidum* and *G. tsugae. G. lucidum* enjoys special veneration in East Asia, where it has been used as a medicinal mushroom in traditional Chinese medicine for more than 2,000 years. It is ranked high in Oriental traditional medicines and has been used as a remedy for many types of chronic diseases (Hobbs, 1995). China is the main producer of *G. lucidum.* Currently, the methods most widely adopted for commercial cultivation are the wood logs, tree stumps, sawdust bags, and bottle procedures (Stamets, 2000). Log cultivation methods include the use of natural logs and tree stumps that are directly inoculated with spawn under natural conditions. The other alternative technique involves the use of sterilized short logs about 12 cm in diameter and approximately 15 cm long that allow for good mycelial running. This method provides for a short growing cycle, higher biological efficiency, and good quality of fruiting bodies.

15.7 Nutritional Value

Mushrooms have been considered a rich food because they contain protein, sugars, vitamins, high proportion of unsaturated fatty acids, crude fibers, and no cholesterol. They also contain important mineral nutrients, which are required for normal functioning of the body. In addition to nutritional components, edible

TABLE 15.7
Nutritive Value of Mushrooms Based on Proximate Composition

Mushroom	Carbohydrates	Fiber	Protein	Fat	Ash	Energy (kcal)
A. bisporus	46.17	20.90	33.48	3.10	5.70	499
Pleurotus sajor-caju	63.40	48.60	19.23	2.70	6.32	412
L. edodes	47.60	28.80	32.93	3.73	5.20	387
Pleurotus ostreatus	57.60	8.70	30.40	2.20	9.80	265
Volvariella volvacea	54.80	5.50	37.50	2.60	1.10	305
C. indica	64.26	3.40	17.69	4.10	7.43	391
F. velutipes	73.10	3.70	17.60	1.90	7.40	378
Auricularia auricula	82.80	19.80	4.20	8.30	4.70	351

Source: Manikandan, K., Nutritional and medicinal values of mushrooms, In: *Mushrooms: Cultivation, Marketing and Consumption*, M. Singh, B. Vijay, S. Kamal, and G.C. Warchaure (eds.), DMR, Chambaghat, Solan, India, 2011, pp. 11–14; Stamets, P., *Mycelium Running: How Mushrooms Can Help Save the World*, Ten Speed Press, Berkeley, CA, 2005 (*A. bisporus, P. sajor-caju, L. edodes*), FAO, *Food Composition Table for Use in East Asia*, FAO, Rome, Italy, 1972 (*P. ostreatus, V. volvacea*); Doshi, A. and Sharma, S.S., Cultivation of white summer mushroom, In *Advances in Horticulture*, K.L. Chadha and S.R. Sharma (eds.), MPH, New Delhi, India, 1995, p. 13 (*C. indica*); Crisan, E.V. and Sands, A., Nutritional value of edible mushrooms, In *The Biology and Cultivation of Edible Mushrooms*, S.T. Chang, and W.A. Hayes (eds.), Academic Press, New York, 1978, pp. 137–168 (*F. velutipes* and *Auricularia* spp.).

mushrooms possess unique characteristics in terms of color, taste, aroma, and texture, which make them attractive for human consumption.

The determination of nutritional value involves chemical analysis of the proximate composition of mushrooms and a study of the spectrum of amino acids, fatty acids, nucleic acids, vitamins, and minerals present. Nutritive value of mushrooms based on proximate composition is given in Table 15.7.

Protein is the most critical component contributing to the nutritional value of food, whereas fats and carbohydrates are rarely deficient in the diet and hence are generally not considered in the nutritional evaluation of food class.

15.7.1 Protein

Protein is an important constituent of dry matter of mushrooms. The protein content of four popular edible mushrooms, *A. bisporus, L. edodes, Pleurotus* spp., and *V. volvacea*, which are commercially cultivated in various countries, ranges from 1.75% to 3.63% of their fresh weight (Chang, 1980). This value can be as high as 5.9% (Flegg and Maw, 1976). However, an average value of 3.5%–4% would appear to be more representative of the earlier said edible mushrooms. On dry weight basis, mushrooms normally contain 19%–35% protein as compared to 7.3% in rice, 13.2% in wheat, 39.1% in soybean, and 25.2% in milk. The protein conversion efficiency of edible mushrooms per unit of land and per unit time is far more superior compared to animal sources of protein (Bano and Rajarathnam, 1988). The quality of a

TABLE 15.8

Contents of EAA of *A. bisporus* and *P. florida* (g/100 g protein)

Amino Acid	Hen's Egg	*A. bisporus*	*P. florida*
Cysteine	2.4	0.86 (0.36)	0.55 (0.23)
Methionine	3.1	0.98 (0.32)	1.84 (0.59)
Lysine	6.4	3.57 (0.56)	3.20 (0.50)
Tryptophan	1.6	1.87 (1.17)	1.08 (0.67)

Values in parentheses are amino acid scores.

TABLE 15.9

EAA Index and BV of *A. bisporus* and *P. florida*

Indicators	*A. bisporus*	*P. florida*
EAA	51.10	46.30
BV	45.09	38.77
Nutritional index	13.69	12.59

protein is judged by its amino acid composition. The most abundant amino acid in *A. bisporus* is the non-essential amino acid (EAA) glutamic acid. The most common EAA is lysine and the most rare amino acids are those containing sulfur (cysteine and methionine). The sulfur amino acids are the limiting amino acids of *Agaricus* proteins (Table 15.8).

EAAs make up between 32% and 43% of the total amino acid contents in *A. bisporus* (Mattila et al., 2002; USDA, 2005). Determination of EAA composition and digestibility of the proteins of mushrooms can give a fair approximation of their protein quality.

The amino acid score, EAA index, and biological value (BV) were calculated as per Oser method (1951, 1959) by Garcha et al. (1993). The BVs and nutritional indices of these mushrooms were reasonably good (see Table 15.9).

Digestibility is another limiting factor for the qualitative assessment of proteins. A good quality protein based on amino acid composition can only be good if it is easily digestible and its amino acids are available for utilization. Table 15.10 compares the in vitro digestibility of sporophores of *Pleurotus* spp. and *A. bisporus.*

15.7.2 Fats/Lipids

Edible mushrooms usually have insignificant lipid level with higher proportion of unsaturated fatty acids. All these result in a low calorific yield in mushroom foods. Mushrooms do not have cholesterol; instead they have ergosterol that acts as a precursor for vitamin D synthesis in human body. Khanna and Garcha (1983) indicated that *Pleurotus* spp. have 3.6%–6.0%

TABLE 15.10

Protein Digestibility, Relative Nutritive Value (RNV), and Protein Efficiency Ratio (PER) of Sporophores of Mushrooms

	Digestibility (%)			
Species	Pepsin	Pepsin + Trypsin	RNV	PER
P. sajor-caju	73.82 ± 2.03	81.38 ± 3.01	86.90	2.29
P. florida	71.32 ± 1.19	79.07 ± 1.04	97.62	2.43
P. ostreatus	70.09 ± 1.79	77.62 ± 1.84	91.67	2.30
A. bisporus	68.51 ± 2.48	76.43 ± 4.26	—	—
Casein			100.00	2.50

—, not determined.

$$\text{RNV} = \frac{\text{Tetrahymena count with sample protein} \times 100}{\text{Tetrahymena count with casein}}$$

Calculated PER = 0.286 + 0.022 (RNV)

TABLE 15.11

Composition of Total Lipids of *Pleurotus* Species

	Dry wt.	Total Lipids	Lipids (%)						
Organism	(g)	(g)	Total	Free	Bound	Volatile	Nonvolatile	Polar	Nonpolar
P. sajor-caju	5.0	0.230	4.6	62.37	37.63	2.97	97.03	62.24	37.76
P. florida	5.0	0.300	6.0	59.92	40.08	3.03	96.97	56.20	43.80
P. sapidus	5.0	0.205	4.1	57.74	42.26	3.84	96.16	59.68	40.32
P. ostreatus	5.0	0.180	3.6	61.38	38.62	2.78	97.22	62.35	37.65

Source: Khanna, P.K. and Garcha, H.S., *Taiwan Mushrooms*, 7(1), 18, 1983.

crude fat on a dry weight basis. More than 50% of the total lipids in all the species of *Pleurotus* occurred as free lipids (Table 15.11).

The fatty acid composition of *A. bisporus* and *P. florida* is given in Table 15.12. *A. bisporus* has most of the common fatty acids with linoleic acid as the main fatty acid followed by palmitic and oleic acids. *A. bisporus* contained about 57% unsaturated fatty acids. On the other hand, *P. florida* has linoleic acid as the main fatty acid followed by oleic acid. These represented nearly 87% of the total unsaturated fatty acids (Garcha et al., 1993).

Triglycerides were the main component (58.72%–63.95%), while free fatty acids and hydrocarbons were also present in significant amount. Partial glycerides, free sterols, and sterol esters were other components of the non-polar lipid fraction, which are present in relatively low concentration (Khanna and Garcha, 1983).

Analysis of phospholipid and glycolipid contents of the polar lipid fraction indicated that phosphatidylcholine (PC) contributed to more than 50% in different species. The other phospholipids detected in high concentration are

TABLE 15.12

Fatty Acid Composition (% Dry Matter) of *A. bisporus* and *P. florida*

Fatty Acid	*A. bisporus*	*P. florida*
Capric acid (10:0)	2.38	—
Lauric acid (12:0)	0.84	Traces
Myristic acid (14:0)	0.25	11.80
Palmitic acid (16:0)	28.12	11.12
Palmitoleic acid (16:1)	4.20	0.36
Stearic acid (18:0)	7.48	—
Oleic acid (18:1)	12.65	13.91
Linoleic acid (18:2)	35.13	72.81
Linolenic acid (18:3)	4.90	—
Arachidonic acid (20:0)	Traces	—
Saturated fatty acids	43.07	12.92
Unsaturated fatty acids	56.83	87.08

Source: Khanna, P.K. and Garcha, H.S., *Taiwan Mushrooms*, 7(1), 18, 1983.

phosphatidylethanolamine (PE) and phosphatidylinositol (PI) (Table 15.12). However, the lyso forms of PC, PE, and phosphatidylserine are present in low concentration.

15.7.3 Carbohydrates and Fibers

Total carbohydrate content in mushrooms varies and can be as high as 82% on dry weight basis. The digestible carbohydrate profile of mushrooms include pentoses, methyl pentoses, hexoses, as well as disaccharides, amino sugars, sugar alcohols, and sugar acids. Mushrooms have low glucose level, more mannitol, which is suitable for diabetics. *Pleurotus* species contain carbohydrates ranging from 46.6% to 81.8% as compared to 60% in *A. bisporus* on a dry weight basis (Bano and Rajarathnam, 1982). Recently, much interest has arisen in characterizing the components of water-soluble polysaccharides obtained from the fruiting bodies of mushrooms because of their ability to inhibit the growth of tumors. The crude fiber composition of the mushroom consists of partially digestible polysaccharides and chitin. The fiber content ranges from 7.4% to 27.6% in *Pleurotus* species (as compared to 10.4% in *A. bisporus*, and 4%–20% in *V. volvacea*). Fiber is considered to be an important ingredient in a balanced and healthy diet.

15.7.4 Minerals and Trace Elements

Mushrooms probably contain every mineral present in their growth substrate. However, the level of minerals and trace elements in mushrooms may also be dependent on strain (Spaulding and Beelman, 2003).

The most common major minerals in mushrooms in general are potassium (K), phosphorous (P), sodium (Na), calcium (Ca), and magnesium (Mg), whereas copper (Cu), zinc (Zn), iron (Fe), manganese (Mn), molybdenum (Mo), and cadmium (Cd) make up their minor mineral elements (Chang and Miles, 2004).

The concentration of K, P, Na, Ca, and Mg constitute about 56%–70% of the total ash content (Li and Chang, 1982). K and P particularly are abundant and account for nearly 45% of the total ash content. The Cu content in *Pleurotus* species varied from 12.2 to 21.9 ppm for the *Pleurotus* species. Ca and lead (Pb) contents varied from 0.3 to 0.5 ppm and from 1.5 to 3.2 ppm, respectively, in all species of *Pleurotus* (Bano et al., 1981). Of all the heavy metals, Zn content was highest in all species of *Pleurotus*. This is especially noteworthy since Zn content of the straw substrate was low. Mushrooms have a high content of selenium, which is regarded as an excellent antioxidant.

While comparing mineral and trace element levels, it is important to compare them in the same part of the mushroom. According to Vetter (1994), sodium is the only mineral element occurring at higher levels in the stipe than in the cap. Other minerals and trace elements are generally found in the stipe at lower or equal levels to those in the cap (van Elteren et al., 1998).

15.7.5 Vitamins

Mushrooms have a high content of several vitamins, which is an important orthomolecular aspect. Cultivated *A. bisporus* seems to be a good source of the B-complex vitamins niacin and folate (Beelman et al., 2003).

TABLE 15.13

Vitamin Content (Expressed per kg Dry Weight) in Cultivated *A. bisporus*

Vitamin	*A. bisporus*
Thiamin (Vit B_1) (mg)	8.8–12
Riboflavin (Vit B_2) (mg)	53–64
Niacin (mg)	476–511
Pantothenic acid (mg)	170–198
Folate (mg)	1.8–2.5
Vitamin B_{12} (μg)	5.3–13
Vitamin C (mg) (ascorbic acid)	277
Vitamin D (μg)	—
α-Tocopherol (Vit E) (mg)	1.3–2.3
Vitamin K (μg)	0.0

Source: USDA, National Nutrient Database for Standard Reference, Release 19 August 2006.

On the other hand, *A. bisporus* contains very low levels of vitamin A, vitamin D, vitamin E, and thiamine (Anderson and Fellers, 1942). The low level of thiamine in *A. bisporus* has been suggested due to the antinutritive thiaminases, which are responsible for the degradation of thiamine in this mushroom (Wakita, 1976). The low level of D vitamins in *A. bisporus* cultivated indoors contrasts with the comparatively high levels in wild mushrooms. Levels of vitamin D_2 are low in spite of relatively high concentrations of the precursor ergosterol.

Rai and Saxena (1989) found 8, 4, and 3 mg vitamin C per 100 g fresh weight in *A. bisporus, P. sajor-caju,* and *P. ostreatus,* respectively. The vitamin content in cultivated *A. bisporus* is given in Table 15.13.

15.8 Medicinal Value

Many edible and non-edible mushrooms have profound medicinal values and are routinely incorporated in foods as health tonics, soups, teas, and herbal formulae in Far East countries. These represent one of the most rapidly growing sources of extractables both for inclusion in food supplements to enhance health, fitness, and prevention of health disorder and for the treatment of human diseases. The edible mushrooms that demonstrate medicinal or functional properties includes *L. edodes, A. auricula, H. erinaceus, G. frondosa, F. velutipes, Pleurotus* spp., and *T. fuciformis* while the others known for their medicinal properties include *G. lucidum* and *Trametes versicolor.* These are non-edible due to their coarse and hard texture or bitter taste. A number of physiologically active substances including high-molecular-weight polysaccharides (β-D-glucans), heteroglucans, chitinous substances, peptidoglucans, proteoglucans, lectins, RNA components, dietary fibers, and low-molecular-weight organic substances such as terpenoids, steroids, and novel phenols have been identified in mushroom species (Chang, 2007). A limited number of partially or highly purified compounds derived from certain medicinal mushrooms are now being used in the Orient as pharmaceutical products in medicine (Table 15.14). These compounds have been extracted from mushroom fruit body, mycelial biomass, and liquid culture broth and act as biological modifiers by stimulating the immune system (Smith et al., 2002). Several purified mushroom compounds, e.g., lentinan, PSK, PSP, AHCC, schizophyllan, and grifolan D, have been shown to enhance or potentiate host resistance in the treatment of various cancers, immunodeficiency diseases, or immune suppression after drug treatments as adjuvants for vaccines and for combination therapy with antibodies.

Ninety-nine percent of all sales of medicinal mushrooms and their derivatives occurred in Asia and Europe, whereas in North America, it was less than 0.1% of the total sales the world over (Khanna, 2011). A large number of clinical evidence demonstrates the beneficial results of mushroom polysaccharides

TABLE 15.14

Medicinal Components of Medicinal Mushrooms and Their Functions

Mushrooms	Medicinal Components	Functions	References
A. bisporus	Lectins	Increase insulin secretion	Ahmad et al. (1984)
		Enhances natural killer cell activity	Wu et al. (2007)
A. aegerita	AG HN-1, AG NH-2	Reduces glycemia	Gunde-Cimerman (1999)
	Agrocybin	Supports cytokine and interleukin production	Ou et al. (2005)
	Immunomodulatory peptides	Restoring immune response	Lee et al. (2009)
A. auricula	1–3-β-glucans, acidic polysaccharides	Antitumor activity, lowers cholesterol, triglycerides, lipid levels, and decrease blood glucose	Misaki and Kishida (1995), Chen (1989), and Yuan et al. (1998)
Cordyceps militaris	Cordycepic acid	Regulating the blood sugar	Kino et al. (1986)
		Anti-metastatic effect	Nakamura et al. (1999)
		Antitumor effect	Bok et al. (1999)
	Polysaccharide fractions CI-P and CI-A	Exhibit substantial antitumor activities in mice with Sarcoma-180	Mizuno (1999)
	β-D-glucans	Increase both innate and cell-mediated immune response	Shin et al. (2003)
	Cordycepin	Cure lung infections	Li et al. (2006)
		Hypoglycemic activity	Ko et al. (2009)
		Antidepression activity	Nishizawa et al. (2007)
Flammulina velutipes	Ergothioneine	Antioxidant activity	Bao et al. (2008)
	Flammulin	Anticancerous, antiviral, and stimulates immune system	Weil (1987)
	FVP (Flammulina polysaccharides)	Antitumor activity	Wasser and Weis (1999)
	Proflamin and prolamin	Antitumor against allogenic and syngeneic tumor by oral administration	Ikekawa et al. (1985)

G. lucidum	β-Glucans	Liver protection	Wang et al. (2007)
		Antibiotic properties	Moradali et al. (2006)
	Ganopoly (GL-P)	Immunomodulating effects. Antitumor and anti-angiogenic activity when used in combination with cytotoxic anticancerous drugs	Lee et al. (1995)
	GL-B7	Antiaging effects by decreasing the production of oxygen free radicals and antagonizes the respiratory burst induced by peripheral mononuclear antioxidant in murine peritoneal macrophages	Li and Lie (2000)
	Triterpenoids (lucidimol, lucidenic acid, ganoderiol, ganoderols, and ganoderic acids)	Possess hypotensive and hypertensive properties	Morigiwa et al. (1986)
		Reduce blood pressure, blood sugar, eliminate cholesterol and inhibit platelet aggregation	Liu (1993), Zhu and Mori (1993), and Mizuno (1995)
		Augments immune system	Lin and Zhang (2004)
	LZ-8	Imparts mitogenic activity in vitro and immunomodulating activity in vivo, thus regulating the immune response and inhibiting anaphylaxis	Kino et al. (1989)
G. frondosa	Polysaccharides (1,3- and 1,6-β-D-glucans)	Anticancerous activity	Ohno et al. (1985), Hishida et al. (1988), and Kurashiga et al. (1997)
		Increase the body immune defense mechanism	Yanaki et al. (1983)
		Prevention and treatment of flu, common infections, AIDS-HIV, diabetes mellitus, hypertension, hypercholesterolemic, and urinary tract infections	Kabir and Kimura (1989), Smith et al. (2002), and Talpur et al. (2002)
	Lectins	Decrease blood glucose	Horio and Ohtsuru (2001)
		Cyclooxygenase inhibitor	Zhang et al. (2002)
		Anticancer and hypoglycemic effects	Konno et al. (2002)

(*continued*)

TABLE 15.14 (continued)

Medicinal Components of Medicinal Mushrooms and Their Functions

Mushrooms	Medicinal Components	Functions	References
L. edodes	Lentinan	Host-mediated anticancerous drug	Ooi and Liv (2000)
		Activate macrophages and T-lymphocytes that modulate the release of cytokines, which in turn accounts for its indirect antitumor and antimicrobial properties	
		Anti-HIV activity	Taguchi and Furue (1985)
		Treatment of chronic hepatitis and viral hepatitis B patients	Amagnase (1987)
	KS-2	Suppresses Sarcoma-180 when administered both orally and intraperitoneally	
	LEM extract	Inhibit HIV infection in vitro and activate the host immune system	Mizuno (1995)
		Lowers blood pressure and free cholesterol in plasma as well as accelerate accumulation of lipids in liver	Kabir and Kimura (1989)
	Eritadenine	Lowers cholesterol	Enman et al. (2007)
Pleurotus spp.	Lovastatin/mevinolin	Hypercholesterolemia, prevents cardiovascular disorders	Mizuno (1999)
	Lovastatin	Lowers cholesterol	Gunde-Cimerman and Cimerman (1995)
		Radical scavenger	Chye et al. (2008)
S. commune	Schizophyllan, sonifillan, sizofiran, and sizofllan	Cytostatic	Wasser and Weis (1999)
	Sulfated schizophyllan	Increase the survival of the patients with head and neck cancers	Kimura et al. (1994)
		Anti-HIV activity	Ito and Sugawara (1990)

	Sulfated polysaccharides (SPG)	Protection against *Staphylococcus* spp., *Pseudomonas*, and *Escherichia coli* infection	Cochran (1978)
		Treatment of chronic hepatitis B, enhances immunological responsiveness to virus, production of interferon "g"	Matsuyama et al. (1992)
T. versicolor	Polysaccharides (PSK, PSP)	Decrease depression	Coles and Toth (2005)
		Anti-HIV	Collins and Ng (1997)
		Inhibits growth of *Candida albicans*	Sakagami et al. (1991)
		Treatment of colorectal, lung, breast, and gastric cancer for stage IV	Kidds (2000)
		Increased delayed type of hypersensitivity on skin tests and enhances chemotactic migration of neutrophils	Kondo and Torisu (1935), Yang et al. (1993)
		PSK in conjugation with chemotherapy is useful in breast cancer patients	Wasser and Weis (1999)
		Amelioration of toxic effects of chemo- and radiotherapy by PSP	Ikekawa et al. (1985)
		Antiviral activity by inhibiting HIV replication	Ikekawa et al. (1985)
	Coreolan (β-glucan protein)	Liver-protective activities	Wasser and Weis (1999)
	Glycoprotein (PSK)	Shows activity against experimental diabetes in animals both in vitro and in vivo	
		Activity against experimental hypertension and thrombosis inhibits blood platelet aggregation, anti-hyperlipidemic, anti-arrhythmic	
	Coriolon	Decrease in LDL cholesterol in hyperlipidemia (stage IIa) patients	
		Antibiotic inhibits Gram-positive and *Trichomonas vaginalis*	

in (1) prevention of oncogenesis by oral consumption of mushrooms or their preparations, (2) direct antitumor activity against various allogeneic and syngeneic tumors, (3) immunopotentiation activity against tumors in conjunction with chemotherapy, and (4) preventive effects on tumor metastasis. The value of mushrooms has recently been promoted by conducting the trials using *G. lucidum*, the medicinal mushroom on the HIV/AIDS patients in Africa, and reporting the encouraging results (Mshigeni et al., 2005).

15.9 Growth Potential and Opportunities

Mushroom production has a special relevance with respect to Indian context. It addresses the twin problems of crop diversification and sustainability as well as utilization of vast reservoir of crop residues available, for mushroom production. Since 2003/2004, India has accounted for an estimated 25% of all U.S. imports of canned mushroom. There is still a vast untapped market that can be exploited under the government policy of liberalization and globalization of trade. In spite of the various reasons of slow growth of mushroom production in the country, there seems enormous potential of developing mushroom industry into a highly remunerative activity because of the following inherent advantages.

15.9.1 Availability of Basic Raw Material

For mushroom cultivation, basic raw materials invariably are cereal straws, other lignocellulosics, and organic residues that are available abundantly and continuously. These residues are generated through agriculture, forest, and food industry operations. Sufficient quantity of crop residues is available with the farmer at his or her door step at no cost. Mushroom cultivation offers a low cost bio-conversion of straw into edible food, which is an important need of the country for meeting the requirement of the exponentially growing population, besides solving the problem of protein energy malnutrition.

15.9.2 Suitable Technology

The development of mushroom production technology suited to varied agroclimatic conditions of the country and its adoption has enabled many states to establish mushroom industry. At present, mushroom cultivation of one variety or the other is being carried out at more than 15 states of the country following low-cost cultivation technology. In addition, technology for controlled mushroom production is also available and is being practiced by some small climate-controlled farms as well as export-oriented units. Many of these farms are producing their own compost, spawn, and mushrooms and processing them as canned product.

15.9.3 Suitable Environment and Diversification of Species

India is endowed with varied climate and, thus, has the inherent advantage for the diversification of mushroom species in different regions and seasons of the country. In India, the mushroom production systems are mixed type comprising both seasonal farming and hi-tech industry. Today, commercially grown species are button and oyster mushrooms, followed by other tropical mushrooms like paddy straw mushroom and milky mushroom. Two to three crops of button mushroom are grown seasonally in temperate regions with minor adjustments of temperature in the growing rooms, while one crop of button mushroom is raised in North Western plains of India seasonally. Oyster, paddy straw, and milky mushrooms are grown seasonally in the tropical/sub-tropical areas from April to October.

15.9.4 Low-Cost Labor

In the developed countries of Europe and America, mushroom farming has attained the status of hi-tech industry, with high levels of mechanization and automation. In India, however, mushroom production is labor intensive. Cheap and skilled labor along with the availability of raw material is the main strength for mushroom growth in the country. According to one analysis of cost break-up in mushroom cultivation, the labor cost averages 7% of total production cost in India compared to 47% in the Western world. Since mushroom cultivation is a labor-intensive process, it simultaneously creates new avenues for employment.

15.9.5 Domestic Consumption and Affordability

The per capita consumption of mushrooms in India is dismally low about 40 g as compared to very high consumption in the United States and Europe. India is a big market and estimates suggest that increase in per capita consumption of even up to 100 g will help growers to market over 1 lakh ton mushrooms within the country (Singh, 2011). Fresh/canned mushrooms are available in almost all the towns and cities. Due to increased urbanization and growth in per capita income of the middle class in developing countries that provide affordability, there has been visible shift in the dietary pattern with preference for fresh mushroom consumption.

15.9.6 Export Potential

Although a highly competitive field, mushroom export is a potential means to earn foreign exchange. There has been a tremendous increase in mushroom export from 1990 to 1991 to date. Earlier the major export from India to Switzerland, France, and Germany was of dried mushroom collections of morels (gucchi) and oyster mushroom from forests. With the inclusion of preserved mushrooms in the exportable quota, the export of canned button mushrooms has increased substantially during the last one-and-a-half decade. Bulk of export in the past years has gone to the United States, where the major competition has been from China, the largest exporter of mushroom from Asia. With further reduction in the cost of production and government intervention in acquiring EEC quota, freight subsidy, etc., there is every possibility of further boosting the export of mushroom from India in order to earn valuable foreign exchange.

India can also enter into a big and lucrative international trade in medicinal mushroom varieties like *Lentinus, Ganoderma,* and *Cordyceps* spp., presently monopolized by some East Asian countries. There is a tremendous scope for diversifying mushroom export by including many other mushroom spp., which are still under experimental cultivation. With the current growth rate of the Indian economy, domestic market for mushrooms too is likely to swell.

15.9.7 Source of Food Protein

Indians, largely a vegetarian population, live on grains, tubers, and small quantities of pulses and vegetables, all of which are a poor quantitative and qualitative source of protein. Mushroom protein, which is highly digestible and has a nutritional score of 6–31, can be of considerable significance in filling the protein gap especially of the rural masses, because of the close location of mushroom production and consumption centers.

Being rich in minerals, their usefulness becomes still more important. Edible mushrooms are apparently innocuous, and this has been confirmed at both cellular and biochemical levels. With the increasing awareness in the production of mushrooms, their availability can be increased, which would in turn improve per capita consumption.

15.9.8 Minimum Land Use/Intensive Use of Indoor Space

Mushrooms are cultivated indoors and do not require arable land. Small farmers and landless workers constitute major fraction of Indian population, and mushroom cultivation is highly suitable for economic upliftment and social security of this group. Comparatively speaking, growing mushrooms as food protein requires less land than producing protein from animal and other plant sources. It has been estimated that many white button mushroom growers throughout the world currently can produce 6,730–7,826 kg of dry protein/ha/year compared with 78 kg of dry animal protein/ha/year and 673 kg of dry protein/ha/year from fish farming. The production of over 620 kg of *Agaricus* mushrooms from 1 m^2 bed area based on 5 crops/year with five shelves in a growing room with a yield of 24.4 kg/m^2 has been recorded (Quimio and Chang, 1990). Thus, less land required for growing mushrooms is very relevant in the context of intensive farming being practiced due to reduction in land holdings.

15.9.9 Recycling Waste Residues

The availability of organic residues has increased tremendously during the last decade, which are partly fed to the cattle, partly used as fuel in rural India, but mostly burned. Mushroom growing represents the only commercially viable process where these wastes can be converted through SSF process into quality food protein (DMR, 2007). India produces about 600 MT of agricultural wastes per annum, and a major part of it is left out to decompose naturally or burned in situ. Guestimates suggest that by just diverting 1% of agro waste toward mushroom production, India can produce 3 MT of mushroom and about 15 MT of spent compost (DMR, 2007).

15.9.10 Self-Employment

Mushroom production is a labor-intensive indoor crop, which is safe from vagaries of weather. It provides immense opportunities for taking it up as self-employment, income-generating avocation, and particularly for the employment of both rural and urban women through cultivation, production of value-added products like soups, soup powder, curies, and nuggets, and their marketing.

15.10 Summary and Future Prospects

All over the world, the last 10–15 years have been a period of strategic research on various aspects of mushroom production like compost formulation and composting, pasteurization and empirical calculations, spawn production, environmental engineering, and post-harvest technology. In India, this has resulted in a phenomenal growth horizontally as well as vertically, both in the productivity and in the overall production, thus helping to establish a sustained system of production both for domestic consumption and for export. At the same time, this spurt in mushroom production has thrown open new challenges, if such high rate of production is to be maintained, which requires a multidimensional strategy to be followed in a number of areas of mushroom research and technology as briefed in the following text:

1. More efforts are required toward the refinement of farm structures with respect to economical size of the growing rooms, insulating materials to be used, environmental control, etc.
2. Shortening of composting period to reduce organic matter-less and standardizing compost formulation for getting consistent yields.
3. A lot of empirical data on SMC needs to be generated to develop productive synthetic compost formulations based on locally available

agri-residues, which will help to predict optimum conditions to carry out phase II of composting.

4. Focus should be shifted toward the production of environmental friendly compost through indoor composting in order to obviate the objectionable smell and pollution caused by malodorous gases.
5. A systematic stream development/improvement programs needs to be taken up for the development of hybrids and strains. The certification of spawn-producing laboratories/centers in the public sector should be made through the imposition of legislative measures of the timely supply of quality spawn.
6. The mushroom portfolio will have to be broadened through diversification and domestication of not only culinary, but medicinal cultivation as well.
7. Mushrooms are perishable commodity and need to be marketed and consumed in a short span. Efforts need to be made to set up suitable growthing units in or cluster near a common collection center to facilitate collective marketing of the produce.

Mushroom production is posed for a phenomenal rise in the coming decade as the knowledge about the nutritional and medicinal qualities of mushrooms is spreading fast, but it is imperative that efforts to improve and refine the existing mushroom production technologies are continuously made. Mushroom producers will have to work with a competitive zeal to make their produce cheaper and qualitatively superior in order to compete in global market. The support of the Government to the commercial and industrial units to facilitate foreign trades, for the opening up of mother spawn and custom composting units, and in providing engineering industry support and post-harvesting processing facility can help a long way in further boosting mushroom production, consumption, and trade.

References

Ahmad, N., Bansal, A.K., and Kidwai, J.R. 1984. Effect of PHA-B fraction of *Agaricus bisporus* lectin on insulin release and $^{45}Ca^{2+}$ uptake by islets of Langerhans in vitro. *Acta Diabetologica* 21: 63–70.

Amagnase, H. 1987. Treatment of Hepatitis B patients with *Lentinus edodes* mycelium. In: *Proceedings of the XII International Conference of Gastroenterology*, Lisbon, Portugal.

Anderson, E.E. and Fellers, C.R. 1942. The food value of mushrooms (*Agaricus campestris*). *Proceedings of American Society for Horticulture Science* 41: 301–303.

Bano, Z., Bhagya, S., and Srinivasan, K.S. 1981. Essential amino acid composition and proximate analysis of the mushroom *Pleurotus eous* and *Pleurotus florida*. *Mushroom Newsletter for the Tropics* 1: 6–10.

Bano, Z. and Rajarathnam, S. 1982. *Pleurotus* mushroom as a nutritious food. In: *Tropical Mushrooms: Biological Nature and Cultivation Methods*, eds. S.T. Chang and T.H. Quimio. The Chinese University Press, Hong Kong, China.

Bano, Z. and Rajarathnam, S. 1988. *Pleurotus* mushroom part II. Chemical composition nutritional value, post harvest physiology, preservation and role as human food. *Critical Reviews in Food Science and Nutrition* 27: 87–158.

Bao, H.N., Ushio, H., and Ohshima, T. 2008. Antioxidative activity and antidiscoloration efficacy of ergothioneine in mushroom (*Flammulina velutipes*) extract added to beef and fish meats. *Journal of Agricultural and Food Chemistry* 56(21): 10032–10040.

Beelman, R.B., Royse, D.J., and Chikthimmah, N. 2003. Bioactive components in button mushroom *Agaricus bisporus* (J. Lge) Imbach (Agaricomycetideae) of nutritional, medicinal, and biological importance (review). *International Journal of Medicinal Mushrooms* 5: 321–337.

Bisht, I.S. and Singh, R.N. 1986. Prospects of active mycelium spawning technique in commercial cultivation of button mushroom, *Agaricus bisporus*. *Indian Journal of Plant Pathology* 16: 301.

Bok, J.W., Lermer, L., Chilton, J., Klingeman, H.G., and Towers, G.H. 1999. Antitumor sterols from the mycelia of *Cordyceps sinensis*. *Photochemistry* 51: 891–898.

Chang, S.T. 1980. Mushrooms as human food. *Bioscience* 30: 399–401.

Chang, S.T. 1993. Mushrooms and mushroom biology. In: *Genetics and Breeding of Edible Mushrooms*. eds. S.T. Chang, J.A. Buswell, and P.G. Miles, pp. 1–13. Gordon Breach Science Publishers, Philadelphia, PA.

Chang, S.T. 2007. Development of the world mushroom industry and its roles in human health: A keynote address. In: *International Conference on Mushroom Biology and Biotechnology*, Souvenir-cum-abstracts, Solan, India, 10–11 February, pp. 3–16.

Chang, S.T. and P.G. Miles. 2004. The nutritional attributes of edible mushrooms. In: *Mushrooms: Cultivation, Nutritional Value, Medicinal Effect, and Environmental Impact*, 2nd edn., pp. 27–36, CRC Press, Boca Raton, FL.

Chen, Q. 1989. Antilipemic effect of polysaccharide from *Auricularia auricula*, *Tremella fuciformis* and *T. fuciformis* spores. *Zhongguo Yaoke Daxue Xuebae* 20: 344–347.

Chye, F.Y., Wong, J.Y., and Lee, J.S. 2008. Nutritional quality and antioxidant activity of selected edible wild mushrooms. *Food Science and Technology International* 14: 375–384.

Cochran, K.W. 1978. Medical effects. In: *The Biology and Cultivation of Edible Mushrooms*, eds. S.T. Chang and W.A. Hayes, pp. 160–187, Academic Press, New York.

Coles, M. and Toth, B. 2005. Lack of prevention of large intestinal cancer by VPS, an extract of *Coriolus versicolor* mushroom. *In Vivo* 19: 867–871.

Collins, R.A. and Ng, T.B. 1997. Polysaccharopeptide from *Coriolus versicolor* has potential for use against human immunodeficiency virus type 1 infection. *Life Science* 60(25): 383–387.

Crisan, E.V. and Sands, A. 1978. Nutritional value of edible mushrooms. In: *The Biology and Cultivation of Edible Mushrooms*, eds. S.T. Chang and W.A. Hayes, pp. 137–168, Academic Press, New York.

Dhar, B.L, Vijay, B., Upadhyay, R.C., and Sohi, H.S. 1985. Effect of compost depth in polythene bags on yield of *Agaricus bisporus*. *Indian Journal of Mycology and Plant Pathology* 43: 74–76.

Directorate of Mushroom Research (DMR). 2007. NRCM perspective plan vision-2025, Directorate of Mushroom Research, Chambaghat, Solan, Himachal Pradesh, India.

Doshi, A. and Sharma, S.S. 1995. Cultivation of white summer mushroom. In: *Advances in Horticulture*, eds. K.L. Chadha and S.R. Sharma, p. 13. MPH, New Delhi, India.

van Elteren, J.T., Woroniecka, U.D., and Kroon, K.J. 1998. Accumulation and distribution of selenium and cesium in the cultivated mushroom *Agaricus bisporus*: A radiotracer-aided study. *Chemosphere* 36: 1787–1798.

Enman, J., Rova, U., and Berglund, K.A. 2007. Quantification of the bioactive compound eritadenine in selected strains of Shiitake mushroom (*Lentinus edodes*). *Journal of Agricultural and Food Chemistry* 55: 1177–1180.

FAO, U.N. 1972. *Food Composition Table for Use in East Asia*. FAO, Rome, Italy.

FAOSTAT. 2009. *World Mushrooms and Truffles: Production 1961–2009*, FAO, United Nations, Rome, Italy.

Flegg, P.B. and Maw, G.A. 1976. Mushrooms and their possible contribution to world protein needs. *The Mushroom Journal* 48: 396–405.

Garcha, H.S. and Khanna, P.K. 2003. *Mushroom Cultivation*, p. 60, Punjab Agricultural University, Ludhiana, India.

Garcha, H.S., Khanna, P.K., and Soni, G.L. 1993. Nutritional importance of mushrooms. In: *Mushroom Biology and Mushroom Products*, eds. S.T. Chang, J.A. Buswell, and S. Chiu, pp. 227–235. The Chinese University Press, Hong Kong, China.

Gaze, R. 2005. The larger world of global 'mushrooms'. *The Mushroom Journal* 670: 11–13.

Gerrits, J.P.G. and Van Griensen, L.J.L.D. 1990. New developments in indoor composting (tunnel process). *The Mushroom Journal* 205: 21–29.

Global Trade Information Service, Inc. (GTIS). 2008. World Trade Atlas Database. GTIS, Columbia, SC.

Gulliver, A., Miller, F.C., Harper, E., and Macauley, B.J. 1991. Environmentally controlled composting on a commercial scale in Australia. *Mushroom Science* 13: 155–164.

Gunde-Cimerman, N. 1999. Medicinal value of the genus *Pleurotus* (Fr.) P. Karst. (Agaricales S.R., Basidiomycetes). *International Journal of Medicinal Mushrooms* 1: 69–80.

Gunde-Cimerman, N. and Cimerman, A. 1995. *Pleurotus* fruiting bodies contain the inhibitor of 3-hydroxy-3-methyl-glutaryl-coenzyme A reductase-lovastatin. *Experimental Mycology* 19: 1–6.

Hawksworth, D.L. 2001. Mushrooms: The extent of the unexplored potential. *International Journal of Medicinal Mushrooms* 3: 333–337.

Hayes, W.A. 1977. Mushroom nutrition and the role of microorganisms in composting process. In: *Composting*, ed. W.A. Hayes, pp. 1–29, University of Aston, Birmingham, U.K.

Hishida, I., Nanba, H., and Kuroda, H. 1988. Antitumor activity exhibited by orally administered extract from fruit body of *Grifola frondosa* (maitake). *Chemical and Pharmaceutical Bulletin* 36: 1819–1827.

Hobbs, C. 1995. *Medicinal Mushrooms: An Exploration of Traditional, Healing and Culture*, Botanica Press, Santa Cruz, CA.

Horio, H. and Ohtsuru, M. 2001. Maitake (*Grifola frondosa*) improve glucose tolerance of experimental diabetic rats. *Journal of Nutritional Science and Vitaminology* 47: 57–63.

Ikekawa, T., Uehara, N., Maeda, Y., Nakamishi, M., and Fukuoka, F. 1985. Proflamin, a new antitumor agent: Preparation, physiochemical properties and antitumor activity. *Japanese Journal of Cancer Research (Gann.)* 76: 142–148.

Ito, W. and Sugawara, I. 1990. Immunopharmacological study of sulphated schizophyllan (SPG) 1, its action as a mitogen anti-HIV agent. *International Journal of Immunopharmacology* 12: 225–233.

Jain, V.B. and Singh, S.P. 1982. Effect of spawning method on the yield of *Agaricus brunnescens* Peck. *Progressive Horticulture* 14(4): 246–248.

Kabir, Y. and Kimura, S. 1989. Dietary mushrooms reduce blood pressure in spontaneously hypertensive rats. *Journal of Nutritional Science and Vitaminology* 35: 91–94.

Kaur, H. and Khanna, P.K. 2001. Physico-chemical and microbiological characteristics of paddy straw based compost for *Agaricus bisporus* production. *Indian Journal of Mushrooms* 19: 15–20.

Khanna, P.K. 2011. Polysaccharides from medicinal mushrooms and their antitumor activities-A review. *International Journal of Food and Fermentation Technology* 1(1): 17–37.

Khanna, P.K. and Garcha, H.S. 1983. Lipid composition of *Pleurotus* spp. (dhingri). *Taiwan Mushrooms* 7(1): 18–23.

Khanna, P.K. and Kapoor, S. 2007. *A Manual on Mushroom Production*. Punjab Agricultural University, Ludhiana, India.

Kidds, P.M. 2000. The use of mushroom glucans and proteoglycans in cancer treatment. *Alternate Medicinal Review* 5: 4–26.

Kimura, Y., Tsuda, K., Arichi, S., and Takahashi, N. 1994. Clinical evaluation of Sizofiran as assistant immunotherapy in treatment of head and neck cancer. *Acta Otolaryngology* 511: 192–195.

Kino, T., Tabata, H., Ukai, S., and Hara, C. 1986. A minor protein containing galactomannan from a sodium carbonate extracted of *Cordyceps lucidum*. *Carbohydrate Research* 156: 189–197.

Kino, K.Y., Yamaoka, K., Watanabe, J., Shimizu, K., and Tsunoo, H. 1989. Isolation and characterization of a newly immunomodulatory protein Ling Zhi-8 (LZ-8) from *Ganoderma lucidum*. *Journal of Biological Chemistry* 264: 472–478.

Ko, W.S., Hsu, S.L., Chyau, C.C., Chen, K.C., and Peng, R.Y. 2009. Compound *Cordyceps* TCM-700C exhibits potent hepatoprotective capacity in animal model. *Fitoterapia* 81(1): 1–7.

Kondo, M. and Torisu, M. 1985. Evaluation of *Lentinus edodes*, *Grifola frondosa* and *Pleurotus ostreatus* administration on cancer outbreaks and activities of macrophages and lymphocytes in mice treated with a carcinogen. *Immunopharmacology and Immunotoxicology* 19: 175–185.

Konno, S., Aynehchi, S., Dolin, D.J., Schwartz, A.M., Choudhury, M.S., and Tazakin, H.N. 2002. Anticancer and hypoglycemic effects of polysaccharides in edible and medicinal maitake mushroom [*Grifola frondosa* (Dicks.:Fr.) S.F. Gray]. *International Journal of Medicinal Mushrooms* 4:185–195.

Kurashiga, S., Akuzawa, Y., and Eudo, F. 1997. Effects of *Lentinus edodes*, *Grifola frondosa* and *Pleurotus ostreatus* administration on cancer outbreaks and activities of macrophages and lymphocytes in mice treated with carcinogen N-butyl-NI-butamolinitreso-amine. *Immunopharmacology and Immunotoxicology* 19: 175–183.

Laborde, J. 1991. Composting: Current and future techniques in France and Abroad. *Mushroom Information* 8: 4–24.

Laborde, J., Lanzi, G., Francesscutti, R., and Giordani, E. 1993. Indoor composting: General principles and large scale development in Italy. In: *Mushroom Biology and Mushroom Products*, eds. S.T. Chang, J.A. Buswell, and S.W. Chiu, 93–114. The Chinese University Press, Hong Kong, China.

Lee, B.R., Kim, S.Y., Kim, D.W. et al. 2009. Agrocybe chaxingu polysaccharide prevent inflammation through the inhibition of CoX-2 and NO production. *Biology and Molecular Biology Reports* 42(12): 794–799.

Lee, S.S., Wei, Y.H., Chen, C.H., Wang, S.Y., and Chen, K.Y. 1995. Antitumor effects of *Ganoderma lucidum*. *Journal of Chinese Medicine* 6: 1–12.

Li, G.S.F. and Chang, S.T. 1982. Nutritive value of *Volvariella volvacea*. In: *Tropical Mushrooms-Biological Nature and Cultivation Methods*, eds. S.T. Chang and T.H. Quimio, 199–219. The Chinese University Press, Hong Kong, China.

Li, M. and Lie, L.S. 2000. Effect of *Ganoderma lucidum* polysaccharides on oxygen-free radicals in murine peritoneal macrophages. *Zhongguc Yadikxue Yu Dulixue Zazhi—Chinese Journal of Pharmacology and Toxicology* 14: 65–68.

Li, S.P., Zhang, G.H., Zeng, Q. et al. 2006. Hypoglycemic activity of polysaccharide with antioxidation isolated from cultured *Cordyceps* mycelia. *Phytomedicine* 13: 428–433.

Lin, Z.B. and Zhang, H.N. 2004. Anti-tumor and immunoregulatory activities of *Ganoderma lucidum* and its possible mechanisms. *Acta Pharmacologica Sinica* 25(11): 1387–1395.

Liu, G.T. 1993. Pharmacology and clinical uses of *Ganoderma*. In: *Mushroom Biology and Mushroom Products*, eds. S.T. Chang, J.A. Buswell, and S.W. Chiu, pp. 267–273. Chinese University Press, Hong Kong, China.

Manikandan, K. 2011 Nutritional and medicinal values of mushrooms. In: *Mushrooms: Cultivation, Marketing and Consumption*, eds. M. Singh, B. Vijay, S. Kamal, and G.C. Warchaure, pp. 11–14, DMR, Chambaghat, Solan, India.

Mao, S.L. 2004. Rich in resource of Chinese edible and medicinal fungi. *Mushroom Market* 3: 7–9.

Matsuyama, H., Mangindaan, R.E.P., and Yano, T. 1992. Protective effect of schizophyllan and scleroglucans against *Streptococcus* sp. infection in yellow tail (*Seriola quinqueradiata*). *Aquaculture* 101: 97–203.

Mattila, P., Salo-Väänänen, P., Känkö, K., Aro, H., and Jalava, T. 2002. Basic composition and amino acid contents of mushrooms cultivated in Finland. *Journal of Agricultural and Food Chemistry* 50: 6419–6422.

Miller, F.C. 1997. Enclosed Phase I Mushroom composting systems considerations of the underlying technology and methods of implementation. In: *Advances in Mushroom Biology and Production*, eds. R.D. Rai, B.L. Dhar, and R.N. Verma, pp. 129–138, MSI, NRCM Solan, India.

Misaki, A. and Kishida, E. 1995. Straw mushroom, Fukurotake, *Volvariella volvacea*. *International Food Review* 11: 219–223.

Mizuno, T. 1995. Bioactive biomolecules of mushrooms: Food function and medicinal effects of mushroom fungi. *International Food Review* 11: 7–12.

Mizuno, T. 1999. The extraction and development of antitumor active polysaccharides from medicinal mushrooms in Japan (review). *International Journal of Medicinal Mushrooms* 1: 19–30.

Moradali, M.F., Mostafavi, H., Hejaroude, G.A., Tehrani, A.S., Abbasi, M., and Ghods, S. 2006. Investigation of potential antibacterial properties of methanol extracts from fungus *Ganoderma applanatum*. *Chemotherapy* 52(5): 241–244.

Morigiwa, A., Kitabatake, K., Fujimoto, Y., and Ikekawa, J. 1986. Angiotensin converting enzyme-inhibiting triterpenes from *Ganoderma lucidum*. *Chemical and Pharmaceutical Bulletin* 34: 3025–3028.

Mshigeni, K.E., Mtango, D., Massele, A. et al. 2005. Intriguing biological treasures more precious than gold: The case of tuberous truffles, and immunomodulating *Ganoderma* mushrooms with potential for HIV/AIDS treatment. *Discovery and Innovation* 17(3, 4) 2005: 105–109.

Nair, N.G. and Price, G. 1991. A composting process to minimize odour pollution. *Mushroom Science* 13: 205–206.

Nakamura, K., Yamaguchi, Y., Kagota, S., Shinozuka, K., and Kunitomo, M. 1999. Activation of Kupffer cell function by oral administration of *Cordyceps sinensis* in rats. *Japanese Journal of Pharmacology* 79: 505–508.

Nishizawa, K., Torii, K., Kawasaki, A., Katada, M., Ito, M., and Terashita, K. 2007. Antidepressant-like effect of *Cordyceps sinensis* in the mouse tail suspension test. *Biological and Pharmaceutical Bulletin* 30(9): 1758–1762.

Ohno, N., Lino, K., Suzuki, I. et al. 1985 Neutral and acidic antitumor polysaccharides extracted from cultured fruit bodies of *Grifola frondosa*. *Chemical and Pharmaceutical Bulletin* 33: 1181–1186.

Ooi, V.E.C. and Liv, F. 2000. Immunomodulation and anti-cancer activity of polysaccharide protein complexes. *Current Medicinal Chemistry* 7: 715–729.

Oser, B.L. 1951. Method for integrating essential amino acid content in nutritional evaluation of proteins. *Journal of American Dietitians Association* 27: 396–402.

Oser, B.L. 1959. An integrated essential amino acid index for predicting the biological value of protein. In: *Protein and Amino acid Nutrition*, ed. A.A. Albanese, pp. 281–295, Academic Press, New York.

Ou, H.T., Shieh, C.J., Chen, J.Y., and Chang, H.M. 2005. The antiproliferative and differentiating effects of human leukemic U 937 cells are mediated by cytokines from activated mononuclear cells by dietary mushrooms. *Journal of Agricultural and Food Chemistry* 53(2): 300–305.

Patil, B.D., and Shinde, P.A. 1983. Effect of spawning rate and method of spawning on yield of white button mushroom (*Agaricus bisporus*). *Journal of Maharashtra Agricultural Universities* 8(1): 82.

Quimio, T.H. and Chang, S.T. 1990. Technical guidelines for mushroom growing in the tropics. FAO Publication No. 106: 6.

Rai, R.D. and Saxena, S. 1989. Suitability of methods of estimation for critical assessment of the vitamin C content in mushrooms. *Mushrooms Journal for the Tropics* 9: 43–46.

Rambelli, A. 1983. *Manual of Mushroom Cultivation*. FAO, Plant Production and Protection, p. 43, Rome, Italy.

RMCU. 2010. Directorate of Mushroom Research (DMR), Solan, India.

Sakagami, H., Aoki, T., Simpson, A., and Tanuma, S. 1991. Induction of immunopotentiation activity by a protein bound polysaccharide, PDS. *Anticancer Research* 11: 993–1000.

Shandilya, T.R., Seth, P.K., Kumar, S., and Munjal, R.L. 1974. Effect of different spawning methods on the productivity of *Agaricus bisporus*. *Indian Journal of Mycology and Plant Pathology* 4:129–131.

Shin, K.H., Lim, S.S., Lee, S., Lee, Y.S., Jung, S.H., and Cho, S.Y. 2003. Antitumor-stimulating activities of the fruiting bodies of *Paecilomyces japonica*, a new type of *Cordyceps* spp. *Phototherapy Research* 17: 830–833.

Shukla, S. and Jaitly, A.K. 2011. Morphological and biochemical characterization of different oyster mushrooms (*Pleurotus* spp.). *Journal of Phytology* 3(8): 18–20.

Sinden, J.W. and Hauser, E. 1950. The short method of composting. *Mushroom Science* 1: 52–59.

Sinden, J.W. and Hauser, E. 1953. The nature of the composting process and its relation to short composting. *Mushroom Science* 2: 123–131.

Singh, M. 2011. Mushroom production: An agribusiness activity. In: *Mushrooms: Cultivation, Marketing and Consumption*, eds. M. Singh, B. Vijay, S. Kamal, and G.C. Warchaure, pp. 1–10. DMR, Chambaghat, Solan, India.

Smith, J., Rowan, N., and Sullivan, R. 2002. *Medicinal Mushrooms: Their Therapeutic Properties and Current Medical Usage with Special Emphasis on Cancer Treatments*, University of Strathclyde, Glasgow, U.K.

Spaulding, T. and Beelman, R. 2003. Survey evaluation of selenium and other minerals in *Agaricus* mushrooms commercially grown in the United States. *Mushroom News* 51: 6–9.

Stamets, P. 2000. *Growing Gourmet and Medicinal Mushrooms*, 3rd edn. Ten Speed Press, Berkeley, CA.

Stamets, P. 2005. *Mycelium Running: How Mushrooms Can Help Save the World*. Ten Speed Press, Berkeley, CA.

Taguchi, T. and Furue, H. 1985. End point result of a randomized controlled study on the treatment of gastrointestinal cancer with a combination of Lentinan and chemotherapeutic agents. *Expert Medica* 12: 151–165.

Talpur, N., Echard, B., Fan, A., Jaffari, O., Bagchi, D., and Preuss, H. 2002. Antihypertensive and metabolic effects of whole maitake mushroom powder and its fractions in two rat strains. *Molecular and Cellular Biochemistry* 237: 129–136.

Tewari, R.P. 2005. Mushrooms, their role in nature and society. In: *Frontiers in Mushroom Biotechnology*, eds. R.D. Rai, R.C. Upadhyay, and S.R. Sharma, pp. 1–8. NRCM, Solan, India.

Trappe, J.M. 1962. Fungus associates of ectotrophic mycorrhizae in nurseries. *Annual Review of Phytopathology* 15: 203–222.

United States International Trade Commission (USITC). 2010. *Mushroom Industry and Trade Summary*. Office of Industries Publication ITS-07, Washington, DC.

Upadhyay, R.C. and Vijay, B. 1988. Optimum compost depth in polythene bags for *Agaricus bisporus* cultivation. *Indian Journal of Agricultural Sciences* 58: 778–779.

U.S. Department of Agriculture (USDA). 2005. *National Nutrient Database for Standard Reference*, Release 18 August 2005. U.S. Government Printing Office, Washington, DC.

U.S. Department of Agriculture (USDA). 2006. *National Nutrient Database for Standard Reference*, Release 19 August 2006. U.S. Government Printing Office, Washington, DC.

Vetter, J. 1994. Mineral elements in the important cultivated mushrooms *Agaricus bisporus* and *Pleurotus ostreatus*. *Food Chemistry* 50: 277–279.

Vijay, B. 2011 Methods of compost preparation for white button mushroom. In: *Mushrooms: Cultivation, Marketing and Consumption*, eds. M. Singh, B. Vijay, S. Kamal, and G.C. Warchaure, pp. 49–68. DMR, Chambaghat, Solan, India.

Vijay, B. and Gupta, Y. 1995. Production technology of *Agaricus bisporus*. In: *Advances in Horticulture*, eds. K.L. Chadha and S.R. Sharma, pp. 63–98. Malhotra Publishing House, New Delhi, India.

Wakchaure, G.C. 2011. Production and marketing of mushrooms: Global and National Scenario. In: *Mushrooms: Cultivation, Marketing and Consumption*, eds. M. Singh, B. Vijay, S. Kamal, and G.C. Warchaure, pp. 15–20. DMR, Chambaghat, Solan, India.

Wakita, S. 1976. Thiamine-destruction by mushrooms. *Science Reports of the Yokohama National University Section* 2(23): 39–70.
Wang, X., Zhao, X., Li, D., Lou, Y.Q., Lin, Z.B., and Zhang, G.L. 2007. Effects of *Ganoderma lucidum* polysaccharide on CYP2E1, CYP1A2 and CYP3A activities in BCG-immune hepatic injury in rats. *Biological and Pharmaceutical Bulletin* 30(9): 1702–1706.
Wasser, S.P. and Weis, A.L. 1999. Medicinal properties of substances occurring in higher Basidiomycetes mushrooms: Current perspectives. *International Journal of Medicinal Mushrooms* 1: 31–62.
Weil, A. 1987. A mushroom a day. *American Health* 12: 129–134.
Wu, D., Pae, M., Ren, Z., Guo, Z., Smith, D., and Meydani, S.N. 2007. Dietary supplementation with white button mushroom enhances natural killer cell activity in C57BL/6 mice. *Nutritional Immunology* 137: 1472–1477.
Yanaki, T., Ito, W., and Kojima, T. 1983. Correlation between the antitumor activity of a polysaccharide schizophyllan and its triple-helical conformation in dilute aqueous solution. *Biophysical Chemistry* 17: 337–342.
Yang, Q.Y., Hu, Y.J., Li, X.Y. et al. 1993. A new biological response modifier-PSP. In: *Mushroom Biology and Mushroom Products*, eds. S.T. Chang, J.A. Buswell, and S.W. Chi, 247–259. The Chinese University Press, Hong Kong, China.
Yuan, Z., He, P., Cui, J., and Takeuchi, H. 1998. Hypoglycemic effect of water-soluble polysaccharide from *Auricularia auricula-judae* Quel. on genetically diabetic KK-Ay Mice. *Bioscience, Biotechnology and Biochemistry* 62: 1898–1903.
Zhang, Y., Mills, G.L., and Nair, M.G. 2002. Cyclooxygenase inhibitory and antioxidant compounds from the mycelia of the edible mushroom *Grifola frondosa*. *Journal of Agricultural and Food Chemistry* 50(26): 7581–7585.
Zhu, S. and Mori, M. 1993. *The Research on Ganoderma lucidum (Part I)*. Shanghai Medical University Press, Shanghai, China.

16

Value Addition of Agro-Industrial Wastes and Residues

Parmjit S. Panesar, Satwinder S. Marwaha, and Reeba Panesar

CONTENTS

16.1 Introduction 558
16.2 Nature of Food Industry Wastes 559
16.2.1 Fruit and Vegetable Industry 560
16.2.2 Dairy Processing Industry 561
16.2.3 Sugar Industry 561
16.2.4 Brewing Industry 562
16.2.5 Wine Industry 562
16.2.6 Distillery 563
16.2.7 Meat and Poultry Processing Industry 563
16.2.8 Seafood Industry 564
16.2.9 Coffee Industry 564
16.2.10 Olive Oil Industry 565
16.3 Production of Value-Added Products 565
16.3.1 Production of Single-Cell Proteins 566
16.3.2 Production of Baker's Yeast 568
16.3.3 Production of Organic Acids 569
16.3.3.1 Lactic Acid 569
16.3.3.2 Citric Acid 571
16.3.3.3 Gluconic Acid 573
16.3.3.4 Acetic Acid 573
16.3.3.5 Butyric Acid 574
16.3.3.6 Pyruvic Acid 574
16.3.3.7 Fumaric Acid 574
16.3.4 Bio-Fuel Production 574
16.3.4.1 Ethanol Production 574
16.3.4.2 Biogas Production 578
16.3.4.3 Biohydrogen Production 581
16.3.4.4 Production of Biodiesel 581
16.3.4.5 Production of Butanol 582
16.3.5 Production of Enzymes 582
16.3.6 Production of Food Additives 585

16.3.6.1 Production of Polysaccharides....585
16.3.6.2 Production of Bio-Pigments....586
16.3.6.3 Production of Bio-Flavors....587
16.3.6.4 Production of Biosurfactants....587
16.3.6.5 Production of Bacteriocins....588
16.3.6.6 Production of Vitamins....589
16.3.6.7 Production of Amino Acids....589
16.3.7 Production of Antibiotics....589
16.3.8 Production of Mushrooms....590
16.4 Summary and Future Prospects....590
References....591

16.1 Introduction

The technological advancement in the field of agriculture has led to the availability of large quantities of surplus agricultural produce and residues. Agro-processing involves a set of techno-economic activities carried out for preservation, handling, and value addition of agricultural produce, which encompasses all the operations from the stage of harvest till the material is transformed in desirable form for use by the consumers (Kachru, 2006). The net impact of all these activities is the rapid growth of food processing industries preparing both the primary and secondary processed products.

Food processing industry engaged in the processing of a variety of agricultural commodities such as fruits, vegetables, fish, meat, poultry, milk, cereal, pulses, oilseeds, and plantation products is one of the largest industries in terms of production, consumption, and growth prospects. Although the growth in food processing industrialization has resulted in a large variety of food products and provided employment and economic benefits, at the same, it is generating a huge quantity of biodegradable wastes/by-products posing a big threat to the environment. There are reports that food processing industry generates 45%–50% of the total organic industrial pollutants. It is estimated that in Europe only, food processing activities produce about 2.5×10^8 ton/year of by-products and waste (Laufenberg et al., 2003; Awarenet, 2004). The generation of wastes not only is an economic loss, but also requires additional cost for their management. The wastes/residues of common food processing industries contain good amounts of biodegradable materials such as carbohydrates, protein, and fat. Thus, most of these waste residues are nutritionally rich, and dumping of these residues contribute a significant loss of potential food and energy source. These wastes are valuable by-products, if bio-processed using appropriate technologies for value addition.

The stringent environmental legislations have significantly contributed to the introduction of sustainable waste management practices throughout the world. During the last few decades, there has been an increasing trend for the bio-utilization of agro-industrial residues for value addition. Biotechnological processes, especially the fermentation, have contributed enormously in the area of agro-waste and residue utilization. The utilization of agro-industrial residues through fermentation will not only provide an alternative substrate but also help in the reduction of environmental pollution as well as upgradation of agro residues and by-products.

16.2 Nature of Food Industry Wastes

Food processing industry can be divided into different subsectors such as fruit and vegetable, cereal/grain, milk, meat, fish and poultry processing, convenience foods, fermented foods, alcoholic beverage, and soft drink (Figure 16.1). The processing of food materials may involve a number of unit

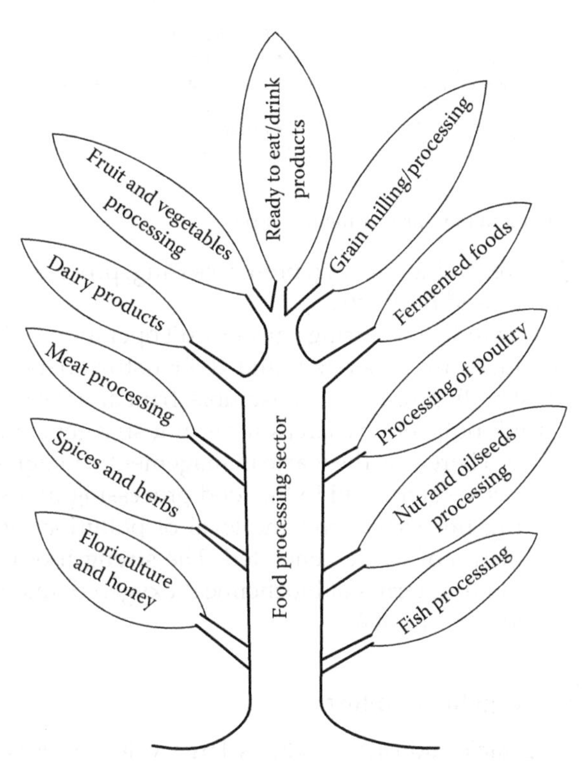

FIGURE 16.1
Major food processing sectors.

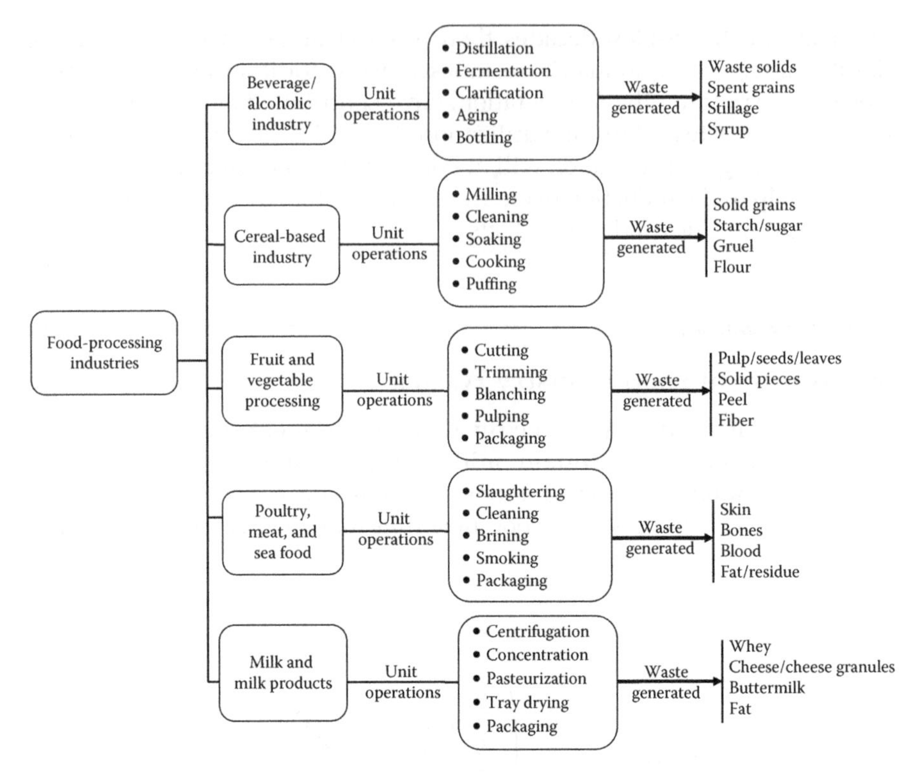

FIGURE 16.2
Type of waste generated in food-processing industries.

operations such as receiving, storing, pre-processing, processing, packaging, and storage of processed products.

In food processing industry during various unit operations, different types of wastes/by-products are generated, and their nature depends upon the type of food industry (Figure 16.2). For example, fruit and vegetable industry generates skin, seed, pulp, etc., whereas meat industry generates hair, hide, blood, etc., as waste. Therefore, the waste management strategies vary in different sectors. The waste generated in the food processing industry depends upon the food being processed, and the level of pollution would depend upon the characteristics of waste generated. The strength of the waste can be primarily estimated in terms of biochemical oxygen demand (BOD) and chemical oxygen demand (COD).

16.2.1 Fruit and Vegetable Industry

The fruit and vegetable industry produces large volumes of waste and by-products (20%–60% w/w) during processing (Awarenet, 2004). The major portion of organic waste is generally pomace and peels of fruits and vegetables. The chemical composition and the amount of organic waste can change during the year due to seasonal variations. Different process operations such as washing of raw materials, peeling, blanching, size reduction, slicing, dicing, juice processing, and filling contribute toward the waste generation (Shah et al., 2011). Thus, fruit and vegetable industry typically generates large volumes of effluents and solid waste. These effluents contain high organic loads, cleansing and blanching agents, salt, organic sugars, starch, and suspended solids.

16.2.2 Dairy Processing Industry

Dairy industry represents a major and important part of food industry and contributes significant volume of liquid wastes. The processing steps of milk and milk products involve the receiving and storing of raw materials, processing into finished products, packaging, storage, etc. All these operations lead to the waste generation in dairy processing industry. The important dairy by-products include whey, butter milk, and ghee residues.

Whey is the main by-product produced in large quantities by the dairy industry throughout the world. It may broadly be defined as the serum or watery part of milk remaining after separation of the curd, which results from the coagulation of milk by acid or proteolytic enzymes. The type and composition of whey mainly depend upon the processing technique applied for casein removal from fluid milk (Fox et al., 2000). Since the rennet-induced coagulation of casein occurs at approximately pH 6.5, this type of whey is referred to as sweet whey. The second type of whey is the acid whey (pH < 5), which results from processes using fermentation or addition of organic or mineral acids to coagulate the casein (Jelen, 2003). The main components of both sweet and acid whey after water are lactose (approx. 70%–72% of the total solids), whey proteins (approx. 8%–10%), and minerals.

Besides whey, large volume of wastewater is also generated during various operations in dairy industry. It has been estimated that 6–10 L of wastewater per liter of milk processed is generated during its processing. Dairy industry wastewater is associated with varying amount of milk constituents such as casein, lactose, fat, and inorganic salts besides detergents and sanitizers used for washings, attributing high BOD and COD to the wastewater (Mohanrao and Subrahmanyam, 1972).

16.2.3 Sugar Industry

In sugar industry, sugarcane and sugar beet are processed to manufacture edible sugar. More than 60% of the world's sugar production is from sugarcane and the remaining is from sugar beet. It is estimated that approximately 20 cubic meters of water per metric ton (m^3/ton) of cane processed is used (MIGA, 2006). Sugar mills use around 2000 L of water and generate about 1000 L of wastewater per ton of cane crushed. The wastewater in sugar industry includes floor washing, condensate water, sugarcane juice, syrup, and molasses (Solomon, 2005).

Molasses is a thick, syrup-like, viscous, and dark-colored liquid produced as a by-product in the production of raw and refined sugar from sugar beet or sugarcane. According to its origin, molasses is known as cane or beet molasses. It is the residual syrup from which no crystalline sucrose can be obtained by simple means (Paturau, 1982). The yield of molasses is approximately 3.0% per ton of sugarcane processed but is influenced by a number of factors. Another important waste generated during sugarcane processing is bagasse, which is fibrous residue of the cane stalk left after crushing and extraction of the juice. However, the utilization of sugarcane bagasse is still limited and is mainly used as a fuel to power the sugar mills (Charles and Shuichi, 2003). Besides bagasse, sugar mills discharge large amount of wastewaters that are acidic in nature and associated with high concentration of suspended solids, etc., contributing toward high dissolved solids, BOD, and COD (Mishra et al., 1987).

16.2.4 Brewing Industry

A brewery generates large quantity of effluent with a very high BOD. These effluents are characterized by rapidly varying pH, temperature, loading, and flow values due to batch-wise operations. The quality and quantity of brewery effluent can vary significantly as it depends on various unit operations that take place within the brewery. The amount of wastewater produced is related to the specific water consumption. In general, water consumption per volume of beer produced attains 4.7 m^3/m^3 (Carlsberg, 2005). A part of the water is disposed with the brewery by-products and a part is lost by evaporation. As a result, the wastewater to beer ratio is often 1.2–2 m^3/m^3, less than the water to beer ratio (Driessen and Vereijken, 2003). The effluent contains suspended organic matter such as spent grain, traub, and waste yeast. Dissolved solids are mainly from beer, wort, and cleaning and sanitizing agents. The excessive use of disinfectants and caustic cleaning solutions can cause toxic shock loadings on biological treatment systems (Vriens et al., 1990).

Brewery wastewater is highly biodegradable as this mainly consists of sugars, soluble starch, ethanol, volatile fatty acids (VFAs), etc., and it has relatively high BOD/COD ratio (Brito et al., 2007). It is estimated that 75% of the fermentable sugars are extracted in brewery plant and the rest enter the by-products (Joshi et al., 2011). Nitrogen and phosphorous levels mainly depend on the handling of raw material and the amount of spent yeast present in the effluent.

16.2.5 Wine Industry

Wineries generate large amount of wastewater, mainly through various washing operations during the crushing and washing of grapes, as well as rinsing of fermentation tanks, barrels, and other vessels used during fermentation (Petruccioli et al., 2000). The volume and pollution load of effluent greatly vary depending upon the technology used and working operations (Brito et al., 2007). Moreover, quantity and quality of effluent generated daily have great variability, which makes the evaluation of pollution more complex.

Typical waste includes unconsumed inorganic and organic acids, microbial cells, wash water containing traces of solvents, alkali, etc., having high BOD and high quantity of solids. The solid waste mostly comprising skin, seed, and pulp is collectively known as pomace. Besides pomace, winery generates pulp and skin sediments rich in yeast. In wastewater, musts and wine constituents are present in variable proportions. Sugars, ethanol, esters, glycerol, organic acids (citric, tartaric, malic, lactic, acetic), phenolic compounds, bacteria, and yeast are the main constituents. All these are easily biodegradable except polyphenols. Effluents contain less nitrogen and phosphorous contents and have a BOD/N/P ratio usually near 100/1/0.3 (Torrijos and Moletta, 1997; Brito et al., 2007).

16.2.6 Distillery

Distillery by-products mainly comprise yeast sludge, spent waste, and carbon dioxide. Out of these, spent waste usually has a very high pollution load (Indira et al., 1980). The spent wash generally discharges 500 ppm of total solids into the environment. Apple-based alcohol beverage industry generates pomace as a waste having high BOD values. Wine distilleries produce large quantities of waste known as "vinasses," which are acidic and organic in nature (Joshi, 2011).

Distiller's dried grain and distiller's dried grain with soluble are by-products from the manufacturing of whiskey or grain fuel alcohol. They contain the nonextracted portions of corn, possibly also some rye and malted barley, and generally contain yeast (Joshi et al., 2011). The COD of effluents of distillery in Wellington, South Africa, ranged from 20,000 to 30,000 mg/L (Wolmarans and De Villiers, 2002), whereas COD values of grape-based distillery effluents have been reported between 22,000 and 48,000 mg/L (Joshi, 2011). The effluents from wine distillery mainly consist of organic acids with high level of soluble biodegradable fractions (Oleszkiewicz and Olthof, 1982). Various phenolic compounds such as gallic acid, p-coumaric acid, and gentisic acid are also other important components of wine distillery wastewaters (Borja et al., 1993a).

16.2.7 Meat and Poultry Processing Industry

Meat and poultry industry consists of slaughterhouses and processing units where meat is prepared, cut into pieces, and is further processed, i.e., either frozen, cooked, cured, smoked, or made into sausages. A major part of the materials handled by the meat and poultry industry originates from slaughterhouses. The wastes coming from these units contain various proportions of blood, fats, residues from the intestine, paunch grass, manure, etc.

(Cournoyer, 1996). The killing and rendering create blood by-products and waste streams, which have very high BOD. Slaughterhouse wastewater is typically high in both moisture (90%–95%) and nitrogen, has a high BOD, and is odorous. The characteristics of waste streams can vary, but these can be generalized as process wastewaters, carcasses and skeleton waste, ejected or unsatisfactory animals, fats, oils and greases, blood, and eviscerated organs (Dessoff, 1997). In general, wastes from meat processing industry contain high concentration of fat and protein, although the composition depends upon the type of production and facilities.

16.2.8 Seafood Industry

Seafood industry waste is a typical characteristic of coastline areas, where large volume of waste is generated by aquaculture and fish-processing industries. In seafood industry, numerous types of seafood such as mollusks (oysters, clams, scallops), crustaceans (crabs and lobsters, saltwater fishes), and freshwater fishes are processed (Tay et al., 2006). Fish processing is an important component of seafood industry. Fish processing waste composed of solid and liquid wastes. During the processing of fish, a large amount of solid waste such as skin, bone, and fin is generated (Shahidi, 1994). The solid waste generation depends upon the type of industry. It has been estimated that fin fish trimming and filleting operations generate 30%–60% waste, shrimp processing generates 40%–80% waste, and crab processing generates 75%–85% waste (Chandra and Shamasundar, 2011).

The liquid waste generally comes from wash water from the fish processing plants. Similar to most processing industries, seafood-processing operations produce wastewater containing substantial contaminants in soluble, colloidal, and particulate forms (Tay et al., 2006). The generated wastewater can be very high in BOD; fat, oil, and grease; and nitrogen content. BOD is derived primarily from the butchering process and general cleaning, and nitrogen originates largely from blood in the wastewater stream (Environment Canada, 1994).

16.2.9 Coffee Industry

Coffee is an important agricultural commodity, which is produced in the tropics and mainly consumed in Europe and the United States. It has been reported that only 9.5% of the weight of the fresh material is used for the preparation of coffee beverage and 90.5% is left as residue (Murthy and Manonmani, 2008). Coffee processing industries generate several residues such as coffee cherry wastes, coffee parchment wastes, sliver skin, coffee spent grounds, and coffee leaves. However, the quantity of coffee wastewater generation varies depending on the processing technology used.

The main pollution in coffee wastewater originates from the organic matter set free during pulping when the mesocarp is removed, and the mucilage texture surrounding the parchment is partly disintegrated (Mburu et al., 1994). Pulping water comprises quickly fermenting sugars from both pulp and mucilage components. Pulp and mucilage contains mainly proteins, sugars, and mucilage in particular of pectins (Avellone et al., 1999). BOD and COD of coffee processing wastewater can be up to 20,000 and 50,000 mg/L, respectively (Von Enden and Calvert, 2002).

16.2.10 Olive Oil Industry

Olive oil industry is a very important economic activity, particularly in Southern Europe. The wastes generated by olive mills represent an important environmental problem since these are generated in huge quantities in short period of time. Olive pulp is a highly polluting semi-solid residue, which is produced during the two-stage extraction processing of olives and is a major environmental issue (Georgieva and Ahring, 2007).

The estimated annual olive oil mill wastewater production in Mediterranean countries is over 3×10^7 m^3 (Borja et al., 1993b). Biological and chemical oxygen demands of this waste can be 100 and 200 g/L, respectively. The organic substances include some sugars, tannins, polyphenols, polyalcohols, pectin, and lipids (Hamdi, 1993). Because of its high content of polyphenol compounds, liquid waste from olive oil industry is a dark-colored juice; however, the color depends on the age and type of olive processed and also the type of technology used (Hamdi and Garcia, 1993). The difficulty of disposing olive oil mill wastewaters is mainly related to its high BOD, COD, and high concentration of organic substances (Saez et al., 1992).

16.3 Production of Value-Added Products

The significant quantity of wastes generated by food processing industry is organic in nature and rich in biodegradable materials, making them suitable substrates for the production of value-added products using conventional and innovative biotechnological tools. The concept of bioconversion of industrial wastes into value-added products has become a major subject of research and process development around the world and is receiving increased attention in view of the fact that there is a need for the best use of valuable resource materials while maintaining the environmentally sustainable processes (Jin et al., 2005).

Application of biotechnological innovations has played a significant role in the production of a variety of value-added products while using agro-industrial wastes as raw material (Figure 16.3). Significant reductions of BOD with concomitant production of useful bio-products such as lactic acid, ethanol, citric acid, enzymes, and single-cell proteins (SCPs) have been

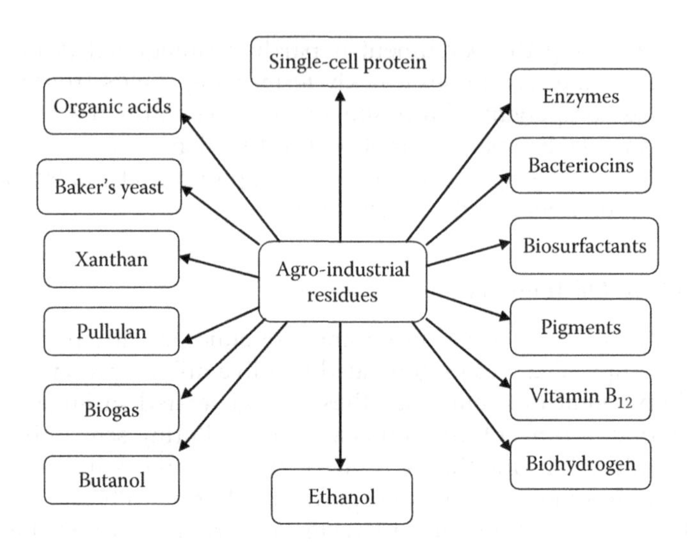

FIGURE 16.3
Value-added products produced from agro-industrial residues through bio-processing.

achieved. Recently, novel products such as exopolysaccharides, biofuels, glycerol, poly-3-hydroxybutyric acid, biosurfactants, and bacteriocins have also been produced using agro-industrial industry waste through biotechnological means. The rest of this chapter focuses on the various approaches being applied for the bio-processing of agro-industrial wastes for value-added products.

16.3.1 Production of Single-Cell Proteins

The term SCP refers to the dried cells of microorganisms such as algae, actinomycetes, bacteria, yeasts, molds, and higher fungi, which can be used as protein source in human food or animal feed (Litchfield, 1983). SCPs are excellent source of proteins that are comparable to other sources. Although the microorganisms are grown primarily for their protein contents in SCP production processes, these also contain carbohydrates, lipids, vitamins, minerals, and nucleic acids. The utilization of agro-industrial waste as the resource material in SCP processes serves two functions, i.e., reduction in pollutants and production of edible protein, which can be either used as a food component or as animal feed. However, each waste material must be assessed for its suitability for conversion to SCP. The main reasons for the increasing interest in SCP production from agro-industrial wastes are escalating cost of conventional protein sources, cheap rather negative values of the substrates, and feasibility of large-scale SCP production by using computer-controlled fermentation processes (Kharatyan, 1978; Litchfield, 1983). The production of SCP at commercial scale is determined by the economic considerations such as availability and cost of raw material, water, labor, capital resources, transportation of finished product, capital investment for fermentation equipment, efficacy of the process, market price of the traditional protein sources, and the disposal of secondary by-products (Arora et al., 2000).

A number of microorganisms including bacteria, yeast, and molds can be employed for the production of SCPs, although each of them has their own advantages and disadvantages. SCP production processes utilizing waste substrates have been reported using various organisms (Table 16.1). The substrates and producer organisms used include sulfite waste liquor (*Paecilomyces variotii, Candida utilis*), molasses (*Saccharomyces cerevisiae*), and cheese whey (*Kluyveromyces fragilis*) while the Symba process developed in Sweden utilizes starchy wastes, e.g., wastewater from potato processing plant by combining two yeasts, *Endomycopsis fibuligera* and *C. utilis* (Skogman, 1976; Litchfield, 1983). Potato peels supplemented with ammonium chloride have also been used for the production of protein using *Pleurotus ostreatus* (Kahlon and Arora, 1986). Another popular process in SCP production is Pekilo process. Pekilo is a fungal protein product that has been produced by continuous fermentation of carbohydrates derived from spent sulfite liquor using *P. variotii* (Litchfield, 1983; Oura, 1983).

SCP production has also been carried out utilizing wastes from fermentation industries, such as winery, brewery, and distillery (Gera and Kramer, 1969; Hang, 1986). *S. cerevisiae, Torulopsis utilis,* and *Rhizopus* sp. have successfully grown on the molasses with good SCP yield, which could be further increased with the addition of corn steep liquor (Shukla and Dutta, 1967).

Different fruit wastes have been utilized for the production of SCP. *Trichoderma viride* and *Aspergillus niger* have successfully been grown on

TABLE 16.1

Utilization of Some Agro-Industrial Waste for Single-Cell Protein Production

Agro-Industrial Residues	Microorganism Employed/ Producer Strain	Reference(s)
Apple pomace	*Saccharomycopsis fibuligera* and *C. utilis*	Fellows and Worgan (1988)
Apple pomace	*T. viride* and *A. niger*	Hang (1986)
Citrus peel waste	*Fusarium* sp.	Bahar and Azuaze (1984)
Molasses	*S. cerevisiae, T. utilis,* and *Rhizopus* sp.	Shukla and Dutta (1967)
Potato processing waste	*E. fibuligera* and *C. utilis*	Skogman (1976)
Potato peels	*P. ostreatus*	Kahlon and Arora (1986)
Sulfite waste liquor	*P. variotii* and *C. utilis*	Oura (1983)
Soybean oil waste	*C. lipolytica*	Takata (1992)
Whey	*Saccharomyces fragilis*	Powell and Robe (1964)
Whey	*K. lactis, K. fragilis, Torulopsis bovina,* and *Candida intermedia*	Moulin and Galzy (1984)
Whey	*K. marxianus* A2	Anvari and Khayati (2011)

dried and pectin-extracted apple pomace using solid and submerged fermentation processes (Hang, 1986). Apple pomace proved to be a good substrate for the growth of *S. fibuligera* and *C. utilis* for use as a source of SCP (Fellows and Worgan, 1988). Bioconversion of apple pomace supplemented with nutrients by *A. niger* resulted in high protein content. Waste citrus peels from orange processing industry have also been utilized for SCP production using *Fusarium* sp. (Bahar and Azuaze, 1984). The suitability of waste soybean oil for biomass production has also been explored and *Candida lipolytica* resulted in highest yield of biomass (Takata, 1992).

Whey is also a suitable substrate for SCP production. Biomass can be produced using whey or ultrafiltration permeate in three ways: direct use of lactose by microorganisms, conversion of lactose into glucose and galactose by enzymatic or chemical hydrolysis, and prior fermentation by lactic bacteria producing a mixture of lactic acid and galactose (Boze et al., 1995). The majority of the studies have been done with lactose-utilizing yeasts, mostly *Kluyveromyces* (*Saccharomyces*) *fragilis* strains, which offer advantages of good growth yields and acceptability as safe microorganisms. The high protein feed production has also been reported by growing lactose fermenting *S. fragilis* on whey (Powell and Robe, 1964). Concentrated whey permeates have also been used for SCP production using *Kluyveromyces* under aerobic and anaerobic conditions (Mahmoud and Kosikowski, 1982).

Depending upon the type of whey used, different supplements may be necessary to complement the nutrient deficiencies. Complete conversion of whey lactose to yeast biomass requires addition of supplementary nitrogen, which can be accomplished by using ammonium salts (Castillo and De Sanchez, 1978). Although whey has significant amounts of vitamins, additions of yeast extract or corn steep liquor increase the specific growth rates. Among the several processes, the Vienna and the Bel processes are the most popular for the production of SCP (Moulin and Galzy, 1984). The production of SCP from deproteinized sweet and sour cheese whey concentrates using *Kluyveromyces marxianus* CBS 6556 indicated that the use of the Crabtree-negative strain yeast was an important factor for gaining high cell concentrations (Schultz et al., 2006). Recently, strain of *K. marxianus* designated as A2 has been isolated, which resulted in the maximum SCP production from whey with the yield of 12.68 g/L (Anvari and Khayati, 2011).

16.3.2 Production of Baker's Yeast

The strains of Baker's yeast (*S. cerevisiae*) have been used all over the world in bread leavening. Earlier, the yeast employed was generally taken from breweries or distilleries. Nowadays, sugarcane or beet molasses are widely used for the production of baker's yeast. However, these raw materials are not sufficiently rich in nitrogenous and phosphorus compounds, and usually the addition of inorganic ammonium compounds, biotin, and phosphoric acid is needed for the good growth of Baker's yeast (Paturau, 1987).

Lactobacillus bulgaricus and *Streptococcus thermophilus* strains have been recommended for the conversion of lactose in whey and its subsequent utilization for baker's yeast multiplication (Moebus and Kiesbye, 1975; Champagne et al., 1990). The industrial process for the production of baker's yeast (*S. cerevisiae*) from whey has been used by Nutrisearch Company at Winchester, Kentucky. The process involves lactose hydrolysis in cheese whey by immobilized lactase followed by glucose–galactose fermentation (Castillo, 1990). The use of nonconventional yeasts as baker's yeast has also been made, but to make this approach workable, such yeasts should show baking properties similar to the baker's yeast. Yeasts like *T. utilis*, *Torula cremoris*, and *Torula casei* have been successfully grown on whey (Vij and Gandhi, 1993).

The use of apple pomace extract as an alternative to molasses (traditionally used as a carbon source) for baker's yeast production has also been explored (Bhushan and Joshi, 2006). Apple pomace extract-based medium in an aerobic fed-batch culture has been employed for the production of baker's yeast. Interestingly, the dough-raising capacity of the baker's yeast grown on the apple pomace extract was apparently the same as that of commercial yeast.

16.3.3 Production of Organic Acids

Different agro-industrial wastes/residues have been employed for the production of organic acids such as lactic, citric, gluconic, and acetic acid. (Table 16.2).

16.3.3.1 Lactic Acid

Lactic acid, its salts, and esters have a wide range of potential applications and are extensively used in diverse fields of food, cosmetic, and pharmaceutical industries (Hofvendalh and Hahn-Hagerdal, 2000; Panesar et al., 2007a). The world market for lactic acid is growing every year, and its current production is about 150 million lb per year worldwide and is likely to grow between 10% and 15% per year in the coming time (Wassewar, 2005).

Lactic acid production by microbial fermentation using refined sugars in submerged fermentation process is an expensive process (Datta et al., 1995). Therefore, cheaply available and renewable agro-industrial waste as raw materials (complex organic sources) has been evaluated and employed for lactic acid production. Molasses after supplementation with nutrients has been used for the production of lactic acid in batch fermentation using bacteria, viz., *Lactobacillus delbrueckii* and *Enterococcus faecalis* (Aksu and Kutsal, 1986; Wee et al., 2004). The continuous production of lactic acid from molasses by *Sporolactobacillus cellulosolvens* has also been reported (Kanwar et al., 1995).

The availability of carbohydrate reservoir of lactose in whey and the presence of other essential nutrients make the whey a potential raw material for

TABLE 16.2

Utilization of Agro-Industrial Residues for the Production of Organic Acids

Agro-Industrial Residues	Microorganism Employed/ Producer Strain	Organic Acid	Reference(s)
Apple pomace	*A. niger*	Citric acid	Shojaosadati and Babaeipour (2002), Dhillon et al. (2011a)
Apple pomace	*L. rhamnosus* CECT-288	Lactic acid	Gullon et al. (2008)
Brewery waste	*A. niger*	Citric acid	Hang et al. (1977)
Cassava bagasse and sugarcane bagasse	*L. delbrueckii*	Lactic acid	John et al. (2006a)
Deproteinized whey	*A. niger*	Gluconic acid	Mukhopadhyay et al. (2005)
Grape must	*A. niger* and *Gluconobacter oxydans*	Gluconic acid	Buzzini et al. (1993)
Grape pomace	*A. niger*	Citric acid	Marwaha et al. (2000)
Molasses	*E. faecalis*	Lactic acid	Wee et al. (2004)
Molasses	*S. cellulosolvens*	Lactic acid	Kanwar et al. (1995)
Molasses	*L. delbrueckii*	Lactic acid	Aksu and Kutsal (1986)
Molasses	*A. niger*	Gluconic acid	Rao and Panda (1994)
Molasses	*A. niger*	Citric acid	Paturau (1987)
Molasses	*C. butyricum*	Butyric acid	Garg et al. (1995)
Molasses	*R. arrhizus*	Fumaric acid	Petruccioli et al. (1996)
Orange peel and molasses	*A. niger*	Citric acid	Hamdy (2012)
Potato waste	*A. niger*	Citric acid	Marwaha et al. (2000)
Salted whey	*K. fragilis, strain* L.	Acetic acid, glycerol	Mostafa (2001)
Wheat bran	*L. casei* and *L. delbrueckii*	Lactic acid	John et al. (2006b)
Whey	*L. Bulgaricus*	Lactic acid	Cox and Macbean (1977)
Whey	*L. delbrueckii*	Lactic acid	Panesar et al. (2007a)
Whey	*A. niger*	Citric acid	El-Samragy et al. (1996); El-Holi and Al-Delaimy (2003)
Whey	*L. casei*	Lactic acid	Panesar et al. (2007b)
Whey permeate	*L. helveticus*	Lactic acid	Panesar et al. (2007a)
Whey permeate	*A. niger*	Citric acid	Hossain et al. (1983)

lactic acid production. However, the supplementation of whey with nutrients like yeast extract, peptones, molasses, corn steep liquor, lactose, vitamins, minerals, and amino acids is desirable for increasing the lactic acid yield (Cox and Macbean, 1977). A three-stage continuous fermentation pilot scale for the production of lactic acid with a productivity of 22 g/L/h has been developed (Boyaval et al., 1987). When the electrodialysis unit was coupled with the earlier process, the production of 85 g/L lactate was recorded.

When reconstituted whey permeate supplemented with yeast extract was used in a continuous stirred tank reactor (CSTR)-membrane recycle bioreactor for lactic acid production using *L. bulgaricus*, the productivity was 6–18 times better than the batch reactor system (Mehaia and Cheryan, 1986). Although, in recent years, the amount of lactic acid produced using biotechnological methods has increased; however, higher costs are associated with the separation steps that are needed to recover and purify the product from the fermentation broth. To decrease the costs of recovery and purification, integrated process for the production of food-grade lactic acid from whey ultrafiltrate (WU) has been suggested. The proposed system consists of fermentation, ultrafiltration, ion exchange, reverse osmosis, and final concentration by vacuum evaporation (González et al., 2007).

The immobilized cell systems have also been tested for the fermentation of whey for lactic acid production. A very stable system has been developed for the bioconversion of whey to L(+)-lactic acid using pectate-entrapped *L. casei* cells (Panesar et al., 2007b). For detailed information on the application of biotechnological techniques for the production of lactic acid from whey, readers may see the review articles published by Panesar et al. (2007a) and Kosseva et al. (2009).

Apple pomace as a potential substrate for lactic acid production has also been evaluated. Apple pomace after enzymatic hydrolysis was used for the fermentation using *Lactobacillus rhamnosus* CECT-288, and lactic acid production up to 32.5 g/L has been obtained (Gullon et al., 2008). Apple pomace has several advantages as a raw material for lactic acid manufacture such as high content of free glucose, fructose, and polysaccharides (cellulose, starch, and hemicelluloses), which can be enzymatically hydrolyzed to give monosaccharides, and presence of metal ions (Mg, Mn, Fe, etc.), which can reduce the nutrient supplementation cost (Hofvendalh and Hahn-Hagerdal, 2000; Kosseva, 2011).

16.3.3.2 Citric Acid

Citric acid is one of the most versatile organic acids, finding increasing applications in the food, beverage, pharmaceuticals, textiles, printing, chemical, and other industries. Solid-state fermentation (SSF) offers many advantages over the submerged fermentation process (Shankaranand and Lonsane, 1994). *A. niger* is the most commonly used mold in citric acid fermentation. Wheat bran, rice bran, and sweet potato fibrous residues have been frequently used as substrates to produce citric acid by SSF. The other wastes including pineapple juice, molasses, beet pulp residues, sweet potatoes residues, sugarcane bagasse, apple pomace, grape pomace, and mandarin orange waste have also been used for citric acid production using *A. niger* (Marwaha et al., 2000).

The fermentation process using molasses and *A. niger* for citric acid production consists of a complex aerobic cycle. More success in citric acid production has been reported with beet molasses as the main raw material than the cane molasses. Aeration and agitation of the medium are essential, and the addition of methanol in cane molasses has been reported advantageous. The citric acid yield is about 65% of total sugar used (Paturau, 1987).

In the SSF process, solid substrates such as wheat or rice bran are moistened, after adjusting the pH, are inoculated with spores of *A. niger,* and are kept in trays or are placed in winrows at 3–5 cm depth, and incubated for citric acid production (Shankaranand and Lonsane, 1994). Citric acid is leached from the fermented solids using hot water, and the extract is subjected to further downstream processing by the conventional technology.

The fermentation of whey permeate has also been attempted for the production of citric acid. In general, high concentration of sugar is required to produce high yields of citric acid. The supplementation of whey permeate with lactose can be useful to increase the yield of citric acid. *A. niger* and its mutants are commonly used for the production of citric acid (Hossain et al., 1983; El-Samragy et al., 1993). The cumulative effect of pH (3.5) of fermentation medium, methanol concentration (4%, v/v), and salt concentration (10%, w/v) during the fermentation of whey supports fourfold increase in citric acid production with *A. niger* CAIM 167 than with *A. niger* CAIM 111 (El-Samragy et al., 1996). Among the different supplementations (sucrose, glucose, fructose, galactose, riboflavin, tricalcium phosphate, and methanol) tested for citric acid production using *A. niger* ATCC 9642, whey with sucrose 15% (w/v) with or without 1% methanol was the most favorable medium (El-Holi and Al-Delaimy, 2003). The studies on utilization of whey and grape must for citric acid production using *Yarrowia lipolytica* NBRC 1658 and *Y. lipolytica* 57 indicated that fructose supplementation can be beneficial for citric acid production (Yalcin et al., 2009). Brewery waste as a substrate for citric acid production using SSF and *A. niger* has also been used (Hang et al., 1977).

Fruit waste, i.e., peels and pineapple pomace, supplemented with 4% methanol has been used for citric acid production using *A. niger* and SSF process (de Lima et al., 1995). Apple pomace has been used for the production of citric acid using *A. niger* via SSF in column reactors (Shojaosadati and Babaeipour, 2002). It has been revealed that the aeration rate and particle size are the most important parameters for maximizing the citric acid yield. The addition of methanol can be beneficial for the fermentation of apple pomace for citric acid production. Apple pomace waste has also been supplemented with rice husk for citric acid production through SSF (Dhillon et al., 2011a). Submerged fermentation has also been investigated for citric acid production using apple pomace ultrafiltration sludge as the substrate and using *A. niger* NRRL567 (Dhillon et al., 2011b). The utilization of inexpensive agro-industrial wastes and their by-products through SSF by natural/wild and genetically engineered strains along with downstream processing strategies has been recently reviewed (Dhillon et al., 2011c). Recently, an orange peel medium supplemented with cane molasses has been used for the citric acid production by *A. niger* (Hamdy, 2012).

16.3.3.3 Gluconic Acid

Gluconic acid and its derivatives such as sodium gluconate have wide range of applications in the formulations of food, pharmaceutical, and hygienic products (Ramachandran et al., 2006). Microbial production of gluconic acid is the preferred method, and *A. niger* is the most widely used microbial culture in fermentation process. Although glucose is a commonly used carbon source for fermentative production of gluconic acid, various agro-industrial wastes have also been explored as substrate. High yield of gluconic acid in media containing glucose or starch hydrolysate as the sole carbon source has been reported (Kundu and Das, 1984). Cane molasses as a source of glucose has also been used for the gluconic acid production (Rao and Panda, 1994).

Deproteinized whey has been used as a fermentation medium for the production of gluconic acid using immobilized *A. niger* (Mukhopadhyay et al., 2005). It has been reported that the addition of a small amount of glucose (0.5%, w/v) in the whey medium can be beneficial in enhancing the production of gluconic acid. Immobilized mycelia showed better results than free mycelia in terms of gluconic acid production. Gluconic acid has also been successfully produced by fermentation from grape must and concentrated rectified grape must using *A. niger* (Buzzini et al., 1993).

16.3.3.4 Acetic Acid

Acetic acid (vinegar) production can be carried by utilizing waste fruit juices or left-over material from juice production. The utilization of orange peel and waste pineapple juice can be made for vinegar production (Gera and Kramer, 1969). Vinegar has been produced by mixing apple pomace extract with molasses (ratio of 2:1) supplemented with ammonium sulfate for alcoholic fermentation followed by acetic acid fermentation (Joshi et al., 1999). Mango peels treated with pectic enzymes and pressing have been utilized for vinegar production. Mango peels and stones have also been utilized for the production of good quality vinegar (with 4.5%–5% acetic acid) with characteristic mango flavor (Ethiraj and Suresh, 1992).

Vinegar has also been produced from whey by fermentation with *Saccharomyces* sp. followed by oxidation with *Acetobacter aceti*. The use of salted whey (liquid by-product from the dairy industry containing 7.5% NaCl) as a substrate for either acetic acid or glycerol production has also been investigated by using two yeast strains (Mostafa, 2001). High percent yield of acetic acid (0.497 g acetic acid/g lactose) has been reported using immobilized cells of isolated yeast strain as compared with *K. fragilis*.

16.3.3.5 Butyric Acid

Butyric acid has several potential applications in the foodstuffs and beverage industries. The strains of *Clostridium* sp. are preferred for butyric acid or butanol production for commercial uses. Although glucose is the common carbon source for butyrate production using *Clostridium* sp., some agro-industrial wastes/by-products such as wheat straw, whey, molasses, and potato wastes have also been exploited (Zigová and Šturdik, 2000). *Clostridium butyricum* has been used in the fermentation of sugar molasses to produce butyric acid (Garg et al., 1995).

16.3.3.6 Pyruvic Acid

Pyruvic acid is used as a reagent for preparation of tyrosine and tryptophan. Citrus peels are used as a carbon source for the production of pyruvic acid (Moriguchi, 1982). Among the different yeasts tested, *Debarormyces coudertii* displayed the production of highest amount of pyruvic acid in peel extract.

16.3.3.7 Fumaric Acid

Fumaric acid is used as an acidulant in food and pharmaceutical preparations. This acid has been produced in repeated batch processes using immobilized *Rhizopus arrhizus* NRRL 1526 in a fluidized-bed reactor using glucose molasses (as glucose equivalent) and ammonium sulfate (Petruccioli et al., 1996). Using SSF, lyophilized orange peels in combination with grape must and concentrated rectified grape must have also been used for fumaric acid production by *R. arrhizus* ATCC 13310 (Buzzini et al., 1990).

16.3.4 Bio-Fuel Production

The continuous depletion of the fossil fuel reserves and consequent escalation in their prices have stimulated an extensive evaluation of alternative technologies and substrates to meet the global energy demand. As a result, alternative sources of energy like ethanol, methane (biogas), and hydrogen are increasingly being considered as potential substitutes. Different agro-industrial wastes have been employed to produce these biochemicals through biotechnological routes.

16.3.4.1 Ethanol Production

Ethanol is an important organic solvent, which has wide range of industrial applications. It is perhaps considered to be the most promising for use as an alternative liquid fuel, since it can be readily produced from a variety of agriculture-based renewable materials like sugarcane juice, molasses, potatoes, corn, and barley. As early as 1890, the potential of ethanol as a substitute for gasoline was conceived (Esser and Karsch, 1984). Currently, yeast (*S. cerevisiae*) is used all over the world as the major ethanol-producing microorganism. Although other microorganisms have also been tested and out of these, *Zymomonas mobilis* has emerged as a potential alternative to presently used yeast for ethanol production (Panesar et al., 2006). Ethanol can be produced from any processing industry waste, rich in sugars. A wide range of food industry wastes have been employed for the production of ethanol (Table 16.3).

Molasses is the traditional raw material for the production of ethanol using *S. cerevisiae* at industrial scale (Maiorella et al., 1981). Batch fermentations of sugarcane blackstrap molasses to ethanol using pressed yeast revealed an exponential relationship between the fermentation time and the initial

TABLE 16.3

Utilization of Agro-Industrial Wastes for Ethanol Production

Agro-Industrial Residues	Microorganism Employed/Producer Strain	Product	Reference(s)
Apple pomace	*S. cerevisiae*	Ethanol	Joshi and Sandhu (1994), Khosravi and Shojaosadati (2003)
Banana peel	*S. cerevisiae*	Ethanol	Tewari et al. (1986)
Citrus peel	*S. cerevisiae*	Ethanol	Grohmann et al. (1994)
Cassava starch	*E. fibuligera* and *Z. mobilis*	Ethanol	Panesar et al. (2006)
Corn starch	*Z. mobilis*	Ethanol	Krishnan et al. (2000b)
Molasses	*S. cerevisiae*	Ethanol	Maiorella et al. (1981)
Molasses	*Z. mobilis*	Ethanol	Rhee et al. (1984), Panesar et al. (2006)
Rice straw and bagasse	*C. shehatae* and *S. cerevisiae*	Ethanol	Palnitkar and Lachke (1990)
Rice straw	*S. cerevisiae* and *Pachysolen tannophilus*	Ethanol	Chadha et al. (1995a)
Rice straw hydrolysate	*C. shehatae*	Ethanol	Abbi et al. (1996)
Rice straw	Recombinant *Z. mobilis*	Ethanol	Yamada et al. (2002)
Rice straw hydrolysate	*Z. mobilis*	Ethanol	Krishnan et al. (2000a)
Whey	Recombinant *S. cerevisiae*	Ethanol	Sreekrishna and Dickson (1985)
Whey	*K. fragilis*	Ethanol	Mawson (1994)
Cheese whey powder	*K. marxianus*	Ethanol	Kargi and Ozmihci (2006)
Whey permeate	*K. pneumoniae*	Butanol	Maddox et al. (1988)
Whey permeate	*C. acetobutylicum*	Butanol	Qureshi and Maddox (1995)
Whey	*C. acetobutylicum*	Butanol	Foda et al. (2010)

concentrations of sugar and the yeast cells (Borzani et al., 1993). Membrane technology can be applied to desalt the molasses for ethanol production using *Z. mobilis*. Fermentation studies using desalted molasses resulted in faster rates of ethanol production and overall higher yields (Rhee et al., 1984). The fermentation of sugarcane syrup and A-molasses by *Z. mobilis* requires addition of small amounts of magnesium sulfate and diammonium phosphate as supplements for their efficient fermentation (Doelle et al., 1990). Ethanol production from molasses has also been successfully scaled up with 91%–95% conversion efficiencies and up to 10% (v/v) ethanol yields with the addition of sucrose/syrup using *Z. mobilis* (Doelle et al., 1991).

Citrus molasses has also been explored for ethanol production by fermentation using yeast in batch process with envisaged usage as neutral spirit for manufacture of alcoholic beverages (Braddock, 1999). It has been reported that 1 ton of citrus molasses can be converted to about 46 gallons of ethanol (96% alcohol), and 40,000 tons of molasses can produce 1.84 million gallons of ethanol. A fixed bed immobilized cell reactor system has been used in continuous fermentation of orange waste (Gera and Kramer, 1969). Citrus peels have also been used for ethanol production using *S. cerevisiae* (Grohmann et al., 1994). The initial saccharification of polysaccharides using commercial cellulase and polygalacturonase is necessary for a successful fermentation.

It is established that the cellulosic and starchy wastes are difficult to ferment as such and need to be broken down to simple sugars by enzymes, acids, etc. Cooking is used to rupture the starch granules followed by enzymatic hydrolysis and subsequent alcoholic fermentation. An SSF process using apple pomace and commercial yeast with fermentation efficiency of 89% has been developed (Khosravi and Shojaosadati, 2003). The amount of ethanol produced depends upon the initial sugar content in the apple pomace, which in turn is influenced by the variety of apple processed, the processing conditions, and the amount of the pressing aids employed (Joshi et al., 1999) The natural microflora present in apple pomace induce fermentation, which can be accelerated by the addition of yeast. The addition of N, P, and trace elements is also beneficial in increasing the fermentation efficiency, and among the different nitrogen sources, ammonium sulfate supported the highest ethanol productivity with *S. cerevisiae* (Joshi and Sandhu, 1994). The utilization of banana peels for ethanol production has also been explored (Tewari et al., 1986). Enzymatic hydrolysis was efficient in maximum saccharification and fermentation.

The lignocellulosic agricultural residues such as rice straw, wheat straw, and sugarcane bagasse are available as low-cost feedstock for fermentative production of ethanol (Palnitkar and Lachke, 1990). A hybrid process has also been developed for the fermentation of rice straw hydrolysates into ethanol to simultaneously utilize cellulose and hemicellulose fractions of the agro residue (Chadha et al., 1995a). Ethanol has also been produced from the lignocellulosic waste by employing recombinant bacterial strains of *Escherichia coli* and *Klebsiella oxytoca* (Katzen and Fowler, 1994). The fermentation of xylose and rice straw hydrolysate to ethanol using *Candida shehatae* NCL-3501 has also been reported. The free or immobilized cells of *C. shehatae* NCL-3501 efficiently utilized sugars present in rice straw hemicellulose hydrolysate (Abbi et al., 1996).

The ethanol production from glucose and xylose using immobilized *Z. mobilis* CP4 (pZB5) has also been carried. The batch fermentation of rice straw hydrolysate containing 76 g/L of glucose and 33.8 g/L of xylose resulted in an ethanol concentration of 44.3 g/L, corresponding to a yield of 0.46 g of ethanol/g of sugars (Krishnan et al., 2000a). By applying the Arkenol process, rice straw was successfully processed and converted into glucose and xylose for fermentation usage in a flash fermentation reactor using recombinant *Z. mobilis* 31821 (pZB5). The recombinant *Z. mobilis* showed the efficient fermentation of glucose and xylose at a relatively high concentration and promises to speed the development of the cellulose-to-ethanol industry (Yamada et al., 2002).

Different strategies have also been employed for the fermentation of whey to ethanol. In one approach, the lactose in whey is hydrolyzed with β-galactosidase and then fermented with non-lactose-utilizing yeasts such as *S. cerevisiae* (O'Leary et al., 1977). A second approach is the direct fermentation of whey using lactose-utilizing yeasts (Bernstein et al., 1977). Another alternative that has been actively explored consists of recombinant DNA techniques, achieving the expression of genes that code for β-galactosidase and lactose permease system of *Kluyveromyces lactis* in *S. cerevisiae* (Sreekrishna and Dickson, 1985). Alcoholic fermentation of cheese whey permeate using a recombinant flocculating *S. cerevisiae* using a continuously operating bioreactor resulted in ethanol productivity near 10 g/L/h (corresponding to 0.45/h dilution rate), which raises new perspectives for the economic feasibility of whey alcoholic fermentation (Domingues et al., 2001).

In batch fermentation, *K. fragilis* utilizes more than 95% of the lactose of unconcentrated whey with a conversion efficiency of 80%–85% of the theoretical value (Mawson, 1994). The production of ethanol from nonconcentrated cheese whey is generally not considered economically feasible (Tin and Mawson, 1993). Therefore, different strains should be selected that have the capability of fermenting concentrated lactose solutions with more than 90% conversion efficiency (Moulin and Galzy, 1984). Costs can be further reduced with the increase in lactose concentration up to about 100–120 g/L lactose (Mawson, 1994).

Cheese whey powder (CWP) solution has been employed for ethanol production using *K. marxianus* NRRL-1195 in batch systems (Kargi and Ozmihci, 2006). The yield coefficient of ethanol varied between 0.35 and 0.54 g ethanol/g sugar and reached the theoretical level at 300 g/L CWP concentration. Recently, deproteinized CWP utilized as fermentation medium for ethanol production by *K. fragilis* has resulted in 80.95 kg/m^3 of ethanol after 44 h of fermentation (Dragone et al., 2011).

16.3.4.2 Biogas Production

The anaerobic digestion of different industrial wastes has attracted the attention of researchers for the safer disposal of wastes including domestic and industrial solids as well as liquids/effluents. This method has one important advantage over the other treatment processes that it produces methane-rich gas (biogas) and thus is an energy-yielding rather than energy-requiring process. Other advantages include low sludge production, low nutrient requirement, and high loading. Besides its role as fuel substitute, it has several social and economic advantages over the existing traditional fuels. Thus, anaerobic waste management has been found to be an attractive alternative to aerobic treatment processes (Webb, 1983). The basic steps involved in the anaerobic digestion of food industry wastes for methane production are depicted in Figure 16.4.

Methane can be produced from agricultural wastes, sewage, and wastes of various industries like distillery, brewery, sugar, dairy, and other food industries (Jain and Mishra, 1989; Viswanath et al., 1992; Panesar et al., 1999; Marwaha et al., 2000; Bouallagui et al., 2005). These waste materials are generally available in large quantities at reasonable rate, thus making the process economically attractive.

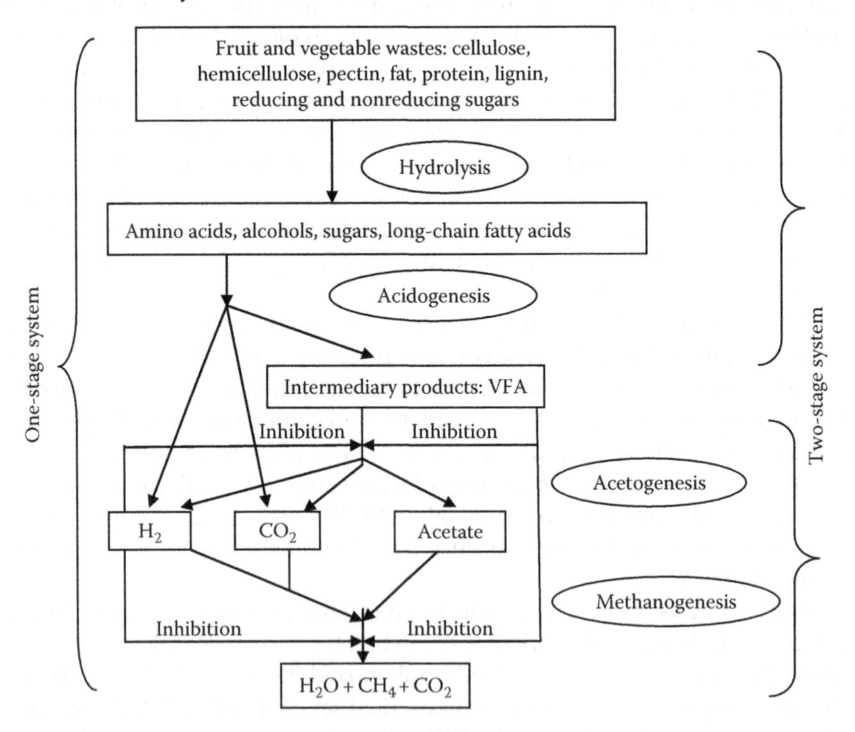

FIGURE 16.4
Anaerobic digestion of particulate organic material of fruit and vegetable wastes. (From Bouallagui, H. et al., *Process Biochem.*, 40, 989, 2005. With permission.)

Sugar mill wastewaters can be treated by anaerobic digestion followed by stabilization in an aerobic oxidation pond (Bhaskaran and Chakraborthy, 1966). The upflow anaerobic sludge blanket (UASB) reactors had been developed for the treatment of sugar mill wastewaters (Heertjes and Vander Meer, 1978).

Brewery wastewater has been successfully treated by UASB process with 90% soluble COD removal at 5–10 kg COD/m^3/day loading rate (Hack, 1985). The methane content of biogas produced approached 80%, and methane yield was observed to be close to the theoretical stoichiometric quantity of 0.35 m^3/kg COD. However, hydraulic loading was the limiting factor of operation of UASB reactor for brewery wastewater treatment. In a two-stage anaerobic unitank system, a BOD and COD reduction of 80%–85% and 84%–90%, respectively, has been reported for combined brewery and malting effluent (Vriens et al., 1990). Various reports are available on the anaerobic treatment of distillery effluents (Jain and Mishra, 1989; Gehlawat, 1998). A typical biogas plant of a distillery having a capacity of 50,000 L alcohol/day resulted in 70% methane production with 89% and 70%, BOD and COD reduction, respectively (Gehlawat, 1988).

Fruit and vegetable processing industry produces large quantities of waste every day. However, physicochemical nature of fruit and vegetable wastes is quite different from the feedstock used in conventional type of digesters. Different anaerobic processes (batch, continuous one-stage, and continuous two-stage systems) with several type of systems like CSTR, tubular reactor, anaerobic sequencing batch reactor, UASB, and anaerobic filters have been used to manage the seasonal availability of fruit and vegetable wastes through anaerobic digestion (Bouallagui et al., 2005). The engineering and economic analysis of methane generation from potato, fruits, tomato, and dairy processing industry waste revealed that the seasonal availability of solid wastes is the major factor for commercialization of the process of anaerobic digestion (Leuschner et al., 1983). The effect of feeding different fruit and vegetable wastes (mango, pineapple, tomato, jackfruit, banana, and orange) in a digester indicated that minor changes in the nutritional and operational parameters could help to operate the digesters for a considerably long time without any noticeable change in the yield of biogas and methane (Viswanath et al., 1992). In India, Central Food Technological Research Institute, Mysore, has developed a viable process for biogas production from mango peel and other fruit and vegetable processing wastes and claimed significant increase in the biogas yield and methane content over the conventionally used process (Nand, 1997). However, the major problem of these wastes is their seasonal availability. Besides this, a major limitation in the anaerobic digestion of fruit and vegetable wastes is the rapid acidification of these wastes (decreasing the pH in the reactor) and production of large quantities of VFAs, which inhibit the activity of methanogenic bacteria (Bouallagui et al., 2005). The continuous two-phase systems have been suggested as more efficient technologies for the anaerobic digestion of fruit and vegetable wastes.

The comparison of conversion rates of different fruit and vegetable wastes revealed that methane yields and methane production rate varied from 0.18 to 0.732 L/g volatile solids (VS) added and 0.016–0.122 per day, respectively, in case of fruit wastes, whereas in case of vegetable wastes, these were 0.19–0.4 L/g VS and 0.053–0.125 per day, respectively (Gunaseelan, 2004). It has also been observed that temperature had no effect on the methane yields of mango peels; however, the conversion kinetics were higher at 35°C than at 28°C.

Hydrogen–methane two-stage fermentation technology has been developed (Nishio and Nakashimada, 2007) and is used for the effective treatment of sugar-rich wastewaters, bread wastes, soybean paste, and brewery wastes. In this process, hydrogen produced in the first stage was used for a fuel cell system to generate electricity, and the methane produced in the second stage was used to generate heat energy to heat the two reactors to meet the heat requirements.

Various attempts have been made to treat the dairy industry wastewater using anaerobic methods. The UASB reactors are becoming more popular for the treatment of dairy effluents due to its lower operation and maintenance cost. The upflow velocity has been the main limiting factor for the design of UASB reactor for dairy wastewater. An upflow velocity of 0.3–0.5 m/h has been considered as optimum for the formation of sludge blanket in the UASB process. The hydraulic retention time (HRT) of 12–18 h has resulted in 85%–90% BOD and 75%–80% COD reduction (Chirlikar, 1998). UASB process has been successfully used at Vasudhara Dairy in Gujarat, India, with a saving of about 50% in terms of power saved, as compared to conventional process (Sayed, 1997).

Different studies have been made for the anaerobic digestion of whey. A two-phase system consisting of two reactors in series for the anaerobic digestion of cheese whey showed higher treatment efficiency in terms of COD reduction than that of the one-stage digestion. However, the two-phase digestion had no advantage over the one-stage digestion in terms of methane production rate (Lo and Liao, 1988). Cheese whey mixed with other wastes (cattle dung, poultry waste) has also been used for methane generation (Desai and Datta, 1994).

Among different support materials (charcoal, gravel, brick pieces, PVC pieces, and pumice stones) used in anaerobic upflow fixed-film reactors for bio-methanation of cheese whey, charcoal fixed film reactor showed the best performance (Patel et al., 1995). The supplementation of sodium lauryl sulfate resulted in 70% increase in gas production with a higher methane content (Patel and Datta, 1998). The nutrient and trace metal supplementation is highly beneficial for the anaerobic treatment of cheese whey, and undiluted cheese whey can be treated anaerobically at relatively short HRT values (2.06–4.95 days) without any significant stability problems (Ergüder et al., 2001). Two-phase anaerobic digestion of cheese whey has been operated in a system consisting of a stirred acidogenic reactor followed by a stirred methanogenic reactor, with methane content of more than 70% (Saddoud et al., 2007). A two-stage continuous process has also been developed for hydrogen and methane production from cheese whey (Venetsaneas et al., 2009).

16.3.4.3 Biohydrogen Production

Biohydrogen (Bio-H_2) from biomass can be a promising sustainable alternative fuel or energy carrier. Hydrogen production from simulated cheese processing wastewater via anaerobic fermentation has been investigated (Yang et al., 2007), which resulted in H_2 yields of 8 and 10 mM/g COD fed in batch system. In continuous system using completely mixed reactor, maximum hydrogen yield between 1.8 and 2.3 mM/g COD fed has been reported with an HRT of 24 h. The effect of initial pH and substrate concentration on both hydrogen molar yield (HMY) and volumetric H_2 production rate using lactose, CWP, and glucose substrates has been studied (Davila-Vazquez et al., 2008), which resulted in a yield of 3.1 mol H_2/mol lactose at pH 6 and 15 g CWP/L and 8.1 mmol H_2/L/h at pH 7.5 with 25 g CWP/L.

Fermentative Bio-H_2 production has also been studied in a CSTR operated with cheese whey as substrate (Davila-Vazquez et al., 2009), which indicated the highest volumetric hydrogen productivity rate (46.61 mmol H_2/L/h) and HMY of 2.8 mol H_2/mol lactose at an organic loading rate of 138.6 g lactose/L/day. Hydrogen (H_2) production from cheese processing wastewater via dark anaerobic fermentation was conducted using mixed microbial communities under thermophilic conditions (Azbar et al., 2009). The biogas contained 5%–82% (45% on average) hydrogen, and the hydrogen production rate ranged from 0.3 to 7.9 L H_2/L/day.

16.3.4.4 Production of Biodiesel

Biodiesel has the advantages of being derived from a renewable, domestic resource, biodegradable, and nontoxic with a more favorable combustion emission profile as compared to petroleum-based diesel. Moreover, carbon dioxide produced by combustion of biodiesel can be recycled by photosynthesis (Agarwal and Das, 2001).

Different continuous processes based on alkaline and acidic conditions for biodiesel production from virgin vegetable oil or waste cooking oil have been developed (Zhang et al., 2003). Two of them were alkali-catalyzed processes, the former using virgin oil and the latter using waste cooking oil. The remaining two processes were acid-catalyzed processes using waste cooking oil as the raw material. Among these four processes, the acid-catalyzed process using waste cooking oil was suggested to be technically feasible and can be a competitive alternative to commercial biodiesel production by the alkali-catalyzed process. Although alkali-catalyzed process using virgin vegetable oil was the simplest process, however, cost of the raw material is the concern.

16.3.4.5 Production of Butanol

Butanol is an important industrial chemical that can be used as a fuel. The feasibility of whey permeate for the production of 2,3-butanediol using immobilized *Klebsiella pneumoniae* in continuous system has been examined (Maddox et al., 1988). The immobilized *Clostridium acetobutylicum* cells have also been tested in a packed-bed or fluidized-bed reactor for continuous production of acetone–butanol–ethanol from whey permeate with productivity values of 3.0–4.0 g/L/h (Qureshi and Maddox, 1995). However, lactose utilization values were low in this process.

The feasibility of butanol production from cheese whey using *C. acetobutylicum* DSM 792 and *C. acetobutylicum* AS 1.224 has also been tested (Foda et al., 2010). Preliminary experiments indicated that *C. acetobutylicum* DSM 792 was better in the solvent production. The studies have demonstrated that the cheese whey can be an excellent substrate for the fermentative production of butanol.

16.3.5 Production of Enzymes

Different enzymes have been produced using various agro-industrial wastes as resource materials (Table 16.4). In enzyme production, pure cultures have been used in most of the studies; however, native microflora has also been used economically for the utilization of cabbage waste for enzyme production like amylase, protease, and cellulases (Krishna and Chandrasekaran, 1995).

Different fruit wastes such as apple pomace, orange pomace, orange peel, lemon pomace, lemon peel, pear peel, banana peel, melon peel, and hazelnut shell have been evaluated as substrate for xylanase production using *Trichoderma harzianum*, and the maximum enzyme activity has been obtained with melon peel as the substrate (Seyis and Aksoz, 2005). Apple pomace has also been used for the production of pectin methylesterase, and the enzyme activity was 2.3 times higher in SSF than by SmF (Joshi et al., 2006).

Dry orange peels have been used for the production of multienzyme preparations containing pectinolytic, cellulolytic, and xylanolytic enzymes by mesophilic fungi *A. niger* BTL, *Fusarium oxysporum* F3, *Neurospora crassa* DSM 1129, and *Penicillium decumbens* in SSF process. Among the various fungi tested, *A. niger* BTL was observed to be the efficient strain for polygalacturonase and pectate lyase production (Mamma et al., 2008).

Wheat bran has been used for the production of extracellular neutral and alkaline amylase, alkaline protease, and alkaline xylanase using SSF process (Joshi et al., 1999). Neutral and alkaline amylases were secreted by *Bacillus coagulans* B49 and *Exiguobacterium aurantiacum*, respectively, whereas *B. licheniformis* A99 and *B. circulans* produced xylanase. Wheat and maize bran supported higher amylase than in mustard oil cake, tapioca, and gram bean. Banana waste has also been employed for amylase production using *B. subtilis*

TABLE 16.4

Utilization of Agro-Industrial Waste for Enzyme Production

Agro-Industrial Residues	Microorganism Employed/ Producer Strain	Enzyme(s)	Reference(s)
Apple pomace	*A. niger*	Pectinase (Pectin methylesterase)	Joshi et al. (2006)
Apple pomace and brewery waste	*P. chrysosporium* BKM-F-1767	Lignin and manganese peroxidase, laccase	Gassara et al. (2010)
Banana waste	*B. subtilis*	Amylase	Krishna and Chandrasekaran (1996)
Banana peel	*Penicillium* sp.	Amylase	Vijayaraghavan et al. (2011)
Dry orange peels	*A. niger* BTL	Polygalacturonase and pectate lyase	Mamma et al. (2008)
Grape pomace	*Aspergillus awamori*	Xylanase and pectinase	Botella et al. (2007)
Melon peel	*T. harzianum*	Xylanase	Seyis and Aksoz (2005)
Rice bran	*Bacillus* THL027	Lipase	Dharmsthiti and Luchai (1999)
Rice bran	*C. rugosa*	Lipase	Rao et al. (1993)
Wheat bran	*Streptomyces lydicus*	Polygalacturonase	Jacob and Prema (2008)
Wheat bran	*Ganoderma* sp.	Laccase	Revankar et al. (2007)
Rice bran, wheat bran, black gram bran, coconut oil cake	*A. niger* BAN3E	Amylase	Suganthi et al. (2011)
Whey	*C. pseudotropicalis*	β-Galactosidase	De Bales and Castillo (1979)
Whey	*A. carbonarius*	β-Galactosidase	El-Gindy (2003)
Whey	*K. marxianus*	β-Galactosidase	Panesar (2008)
Whey	*Bacillus* sp.	α-Amylase	Bajpai et al. (1991)
Whey	*S. marcescens*	Proteases	Romero et al. (2001)
Whey	Recombinant *E. coli*	Penicillin acylase	León-Rodríguez et al. (2006)

(Krishna and Chandrasekaran, 1996). Recently, *Penicillium* sp. has also been used for SSF of banana peel for the production of amylase (Vijayaraghavan et al., 2011). Rice bran has been used for lipase production in SSF process in a tray fermenter using *Candida rugosa* (Rao et al., 1993). A low-cost medium containing rice bran oil and rice bran besides other constituents has also been developed for lipase production using *Bacillus* THL027 (Dharmsthiti and Luchai, 1999).

The production of cellulase enzymes has gained considerable importance in view of their application in the hydrolysis of cellulose for subsequent conversion to ethanol. In view of economic feasibility, production of enzymes like cellulase, hemicellulases, and pectinase in situ using SSF system is considered most beneficial in the processing of agro-industrial wastes (Tengerdy, 1996). Different agro-industrial wastes (wheat bran, sugar beet, rice straw, bagasse, etc.) have been widely employed for the cellulase production (Nirmala and Ramakrishna, 1994). Among the molds, *Trichoderma*, *Aspergillus*, and *Penicillium* have been used for this purpose. The activity of the cellulase, however, varied with the nature of substrate and type of organisms used. Higher yields of cellulases and β-glucosidase on wheat bran and rice bran have been observed in SSF than SmF process (Gupta and Madamwar, 1994). The pretreatment of cellulosic material using alkali improved the enzyme yield. *Phanerochaete chrysosporium* BKM-F-1767 has been investigated for ligninolytic enzyme production by SSF technique using different agro-industrial wastes, such as fishery residues, brewery waste, apple waste (pomace), and pulp and paper industry sludge (Gassara et al., 2010). Brewery wastes and pomace served as excellent sources for the production of manganese peroxidase (MnP), lignin peroxidase (LiP), and laccases.

Most of the work has been carried on the use of whey for β-galactosidase production; however, whey has also been explored as a substrate for the production of different enzymes. Different yeasts, which are able to utilize lactose, can be grown on whey and subsequently been used for β-galactosidase production. *Candida pseudotropicalis*, *K. marxianus*, *K. fragilis*, and *K. lactis* have been used by different researchers for the production of this enzyme from whey-based medium. However in most cases, there is need for supplementation of whey with different nutrients for optimal production of the enzyme (Panesar et al., 2006; Panesar, 2008). The optimal process conditions for the production of the enzyme vary considerably for different yeast strains. The supplementation whey with yeast extract, ammonium sulfate, and potassium dihydrogen orthophosphate had been useful for lactase production by *C. pseudotropicalis* (De Bales and Castillo, 1979). *Aspergillus carbonarius* has also been used for β-galactosidase production grown on deproteinized cheese whey (El-Gindy, 2003). *K. marxianus* has also been tested for the β-galactosidase production from whey (Bansal et al., 2008). Among the four methods tested for extraction of β-galactosidase, SDS-chloroform method was found to be best followed by toluene–acetone, sonication, and homogenization with glass beads. SDS-chloroform method is economical and simple for enzyme extraction from *Kluyveromyces* cells and results in higher enzyme activity as compared to the other extraction methods.

Whey has also been employed as substrate for the production of protease, amylase, and polygalacturonase enzymes; however, its supplementation with different nutrients was essential. The pattern and the extent of protease production and secretion differ with the fungus, age, and the nature of the co-supplement (Ashour et al., 1996). It has been reported that the addition of yeast extract induced the best yield of proteases from both *A. niger* and *A. terreus*. Use of *Serratia marcescens* ATCC 25419 for the production of proteases on reconstituted whey has also been reported (Romero et al., 2001). Cheese whey has been utilized as a carbon source for the production of α-amylase by *Bacillus* sp., and medium and culture conditions have been optimized for maximum enzyme yield (Bajpai et al., 1991). A fermentation process for the production of penicillin acylase by a recombinant *E. coli* and using cheese whey as unique carbon source and inducer has also been developed (León-Rodríguez et al., 2006).

16.3.6 Production of Food Additives

Agro-industrial by-products and residues have also been used in the production of different types of food additives, which have been discussed in the following text.

16.3.6.1 Production of Polysaccharides

Polysaccharides/gums are important molecules, which are mainly used as thickeners and stabilizers besides other applications. Xanthan gum is the one of the most important microbial polysaccharides produced by *Xanthomonas campestris* and has widespread commercial applications as a viscosity enhancer and stabilizer in different industries (Papagianni et al., 2001). Since a number of wastes especially from fruit industry contain sucrose besides other carbohydrates, they can serve as cheaper substrates for the production of polysaccharides. The comparison of xanthan production from citrus waste fractions by *X. campestris* has been made (Bilanovic et al., 1984). Whole waste has been reported to be good substrate as compared to cellulose and hemicellulose. However, pectin fractions were more suitable for xanthan production and with the water-soluble fractions contributing to the maximum. The use of SSF as an alternative strategy for the production of xanthan by *X. campestris* has been proposed, which can overcome problems connected with broth viscosity (Stredansky and Conti, 1999).

The watery extract and pressed apple pomace have been studied as the substrates for chitosan production using SmF and SSF techniques, respectively (Streit et al., 2004). The fungus *Gongronella butleri* yielded the best results and the highest productivity (0.091 g/L/h) of chitosan content and biomass production (0.1783 g/g of apple pomace). An external loop airlift bioreactor for chitosan production using *G. butleri* CCT 4274 on the watery extract of apple pomace has also been used (Vendruscolo, 2005). The studies revealed that higher levels of aeration (0.6 vvm) resulted in greater concentrations of biomass and chitosan.

The partially deproteinized cheese whey has been used for the production of extracellular polysaccharide using *Xanthomonas cucurbitae* PCSIR B-52 (Baig et al., 1990). The bacterium efficiently consumed cheese whey, particularly in the presence of corn steep liquor and penicillin waste mycelium. The feasibility of using cheese whey as carbon source for xanthan gum production was investigated using two strains of *X. campestris,* and maximum xanthan gum productions were observed after 72 h (Silva et al., 2009). The production of pullulan from deproteinized whey by *Aureobasidium pullulans* P56 was investigated using an adaptation technique and a mixed culture system (Roukas, 1999). Enzyme-hydrolyzed whey supplemented with other ingredients (K_2HPO_4, L-glutamic acid, olive oil, Tween-80) resulted in maximum polysaccharide concentration. Pullulan production has also been carried from beet molasses using *A. pullulans* (Lazaridou et al., 2002). Combined pretreatment of molasses with sulfuric acid and activated carbon resulted in a maximum pullulan concentration of 24 g/L, with pullulan yield of 52.5%, and a sugar utilization of 92% under optimum fermentation conditions.

16.3.6.2 Production of Bio-Pigments

The color additives play a very important role in food industry since these impart attractive color to different food products. Bio-pigments produced by fermentation offer immense potential for food applications, because they are natural and can be produced throughout the year (Carvalho et al., 2007).

A number of microorganisms (*Rhodotorula, Xanthophyllomyces, Monascus, Ashbya,* and *Bacillus* sp.) are known to produce pigments through fermentation (Dufossé, 2006). Apple pomace has been employed for carotenoid production using *Rhodotorula* sp. (Sandhu and Joshi, 1997). *Micrococcus* and *Chromobacter* spp. have also been employed for the production of bio-pigments from apple pomace (Attri and Joshi, 2005, 2006). The production of bio-pigments in apple pomace-based medium using SSF resulted in better yield of biomass and carotenoids. The studies on the bio-pigment production using *Rhodotorula mucilaginosa* from different carbon sources (glucose, molasses, sucrose, and whey lactose sugars) indicated highest carotenoid production (89.0 mg/L) with molasses sucrose (20 g/L) and highest product yield (35.0 mg/gm of dry cells) with whey lactose (13.2 g/L) as the carbon source (Aksu and Eren, 2005).

Fermented radish is one of fermented vegetables popularly consumed in the Oriental countries. Its production generates a considerable amount of brine as a waste, which was also tested as a substrate for β-carotene production using *R. glutinis* DM28, and its different cultivation conditions for pigment production were optimized (Malisorn and Suntornsuk, 2008). Tomato processing waste has also been utilized for the production of carotenoid using *R. glutinis* (Chandi et al., 2010).

The suitability of whey for bio-pigment production has also been explored. The strains of the lactose-negative yeast *R. rubra* GED2 and the homofermentative *Lactobacillus casei* subsp. *casei* Ha1 were used for co-cultivation in cheese WU and active synthesis of carotenoids (Frengova et al., 2003). Carotenogenesis of the lactose-negative yeast *Rhodotorula rubra* GED5 was also investigated by co-cultivation with *K. lactis* MP11 in WU. Maximum yields of biomass (24.3 g/L) and carotenoids (10.2 mg/L of culture fluid or 0.421 μg/g of dry cells) have been reported in WU at a temperature of 30°C (Frengova et al., 2004).

16.3.6.3 Production of Bio-Flavors

Flavors and fragrances are extremely important for the food, feed, cosmetic, chemical, and pharmaceutical industries (Janssens et al., 1992). The use of biotechnology for the production of aroma compounds by fermentation/bioconversion can be an economic alternative to the difficult and expensive extraction from raw materials such as plants (Kosseva et al., 2011). Among the eight yeast strains tested for their aromatic potential, *Geotrichum candidum* ATCC 62217 resulted in the production of fruity aroma compounds (pineapple-like) on fermented waste bread (Daigle et al., 1999).

Coffee husk was used as raw material for the production of volatile compounds by SSF using *Ceratocystis fimbriata*. It has been reported that the production of volatile compounds was significantly higher in horizontal drum bioreactor as compared to column bioreactors, and this model of bioreactor presents good perspectives for scale-up and application in industrial-scale production process (Medeiros et al., 2006). The application of microbial enzymes like lipases for flavor generation is gaining acceptance, but the industrial viability of such products would be determined by economics of the process that can be improved by using cheap substrate (Nair, 1994).

Different agro-industrial wastes such as cassava bagasse, apple pomace, amaranth, and soybean have been used for the production of aroma compounds by *C. fimbriata* (Bramorski et al., 1998). The aroma production was growth dependent, and the medium containing apple pomace produced a strong fruity aroma. *Rhizopus* sp. has also been used for the production of volatile compounds using this medium (Christen et al., 2000). It was revealed that the production of volatile compounds was mainly related to the medium used.

16.3.6.4 Production of Biosurfactants

Biosurfactants are amphiphilic molecules produced by a wide variety of microorganisms (bacteria, yeast, and fungi). Although much attention has been focused on the biosurfactant production due to their versatile applications, however, the production economics is the major bottleneck in the process. The raw material can contribute 10%–30% of the production cost in most biotechnological processes (Muthusamy et al., 2008). To reduce the production cost, an alternate approach is the use of low-cost agro-based raw material as substrates for biosurfactant production. A variety of potential inexpensive raw materials, viz. plant-derived oils, oil waste, ground nut oil cake, starchy substance, wheat bran, rice bran, whey, distillery waste, and molasses, have been tested for biosurfactant production (Makkar et al., 2011; Panesar et al., 2011).

Molasses and corn steep liquor have been used as the primary carbon and nitrogen source to produce rhamnolipid biosurfactant using *Pseudomonas aeruginosa* GS3 (Patel and Desai, 1997). Similarly, *P. aeruginosa* BS2 has been cultivated on whey to produce biosurfactants (Dubey and Juwarka, 2004). *Bacillus subtilis* ATCC 21332 cultivated in potato process effluent produced surfactin, and it was reported that the process is oxygen limited and that recalcitrant indigenous bacteria in the potato process effluent hamper continuous surfactin production (Noah et al., 2005). Studies have also suggested that the plant-derived oils can act as effective inexpensive raw materials for biosurfactant production. Rapeseed oil has been reported as the good substrate for the production of rhamnolipids and L(+)-rhamnose using *Pseudomonas* sp. DSM 2874 (Trummler et al., 2003). It has also been reported that *P. aeruginosa* MTCC 2297 cultivated on various inexpensive waste materials such as orange peelings, carrot peel waste, lime peelings, coconut oil cake, and banana wastes produced a surface-active compound rhamnolipid by submerged fermentation (George and Jayachandran, 2009).

16.3.6.5 Production of Bacteriocins

Different agro-industrial wastes like whey, molasses, and barley extract are used for the production of bacteriocins. Low-cost soy waste has been used for the production of nisin. The supplementation of the wastes given earlier with the nutrients resulted in 3% and 7% increase in biomass and nisin production, respectively (Mitra et al., 2010). Nisin A production by *Lactococcus lactis* subsp. *lactis* ATCC 11454 has been reported using fermented barley extract (Furuta et al., 2008). Soybean peptone and sweet whey have also been used for the production of nisin using *L. lactis* UQ2 (González-Toledo et al., 2011).

Different studies have been carried out to use whey for the production of antibacterial substances (Cladera-Olivera et al., 2004; Liu et al., 2004). Both *L. lactis* subsp. *lactis* CECT 539 and *Pediococcus acidilactici* NRRL B-5627 produced higher bacteriocin titers in diluted whey than concentrated whey (Guerra et al., 2001). The recombinant strains of *L. lactis* ssp. *lactis, S. Thermophilus,* and *E. faecalis* have been successfully used for the pediocin production in skim milk and cheese whey (Somkuti and Steinberg, 2003). The studies on pediocin production by *P. acidilactici* NRRL B-5627 revealed that the use of feeding substrates containing glucose instead of lactose could be an appropriate alternative for increasing fed-batch production of pediocin in diluted whey supplemented with yeast extract (Guerra et al., 2007).

The optimization of enterocin AS-48 production by *E. faecalis* A-48-32 strain using a whey-derived substrate has been carried out (Ananou et al., 2008). A maximum activity of 360 AU/mL has been achieved under optimal conditions after 18 h. Enterocin A has also been produced by using cheese whey as the substrate using *Enterococcus faecium* strains (Mirhosseini and Emtiazi, 2011).

16.3.6.6 Production of Vitamins

The suitability of whey for the production of vitamin B_{12} has been explored. The growth pattern of *Propionibacterium shermanii* and vitamin B_{12} production from cheese whey has been investigated (Marwaha and Sethi, 1983; Marcoux et al., 1992). Vitamin B_{12} biosynthesis in whey permeate using 5,6-dimethylbenzimidazole (DMB) as a precursor has been achieved (Marwaha et al., 1983a). The studies have indicated that the addition of precursor DMB is required during the biosynthesis of vitamin B_{12}. Moreover, the addition of betaine and L-glutamic to the whey permeate can also be beneficial in stimulating the vitamin B_{12} production (Marwaha et al., 1983b). Sago waste has also been used for the production of vitamin B_{12} using *Propionibacterium freudenreichii* and other bacterial isolates (Manjunathan et al., 2010).

16.3.6.7 Production of Amino Acids

The amino acids like lysine and glutamic acid have been produced by the fermentation of fruit wastes. Fruit waste obtained from the preparation of fruit juices is pressed, centrifuged, and filtered by ultrafiltration to obtain clear liquid that has been used to manufacture lysine by fermentation (Sato, 1992). The citrus molasses has been successfully used as a substrate for the production of amino acids. After suitable pre-treatment, i.e., liming, pressing, or heating, and centrifugation, the peel extract was concentrated and the molasses was used as a substrate for amino acid production (Tsugawa et al., 1981; Joshi et al., 1999).

The amino acid in the form of monosodium glutamate is employed as a flavor-enhancing agent. The studies on the production of glutamic acid using *Brevibacterium* sp. in SSF technique resulted in the maximum yield when sugarcane bagasse (80%–90% moisture level) containing 10% glucose was used (Nampoothiri and Pandey, 1996).

16.3.7 Production of Antibiotics

Sweet potato residue has been used to produce tetracycline by *Streptomyces viridifaciens* ATCC 11989 in SSF (Yang and Ling, 1989). Corncob, a cellulosic waste, has also been successfully used as a substrate for the production of oxytetracycline using *Streptomyces rimosus* (Yang, 1996). The production of neomycin by *Streptomyces marinensis* under SSF using wheat rawa as a support substrate has also been reported (Adinarayana et al., 2003a). Several substrates (wheat bran, wheat rawa, Bombay rawa, barley, and rice bran) have been evaluated to produce cephalosporin C by *Acremonium chrysogenum* under SSF. A maximum productivity of cephalosporin C was achieved using wheat rawa and soluble starch (1%) and yeast extract (1%) as additives under optimized conditions (Adinarayana et al., 2003b). The accumulation of neomycin was higher with SSF than with SmF. Different strains of *Streptomyces* sp. has been employed to produce tetracycline under SSF conditions using peanut shells, corn cob, corn pomace, and cassava peels as substrates, and results indicated that peanut shells were the most effective substrate for the production of said antibiotics (Asagbra et al., 2005).

Different agro-industrial residues such as apple pomace, cotton seed meal, soybean powder, and wheat bran were evaluated for the production of neomycin by *Streptomyces fradiae* 2418 (Vastrad and Neelagund, 2011). Among the various agro-industrial residues studied, highest production of neomycin was observed with apple pomace in SSF.

16.3.8 Production of Mushrooms

Various lignocellulosic agricultural residues such as wheat straw, paddy straw, and sugarcane bagasse are readily available source of renewable biomass. These materials can be used for the cultivation of edible mushrooms (Kaur and Khanna, 2001). In the mushroom production, fruiting bodies serve as delicious food, and the spent substrate can be used as feed or as humus fertilizer. The technical feasibility of using other agricultural wastes as substrate for the cultivation of *Pleurotus ostreatus* NRRL-0366 has also been evaluated. When comparing the biological efficiency of mushroom production, the highest yield of fruiting bodies was obtained using a mixture of rice waste and date straw in 1:1 ratio. *P. ostreatus* NRRL-0366 can also be cultivated on rice straw supplemented with fruit industry waste in different ratios (Jwanny et al., 1995).

Food industry waste like apple pomace has also found applications in compost making (Upadhyay and Sohi, 1980). Submerged production of mushrooms using citrus waste streams has also been explored. Orange juice, citrus press liquor, peel extract, and synthetic media have been used for the production of fungal mycelium (Block et al., 1953; Labaneih et al., 1979). See Chapter 15 for information on mushroom production.

16.4 Summary and Future Prospects

With the rapid growth and development of food processing industry, the nature and quantum of waste/by-products generation has also increased considerably during the last two to three decades. Industries are facing increasing constraints in disposal of their waste and by-products due to stringent regulation laid down by regulatory agencies. Therefore, industrial organizations are always looking forward for effective and efficient systems for the management of their by-products and residues. The waste generated by food industry is a rich source of biodegradable components, which can be utilized using various microorganisms. Thus, food industry waste and by-products can be of considerable value and improve the economics of the process, if these are efficiently utilized as raw materials for the production of value-added products. During the last few decades, there has been increasing interest in biotechnology all over the world to utilize its techniques/tools in the bio-processing of wastes and by-products for the production of food, fuel, bio-fertilizers, and other value-added products.

A strong agro-based economy demands greater efficiency in production systems, the avoidance of losses through improvement in agro-processing, enhanced nutritional quality, and safety of processed products. Biotechnology could add value not only to agro/food processing, but also to their by-products/residues through production of wide range of products. An integrated approach comprising stripping off the useful recoverable products followed by waste/by-products management technologies is required. Moreover, application of novel biotechnological techniques such as immobilization, genetic engineering, and enzyme engineering technology can further improve the nutritive quality of products produced and consequently the economics of these processes. Therefore, much can be gained through biotechnological strategies in the near future.

References

Abbi, M., Kuhad, R.C., and Singh, A. 1996. Fermentation of xylose and rice straw hydrolysate to ethanol by *Candida shehatae* NCL-3501. *Journal of Industrial Microbiology* 17: 20–23.

Adinarayana, K., Ellaiah, P., Srinivasulu, B. et al. 2003a. Response surface methodological approach to optimize the nutritional parameters for neomycin production by *Streptomyces marinensis* under solid-state fermentation. *Process Biochemistry* 38: 1565–1572.

Adinarayana, K., Prabhakar, T., Srinivasulu, V., Rao, A.M. et al. 2003b. Optimization of process parameters for cephalosporin C production under solid state fermentation from *Acremonium chrysogenum*. *Process Biochemistry* 39: 171–177.

Agarwal, A.K. and Das, L.M. 2001. Biodiesel development and characterization for use as a fuel in compression ignition engines. *Journal of Engineering Gas Turbines Power* 123: 440–447.

Aksu, Z. and Eren, A.T. 2005. Carotenoids production by the yeast *Rhodotorula mucilaginosa*: Use of agricultural wastes as a carbon source. *Process Biochemistry* 40: 2985–2991.

Aksu, Z. and Kutsal, T. 1986. Lactic acid production from molasses utilizing *Lactobacillus delbrueckii* and invertase together. *Biotechnology Letters* 8: 157–160.

Ananou, S., Muñoz, A., Gálvez, A., Martínez-Bueno, M., Maqueda, M., and Valdivia, E. 2008. Optimization of enterocin AS-48 production on a whey-based substrate. *International Dairy Journal* 18: 923–927.

Anvari, M. and Khayati, G. 2011. Submerged yeast fermentation of cheese whey for protein production and nutritional profile analysis. *Advance Journal of Food Science and Technology* 3: 122–126.

Arora, J.K, Marwaha, S.S., and Bakshi, A. 2000. Biotechnological advancements in food processing: An overview. In: *Food Processing: Biotechnological Applications*, eds. S.S. Marwaha and J.K. Arora, pp. 1–23, Asiatech Publishers, New Delhi, India.

Asagbra, A.E., Sanni, A.I., and Oyewole, O.B. 2005. Solid-state fermentation production of tetracycline by *Streptomyces* strains using some agricultural wastes as substrate. *World Journal of Microbiology and Biotechnology* 21: 107–114.

Ashour, S.A., el-Shora, H.M., Metwally, M., and Habib, S.A. 1996. Fungal fermentation of whey incorporated with certain supplements for the production of proteases. *Microbios* 86: 59–69.

Attri, D. and Joshi, V.K. 2005. Optimization of apple pomace based medium and fermentation conditions for pigment production by *Micrococcus* species. *Journal of Scientific and Industrial Research* 64: 598–601.

Attri, D. and Joshi, V.K. 2006. Optimization of apple pomace based pigment production medium and fermentation conditions for pigment production by *Chromobacter* species. *Journal of Food Science and Technology* 43: 484–487.

Avellone, S., Guyot, B., Michaux-Ferriere, J.P.G., Palacios, E.O., and Brillouet, J.M. 1999. Cell wall polysaccharides of coffee bean mucilage. Histological characterisation during fermentation. In: *Proceedings of the 18th International Scientific Colloquium on Coffee Science (ASIC)*, pp. 463–470. Helsinki, Finland.

Awarenet. 2004. *Handbook for the Prevention and Minimisation of Waste and Valorisation of By-Products in European Agro-Food Industries*. Fundacién Gaiker, Bilbao, Spain.

Azbar, N., Dokgöz, F.T.Ç., Keskin, T., Korkmaz, K.S., and Syed, H.M. 2009. Continuous fermentative hydrogen production from cheese whey wastewater under thermophilic anaerobic conditions. *International Journal of Hydrogen Energy* 34: 7441–7447.

Bahar, S. and Azuaze, J.T. 1984. Studies on the growth of *Fusarium* sp. on citrus waste for the production of single cell proteins. *Journal of Food Science and Technology* 21: 63–67.

Baig, S., Qadeer, M.A., Akhtar, M.S., and Ahmed, T. 1990. Utilization of unhydrolyzed cheese whey for the production of extracellular polysaccharide by *Xanthomonas cucurbitae* PCSIR B-52. *Journal of Fermentation and Bioengineering* 69: 345–349.

Bajpai, P., Verma, N., Neer, J., and Bajpai, P.K. 1991. Utilization of cheese whey for production of α-amylase enzyme. *Journal of Biotechnology* 18: 265–270.

Bansal, S., Oberoi, H.S., Dhillon, G.S., and Patil, R.T. 2008. Production of β-galactosidase by *Kluyveromyces marxianus* MTCC 1388 using whey and effect of four different methods of enzyme extraction on β-galactosidase activity. *Indian Journal of Microbiology* 48: 337–341.

Bernstein, S., Tzeng, C.H., and Sisson, D. 1977. The commercial fermentation of cheese whey for the production of protein and/or alcohol. *Biotechnology and Bioengineering* 7: 1–9.

Bhaskaran, T.R. and Chakraborthy, R.N. 1966. Pilot plant for treatment of cane sugar wastes. *Journal of Water Pollution Control* 38: 1060–1169.

Bhushan, S. and Joshi, V.K. 2006. Baker's yeast production under fed batch culture from apple pomace. *Journal of Scientific and Industrial Research* 65: 72–76.

Bilanovic, D., Shelef, G., and Green, M. 1984. Xanthan fermentation of citrus waste. *Bioresource Technology* 48: 169–172.

Block, S.S., Steams, T.W., Stephens, P.L., and McCandless, R.F.L. 1953. Mushroom mycelium experiments with submerged culture. *Journal of Agricultural and Food Chemistry* 1: 890–893.

Borja, R., Garrido, E.S., Martinez, L., Cormenzana, R.A., and Martin, A. 1993b. Kinetic study of anaerobic digestion of olive mill wastewater previously fermented with *Aspergillus terreus*. *Process Biochemistry* 28: 397–404.

Borja, R., Martin, A., Maestro, R., Luque, M., and Duran, M.M. 1993a. Enhancement of anaerobic digestion of wine-distillery waste water by removal of phenolic inhibitor. *Bioresource Technology* 45: 99–104.

Borzani, W., Gerb, A., Delahiguera, G.A., Pires, M.H., and Piolovic, R. 1993. Batch ethanol fermentation of molasses-A correlation between the time required to complete fermentation and the initial concentration of sugar and yeast cells. *World Journal of Microbiology and Biotechnology* 9: 265–268.

Botella, C., Diaz, A., Ory, I., Webb, C., and Blandino, A. 2007. Xylanase and pectinase production by *Aspergillus awamori* on grape pomace in solid state fermentation. *Process Biochemistry* 42: 98–101.

Bouallagui, H., Touhami, Y., Ben, Cheikh, R., and Hamdi, M. 2005. Bioreactor performance in anaerobic digestion of fruit and vegetable wastes. *Process Biochemistry* 40: 989–995.

Boyaval, P., Corre, C., and Terre, S. 1987. Continuous lactic acid fermentation with concentrated product recovery by ultrafiltration and electrodialysis. *Biotechnology Letters* 9: 207–212.

Boze, H., Moulin, G., and Galzy, P. 1995. Production of microbial biomass. In: *Biotechnology*, eds. H.J. Rehm and G. Reed, Vol. 9, pp. 167–220. VCH Verlagsgesellschaft mbH, Weinheim, Germany.

Braddock, R.J. 1999. *Handbook of Citrus By-Products and Processing Technology*. John Wiley & Sons, Inc., New York.

Bramorski, A., Soccol, C.R., Christen, P., and Revah, S. 1998. Fruit aroma production by *Ceratocystis fimbriata* in solid cultures from agro-industrial wastes. *Revista de Microbiologia* 29: 208–212.

Brito, A.G., Peixoto, J., Oliveira, J.M. et al. 2007. Brewery and winery wastewater treatment: Some focal points of design and operation. In: *Utilization of By-Products and Treatment of Waste in the Food Industry*, eds. V. Oreopoulou and W. Russ, Vol. 3, pp. 109–131. ISEKI Food Series, Springer, New York.

Buzzini, P., Gobbetti, M., and Rossi, J. 1990. Fumaric acid production by *Rhizopus arrhizus* in solid state fermentation on lyophilized orange peels. *Annali della Facolta' di Agraria* 44: 661–673.

Buzzini, P., Gobbetti, M., Rossi, J., and Ribaldi, M. 1993. Utilization of grape must and concentrated rectified grape must to produce gluconic acid by *Aspergillus niger* in batch fermentations. *Biotechnology Letters* 15: 151–156.

Carlsberg/Carlsberg Breweries A/S Environmental Report 2003 and 2004, 2005, Copenhagen, Denmark, February 1, 2005. www.carlsberg.com

Carvalho, J.C., Oishi, B.O., Woiciechowski, A.L., Pandey, A., Babitha, S., and Soccol, C.R. 2007. Effect of substrates on the production of *Monascus* biopigments by solid-state fermentation and pigment extraction using different solvents. *Indian Journal of Biotechnology* 6: 194–199.

Castillo, F.J. 1990. Lactose metabolism by yeasts. In: *Yeast Biotechnology and Biocatalysis*, eds. H. Verachtert and R. De Mot, pp. 297–320. Marcel Dekker, New York.

Castillo, F.J. and De Sanchez, S.B. 1978. Studies on the growth of *Kluyveromyces fragilis* in whey for the production of yeast protein. *Acta Cient Venez* 29: 113–118.

Chadha, B.S., Kanwar, S.S., Saini, H.S., and Garcha, H.S. 1995a. Hybrid process for ethanol production from rice straw. *Acta Microbiologica et Immunologica Hungarica* 42: 53–59.

Champagne, C.P., Goulet, J., and Lachance, R.A. 1990. Production of baker's yeast in cheese whey ultrafiltrate. *Applied and Environmental Microbiology* 56: 425–430.

Chandi, G.K., Singh, S.P., Gill, B.S., Sogi, D.S., and Singh, P. 2010. Optimization of carotenoids by *Rhodotorula glutinis*. *Food Science and Biotechnology* 19: 881–887.

Chandra, M.V. and Shamasundar, B.A. 2011. Fish processing waste treatment. In: *Food Processing Waste Management: Treatment and Utilization Technology*, eds. V.K. Joshi and S.K. Sharma, pp. 161–194. New India Publishing Agency, New Delhi, India.

Charles, M. and Shuichi, F. 2003. Electricity from bagasse in Zimbabwe. *Biomass and Bioenergy* 25: 197–207.

Chirlikar, S. 1998. Treatment of dairy wastewater. In: *Advances in Wastewater Treatment Technologies*, ed. R.K. Trivedy, pp. 94–98. Global Science Publishing Ltd., Aligarh, India.

Christen, P., Bramorski, A., Revah, S., and Soccol, C.R. 2000. Characterization of volatile compounds produced by *Rhizopus* strains grown on agro-industrial solid wastes. *Bioresource Technology* 71: 211–215.

Cladera-Olivera, F., Caron, G.R., and Brandelli, A. 2004. Bacteriocin production by *Bacillus licheniformis* strain P40 in cheese whey using response surface methodology. *Biochemical Engineering Journal* 21: 53–58.

Cournoyer, M.S. 1996. Sanitation and stabilization of slaughter-house sludges through composting. In: *Proceedings of the Canadian Meat Research Institute Technology Symposium*, pp. 1–7. Canadian Meat Research Institute, Ottawa, Ontario, Canada.

Cox, G.C. and Macbean, R.D. 1977. Lactic acid production by *Lactobacillus bulgaricus* in supplemented whey ultrafiltrate. *Australian Journal of Dairy Technology* 32: 19–22.

Daigle, P., Gelinas, P., Leblanc, D., and Morin, A. 1999. Production of aroma compounds by *Geotrichum candidum* on waste bread crumb. *Food Microbiology* 16: 517–522.

Datta, R., Tsai, S.P., Bonsignore, P., and Moon, S.H. 1995. Technological and economic potential of poly(lactic acid) and lactic acid derivatives. *FEMS Microbiology Reviews* 16: 221–231.

Davila-Vazquez, G., Alatriste-Mondragón, F., de León-Rodríguez, A., and Razo-Flores, E. 2008. Fermentative hydrogen production in batch experiments using lactose, cheese whey and glucose: Influence of initial substrate concentration and pH. *International Journal of Hydrogen Energy* 33: 4989–4997.

Davila-Vazquez, G., Cota-Navarro, C.B., Rosales-Colunga, L.M., de León-Rodríguez, A., and Razo-Flores, E. 2009. Continuous biohydrogen production using cheese whey: Improving the hydrogen production rate. *International Journal of Hydrogen Energy* 34: 4296–4304.

De Bales, S.A. and Castillo, F.J. 1979. Production of lactase by *Candida pseudotropicalis* grown in whey. *Applied and Environmental Microbiology* 37: 1201–1205.

Desai, M. and Datta, M. 1994. Anaerobic digestion of a mixture of cheese whey, poultry waste and cattle dung: A study of the use of adsorbents to improve digester performance. *Environmental Pollution* 86: 337–340.

Dessoff, A. 1997. New meat sanitation regs kick in: Industry keeps an eye on plant operations. *Food Processing* 58: 60–66.

Dharmsthiti, S. and Luchai, S. 1999. Production, purification and characterization of thermophilic lipase from *Bacillus* sp. THL027. *FEMS Microbiology Letters* 179: 241–246.

Dhillon, G.S., Brar, S.K., Verma, M., and Tyagi, R.D. 2011a. Enhanced solid-state citric acid bio-production using apple pomace waste through surface response methodology. *Journal of Applied Microbiology* 110: 1045–1055.

Dhillon, G.S., Brar, S.K., Verma, M., and Tyagi, R.D. 2011b. Apple pomace ultrafiltration sludge—A novel substrate for fungal bioproduction of citric acid: Optimisation studies. *Food Chemistry* 128: 864–871.

Dhillon, G.S., Brar, S.K., Verma, M., and Tyagi, R.D. 2011c. Recent advances in citric acid bio-production and recovery. *Food and Bioprocess Technology* 4: 505–529.

Doelle, M.B., Greenfield, P.F., and Doelle, H.W. 1990. Effect of mineral ions on ethanol formation during sugar cane molasses fermentation using *Zymomonas mobilis* ATCC 39676. *Process Biochemistry International* 25: 151–156.

Doelle, H.W., Kennedy, L.D., and Doelle, M.B. 1991. Scale up of ethanol production from sugarcane using *Zymomonas mobilis. Biotechnology Letters* 13: 131–136.

Domingues, L., Lima, N., and Teixeira, J.A. 2001. Alcohol production from cheese whey permeate using genetically modified flocculent yeast cells. *Biotechnology and Bioengineering* 72: 507–514.

Dragone, G., Mussatto, S.I., Almeida e Silva, J.B., and Teixeira, J.A. 2011. Optimal fermentation conditions for maximizing the ethanol production by *Kluyveromyces fragilis* from cheese whey powder. *Biomass Bioenergy* 35: 1977–1982.

Driessen, W. and Vereijken, T. 2003. *Recent Developments in Biological Treatment of Brewery Effluent*. The Institute and Guild of Brewing Convention, Livingstone, Zambia.

Dubey, K. and Juwarka, A. 2004. Determination of genetic basis for biosurfactant production in distillery and curd whey wastes utilizing *Pseudomonas aeruginosa* strain BS2. *Indian Journal of Biotechnology* 3: 74–81.

Dufossé, L. 2006. Microbial production of food grade pigments. *Food Technology and Biotechnology* 44: 313–321.

El-Gindy, A. 2003. Production, partial purification and some properties of β-galactosidase from *Aspergillus carbonarius. Folia Microbiologica* 48: 581–584.

El-Holi, M.A. and Al-Delaimy, K.S. 2003. Citric acid production from whey with sugars and additives by *Aspergillus niger. African Journal of Biotechnology* 2: 356–359.

El-Samragy, Y.A., Khorshid, M.A., Foda, M.I., and Shehata, A.E. 1996. Effect of fermentation conditions on the production of citric acid from cheese whey by *Aspergillus niger. International Journal of Food Microbiology* 29: 411–416.

El-Samragy, Y.A., Shehata, A.E., Foda, M.I., and Khorshid, M.A. 1993. Suitability of strains and mutants of *Aspergillus niger* for the production of citric acid from cheese whey. *Milchwissenschaft* 48: 498–501.

Environment Canada. 1994. Canadian biodiversity strategy: Canadian response to the convention on biological diversity. Report of the Federal Provincial Territorial Biodiversity Working Group Environment. Ottawa, Ontario, Canada.

Ergüder, T.H., Tezel, U., Güven, E., and Demirer, G.N. 2001. Anaerobic biotransformation and methane generation potential of cheese whey in batch and UASB reactors. *Waste Management* 21: 643–650.

Esser, K. and Karsch, T. 1984. Bacterial ethanol production: Advantages and disadvantages. *Process Biochemistry* 17: 116–121.

Ethiraj, S. and Suresh, E.R. 1992 Studies on the utilization of mango processing wastes for production of vinegar. *Journal of Food Science and Technology* 29: 48–50.

Fellows, P.J. and Worgan, J.T. 1988. Growth of *Saccharomycopsis fibuliger* and *Candida utilis* in mixed culture on apple processing wastes. *Enzyme and Microbial Technology* 9: 434–437.

Foda, M.I., Dong, H., and Li, Y. 2010. Study the suitability of cheese whey for biobutanol production by Clostridia. *American Journal of Science* 6: 39–46.

Fox, P.F., Guinee, T.P., Cogan, T.M., and McSweeney, P.L.H. 2000. *Fundamentals of Cheese Science*. Aspen Publishers, Inc., Gaithersburg, MD.

Frengova, G.I., Emilina, S.D., and Beshkova, D.M. 2003. Carotenoid production by lactose-negative yeast co-cultivated with lactic acid bacteria in whey ultrafiltrate. *Z Naturforsch* 58: 562–567.

Frengova, G., Simova, E., and Beshkova, D. 2004. Use of whey ultrafiltrate as a substrate for production of carotenoids by the yeast *Rhodotorula rubra*. *Applied Biochemistry and Biotechnology* 112: 133–141.

Furuta, Y., Maruoka, N., Nakamura, A., Omori, T., and Sonomoto, K. 2008. Utilization of fermented barley extract obtained from a by-product of barley *Shochu* for nisin production. *Journal of Bioscience and Bioengineering* 106: 393–397.

Garg, K., Nayak, K.K., and Sinkar, V.P. 1995. Butyric and production by *Clostridium butyrium* in high density fermentation. In: *MICON-95*, Abstract no. IMP-31, Mysore, India.

Gassara, F., Brar, S.K., Tyagi, R.D., Verma, M., and Surampalli, R.Y. 2010. Screening of agro-industrial wastes to produce ligninolytic enzymes by *Phanerochaete chrysosporium*. *Biochemical Engineering Journal* 49: 388–394.

Gehlawat, J.K. 1998. Tapping wealth from effluents of agrobased industries. In: *Advances in Wastewater Treatment Technologies*, ed. R.K. Trivedy, pp. 155–156. Global Science Publishing Ltd., Aligarh, India.

George, S. and Jayachandran, K. 2009. Analysis of rhamnolipid biosurfactant produced through submerged fermentation using orange fruit peelings as sole carbon source. *Applied Biochemistry and Biotechnology* 158: 694–705.

Georgieva, T.I. and Ahring, B.K. 2007. Potential of agroindustrial waste from olive oil industry for fuel ethanol production. *Biotechnology Journal* 2: 1547–1555.

Gera, I.B. and Kramer, A. 1969. The utilization of food industries wastes. In: *Advances in Food Research*, eds. C.O. Chichester, E.M. Mkak, and G.F. Stewart, pp. 78–135. Academic Press, New York.

González, M.I., Álvarez, S., Riera, F., and Álvarez, R. 2007. Economic evaluation of an integrated process for lactic acid production from ultrafiltered whey. *Journal of Food Engineering* 80: 553–561.

González-Toledo, S.Y., Domínguez-Domínguez, J., García-Almendárez, B.E., Prado-Barragán, L.A., and Regalado-González, C. 2011. Optimization of nisin production by *Lactococcus lactis* UQ2 using supplemented whey as alternative culture medium. *Journal of Food Science* 75: 347–353.

Grohmann, K., Baldwin, E.A., and Buslig, B.S. 1994. Production of ethanol from enzymatic hydrolyzed orange peel by the yeast *Saccharomyces cerevisiae*. *Applied Biochemistry and Biotechnology* 45: 315–327.

Guerra, N.P., Bernárdez, P.F., and Castro, L.P. 2007. Fed-batch pediocin production on whey using different feeding media. *Enzyme and Microbial Technology* 41: 397–406.

Guerra, N.P., Rua, M.L., and Pastrana, L. 2001. Nutritional factors affecting the production of two bacteriocins from lactic acid bacteria on whey. *International Journal of Food Microbiology* 70: 267–281.

Gullon, B., Yanez, R., Alonso, J.L., and Parajo, J.C. 2008. L-Lactic acid production from apple pomace by sequential hydrolysis and fermentation. *Bioresource Technology* 99: 308–319.

Gunaseelan, V.N. 2004. Biochemical methane potential of fruits and vegetable solid waste feedstocks. *Biomass and Bioenergy* 26: 389–399.

Gupta, A. and Madamwar, D. 1994. High strength cellulase and β-glucosidase formation from *Aspergillus* spp. under solid state fermentation. In: *Solid State Fermentation*, ed. A. Pandey, pp. 130–133. Wiley Eastern Limited, New Delhi, India.

Hack, P.J.F. 1985. Application of the UASB reactor for anaerobic treatment of brewery effluent. *Water Science and Technology* 17: 1489–1490.

Hamdi, M. 1993. Future prospect and constraints of olive mill wastewater use and treatment: A review. *Bioprocess Engineering* 8: 209–214.

Hamdi, M. and Garcia, L.J. 1993. Anaerobic digestion of olive mill wastewaters after detoxification by prior culture of *Aspergillus niger*. *Process Biochemistry* 28: 155–159.

Hamdy, H.S. 2012. Citric acid production by *Aspergillus niger* grown on orange peel medium fortified with cane molasses. *Annals of Microbiology* 63: 267–278.

Hang, Y.D. 1986. Production of single cell proteins from food processing wastes. In: *Food Processing Waste*, eds. J.H. Green and A. Kramer, p. 442, AVI Publ. Co., Westport, CT.

Hang, Y.D., Splittstoesser, O.F., Woodams, E.E., and Sherman, R.M. 1977. Citric acid fermentation of brewery waste. *Journal of Food Science* 42: 383–384.

Heertjes, P.M. and Vander Meer, R.R. 1978. Dynamics of liquid flow in an upflow reactor used for anaerobic treatment of wastewaters. *Biotechnology and Bioengineering* 20: 1577–1594.

Hofvendalh, K. and Hahn-Hagerdal, B. 2000. Factors affecting the fermentative lactic acid production from renewable resources. *Enzyme and Microbial Technology* 26: 87–107.

Hossain, M., Brooks, J.D., and Maddox, I.S. 1983. Production of citric acid from whey permeate by fermentation using *Aspergillus niger*. *New Zealand Journal of Dairy Science and Technology* 18: 161–68.

Indira, A.S., Murthi malathi, H.N., Rajalakshmi, D., and Potty, V.H. 1980. Survey of industrial wastes and by-products from distilleries and breweries. In: *Proceedings of the Symposium on By-Products from Food Industries: Utilization and Disposal*, eds. K.T. Acharya et al., p. 7. Association of Food Scientists and Technologists (India), CFTRI, Mysore, India.

Jacob, N. and Prema, P. 2008 Novel process for the simultaneous extraction and degumming of banana fibers under solid-state cultivation. *Brazilian Journal of Microbiology* 39: 115–121.

Jain, N.K. and Mishra, R.B. 1989. Methane from distillery effluent: Experiments with two stage packed bed reactors. *Journal of Microbiology and Biotechnology* 4: 21–28.

Janssens, L., De Pooter, H.L., Schamp, N.M., and Vandamme, E.J. 1992. Production of flavours by microorganisms. *Process Biochemistry* 27: 195–215.

Jelen, P. 2003 Whey processing. In: *Encyclopedia of Dairy Sciences*, eds. H. Roginski, J.W. Fuquay, and P.F. Fox, pp. 2739–2751. Academic Press, London, U.K.

Jin, B., Yin, P., Ma, Y., and Zhao, L. 2005. Production of lactic acid and fungal biomass by Rhizopus fungi from food processing waste streams. *Journal of Industrial Microbiology Biotechnology* 32: 678–686.

John, R.P., Nampoothiri, K.M., and Pandey, A. 2006a. Solid-state fermentation for L-lactic acid production from agro wastes using *Lactobacillus delbrueckii. Process Biochemistry* 41: 759–763.

John, R.P., Nampoothiri, K.M., and Pandey, A. 2006b Simultaneous saccharification and L-(+)-lactic acid fermentation of protease-treated wheat bran using mixed culture of lactobacilli. *Biotechnology Letters* 28: 1823–1826.

Joshi, C. 2011. Food processing waste treatment technology. In: *Food Processing Waste Management: Treatment and Utilization Technology*, eds. V.K. Joshi and S.K. Sharma, pp. 357–411. New India Publishing Agency, New Delhi, India.

Joshi, V.K., Pandey, A., and Sandhu, D.K. 1999. Fermentation technology for food industry waste utilization. In: *Biotechnology: Food Fermentation*, Vol. 2, pp. 1291–1348. Asiatech Publishers, Inc., Delhi, India.

Joshi, V.K., Parmar, M., and Rana, N.S. 2006. Pectin esterase production from apple pomace in solid-state and submerged fermentations. *Food Technology and Biotechnology* 44: 253–256.

Joshi, V.K., Raj, D., and Joshi, C. 2011. Utilization of waste from food fermentation industry. In: *Food Processing Waste Management: Treatment and Utilization Technology*, eds. V.K. Joshi and S.K. Sharma, pp. 295–356. New India Publishing Agency, New Delhi, India.

Joshi, V.K. and Sandhu, D.K. 1994. Solid state fermentation of apple pomace for production of ethanol and animal feed. In: *Solid State Fermentation*, ed. A. Panday, pp. 93–98. Wiley Eastern Limited, New Delhi, India.

Jwanny, E.W., Rashad, M.M., and Abdu, H.M. 1995. Solid state fermentation of agricultural wastes into food through *Pleurotus* cultivation. *Applied Biochemistry and Biotechnology* 50: 71–80.

Kachru, R.P. 2006. Agro-processing industries in India—Growth, status and prospects. www.agricoop.nic.in (accessed August 14, 2011).

Kahlon, S.S. and Arora, M. 1986. Utilization of potato peels by fungi for protein production. *Journal of Food Science and Technology* 23: 264–267.

Kanwar, S.S., Chadha, B.S., Tewari, H.K., and Sharma, V.K. 1995. Continuous production of lactic acid from molasses by free and immobilized *Sporolactobacillus cellulosolvens. World Journal of Microbiology and Biotechnology* 11: 687–688.

Kargi, F. and Ozmihci, S. 2006. Utilization of cheese whey powder (CWP) for ethanol fermentations: Effects of operating parameters. *Enzyme and Microbial Technology* 38: 711–718.

Katzen, R. and Fowler, D.E. 1994. Ethanol from lignocellulosic waste with utilization of recombinant bacteria. *Applied Biochemistry and Biotechnology* 45/46: 697–707.

Kaur, H. and Khanna, P.K. 2001. Physico-chemical and microbiological characteristics of paddy straw based compost for *Agaricus bisporus* production. *Indian Journal of Mushrooms* 19: 15–20.

Kharatyan, S.G. 1978. Microbes as foods for humans. *Annual Review of Microbiology* 32:301–327.

Khosravi, K. and Shojaosadati, S.A. 2003. A solid state of fermentation system for production of ethanol from apple pomace. *Fanni va Muhandisi-i Mudarris* 10: 55–60.

Kosseva, M.R. 2011. Management and processing of food wastes. In: *Comprehensive Biotechnology*, 2nd edn., ed. M. Moo-Young, Vol. 6, pp. 557–593, Elsevier, New York.

Kosseva, M.R., Panesar, P.S., Kaur, G., and Kennedy, J.F. 2009. Use of immobilised biocatalysts in the processing of cheese whey. *International Journal of Biological Macromolecules* 45: 437–447.

Krishna, C. and Chandrasekaran, M. 1995. Economic utilization of cabbage waste through solid state fermentation by native micro flora. *Journal of Food Science and Technology* 32: 199–201.

Krishna, C. and Chandrasekaran, M. 1996 Banana waste as substrate for α-amylase production by *Bacillus subtilis* under solid state fermentation. *Applied Microbiology and Biotechnology* 46: 106–111.

Krishnan, M.S., Blanco, M., Shattuck, C.K., Nghiem, N.P., and Davison, B.H. 2000a. Ethanol production from glucose and xylose by immobilized *Zymomonas mobilis* CP4(pZB5). *Applied Biochemistry and Biotechnology* 84–86: 525–541.

Krishnan, M.S., Taylor, F., Davison, B.H., and Nghiem, N.P., 2000b. Economic analysis of fuel ethanol production from corn starch using fluidized-bed bioreactors. *Bioresource Technology* 75: 99–105.

Kundu, P. and Das, A. 1984. Utilization of cheap carbohydrate sources for calcium gluconate production by *Penicillium funiculosum* mutant MN 238. *Indian Journal of Experimental Biology* 22: 279–281.

Labaneih, M.E.O., AbouDonia, S.A., Mohamed, M.S., and Eizalak, E.M. 1979. Utilization of citrus waste for the production of fungal protein. *Journal of Food Technology* 14: 95–100.

Laufenberg, G., Kunz, B., and Nystroem, M. 2003. Transformation of vegetable waste into value added products: (A) the upgrading concept; (B) practical implementations. *Bioresource Technology* 87: 167–198.

Lazaridou, A., Biliaderis, C.G., Roukas, T., and Izydorczyk, M. 2002. Production and characterization of pullulan from beet molasses using a nonpigmented strain of *Aureobasidium pullulans* in batch culture. *Applied Biochemistry and Biotechnology* 97: 1–22.

León-Rodríguez, A.D., Rivera-Pastrana, D., Medina-Rivero, E. et al. 2006. Production of penicillin acylase by a recombinant *Escherichia coli* using cheese whey as substrate and inducer. *Biomolecular Engineering* 23: 299–305.

Leuschner, A.P., West, C.E., and Ashare, E. 1983. Assessment of secondary agricultural residues. Part-II. Engineering and economic analysis for conversion to methane and/or alcohol fuels. In: *Fuel Gas Systems*, ed. D.L. Wise. *CRC Series in Bioenergy Systems*. CRC Press, Boca Raton, FL.

de Lima, V.A.G., Stamford, T.M., and Salgueiro, A.A. 1995. Citric acid production from pineapple waste by solid state fermentation using *Aspergillus niger. Brazilian Archives of Biology and Technology* 38: 773–783.

Litchfield, J.H. 1983. Single-cell proteins. *Science* 219: 740–746.

Liu, C., Liu, Y., Liao, W., Wen, Z., and Chen, S. 2004. Simultaneous production of nisin and lactic acid from cheese whey: Optimization of fermentation conditions through statistically based experimental designs. *Applied Biochemistry and Biotechnology* 114: 627–638.

Lo, K.V. and Liao, P.H. 1988. Laboratory scale studies on the mesophilic anaerobic digestion of cheese whey in different digester configurations. *Journal of Agricultural Engineering Research* 39: 99–105.

Maddox, I.S., Qureshi, N., and McQueen, J. 1988. Continuous production of 2,3-butanediol from whey permeate using cells of *Klebsiella pneumoniae* immobilized on to bonechar. *New Zealand Journal of Dairy Science and Technology* 23:127–132.

Mahmoud, M.M. and Kosikowski, F.V. 1982. Alcohol and SCP production by *Kluyveromyces* in concentrated whey permeates with reduced ash. *Journal of Dairy Science* 65: 2082–2087.

Maiorella, B.L., Wilke, C.R., and Blank, H.W. 1981. Alcohol production and recovery. *Advances in Biochemical Engineering* 20: 44–92.

Makkar, R.S., Cameotra, S.S., and Banat, I.M. 2011. Advances in utilization of renewable substrates for biosurfactant production. *AMB Express* 1: 5.

Malisorn, C. and Suntornsuk, W. 2008. Optimization of β-carotene production by *Rhodotorula glutinis* DM28 in fermented radish brine. *Bioresource Technology* 99: 2281–2287.

Mamma, D., Kourtoglou, E., and Christakopoulos, P. 2008. Fungal multienzyme production on industrial by-products of the citrus-processing industry. *Bioresource Technology* 99: 2373–2383.

Manjunathan, J., Raja, S.S.S., and Kaviyarasan, V. 2010. Production of vitamin B_{12} by *Pseudomonas* isolated from sago waste. *International Journal of Medicobiological Research* 1: 68–72.

Marcoux, V., Beaulieu, Y., Champagne, C.P., and Goulet, J. 1992. Production of *Propionibacterium freudenreichii* subsp. *shermanii* in whey-based media. *Journal of Fermentation Bioengineering* 74: 95–99.

Marwaha, S.S., Arora, J.K., and Grover, R. 2000. Biomanagement of food industry waste. In: *Food Processing: Biotechnological Applications*, eds. S.S. Marwaha and J.K. Arora, pp. 295–347. Asiatech Publishers, New Delhi, India.

Marwaha, S.S. and Sethi, R.P. 1983. Utilization of dairy waste for vitamin B_{12} fermentation. *Agricultural Wastes* 9: 111–130.

Marwaha, S.S., Sethi, R.P., and Kennedy, J.F. 1983a. Influence of 5,6-dimethylbenzimidazole (DMB) on vitamin B_{12} biosynthesis by strains of *Propionibacterium*. *Enzyme and Microbial Technology* 5: 361–364.

Marwaha, S.S., Sethi, R.P., and Kennedy, J.F. 1983b. Role of amino acids, betaine and choline on vitamin B_{12} biosynthesis by strains of *Propionibacterium*. *Enzyme and Microbial Technology* 5: 454–464.

Mawson, A.J. 1994. Bioconversions for whey utilization and waste abatement. *Bioresource Technology* 47: 195–203.

Mburu, J.K., Thuo, J.T., and Marder. R.C. 1994. The characterization of coffee waste water from coffee processing factories in Kenya. *Kenya Coffee* 59: 1757–1761.

Medeiros, A.B.P., Pandey, A., and Vandenberghe, L.P.S. 2006. Production and recovery of aroma compounds produced by solid-state fermentation using different adsorbents. *Food Technology and Biotechnology* 44: 47–51.

Mehaia, M.A. and Cheryan, M. 1986. Lactic acid from acid whey permeate in a membrane recycle bioreactor. *Enzyme and Microbial Technology* 8: 289–292.

Mirhosseini, M. and Emtiazi, G. 2011. Optimization of enterocin A production on a whey-based substrate. *World Applied Science Journal* 14: 1493–1499.

Mishra, I.M., Deepak, D., Panesar, P.S., and Saraf, S.K. 1987. Environmental pollution due to sugar industry. *Environmental Management* 129–140.

Mitra, D., Pometto, A.L., Khanal, S.K., Karki, B., Brehm-Stecher, B.F., and van Leeuwen, J.H. 2010. Value-added production of nisin from soy whey. *Applied Biochemistry and Biotechnology* 162: 1819–1833.

Moebus, O. and Kiesbye, P. 1975. Continuous process for producing yeast protein and baker's yeast. G.F.R. *Patent Application* 2410349.

Mohanrao, G.J. and Subrahmanyam, P.V.R. 1972. Sources, flows and characteristics of dairy wastes. *Indian Journal of Environmental Health* 14: 207–217.

Moriguchi, M. 1982. Fermentative production of pyruvic acid from citrus peel extract by *Debaryomyces coudertii*. *Agricultural and Biological Chemistry* 46: 955–961.

Mostafa, N.A. 2001. Production of acetic acid and glycerol from salted and dried whey in a membrane cell recycle bioreactor. *Energy Conversion and Management* 42: 1133–1142.

Moulin, G. and Galzy, P. 1984. Whey, a potential substrate for biotechnology. *Biotechnology & Genetic Engineering Reviews* 1: 347–374.

Mukhopadhyay, R., Chatterjee, S., Chatterjee, B.P., Banerjee, P.C., and Guha, A.K. 2005. Production of gluconic acid from whey by free and immobilized *Aspergillus niger*. *International Dairy Journal* 15: 299–303.

Multilateral Investment Guarantee Agency (MIGA). 2006. World Bank Group. Environmental Guidelines for Sugar Manufacturing. www.miga.org/documents/Sugar Manufacturing.pdf

Murthy, P.S. and Manonmani, H.K. 2008. Bioconversion of coffee industry wastes with white rot fungus *Pleurotus florida*. *Research Journal of Environmental Sciences* 2: 145–150.

Muthusamy, K., Gopalkrishnan, S., Ravi, T.K., and Sivachidambaram, P. 2008. Biosurfactant: Properties, commercial production and application. *Current Science* 94: 736–747.

Nair, P.M. 1994. Biotechnology and hi-technology in food production, processing and preservation—Industries and export opportunities. *Indian Food Industry* 13: 18–24.

Nampoothiri, M.K. and Pandey, A. 1996. Solid state fermentation for L-glutamic acid production using *Brevibacterium* sp. *Biotechnology Letters* 18: 199–204.

Nand, K. 1997. CFTRI Initiatives for energy generation from fruit and vegetable processing waste. *Bioenergy News* 2: 8–9.

Nirmala, P.J. and Ramakrishna, M. 1994. Production of cellulases by solid state fermentation. Recent development. In: *Solid State Fermentation*, ed. A. Pandey, p. 145. New Age Publisher, New Delhi, India.

Nishio, N. and Nakashimada, Y. 2007. Recent development of anaerobic digestion processes for energy recovery from wastes. *Journal of Bioscience and Bioengineering* 103: 105–112.

Noah, K., Bruhn, D.F., and Bala, G.A. 2005. Surfactin production from potato process effluent by *Bacillus subtilis* in a chemostat. *Applied Biochemistry and Biotechnology* 122: 465–474.

O'Leary, V.S., Green, R., Sullivan, B.C., and Holsinger, V.H. 1977. Alcohol production by selected yeast strains in lactase hydrolyzed acid whey. *Biotechnology and Bioengineering* 19: 1019–1035.

Oleszkiewicz, J.A. and Olthof, M. 1982. Anaerobic treatment of food industry wastewater. *Food Technology* 36: 78–82.

Oura, E. 1983. Biomass from carbohydrates. In: *Biotechnology*, eds. H.J. Rehm and G. Reed, Vol. 3, pp. 3–42. Verlag Chemie, Weinheim, Germany.

Palnitkar, S.S. and Lachke, A.H. 1990. Efficient simultaneous saccharification and fermentation of agricultural residues by *Saccharomyces cerevisiae* and *Candida shehatae*. The D-xylose fermenting yeast. *Applied Biochemistry and Biotechnology* 26: 151–158.

Panesar, P.S. 2008. Production of β-D-galactosidase from whey using *Kluyveromyces marxianus*. *Research Journal of Microbiology* 3: 24–29.

Panesar, P.S., Kennedy, J.F., Gandhi, D.N., and Bunko, K. 2007a. Bioutilisation of whey for lactic acid production. *Food Chemistry* 105: 1–14.

Panesar, P.S., Kennedy, J.F., Knill, C.J., and Kosseva, M. 2007b. Applicability of pectate entrapped *Lactobacillus casei* cells for L(+) lactic acid production from whey. *Applied Microbiology and Biotechnology* 74: 35–42.

Panesar, R., Panesar, P.S., and Bera, M.B. 2011. Development of low cost medium for the production of biosurfactants. *Asian Journal of Biotechnology* 3: 388–396.

Panesar, P.S., Panesar, R., Singh, R.S., and Kennedy, J.F. 2006. Microbial production, immobilization and applications of β-D-galactosidase. *Journal of Chemical Technology and Biotechnology* 81: 530–543.

Panesar, P.S., Rai, R., and Marwaha, S.S. 1999. Biological treatment of dairy industry wastewater. *Asian Journal of Microbiology, Biotechnology and Environmental Sciences* 1: 67–72.

Papagianni, M., Psomas, S.K., Batsilas, L. et al. 2001. Xanthan production by *Xanthomonas campestris* in batch cultures. *Process Biochemistry* 37: 73–80.

Patel, P. and Datta, M. 1998. Surfactants in anaerobic digestion of salty cheese whey using upflow fixed film reactor for improved biomethanation. *Process Biochemistry* 33: 199–203.

Patel, R.M. and Desai, A.J. 1997. Biosurfactant production by *Pseudomonas aeruginosa* GS3 from molasses. *Letters in Applied Microbiology* 25: 91–94.

Patel, P., Desai, M., and Datta, M. 1995. Biomethanation of cheese whey using anaerobic upflow fixed film reactor. *Journal of Fermentation and Bioengineering* 79: 398–399.

Paturau, J.M. 1982. *By-Products of the Cane Sugar Industry*, 2nd edn, p. 365. Elsevier, Amsterdam, the Netherlands.

Paturau, J.M. 1987. Alternative uses of sugarcane and its byproducts in agroindustries. Food and Agricultural Organization of the United Nations (FAO), Rome, Italy.

Petruccioli, M., Angiani, E., and Federici, F. 1996. Semi-continuous fumaric acid production by *Rhizopus arrhizus* immobilized in polyurethane sponge. *Process Biochemistry* 5: 463–469.

Petruccioli, M., Duarte, J.C., and Federici, F. 2000. High rate aerobic treatment of winery wastewater using bioreactors with free and immobilized activated sludge. *Journal of Bioscience and Bioengineering* 90: 381–386.

Powell, M.E. and Robe, K. 1964. High protein feed production by lactose fermenting yeast on whey. *Food Processor* 25: 80–95.

Qureshi, N. and Maddox, I.S. 1995. Continuous production of acetone-butanol-ethanol using immobilized cells of *Clostridium acetobutylicum* and integration with product removal by liquid–liquid extraction. *Journal of Fermentation Bioengineering* 80: 185–189.

Ramachandran, S., Fontanille, P., Pandey, A., and Larroche, C. 2006. Gluconic acid: Properties, applications and microbial production. *Food Technology and Biotechnology* 44: 185–195.

Rao, P.V., Jayaram, K., and Lakshmanan, C.M. 1993. Production of lipase by *Candida rugosa* in solid fermentation. 2: Medium optimization and effect of aeration. *Process Biochemistry* 28: 391–395.

Rao, S. and Panda, T. 1994. Critical analysis of the metal ions on gluconic acid production by *Aspergillus niger* using a treated Indian cane molasses. *Bioprocess Engineering* 10: 99–107.

Revankar, M.S., Desai, K.M., and Lele, S.S. 2007. Solid-state fermentation for enhanced production of laccase using indigenously isolated *Ganoderma* sp. *Applied Biochemistry and Biotechnology* 143:16–26.

Rhee, S.K., Pagan, R.J., Lefebvre, M.F., Wong, L., and Rogers, P.L. 1984. Ethanol production from desalted molasses using *Saccharomyces uvarum* and *Zymomonas mobilis*. *Journal of Fermentation Technology* 62: 297–300.

Romero, F.J., García, L.A., Salas, J.A., Díaz, M., and Quirós, L.M. 2001. Production, purification and partial characterization of two extracellular proteases from *Serratia marcescens* grown in whey. *Process Biochemistry* 36: 507–515.

Roukas, T. 1999. Pullulan production from deproteinized whey by *Aureobasidium pullulans*. *Journal of Industrial Microbiology and Biotechnology* 22: 617–621.

Saddoud, A., Hassaïri, I., and Sayadi, S. 2007. Anaerobic membrane reactor with phase separation for the treatment of cheese whey. *Bioresource Technology* 98: 2102–2108.

Saez, L., Perez, J., and Martinez, J. 1992. Low molecular weight phenolics attenuation during simulated treatment of wastewaters from olive oil mill in evaporation ponds. *Water Research* 26: 1261–1266.

Sandhu, D.K. and Joshi, V.K. 1997. Development of apple pomace medium, optimization of conditions for pigment production by *Rhodotorula*. *Advances in Food Research* 19: 31–34.

Sato, A. 1992. Lysine manufacture from fruit waste. Japanese Patent 92-214004.

Sayed, M.S. 1997. Pollution control in dairy plants. In: *Dairy India*, ed. P.R. Gupta, pp. 351–353. NDDB, New Delhi, India.

Schultz, N., Chang, L., Hauck, A., Reuss, M., and Syldatk, C. 2006. Microbial production of single-cell protein from deproteinized whey concentrates. *Applied Microbiology Biotechnology* 69: 515–520.

Seyis, I. and Aksoz, N. 2005. Xylanase production from *Trichoderma harzianum* 1073 D3 with alternative carbon source and nitrogen sources. *Food Technology and Biotechnology* 43: 37–40.

Shah, S., Ramnan, V.V., and Sharma, S.K. 2011. Impact of food industry waste on environment. In: *Food Processing Waste Management: Treatment and Utilization Technology*, eds. V.K. Joshi and S.K. Sharma, pp. 31–52. New India Publishing Agency, New Delhi, India.

Shahidi, F. 1994. Seafood processing by-products. In: *Seafoods Chemistry, Processing, Technology and Quality*, eds. F. Shahidi and J.R. Botta, pp. 320–334. Chapman & Hall, London, U.K.

Shankaranand, V.S. and Lonsane, B.K. 1994. Citric acid by solid state fermentation: A case study for commercial exploitation. In: *Solid Fermentation*, ed. A. Pandey, pp. 149–152. Wiley Eastern Limited, New Delhi, India.

Shojaosadati, S.A. and Babaeipour, V. 2002. Citric acid production from apple pomace in multi-layer packed bed solid-state bioreactor. *Process Biochemistry* 37: 909–914.

Shukla, J.P. and Dutta, S.M. 1967. Production of fungal protein from waste molasses. *Indian Journal of Technology* 5: 27.

Silva, M.F., Fornari, R.C.G., Mazutti, M.A. et al. 2009. Production and characterization of xanthan gum by *Xanthomonas campestris* using cheese whey as sole carbon source. *Journal of Food Engineering* 90: 119–123.

Skogman, H. 1976. Production of symba-yeast from potato waste. In: *Food from Waste*, eds. G.G. Birch, K.J. Parkar, and J.T. Worgan, pp. 167–179. Applied Science Publishers, London, U.K.

Solomon, S.K. 2005. Environmental pollution and its management in sugar industry in India: An appraisal. *Sugar Technology* 7: 77–81.

Somkuti, G.A. and Steinberg, D.H. 2003. Pediocin production by recombinant acetic acid bacteria. *Biotechnology Letters* 25: 473–477.

Sreekrishna, K. and Dickson, R.C. 1985. Construction of strains of *Saccharomyces cerevisiae* that grow on lactose. *Proceedings of the National Academy of Sciences United States of America* 82: 7909–7913.

Stredansky, M. and Conti, E. 1999. Xanthan production by solid state fermentation. *Process Biochemistry* 34: 581–587.

Streit, F., Koch, F., Trossini, T.G., Laranjeira, M.C.M., and Ninow, J.L. 2004. An alternative process for the production of an additive for the food industry. In: *Proceedings of the Chitosan International Conference Engineering and Food—ICEF 9*, Montpellier, France.

Suganthi, R., Benazir, J.F., Santhi, R. et al. 2011 Amylase production by *Aspergillus niger* under solid state fermentation using agroindustrial wastes. *International Journal of Engineering Science and Technology* 3: 1756–1763.

Takata, Y. 1992. Production of biomass from soybean oil and waste oil. *Journal of the Japanese Society for Food Science and Technology* 39: 253–256.

Tay, J.H., Show, K.Y., and Hung, Y.T. 2006. Seafood processing wastewater treatment. In: *Waste Treatment in the Food Processing Industry*, eds. L.K. Wang, Y.T. Hung, H.H. Low, and C. Yapijakis, pp. 29–66. Taylor & Francis, Boca Raton, FL.

Tengerdy, R.P. 1996. Cellulase production by solid state substrate fermentation. *Journal of Scientific and Industrial Research* 55: 313–316.

Tewari, H.K., Marwaha, S.S., and Rupal, K. 1986. Ethanol from banana peels. *Agricultural Wastes* 16: 135–146.

Tin, C.S.F. and Mawson, A.J. 1993. Ethanol production from whey in a membrane recycle bioreactor. *Process Biochemistry* 28: 217–221.

Torrijos, M. and Moletta, R. 1997. Winery wastewater depollution by sequencing batch reactor. *Water Science and Technology* 35: 249–257.

Trummler, K., Effenberger, F., and Syldatk, Z.C. 2003. An integrated microbial/enzymatic process for production of rhamnolipids and L-(+)-rhamnose from rapeseed oil with *Pseudomonas* sp. DSM 2874. *European Journal of Lipid Science and Technology* 105: 563–571.

Tsugawa, R., Nakamura, A., Takeyama, T., Ueno, S., Soejima, T., and Minematsu, T. 1981. Production of purified citrus molasses for the raw materials of amino acid fermentation. *Abstract at International Citrus Congress*, Tokyo, Japan.

Upadhyay, R.C. and Sohi, H.S. 1980. Apple pomace—A good substrate for the cultivation of edible mushrooms. *Current Science* 57: 1189–1190.

Vastrad, B.M. and Neelagund, S.E. 2011. Optimization and production of neomycin from different agro industrial wastes in solid state fermentation. *International Journal of Pharmaceutical Sciences and Drug Research* 3: 104–111.

Vendruscolo, F. 2005. Cultivo em meio solido e submerso do bagaco de maca por *Gongronella butleri* e avaliacao do seu potencial biotecnologico. Master's dissertation, Universidade Federal de Santa Catarina, Florianopolis, Brazil.

Venetsaneas, N., Antonopoulou, G., Stamatelatou, K., Kornaros, M., and Lyberatos, G. 2009. Using cheese whey for hydrogen and methane generation in a two-stage continuous process with alternative pH controlling approaches. *Bioresource Technology* 100: 3713–3717.

Vij, S. and Gandhi, D.N. 1993. Isolation and identification of lactose fermenting yeasts from various dairy products. *Journal of Food Science and Technology* 30: 222–223.

Vijayaraghavan, P., Devi, V.S.L., and Vincent, S.G.P. 2011. Bio-processing of banana peel for amylase production by *Penicillium* sp. *Asian Journal of Experimental Sciences* 2: 257–264.

Viswanath, P., Devi, S.S., and Nand, K. 1992. Anaerobic digestion of fruit and vegetable processing wastes for biogas production. *Bioresource Technology* 40: 43–48.

Von Enden, J.C. and Calvert, K.C. 2002. Review of coffee waste water characteristics and approaches to treatment. www.coffee.20m.com/CoffeeProcessing/CoffeeWasteWater.pdf (accessed October 4, 2011).

Vriens, L., Van Soest, H., and Verachtert, H. 1990. Biological treatment of malting and brewing effluents. *Critical Reviews in Biotechnology* 10: 1–27.

Wassewar, K.L. 2005. Separation of lactic acid: Recent advances. *Chemical and Biochemical Engineering Quarterly* 19: 159–172.

Webb, L.J. 1983. In: *Proceedings of Conference on Biotechnology in the Pulp and Paper Industries*, London, U.K., September 12–14, 1983, pp. 1–16.

Wee, Y.J., Kim, J.N., Yun, J.S., and Ryu, H.W. 2004. Utilization of sugar molasses for economical L(+)-lactic acid production by batch fermentation of *Enterococcus faecalis*. *Enzyme and Microbial Technology* 35: 568–573.

Wolmarans, B. and De Villiers, G.H. 2002. Start up of a UASB effluent treatment plant on distillery wastewater. *Water SA* 28: 63–68.

Yalcin, S.K., Bozdemir, M.T., and Ozbas, Z.Y. 2009. Utilization of whey and grape must for citric acid production by two *Yarrowia lipolytica* strains. *Food Biotechnology* 23: 266–283.

Yamada, T., Fatigati, M.A., and Zhang, M. 2002 Performance of immobilized *Zymomonas mobilis* 31821 (pZB5) on actual hydrolysates produced by Arkenol technology. *Applied Biochemistry and Biotechnology* 98–100: 899–907.

Yang, S.S. 1996. Antibiotics production of cellulosic wastes with solid state fermentation by *Streptomyces*. *Renewable Energy* 9: 976–979.

Yang, S.S. and Ling, M.Y. 1989. Tetracycline production with sweet potato residue by solid state fermentation. *Biotechnology and Bioengineering* 33: 1021–1028.

Yang, P., Zhang, R., McGarvey, J.A., and Benemann, J.R. 2007. Biohydrogen production from cheese processing wastewater by anaerobic fermentation using mixed microbial communities. *International Journal of Hydrogen Energy* 32: 4761–4771.

Zhang, Y., Dube, M.A., McLean, D.D., and Kates, M. 2003. Biodiesel production from waste cooking oil: 1. Process design and technological assessment. *Bioresource Technology* 89: 1–16.

Zigová, J. and Šturdik, E. 2000. Advances in biotechnological production of butyric acid. *Journal of Industrial Microbiology and Biotechnology* 24: 153–160.

Index

A

α-Amylases, 340
Abiotic stress tolerance, 160–161
Acesulfame-K, 395
Acetan, *see* Xylinan
Acetic acid production, 224–225
Acetobacter species, 224–225
Aerobes and anaerobes, 228
Agrobacterium transformation
  tissue culture–dependent, 142–143
  tissue culture–independent, 143–144
Agrocybe aegerita, 532
Agro-industrial wastes and residues
  environmental legislations, 558–559
  food processing
    brewing industry, 562
    coffee industry, 564–565
    dairy industry, 561
    distillery, 563
    fruit and vegetable industry, 560–561
    meat and poultry industry, 563–564
    olive oil industry, 565
    seafood industry, 564
    sugar industry, 561–562
    wine industry, 562–563
  utilization, 559
  value-added products production
    antibiotics, 589–590
    Baker's yeast *(S. cerevisiae),* 568–569
    bio-fuel, 574–582
    enzymes, 582–585
    food additives, 585–589
    mushrooms, 590
    organic acids, 569–574
    single-cell proteins, 566–568
Alcoholic beverages production, 31–32
Alitame, 395
Alkaline fermentations, 225
Amino acids production, 589
Amylopectin, 397–398
Amylose, 397–398
Androgenesis, *see* Anther culture and haploid production
Anther culture and haploid production
  applications
    early release of varieties, 100
    hybrid development, 99
    mutation induction, 99
    pure homozygous lines development, 99
  case study, 100–101
  culture media and nutritional requirements
    direct androgenesis, 98
    growth regulators, 96–97
    indirect androgenesis, 99
    modes of, 97
  optimum condition, 95
  techniques, 96
Anthraquinones, biocolors production
  emodin pigment, 426
  factors influence, 428–429
  isolated and identification, 428
  P. oxalicum, arpink red, 425, 427
  red pigment, 426
Antibiotics production, 589–590
Antimicrobial substances, fermentation, 227
Apomixis, micropropagation, 86
Aspartame, 395–396
Astaxanthin production
  Ag. aurantiacum, 438
  X. dendrorhous (Pha. rhodozyma), 437–438
Auricularia polytricha, 532
Azaphilones, 421
Azospirillum, 173
Azotobacter, 173

B

Bacillus thuringiensis (Bt), 192, 194–203, 205–206, 208; *see also* Biopesticides
Bacteria, food fermentation; *see also Lactobacillus* bacteria, therapeutic agents
  acetic acid, 224–225
  alkaline fermentations, 225
  lactic acid, 224
Bacteriocins production, 588–589
Baker's yeast *(S. cerevisiae)* production, 568–569
Baking industry, enzymes
  α-amylases, 340
  glucose oxidase, 340
  lipases, 341
  proteases, 341
  use, 339
  xylanases, 340–341
Bergmann's cell plating technique, 77–78
Biocolors production
  anthraquinones
    factors influence, 428–429
    isolated and identification, 428
    pigment emodin, 426
    P. oxalicum, arpink red, 425, 427
    red pigment, 426
  B. trispora, 432–436
  canthaxanthin, 439–440
  carotenoid production, *Ag. aurantiacum*, 438
  fermentative food-grade pigments, 420
  F. sporotrichioides, 436–437
  microbial food-grade pigments, 419
  Monascus pigments
    cultivation, 421
    fungal metabolites, 421–422
    mycotoxin control production methods, 423–424
    nontoxigenic *Penicillium* species, 424
    production methods, 422–423
  Mu. circinelloides, 434–435
  Ph. blakesleeanus, 434
  phycocyanin, 430–432
  pigmented microorganisms, 418
  red microalga *Porphyridium*, phycobiliproteins, 429–430
  riboflavin, 429
  torulene and torularhodin, 440
  X. dendrorhous (Pha. rhodozyma), 437–438
  zeaxanthin, 438–439
Biocontrol agents
  awareness, 13–14
  biofertilizers, 15 (*see also* Biofertilizers)
  bioherbicides, 14
  biopesticides, 14 (*see also* Biopesticides)
  disease control agents, 14–15
Biodiesel production, 581; *see also* Bio-fuel production
Biofertilizers
  benefits, 177–178
  bioinoculants, 170
  constraints, 185–186
  global status
    commercial scale producers, 184–185
    nitrogenous, 184
  microbial cultures
    microbial consortia, 176–177
    microorganisms used, 170–171
    nitrogen-fixing microorganisms, 170–174
    PGPR, 176
    phosphate-solubilizing bacteria, 174–175
    vesicular arbuscular mycorrhiza, 175
  production technology
    bacterial cultures, carrier materials, 181–182
    mass multiplication, 181
    media composition, 180
    microbial strain selection, 179
    mother culture preparation, 179
    quality control, 182–183
    raw material and equipment, 179–181
    trace metal mix composition, 181

Bioflavors production
biotechnological method
enzymatic method, 460, 463
microbial method, 463–469
plant tissue culture method, 460–462
characteristics, 448
classification
carboxylic acids, 457–459
diacetyl, 450
esters, 456–457
lactones, 450, 452–454
pyrazines, 450–451
terpenes, 451, 455–456
and fragrances production, 587
fruit and vegetable biogenesis, 448–449
natural flavor, 448
safety and regulatory aspects
Australia New Zealand Food Standards Code, 471
FEMA, 470–471
food sanitation law, 471–472
GMO labeling, 471
joint FAO/WHO food standards, 472
Bio-fuel production
biodiesel, 581
biogas, 578–581
biohydrogen (Bio-H_2), 581
butanol, 582
ethanol, 574–577
Bioinsecticide production, 195, 198
Biopesticides
advantages
residual effect, 208
resistance management, 208
restricted-entry intervals, 208
role, 207
Bacillus thuringiensis (Bt)
characteristics, 192, 194
formulations, 195–197
guidelines, 194
identification, 198
isolation, 197
nutritional requirements, 199
production flow sheet, 198
production technology, 199–200
strains, 194
biocontrol potential, 193
Biopesticide and Pollution Prevention Division (BPPD), 208
categories, 192
compound annual growth rate, 192
constraints/challenges, 206
crystal proteins, mode of action, 205–206
disadvantages, 207–208
features, 192
and integrated pest management, 209
production technology
downstream process, 200–205 (*see also* Downstream process)
strain selection, 195, 197–199
serotypes, 194
Bio-pigments production, food additives, 586–587
Biosurfactants production, food additives, 587–588
Biotechnology role
agriculture, techniques
biocontrol agents, 13–15
genetically modified crops, 9–13
plant cell and tissue culture, 8–9
concept, 4–5
crops (*see also* Genetically modified (GM) foods)
corn, 499–500
cotton, 501–502
potato, 500–501
soybean, 499
food-processing industry
alcoholic beverages production, 31–32
cereal processing, 18–19
food additives production, 22
fruit and vegetable processing, 19–20
meat and fish processing, 20–21
milk processing, 16–18
single cell protein production, 21–22
waste management, 32
genomics impact, 32–34
historical developments, 5–6
interdisciplinary nature, 6–7

Blakeslea trispora
β-carotene, 432–434
lycopene, 435–436
Blue-green algae, 173–174, 368
Brewing industry, 562
Bt.-based biopesticides, *see* Biopesticides
Butanol production, 582
Button mushroom production, compost
Anglo-Dutch method, 526
base materials, 520
concentrate meals, 520–521
formulations, 521, 523
long method, 521, 523–524
recent trends, 525–527
rectifying mineral deficiencies supplements, 521
short method, 524–525
supplements, activating fermentation, 520

C

Calocybe indica, 532
Carboxylic acids bioflavors, 457–459
Carotenoid production
astaxanthin production
Ag. aurantiacum, 438
X. dendrorhous (Pha. rhodozyma), 437–438
β-carotene
B. trispora, 432–434
Mu. circinelloides, 434–435
Ph. blakesleeanus, 434
lycopene
B. trispora, 435–436
F. sporotrichioides, 436–437
Cereal processing
cereal-legume products, indigenous foods
adai, 262–263
dhokla, 262
dosa, 261–262
idli, 260–261
food-processing industry, 18–19
indigenous fermented foods
bhatura, 247
bread, 245–246
bushera, 251
dolo and pito, 253
injera, 249
jalebis, 247–248
kaffir beer and chibuku, 253
kenkey, 249
kishk, 251
kisra, 250–251
mahewu, 249–250
nan, 247
ogi, 248
pasta, 246–247
togwa, 51–253
uji, 248–249
proteins and peptides, 313
Chemotherapy, meristem culture, 84
Coffee industry, 564–565
Corn/glucose syrups
enzymatic processes, 401
hydrolysis processes, 399–400
purification steps, 400
Crop improvement programs; *see also* Tissue culture
genome-wide association studies, mapping, 53–54
genomic marker identification, 51–52
genomic selection, 54–55
genotyping by sequencing, 52–53
marker-assisted selection, plant breeding, 48
next generation sequencing-facilitated crop improvement, 48–51
Crown galls, 482
Crystal proteins, 205–206
Curvularia lunata, 426
Cyanobacteria, *see* Blue-green algae
Cybridization, 108–109; *see also* Protoplast culture
Cyclamate, 396
Cytoplasmic male sterility, 111

D

Dairy industry, 561; *see also* Milk processing
Debranching enzymes, 401
Dextrans, 373–374
Diacetyls
bioflavors production, 450
biotransformation, 467–468
de novo synthesis, 465

Downstream process
Bt.
disk–stack centrifuge/rotary vacuum filter, 202–203
ISPR technique, 204
issues, 201
MEUF, 203–204
microfiltration membranes function, 204–205
spray dying method, 201–202
enzymes production and purification, 335

E

Edible vaccines, GM crops, 12–13
Electroporation-mediated transformation, 145
Embryo culture
applications
embryo abortion prevention, 103
interspecific crosses, 104
nutritional requirements, 103–104
precocious germination, 103
seed dormancy, 103
case study, 104
description, 101
factors affecting, 102
types, 102
uses, 101
Endomycorrhiza, *see* Vesicular arbuscular mycorrhiza (VAM)
δ-Endotoxins, 205, 487
Enokitake, *see Flammulina velutipes*
Environmental Protection Agency (EPA)
biopesticides, 194
genetic transformation and crop improvement, 149
residual effect, guidelines, 208
Enzymes
advantages, 330
agro-industrial wastes and residues, production, 582–585
applications
baking industry, 339–341
fruit juice industry, 341–342
meat processing, 337–338
milk and cheese industries, 338–339
wine production, 342–344
bioflavors, 460, 463
biosensors
applications, 346–347
biochemical and physical transducer, 345
engineering studies in, 332–333
immobilization techniques
advantages, 335
methods, 335–336
potential applications, 337
mechanism of action, 330–332
production and purification
downstream process, 335
fermentations, 333–334
growth and nongrowth-associated, 333
steps used, 334
recombinant DNA technology, 332–333
Erucin (ER), 310
Esters
bioflavors, 456–457
biotransformation, 467
de novo synthesis, 466
Ethanol production, 574–577
Exopolysaccharides (EPS), *see* Polysaccharides

F

Farmaceuticals, 296
Fermentation
enzymes production and purification, 333–334
factors affecting
antimicrobial substances, 227
biological structures, 227
gases, 228
moisture content, 226–227
nutrient content, 227
osmotic pressure, 227
pH, 226
temperature, 228
indigenous fermented foods
cereal–legume products, 260–263
cereal products, 245–253
legume products, 253–260
meat and fish products, 265–267
milk products, 230–245
vegetable products, 263–265

methods
solid-state, 222–223, 422–423
submerged state, 222, 422
surface, 223
microorganisms
bacteria, 223–225
molds, 226
yeasts, 225
nutritional and health benefits, 267–268
safety aspects, 268–269
starter cultures
benefits, 229
preparation, 229–230
role, 228
wine, 221
Fiber, 314–315
Fish processing
fish paste, 266–267
fish sauce, 267
sausages, 265–266
Fission yeast, *see Schizosaccharomyces pombe*
Flammulina velutipes, 532
Flavonoids, 306–307
Food-processing industry
agro-industrial wastes and residues
brewing industry, 562
coffee industry, 564–565
dairy industry, 561
distillery, 563
fruit and vegetable industry, 560–561
meat and poultry industry, 563–564
olive oil industry, 565
seafood industry, 564
sugar industry, 561–562
wine industry, 562–563
alcoholic beverages, 31–32
amino acids, 28, 589
bacteriocins, 29–30, 588–589
bio flavors and fragrances, 25, 587
bio-pigments, 22, 24–25, 586–587
biosurfactants, 30, 587–588
cereal processing, 18–19
fruit processing, 19–20
gums production, 27–28
meat and fish processing, 20–21
milk processing, 16–18
oils and fats, 30–31
organic acids, 25–27
polysaccharides/gums, 585–586
single cell protein production, 21–22
vitamin B_{12}, 28–29, 589
waste management, 32
Food Sanitation Law
natural flavoring agent, 471
recombinant DNA technique, 472
Fruit processing industry
agro-industrial wastes and residues, 560–561
bioflavors production, 448–449
biotechnology role, 19–20
juice industry, enzymes applications, 341–342
Functional foods
definition, 281–282
fiber, 314–315
health benefits
capsaicin, 310
carotenoids, 307–308
curcumin, 309
flavonoids, 306–307
isoflavones, 305–306
isothiocyanates, 310–311
limonoids, 308–309
oleuropein, 311
omega-3 fatty acids, 302, 305
organosulfur compounds, 309
phytosterols and phytostanols, 311–312
proteins and peptides, 312–314
tocopherols and tocotrienols, 306
xylitol, 311
Lactobacillus bacteria
allergies/eczema, 291–292
anticarcinogenic activity, 291
immune system enhancement, 290
intestinal microflora, 290
lactose intolerance reduction, 290
serum cholesterol levels reduction, 291
ulcer protection, 291
nutraceuticals
bioavailability, 301
bioproduction, 298–300

classification, 296
and human health, 297
safety, 301
prebiotics
classification, 284–286
definition, 282–283
human health role, 286–287
properties, 283–284
probiotics
criteria, 289
microorganisms, 288
properties, 289
synbiotics
clinical applications, 293–294
description, 282
mode of action, 294–295
prebiotics and probiotics, 292–294
Functional ingredients, 281
Fusogens, 106; *see also* Protoplast culture

G

Gamma-aminobutyric acid (GABA), 314
Ganoderma lucidum, 533
Gases, fermentation, 228
Gellan, 366–367
Genetically modified (GM) foods
benefits of, 488–492
biotechnological methods, 480
commercialized crops
corn, 499–500
cotton, 501–502
potato, 500–501
soybean, 499
crops
benefits, 11–12
commercialized, 498–502
country-wise area, 485–486
edible vaccines, crops, 12–13
Food and Drug Administration (FDA), 9
list of, modified genes and benefits, 10–11
in Russia, 488
safety concerns, 13
types, 486–488
first, second, and future generations, 480–481
gene expression control, 484–485
genomic analysis, 484
plant transformation vectors
ballistic method, 482–483
development techniques, 481–482
multistep process, 481
safety concerns
assessment system, Russian federation, 494–496
identification and quantification methods, 497
labeling issues, 496–497
principles, 489, 492
recombinant DNA, 492–493
world practices in, assessment, 493
selection and regeneration, transformants, 483–484
Genetic transformation
advancements, 134
advantages and limitations, 134–135
Agrobacterium-based T-DNA, 134
crop improvement, different means, 135
green revolution, 134
history, 136–138
importance
applied plant biotechnology, 140–141
understanding technology, 138–140
plant trait modification
abiotic stress tolerance, 160–161
applications, 149–157
herbicide tolerance, 157–158
insect and disease resistance, 158–159
quality improvement, 159–160
purpose, 133–134
totipotent nature, 138
transgenic plants development
Agrobacterium transformation, 142–144
biolistic transformation and chloroplast genetic engineering, 144–145
DNA transfer, 146
electroporation-mediated transformation, 145
microinjection, 146

PEG-mediated genetic transformation, 145–146
post-transformation steps, 146–149
Genomic estimated breeding values (GEBVs), 54
Genomics impact
agriculture and food processing, 32–34, 47
crop improvement programs
genome-wide association studies, mapping, 53–54
genomic marker identification, 51–52
genomic selection, 54–55
genotyping by sequencing, 52–53
marker-assisted selection, plant breeding, 48
next generation sequencing-facilitated crop improvement, 48–51
malnutrition tackling, 46
nutritional traits improvement
efficient food processing, microbial genomics, 60–61
and food-processing industry, 59
food safety, microbial genomics, 61–62
genome editing, 59
GMOs, biological species concept, 56–57
humanitarian golden rice initiative, 55–56
induced local lesions targeting, 58–59
mutagenesis, nutritional quality improvement, 57
seed quality improvement, 46–47
storage and food processing, 59–60
Glucose oxidase, 340
Glycemic index (GI), 391–392
Glyphosate, 486–487
GM crops, *see* Genetically modified (GM) foods; Plant transformation

H

Haber–Bosch process, 170; *see also* Nitrogen-fixing microorganisms
Health benefits, functional foods
capsaicin, 310
carotenoids, 307–308
curcumin, 309
flavonoids, 306–307
isoflavones, 305–306
isothiocyanates, 310–311
limonoids, 308–309
oleuropein, 311
omega-3 fatty acids, 302, 305
organosulfur compounds, 309
phytosterols and phytostanols, 311–312
proteins and peptides, 312–314
tocopherols and tocotrienols, 306
xylitol, 311
Herbicide tolerance, 157–158
High-fructose corn syrup (HFCS)
activated carbon and ion-exchange steps, 402
description, 388–389
syrups characteristics, 403
Hyaluronic acid (HA), 370–372

I

Immobilization techniques, enzymes
advantages, 335
methods, 335–336
potential applications, 337
Indigenous fermented foods; *see also* Fermentation
cereal-legume products
adai, 262–263
dhokla, 262
dosa, 261–262
idli, 260–261
cereal products
bhatura, 247
bread, 245–246
bushera, 251
dolo and pito, 253
injera, 249
jalebis, 247–248
kaffir beer and chibuku, 253
kenkey, 249
kishk, 251
kisra, 250–251
mahewu, 249–250

nan, 247
ogi, 248
pasta, 246–247
togwa, 51–253
uji, 248–249
legume products
bhallae, 259
dawadawa, 258
miso, 255–256
natto, 255
oncom, 257–258
papadam, 259–260
soy sauce, 256–257
sufu, 256–257
tempeh, 254–255
vada/vadai, 259
warri/wadiyan, 258–259
meat and fish products
fish paste, 266–267
fish sauce, 267
sausages, 265–266
milk products
acidophilous milk, 242–243
butter and ghee, 243
cheese, 230, 239–240
cultured buttermilk and cream, 242
kefir, 244–245
koumiss, 244
shrikhand, 243
yogurt, 240–241
vegetable products
pickles, 264–265
sauerkraut, 263–264
In situ product removal (ISPR) technique, 204
Insect and disease resistance, 158–159
Integrated pest management (IPM), 209
Inulin syrup, sweeteners
high-fructose syrups production, 407–408
hydrolysis process, 406–407
immobilized inulinases, 409
structure, 406
Invert sugar/syrup
hydrolysis of, 404
microorganisms uses, 404
production of, 405
Isoflavones, 305–306
Isothiocyanates (ITCs)
erucin, 310
sulforaphane, 310–311

L

Lactic acid bacteria (LAB), 224, 369–370
Lactobacillus bacteria, therapeutic agents
allergies/eczema, 291–292
anticarcinogenic activity, 291
immune system enhancement, 290
lactose intolerance reduction, 290
normal intestinal microflora, 290
serum cholesterol levels reduction, 291
ulcer protection, 291
Lactones
bioflavors, 450, 452–454
de novo synthesis, 465–466
Lentinus edodes, 531
Limonoids, 308–309
Lipases, 341
Long-chain polyunsaturated fatty acids, 299

M

Maceration enzymes, 344
Marker genes, 483–484
Meat processing
agro-industrial wastes and residues, 563–564
biotechnology role, 20–21
endogenous enzymes, 337
enzymes applications, 337–338
indigenous fermented foods, 265–267
proteolytic enzymes, 337–338
Mellowing, 239
Meristem culture
purpose, 82–83
technique, 83
virus elimination methods
chemotherapy, 84
potato virus X, 85
research work on, 84
thermotherapy/heat treatment, 83–84
tobacco mosaic virus, 84–85
virus indexing, 85

Micellar-enhanced ultrafiltration (MEUF), 203–204
Microaerophiles, 228
Microbial method
  biotransformation
    diacetyls, 467–468
    esters, 467
    terpenes, 466–467
    vanillin production, 467–469
  consortia
    IAA production, 176
    lacZ and gusA, 177
  de novo synthesis
    aromatic compounds production, 464
    diacetyl, 465
    esters, 466
    pyrazines, 465
    terpenes, 466
  genomics (*see* Genomics)
Microinjection, transgenic plants development, 146
Micronutrient malnutrition (MNM), 45–46
Micropropagation
  advantage, 85–86
  adventitious buds/shoots regeneration, 87
  apomixis, 86
  axillary branching, 86–87
  case study, 93
  horticultural and forest tree species, progress, 93–95
  single-node culture, 86
  stages
    aseptic culture establishment, 87–88
    explants, preparation and pretreatment, 87
    genetic stability molecular analysis, 92–93
    root regeneration, in vitro developed shoots, 90, 92
    shoot multiplication, 88–91
    subsequent field planting hardening, 92
Milk processing
  biotechnology role, 16–18
  and cheese production
    rennet, 338
    whey proteins and lipases, 339
  enzymes applications, 338–339
  indigenous fermented foods
    acidophilous milk, 242–243
    butter and ghee, 243
    cheese, 230, 239–240
    cultured buttermilk and cream, 242
    kefir, 244–245
    koumiss, 244
    shrikhand, 243
    yogurt, 240–241
  proteins and peptides, 312–313
Molds, 226
Monascidin A, 422–423
Monascus pigments
  biocolors production
    cultivation, 421
    fungal metabolites, 421–422
    mycotoxin control production methods, 423–424
    nontoxigenic *Penicillium* species, 424
    production methods, 422–423
Monosodium glutamate (MSG), 25
Mushrooms production
  agro-industrial wastes and residues, 590
  button variety
    compost, 519–521
    compost formulations, 521–522
    methods, compost preparation, 521, 523–527
  crop
    casing materials and treatment, 529
    casing soil, 529
    environmental conditions, 529–530
    growing system, 527–528
    spawning, 528–529
  cultivation history, 510–511
  growth potential
    basic raw material availability, 544
    development, 544
    domestic consumption and affordability, 545
    environment and diversification, 545

export potential, 545–546
food protein source, 546
low-cost labor, 545
minimum land use/intensive use, indoor space, 546
recycling waste residues, 547
self-employment, 547
important cultivated types portfolio
Agrocybe aegerita, 532
Auricularia polytricha, 532
Calocybe indica, 532
Flammulina velutipes, 532
Ganoderma lucidum, 533
Lentinus edodes, 531
Pleurotus spp., 530–531
Volvariella volvacea, 533
medicinal values, 539–544
nutritional values
carbohydrates and fibers, 537
fats/lipids, 535–537
minerals and trace elements, 537–538
protein, 534–536
proximate composition, 534
vitamins, 538–539
trade and consumption
European Union production, 513
global imports, canned and fresh, 513–515
global production, 511–512
indian scenario, 515–517
types, 518–519

N

N-acetyl neuraminic acid, 376
Nature's unique plant, 501–502
Neotame, 396–397
Nitrogen-fixing microorganisms
azospirillum, 173
azotobacter, 173
blue-green algae, 173–174
microbial consortia, 176–177
phosphate-solubilizing bacteria, 174–175
rhizobacteria, 176
rhizobium, 172
vesicular *arbuscular mycorrhiza,* 175
Nonsaccharides
natural high-intensity sweeteners, 393–395
sugar alcohols
biologic agents use, 393
GI, 391–392
synthetic high-intensity sweeteners, 395–397
Nutraceuticals
bioavailability, 301
bioproduction
large-molecule, 299–300
lipid-based, 299
long-chain polyunsaturated fatty acids, 299
polar lipids, 298–299
small-molecule, 300
classification, 296
description, 295–296
dietary supplements, 296
farmaceuticals, 296
functional foods, 296
and human health, 297
medical foods, 296
safety, 301
Nutritional traits improvement, genomics
efficient food processing, microbial genomics, 60–61
food processing and storage, 59–60
and food-processing industry, 59
food safety, microbial genomics, 61–62
genome editing, 59
GMOs, biological species concept, 56–57
humanitarian golden rice initiative, 55–56
induced local lesions targeting, 58–59
mutagenesis, nutritional quality improvement, 57

O

Oleuropein, 311
Olive oil industry, 565
Omega-3 fatty acids, 302, 305

Organic acids production
acetic acid, 573
butyric acid, 574
citric acid, 571–573
fumaric acid, 574
gluconic acid, 573
lactic acid, 569–571
pyruvic acid, 574
Osmotic pressure, fermentation, 227

P

Pectinase, 341–343
Pectinolytic enzymes, 343–344
PEG-mediated genetic transformation, 145–146
Pest management, 209; *see also* Biopesticides
PGPR, *see* Plant growth-promoting rhizobacteria (PGPR)
pH, fermentation, 226
Phosphate-solubilizing bacteria, 174–175
Phosphinothricin, 487
Phytostanols, 311–312
Phytosterols, 311–312
Plant cell culture, 75; *see also* Tissue culture
Plant growth-promoting rhizobacteria (PGPR), 176; *see also* Biofertilizers
Plant tissue culture method, 460–462
Plant trait modification, genetic transformation
abiotic stress tolerance, 160–161
applications, yield and quality improvement, 149–157
herbicide tolerance, 157–158
insect and disease resistance, 158–159
quality improvement, 159–160
Plant transformation; *see also* Transgenic plants development
gene expression control, 484–485
genomic analysis, 484
selection and regeneration, 483–484
vectors
ballistic method, 482–483
development techniques, 481–482
multistep process, 481
Polar lipids, 298–299
Polyphenol extraction, *see* Pectinase
Polysaccharides
bacterial
dextrans, 373–374
pullulan, 374
scleroglucan and schizophyllan, 375
xylinan, 372–373
cyanobacterial, 368–369
food additives, gums production, 585–586
gellan, 366–367
hyaluronic acid
applications, 371–372
β-(1→4) N-acetyl-d-glucosamine and β-(1→3) d-glucuronic acid, 370
lactic acid bacteria, 369–370
microbes, 362
pyranose/furanose carbohydrate ring structures, 355–361
vaccines, 375–377
xanthan
gum, 365–366
microbial production, 364
polytetramer and polytrimer, 365
trisaccharide chain, 363
Pomace, 563
Porphyridium sp.
fluorescent pink color, 430
Po. aerugineum, 430–432
red microalga, 429–430
Poultry industry, 563–564; *see also* Meat processing
Prebiotics
classification, 284
fructo-oligosaccharides, 285–286
galacto-oligosaccharides, 285
soybean oligosaccharides, 286
xylo-oligosaccharides, 286
definition, 282–283
functions, 287
human health role, 286–287
properties, 283–284
Probiotics
description, 287
microorganism lists, 288
properties, 289
role, 288
strains criteria, 289

Proteases, 341
Proteins and peptides
cereals, 313
egg and meat, 314
GABA, 314
legumes, 313
milk and milk products, 312–313
Protoplast culture
applications
disease resistance genes, transfer, 111
male sterility, transfer, 111
somatic hybrids, gene transfer, 110–111
case study, 112
cybridization, 108–109
description, 104–105
heterokaryons/hybrid cells, identification and selection
auxotrophic mutants, 109
density gradient centrifugation, 109
drug sensitivity, 109
flow cytometry, 110
isolation and culture
mixed enzymatic treatment, 106
sequential enzymatic treatment, 105
protoplast fusion, 105–108
and regeneration, 106–107
somatic hybrid plants characterization, 110
Pullulan, 374
Pyrazines
bioflavors production, 450–451
de novo synthesis, 465

R

Recombinant DNA technology, 332
Rhizobium, 172
Riboflavin (vitamin B_2), 429

S

Saccharides
oligosaccharides, 390
sugars, 389–390
Saccharin, 397
Saccharomyces cerevisiae, see Yeasts
Schizophyllan, 375
Schizosaccharomyces pombe, 464
Scleroglucan, 375
Seafood industry, 564; *see also* Fish processing
Secondary metabolites
commercially viable products, 79
in vitro production, 80–82
production, 78–79
Shiitake, *see Lentinus edodes*
Single-cell proteins (SCP) production
agro-industrial wastes and residues, 566–568
definition, 566
description, 21–22
SmF, *see* Submerged state fermentation (SmF)
Solid-state fermentation (SSF), 222–223
Somaclonal variations
applications, 116–117
case study, 117
chromosomal rearrangements, 113
description, 112–113
detection and isolation
in vitro cell selection, 115–116
screening, 115
disease resistance and herbicide resistance, 118
factors affecting
biochemical causes, 115
genetic causes, 114
physiological causes, 114
Somatic cell hybridization, *see* Protoplast culture
Spray dying method, 201–202
SSF, *see* Solid-state fermentation (SSF)
Starch, sweeteners production
corn/glucose syrups, 399–402
HFCS syrups, 402–404
major components, 397–398
schematic overview, 398–399
Starter cultures
benefits, 229
preparation, 229–230
Submerged state fermentation (SmF), 222, 422
Sucralose, 397

Sugar industry, 561–562
Sulforaphane, 310–311
Surface fermentation, 223
Sweeteners production
  developments, 387
  division of, 388
  HFCS, 388–389
  inulin syrup
    high-fructose syrups production, 407–408
    hydrolysis process, 406–407
    immobilized inulinases, 409
    structure, 406
  invert sugar, 404–405
  starch
    corn/glucose syrups, 399–402
    HFCS syrups, 402–404
    major components, 397–398
    schematic overview, 398–399
  sweetening agents, 387
  types
    nonsaccharides, 391–397
    saccharides, 389–391
Synbiotics
  clinical applications, 293–294
  description, 282
  mode of action, 294–295
  prebiotics and probiotics, 292–294

T

Terpenes
  bioflavors, 451, 455–456
  biotransformation, 466–467
  de novo synthesis, 466
Therapeutic agents, *see Lactobacillus* bacteria, therapeutic agents
Thermotherapy/heat treatment, meristem culture, 83–84
Third-generation sequencing (TGS) technologies, 51
Tissue culture
  anther culture and haploid production
    applications, 99–100
    case study, 100–101
    culture media and nutritional requirements, 96–99
    techniques, 96
  applications
    biotransformation, 79
    mutant selection, 78
    secondary metabolites production, 78–82
  description, 75–76
  embryo culture
    applications, 102–104
    case study, 104
    description, 101
    factors affecting, 102
    types, 102
    uses, 101
  meristem culture
    purpose, 82–83
    virus elimination methods, 83–85
    virus indexing, 85
  micropropagation
    adventitious buds/shoots regeneration, 87
    axillary branching, 86–87
    progress, horticultural and forest tree species, 93–95
    single-node culture, 86
    stages, 87–93
  nutrient medium, 76–77
  protoplast culture
    applications, 110–111
    case study, 112
    cybridization, 108–109
    description, 104–105
    heterokaryons/hybrid cells, identification and selection, 109–110
    isolation and culture, 105–106
    protoplast fusion, 105–108
    and regeneration, 106
    somatic hybrid plants characterization, 110
  somaclonal variations
    applications, 116–117
    case study, 117
    chromosomal rearrangements, 113
    description, 112–113
    detection and isolation, 115–116
    disease resistance and herbicide resistance, 118
    factors affecting, 114–115
  techniques, 77–78

Tocopherols, 306
Tocotrienols, 306
Totipotency, 75; *see also* Tissue culture
Transgenic plants development
  Agrobacterium transformation
    tissue culture–dependent, 142–143
    tissue culture–independent, 143–144
  biolistic transformation and chloroplast genetic engineering, 144–145
  description, 141–142
  DNA transfer, 146
  electroporation-mediated transformation, 145
  genetic transformation, 141
  microinjection, 146
  PEG-mediated genetic transformation, 145–146
  post-transformation steps
    crop improvement and commercialization, 146–147
    evaluation and characterization, 148
    modified plant cells and regeneration selection, 147–148
    transgene regulation, commercialization, 148–149

V

Vaccines
  GM crops, edible, 12–13
  polysaccharides, 375–377
Value-added products production
  antibiotics, 589–590
  Baker's yeast *(S. cerevisiae)*, 568–569
  bio-fuel, 574–582
  enzymes, 582–585
  food additives, 585–589
  mushrooms, 590
  organic acids, 569–574
  single-cell proteins, 566–568
Vesicular arbuscular mycorrhiza (VAM), 175
Vinasses, 563
Vinification, 343–344
Virus elimination methods, meristem culture
  chemotherapy, 84
  potato virus X, 85
  research work on, 84
  thermotherapy/heat treatment, 83–84
  tobacco mosaic virus, 84–85
Virus indexing, meristem culture, 85
Vitamin B_{12} production, 589
Volvariella volvacea, 533

W

Waste management, *see* Agro-industrial wastes and residues
Wastewater sludge, 199–200
Wine
  fermentation, 221
  industry, 562–563

X

Xanthan
  gum, 365–366
  microbial production, 364
  polytetramer and polytrimer, 365
  trisaccharide chain, 363
Xylanases, 340–341
Xylinan, 372–373
Xylitol, 311

Y

Yeasts, 25–26, 225; *See also* Fermentation
Yellow food colorant, *see* Riboflavin (vitamin B_2)
Yield and quality improvement, applications, 149–157